FUNDAMENTAL PHYSICOCHEMICAL PROPERTIES OF GERMANENE-RELATED MATERIALS

FUNDAMENTAL PHYSICOCHEMICAL PROPERTIES OF GERMANENE-RELATED MATERIALS

A THEORETICAL PERSPECTIVE

SHIH-YANG LIN
Department of Physics, National Cheng Kung University, Tainan, Taiwan

VO KHUONG DIEN
Department of Physics, National Cheng Kung University, Tainan, Taiwan

THI DIEU HIEN NGUYEN
Department of Physics, National Cheng Kung University, Tainan, Taiwan

CHI-HSUAN LEE
Department of Physics, National Cheng Kung University, Tainan, Taiwan

MING-FA LIN
Department of Physics, National Cheng Kung University, Tainan, Taiwan

ELSEVIER

Elsevier
Radarweg 29, PO Box 211, 1000 AE Amsterdam, Netherlands
The Boulevard, Langford Lane, Kidlington, Oxford OX5 1GB, United Kingdom
50 Hampshire Street, 5th Floor, Cambridge, MA 02139, United States

ISBN: 978-0-443-15801-8

For information on all Elsevier Science publications visit our website at https://www.elsevier.com/books-and-journals

Publisher: Candice G. Janco
Acquisitions Editor: Charles Bath
Editorial Project Manager: Anthony Marvullo
Production Project Manager: Erragounta Saibabu Rao
Cover Designer: Matthew Limbert

Typeset by TNQ Technologies

Contents

7. Hydrogen-adsorbed group-IV materials

Vo Khuong Dien, Shih-Yang Lin, Chi-Hsuan Lee, Hsin-Yi Liu, Thi My Duyen Huynh, Nguyen Thi Han, Ngoc Thanh Thuy Tran, Thi Dieu Hien Nguyen, Wei-Bang Li, and Ming-Fa Lin

8. Fundamental properties of transition-metal-adsorbed germanene: a DFT study

Hai Duong Pham, Shih-Yang Lin, Vo Khuong Dien, Chi-Hsuan Lee, Hsin-Yi Liu, Thi My Duyen Huynh, Nguyen Thi Han, Ngoc Thanh Thuy Tran, Thi Dieu Hien Nguyen, Wei-Bang Li, and Ming-Fa Lin

9. Geometric structures, electronic, and excitonic optical properties of germanium dioxide α – quartz GeO_2

Thi Dieu Hien Nguyen, Shih-Yang Lin, Vo Khuong Dien, Chi-Hsuan Lee, and Ming-Fa Lin

10. Spin-diversified quasiparticle behaviors in rare-rare-earth La- and Eu-adsorbed germanene materials

Vo Khuong Dien, Shih-Yang Lin, Chi-Hsuan Lee, Hsin-Yi Liu, Thi My Duyen Huynh, Nguyen Thi Han, Ngoc Thanh Thuy Tran, Thi Dieu Hien Nguyen, Wei-Bang Li, and Ming-Fa Lin

11. Germanene on Ag (111)

Shih-Yang Lin, Thi Dieu Hien Nguyen, Vo Khuong Dien, Chi-Hsuan Lee, Hsin-Yi Liu, Hai Duong Pham, Thi My Duyen Huynh, Nguyen Thi Han, Ngoc Thanh Thuy Tran, Wei-Bang Li, and Ming-Fa Lin

12. Bilayer germanene/germanene composites

Shih-Yang Lin, Thi Dieu Hien Nguyen, Vo Khuong Dien, Chi-Hsuan Lee, Hsin-Yi Liu, Hai Duong Pham, Thi My Duyen Huynh, Nguyen Thi Han, Ngoc Thanh Thuy Tran, Wei-Bang Li, and Ming-Fa Lin

13. Germanene nanotubes

Hsin-Yi Liu, Shih-Yang Lin, Vo Khuong Dien, Chi-Hsuan Lee, Hai Duong Pham, Thi My Duyen Huynh, Nguyen Thi Han, Ngoc Thanh Thuy Tran, Thi Dieu Hien Nguyen, Wei-Bang Li, and Ming-Fa Lin

List of contributors

Vo Khuong Dien Department of Physics, National Cheng Kung University, Tainan, Taiwan

Thi My Duyen Huynh Department of Physics, National Cheng Kung University, Tainan, Taiwan

Nguyen Thi Han Department of Physics, National Cheng Kung University, Tainan, Taiwan

Chi-Hsuan Lee Department of Physics, National Cheng Kung University, Tainan, Taiwan

Wei-Bang Li Department of Physics, National Cheng Kung University, Tainan, Taiwan

Shih-Yang Lin Department of Physics, National Cheng Kung University, Tainan, Taiwan

Ming-Fa Lin Department of Physics, National Cheng Kung University, Tainan, Taiwan; Hierarchical Green-Energy Materials (Hi-GEM) Research Center, National Cheng Kung University, Tainan, Taiwan; Center for Micro/Nano Science and Technology, National Cheng Kung University, Tainan, Taiwan

Hsin-Yi Liu Department of Physics, National Cheng Kung University, Tainan, Taiwan; Center for Micro/Nano Science and Technology, National Cheng Kung University, Tainan, Taiwan

Thi Dieu Hien Nguyen Department of Physics, National Cheng Kung University, Tainan, Taiwan

Hai Duong Pham Department of Physics, National Cheng Kung University, Tainan, Taiwan

Jheng-Hong Shih Department of Physics, National Cheng Kung University, Tainan, Taiwan

Ngoc Thanh Thuy Tran Hierarchical Green-Energy Materials (Hi-GEM) Research Center, National Cheng Kung University, Tainan, Taiwan

Yu-Ming Wang Department of Physics, National Cheng Kung University, Tainan, Taiwan

CHAPTER

1

Introduction

Shih-Yang Lin[1], *Thi Dieu Hien Nguyen*[1], *Vo Khuong Dien*[1], *Chi-Hsuan Lee*[1], *Hsin-Yi Liu*[1], *Hai Duong Pham*[1], *Thi My Duyen Huynh*[1], *Nguyen Thi Han*[1], *Ngoc Thanh Thuy Tran*[2], *Wei-Bang Li*[1] *and Ming-Fa Lin*[1]

[1]Department of Physics, National Cheng Kung University, Tainan, Taiwan; [2]Hierarchical Green-Energy Materials (Hi-GEM) Research Center, National Cheng Kung University, Tainan, Taiwan

Very interesting, the germanium-based materials play the critical roles in the basic and applied sciences [1–8,9], as clearly revealed in other group-IV [10–15] and group-V [16–18] condensed-matter systems. Their atomic configurations are rather suitable in creating the active chemical bondings among the identical and/or different nearest-neighboring atoms [3,13], leading to the diverse physical/chemical/material environments [19–25]. A lot of mom-element crystals and multi-component compounds, which are mainly driven by dimensions [26–33], stackings [34–41], layer numbers [3,42–46], cylindrical/curved/folded structures [47–52], open/closed edges [37,53–58], different atomic compositions [59–63], composites of two subsystems, and heterojunctions, have been successfully synthesized/generated by the various experimental growth manners. Obviously, such main-stream materials include emergent systems (e.g., layered graphenes/silicenes/germaneness/tinenes/plumbenes), semiconductors (II-V/II-VI/III-V/III-VI), and green energies (batteries, solar cells; hydrogen storages). The same-component allotropes are greatly developed by through much progress about the high-resolution measurements and the accurate predictions. The current book is mainly focused on the first-principles simulations (details discussed later in Chapter 3) and their close relations with the phenomenological models (details in Chapter 2), in which both approaches are thoroughly consistent under the quasiparticle framework [3,40,64]. The former are capable of providing the enough information in evaluating whether the reliable former could be well established under the unusual crystal symmetries and complicated orbital hybridizations [12,65,66]. Moreover, the detailed comparisons

Fundamental Physicochemical Properties of Germanene-related Materials
https://doi.org/10.1016/B978-0-443-15801-8.00019-0

between theoretical predictions and examined results are illustrated thoroughly in Chapters 4–20, covering the various merits and drawbacks of experimental observations [59,67–74].

Both numerical simulations and model calculations are strong required to explore the diverse phenomena due to the natural and/or artificial operations. The long-term theoretical developments clearly show that these two approaches have created the strong modifications of their manners and techniques. For example, the up-to-date VASP results present the reliable optical spectra in the presence/absence of excitonic effects using the efficient computer evaluations, while the opposite is true for the Coulomb-field excitations [31,40,75–77]. Most important, how to directly link the geometric symmetry [78,79], the featured electronic properties [42,80–82], the prominent absorption channels [83–85], and the unusual spin configurations/the net magnetic moments [86–88] needs to be thoroughly examined under an unified framework. This has been successfully investigated by using the quasiparticle viewpoints, e.g., the systematic investigations about the fundamental properties in graphene- [37,89–91], graphene-nanoribbon- [55,89,92,93], carbon nanotube- [48,69,94–100], silicene-, and battery-related systems [9,11,40,101,102]. The calculated physical & chemical quantities fully agree with one another through the quasiparticle framework. Specifically, the first-principles method cannot deal with the carrier responses of magnetic-field perturbations, since the vector potential leads to a quite large period. More quasiparticles, such as, phonons [85,103–105], photons [106–108], polarons [109] and polaritons [110–112], could largely extend the electron-based scopes through the significant couplings of the identical an distinct quasiparticles. On the model side, the generalized tight-binding model (external fields), the modified random-phase approximations (RPA in Refs. [113–117]), and the corrected self-energy method, which are, respectively, developed for the rich magnetic quantization [3,118,119], the diverse (momentum, frequency)-excitation phase diagrams, and the unique Coulomb decay rates for the layered/coaxial materials. Furthermore, their direct combinations with the well-known formulas will become highly efficient and reliable in comprehending the unusual quasiparticle behaviors, e.g., the well linking of the first one with the static/dynamic Kubo formula in the clear evidences of quantum Hall conductivities [40,119]/the magneto-optical selection rules [40,120–124]. Of course, the suitable parameters of the tight-binding model, which correspond to the orbital- and spin-dependent intrinsic interactions are achieved from a good fitting with the main features of the VASP results. Their values are very difficult to get for Moire superlattices, in which large unit cells generate very complex orbital-hybridization environments and position-dependent spin moments [53,90,118,125–129].

Very obviously, the low-dimensional materials, as clearly revealed by the graphene-related sp^2-bonding crystals, have driven much progress from of the initial growths, the consistent examinations among the essential properties, and the commercial applications. For example, the C_{60} fullerenes [130–133], carbon nanotubes [48,69,95,134,135], and few-layer graphenes, are produced from laboratories, respectively, 1985, 1991 [132], and 2004 [136], through a graphite vaporization by laser irradiation, an arc-discharge evaporation method, and a mechanical exfoliation. The various synthesis methods are successfully developed for the large-scaled quantities, being very helpful in getting the sufficiently strong measurement responses. And then, the verifications of nanoscale crystal symmetries have been conducted by the high-resolution scanning tunneling microscopy (STM; details in Chapter 3 [9,137,138]), tunneling electron microscopy (TEM [67,87,139,140]), and low-energy electron diffraction (LEED [9,141–145]). The STM measurements have observed a uniform hexagon with an achiral/chiral distribution on a cylindrical nanotube surface, a planar carbon-honeycomb crystal of monolayer graphene, and a finite-width graphene nanoribbon with achiral/chiral edge structure. The simultaneous scanning tunneling spectroscopy (SS) are able to clearly identify the narrow-/middle-gap/metallic nanotubes, a zero-gap semiconducting graphene, and the semiconducting graphene nanoribbons due to the unique widths and open edges. The rich quasiparticle phenomena are directly reflected in band structures, Coulomb-field excitations, optical vertical transitions, and quantum Hall effects, in which such properties can be fully examined through the angle-resolved photoemission spectroscopy (ARPES [30,146,147]), electron energy loss spectroscopy (EELS [130,148—152]), reflectance/absorption/transmission/photoluminescence spectroscopies [153—161], and transport apparatus [162]. A plenty of diversified quasiparticle phenomena have been successfully observed by examining the featured responses. e.g., a Dirac-cone structure (ARPES in Refs. [30,146,147]) and the alkali-doping-induced free carrier plasmon/the unusual Coulomb decay rate of monolayer graphene (EELS in Refs. [130,148—152]/ARPES in Refs. [30,146,147]). The direct combination of different equipment and techniques could investigate the unusual more efficiently, e.g., the combined (STM, ARPES) and (LEED, EELS). Most of them are consistent with the theoretical predictions within the quasiparticle framework However, certain experimental measurements might become meaningless under the intrinsic limits, covering those of STM on 3D materials/thin films, ARPES on a large Moire superlattice/a semiinfinite crystal with a nonnegligible surface/a curved axis without a well-defined momentum, and optical spectroscopies/EELS on wide-frequency excitations [106,130,148—150,163].

Other group-IV-based materials quite differ from carbon-related systems in terms of geometric, electronic, optical, Coulomb-excitations, magnetic, transport and mechanical properties. Few-layer silicene/germanene/tinene/plumbene systems ([40,101,102]/[1—3,8,9,40, 164—166]/[1,2,101,167,168]/[10,90,169,170]) are frequently synthesized on the different substrates through the molecular beam epitaxy (MBE [171—173]). Very interesting, the first-step VASP simulations show that the significant orbital hybridizations, as well as, the ferromagnetic spin configurations, in honeycomb lattices, substrate crystal and heterojunction will greatly diversify quasiparticle behaviors. These systems exhibit the buckled structures under the high-resolution observations, indicating the nonperpendicular π and σ bondings. Specifically, the thoroughly experimental examinations on the germanene-/germanium-related materials are urgently required in the current science researches, since they could reveal and identify a lot of unusual physical/chemical/material phenomena. For example, STM/TEM/LEED [9,137,138,141,142], STS [18,138,174], ARPES [30,39,146], EELS [130,148,149], and optical spectroscopies, are, respectively, rather powerful in verifying the top & side surface morphologies, the featured van Hove singularities, the low-lying occupied energy subbands, the π & σ plasmon modes, and the prominent absorption peaks in th presence/absence of excitonic effects for layered germanene systems (the details of all Ge-associated systems in Chapters 4—20). It should be noticed that most of significant measurements do not finish up to now because of many intrinsic and extrinsic factors, e.g., the direct tests of the spin-orbital coupling strengths/the sp^3-multi-orbital hybridizations/the optical absorption structures/the chemical modifications in silicene, germanene, tinene and plumbene. Maybe, a well-established quasiparticle framework is very helpful to persue the simultaneous development of the equipment designs. Their specific characteristics could be further developed to become the realized devices with the special electronic and optical functionalities [175], as clearly illustrated in Chapter 22 for the highly potential applications.

Monolayer materials were discovered originally in 2004, graphene, which is exfoliated from a graphite surface via a mechanical method [176]. After that, other group-IV materials such as silicene and germanene have been synthesized successfully on specific substrates, for example, silicene on Ag(111), Ir(111) and ZrB_2 surfaces [177–179], and germanene on Pt(111), Au(111) and Al(111) surfaces [137,180,181]. The group-IV materials, which are cataloged from C, Si, Ge, Sn and Pb atoms, present four half-occupied orbitals in condensed-matter systems, respectively (ns, np_x, np_y, np_z), where n = 2–5. They can generate π, σ and sp^3 chemical bonding and their atomic configurations are quite different from those in an isolated state. This will illustrate clearly in this chapter that monolayer surfaces can provide very active chemical environments by results of charge density. Such its novel electronic properties [182] arouse everyone's attention. It is not only for theoretical research, but also for applications of electronic devices. In this chapter, we will put more details about counterparts of graphene, germanene. It possesses also a honeycomb structure like graphene and shows a large spin-orbit coupling (SOC), which are suitable for applications of topological insulators (TI) and spintronic devices [183]. Production of germanene requires a metal substrate and the interaction between germanene and substrate is very strong. A relative weak interaction of germanene on a semiconducting GaAs(0001) substrate is reduced by intercalation of H at interface [184] to formation of the quasi free-standing germanene. In this chapter we focus on the diversified electronic properties of monolayer, bilayer and different sliding configurations of bilayers, which their spin-orbit coupling are taken into consideration, investigated by first-principles calculations. A relativistic interaction of a carbon atom's spin with its motion inside a potential is normally weak such as graphene. However, it is significant in a germanium atom, buckled monolayer germanene shows a band gap opening around 22.9–23.9 meV at K point by DFT calculations [13,183,185]. A large spin orbital coupling leads to observe the Quantum Spin Hall effect (QSHE). Optimal geometric structure, bond lengths, ground state energy, buckling distance, energy band structure, energy gap (E_g), charge density (ρ) and orbital-projected density of states (PDOSs) will be fully included in the calculations.

The planar/buckled/curved/folded honeycomb lattices provide very outstanding chemical environments [186] in creating the various compounds through the active chemisorptions [84], greatly diversifying the essential quasiparticle properties [96], and widely promoting the potential applications. For example, group-IV elements have been successfully synthesized under the layered structures, clearly showing the dangling bonds of C-$2p_z$ and other (s, px, py, pz) orbitals in graphene and other systems, respectively [187]. The optimal geometric symmetries are consistent with each other from the experimental measurements (STM and TEM; details in Chapter 3 [188]) and the first-principles simulations [189]. In Chapter 5, germanene, with a buckled structure, serve as a host material, in which three kinds of adsorption adatoms, borons, carbons and nitrogens, are able to largely enrich the quasiparticle behaviors, e.g., the modulation of band gaps [190], and the semiconductor/metal & transitions [191] during the variation of chemisorption cases. The atomic configurations of adatoms and host ones are predicted to play the critical roles. Certain important differences, corresponding to the diverse orbital hybridization [192] and spin-dependent magnetisms [193] will be delicately determined from all calculated results. The similar chemical modifications could be achieved from hydrogenation [194], alkalization [195], oxidization [93], and halogenation [196], as revealed in graphene-related compounds. It is very interesting to fully comprehend the diversified quasiparticle phenomena among the significant chemisorptions of distinct adatoms [197].

Layered materials are capable of providing the planar/buckled/deformed surfaces with the very active dangling bonds, especial for group-IV ones (graphene/silicene/germanene/

tinene/plumbene with four half-occupied in each atom [136]/[198]/[180]/[2]/[169]). Such chemical environments are very powerful in generating the various modifications through the modulated chemisorption [187,199] and substitution [200,201]. For example, the alkali-adatom adsorption on graphene has successfully led to a semiconductor-metal transition using the ARPES measurements [202,203], or a blue shift of the Fermi level from the Dirac point to the linear conduction band. The chemisorption reactions, e.g., hydrogenations [12], alkalizations [13], oxidizations [14] and halogenation [6], are expected to be rather powerful in widening the diversities of fundamental quasiparticle properties [7], and enhancing the potential applications [15,16]. The hydrogenated group-IV materials are outstanding candidates in greatly enhancing the theoretical and experimental progresses. The unusual hydrogenation -effects are able to stir the establishment of phenomenological models [17], the thorough simulations of first-principles calculations [18], the high-precision measurements [19,20], and the specific functionalities of highly potential applications [15,21]. These are completely consistent with the development of quasiparticle framework in the current book [7,22]. The simulation predictions, which are conducted on graphene [17] and graphone [18] under the first-principles evaluations, are earlier than the experimental syntheses about the latter. Both of them, respectively, correspond the full hydrogen-adatom chemisoprptions of the double- and single-side configurations. The similar VASP methods have been generalized to other group-IV materials [23,24]. However, the critical mechanisms (the concise physical and chemical pictures) cannot be achieved under such studies. As for the experimental growths [25] and examinations [12] for the nongraphene-related hydrogenations are almost absent in the previous works, being urgently required in the near-future searches. This work, the systematic investigations on the geometric, electronic and optical properties of hydrogenated group-IV monolayer systems, will present the diverse quasiparticle behaviors, e.g., the critical factors due to the various atoms, orbitals, spin configurations, and chemisorption cases. Moreover, the significant mechanisms, the perpendicular $\pi/\sigma/sp^3/ssp^3$ chemical bondings [26]/[12]/[17], as well as, the observable spin-orbital couplings [27] and/or on-site electron-electron Coulomb interactions (the Hubbard-like spin-based ones [28]), are completely identified from the relevant physical/chemical quantities. The main features of crystal symmetries, electronic energy spectra & wave functions, and optical excitations are directly linked together only under this unified viewpoint. Specifically, it is very interesting in developing the excitonic theories of layered materials, since such systems might possess the unusual charge screening effects [29]. How to solve the numerical calculations would become an open issue. This study can provide the sufficient information in comprehending whether the above-mentioned materials are suitable in serving as the hydrogen energy storages [30].

Chapter 8 studies the electronic, magnetic properties of transition metal (T)-doped germanene systems using the first-principles method. Dedicated calculations include the bond length, the buckling, preferable adsorption sides and height of adatoms, the adsorbed energy, atom-dominated band structure, adatom-induced free carrier density along with energy gap, spin-density distributions, spatial charge distribution, and atom-, orbital- and spin-projected density of states (DOS). All absorption systems show optimized structure on the hollow side. The hexagonal honeycomb lattices were slightly deformed after the TMs were absorbed together with the new bonds between the TM-Ge which created a nonuniform environment thus being responsible for the emergence of unusual properties. Out of three types of transition metal-adatom adsorption, Fe and Coplay important roles in inducing magnetic properties, while the Ni−Ge system exhibits nonmagnetic behavior. The magnetic configurations are well determined from the total magnetic moment, spin split energy band, and spin density distribution. In this work, the theoretical predictions presented can be verified by experimental measurements. In addition, a comparison of significant similarities and differences with systems involving graphene, silicene is also considered. Finally, potential applications of adsorption systems will be proposed and discussed.

The research of $\alpha - GeO_2$ attracts much attention due to broad applications in industries, especially optoelectronic devices [204−207] Chapter 9 provides the full comprehending of binary $\alpha - GeO_2$ compound. Germanium dioxide is successfully synthesized by a chemical precipitation method [208]. The featured quasiparticle properties are investigated under the first-principles calculations [206]. The main characteristics cover geometric structure, band structures, atom dominances, orbital-decomposed density of states, charge densities, dielectric functions, absorption coefficient, energy loss spectra, and reflectivity index. All these are combined in the unified theoretical framework of orbital hybridizations. The featured phenomena also include excitonic effects, which frequently apply in the wide band gap materials [209]. The strong cooperation of optical properties and available channels in band edge stages satisfied the joint density of states, the conservation of energy, and Pauli exclusive. These properties will contribute a great knowledge to fully explore the physical, chemical, and material characteristics of salt of germanium.

Rare-earth (RE) atoms possess an intrinsic large spin-orbit interaction and their magnetic states are well decoupled from the environment [1,2]. These properties have been exploited to design single ion molecular magnets with magnetic bistability on a long timescale. Magnetic remanence and long magnetic lifetimes have also been achieved in single rare-earth (RE) adatoms. To date, there are much less studies of interaction between RE atoms and 2D materials [3,4]. The main challenge is complicated of the sp-f chemical bonding between group IV elements and RE adatoms [5]. The previous theoretical and experimental have been investigated on the RE-related graphene [4,6], however, comprehensive and comparative work to fully understanding the orbital bonding-enrich phenomena is rather limited up to now. The theoretical framework based on first-principles calculations was adopted for the germanene adsorbed rare-earth adatoms (La and Eu). Similar to that of graphene [6], the favorite adsorption side for rare-earth atoms is the hollow side. All rare-earth atoms doped germanene present n-type doping with the significant blue-shift of the Fermi, and thus, the electric conductivity can be significantly improved. The optical properties of germanene-based systems are also changed with the adatom's kinds and concentrations. The tunable electronic and optical properties clearly indicate the potential applications in various fields, e.g., lithium batteries [7], suppercapacitors [9], electronic and optical devices [8]. Especially, germanene adsorbed Eu atoms might exhibit the splitting of the spin-up and spin-down energy bands. Such electronic states in transport measurements are expected to produce the spin-polarized current, and thus, suit for spintronic applications [10,11].

The atomic interactions between a germanene layers and substrate can create diversified phenomena. The monolayer and few-layer materials are frequently grown on various through the molecular beam epitaxy (MBE) or chemical vapor deposition (CVD), which is sensitive to the lattice constant. For instance, 2D few-layer (C, Si, Ge, Sn, Pb) materials are, synthesized on Ag(111)/Ir(111)/ZrB2R(0001) [210−212], Pt(111)/Al(111)/Ag(111) [125,181,210,213,214], and $Bi_2Te_3(111)$ [73,215−217]. The experimental observations of STM and LEED/RHEED/TEM are conduct to explore the surface morphologies. The direct evidences of the top and side views can be confirmed from these measurements. In most of cases, the surface reconstructions it clearly revealed in the adlayer and substrate, but only a few of them remains the original lattice parameters. Specially, the germanene adlayer and substrate atoms are capable of producing the nonnegligible orbital hybridization, resulting in the diversified lattice parameters and electronic properties. The Moire superlattice and the sufficiently thick substrate are fully explore under the substrate effects by using the first-principles calculations. The monolayer/bilayer germanene on Ag(111) is chosen for models study. The critical roles of Ge—Ag, Ge—Ge, and Ag—Ag chemical bonds being main focused of research. The magnetic and nonmagnetic configurations through critical orbital hybridization of Ge-(4s, 4px, 4py, 4pz) and Ag-($3d^2_z$, $3d^2_{x}$-^2_y, $3d_{xz}$, $3d_{zy}$, $3d_{yx}$) are well simulated and could be further test by future experiments.

Tubular nanostructures have been always attracted highly research attention. Carbon nanotubes (CNTs), which are created experimentally [54,132,210—214], belong to one dimensional semiconducting nanostructures. Basically, their low dimensionality and the quantum confinement effect are with potentials for various technology Products such as nanocomposites, field emission sources, chemical sensor, photonics, Lithium ion batteries, supercapacitors and actuators because of its distinctive mechanical, electronic and optical characteristics [215—220]. The anode of Lithium ion batteries which have been commercialized exhibits limited capacity of graphite is 372 mA h/g theoretically. When a graphitic carbon anode is replaced by tubular nanostructures of C, Si and Ge may increase the anode capacity about two times to 10 times. Carbon nanotubes exhibit reversible capacities of 573 mAh/g at a small current of 0.2 C and 520 mAh/g at a large current of 2 C [221]. Lithium-ion diffusivity of Germanium is 400 times faster than Si and electrical conductivity of Germanium is 10^4 times higher than Si. Besides, theoretical specific capacity of Germanium is very large about 1600 mAh/g, Germanium is also as a promising anode material for lithium-ion batteries. The metals of Si, Ge and Sn exhibit multiple times of capacities when a graphitic anode is replaced. However, the large volume change of IV-grope metals has caused rapid capacity fading during cycling. The growth of germanium nanostructures, which are nanowire (NW) and nanotube (NT) can be controlled by methods of chemical vapor deposition (CVD). Both diameters are with variable diameters from 10 to 500 nm by use of a porous anodic aluminum oxide (AAO) template nanochannels and a wall thickness is controlled between 6 and 18 nm [222]. There are a lot of theoretical studies on the electronic properties of carbon and silicon nanotubes. They are analyzed by the tight-binding model and first-principle method [80,223—228]. However, simulated calculation of germanium nanotubes published quite a few. Bayani et al. performed carbon and germanium nanotubes applied to field effect transistors (FETs) by simulation of first-principle. The results showed better transistor performance of germanium nanotubes, which output characteristic voltage reached to 0.6 V, compared to carbon nanotubes (0.4 V) [229]. Therefore, it is very significant to realize the electronic properties. In this chapter we focus on the diversified electronic properties of single wall germanium nanotubes, investigated by first-principles calculations. Achiral germanium nanotubes are taken into consideration to reveal that there are two basic kinds of geometric structures (armchair and zigzag), which can be further modulated by various diameters and chiral vectors. This work presents bond lengths b_1 and b_2, ground state energy as well as total energy E_0, buckling distance Δh, energy band structure, energy gap (E_g), charge density (ρ) and projected density of states (PDOSs) and the both structures dependent on chiral vectors and diameters are fully included in the calculations.

Generally, quantum confinement plays a critical role in diversifying the physical/chemical/material phenomena, while any condensed-matter system is reduced from 3D dimension to 2D/1D/0D. For example, the dramatic transformation is clearly revealed from a bulk semiconductor [1] into a quantum well/wire/dot [2,3]. The close relations are obviously shown among the AA-/AB-/ABC-stacked graphite systems [4,5] and the few-layer graphene systems [6–10]/graphene nanoribbons [11–13]/carbon nanotubes [14–18]/carbon tori [19–21]/carbon disks [22]/carbon onions [23,24]/carbon fullerenes [25,26]. The quantum-confinements are fully assisted by the different dimensionalities and crystal symmetries, Very interesting, there exist a lot of theoretical [6,8–10,27–31] and experimental researches [7,27,29,32] on carbon-related materials in terms of the geometric [18,33,34], electronic [5,12,21,24,35–41], Coulomb-field excitation [4], optical [35,39], magnetic [2,20,35,42–46], transport [12,28,47–49], and mechanical properties [38,50,51] in the absence/presence of chemical adsorptions/substitutions [37,49,52–55], especially for the systematic investigations in the previous books [56] under a unified quasiparticle framework. The simultaneous measurements of STM and STS on a graphene nanoribbon (details in Chapter 3) can identify the width/chirality of edge structure and band gap. E.g., quickly declines in the increment of width; furthermore, there are three kinds of bandgap approaches for armchair systems with the dimer-line numbers of $NA = 3I$, $3I + 1$; $3I + 2$ (I: an integer [56]). The low-lying valence subbands, which correspond to parabolic energy dispersions, have been successfully verified from the ARPES examinations (Chapter 3). Specifically, the first-principles simulations have shown that zigzag graphene nanoribbons possess the antiferromagnetic/ferromagnetic spin configuration across the ribbon center/on the open edges [11–13,36,47]. However, the directly experimental evidence is absent up to now. The above-mentioned quasiparticle behaviors are expected to be greatly diversified in germanene nanoribbons, since the significant couplings among the highly nonuniform geometries (width, chirality and buckling [36,42,43,57–59]), the nonperpendicular π and σ bondings [59], and the spin-orbital interactions/the Hubbard-like on-site Coulomb interactions, are responsible for the rich phenomena. The detailed VASP calculations, analyses and critical mechanisms are clearly illustrated in Chapter 14.

Group I materials in a periodic table, alkali atoms (lithium/sodium/potassium/rubidium/cesium), are one of the most important elements in creating the various pure crystals and multi-component compounds. The atomic configuration of each atom possesses a very active half-occupied bond corresponding to the outmost s-orbital [1], being responsible for the critical bonding in alkali-related condensed-matter systems. Very interesting, the systematic first-principles investigations, which are conducted on the highly efficient cathode/electrolyte/anode materials of lithium-ion-based batteries, clearly identify the concise orbital hybridizations by examining all Li-related bonds in a unit cell. The nonuniform chemical environment, the atom-dominated band structures and wave functions, the spatial charge densities, and the atom-/orbital-decomposed van Hove singularities are unified together to achieve the critical mechanism, such as, the multi-orbital hybridizations of 2s-(3d_(z2), 3d_(x2-y2), 3dxy, 3dyz, 3dzx) and 2s-(4s, 4px, 4py, 4pz) and 2s-(2s, 2px, 2py, 2pz)/2s-(3d_(z2), 3d_(x2-y2), 3dxy, 3dyz, 3dzx) and 2s-(2s, 2px, 2py, 2pz) for the LiTiO [2]/LiGeO [3,4]/LiFeO [5] materials. Moreover, alkali atoms are frequently utilized as guest atoms for the chemical modifications, e.g., graphite [6,7]/carbon nanotubes [9] intercalation compounds and chemically absorbed graphene [8,10–12]/graphene nanoribbons [8,13–15], with the high carrier densities of conduction electrons. The metallic compounds are thoroughly examined to arise from a large charge transfer phenomenon of the outmost-s orbital in alkali atoms. For the planar carbon-honeycomb crystal, a 1D graphene nanoribbon is an outstanding candidate in evaluating the great enhancement of free carriers per length, where a linear relation between their density and alkali-atom concentration is verified from a lot of chemisorption cases. In addition, a 2D graphene is rather difficult to prove the 100% charge transfer of the alkali-C bonds because of anisotropic energy spectra near the blue-shift Fermi level [16–18]. The rich quasiparticle phenomena will be revealed in the alkali-absorbed germanene compounds through the various adsorption modulations, since there exist the significant geometry couplings among the finite-width quantum confinements [19–21], open edge boundaries [22], buckled honeycomb lattices [23–25], and guest-adatom distribution configurations (height and concentrations [14,26–29]). The free conduction electrons, being created by guest alkali atoms, are directly reflected in the main featured of electronic properties. Whether their prominent effects could be understood from the establishment of the tight-binding model/the generalized tight-binding model [30,31] is under the current investigation.

The tetrahedral structure of Li-Ge-O compound exhibits high optical transparency, the strong Stokes shift of self-trap excitons [230], the polar orthorhombic symmetry in the crystal [231,232] strongly demonstrate that it is a remarkable interest material for piezoelectric, pyroelectric, and optoelectronic applications [232]. This material also presents a reliable ionic conductivity $(5 \times 10^{-5} \Omega .cm^{-1})$ [233], wide cycling stability with a reserved charge capacity of 725 $mAhg^{-1}$ after 300 cycles at 50 mAg^{-1} [234] and thus, suitable for the selection of anode/electrolyte in LIBs [79,234]. In general, the physical/chemical environment of this ternary compound are rather complicated and directly related to the electronic and optical properties of the system. How to get the concise quasi-particle features, the orbital hybridizations, and the connections with the final & initial states are the main challenge of the current studies [235,236]. The theoretical framework is based on the numerical calculations and delicate analyses that were developed and applied for the three-dimensional ternary Li_2GeO_3 compound, a candidate for the electrolyte materials. The fundamental features, the critical quasiparticle properties, and the significant orbital hybridizations in various chemical bonds are thoroughly examined from the optimal geometric structure, the energy spectrum with atom-dominated, the special van Hove singularities and the charge density distributions related to different orbital hybridizations. In addition, the unique optical excitations cover the single-particle & collective excitations, the reflection, adsorption spectrums and the related of the prominence peaks with specific orbital hybridizations. Our predictions provide certain meaningful information about the critical physical/chemical pictures in LIBs. Such states-of-the-art analysis is very useful for fully comprehending the diversified properties in binary/ternary/quaternary germanium-based compounds.

Perovskite solar cell materials have become common basic materials for science and engineering applications. They can provide eco-friendly energy, their efficiency and a special route to convert photon energy into electricity [237–245]. The principle architecture of a perovskite solar cell includes the absorber layer sandwich between the electron-transporting and the hole-transporting layers [94,246,247]. In which, the first component is very important since it absorbs the electromagnetic wave (EM) and generates the electron-hole pairs [248]. Apparently, the geometric, electronic properties and optical excitation mechanisms of these components are regarded as very important common characteristics. These factors should be critical conditions in determining the efficiency of light-generated current and searching for optimal match of three kinds of core components.

Recently, the metal halide perovskites with the general formula ABX_3 (where, A = cation, B = metal ion, and X = halogen anion) have been shown to be a novel class of promising semiconductor materials in photovoltaic applications [249,250]. For example, some pre sentative systems for this kind of materials, named $CaTiO_3$, $BiVO_3$, $CsSnCl_3$, $CsPbBr_3$ and so on are investigated by many methods [153,251–256]. Among them, $CsGeI_3$ has attracted much more and more attention from researchers [257–260]. Although a lot of theoretical and experimental studies have been conducted on the geometric and electronic properties of metal halide perovskites [257–260]; however, the delicate results and analyses are thoroughly absent up to now. That is to say, the calculated results are insufficient and there are no critical mechanisms (concise physical/chemical pictures) in comprehending the diversified phenomena.

In this Chapter 18, we have exposed germanium-based halide perovskite $CsGeI_3$ by using the first-principles calculations [261,262]. The sophisticated analysis has been conducted on the highly accurate results of optimized geometric structure, the atom dominated energy spectrum at various energy ranges, the spatial charge density distribution due to different orbitals and density of states are associated with the overlap of orbitals, the single-particle excitations under the absent/present of excitonic effects, the collective responses and various optical properties. As a result, the good connections of the electronic and optical properties are successfully achieved. The presented theoretical framework will fully comprehend the diverse phenomena and potential applications for other solar cell and optoelectronic materials [263–265].

An antimony film is known to exhibit topological surface states depending on the thickness of the film [266]. The surface states near Γ point own the linear crossing dispersion and possess time-reversal symmetry. If the thickness of the film is reduced to as low as four bilayers, for example, Dirac cones disappear as a result of quantum tunneling. The chemical adsorption on the one-side pristine surface could induce the large asymmetry in this film and suppress quantum tunneling. We use density functional calculation to investigate the electronic structure of the four-bilayer Sb film and find that adsorptions of impurity atoms on the film. The binding energies of the impurity atoms on the Sb film are also calculated and they are generally associated with charge transfer between the TM and Sb film. The magnetic moment of the impurity atom also play an important role in the electronic properties.

A free-standing four-bilayer Sb (111) film exists a large energy gap due to the quantum tunneling between the two surfaces. The gap can vanish, and the spin-split surface states can be induced as a result of the suppression of the tunneling-induced coupling when the thin film is deposited on a substrate [266]. We investigate Sb (111) films in contact with a single layer of Bi with and without defects by using density functional calculations, Furthermore, the electronic properties of Sb films on pure and boron-doped graphene are calculated. The stacking configurations between Sb films and substrates are also taken into account. The binding energy has a certain relation to the charge transfer of Sb films and substrates. Finally, we choose the h-BN as the substrate to compare the results of pure and boron-doped graphene.

The current book is organized as follows. The concise motivations for all chapters are clearly expressed in the above paragraphs. Chapter 2 illustrates the development of quasiparticle frameworks based on the first-principles simulations [267] and the phenomenological models [268]. Its scope obviously shows a great enhancement, compared with the previously published books [12]. The different experimental techniques, which cover X-ray/STM/SP-STM/LLED/TEM, STS/SP-STS, ARPES/SP-ARPES and reflectance/absorption/transmission/photoluminescence spectroscopies, are thoroughly characterized in Chapter 3. Their merits and drawbacks, as well as, the measurement examinations on the theoretical predictions, are explored in detail. Chapter 4 presents the main features of geometric symmetries and electronic properties for monolayer and monolayer/AA-bt/AA-bb/AB-bt/AB-bb/sliding bilayer germanene systems. The intralayer & interlayer sp^3 bondings and the spin-dependent interactions are thoroughly examined to propose the delicately concise mechanisms for the diversified quasiparticle behaviors. As for Chapter 4, the chemical modifications of boron-, carbon- and nitrogen-adatoms on monolayer germanene are conducted on the double- and single-side chemisorptions through the different concentration and distribut6ion cases. Whether the pure/hybridized π, σ and sp^3 bondings are well defined through the chemical bonds of the identical/different neighboring atoms is completely identified from the highly nonuniform chemical environments, as well as, the featured band structures, wave functions, charge/spin density distributions and orbital-/spin0decomposed van Hove singularities. The similar researches are generalized to the substitution cases in Chapter 6, while such two chemical modifications totally differ from each other in terms of enlarged unit cells with/without buckled structures, significant orbital hybridizations and spin-induced magnetic configurations/interactions. The diverse phenomena are clearly revealed by them. A charge impurity, corresponding to a very dilute substitution is also discussed in quasi-bound states and residual transports.

The hydrogenation effects of Chapter 7, being important in hydrogen energy storages [269], are widely investigated for graphene/silicene/germanene/tinene/plumbene materials through the 15 cases pristine, double- and single-side adsorptions on five group-IV materials. The geometric, electronic and optical properties are fully included in the VASP simulations to achieve the critical mechanisms, especially for the presence/absence of excitonic effects. The different absorption peaks, which are completely assisted by the orbital-/spin-projected van Hove singularities, can be assigned the specific orbital hybridizations associated with the initial and final band-edge states. The existences of stable/quasi-stable excitons might be determined by the band properties near the Fermi level. The establishment of 2D excitonic theories is also evaluated from the phenomenological viewpoint. The prominent iron-/cobalt-/nickel-adatom chemisorptions on monolayer germanene systems, as clearly indicated in Chapter 8, can display the magnetic and nonmagnetic configurations through the 10 3d-orbitals ($3d_z^2$, $3d_{x^2-y^2}$, $3d_{xz}$, $3d_{zy}$, $3d_{yx}$) [270]. The strong cooperation/competition between the complicated multi-orbital hybridizations in Ge–Ge/Ge–Fe/Ge–Co/Ge–Ni bonds and the spin-dependent on-site Coulomb interactions [271] need to be thoroughly examined from the charge- and spin-related physical quantities, covering the net magnetic moments, the spin-split band structures, the spatial spin density distributions, and the spin-projected van Hove singularities. How to delicately measure them is also discussed in detail. Chapter

9 presents the oxidization effects on 2D germanene and its 3D allotrope, where both binary compounds are outstanding candidates in showing the chemisorption- and dimensionality-enriched quasipaticle phenomena. The critical roles of Ge–O and Ge–Ge bonds will be completely clarified from the specific multi-orbital hybridizations in terms of the fundamental properties and potential applications [272]. The rare-earth metal adatoms on a germanene surface is capable of tailoring the unique configurations of ferromagnetic spin distributions through the 14 *4f/5f*-orbitals (f_{z^3}, f_{xz^2}, $f_{yz^2}$$f_{xyz}$, $f_{z(x^2-y^2)}$, $f_{x(x^2-3y^2)}$, $f_{y(y^2-3x^2)}$) [273] with the significant magnetic moments. In addition to the spin-enriched physical environments and electronic properties, the magneto-optical transitions are expected to exhibit the unusual absorption structures. Whether the spin-split optical spectra could be clearly observed within the VASP simulations is test in the presence/absence of excitonic effects. Certain important differences Compared with transition metal chemisorptions of Chapter 8, certain important differences obviously illustrate the diverse spin configurations.

Very interesting, Chapter 11 is mainly focused on the heterojunction effects, mainly arising from the creation of complicated chemical bondings at the boundary of two different subsystems [274], e.g., the clear evidences for germanene grown on Al(111) through the MBE synthesis [274]. The suitable substrate width is delicately examined using the VASP calculations, being consistent with the necessary requirements of the high-precision STS and ARPES measurements (details in Chapter 3). Three kinds of multi-orbital hybridizations, which survive in Ge–Ge, Ge–Al and Al–Al bonds, are fully identified from a Moire superlattice, the featured energy spectra & atom dominances, the highly nonuniform charge density distributions, and the atom- & orbital-decomposed van Hove singularities. Such concise mechanisms are greatly diversified by the distinct 2D group-IV/group-V and substrates, so that they are responsible for the various measurements. The composite bilayer group-IV material, silicene & germanene compound, is chosen in Chapter 12 to explore the critical roles of the enhanced interlayer atomic interactions (impure orbital hybridizations). Their close relations with the enlarged unit cells under the various commensurate couplings, the buckled structures, the spin-orbital couplings, and the intralayer sp^3 bondings need to be clarified from all the calculated results. Compared with graphene-germanene compounds, more complicated environment are expected to strongly modify and thus diversify the low-energy essential properties. The curvature effects will be obviously shown in germanene nanotubes through the delicate analyses of Chapter 13. The significant couplings of the cylindrical surface and the buckled structures can lead to the great enhancement of the sp^3-orbital hybridizations in neighboring Ge atoms. The periodical boundary condition is thoroughly examined in explaining the featured energy spectra, e.g., the sensitive dependences of band gaps on the nanotube radii and chiral angles. The intralayer and interlayer atomic interactions, with four-orbital mis-orientations, are proposed to illustrate the featured results. The important difference among germanium, silicon and carbon nanotubes are explored in details from the charge- and spin-related viewpoints. On the other side, germanium nanoribbons in Chapter 14 can display the quantum-confinement effects (open boundaries), being fully assisted by the edges structures. Whether the essential properties could be divided into three-width types is required to test for achiral armchair systems by the accurate VASP simulations. Furthermore, the antiferromagnetic spin configurations on the specific zigzag edges is one of the studying focuses. The diversified quasiparticle behaviors should be obviously observed in group-IV nanoribbons and nanotubes. The rich surface morphology of a germanium nanoribbon, as indicated in Chapter 15, is easily decorated with the distinct adatoms because of dangling bonds at honeycomb lattices and chiral/achiral edges. The composite effects, which principally come from the finite width, the open edges, and the buckled structure, account for the main features of quasiparticle properties. Their prominent relations with the unusual multi-orbital mixings and the spin-related interactions are accurately achieved from all the consistently physical quantities.

Chapter 16 fully explores the diversified quasiparticle phenomena in ternary RE-T-silicon/-germanium intermetallic materials [275,276], mainly owing to the strong coupling effects of orbital- and spin-dependent interactions. Such systems cover fourteen-*f*-, 10-*d*-, and sp^3-chemical bondings and the intrinsic magnetic moments of two formers. Very complicated chemical and physical environments are capable of creating the various multi-orbital hybridizations and the specific ferromagnetic spin distributions simultaneously, leading to the diverse features of geometric, electronic and magnetic properties. Certain important differences among the distinct compounds are illustrated in detail, i.e., the composite effects due to the distinct rare-earth and transition-metal atoms. The theoretical predictions are thoroughly examined by the previously experimental measurements. A 3D Li-Ge-O ternary compound, as thoroughly investigated in Chapter 17, frequently serves as electrolyte of Li^+-based batteries. How to directly link a Moire superlattice, electronic properties and optical excitations is the studying focus. Both Li—O and Ge—O bonds, with significant fluctuations in their lengths, are available in fully exploring the featured valence and conduction bands, the spatial deformations of the atom-related orbitals, and the merged van Hove singularities of the orbital-decomposed density of states. The critical pictures of the multi-orbital hybridizations are achieved by the delicate analyses, being very useful for specifying each prominent absorption peak in the presence/absence of excitonic effects. Chapter 18 is mainly conducted on the standard solar-cell materials of $CsGeI_3$-related ternary perovskite systems [67,83,260,277]. After the VASP optimization, their crystal structures clearly show the unusual bonding environments and thus account for the rich quasiparticle behaviors of electronic properties and optical excitations (with many-body effects). Whether the critical mechanisms of unusual multi-orbital hybridizations in Ge—I and Cs—I bonds is available in qualitatively comprehending the high light-electricity conversion efficiency is investigated thoroughly. Moreover, the diverse phenomena, which are driven by the distinct germanium-based ternary 3D compounds, are thoroughly examined through the accurate analyses. Chapter 19 study an antimony film, which is known to exhibit topological surface states depending on the thickness of the film. The density functional calculation is used to investigated the electronic structure of the four-bilayer Sb film and find that adsorptions of impurity atoms on the film. Chapter 20 presents a free-standing four-bilayer Sb (111) film exists a large energy gap due to the quantum tunneling between the two surfaces.

Concluding remarks in Chapter 21 present the theoretical and experimental progresses for germanium-based and germanium-related 3D/2D/1D materials. Their strong cooperation, as well as, the close relation between two approaches of simulations and phenomenological models, are clearly illustrated through the up-to-date development. Very extensive researches have provided the diverse physical/chemical/material properties, being very useful in exploring the full quasiparticle framework [40,64]. This book will greatly modify the previous strategies, extend the current quasiparticle framework [40,64], and open the near-future issues. In addition to the basic sciences, the commercial and potential applications are thoroughly illustrated in Chapter 22, especially for the semiconductor industry. Such main-stream-main systems will play the critical roles in the next-generation nano-devices, such as, various functional chips. The unsolved and emergent research topics are clearly presented in Chapter 23. Certain of them will be worthy of the systematic investigations, such as, the germanium-/silicon-related solar cells [278–281], batteries [61,282–285] and intermetallics [286–288], and the reliable establishment of the tight-binding models/the generalized tight binding model in the absence/presence of external fields. Finally, the concise problems, which are closely related to the main-stream materials and the suitable calculation methods, are proposed in Chapter 24. Specifically, they are designed for the junior researchers and the graduate-level physics courses, being consistent with the quasiparticle behaviors in Chapters 4–20.

References

[1] Bhimanapati GR, Lin Z, Meunier V, Jung Y, Cha J, Das S, Xiao D, Son Y, Strano MS, Cooper VR, Liang L, Louie SG, Ringe E, Zhou W, Kim SS, Naik RR, Sumpter BG, Terrones H, Xia F, Wang Y, Zhu J, Akinwande D, Alem N, Schuller JA, Schaak RE, Terrones M, Robinson JA. Recent advances in two-dimensional materials beyond graphene. ACS Nano 2015;9:11509–39.

[2] Zhu FF, Chen WJ, Xu Y, Gao CL, Guan DD, Liu CH, Qian D, Zhang SC, Jia JF. Epitaxial growth of two-dimensional stanene. Nat Mater 2015;14:1020–5.

[3] Lin C-Y, Wu J-Y, Chiu C-W, Lin M-F. Coulomb excitations and decays in graphene-related systems. CRC Press; 2019.

[4] Molle A, Grazianetti C, Tao L, Taneja D, Alam MH, Akinwande D. Silicene, silicene derivatives, and their device applications. Chem Soc Rev 2018;47:6370–87.

[5] Ni Z, Liu Q, Tang K, Zheng J, Zhou J, Qin R, Gao Z, Yu D, Lu J. Tunable bandgap in silicene and germanene. Nano Lett 2012;12:113–8.

[6] Wang Y, Zheng J, Ni Z, Fei R, Liu Q, Quhe R, Xu C, Zhou J, Gao Z, Lu J. Half-metallic silicene and germanene nanoribbons: towards high-performance spintronics device. Nano 2012;7:1250037.

[7] Roome NJ, Carey JD. Beyond graphene: stable elemental monolayers of silicene and germanene. ACS Appl Mater Interfac 2014;6:7743–50.

[8] Monshi MM, Aghaei SM, Calizo I. Edge functionalized germanene nanoribbons: impact on electronic and magnetic properties. RSC Adv 2017;7:18900–8.

[9] Meng L, Wang Y, Zhang L, Du S, Wu R, Li L, Zhang Y, Li G, Zhou H, Hofer WA, Gao HJ. Buckled silicene formation on Ir(111). Nano Lett 2013;13:685–90.

[10] Lee KW, Lee CE. Spin-orbit coupling-induced band inversion and spin Chern insulator phase in plumbene and stanene. Curr Appl Phys 2020;20:413–8.

[11] Vacacela Gomez C, Pisarra M, Gravina M, Riccardi P, Sindona A. Plasmon properties and hybridization effects in silicene. Phys Rev B 2017;95:85419.

[12] Lin S-Y, Liu H-Y, Nguyen DK, Tran NTT, Pham HD, Chang S-L, Lin C-Y, Lin M-F. Silicene-based layered materials. IOP Publishing; 2020.

[13] Liu HY, Lin SY, Wu JY, Wu JY. Stacking-configuration-enriched essential properties of bilayer graphenes and silicenes. J Chem Phys 2020;153.

[14] Cornet C, Charbonnier S, Lucci I, Chen L, Létoublon A, Alvarez A, Tavernier K, Rohel T, Bernard R, Rodriguez JB, Cerutti L, Tournié E, Léger Y, Bahri M, Patriarche G, Largeau L, Ponchet A, Turban P, Bertru N. Zinc-blende group III-V/group IV epitaxy: importance of the miscut. Phys Rev Mater 2020;4.

[15] Hillerich K, Dick KA, Wen CY, Reuter MC, Kodambaka S, Ross FM. Strategies to control morphology in hybrid group III-V/group IV heterostructure nanowires. Nano Lett 2013;13:903–8.

[16] Pumera M, Sofer Z. 2D monoelemental arsenene, antimonene, and bismuthene: beyond black phosphorus. Adv Mater 2017;29.

[17] Hsu CH, Ozolins V, Chuang FC. First-principles study of Bi and Sb intercalated graphene on SiC(0001) substrate. Surf Sci 2013;616:149–54.

[18] Zheng X, Zhang JF, Tong B, Du RR. Epitaxial growth and electronic properties of few-layer stanene on InSb (1 1 1). 2D Mater 2020;7:011001.

[19] Kim K, Sussman A, Zettl A. Graphene nanoribbons obtained by electrically unwrapping carbon nanotubes. ACS Nano 2010;4:1362–6.

[20] Ong SP, Wang L, Kang B, Ceder G. Li - Fe - P - O2 phase diagram from first principles calculations. Chem Mater 2008;20:1798–807.

[21] Kwak K, Choi W, Tang Q, Kim M, Lee Y, Jiang D, Lee D. A molecule-like PtAu24(SC6H13)18 nanocluster as an electrocatalyst for hydrogen production. Nat Commun 2017;8:14723.

[22] Červenka J, Katsnelson MI, Flipse CFJ. Room-temperature ferromagnetism in graphite driven by two-dimensional networks of pointdefects. Nat Phys 2009;5:840–4.

[23] Zhao H, Deng N, Yan J, Kang W, Ju J, Ruan Y, Wang X, Zhuang X, Li Q, Cheng B. A review on anode for lithium-sulfur batteries: progress and prospects. Chem Eng J 2018;347:343–65.

[24] Lee S-J, Eom A-H, Ryu S, Won J-P. Resistance of an eco-friendly nano-polymer fireproof cementitious composite to physical and chemical environments. Compos Struct 2019;222:110901.

[25] Awater RHP, Dorenbos P. The Bi3+ 6s and 6p electron binding energies in relation to the chemical environment of inorganic compounds. J Lumin 2017;184:221–31.

[26] Yera R, Rossi N, Méndez CG, Huespe AE. Topology design of 2D and 3D elastic material microarchitectures with crystal symmetries displaying isotropic properties close to their theoretical limits. Appl Mater Today 2020;18:100456.

[27] Yang F, Miao L, Wang ZF, Yao MY, Zhu F, Song YR, Wang MX, Xu JP, Fedorov AV, Sun Z, Zhang GB, Liu C, Liu F, Qian D, Gao CL, Jia JF. Spatial and energy distribution of topological edge states in single Bi(111) bilayer. Phys Rev Lett 2012;109:16801.

[28] Du H, Tu M, Luo S, Liu Y, Qiu X, Lu H, Li S, Yuan S, Huang W, Jie W, Hao J. Reversible transition between bipolar resistive switching and threshold switching in 2D layered III–VI semiconductor GaSe. Appl Phys Lett 2020;116:253102.

[29] Chen HH, Su SH, Chang SL, Cheng BY, Chen SW, Chen HY, Lin MF, Huang JCA. Tailoring low-dimensional structures of bismuth on monolayer epitaxial graphene. Sci Rep 2015;5.

[30] Senkovskiy BV, Fedorov AV, Haberer D, Farjam M, Simonov KA, Preobrajenski AB, Mårtensson N, Atodiresei N, Caciuc V, Blügel S, Rosch A, Verbitskiy NI, Hell M, Evtushinsky DV, German R, Marangoni T, van Loosdrecht PHM, Fischer FR, Grüneis A. Semiconductor-to-Metal transition and quasiparticle renormalization in doped graphene nanoribbons. Adv Electron Mater 2017;3:1600490.

[31] Shyu FL, Lin MF. Π plasmons in two-dimensional arrays of aligned carbon nanotubes. Phys Rev B Condens Matter 1999;60:14434–40.

[32] Umemoto Y, Sugawara K, Nakata Y, Takahashi T, Sato T. Pseudogap, Fermi arc, and Peierls-insulating phase induced by 3D–2D crossover in monolayer VSe2. Nano Res 2019;12:165–9.

[33] Averyanov DV, Sokolov IS, Platunov MS, Wilhelm F, Rogalev A, Gargiani P, Valvidares M, Jaouen N, Parfenov OE, Taldenkov AN, Karateev IA, Tokmachev AM, Storchak VG. Competing magnetic states in silicene and germanene 2D ferromagnets. Nano Res 2020;13:3396–402.

[34] Fayazi M, Liu B, Lei L, Shuai G, Odunmbaku O, Wang S, He Y, Boi FS. Ferromagnetic hysteresis and structural recrystallization in turbostratic graphite. Mater Res Express 2019;6:105612.

[35] Hattendorf S, Georgi A, Liebmann M, Morgenstern M. Networks of ABA and ABC stacked graphene on mica observed by scanning tunneling microscopy. Surf Sci 2013;610:53–8.

[36] Ohta T, Bostwick A, McChesney JL, Seyller T, Horn K, Rotenberg E. Interlayer interaction and electronic screening in multilayer graphene investigated with angle-resolved photoemission spectroscopy. Phys Rev Lett 2007;98:206802.

[37] Lee CH, Chen SC, Yang CK, Su WS, Lin MF. Low-energy electronic structures of nanotube-nanoribbon hybrid systems. Comput Phys Commun 2011;182:68–70.

[38] Bhattacharyya S, Singh AK. Lifshitz transition and modulation of electronic and transport properties of bilayer graphene by sliding and applied normal compressive strain. Carbon N. Y 2016;99:432–8.

[39] Coletti C, Forti S, Principi A, Emtsev KV, Zakharov AA, Daniels KM, Daas BK, Chandrashekhar MVS, Macdonald AH, Polini M, Starke U. Revealing the electronic band structure of quasi-free trilayer graphene on SiC(0001). Mater Res Soc Symp Proc 2014;1693:155439.

[40] Lin C-Y, Ho C-H, Wu J-Y, Do T-N, Shih P-H, Lin S-Y, Lin M-F. In: Lin C-Y, Ho C-H, Wu J-Y, Do T-N, H Shih P-, Lin S-Y, Lin M-F, editors. Diverse quantization phenomena in layered materials. CRC Press; 2019.
[41] de la Figuera J, González M, García-Martínez R, Rojo J, Hernán O, Vázquez de Parga A, Miranda R. STM characterization of extended dislocation configurations in Au(001). Phys Rev B Condens Matter 1998;58:1169–72.
[42] Zhou SY, Gweon GH, Fedorov AV, First PN, De Heer WA, Lee DH, Guinea F, Castro Neto AH, Lanzara A. Substrate-induced bandgap opening in epitaxial graphene. Nat Mater 2007;6:770–5.
[43] Stéphan O, Taverna D, Kociak M, Suenaga K, Henrard L, Colliex C. Dielectric response of isolated carbon nanotubes investigated by spatially resolved electron energy-loss spectroscopy: from multiwalled to single-walled nanotubes. Phys Rev B Condens Matter 2002;66:1–6.
[44] Masubuchi S, Ono M, Yoshida K, Hirakawa K, MacHida T. Fabrication of graphene nanoribbon by local anodic oxidation lithography using atomic force microscope. Appl Phys Lett 2009;94:82107.
[45] Liu Z, Suenaga K, Harris PJF, Iijima S. Open and closed edges of graphene layers. Phys Rev Lett 2009;102:15501.
[46] Katkov MV, Sysoev VI, Gusel'Nikov AV, Asanov IP, Bulusheva LG, Okotrub AV. A backside fluorine-functionalized graphene layer for ammonia detection. Phys Chem Chem Phys 2015;17:444–50.
[47] Wang Z, Ratvik AP, Grande T, Selbach SM. Diffusion of alkali metals in the first stage graphite intercalation compounds by vdW-DFT calculations. RSC Adv 2015;5:15985–92.
[48] Feng J, Qi L, Huang JY, Li J. Geometric and electronic structure of graphene bilayer edges. Phys Rev B Condens Matter 2009;80:165407.
[49] Chang SL, Lin SY, Lin SK, Lee CH, Lin MF. Geometric and electronic properties of edge-decorated graphene nanoribbons. Sci Rep 2014;4:1031–9.
[50] De Jong S, Van Heumen E, Thirupathaiah S, Huisman R, Massee F, Goedkoop JB, Ovsyannikov R, Fink J, Dürr HA, Gloskovskii A, Jeevan HS, Gegenwart P, Erb A, Patthey L, Shi M, Follath R, Varykhalov A, Golden MS. Droplet-like Fermi surfaces in the anti-ferromagnetic phase of EuFe 2As2, an Fe-pnictide superconductor parent compound. EPL 2010;89:27007.
[51] Zhang J, Xiao J, Meng X, Monroe C, Huang Y, Zuo JM. Free folding of suspended graphene sheets by random mechanical stimulation. Phys Rev Lett 2010;104:166805.
[52] Gülseren O, Yildirim T, Ciraci S. Systematic ab initio study of curvature effects in carbon nanotubes. Phys Rev B Condens Matter 2002;65:1–4.
[53] Lin YT, Lin SY, Chiu YH, Lin MF. Alkali-created rich properties in graphene nanoribbons: chemical bondings. Sci Rep 2017;7.
[54] Thess A, Lee R, Nikolaev P, Dai H, Petit P, Robert J, Xu C, Lee YH, Kim SG, Rinzler AG, Colbert DT, Scuseria GE, Tománek D, Fischer JE, Smalley RE. Crystalline ropes of metallic carbon nanotubes. Science 1996;273:483–7.
[55] Gorjizadeh N, Farajian AA, Esfarjani K, Kawazoe Y. Spin and band-gap engineering in doped graphene nanoribbons. Phys Rev B Condens Matter 2008;78:155427.
[56] Corso M, Carbonell-Sanromà E, De Oteyza DG. Bottom-up fabrication of atomically precise graphene nanoribbons. arXiv 2017;466:470–3.
[57] Xie L, Wang H, Jin C, Wang X, Jiao L, Suenaga K, Dai H. Graphene nanoribbons from unzipped carbon nanotubes: atomic structures, Raman spectroscopy, and electrical properties. J Am Chem Soc 2011;133:10394–7.
[58] Lee G, Cho K. Electronic structures of zigzag graphene nanoribbons with edge hydrogenation and oxidation. Phys Rev B Condens Matter 2009;79:165440.
[59] García-Vidal FJ, Pitarke JM. Optical absorption and energy-loss spectra of aligned carbon nanotubes. Eur Phys J B 2001;22:257–65.
[60] Park C, Crooks RE, Siochi EJ, Harrison JS, Evans N, Kenik E. Adhesion study of polyimide to single-wall carbon nanotube bundles by energy-filtered transmission electron microscopy. Nanotechnology 2003;14:L11.
[61] Wang Y, Song S, Xu C, Hu N, Molenda J, Lu L. Development of solid-state electrolytes for sodium-ion battery—A short review. Nano Mater Sci 2019;1:91–100.
[62] Büyükyazi M, Mathur S. 3D nanoarchitectures of α-$LiFeO_2$ and α-LiFeO2/C nanofibers for high power lithium-ion batteries. Nano Energy 2015;**13** 28:35.
[63] Lin S-Y, Liu H-Y, Nguyen DK, Tran NTT, Pham HD, Chang S-L, Lin C-Y, Lin M-F. Silicene—, germanene— and stanene—graphene composites Silicene-Based Layered Materials. IOP Publishing); 2020.
[64] Chen SC, Wu JY, Lin CY, Lin MF. Theory of magnetoelectric properties of 2D systems. IOP Publishing); 2017.

[65] Lin SY, Tran NTT, Fa-Lin M. Structure- and adatom-enriched essential properties of graphene nanoribbons. CRC press; 2018.

[66] Tran NTT, Lin SY, Lin CY, Lin MF. Geometric and electronic properties of graphene-related systems: chemical bonding schemes. CRC Press; 2017.

[67] Sugawara E, Nikaido H. Properties of AdeABC and AdeIJK efflux systems of Acinetobacter baumannii compared with those of the AcrAB-TolC system of *Escherichia coli*. Antimicrob Agents Chemother 2014;58:7250–7.

[68] Cuk T, Lu DH, Zhou XJ, Shen ZX, Devereaux TP, Nagaosa N. A review of electron-phonon coupling seen in the high-Tc. Superconductors by angle-resolved photoemission studies (ARPES). Phys Status Solidi Basic Res 2005;242:11–29.

[69] Mohr M, Maultzsch J, Dobardžić E, Reich S, Milošević I, Damnjanović M, Bosak A, Krisch M, Thomsen C. Phonon dispersion of graphite by inelastic x-ray scattering. Phys Rev B Condens Matter 2007;76:35439.

[70] Richter K, Waldmann T, Paul N, Jobst N, Scurtu RG, Hofmann M, Gilles R, Wohlfahrt-Mehrens M. Low-Temperature charging and aging mechanisms of Si/C composite anodes in Li-ion batteries: an operando neutron scattering study. ChemSusChem 2020;13:529–38.

[71] Kugler M, Renner C, Fischer MV, Batey G. A 3He refrigerated scanning tunneling microscope in high magnetic fields and ultrahigh vacuum. Rev Sci Instrum 2000;71:1475–8.

[72] Kuzmenko AB, Crassee I, Van Der Marel D, Blake P, Novoselov KS. Determination of the gate-tunable band gap and tight-binding parameters in bilayer graphene using infrared spectroscopy. Phys Rev B Condens Matter 2009;80:165406.

[73] Li J, Lei T, Wang J, Wu R, Zhao J, Zhao L, Guo Y, Qian H, Ibrahim K. Anisotropic electronic structure and interfacial chemical reaction of stanene/Bi2Te3. J Phys Chem C 2020;124:4917–24.

[74] Liu D, Chu XQ, Lagi M, Zhang Y, Fratini E, Baglioni P, Alatas A, Said A, Alp E, Chen SH. Studies of phononlike low-energy excitations of protein molecules by inelastic X-ray scattering. Phys Rev Lett 2008;101:135501.

[75] Berman OL, Gumbs G, Lozovik YE. Magnetoplasmons in graphene structures. Prog Electromagn Res Symp 2008;2:1210–4.

[76] Ho JH, Lu CL, Hwang CC, Chang CP, Lin MF. Coulomb excitations in AA- and AB-stacked bilayer graphites. Phys Rev B Condens Matter 2006;74:85406.

[77] Lin C-Y, Wu J-Y, Chiu C-W, Lin M-F. Coulomb excitations and decays in graphene-related systems. 2019.

[78] Chen RB, Lee CH, Chang CP, Lin MF. Electronic and optical properties of finite carbon nanotubes in an electric field. Nanotechnology 2007;18:508–15.

[79] Khuong Dien V, Thi Han N, Nguyen TDH, Huynh TMD, Pham HD, Lin MF. Geometric and electronic properties of Li_2GeO_3. Front Mater 2020;7.

[80] Behzad S, Chegel R. Magnetic field-induced splitting of optical spectra in silicon nanotubes: tight binding calculations. Silicon 2016;8:43–55.

[81] Lee RS, Kim HJ, Fischer JE, Thess A, Smalley RE. Conductivity enhancement in single-walled carbon nanotube bundles doped with K and Br. Nature 1997;388:255–7.

[82] Berdiyorov GR, Neek-Amal M, Peeters FM, Van Duin ACT. Stabilized silicene within bilayer graphene: a proposal based on molecular dynamics and density-functional tight-binding calculations. Phys Rev B Condens Matter 2014;89:24107.

[83] Fujiwara H, Fujimoto S, Tamakoshi M, Kato M, Kadowaki H, Miyadera T, Tampo H, Chikamatsu M, Shibata H. Determination and interpretation of the optical constants for solar cell materials. Appl Surf Sci 2017;421:276–82.

[84] Papagno M, Rusponi S, Sheverdyaeva PM, Vlaic S, Etzkorn M, Pacilé D, Moras P, Carbone C, Brune H. Large band gap opening between graphene Dirac cones induced by Na adsorption onto an Ir superlattice. ACS Nano 2012;6:199–204.

[85] Ruckert T, Keune W, Sahoo B, Sturhahn W, Toellner TS, Alp EE, Röhlsberger R. Atomic vibrational dynamics of thin films studied by nuclear resonant inelastic X-ray scattering: amorphous Tb1-xFex alloys. Hyperfine Interact 2002;144–145:65–76.

[86] Friedman AL, van 't Erve OMJ, Robinson JT, Whitener KE, Jonker BT. Hydrogenated graphene as a homoepitaxial tunnel barrier for spin and charge transport in graphene. ACS Nano 2015;9:6747–55.

[87] He Z, He K, Robertson AW, Kirkland AI, Kim D, Ihm J, Yoon E, Lee GD, Warner JH. Atomic structure and dynamics of metal dopant pairs in graphene. Nano Lett 2014;14:3766–72.

[88] Magda GZ, Jin X, Hagymási I, Vancsó P, Osváth Z, Nemes-Incze P, Hwang C, Biró LP, Tapasztó L. Room-temperature magnetic order on zigzag edges of narrow graphene nanoribbons. Nature 2014;514:608–11.
[89] Lai L, Lu J, Wang L, Luo G, Zhou J, Qin R, Chen Y, Li H, Gao Z, Li G, Mei WN, Maeda Y, Akasaka T, Sanvito S. Magnetism in carbon nanoscrolls: quasi-half-metals and half-metals in pristine hydrocarbons. Nano Res 2009;2:844–50.
[90] Bihlmayer G, Sassmannshausen J, Kubetzka A, Blügel S, von Bergmann K, Wiesendanger R. Plumbene on a magnetic substrate: a combined scanning tunneling microscopy and density functional theory study. Phys Rev Lett 2020;124:126401.
[91] Amelinckx S, Lucas A, Lambin P. Electron diffraction and microscopy of nanotubes. Rep Prog Phys 1999;62:1471–524.
[92] Lee J-K, Kim J-G, Hembram KPSS, Kim Y-I, Min B-K, Park Y, Lee J-K, Moon DJ, Lee W, Lee S-G, John P. The nature of metastable AA′ graphite: low dimensional nano- and single-crystalline forms. Sci Rep 2016;6:39624.
[93] Wang Y, Ding Y. Mechanical and electronic properties of stoichiometric silicene and germanene oxides from first-principles. Phys Status Solidi Rapid Res Lett 2013;7:410–3.
[94] Hussain I, Tran HP, Jaksik J, Moore J, Islam N, Uddin MJ. Functional materials, device architecture, and flexibility of perovskite solar cell. Emergent Mater 2018;1:133–54.
[95] Lin J, Raji ARO, Nan K, Peng Z, Yan Z, Samuel ELG, Natelson D, Tour JM. Iron oxide nanoparticle and graphene nanoribbon composite as an anode material for high-performance li-ion batteries. Adv Funct Mater 2014;24:2044–8.
[96] Bostwick A, Ohta T, Seyller T, Horn K, Rotenberg E. Quasiparticle dynamics in graphene. Nat Phys 2007;3:36–40.
[97] Dubois SMM, Zanolli Z, Declerck X, Charlier JC. Electronic properties and quantum transport in Graphene-based nanostructures. Eur Phys 2009;72:1–24.
[98] Tapasztó L, Dobrik G, Lambin P, Biró LP. Tailoring the atomic structure of graphene nanoribbons by scanning tunnelling microscope lithography. Nat Nanotechnol 2008;3:397–401.
[99] Cervantes-Sodi F, Csányi G, Piscanec S, Ferrari AC. Edge-functionalized and substitutionally doped graphene nanoribbons: electronic and spin properties. Phys Rev B Condens Matter 2008;77:165427.
[100] Singh K, Raj B. Temperature-dependent modeling and performance evaluation of multi-walled CNT and single-walled CNT as global interconnects. J Electron Mater 2015;44:4825–35.
[101] Lin CH, Huang A, Pai WW, Chen WC, Chen TY, Chang TR, Yukawa R, Cheng CM, Mou CY, Matsuda I, Chiang TC, Jeng HT, Tang SJ. Single-layer dual germanene phases on Ag(111). Phys Rev Mater 2018;2:24003.
[102] Dávila ME, Le Lay G. Few layer epitaxial germanene: a novel two-dimensional Dirac material. Sci Rep 2016;6:20714.
[103] Saito R, Dresselhaus G, Dresselhaus MS. Physical properties of carbon nanotubes. Published by Imperial College Press and Distributed by World Scientific Publishing Co.); 1998.
[104] Li M, Hong M, Tang X, Sun Q, Lyu W-Y, Xu S-D, Kou L-Z, Dargusch M, Zou J, Chen Z-G. Crystal symmetry induced structure and bonding manipulation boosting thermoelectric performance of GeTe. Nano Energy 2020;73:104740.
[105] Ko W, Hus SM, Li X, Berlijn T, Nguyen GD, Xiao K, Li A-P. Tip-induced local strain on MoS2/graphite detected by inelastic electron tunneling spectroscopy. Phys Rev B 2018;97:125401.
[106] Strocov VN, Schmitt T, Flechsig U, Schmidt T, Imhof A, Chen Q, Raabe J, Betemps R, Zimoch D, Krempasky J, Wang X, Grioni M, Piazzalunga A, Patthey L. High-resolution soft X-ray beamline address at the swiss light source for resonant inelastic X-ray scattering and angle-resolved photoelectron spectroscopies. J Synchrotron Radiat 2010;17:631–43.
[107] De Wild J, Meijerink A, Rath JK, Van Sark WGJHM, Schropp REI. Upconverter solar cells: materials and applications. Energy Environ Sci 2011;4:4835–48.
[108] Dakovski GL, Li Y, Durakiewicz T, Rodriguez G. Tunable ultrafast extreme ultraviolet source for time- and angle-resolved photoemission spectroscopy. Rev Sci Instrum 2010;81:73108.
[109] Axmann P, Stinner C, Wohlfahrt-Mehrens M, Mauger A, Gendron F, Julien CM. Nonstoichiometric $LiFePO_4$: defects and related properties. Chem Mater 2009;21:1636–44.
[110] Tan LB, Cotlet O, Bergschneider A, Schmidt R, Back P, Shimazaki Y, Kroner M, İmamoǧlu A. Interacting polaron-polaritons. Phys Rev 2020;X.

[111] Dai S, Fang W, Rivera N, Stehle Y, Jiang BY, Shen J, Tay RY, Ciccarino CJ, Ma Q, Rodan-Legrain D, Jarillo-Herrero P, Teo EHT, Fogler MM, Narang P, Kong J, Basov DN. Phonon polaritons in monolayers of hexagonal boron nitride. Adv Mater 2019.

[112] Luo C, Guo X, Hu H, Hu D, Wu C, Yang X, Dai Q. Probing polaritons in 2D materials. Adv Opt Mater 2020.

[113] Dinh Dang N, Zelevinsky V. Improved treatment of ground-state correlations: modified random phase approximation. Phys Rev C: Nucl Phys 2001;64:643191−6431910.

[114] Rani P, Dubey GS, Jindal VK. DFT study of optical properties of pure and doped graphene. Phys E Low-Dimens Syst Nanostruct 2014;62:28−35.

[115] Kusakabe K, Wake A, Nagakubo A, Murashima K, Murakami M, Adachi K, Ogi H. Interplanar stiffness in defect-free monocrystalline graphite. Phys Rev Mater 2020;4:043603.

[116] Mohammadi Y, Moradian R, Shirzadi Tabar F. Effects of doping and bias voltage on the screening in AAA-stacked trilayer graphene. Solid State Commun 2014;193:1−5.

[117] Berciu M, Bhatt RN. Spin waves in disordered III-V diluted magnetic semiconductors by a modified RPA approach. Phys Rev B Condens Matter 2002;66:852071−8520713.

[118] Cui Y, Liang F, Xu S, Ding Y, Lin Z, Liu J. Interfacial wetting behaviors of liquid Ga alloys/FeGa 3 based on metallic bond interaction. Coll Surface A Physicochem Eng Asp 2019;569:102−9.

[119] Gusynin VP, Sharapov SG, Carbotte JP. Magneto-optical conductivity in graphene. J Phys Condens Matter 2007;19:26222.

[120] Huang YK, Chen SC, Ho YH, Lin CY, Lin MF. Feature-rich magnetic quantization in sliding bilayer graphenes. Sci Rep 2014;4:7509.

[121] Lin C-Y, Do T-N, Huang Y-K, Lin M-F. Optical properties of graphene in magnetic and electric fields. IOP Publishing, e-book); 2017.

[122] Saroka VA, Shuba MV, Portnoi ME. Optical selection rules of zigzag graphene nanoribbons. arXiv 2017;95:155438.

[123] Liao W, Zhou G, Xi F. Optical properties for armchair-edge graphene nanoribbons. J Appl Phys 2008;104.

[124] Ho YH, Chiu YH, Lin DH, Chang CP, Lin MF. Magneto-optical selection rules in bilayer bernal graphene. ACS Nano 2010;4:1465−72.

[125] Rajasekaran S, Kaya S, Anniyev T, Ogasawara H, Nilsson A. Probing substrate effects in the carbon-projected band structure of graphene on Pt(111) through resonant inelastic x-ray scattering. Phys Rev B Condens Matter 2012;85:45419.

[126] Konschuh S, Gmitra M, Fabian J. Tight-binding theory of the spin-orbit coupling in graphene. Phys Rev B Condens Matter 2010;82:245412.

[127] Lin SY, Chang SL, Thuy Tran NT, Yang PH, Lin MF. H-Si bonding-induced unusual electronic properties of silicene: a method to identify hydrogen concentration. Phys Chem Chem Phys 2015;17:26443−50.

[128] Hsu PJ, Kubetzka A, Finco A, Romming N, Von Bergmann K, Wiesendanger R. Electric-field-driven switching of individual magnetic skyrmions. Nat Nanotechnol 2017;12:123−6.

[129] Volpe M, Cleri F. Chemisorption of atomic hydrogen in graphite and carbon nanotubes. Surf Sci 2003;544:24−34.

[130] Kuzuo R, Terauchi M, Tanaka M. Electron energy-loss spectra of carbon nanotubes. Jpn J Appl Phys 1992;31:L484−1487.

[131] Lobach AS, Shul'ga YM, Roshchupkina OS, Rebrov AI, Perov AA, Morozov YG, Spector VN, Ovchinnikov AA. C60H18, C60H36 and C70H36 fullerene hydrides: study by methods of IR, NMR, XPS, EELS and magnetochemistry. Fullerene Sci Technol 1998;6:375−91.

[132] Iijima S. Microtubules of graphitic carbon. Nature 1991 Helical;354:56−8.

[133] Tomaânek D, Zhong W, Krastev E. Stability of multishell fullerenes. Phys Rev B 1993;48:15461−4.

[134] Yoo EJ, Kim J, Hosono E, Zhou HS, Kudo T, Honma I. Large reversible Li storage of graphene nanosheet families for use in rechargeable lithium ion batteries. Nano Lett 2008;8:2277−82.

[135] Hashimoto A, Suenaga K, Gloter A, Urita K, Iijima S. Direct evidence for atomic defects in graphene layers. Nature 2004;430:870−3.

[136] Geim AK, Novoselov KS. The rise of graphene Nanoscience and technology: a Collection of Reviews from nature journals. World Scientific; 2009. p. 11−9.

[137] Zhuang J, Gao N, Li Z, Xu X, Wang J, Zhao J, Dou SX, Du Y. Cooperative electron-phonon coupling and buckled structure in germanene on Au(111). ACS Nano 2017;11:3553−9.

[138] Pierucci D, Sediri H, Hajlaoui M, Girard JC, Brumme T, Calandra M, Velez-Fort E, Patriarche G, Silly MG, Ferro G, Soulière V, Marangolo M, Sirotti F, Mauri F, Ouerghi A. Evidence for flat bands near the Fermi level in epitaxial rhombohedral multilayer graphene. ACS Nano 2015;9:5432–9.

[139] Egerton RF. Prospects for vibrational-mode EELS with high spatial resolution. Microsc Microanal 2014;20:658–63.

[140] Sato Y, Terauchi M. High-energy resolution electron energy-loss spectroscopy study of interband transitions characteristic to single-walled carbon nanotubes. Microsc Microanal 2014;20:807–14.

[141] Hofmann P. The surfaces of bismuth: structural and electronic properties. Prog Surf Sci 2006;81:191–245.

[142] Pfnür H, Tegenkamp C, Vattuone L. Plasmons in one and two dimensions. Springer Handbooks; 2020. p. 557–84.

[143] Gensterblum G, Li-Ming Y, Pireaux JJ, Thiry PA, Caudano R, Lambin P, Lucas AA, Krätschmer W, Fischer JE. High resolution electron energy loss spectroscopy of epitaxial films of C60 grown on GaSe. J Phys Chem Solid 1992;53:1427–32.

[144] Bobaru S, Gaudry É, De Weerd MC, Ledieu J, Fournée V. Competing allotropes of Bi deposited on the Al13Co 4(100) alloy surface. Phys Rev B Condens Matter 2012;86:214201.

[145] Moritz W, Wang B, Bocquet ML, Brugger T, Greber T, Wintterlin J, Günther S. Structure determination of the coincidence phase of graphene on Ru(0001). Phys Rev Lett 2010;104:136102.

[146] Usachov D, Fedorov A, Otrokov MM, Chikina A, Vilkov O, Petukhov A, Rybkin AG, Koroteev YM, Chulkov EV, Adamchuk VK, Grüneis A, Laubschat C, Vyalikh DV. Observation of single-spin Dirac Fermions at the graphene/ferromagnet interface. Nano Lett 2015;15:2396–401.

[147] Dávila ME, Xian L, Cahangirov S, Rubio A, Le Lay G. Germanene: a novel two-dimensional germanium allotrope akin to graphene and silicene. New J Phys 2014;16:95002.

[148] Erni R, Browning ND. Valence electron energy-loss spectroscopy in monochromated scanning transmission electron microscopy. Ultramicroscopy 2005;104:176–92.

[149] Servanton G, Pantel R. Arsenic dopant mapping in state-of-the-art semiconductor devices using electron energy-loss spectroscopy. Micron 2010;41:118–22.

[150] Xu Z, Bando Y, Liu L, Wang W, Bai X, Golberg D. Electrical conductivity, chemistry, and bonding alternations under graphene oxide to graphene transition as revealed by in situ TEM. ACS Nano 2011;5:4401–6.

[151] Fultz B, Howe JM. Transmission electron microscopy and diffractometry of materials. Springer Science & Business Media; 2008.

[152] Strydom I, Wells A, Ferreira NP. Surface layers on fe formed during plasma nitriding studied by XPS and reflection EELS. Surf Interface Anal 1988;12:211–7.

[153] Diallo PT, Jeanlouis K, Boutinaud P, Mahiou R, Cousseins JC. Improvement of the optical performances of Pr3+ in $CaTiO_3$. J Alloys Compd 2001;323–324:218–22.

[154] Diallo PT, Jeanlouis K, Boutinaud P, Mahiou R, Cousseins JC. Improvement of the optical performances of Pr3+ in $CaTiO_3$. J Alloys Compd 2001;323–324:218–22.

[155] Wolf O, Dasog M, Yang Z, Balberg I, Veinot JGC, Millo O. Doping and quantum confinement effects in single Si nanocrystals observed by scanning tunneling spectroscopy. Nano Lett 2013;13:2516–21.

[156] Sofer Z, Bouša D, Luxa J, Mazanek V, Pumera M. Few-layer black phosphorus nanoparticles. Chem Commun 2016;52:1563–6.

[157] Zaric S, Ostojic GN, Kono J, Shaver J, Moore VC, Strano MS, Hauge RH, Smalley RE, Wei X. Optical signatures of the Aharonov-Bohm phase in single-walled carbon nanotubes. Science 2004;304:1129–31.

[158] Uematsu T, Wajima K, Sharma DK, Hirata S, Yamamoto T, Kameyama T, Vacha M, Torimoto T, Kuwabata S. Narrow band-edge photoluminescence from AgInS2 semiconductor nanoparticles by the formation of amorphous III–VI semiconductor shells. NPG Asia Mater 2018;10:713–26.

[159] Guo F, Lyu Y, Jedrzejczyk MB, Zhao Y, Io WF, Bai G, Wu W, Hao J. Piezoelectric biaxial strain effects on the optical and photoluminescence spectra of 2D III–VI compound α -In 2 Se 3 nanosheets. Appl Phys Lett 2020;116:113101.

[160] Mortimer IB, Nicholas RJ. Role of bright and dark excitons in the temperature-dependent photoluminescence of carbon nanotubes. Phys Rev Lett 2007;98:27404.

[161] Gangopadhyay P, Kesavamoorthy R, Bera S, Magudapathy P, Nair KGM, Panigrahi BK, Narasimhan SV. Optical absorption and photoluminescence spectroscopy of the growth of silver nanoparticles. Phys Rev Lett 2005;94.

[162] Nikolic BK. Quantum transport through mesoscopic disordered interfaces, junctions, and multilayers. J Phys Condens Matter 2002;14:7871–99.

[163] Chiu CW, Shyu FL, Chang CP, Chuu DS, Lin MF. Coulomb scattering rates of excited carriers in moderate-gap carbon nanotubes. Phys Rev B Condens Matter 2006;73:235407.

[164] Hanf MC, Marjaoui A, Stephan R, Zanouni M, Diani M, Sonnet P. Undulated silicene and germanene free-standing layers: why not? J Phys Condens Matter 2020;32.

[165] Fan Y, Liu X, Wang J, Ai H, Zhao M. Silicene and germanene on InSe substrates: structures and tunable electronic properties. Phys Chem Chem Phys 2018;20:11369–77.

[166] Matthes L, Pulci O, Bechstedt F. Massive Dirac quasiparticles in the optical absorbance of graphene, silicene, germanene, and tinene. J Phys Condens Matter 2013;25.

[167] Kumbhakar P, Chowde Gowda C, Mahapatra PL, Mukherjee M, Malviya KD, Chaker M, Chandra A, Lahiri B, Ajayan PM, Jariwala D, Singh A, Tiwary CS. Emerging 2D metal oxides and their applications. Mater Today 2021;45:142–68.

[168] Ogikubo T, Shimazu H, Fujii Y, Ito K, Ohta A, Araidai M, Kurosawa M, Le Lay G, Yuhara J. Continuous growth of germanene and stanene lateral heterostructures. Adv Mater Interfac 2020;7.

[169] Yuhara J, He B, Matsunami N, Nakatake M, Le Lay G. Graphene's latest cousin: plumbene epitaxial growth on a "nano WaterCube". Adv Mater 2019;31:1901017.

[170] Yuhara J, Le Lay G. Beyond silicene: synthesis of germanene, stanene and plumbene. Jpn J Appl Phys 2020;59:SN0801.

[171] Moreau E, Godey S, Wallart X, Razado-Colambo I, Avila J, Asensio MC, Vignaud D. High-resolution angle-resolved photoemission spectroscopy study of monolayer and bilayer graphene on the C-face of SiC. Phys Rev B Condens Matter 2013;88:75406.

[172] Haeni JH, Theis CD, Schlom DG. RHEED intensity oscillations for the stoichiometric growth of SrTiO3 thin films by reactive molecular beam epitaxy. J Electroceram 2000;4:385–91.

[173] Braun W. Applied RHEED reflection high-energy electron diffraction during crystal growth, vol. 154; 1999.

[174] Jałochowski M, Krawiec M. Antimonene on Pb quantum wells. 2D Mater 2019;6.

[175] Bent SF. Organic functionalization of group IV semiconductor surfaces: principles, examples, applications, and prospects. Surf Sci 2002;500(1–3):879–903.

[176] Morozov SV, Jiang D, Zhang Y, Dubonos Sv, Novoselov KS, Geim AK, Morozov SV, Jiang D, Zhang Y, Dubonos Sv, et al. Electric field effect in atomically thin carbon films. Science 2004;306(5296):666–9.

[177] Kiraly B, Iski EV, Mannix AJ, Fisher BL, Hersam MC, Guisinger NP. Solid-source growth and atomic-scale characterization of graphene on Ag(111). Nat Commun 2013;4:2804.

[178] Coraux J, N'Diaye AT, Busse C, Michely T. Structural coherency of graphene on Ir(111). Nano Lett 2008;8:565–70.

[179] Walker LS, Marotto VR, Rafiee MA, Koratkar N, Corral EL. Toughening in graphene ceramic composites. ACS Nano 2011;5:3182–90.

[180] Li L, Lu SZ, Pan J, Qin Z, Wang YQ, Wang Y, Cao GY, Du S, Gao HJ. Buckled germanene formation on Pt(111). Adv Mater 2014;26:4820–4.

[181] Derivaz M, Dentel D, Stephan R, Hanf MC, Mehdaoui A, Sonnet P, Pirri C. Continuous germanene layer on Al(111). Nano Lett 2015;15:2510–6.

[182] Castro Neto AH, Guinea F, Peres NMR, Novoselov KS, Geim AK. The electronic properties of graphene. Rev Mod Phys 2009;81:109–62.

[183] Liu CC, Feng W, Yao Y. Quantum spin Hall effect in silicene and two-dimensional germanium. Phys Rev Lett 2011;107.

[184] Kaloni TP, Schwingenschlögl U. Weak interaction between germanene and GaAs(0001) by H intercalation: a route to exfoliation. J Appl Phys 2013;114.

[185] Kaloni TP, Modarresi M, Tahir M, Roknabadi MR, Schreckenbach G, Freund MS. Electrically engineered band gap in two-dimensional GE, SN, and PB: a first-principles and tight-binding approach. J Phys Chem C 2015;119:11896–902.

[186] Peng Q, Wen X, De S. Mechanical stabilities of silicene. RSC Adv 2013;3:13772–81.

[187] Ngoc Thanh Thuy T, Lin S-Y, Lin C-Y, Lin M-F. Geometric and electronic properties of graphene-related systems. CRC Press; 2017.

[188] Marchini S, Günther S, Wintterlin J. Scanning tunneling microscopy of graphene on Ru(0001). Phys Rev B Condens Matter 2007;76.
[189] Osborn TH, Farajian AA, Pupysheva OV, Aga RS, Lew Yan Voon LC. Ab initio simulations of silicene hydrogenation. Chem Phys Lett 2011;511:101–5.
[190] Gao H, Wang L, Zhao J, Ding F, Lu J. Band gap tuning of hydrogenated graphene: H coverage and configuration dependence. J Phys Chem C 2011;115:3236–42.
[191] Chen W, Li Y, Yu G, Li CZ, Zhang SB, Zhou Z, Chen Z. Hydrogenation: a simple approach to realize semiconductor-half-metal-metal transition in boron nitride nanoribbons. J Am Chem Soc 2010;132:1699–705.
[192] Lin S-Y, Lin Y-T, Tran NTT, Su W-P, Lin M-F. Feature-rich electronic properties of aluminum-adsorbed graphenes. Carbon N. Y 2017;120:209–18.
[193] Nguyen DK, Tran NTT, Chiu YH, Lin MF. Concentration-diversified magnetic and electronic properties of halogen-adsorbed silicene. Sci Rep 2019;9.
[194] Elias DC, Nair RR, Mohiuddin TMG, Morozov SV, Blake P, Halsall MP, Ferrari AC, Boukhvalov DW, Katsnelson MI, Geim AK, Novoselov KS. Control of graphene's properties by reversible hydrogenation: evidence for graphane. Science 2009;323:610–3.
[195] Sahin H, Peeters FM. Adsorption of alkali, alkaline-earth, and 3d transition metal atoms on silicene. Phys Rev B Condens Matter 2013;87.
[196] Karlický F, Kumara Ramanatha Datta K, Otyepka M, Zbořil R. Halogenated graphenes: rapidly growing family of graphene derivatives. ACS Nano 2013;7:6434–64.
[197] Miura Y, Kasai H, Diño WA, Nakanishi H, Sugimoto T. Effective pathway for hydrogen atom adsorption on graphene. J Phys Soc Japan 2003;72:995–7.
[198] Lalmi B, Oughaddou H, Enriquez H, Kara A, Ś V, Ealet B, Aufray B. Epitaxial growth of a silicene sheet. Appl Phys Lett 2010;97:223109.
[199] Tran NTT, Nguyen DK, Glukhova OE, Lin MF. Coverage-dependent essential properties of halogenated graphene: a DFT study. Sci Rep 2017;7:1–13.
[200] Nguyen DK, Tran NTT, Chiu YH, Gumbs G, Lin MF. Rich essential properties of Si-doped graphene Scientific Reports, vol. 10. CRC Press; 2020. p. 83–107.
[201] Shih P-H, Do T-N, Gumbs G, Huang D, Pham TP, Lin M-F. Magneto-transport properties of B-, Si-and N-doped graphene. Carbon N. Y 2020;160:211–8.
[202] Bianchi M, Rienks EDL, Lizzit S, Baraldi A, Balog R, Hornekær L, Hofmann P. Electron-phonon coupling in potassium-doped graphene: angle-resolved photoemission spectroscopy. Phys Rev B Condens Matter 2010;81:41403.
[203] Bisti F, Profeta G, Vita H, Donarelli M, Perrozzi F, Sheverdyaeva PM, Moras P, Horn K, Ottaviano L. Electronic and geometric structure of graphene/SiC(0001) decoupled by lithium intercalation. Phys Rev B Condens Matter 2015;91:245411.
[204] Su Y, Liang X, Li S, Chen Y, Zhou Q, Yin S, Meng X, Kong M. Self-catalytic VLS growth and optical properties of single-crystalline GeO_2 nanowire arrays. Mater Lett 2008;62:1010–3.
[205] Terakado N, Tanaka K. Photo-induced phenomena in sputtered GeO_2 films. J Non-Cryst Solids 2005;351:54–60.
[206] Liu QJ, Liu ZT, Feng LP, Tian H. First-principles study of structural, elastic, electronic and optical properties of rutile GeO_2 and α-quartz GeO_2. Solid State Sci 2010;12:1748–55.
[207] Walker B, Dharmawardhana CC, Dari N, Rulis P, Ching WY. Electronic structure and optical properties of amorphous GeO_2 in comparison to amorphous SiO_2. J Non-Cryst Solids 2015;428:176–83.
[208] Ramana CV, Troitskaia IB, Gromilov SA, Atuchin VV. Electrical properties of germanium oxide with α-quartz structure prepared by chemical precipitation. Ceram Int 2012;38:5251–5.
[209] Franz M. Solid-state physics: U-turns strictly prohibited. Nature 2010;466:323–4.
[210] Journet C, Maser WK, Bernier P, Loiseau A, Lamy de la Chapelle M, Lefrant S, Deniard P, Lee R, Fischer JE. Large-scale production of single-walled carbon nanotubes by the electric-arc technique. Nature 1997;388:756–8.
[211] Yudasaka M, Komatsu T, Ichihashi T, Iijima S. Single-wall carbon nanotube formation by laser ablation using double-targets of carbon and metal. Chem Phys Lett 1997;278:102–6.
[212] Karthikeyan S, Mahalingam P, Karthik M. Large scale synthesis of carbon nanotubes. E-J Chem 2009;6:1–12.
[213] Ebbesen TW, Hiura H, Fujita J, Ochiai Y, Matsui S, Tanigaki K. Patterns in the bulk growth of carbon nanotubes. Chem Phys Lett 1993;209:83–90.

[214] Seraphin S, Zhou D, Jiao J, Withers JC, Loutfy R. Effect of processing conditions on the morphology and yield of carbon nanotubes. Carbon N. Y 1993;31:685–9.

[215] Terrones M. Science and technology of the twenty-first century: synthesis, properties, and applications of carbon nanotubes. Annu Rev Mater Res 2003;33:419–501.

[216] Yue GZ, Qiu Q, Gao B, Cheng Y, Zhang J, Shimoda H, Chang S, Lu JP, Zhou O. Generation of continuous and pulsed diagnostic imaging x-ray radiation using a carbon-nanotube-based field-emission cathode. Appl Phys Lett 2002;81:355–7.

[217] Baughman RH, Zakhidov AA, De Heer WA. Carbon nanotubes - the route toward applications. Science 2002;297:787–92.

[218] Crompton KR, Hladky MP, Park HH, Prokes SM, Love CT, Landi BJ. Lithium-ion cycling performance of multi-walled carbon nanotube electrodes and current collectors coated with nanometer scale Al_2O_3 by atomic layer deposition. Electrochim Acta 2018;292:628–38.

[219] Frackowiak E, Béguin F. Electrochemical storage of energy in carbon nanotubes and nanostructured carbons. Carbon N. Y 2002;40:1775–87.

[220] Memarian F, Fereidoon A, Khodaei S, Mashhadzadeh AH, Ganji MD. Molecular dynamic study of mechanical properties of single/double wall SiCNTs: consideration temperature, diameter and interlayer distance. Vacuum 2017;139:93–100.

[221] Chen S, Chen P, Wang Y. Carbon nanotubes grown in situ on graphene nanosheets as superior anodes for Li-ion batteries. Nanoscale 2011;3:4323–9.

[222] Li X, Meng G, Xu Q, Kong M, Zhu X, Chu Z, Li AP. Controlled synthesis of germanium nanowires and nanotubes with variable morphologies and sizes. Nano Lett 2011;11:1704–9.

[223] Chegel R, Behzad S. Bandstructure modulation for Si-h and Si-g nanotubes in a transverse electric field: tight binding approach. Superlattice Microst 2013;63:79–90.

[224] Ahmadi N, Shokri AA, Elahi SM. Optical transition of zigzag silicon nanotubes under intrinsic curvature effect. Silicon 2016;8:217–24.

[225] Chegel R, Behzad S. Electronic properties of SiNTs under external electric and magnetic fields using the tight-binding method. J Electron Mater 2014;43:329–40.

[226] Zhang RQ, Lee HL, Li WK, Teo BK. Investigation of possible structures of silicon nanotubes via density-functional tight-binding molecular dynamics simulations and ab initio calculations. J Phys Chem B 2005;109:8605–12.

[227] Seifert G, Köhler T, Urbassek HM, Hernández E, Frauenheim T. Tubular structures of silicon. Phys Rev B Condens Matter 2001;63.

[228] Yang X, Ni J. Electronic properties of single-walled silicon nanotubes compared to carbon nanotubes. Phys Rev B Condens Matter 2005;72.

[229] Bayani AH, Dideban D, Akbarzadeh M, Moezi N. Benchmarking performance of a gate-all-around germanium nanotube field effect transistor (GAA-GeNTFET) against GAA-CNTFET. ECS J Solid State Sci Technol 2017;6:M24–8.

[230] Trukhin AN, Rogulis U, Spingis M. Self-trapped exciton in Li_2GeO_3. J Lumin 1997;72–74:890–2.

[231] Bhalla AS, Cross LE, Vedam K. Unusual white light conoscopic figure in single crystal lithium metagermanate Li_2GeO_3. Appl Opt 1978;17:33391.

[232] Haussühl S, Liebertz J, Stähr S. Single crystal growth and pyroelectric, dielectric, piezoelectric, elastic, and thermoelastic properties of orthorhombic Li_2SiO_3, Li_2GeO_3, and Na_2GeO_3. Cryst Res Technol 1982;17:521–6.

[233] Liebert BE, Huggins RA. Ionic conductivity of Li_4GeO_4, Li_2GeO_3 and $Li_2Ge_7O_{15}$. Mater Res Bull 1976;11:533–8.

[234] Rahman MM, Sultana I, Yang T, Chen Z, Sharma N, Glushenkov AM, Chen Y. Lithium germanate (Li_2GeO_3): a high-performance anode material for lithium-ion batteries. Angew Chem Int Ed 2016;55:16059–63.

[235] Han NT, Dien VK, Thuy Tran NT, Nguyen DK, Su WP, Lin MF. First-principles studies of electronic properties in lithium metasilicate (Li_2SiO_3). RSC Adv 2020;10:24721–9.

[236] Han NT, Dien VK, Lin MF. Excitonic effects in the optical spectra of Li_2SiO_3 compound. Sci Rep 2021;11.

[237] Lewis NS. Research opportunities to advance solar energy utilization. Science 2016:351.

[238] Rong Y, Hu Y, Mei A, Tan H, Saidaminov MI, Seok SI, McGehee MD, Sargent EH, Han H. Challenges for commercializing perovskite solar cells. Science 2018;361.

[239] Wang R, Mujahid M, Duan Y, Wang ZK, Xue J, Yang Y. A review of perovskites solar cell stability. Adv Funct Mater 2019;29.

[240] Lewis NS. Toward cost-effective solar energy use. Science 2007;315:798–801.
[241] Atwater HA, Polman A. Plasmonics for improved photovoltaic devices. Mater Sustain Energy A Collect Peer-Reviewed Res Rev Artic from Nat Publ Gr 2010:3–11.
[242] Chopra KL, Paulson PD, Dutta V. Thin-film solar cells: an overview. Prog Photovoltaics Res Appl 2004;12:69–92.
[243] Polman A, Knight M, Garnett EC, Ehrler B, Sinke WC. Photovoltaic materials: present efficiencies and future challenges. Science 2016:352.
[244] Yang G, Tao H, Qin P, Ke W, Fang G. Recent progress in electron transport layers for efficient perovskite solar cells. J Mater Chem 2016;4:3970–90.
[245] Pham HD, Yang TCJ, Jain SM, Wilson GJ, Sonar P. Development of dopant-free organic hole transporting materials for perovskite solar cells. Adv Energy Mater 2020;10.
[246] Grånäs O, Vinichenko D, Kaxiras E. Establishing the limits of efficiency of perovskite solar cells from first principles modeling. Sci Rep 2016;6:1–6.
[247] Fujiwara H, Kato M, Tamakoshi M, Miyadera T, Chikamatsu M. Optical characteristics and operational principles of hybrid perovskite solar cells. Phys Status Solidi Appl Mater Sci 2018;215:1700730.
[248] Gregg BA. Excitonic solar cells. J Phys Chem B 2003;107:4688–98.
[249] Rahul, Singh PK, Singh R, Singh V, Bhattacharya B, Khan ZH. New class of lead free perovskite material for low-cost solar cell application. Mater Res Bull 2018;97:572–7.
[250] Yusoff ARBM, Nazeeruddin MK. Organohalide lead perovskites for photovoltaic applications. J Phys Chem Lett 2016;7:851–66.
[251] Lemanov VV, Sotnikov AV, Smirnova EP, Weihnacht M, Kunze R. Perovskite $CaTiO_3$ as an incipient ferroelectric. Solid State Commun 1999;110:611–4.
[252] Dutta DP, Jayakumar OD, Tyagi AK, Girija KG, Pillai CGS, Sharma G. Effect of doping on the morphology and multiferroic properties of BiFeO3 nanorods. Nanoscale 2010;2:1149–54.
[253] Yang JC, He Q, Suresha SJ, Kuo CY, Peng CY, Haislmaier RC, Motyka MA, Sheng G, Adamo C, Lin HJ, Hu Z, Chang L, Tjeng LH, Arenholz E, Podraza NJ, Bernhagen M, Uecker R, Schlom DG, Gopalan V, Chen LQ, Chen CT, Ramesh R, Chu YH. Orthorhombic $BiFeO_3$. Phys Rev Lett 2012;109.
[254] Huang LY, Lambrecht WRL. Electronic band structure, phonons, and exciton binding energies of halide perovskites CsSnCl3, CsSnBr3, and CsSnI3. Phys Rev B Condens Matter 2013;88.
[255] Huang LY, Lambrecht WRL. First-principles calculations of phonons and Raman spectra in monoclinic CsSnCl3. Phys Rev B Condens Matter 2015;91.
[256] Shen ZX, Loo WL, Kuok MH, Tang SH. High pressure phase transition studies of CsSnCl3. J Mol Struct 1994;326:73–80.
[257] Raj A, Kumar M, Singh PK, Chandra Singh R, Bherwani H, Gupta A, Anshul A. A computational approach to investigate the suitable ETL for lead-free CsGeI3 based perovskite solar cell. Mater Today Proc 2021.
[258] Kopacic I, Friesenbichler B, Hoefler SF, Kunert B, Plank H, Rath T, Trimmel G. Enhanced performance of germanium halide perovskite solar cells through compositional engineering. ACS Appl Energy Mater 2018;1:343–7.
[259] Ming W, Shi H, Du MH. Large dielectric constant, high acceptor density, and deep electron traps in perovskite solar cell material CsGeI3. J Mater Chem 2016;4:13852. 8.
[260] Jonga U-G, Yu C-J, Kye Y-H, Choe Y-G, Wei H, Li S. Structural and optoelectronic properties of the inorganic perovskites AGeX3 (A = Cs, Rb; X = I, Br, Cl) for solar cell application arXiv Prepr. arXiv1810.02543. 2018.
[261] Vuckovic S, Song S, Kozlowski J, Sim E, Burke K. Density functional analysis: the theory of density-corrected DFT. J Chem Theor Comput 2019;15:6636–46.
[262] Lan TN, Shee A, Li J, Gull E, Zgid D. Testing self-energy embedding theory in combination with GW. Phys Rev B 2017;96:155106.
[263] Bucher E. Solar cell materials and their basic parameters. Appl Phys 1978;17:1–26.
[264] Ostroverkhova O. Organic optoelectronic materials: mechanisms and applications. Chem Rev 2016;116:13279–412.
[265] Pazoki M, Edvinsson T. Metal replacement in perovskite solar cell materials: chemical bonding effects and optoelectronic properties. Sustain Energy Fuels 2018;2:1430–45.
[266] Bian G, Wang X, Liu Y, Miller T, Chiang TC. Interfacial protection of topological surface states in ultrathin Sb films. Phys Rev Lett 2012;108.

[267] Kresse G, Hafner J. Ab initio molecular dynamics for liquid metals. Phys Rev B 1993;47(1):558.
[268] Kresse G, FURTHMÜLLER J. Efficient iterative schemes for ab initio total-energy calculations using a plane-wave basis set. Phys Rev B 1996;54(16):11169.
[269] Abe JO, Popoola API, Ajenifuja E, Popoola OM. Hydrogen energy, economy and storage: review and recommendation. Int J Hydrog Energy 2019;44(29):15072–15086.
[270] Liao M-S, Scheiner S. Electronic structure and bonding in metal phthalocyanines, metal= Fe, Co, Ni, Cu, Zn, Mg. J Chem Phys 2001;114(22):9780–9791.
[271] Neroni A, Sasıoğlu E, Hadipour H, Friedrich C, Blügel S, Mertig I, et al. First-principles calculation of the effective on-site Coulomb interaction parameters for Sr_2ABO_6 (A= Cr, Mn, Fe, Co, Ni, and B= Mo, W) double perovskites. Phys Rev B 2019;100(11):115113.
[272] Vaughn II, DD, Schaak RE. Synthesis, properties and applications of colloidal germanium and germanium-based nanomaterials. Chem Soc Rev 2013;42(7):2861–2879.
[273] Dolg M, Stoll H, Preuss H. Energy-adjusted abinitio pseudopotentials for the rare earth elements. J Chem Phys 1989;90(3):1730–1734.
[274] Derivaz M, Dentel D, Stephan R, Hanf MC, Mehdaoui A, Sonnet P, et al. Continuous germanene layer on Al (111). Nano Lett 2015;15(4):2510–2516.
[275] Bugaris DE, Malliakas CD, Han F, Calta NP, Sturza M, Krogstad MJ, Osborn R, Rosenkranz S, Ruff JPC, Trimarchi G, Bud'ko SL, Balasubramanian M, Chung DY, Kanatzidis MG. Charge density wave in the new polymorphs of RE2Ru3Ge5 (RE = Pr, Sm, Dy). J Am Chem Soc 2017;139:4130–43.
[276] Morozkin AV, Seropegin YD. Sm-Ru-Ge system at 1070 K. J Alloy Compd 2004;365:168–72.
[277] Benrekia AR, Benkhettou N, Nassour A, Driz M, Sahnoun M, Lebègue S. Structural, electronic and optical properties of cubic $SrTiO_3$ and KTaO3: ab initio and GW calculations. Phys B Condens Matter 2012;407:2632–6.
[278] Yang Z, Hao J. Recent progress in 2D layered III–VI semiconductors and their heterostructures for optoelectronic device applications. Adv Mater Technol 2019;4:1900108.
[279] Goodenough JB, Park K-S. The Li-ion rechargeable battery: a perspective. J Am Chem Soc 2013;135:1167–76.
[280] Lee J, Tian WC, Wang WL, Yao DX. Two-dimensional pnictogen honeycomb lattice: structure, on-site spin-orbit coupling and spin polarization. Sci Rep 2015;5:11512.
[281] Zhao J, Zeng H, Yao G. Computational design of a polymorph for 2D III-V orthorhombic monolayers by first principles calculations: excellent anisotropic, electronic and optical properties. Phys Chem Chem Phys 2021;23:3771–8.
[282] Li L, Jacobs R, Gao P, Gan L, Wang F, Morgan D, Jin S. Origins of large voltage hysteresis in high-energy-density metal fluoride lithium-ion battery conversion electrodes. J Am Chem Soc 2016;138:2838–48.
[283] Wang L, He X, Sun W, Wang J, Li Y, Fan S. Crystal orientation tuning of LiFePO4 nanoplates for high rate lithium battery cathode materials. Nano Lett 2012;12:5632–6.
[284] Lin MC, Gong M, Lu B, Wu Y, Wang DY, Guan M, Angell M, Chen C, Yang J, Hwang BJ, Dai H. An ultrafast rechargeable aluminium-ion battery. Nature 2015;520:325–8.
[285] Nitta N, Wu F, Lee JT, Yushin G. Li-ion battery materials: present and future. Mater Today 2015;18:252–64.
[286] Lekhal A, Benkhelifa FZ, Mecabih S, Abbar B, Bouhafs B. Structural and electronic properties of non-magnetic intermetallic YAuX (X = Ge and Si) in hexagonal and cubic phases. Bull Mater Sci 2016;39:195–200.
[287] Sasaki T, Kato T, Fukushima J, Hayashi Y, Takizawa H. High-pressure synthesis and crystal structure of a novel intermetallic compound Mn(Al,Ge)5. J Alloys Compd 2019;806:58–62.
[288] Saha P, Datta MK, Velikokhatnyi OI, Manivannan A, Alman D, Kumta PN. Rechargeable magnesium battery: current status and key challenges for the future. Prog Mater Sci 2014;66:1–86.

CHAPTER

2

Quasiparticle framework

Vo Khuong Dien[1], *Shih-Yang Lin*[1], *Chi-Hsuan Lee*[1], *Hsin-Yi Liu*[1], *Hai Duong Pham*[1], *Thi My Duyen Huynh*[1], *Nguyen Thi Han*[1], *Ngoc Thanh Thuy Tran*[2], *Thi Dieu Hien Nguyen*[1], *Wei-Bang Li*[1] *and Ming-Fa Lin*[1,2]

[1]Department of Physics, National Cheng Kung, University; [2]Hierarchical Green-Energy Materials (Hi-GEM) Research Center, National Cheng Kung University, Tainan, Taiwan

Two approaches of theoretical developments, first-principles simulations [1,2] and phenomenological models [3,4], will be unified together under the same quasiparticle framework [3] using the full support of each other. The former is very suitable for the germanene-/germanium-based emergent materials (details in Chapters 4–20) in identifying the diverse physical/chemical/material phenomena. Their geometric, electronic, magnetic and optical properties are thoroughly explored to clearly illustrate the various orbital hybridizations and spin-dependent configurations. The concise mechanisms are very useful in further understanding the intrinsic interactions for neighboring atoms and thus developing the tight-binding model/the generalized tight-binding model in the absence/presence of field perturbations [5,6]. This is worthy of near-future systematic investigations. The simultaneous developments of different methods will be finished under the thorough analyses.

2.1 The first-principles simulations

The accurate VASP simulations are available in fully exploring the optimal crystal structures, wave-vector-dependent band structures & atom-decomposed wave functions, spatial charge density distributions, atom-/orbital-/spin-projected density of states, net magnetic moments, spin configurations and optical absorption peaks in the presence/absence of excitonic effects. All the relevant quantities are consistent with one another in achieving the

Fundamental Physicochemical Properties of Germanene-related Materials
https://doi.org/10.1016/B978-0-443-15801-8.00012-8

concise pictures of critical mechanisms. This strategy could be generalized to a lot of mainstream materials with the unusual quasiparticle phenomena. On the other side, such numerical calculations are very difficult in dealing with very large Moiré superlattices due to the external-field modulations and the chemical modifications.

2.1.1 Optimal crystal

Using first-principle simulations, this book is depicted to theoretical framework to combine various physical quantities. The succinct physical/chemical/material phenomena are exposed to comprehend geometric, electronic, magnetic and optical properties in the 3D germanium crystal/2D germanene/1D germanium nanotubes/germanium nanoribbons, bilayer germanene systems with different stacking configurations and buckling, chemical adsorption- and substituted-enriched germanium compounds, alkalizations and hydrogenations, transition-metal and rare-earth-created ferromagnetic materials, the heterojunction-generated more complicated Moire superlattices, the composite bilayer crystals, the battery electrolytes of Li_2GeO_3, the $CsGeI_3$ perovskite solar cell, and few-layer antimony systems.

Germanium- and germanene-based materials are studied as the model material to critical mechanisms with the unusual quasiparticle phenomena. Computer modeling and calculations are necessary methods in basic science research. Numerical simulations demand to resolve the energy spectra and wave functions of quasiparticles in condensed-matter systems exactly and simultaneously. All quasiparticles frequently experience specific scattering events through wave pocket forms, e.g., the electron-crystal-potential, electron–electron Coulomb interactions and hole-electron pair interactions. It is very difficult to deal with a Schrödinger equation [7] with single-/many-particle interactions. Some approximate models have been developed to obtain trustworthy solutions. Among them, first-principles simulations are an accurate method in the quantum-mechanical calculations [8] of periodic systems.

So far, the first-principles simulations become a dominant method in getting the quantum states of periodic systems and thus reliable geometric, electronic, magnetic and optical properties. Such software packages as the Vienna ab initio simulation (VASP) [9], WIEN2k [10], EXCITING [11], Quantum Espresso [12] are very efficient for studying the featured physical properties. Especially, VASP package is very powerful to calculate an approximate solution to the many-body Schrödinger equation within the density functional theory (DFT) [1] by solving the Kohn– Sham equations [13]. All quasiparticle properties, such as energy bands, electronic and optical excitations, screening charges and effective Coulomb potentials, are confirmed mainly by their spins and charges. Most of these could be resolved using the static and time-dependent first-principles methods [14]. The calculated results consist the optimal geometries, energy bands, wave-vector-dependent wave functions, spatial charge densities, van Hove singularities in the density of states, net magnetic moments, and ferromagnetic, antiferromagnetic and nonmagnetic spin configurations, optical responses. In addition, other numerical methods are very powerful for specific topics, e.g., the molecular dynamics (MD) [15] approach in simulating dynamic growth and structural transformation, and quantum Monte Carlo [16] approaches in determining the charge-density-induced exchange and correlation interactions of emergent materials.

In this book, we have exhibited the density functional theory (DFT) method via the Vienna Ab-initio Simulation Package (VASP) to perform the optimization of the structural crystals for the fundamental properties of Germanene-related materials at zero temperature. The projector augmented wave (PAW) method [17] was utilized for characterization of the ion and valence electron interactions. The convergence condition of the ground-state is set to be 10^{-8} eV between two consecutive simulation steps and all atoms could fully relax during the geometric optimization until the Hellmann-Feynman force acting on each atom was smaller than 0.01 eV. The Perdew-Burke-Ernzerhof (PBE) [18] generalized gradient approximation was used for the exchange-correlation functional. Specifically, the spin-dependent single-/many-particle interactions need to be taken into account for the significant materials and adatoms, such as the spin–orbit coupling [19] in C/Ge/Sn/Pb-monolayer systems. The Hubbard-like onsite Coulomb interactions might play important roles in determining the total ground state energies, optimal geometries, band structures, Van Hove singularities, atom- and boundary-induced magnetic configurations, magnetic moments and optical absorption peaks in the presence/absence of excitonic effects. For instance, monolayer germanene, a vacuum space of 15 Å is inserted between periodic images to avoid the interactions of neighboring cells. The cutoff energy of the Block wave functions, which are built from complete plane waves, is set to be ~500 eV. The first Brilouin zone is sampled in a Gamma scheme [20] along the two-dimension periodic direction by 20 × 20 × 1 k-points for structure relaxations, and by 200 × 200 × 1 for accurate evalutions on electronic properties of two atoms in unit cell. To get the optimal geometry with the lowest ground state energy, all the atoms of a condensed-matter system will experience relaxation processes by adjusting their positions. Then, the most stable configuration is utilized as an initial input for further simulations of the physical properties. The sketching chart of the first-principles calculations is exposed in Fig. 2.1.

2.1.2 Electronic properties

To date, first-principles calculations have proven effective in solving very complex intrinsic interactions under a series of approximations and simplifications to obtain eigenvalues and eigenfunctions thus obtain reliable geometric, electronic, and magnetic properties. Most prominently, the Vienna ab initio simulation package (VASP) [21,22] computes an approximation to the multi-body Schrodinger equation, using density functional theory (DFT), solve the Kohn-Sham [23] and Roothaan equations [24], or in the Hartree-Fock approximation [25]. Besides, a number of other effective methods can be applied to the specific topic, for example, molecular dynamics [26] and the quantum Monte Carlo approach [27], in the simulation of dynamic growth and structural transformation and exchange induced by charge density and interaction correlation of the supernatant.

In this book, based on the quasiparticle framework and multi-orbital hybridizations in chemical bonds, a systematic study including the geometric symmetries, band structures/wave packets, spatial charge densities, and atom- & orbital-decomposed can Hove singularities and the strong optical responses are totally investigated in various systems as stacking configuration (mono/bilayer germanene), quantum confinement boundaries (hydrogenation), nanotube, adatom absorptions, host-atom substitutions, heterojunction (substrate effect), and plenty of functionalized germanene systems. All rich essential properties of germanene-related materials at zero temperature are thoroughly explored by DFT which implemented VASP code. Besides, the exchange and correlation energies were calculated with the use of the Perdew–Burke–Ernzerhof (PBE) functional under the generalized gradient approximation [28]. The projector augmented wave approach (PAW) [29] is available in characterizing the electron-ion crystal potential. In particular, spin-dependent single- and multiple-particle interactions (spin polarization and spin-orbital coupling) need to be taken into account for important materials [30,31].

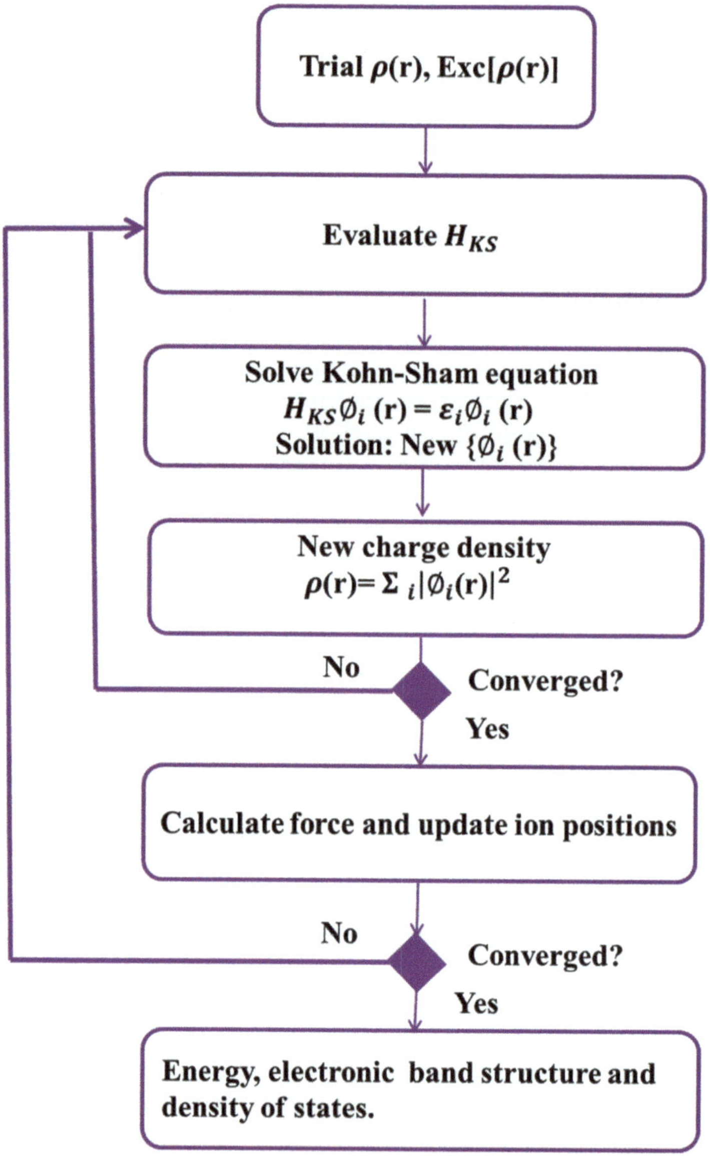

FIG. 2.1 The sketching chart of the first-principles calculations.

For 1D (nanoribbon) and 2D (mono/bilayer) germanene [32,33], a vacuum space of 15 Å is added between the periodic lattices to avoid the interaction of neighboring cells. To reach the optimal geometry with the lowest ground-state energy, entire atoms of a condensed matter system would undergo a delicate relaxation process by adjusting their positions. And then, the most stable configuration is used as the initial input for further calculations of the underlying physical properties. Clearly, the treatment of orbital hybridization and analysis sophisticated spin configuration will determine the type of electronic (finite-gap/zero-gap semiconductors, semimetals, or metals) and magnetic properties (ferromagnetism, antiferromagnetism, or non-magnetism) that the material possesses. This work will be even more difficult when we have to deal with the chemical bonding in a system with nonuniform environments. Such chemical bondings play an important role in fundamental properties, giving rise to rich and unique electronic structures and geometries. In addition to orbital hybridization, spin density configurations are important for generating diverse magnetic configurations. By using spin-polarized calculations, magnetic behavior can be distinguished using spin split band structure, net magnetic moment, spin density distribution, and spin decay DOS.

In the task boosts germanene become one of the leading potential materials in the field of nanotechnology, enhancing the free carrier density and opening the band gap is considered an urgent mission to improve the practical applications. Various methods have been proposed to modify the intrinsic feature of materials including chemical adatom adsorption [34], guest atom substitutions [35], defects/vacancies [36], quantum confinement and edge structures [37], heterojunctions [38], gate voltages [39], magnetic fields [40], and mechanical strain [41]. Among them, chemical modifications are very effective in generating completely different condensed matter systems of pristine lattice through host-guest multi-orbital hybridizations and significant atom-induced spin distributions. For example, chemisorption of alkali, hydrogen, and oxygen atoms in hollow, top, and bridge sites on graphene surfaces [42–44] present the complex orbital-hybridizations in form of 1s-2p_z, sp^3-sp^3 and 1s-sp^3. The first type of hybridization barely changes the honeycomb lattice and σ bondings, but the rigid blue-shift of the Fermi level (creates high-density conduction electrons). However, the other two types cause buckling crystals thus creating destroyed or serious distortion of Dirac-cone bands, finite gaps, more energy dispersion and more van Hove singularities with distinct forms, especially for high concentration absorption cases.

The position-dependent chemical bonds, namely the fluctuations of the bond lengths, create a highly heterogeneous environment that is responsible for the important orbital hybridization and thus determines the fundamentally different properties. Apparently, the active surface environment of the material can be easily achieved in graphene, silicene, germanene and tinene systems due to possessing diverse dangling bonds [45,46]. Apparently, such 2D material properties are mainly determined by the outermost p_z orbitals. The chemical modification could slightly be modified or even totally destroyed π-bondings, consequently and the creation of the single-/multi-orbital hybridizations often appears in specific chemical bonds. We have had many previous systematic studies about chemical modifications in both graphene and silicene systems [47–49]. For example, when boron, carbon, and nitrogen are substituted for silicene, systems can significantly diversify the geometric, electronic, and magnetic properties of silicene-related binary compounds, since the guest-atoms can provide three, four, and five valence electrons. For the two Bo-/C-substituted silicene fields, the magnetic spin configuration completely disappears through the well-defined and σ chemical bonds. In contrast, the third case exhibits a four-orbit complex hybridization and a semimetallic/semiconductor ferromagnetic configuration under certain nitrogen substitution conditions. Of the three types of guest atoms, the bond length relative to carbon/boron is the shortest/longest and, therefore, corresponds to the lowest/highest ground state energy (the most stable systems). Furthermore, the C, B, and N guest atom substitutions produce a diverse electronic band structure near the Fermi level that exhibits semiconductor, p-type, and semimetallic, semiconductor behaviors, respectively.

2.1.3 Magnetic properties

The magnetic properties, covering spin configurations of electron/hole quasiparticles in the various group-IV-related materials, are greatly diversified by the number of layers, configuration of stackings, chemisorption of adatoms, guest atom substitutions, and hetero junctions. Based on the VASP calculations, the relativistic spin-orbital couplings are characterized by the single-particle form, and the spin-created electron-electron Coulomb interactions are evaluated within the Hubbard-like model using the mean-field approach (the standard strength U; [50,51]).

As for all VASP calculations on magnetic systems, spin spirals may be conveniently modeled using a generalization of the Bloch condition. The initial magnetic moments order of the magnetic subsystem will be ferromagnetic in case one does not specify, and even if the ground state of your system is not ferromagnetic, the initial ferromagnetic state might very well be a local minimum in which the system can remain stuck during the electronic minimization. Generally speaking, spin-spiral calculations for the magnetic subsystem is not completely characterized by the magnetic configuration within the unit cell, but will additionally depend on how the magnetization density changes from one unit cell to the next as determined by the spiral propagation vector.

The noncolinear calculation in the latest version of VASP can provide much detail information about the spin axis orientation. This option works only for PAW potentials and might not supported by ultrasoft pseudopotentials. If spin-orbit coupling is not included in the calculations, the ground state energy does not depend on the direction of the magnetic moment, i.e., rotating all magnetic moments for each element by the same angle results in principle exactly in the same energy. Hence, there is no need to define the spin quantization axis, as long as spin-orbit coupling is not included. However, spin-orbit coupling couples the spin to the crystal structure.

The ferromagnetic, antiferromagnetic, and nonmagnetic spin configurations are thoroughly studied through the spin-split energy band, the existence of magnetic moments, the spin density distributions, and the spin-decomposed density of states. It is noticed that the former two are distinguished from the spin distributions. The spin configurations are one of the studying focuses, which are driven by the host atoms, guest ones, edge structures, or hetero junctions of substrates. The spin-induced essential properties could be further verified by the spin-polarized ARPES [52–55], STM [56–59], STS [60,61], optical spectroscopies [54,62], and quantum transport apparatuses [63]. In addition, the spin configurations could also be studied by the phenomenological Hubbard model (the tight-binding model with the on-site spin-dependent Coulomb interactions). For instance, the antiferromagnetic configurations induced by the boundary-carbons of zigzag graphene nanoribbons are exhibited in the spin-degenerate energy bands and density of states.

2.1.4 Optical transitions in the presence/absence of excitonic effects

The special optical properties, which appear in various condensed-matter materials, are frequently utilized in the highly potential applications, e.g., optical sensors [64], modulators [65] and solar cells [66]. The main characteristics are clearly revealed in the optical gap (threshold excitation frequency), prominent absorption structures, absorption reflection coefficient and energy loss function with charge screenings. The position-dependent chemical environment can create the strongly anisotropic excitation spectra, being sensitive to the directions of electric polarizations. Moreover, the many-body effects, the excitonic ones [67,68], might play an important role in determining the featured optical properties, e.g., the red-shift of initial absorption frequencies due to the stable excions. How to use the first-principles results for examining the critical orbital hybridizations is the main studying focus. The Germanium and its related materials are chosen as the typical systems to test the developed theoretical framework.

A condensed-matter system is easily modulated by a time- and position-dependent electromagnetic field, in which there exist the significant couplings of between charges and propagating waves (electrons and photons). The carrier dynamic and wave dispersion relation, which are, respectively, solved by quantum scattering theories and the Maxwell equations, determine all the features of optical properties. Electrons are vertically promotion from the occupied states to the unoccupied ones with the same wave vector, since momentum of photons are usually insignificant compare with the electron one (the long wavelength limit). Such carriers effectively screen the perturbed electric field, create the induced current density, and thus established the electric polarization. The charge screening ability is fully characterized by the transverse dielectric function (a dimensionless $\varepsilon(q \rightarrow 0, \omega)$), as done for the macroscopic dielectric constant due to the static Coulomb-field perturbation [69]. That is, its definition is the specific ration of the electric fields before and after the charge screenings under the (momentum, frequency) spaces.

As a result, its real and imaginary parts, $Re[\varepsilon(\omega)]$ and $Im[\varepsilon(\omega)]$, can define all the optical physical quantities through the long wavelength limit (the further discussions; [70]). Very obviously, how to calculate and get the suitable and reliable dielectric function in the absence/presence of many-body effects should be the most important focus.

To evaluate the complex dielectric function, experimental measurements and theoretical calculations present the different manners. The former is able to examine the frequency-dependent absorption/reflection/transmission spectra, the measured results could be supported by the following Kramers-Kronig relations to get the real- and imaginary-part response functions simultaneously.

Eq. (2.1); relations using $Re[\varepsilon - 1]$ & $Im[\varepsilon - 1]$

$$Re\frac{\varepsilon(\omega)}{\varepsilon_0} = 1 + \frac{2}{\pi}P\int_0^\infty \frac{Im\frac{\varepsilon(\omega')}{\varepsilon_0}}{\omega' - \omega}d\omega' \tag{2.1}$$

$$Im\frac{\varepsilon(\omega)}{\varepsilon_0} = -\frac{1}{\pi}P\int_0^\infty \frac{\left[Re\frac{\varepsilon(\omega')}{\varepsilon_0} - 1\right]}{\omega' - \omega}d\omega'$$

Any physical responses come to exist after the external perturbations. therefore, they are well-behaved functions of excitation frequencies in the presence of deexcitation mechanisms [47,70]. While the absorption frequency is further regarded as a complex variable, the real- and imaginary-part response functions are examined to be associated with each other. That is, doing the principle-value integration of $Re[\varepsilon -1]/Im[\varepsilon -1]$ can achieve $Im[\varepsilon -1]/Re[\varepsilon -1]$. Furthermore, the $\int_0^\infty$ integration is enough because of the even/odd property in $Re[\varepsilon -1]/Im[\varepsilon -1]$.It should be noticed that very wide frequency ranges are required in the various optical spectroscopies. This might induce the main drawbacks for estimating the frequency-dependent dielectric function. How to overcome such frequent issues could utilize the full assistance of the empirical models, e.g., the high-frequency plasmon-mode approximation [71].

The dynamic Kubo formula [72], without/with the many-body effects, is frequently utilized to explore the optical excitations. To satisfy the carrier dynamics under an electromagnetic wave, the Canonical momentum of electron consists of the kinetic and magnetic-field ones, in which the latter is proportional to field strength and charge quantity [73]. The dominating Hamiltonian perturbation belongs to the electric-dipole moment. Its square of the inner product between the final and initial states will dominate the scattering probability of each available transition channel and thus even determine the well-behaved optical selection rules [47,74]. In addition to absorption intensity, the joint density of states, being generated by the occupied and unoccupied electronic states, represents the available number of excitation channels. These two critical factors are responsible for the dielectric function under the single-particle pictures [47,74], as revealed in the previously many researches. But for semiconducting or insulation materials, the excitonic effects come to exist very often under the suitable environment, e.g., the great reduce of threshold excitation frequency in a wide-gap system [75,76], and certain prominent absorption peaks at specific frequencies lower than the single-particle ones [76]. Excitons are stable/meta-stable bound states of the excited electrons and holes, since there exist the significant attractive Coulomb potential between them. Their binding strength is rather sensitive to the charge screenings of the other carriers. The creation of two-particle states and the dielectric screenings need to be included in all the scattering Feynman diagrams [77]. This corresponds to the standard Bethe-Salpeter equation [78], as done in the current VASP calculations (the flowing chart in Fig. 2.2).

The various optical properties are clearly characterized under the 4 Maxwell equations of electric & magnetic fields and the related boundary conditions [69]. Under the Coulomb gauge of $\nabla . A = 0$, the electromagnetic and Coulomb fields, respectively, possess the transverse and longitudinal electric components. As for condensed-matter systems without free current densities and magnetic configurations, the significant couplings between an electromagnetic wave and charge carriers are well described through the Helmhortz equation

$$c^2\nabla^2 E = \frac{\partial^2 D}{\partial t^2} \tag{2.2}$$

Apparently, this equation corresponds to the strong dependences of position vectors and time. By the further Fourier transforms into the (momentum, frequency)-space, one can achieve the dispersion relation

$$c^2k^2 = \varepsilon(\omega)\omega^2 \tag{2.3}$$

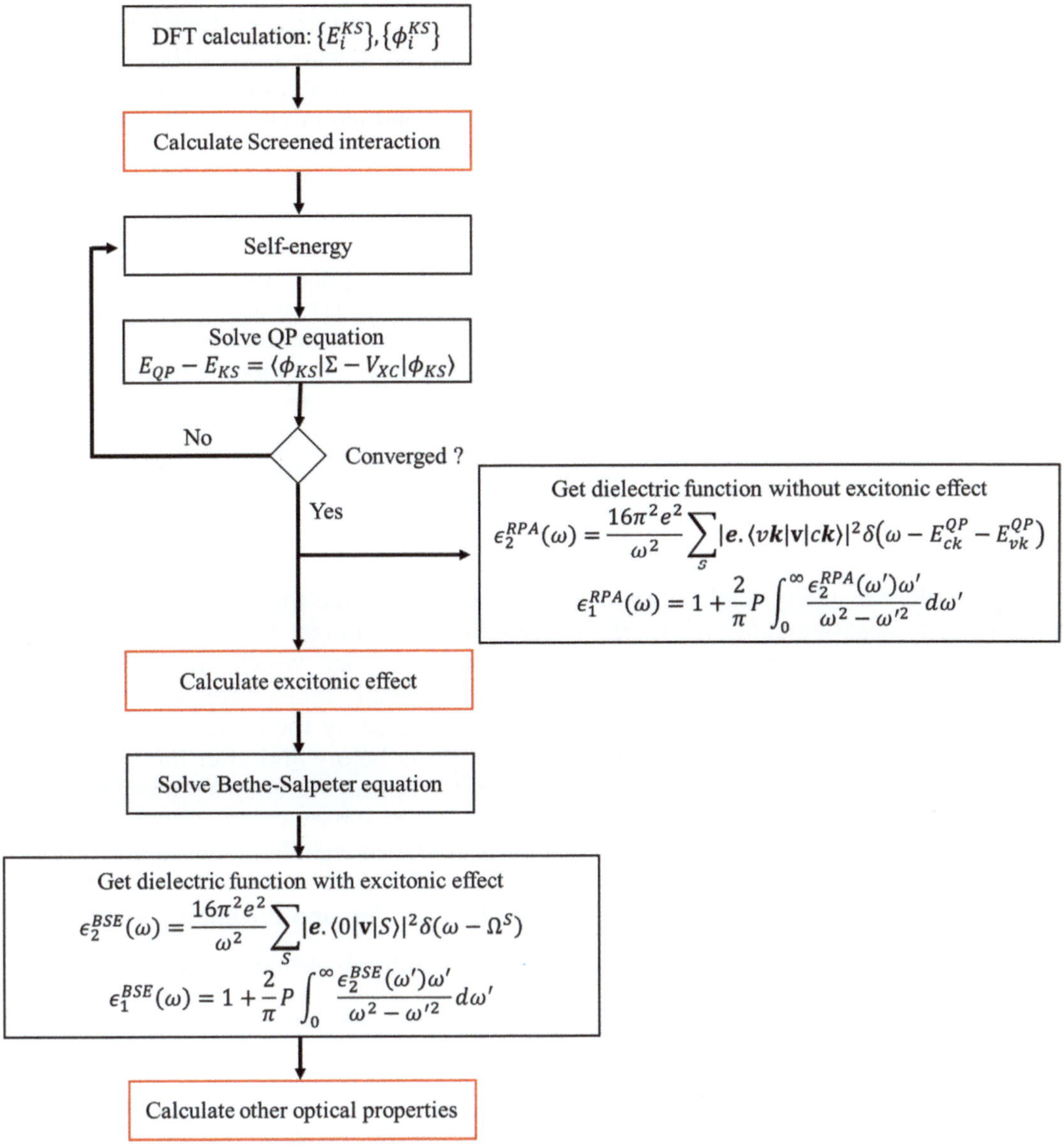

FIG. 2.2 The first-principles calculations on the optical properties in the presence/absence of excitonic effects, covering the real- and imaginary-part dielectric functions, absorption coefficients, reflectance spectra, and energy loss functions.

The wave vector of an electromagnetic wave agrees with the close relation of $k = \frac{\omega\sqrt{\varepsilon(\omega)}}{c}$. Its value might be real or complex, being mainly determined by the imaginary-part dielectric function (being sensitive to the available excitation channels).

In general, $\sqrt{\varepsilon(\omega)} = n + i\frac{c}{\omega}\alpha$; therefore, the complex k is utilized to define the refractive index and the absorption coefficient simultaneously. The inverse of α is the characteristic propagation length which an electromagnetic wave can penetrate a condensed-matter systems. Very interesting, the various kinds of valence/conduction electrons are able to efficiently screen the electromagnetic wave at the different frequency ranges, or such carriers can generate the diversified optical excitations. The significant couplings, with an electromagnetic wave absorption, directly determine the reflectance, absorption and transmission spectra [70].

When an electromagnetic wave is incident on a sample along the normal direction, some of the incident photon will be reflected, while the other is panentrate into the medium and will be adsorbed within a specific range. There exist two continuous boundary conditions for the tangential electric and magnetic fields. By using the Faraday's and modified Ampere's laws, the reflectance, being characterized the square ratio of the reflected versus incident electric fields, is expressed as

$$R(\omega) = \left|\frac{\sqrt{\varepsilon(\omega)} - 1}{\sqrt{\varepsilon(\omega)} + 1}\right|^2 \tag{2.4}$$

Apparently, the nearly total reflectance comes to exist, while the dielectric function presents a purely negative value within the specific frequency range. The very large R is very useful in fully understanding the forbidden frequency range which an electromagnetic wave only declines quickly. The similar behavior is revealed in the low-frequency reflectance due to the conduction electrons, being frequently described by the Drude model [70]. The rapid decrease of reflectance, the plasmon edge, appears before and after the frequency of coherent charge oscillations [70]. On the other hand, the optical excitations thoroughly disappear for any semiconductor within an optical gap. The imaginary-part dielectric function is vanishing, and its real-part one can provide a low reflectance, mainly owing to free propagation in the absence of energy loss.

It is well known that the imaginary-part dielectric function and the energy loss function, $Im[\varepsilon]$ and $Im\left[\frac{-1}{\varepsilon}\right]$, respectively, the available optical/Coulomb excitations under the single-particle and collective pictures. The former and the latter are responsible for the electron-hole pair excitations and the plasmon modes at the various frequency ranges. That is, all screening behaviors due to all the kinds of valence/conductions are obviously covered in the electron energy loss spectra corresponding to the directly measured results (EELS in Ref. [47]). Such important differences have been very successfully clarified from a lot of experimental [79,80] and theoretical researches [80]. For example, Taft and Philipp can measure the high-resolution reflectance spectrum of the semimetallic Bernal graphite within a broad frequency range of $\omega < 26$ eV [79]. Such wide absorption range is very sufficient and reliable in fully understanding the optical excitations associated the free conduction electrons, the valence π electrons (the carbon-$2p_z$-orbitals], and the deeper-energy σ ones (the C-(2s, $2p_x$, $2p_y$)-orbitals)). These quasiparticle behaviors directly reflect the energy-dependent band structure. That is, they are respectively, dominated by the weak, but significant interlayer van der Waals interactions (the interlayer $2p_z$-orbital hybridizations), the important intralayer nearest-neighbor hopping integrals (the intralayer π bondings), and the intralayer σ-electron multi-orbital hybridizations. Using the delicate analyses on $Im[\varepsilon]$ and $Im\left[\frac{-1}{\varepsilon}\right]$, the first, second and third kinds of quantum quasiparticles are clearly identified to, respectively, appear in (I) $\omega < 0.8$ eV (II) 0.8 eV $< \omega < 9$ eV, and (III). $\omega > 9$ eV Furthermore, the π and σ plasmon modes under long wavelengths come to exist at 7.0 and 25.0 eVs, respectively.

In this book, one of the main focuses is to fully explore the optical properties in germanium- and germanene-based materials through the first-principles calculations. By using the nonlocal Kohn-Sham equation with the biparticle Coulomb interactions (details in Chapter 2), the wave-vector-dependent energy spectra and wave functions are available in further calculating the dielectric function in the absence of excitonic effects. But when the significant Coulomb attractions between the photon-induced electron and hole pairs take part in all the intermediate excitation processes, the standard Bethe and Salpeter equation, as clearly indicated in Fig. 2.2, needs to be solved for the excitonic effects. How to analyze their strong modifications on the various optical properties might be relatively easy for condensed-matter systems with few valence and conduction energy subbands, e.g., the excitonic effects in germanene and its hydrogenated.

The theoretical and experimental analyses, which are mainly conducted on the important absorption structures of Germanium- and germanene-based materials. In general, all the prominent absorption structures might be due to the stable bound excitons and the band-edge states with the joint van Hove singularities. The former frequently appear before the single-particle threshold absorption frequency, e.g., the initial strong peaks of the 1, 2 and 3s excitons [77]. Very interesting, the complicated optical excitation processes, which directly determine two kinds of pronounced absorption structures, might become open issues in identifying the specific transition channels. How to reach the excellent predictions on the unusual optical properties will be solved in the different chapters under the suitable physical pictures, e.g., the delicate analyses on unique absorption spectra of group-IV monolayers and its hydrogenated systems in the absence and presence of excitonic effects (details in Chapter 7). Most importantly, the critical orbital hybridizations in chemical bonds, being associated with the band-edge states, are found to be meaningful in dominating the various absorption peaks through the specific absorption frequencies. A broader theoretical framework, which consists of the geometric, electronic, magnetic and optical properties, will show a successful development under the unified viewpoints. Furthermore, the diversified chemical bondings are clearly revealed in the germanium- and germanene-based materials.

2.2 The phenomenological models and their relations with VASP simulations

The phenomenological models are very useful in providing the sufficient information for the rich quasiparticle behaviors. Their systematic investigations have been successfully finished on most of essential properties in graphene-related sp^2-bonding systems (layered graphenes/carbon nanotubes/graphene nanoribbons; [81–83]), as clearly illustrated in several books [5,6,40,84]. That the π bondings of C-$2p_z$ orbitals dominate the low-energy physics is the main reason. This strategy is also suitable for silicene-/geremanene/-tinene/

plumbene-based materials under the concise physical/chemical/material pictures of th intrinsic interactions [85–88]. In general, the reliable establishment of quasiparticle models will require the full supports from the first-principles calculations (details in Chapters 1 and 2), while there exist the complicated orbital hybridizations and spin-dependent interactions. Both clear mechanisms and delicate analyses become the necessary conditions in developing the various phenomenological models [3,4].

2.2.1 The tight-binding model/the generalized tight-bonding model

The previously published books have successfully presented the tight-binding models for few-layer graphenes with AAA [89,90], ABA [91,92], ABC [92–94] and AAB [95–97] stackings, simple hexagonal/Bernal/rhombohedral bulk graphites [84,98–103], single-/multi-walled carbon nanotubes [104,105], graphene nanoribbons [83], monolayer silicene/germanene/tinene/plumbene [46,106–108], and layered phosphorene systems [109]. As for bilayer/trilayer nongraphene group-IV materials, the reliable Hamiltonians are not well established up to now, since the coupling effects due the buckling, stacking, nonperpendicular orbital hybridization and spin-orbital interaction might be too strong to clearly identify each term. Apparently, orbital mixings and spin configurations play the critical roles in determining all the intrinsic interactions. Their clear identifications for germanium-/germanene-related systems in this book (details in Chapters 4–20, or the second paragraph of Chapter 21) will be very useful in the further model developments.

Theoretical predictions [110] and experimental examinations [111] on graphene-related systems clearly show that most of essential properties are dominated by the intralayer & interlayer atomic interactions of carbon-$2p_z$ orbitals, such as, those in layered graphene systems, three kinds of bulk graphites, single-/multi-walled carbon nanotubes and graphene nanoribbbons (reliable quasiparticle frameworks in books of Ref. [5]). The $2p_z$-based Hamiltonians are well developed in the previous works. For example, the AAA- [109], ABA- [112], ABC- [113] and AAB-stacked (10 parameters in Ref. [96] trilayer graphene systems (Figs. 2.3A–D)) possess the stacking-enriched hooping integrals and thus exhibit the rich and unique energy dispersions. As a result, their magneto-electronic [95,114], Coulomb-excitation [5,115], optical/magneto-optical [114] and quantum transport properties excitations [116] are totally different from one another. However, the curvature effects of $2p_z$-mis-orientation and sp^3 hybridizations need to be taken into account for very small carbon nanotubes [82], and the antiferromagnetic spin configurations come to exist in open zigzag edges of graphene nanoribbons [83]. When the nonperpendicular π and σ bondings, as well as the spin-orbital interactions, are included in the buckled monolayer silicene/germanene/stanine/pumbene, the strongly modified valence and conduction subbbands appear at the highly asymmetric energy spectra about the Fermi level, e.g., the obvious modifications about energy gap, band widths, energy dispersions, band-edge states, and noncrossing/crossing/anticrossing behaviors. The above-mentioned normal materials, without deformation modulations, possess small primitive unit cells, being very suitable in thoroughly exploring the diverse magnetic quantization phenomena. The concise pictures of all orbital- and spin-dependent interactions are available for covering the magnetic-/electric-field effects. The vector potential leads to another period associated with the Peierls phases and

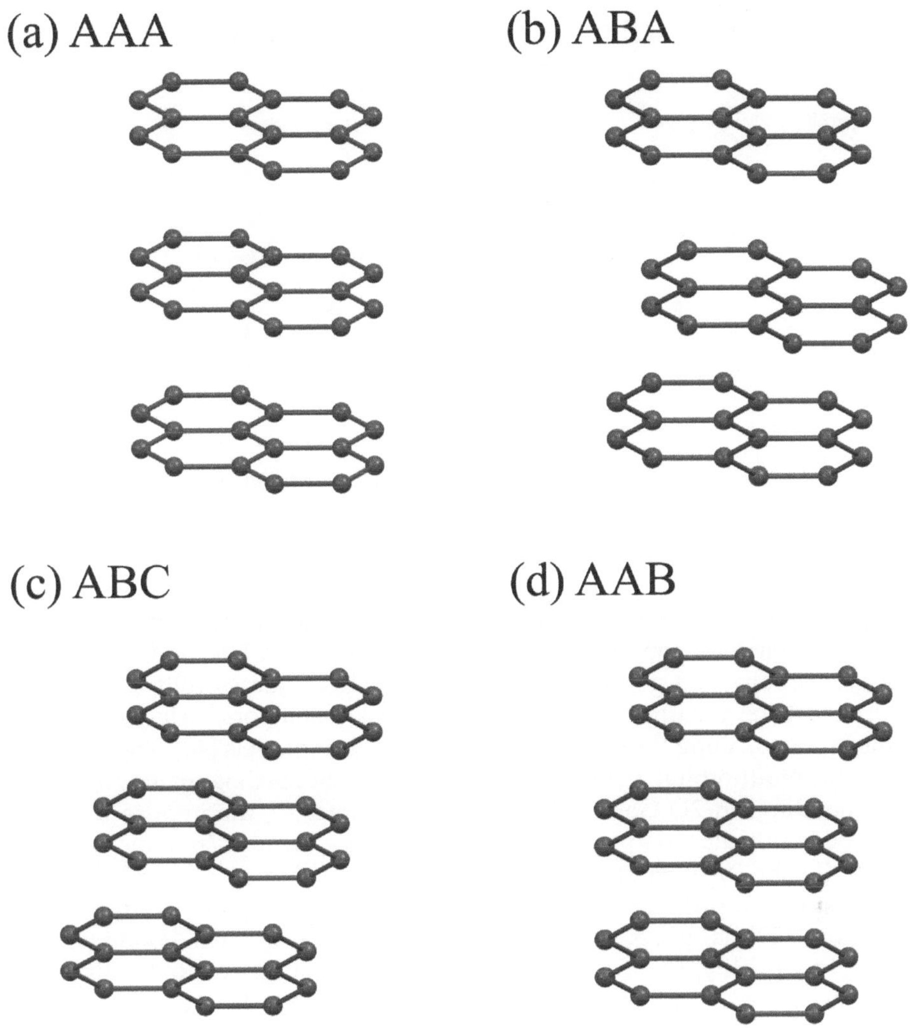

FIG. 2.3 The geometric structures for trilayer graphene systems under (A) AAA, (B) ABA, (C) ABC, and (D) AAB stackings.

thus creates a giant Moire superlattice (Fig. 2.4). Very interesting, the generalized tight-binding models, which are able to deal with the uniform and/or spatially modulated external fields, have been successfully developed for a lot of 1D, 2D and 3D, especially the systematic investigations about the graphene-based materials. Such strategy is built from the tight-binding functions of the layer-dependent sublattices; that is, the field effects and the intra-layer/interlayer atomic interactions can be included in Hamiltonians simultaneously. This is totally different from the effective-mass approximations. The latter needs to solve the approximate energy dispersions near the High-symmetry points and then do the magnetic quantization. Two-step distinct processes would induce much trouble, when there exist a

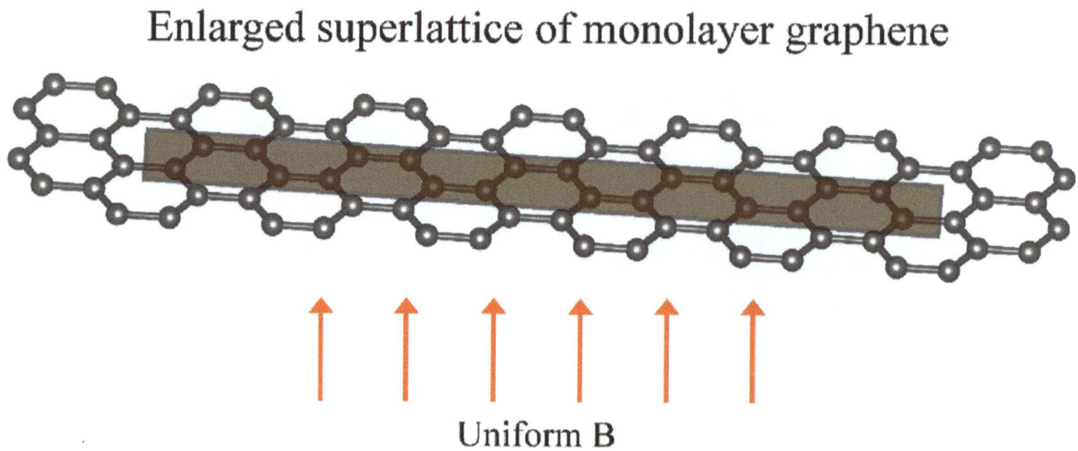

FIG. 2.4 Monolayre graphene in a perpendicular magnetic field leading to a giant moire superlattice.

lot of low-lying energy subbands [117], nonmonotonous energy dispersions [118], and degenerate constant-energy loops [5]. Obviously, only the former is suitable for the magnetically diversified phenomena in condensed-matter systems with the various orbital hybridizations [49,119] and spin-enriched configurations [49]. being discussed later in Chapters 2–4.

A lot of germanium- and germanene-based materials, as successfully identified in Chapters 4–20, present the greatly diversified quasiparticle behaviors, belonging to the critical mechanisms for establishing the reliable phenomenological models [40]. The concise chemical pictures for the multi-orbital hybridizations cover the nonorthogonality/orthogonality of pure π & σ bondings in 2D Ge–Ge/C–C bonds (Chapter 7), the impure π & σ bondings with the nonperpendicular/perpendicular relations in the buckled/nonbuckled binary compounds of B-, C-, and N-substituted germanene systems (Chapter 6), s-sp^3 in H–Ge/alkali-Ge bonds (Chapter 7), pure sp^3-sp^3 in 3D Ge–Ge bonds (Chapter 13), hybridized sp^3-sp^3 in Ge–O bonds [120], {s, p_x, p_y, p_z}-{d_{z^2}, $d_{x^2-y^2}$, d_{xz}, d_{zy}, d_{yx}} in Ge-T bonds (Chapter 8), as well as, {s, p_x, p_y, p_z}-{f_{z^2}, f_{xz^2}, f_{yz^2} f_{xyz}, $f_{z(x^2-y^2)}$, $f_{x(x^2-3y^2)}$, $f_{y(y^2-3x^2)}$} in Ge–Re bonds (Chapter 10). Very interesting, the rich and unique chemical bondings are rather sufficient for the further development of a unified quasiparticle framework. For example, the hopping integrals, on-side Coulomb potentials, spin-orbital couplings, and spin-induced electron-electron interactions, which frequently appear in the Hubbard-like tight-binding models [50,121–123], are quite sensitive to the various orbitals, especially for a lot of magnetic *d*- and *f*-ones. The suitable parameters can match the low-lying valence and conduction energy subbands from the VASP simulations [2] through the delicate calculations and analyses. Of course, their scientific values are worthy of the systematic investigations and are able to provide useful pictures in understanding the diverse phenomena. Also noticed that the reliable establishment of the generalized tight-binding model under the modulations of external fields is relatively easy for the man-stream materials with the concise physical/chemical/material mechanisms in a small unit cell, e.g., the Ge–B/Ge–C/Ge–N saturated binary compounds (details in Chapter 6) in the presence of a uniform perpendicular magnetic field.

2.2.2 RPA/modified RPA & self-energy method/corrected one

The Coulomb-field perturbations, which belong to the natural charge screening behaviors, are able to generating a plenty of greatly diversified excitation/deexcitation phenomena, e.g., the various electron-hole pair excitations with the specific boundaries [124], as well as, the distinct plasmon modes with the strong energy dispersion relations for graphene-related materials [125]. Both classical and quantum pictures are available in fully comprehending the screening/response abilities of charge carriers, e.g., the identical pictures for the coherent charge oscillation modes and the quasiparticle plasmons [115]. All charges in a condensed-matter system, which cover conductions and/or valence electrons, can display the distinct single-particle and collective excitations modes at the specific energy ranges (discussed later). Very interesting, the whole excitation processes are thoroughly revealed in the (moment, frequency)-excitation phase diagrams, in which the screened response abilities are clearly illustrated by the various intensities of energy loss functions (much difficulty in achieving many data; details in Books [5]). These up-to-date predictions, belonging to a part of quasiparticle framework, will require the urgent EELS examinations [126]. In addition to electronic excitations, the inelastic Coulomb scatterings will provide the rather efficient decay rates [5]; furthermore, they are assisted by the electron-phonon interactions at finite temperatures [127]. The close relations between Coulomb excitations and deexcitations have been successfully identified for many layered emergent materials, mainly owing to viewpoint modifications of many-particle theories.

Generally speaking, the random-phase approximation (the self-consistent-field approach; Ref. [128]), is frequently utilized to explore the charge screening abilities valence and conduction electrons in the different energy subbands, in which the significant effects due to energy spectra and wave functions are included in bare polarization functions [129]. While the intralayer & interlayer orbital hybridizations of layered systems, the intralayer & interlayer Coulomb interactions, as well as, the electric and magnetic modulations need to consider simultaneously under a unified framework, a thorough generalization of RPA is made by the projection of Bloch wave function about the layer & sublattice dependence. The similar decomposition is able to deal with the screened exchange energy (the modified self-energy method; book; under the current investigations for few-layer graphene systems). The previous books have clearly shown the reliable results for layered graphenes/monolayer silicene & germanene with/without n-/p-type dopings [130–133] in the presence/absence of magnetic/electric fields [134–137]. For example, trilayer graphenes of AAA, ABA, ABC and AAB stackings, clearly reveal the diverse behaviors of excitation-phase diagrams in terms of electron-hole regions/boundaries [112,113,138], featured plasmon/magnetoplasmon energy dispersion relations (the number, strength and frequency of collective excitations [96,113,139]), and their coexistent characteristics (the existence/absence of Landau dampings). The unusual Coulomb excitation phenomena is further driven by the magnetic quantization, since there exists a prominent competition of the longitudinal Coulomb-force charge oscillations and the transverse cyclotron motion, such as, the nonmonotonous momentum-dependence for group velocity of magneto-plasmon. The magnetically Coulomb excitations can be explored exactly and efficiently, being attributed to a unified quasiparticle framework based on layer- and sublattice-dependences.

Within the modified RPA for 2D systems [140–142], the bare electric polarizations belong to layer-dependent tensor functions [125,143], in which the excited electrons and holes could exist on the same or different layers. Both intralayer and interlayer charge screening abilities, as revealed by all valence and conduction electrons in any layers, are taken into consideration simultaneously. As a result, the screened response function (the effective energy loss functions [97,106]); is achieved from the detailed calculations and thus is responsible for the measured EELS probability about the inelastic scattering events [144]. The simultaneous combinations of induced charge density fluctuations each layer have created the diverse quasiparticle phenomena, e.g., the static carrier screening of few-layer graphene/carbon nanotubes for a charged impurity [145–147], the dynamics of plasma waves [148–150], and the Coulomb decays of excited quasiparticles [5]. Free electrons or holes in metallic graphene-related systems can rapidly terminate the Coulomb potential with the long-range dependence of 1/r [150]; that is, their effective Coulomb potentials behave as the Yukawa's form within short distances. The total/induced potentials and screening charge density display the Friedel oscillations at sufficiently long distances, respectively, revealing the sin $(2k_{Fr})/r^2$ and sin $(2k_{Fr})/r$ behaviors for 2D (k_F the Fermi momentum [5]); and 1D metals. Very interesting, the superposition of two normal oscillation modes with two distinct Fermi moments (the beating phenomena) is predicted to exist in metallic bilayer graphene systems/double-walled carbon nanotubes [140,151–153]. However, the semiconducting ones display the original distance dependence through the observable decline of potential/screening charge amplitude. Certain important differences between them are thoroughly investigated for 2D graphene and 1D carbon nanotubes. As for the time-dependent fluctuations of screening charge densities and total/induced Coulomb potentials, they behave as the propagating plasma waves with the Friedel oscillation form. Such simulations are very useful in understanding the dynamic carrier response of a whole system due to a perturbed electron/ion beam.

The quasiparticle self-energy, which covers energy spectrum and lifetime [154], needs to be thoroughly modified for layered systems, as done for the random-phase approximation [106,140,155]. The layer- and sublattice-decomposed viewpoint is available in solving the complicated intralayer and interlayer orbital hybridizations/Coulomb interactions/excitations/deexcitations, as well as, the external-field-enriched coupling effects [156]. This strategy has been successfully for exploring the Coulomb decay rates of monolayer graphene/silicene/germanene, single-walled carbon nanotubes, and double-walled ones. Furthermore, bilayer AA-/AB-stacked ad trilayer AAA-/ABA-/ABC-/AAB-stacked graphene systems are under the current investigations. Very interesting, as for extrinsic *n*-type dopings (intrinsic cases), there exist excited quasiparticles of conduction electrons, conduction holes and valence holes (conduction electrons and valence holes). While they decay by the inelastic electron-electron interactions, the sufficient momentum and energy transfers under the Pauli exclusion principle play the critical roles in determining how many available deexcitation channels, especially those due to the electron-hole excitation boundaries and the plasmon modes. In general, the quasiparticle lifetimes in graphene-related systems exhibit the highly anisotropic behaviors, the largest energy uncertainty below the Dirac point, and the unusual Fermi-momentum dependence near the Fermi level at zero temperature (details in books; [5]). These are totally absent in 3D/2D/1D electron gases. How to generalize the layer- and sublattice-decomposed viewpoints in any layered condensed-matter systems with the various orbital hybridizations and spin-dependent configurations is one of the near-future research focuses, such as, the greatly enriched Coulomb deexcitations/excitations due to hydrogenations of graphene/silicene/germanene/tinene/plumbene systems (details in Chapter 7).

2.2.3 Optical & magneto-excitation theories with/without excitonic effects

Optical properties, which correspond to the vertical excitations between the initial occupied and final unoccupied states [56,157,158], present the featured energy spectra and wave functions [143,152]. In general, they cover the dielectric function (the bare screening ability; [159]), energy loss function (the screened response function; [106,124,160,161]), reflectance [162] and absorption coefficient [118,163]. While charge carriers experience a perturbation of electromagnetic wave, their time- and position-dependent moving can create an induced current density and thus change the effective electric field, in which a linear relation between them is just the Ohm's law. The Kubo formula within the linear response is sufficient in understanding the main features of optical excitations in the absence of excitonic effects, especially for semimetallic [51,63], metallic [51,63,164,165] and narrow-gap materials [166,167]. There exist the available excitation channels (the joint density of states [168,169]); and the square of electric dipole matrix element in determining the number, frequency, form and intensity of prominent absorption structures. The latter could be evaluated from the gradient approximations, as very successfully done for graphene-related sp^2-bonding systems. This approximate method is fully consistent with the layer- and sublattice-associated bases in the generalized tight-binding model. The direct combinations between the electronic and optical models are very suitable for thoroughly exploring optical and magneto-optical excitations simultaneously. According to the previous theoretical predictions (books), this strategy has successfully diversified the vertical transition/magneto-transition phenomena. For example, the sliding bilayer graphene systems clearly display the stacking-enriched configurations [170–172], induce the dramatic transformations about energy spectra/wave functions, and create three kinds of Landau levels (the well-behaved, perturbed and undefined LLs; [170]), thus leading to the standard magneto-optical selection rules [40,173], the extra ones [40], and the total absence. Moreover, the trilayer graphene systems, with AAA, ABA, ABC and AAB stacking, possess the unusual energy dispersions in three pairs of low-lying valence and conduction energy subbands. By using the generalized tight-binding model, the rich and unique magneto-electronic states are clearly revealed through the magnetic-field-dependent LL energies [40,118,174,175] and the localized oscillation functions [176], such as, the close relation between LL energies and field strength, the layer-/sublattice-related probability distributions, their normal/perturbed/irregular zero-point numbers, and the noncrossing/crossing/anticrossing behaviors [40,170,177]. The further combination with the Kubo formula is able to show a lot of prominent interLL absorption peaks, in which they could be classified into several magneto-excitation subgroups, based on the spatial oscillation features of each LL subgroup. For example, there exist nine interLL transition subgroups for ABC [115,178] and AAB stackings [40,96]. The complicated interlayer $2p_z$-orbital hybridizations are responsible for the featured absorption/magneto-absorption spectra in terms of forms, intensities, numbers, and frequencies. The delicate simulations and analyses are also successfully applied in bulk graphite systems [179,180], carbon nanotubes [181,182], graphene nanoribbons [183,184], and silicene/germanene/tinen/phosphorene [61,185–190].

The layered materials are outstanding candidates in observing the dimensionality crossover behaviors [191–193] through the optical transitions. Reflectance, absorption and transmittance, which frequently come to exist in the layer-dependent crystal structures, need to be examined thoroughly. The preciously optical works are only focused on a semiinfinite sample, in which the significant coupling effects, being closely related to the intralayer/interlayer atomic interactions, as well as, the modulations of external electric/magnetic fields, are totally absent. Also noticed that all intrinsic mechanisms have been successfully included in the current theory of optical scatterings. The reliable calculations of the dielectric function, the suitable boundary conditions, and the multi-optical processes on each interface are thoroughly required during the numerical simulations. Apparently, few-layer graphene systems [40,118,194] and bulk graphites [177,179,195] exhibit the quite different absorption spectra, even if both of them belong to semimetals. The up-to-date results show the sensitive dependences on the layer number and stacking configuration, such as, the rich optical properties of AAA- [113], ABA- [196], ABC- [113] and AAB-stacked [96] few-/multi-layer graphene systems. The important effects from the modulations of magnetic fields and gate voltages could be directly explored under the analytic formulas of all optical properties. That is, the optical scattering theory of layered thin films and the generalized tight-binding is consistently unified under the quasiparticle framework. However, whether the excitonic effects must be taken into consideration, depends on the gap magnitudes of the semiconducting materials.

The stable/quasi-stable excitons, which are formed by the bound states of the excited valence holes and conduction electrons [159,197], can strongly modified the initial certain absorption spectra [118,198], and provide the photoluminescence spectra for potential applications [181,199,200]. In general, the optical threshold frequency exhibits a red shift, i.e., an optical gap is smaller than band one. Furthermore, there exist few prominent exciton-related absorption structures below/near the characteristic frequency. The up-to-date results that the VASP simulations are able to distinguish the single-particle and many-body optical transitions (detail in Chapters 1, 2 and 4). However, it is very difficult to achieve the suitable physical pictures in fully understanding the diverse phenomena of excitonic effects. The low-dimensional excitonic theories are urgently required for the emergent layered semiconductors, e.g., the graphene-/gremanene-/silicene-/tinene-/plumbene-hydrogenated systems in the current book (details in Chapter 7; Ref. [201]). How to simultaneously cover the band-structure effects, the charge screenings on the electron-hole Coulomb potentials and the pairing-enriched optical scatterings become open issues, since these three quasiparticle ehaviors belong to the many-particle dynamic processes. May be, the 2D excitonic theory could be developed within the self-consistent manner, in which the numerical calculations might be full of challenges.

2.2.4 Hall quantum transports

Very interesting, layered materials clearly reveal the diverse magneto-quantum transport phenomena, as extensively shown in graphene systems experimentally [202,203] and theoretically [114,118]. Under the direct combination of the static Kubo formula [63,114] and the generalized tight-binding model [40,114,170,204], the various quantum behaviors of electrical conductivities have been successfully attributed to the rich field-dependent LL energy spectra [96,178,185], the highly degenerate magneto-electronic states [118,184,188], the localized oscillations on layer-related sublattices through the symmetric/antisymmetric/irregular probability distributions, and the specific scattering rules associated with the initial- and final-state quantum numbers. This strategy is capable of fully understanding each significant contribution from the available interLL scattering channel. The delicate simulations and analyses are finished for certain outstanding systems, such as, monolayer graphene/silicene, the sliding bilayer graphenes with the distinct stacking configurations, bilayer silicene under a specific stacking and buckling [205–207], and the trilayer AAA-, ABA-, ABC- and AAB-stacked trilayer graphene materials. Very interesting, their Hall transport properties are rather sensitive to the magnetic-field and Fermi-level dependences (free carrier density [208,209]); under the heights, widths, and increasing/decreasing approaches of quantum plateaus. Apparently, a lot of unusual transport directly reflect the featured Landau levels. Most important, the electronic, optical and transport properties of magnetic quantization have been unified together by the quasiparticle framework in the current and previous books [6,40,109]. On the other side, the effective-mass approximation [103,210] cannot deal with the low-lying Landau levels related to the nonmonotonous energy bands, so it is not suitable for exploring the static and dynamic interLL scatterings.

The germanene-related 2D materials, being fully explored in Chapters 4–20, are expected to display the unusual quantum transport properties [33,211], mainly owing to the various intrinsic interactions. The coupling effects, which arise from the impure/pure multi-orbital hybridizations, the spin-orbital couplings/the spin-induced on-site electron-electron interactions, the vector-potential-created modifications on the nearest-neighbor hopping integrals, and the Coulomb-potential differences of the different-height sublattices, will play the critical roles in diversifying the magnetic quantization phenomena [109]. For example, the boron-/carbon-/nitrogen-substituted germanium compounds, with a saturated concentration, has the concise orbital hybridizations in the smallest unit cell (details in Chapter 6). By the VASP simulations and thorough analyses, a perpendicular relation between the impure π and σ bondings is responsible for the essential properties. The specific multi-orbital hybridizations can provide sufficient information in establishing the orbital-dependent hopping integrals and the on-site Coulomb potentials for the generalized tight-binding model. The systematic investigations would be conducted on the main features of Landau levels [174,212–214], magneto-electronc excitations (inter-LL single-particle and collective excitations; [114,186,188]), magneto-optical selection rules [40,114,173], and quantum Hall conductivities [109,215,216]. The similar researches could be generalized to the prominent chemsorption cases, such as, hydrogenation, alkaliation, oxidization, and halogenation.

References

[1] Orio M, Pantazis DA, Neese F. Density functional theory. Photosynth Res 2009;102:443–53.

[2] Arias MM. Vienna Ab-initio simulation package. Edinburgh, UK: The University of Edinburgh; 2009.

[3] Foulkes WMC, Haydock R. Tight-binding models and density-functional theory. Phys Rev B 1989;39:12520–36.

[4] Goringe CM, Bowler DR, Hernández E. Tight-binding modelling of materials. Rep Prog Phys 1997;60:1447–512.

[5] Lin C-Y, Wu J-Y, Chiu C-W, Lin M-F. Coulomb excitations and decays in graphene-related systems. CRC Press; 2019.

[6] Chen SC, Wu JY, Lin CY, Lin MF. Theory of magnetoelectric properties of 2D systems. IOP Publishing); 2017.

[7] Chinesta F, Abisset-Chavanne E. The Schrödinger equation. SpringerBriefs Appl Sci Technol 2018:1–27.

[8] Ratcliff LE, Mohr S, Huhs G, Deutsch T, Masella M, Genovese L. Challenges in large scale quantum mechanical calculations. Wiley Interdiscip Rev Comput Mol Sci 2017;7.

[9] Sun G, Kürti J, Rajczy P, Kertesz M, Hafner J, Kresse G. Performance of the Vienna ab initio simulation package (VASP) in chemical applications. J Mol Struct Theochem 2003;624:37–45.

[10] Blaha P, Schwarz K, Madsen GKH. WIEN2K, an augmented plane Wave+ local orbitals program for calculating crystal properties. Austria: TU Wien; 2001. 2001.

[11] Gulans A, Kontur S, Meisenbichler C, Nabok D, Pavone P, Rigamonti S, Sagmeister S, Werner U, Draxl C. Exciting: a full-potential all-electron package implementing density-functional theory and many-body perturbation theory. J Phys Condens Matter 2014;26.
[12] Giannozzi P, Baroni S, Bonini N, Calandra M, Car R, Cavazzoni C, Ceresoli D, Chiarotti GL, Cococcioni M, Dabo I, Dal Corso A, De Gironcoli S, Fabris S, Fratesi G, Gebauer R, Gerstmann U, Gougoussis C, Kokalj A, Lazzeri M, Martin-Samos L, Marzari N, Mauri F, Mazzarello R, Paolini S, Pasquarello A, Paulatto L, Sbraccia C, Scandolo S, Sclauzero G, Seitsonen AP, Smogunov A, Umari P, Wentzcovitch RM. Quantum ESPRESSO: a modular and open-source software project for quantum simulations of materials. J Phys Condens Matter 2009;21.
[13] Castro A, Marques MAL, Rubio A. Propagators for the time-dependent Kohn-Sham equations. J Chem Phys 2004;121:3425–33.
[14] Zhang Y, Chen S, Chen G. First-principles time-dependent quantum transport theory. Phys Rev B Condens Matter 2013;87.
[15] Hansson T, Oostenbrink C, van Gunsteren W. Molecular dynamics simulations. Curr Opin Struct Biol 2002;12:190–6.
[16] Koch KR. Monte Carlo methods. GEM - Int J Geomathemat 2018;9:117–43.
[17] Blöchl PE. Projector augmented-wave method. Phys Rev B 1994;50:17953.
[18] Perdew JP, Burke K, Ernzerhof M. Perdew, burke, and ernzerhof reply. Phys Rev Lett 1998;80:891.
[19] Elliott RJ. Theory of the effect of spin-Orbit coupling on magnetic resonance in some semiconductors. Phys Rev 1954;96:266–79.
[20] Shim J, Lee EK, Lee YJ, Nieminen RM. Density-functional calculations of defect formation energies using the supercell method: brillouin-zone sampling. Phys Rev B Condens Matter 2005;71.
[21] Kresse G, Furthmüller J. Efficient iterative schemes for ab initio total-energy calculations using a plane-wave basis set. Phys Rev B Condens Matter 1996;54:11169–86.
[22] Joubert D. From ultrasoft pseudopotentials to the projector augmented-wave method. Phys Rev B Condens Matter 1999;59:1758–75.
[23] Hu G, Xie H, Xu F. A multilevel correction adaptive finite element method for Kohn–Sham equation. J Comput Phys 2018;355:436–49.
[24] Rodriguez-Bautista M, Díaz-García C, Navarrete-López AM, Vargas R, Garza J. Roothaan's approach to solve the Hartree-Fock equations for atoms confined by soft walls: basis set with correct asymptotic behavior. J Chem Phys 2015;143(3):034103.
[25] Duguet T, Bally B, Tichai A. Zero-pairing limit of Hartree-Fock-Bogoliubov reference states. Phys Rev C 2020;102(5):054320.
[26] Zhao C, Liu F, Kong X, Yan T, Ding F. The wrinkle formation in graphene on transition metal substrate: a molecular dynamics study. Int J Smart Nano Mater 2020;11(3):277–87.
[27] Nagy A, Savona V. Variational quantum Monte Carlo method with a neural-network ansatz for open quantum systems. Phys Rev Lett 2019;122(25):250501.
[28] Perdew JP, Burke K, Ernzerhof M. Generalized gradient approximation made simple. Phys Rev Lett 1996;77:3865–8.
[29] Blanc X, Cancès É, Dupuy M-S. Variational projector augmented-wave method. Compt Rendus Math 2017;355(6):665–70.
[30] Stroppa A, Kresse G, Continenza A. Spin polarization tuning in Mn5-x Fex Ge3. Appl Phys Lett 2008;93(9):092502.
[31] Chen RB, Chen SC, Chiu CW, Lin MF. Optical properties of monolayer tinene in electric fields. Sci Rep 2017;7(1):1–9.
[32] Monshi MM, Aghaei SM, Calizo I. Edge functionalized germanene nanoribbons: impact on electronic and magnetic properties. RSC Adv 2017;7(31):18900–8.
[33] Ni Z, Liu Q, Tang K, Zheng J, Zhou J, Qin R, Gao Z, Yu D, Lu J. Tunable bandgap in silicene and germanene. Nano Lett 2012;12:113–8.
[34] Karlický F, Lepetit B, Lemoine D. Quantum modelling of hydrogen chemisorption on graphene and graphite. J Chem Phys 2014;140(12):124702.
[35] Pham HD, Su WP, Nguyen TD, Tran NT, Lin MF. Rich p-type-doping phenomena in boron-substituted silicene systems: boron-substituted silicene. R Soc Open Sci 2020;7(12):200723.

[36] Banhart F, Kotakoski J, Krasheninnikov AV. Structural defects in graphene. ACS Nano 2011;5(1):26–41.
[37] Kuzmany H, Shi L, Martinati M, Cambré S, Wenseleers W, Kürti J, et al. Well-defined sub-nanometer graphene ribbons synthesized inside carbon nanotubes. Carbon N. Y 2021;171:221–9.
[38] Dreher M, Gunder D, Zorb S, Witte G. Van der Waals bound organic semiconductor/2D-material hybrid heterosystems: intrinsic epitaxial alignment of perfluoropentacene films on transition metal dichalcogenides. Chem Mater 2020;32(20):9034–43.
[39] Katayama I, Inoue KI, Arashida Y, Wu Y, Yang H, Inoue T, et al. Ultrafast optical modulation of Dirac electrons in gated single-layer graphene. Phys Rev B 2020;101(24):245408.
[40] Lin C-Y, Do T-N, Huang Y-K, Lin M-F. Optical properties of graphene in magnetic and electric fields. IOP Publishing, e-book); 2017.
[41] Sharma A, Harnish P, Sylvester A, Kotov VN, Neto AHC. Van der Waals forces and electron-electron interactions in two strained graphene layers. Phys Rev B Condens Matter 2014;89(23):235425.
[42] Wang Z, Yang L, Liu S, Yu H, Li X, Xu Y, et al. ZIF-derived porous carbon composites coated on $NiCo_2S_4$ nanotubes array toward efficient water splitting. Nanotechnology 2020;31(19):195402.
[43] Huang HC, Lin SY, Wu CL, Lin MF. Configuration- and concentration-dependent electronic properties of hydrogenated graphene. Carbon N. Y 2016;103:84–93.
[44] Lin YT, Lin SY, Chiu YH, Lin MF. Alkali-created rich properties in graphene nanoribbons: chemical bondings. Sci Rep 2017;7.
[45] Lima MP, Fazzio A, Da Silva AJR. Silicene-based FET for logical technology. IEEE Electron Device Lett 2018;39(8):1258–61.
[46] Yuhara J, Le Lay G. Beyond silicene: synthesis of germanene, stanene and plumbene. J Appl Phys 2020;59:SN0801.
[47] Lin CY, Chen RB, Ho YH, Lin MF. Electronic and optical properties of graphite-related systems. 2017.
[48] Lin S-Y, Liu H-Y, Nguyen DK, Tran NTT, Pham HD, Chang S-L, Lin C-Y, Lin M-F. Silicene-based layered materials. IOP Publishing); 2020.
[49] Lin SY, Tran NTT, Fa-Lin M. Structure- and adatom-enriched essential properties of graphene nanoribbons. CRC press; 2018.
[50] Fujita M, Wakabayashi K, Nakada K, Kusakabe K. Peculiar localized state at zigzag graphite edge. J Phys Soc Japan 1996;65:1920–3.
[51] Ohki D, Omori Y, Kobayashi A. Domain wall conductivity with strong Coulomb interaction of two-dimensional massive Dirac electrons in the organic conductor α-(BEDT-TTF)2 I3. Phys Rev B 2019;100.
[52] Usachov D, Fedorov A, Otrokov MM, Chikina A, Vilkov O, Petukhov A, Rybkin AG, Koroteev YM, Chulkov EV, Adamchuk VK, Grüneis A, Laubschat C, Vyalikh DV. Observation of single-spin Dirac Fermions at the graphene/ferromagnet interface. Nano Lett 2015;15:2396–401.
[53] Pogrebna A, Vujičić N, Mertelj T, Borzda T, Cao G, Xu ZA, Chu JH, Fisher IR, Mihailovic D. Spectrally resolved femtosecond reflectivity relaxation dynamics in undoped spin-density wave 122-structure iron-based pnictides. Phys Rev B Condens Matter 2014;89:165131.
[54] Cattelan M, Fox NA. A perspective on the application of spatially resolved ARPES for 2D materials. Nanomaterials 2018;8:284.
[55] De Jong S, Van Heumen E, Thirupathaiah S, Huisman R, Massee F, Goedkoop JB, Ovsyannikov R, Fink J, Dürr HA, Gloskovskii A, Jeevan HS, Gegenwart P, Erb A, Patthey L, Shi M, Follath R, Varykhalov A, Golden MS. Droplet-like Fermi surfaces in the anti-ferromagnetic phase of EuFe 2As2, an Fe-pnictide superconductor parent compound. EPL 2010;89:27007.
[56] Bihlmayer G, Sassmannshausen J, Kubetzka A, Blügel S, von Bergmann K, Wiesendanger R. Plumbene on a magnetic substrate: a combined scanning tunneling microscopy and density functional theory study. Phys Rev Lett 2020;124:126401.
[57] Wiesendanger R, Gntherodt HJ, Gntherodt G, Gambino RJ, Ruf R. Observation of vacuum tunneling of spin-polarized electrons with the scanning tunneling microscope. Phys Rev Lett 1990;65:247–50.
[58] Heinze S, Blügel S. Prediction of bias-voltage-dependent corrugation reversal for STM images of bcc (110) surfaces: W(110), Ta(110), and Fe(110). Phys Rev B Condens Matter 1998;58:16432–45.
[59] Wiesendanger R. Spin mapping at the nanoscale and atomic scale. Rev Mod Phys 2009;81:1495–550.
[60] Hanaguri T. Development of high-field STM and its application to the study on magnetically-tuned criticality in $Sr_3Ru_2O_7$. J Phys Conf 2006;51:514–21.

[61] Zheng X, Zhang JF, Tong B, Du RR. Epitaxial growth and electronic properties of few-layer stanene on InSb (1 1 1). 2D Mater 2020;7:011001.
[62] Cammarata M, Bertoni R, Lorenc M, Cailleau H, Di Matteo S, Mauriac C, Matar SF, Lemke H, Chollet M, Ravy S, Laulhé C, Létard JF, Collet E. Sequential activation of molecular breathing and bending during spin-crossover photoswitching revealed by femtosecond optical and X-Ray absorption spectroscopy. Phys Rev Lett 2014;113:227402.
[63] Nikolic BK. Quantum transport through mesoscopic disordered interfaces, junctions, and multilayers. J Phys Condens Matter 2002;14:7871–99.
[64] Russell AP, Fletcher KS. Optical sensor for the determination of moisture. Anal Chim Acta 1985;170:209–16.
[65] Liu M, Yin X, Ulin-Avila E, Geng B, Zentgraf T, Ju L, Wang F, Zhang X. A graphene-based broadband optical modulator. Nature 2011;474:64–7.
[66] Neville RC, Loferski JJ. Solar energy conversion: the solar cell. Phys Today 1982;35. 65–65.
[67] Rohlfing M, Louie SG. Electron-hole excitations in semiconductors and insulators. Phys Rev Lett 1998;81:2312–5.
[68] Rohlfing M, Louie SG. Excitonic effects and the optical absorption spectrum of hydrogenated Si clusters. Phys Rev Lett 1998;80:3320–3.
[69] Jackson JD, Fox RF. Classical electrodynamics. Am J Phys 1999;67:841–2.
[70] Fox M, Bertsch GF. Optical properties of solids. Am J Phys 2002;70:1269–70.
[71] Surh MP, Louie SG, Cohen ML. Quasiparticle energies for cubic BN, BP, and BAs. Phys Rev B 1991;43:9126–32.
[72] Ziegler K. Minimal conductivity of graphene: nonuniversal values from the kubo formula. Phys Rev B Condens Matter 2007;75.
[73] Sakurai JJ, Commins ED. Modern quantum mechanics, revised edition. Am J Phys 1995;63:93–5.
[74] Lin C-Y, Wu J-Y, Chiu C-W, Lin M-F. Coulomb excitations and decays in graphene-related systems. 2019.
[75] Han NT, Dien VK, Lin MF. Excitonic effects in the optical spectra of Li_2SiO_3 compound. Sci Rep 2021;11.
[76] Dien VK, Pham HD, Tran NTT, Han NT, Huynh TMD, Nguyen TDH, Fa-Lin M. Orbital-hybridization-created optical excitations in Li_2GeO_3. Sci Rep 2021;11.
[77] Mahan GD. Many-particle physics. 2000.
[78] Cutkosky RE. Solutions of a bethe-salpeter equation. Phys Rev 1954;96:1135–41.
[79] Taft EA, Philipp HR. Optical properties of graphite. Phys Rev 1965;138.
[80] Yang L, Deslippe J, Park CH, Cohen ML, Louie SG. Excitonic effects on the optical response of graphene and bilayer graphene. Phys Rev Lett 2009;103.
[81] Geim AK, Novoselov KS. The rise of graphene. Nanosci Technol A Collect Rev from Nat J 2009;6:11–9.
[82] Iijima S. Microtubules of graphitic carbon. Nature 1991 Helical;354:56–8.
[83] Gorjizadeh N, Farajian AA, Esfarjani K, Kawazoe Y. Spin and band-gap engineering in doped graphene nanoribbons. Phys Rev B Condens Matter 2008;78:155427.
[84] Chiu CW, Lee SH, Chen SC, Shyu FL, Lin MF. Absorption spectra of AA-stacked graphite. New J Phys 2010;12:083060.
[85] Meng L, Wang Y, Zhang L, Du S, Wu R, Li L, Zhang Y, Li G, Zhou H, Hofer WA, Gao HJ. Buckled silicene formation on Ir(111). Nano Lett 2013;13:685–90.
[86] Li L, Lu SZ, Pan J, Qin Z, Wang YQ, Wang Y, Cao GY, Du S, Gao HJ. Buckled germanene formation on Pt(111). Adv Mater 2014;26:4820–4.
[87] Zhu F, Chen W, Xu Y, Gao C, Guan D, Liu C, Qian D, Zhang S-C, Jia J. Epitaxial growth of two-dimensional stanene. Nat Mater 2015;14:1020–5.
[88] Yuhara J, He B, Matsunami N, Nakatake M, Le Lay G. Graphene's latest cousin: plumbene epitaxial growth on a "nano WaterCube". Adv Mater 2019;31:1901017.
[89] Mohammadi Y. Integer quantum Hall effect in AAA-stacked trilayer graphene. Phys B Condens Matter 2021;613.
[90] Mohammadi Y, Moradian R, Shirzadi Tabar F. Effects of doping and bias voltage on the screening in AAA-stacked trilayer graphene. Solid State Commun 2014;193:1–5.
[91] Huang M, Bakharev PV, Wang ZJ, Biswal M, Yang Z, Jin S, Wang B, Park HJ, Li Y, Qu D, Kwon Y, Chen X, Lee SH, Willinger MG, Yoo WJ, Lee Z, Ruoff RS. Large-area single-crystal AB-bilayer and ABA-trilayer graphene grown on a Cu/Ni(111) foil. Nat Nanotechnol 2020;15:289–95.

[92] Sugawara K, Yamamura N, Matsuda K, Norimatsu W, Kusunoki M, Sato T, Takahashi T. Selective fabrication of free-standing ABA and ABC trilayer graphene with/without Dirac-cone energy bands. NPG Asia Mater 2018;10:e466.

[93] Gao Z, Wang S, Berry J, Zhang Q, Gebhardt J, Parkin WM, Avila J, Yi H, Chen C, Hurtado-Parra S, Drndić M, Rappe AM, Srolovitz DJ, Kikkawa JM, Luo Z, Asensio MC, Wang F, Johnson ATC. Large-area epitaxial growth of curvature-stabilized ABC trilayer graphene. Nat Commun 2020;11.

[94] Warner JH, Mukai M, Kirkland AI. Atomic structure of ABC rhombohedral stacked trilayer graphene. ACS Nano 2012;6:5680–6.

[95] Do TN, Lin CY, Lin YP, Shih PH, Lin MF. Configuration-enriched magneto-electronic spectra of AAB-stacked trilayer graphene. Carbon N. Y 2015;94:619–32.

[96] Do TN, Shih PH, Chang CP, Lin CY, Lin MF. Rich magneto-absorption spectra of AAB-stacked trilayer graphene. Phys Chem Chem Phys 2016;18:17597–605.

[97] Lin CY, Huang BL, Ho CH, Gumbs G, Lin MF. Geometry-diversified Coulomb excitations in trilayer AAB stacking graphene. Phys Rev B 2018;98.

[98] Lee JK, Lee SC, Ahn JP, Kim SC, Wilson JIB, John P. The growth of AA graphite on (111) diamond. J Chem Phys 2008;129.

[99] Warner JH, Rümmeli MH, Gemming T, Büchner B, Briggs GAD. Direct imaging of rotational stacking faults in few layer graphene. Nano Lett 2009;9:102–6.

[100] Lee J-K, Kim J-G, Hembram KPSS, Kim Y-I, Min B-K, Park Y, Lee J-K, Moon DJ, Lee W, Lee S-G, John P. The nature of metastable AA' graphite: low dimensional nano- and single-crystalline forms. Sci Rep 2016;6:39624.

[101] Taut M, Koepernik K. Electronic structure of interfaces between hexagonal and rhombohedral graphite. Phys Rev B 2016;94:035446.

[102] Taut M, Koepernik K, Richter M. Electronic structure of stacking faults in rhombohedral graphite. Phys Rev B 2014;90:085312.

[103] Kopnin NB, Ijäs M, Harju A, Heikkilä TT. High-temperature surface superconductivity in rhombohedral graphite. Phys Rev B Condens Matter 2013;87.

[104] Dresselhaus MS, Dresselhaus G, Saito R. Physics of carbon nanotubes. Carbon N. Y 1995;33:883–91.

[105] Wildöer JWG, Venema LC, Rinzler AG, Smalley RE, Dekker C. Electronic structure of atomically resolved carbon nanotubes. Nature 1998;391:59–62.

[106] Vacacela Gomez C, Pisarra M, Gravina M, Riccardi P, Sindona A. Plasmon properties and hybridization effects in silicene. Phys Rev B 2017;95:85419.

[107] Derivaz M, Dentel D, Stephan R, Hanf MC, Mehdaoui A, Sonnet P, Pirri C. Continuous germanene layer on Al(111). Nano Lett 2015;15:2510–6.

[108] Bhimanapati GR, Lin Z, Meunier V, Jung Y, Cha J, Das S, Xiao D, Son Y, Strano MS, Cooper VR, Liang L, Louie SG, Ringe E, Zhou W, Kim SS, Naik RR, Sumpter BG, Terrones H, Xia F, Wang Y, Zhu J, Akinwande D, Alem N, Schuller JA, Schaak RE, Terrones M, Robinson JA. Recent advances in two-dimensional materials beyond graphene. ACS Nano 2015;9:11509–39.

[109] Lin CY, Do TN, Wu JY, Shih PH, Lin SY, Ho CH, Lin MF. Diverse quantization phenomena in layered materials. CRC Press; 2019.

[110] Gülseren O, Yildirim T, Ciraci S. Systematic ab initio study of curvature effects in carbon nanotubes. Phys Rev B Condens Matter 2002;65:1–4.

[111] Li J, Lei T, Wang J, Wu R, Zhao J, Zhao L, Guo Y, Qian H, Ibrahim K. Anisotropic electronic structure and interfacial chemical reaction of stanene/Bi_2Te_3. J Phys Chem C 2020;124:4917–24.

[112] Lin CY, Ho CH, Wu JY, Lin MF. Unusual electronic excitations in ABA trilayer graphene. Sci Rep 2020;10.

[113] Lin YP, Lin CY, Ho YH, Do TN, Lin MF. Magneto-optical properties of ABC-stacked trilayer graphene. Phys Chem Chem Phys 2015;17:15921–7.

[114] Lin C-Y, Ho C-H, Wu J-Y, Do T-N, Shih P-H, Lin S-Y, Lin M-F. In: Lin C-Y, Ho C-H, Wu J-Y, Do T-N, H Shih P-, Lin S-Y, Lin M-F, editors. Diverse quantization phenomena in layered materials. CRC Press; 2019.

[115] Chiu CW. Coulomb excitations in rhombohedral graphite arXiv. 2018.

[116] Huang B, Li Z, Liu Z, Zhou G, Hao S, Wu J, Gu BL, Duan W. Adsorption of gas molecules on graphene nanoribbons and its implication for nanoscale molecule sensor. J Phys Chem C 2008;112:13442–6.

[117] Ueda A, Matsuda K, Tayagaki T, Kanemitsu Y. Carrier multiplication in carbon nanotubes studied by femtosecond pump-probe spectroscopy. Appl Phys Lett 2008;92:233105.

[118] Huang YC, Chang CP, Lin MF. Magnetic and quantum confinement effects on electronic and optical properties of graphene ribbons. Nanotechnology 2007;18:495401.
[119] Tran NTT, Lin SY, Lin CY, Lin MF. Geometric and electronic properties of graphene-related systems: chemical bondings. Boca Raton, FL: CRC Press; 2017.
[120] Liu G, Liu SB, Xu B, Ouyang CY, Song HY. First-principles study of the stability of free-standing germanene in oxygen atmosphere. J Appl Phys 2015;118.
[121] Rhim JW, Moon K. Edge states of zigzag bilayer graphite nanoribbons. J Phys Condens Matter 2008;20:365202.
[122] Hasan MZ, Isaacs ED, Shen ZX, Miller LL, Tsutsui K, Tohyama T, Maekawa S. Electronic structure of Mott insulators studied by inelastic X-ray scattering. Science 2000;288:1811–4.
[123] Alfonsi J, Meneghetti M. Excitonic properties of armchair graphene nanoribbons from exact diagonalization of the Hubbard model. New J Phys 2012;14:53047.
[124] Wu JY, Chen SC, Roslyak O, Gumbs G, Lin MF. Plasma excitations in graphene: their spectral intensity and temperature dependence in magnetic field. ACS Nano 2011;5:1026–32.
[125] Lin M, Chuu D, Huang C, Lin Y, Shung KWK. Collective excitations in a single-layer carbon nanotube. Phys Rev B Condens Matter 1996;53:15493–6.
[126] Erni R, Browning ND. Valence electron energy-loss spectroscopy in monochromated scanning transmission electron microscopy. Ultramicroscopy 2005;104:176–92.
[127] Lopatin S, Cheng B, Liu WT, Tsai ML, He JH, Chuvilin A. Optimization of monochromated TEM for ultimate resolution imaging and ultrahigh resolution electron energy loss spectroscopy. Ultramicroscopy 2018;184:109–15.
[128] Matsen MW. Self-consistent field theory and its applications. Soft Matter 2006:1.
[129] Yi KS, Kim HJ. Dielectric response functions of multi-component hot carrier plasmas. Phys B Condens Matter 2015;457:149–55.
[130] Ni Z, Zhong H, Jiang X, Quhe R, Luo G, Wang Y, Ye M, Yang J, Shi J, Lu J. Tunable band gap and doping type in silicene by surface adsorption: towards tunneling transistors. Nanoscale 2014;6:7609–18.
[131] Guo Y, Zhou S, Bai Y, Zhao J. Tunable thermal conductivity of silicene by germanium doping. J Supercond Nov Magnet 2016;29:717–20.
[132] Hernández Cocoletzi H, Castellanos Águila JE. DFT studies on the Al, B, and P doping of silicene. Superlattice Microst 2018;114:242–50.
[133] Pablo-Pedro R, Magaña-Fuentes MA, Videa M, Kong J, Li M, Mendoza-Cortes JL, Van Voorhis T. Understanding disorder in 2D materials: the case of carbon doping of silicene. Nano Lett 2020;20:6336–43.
[134] Hoat DM, Nguyen DK, Ponce-Pérez R, Guerrero-Sanchez J, On V Van, Rivas-Silva JF, Cocoletzi GH. Opening the germanene monolayer band gap using halogen atoms: an efficient approach studied by first-principles calculations. Appl Surf Sci 2021;551.
[135] Wang Y, Ding Y. Strain-induced self-doping in silicene and germanene from first-principles. Solid State Commun 2013;155:6–11.
[136] Xia W, Hu W, Li Z, Yang J. A first-principles study of gas adsorption on germanene. Phys Chem Chem Phys 2014;16:22495–8.
[137] Zhou Q, Ju W, Yong Y, Liu Y, Li J. Quantum capacitance of supercapacitor electrodes based on germanene influenced by vacancy and co-doping: a first-principles study. Comput Mater Sci 2021;188.
[138] Guinea F, Castro Neto AH, Peres NMR. Electronic states and Landau levels in graphene stacks. Phys Rev B Condens Matter 2006;73.
[139] Lin CY, Huang BL, Gumbs G, Lin MF. Geometry-diversified Coulomb excitations in trilayer AAB stacking graphene. arXiv 2018;98:195442.
[140] Berman OL, Gumbs G, Lozovik YE. Magnetoplasmons in graphene structures. Prog Electromagn Res Symp 2008;2:1210–4.
[141] Magda GZ, Jin X, Hagymási I, Vancsó P, Osváth Z, Nemes-Incze P, Hwang C, Biró LP, Tapasztó L. Room-temperature magnetic order on zigzag edges of narrow graphene nanoribbons. Nature 2014;514:608–11.
[142] Berciu M, Bhatt RN. Spin waves in disordered III-V diluted magnetic semiconductors by a modified RPA approach. Phys Rev B Condens Matter 2002;66:852071–8520713.
[143] Brey L, Fertig HA. Elementary electronic excitations in graphene nanoribbons. Phys Rev B Condens Matter 2007;75:125434.

[144] Kuzuo R, Terauchi M, Tanaka M. Electron energy-loss spectra of carbon nanotubes. Jpn J Appl Phys 1992;31:L484–1487.

[145] Ophus C, Rasool HI, Linck M, Zettl A, Ciston J. Automatic software correction of residual aberrations in reconstructed HRTEM exit waves of crystalline samples. Adv Struct Chem Imaging 2016;2:15.

[146] Rao AM, Eklund PC, Bandow S, Thess A, Smalley RE. Evidence for charge transfer in doped carbon nanotube bundles from Raman scattering. Nature 1997;388:257–9.

[147] Carvalho A, Wang M, Zhu X, Rodin AS, Su H, Castro Neto AH. Phosphorene: from theory to applications. Nat Rev Mater 2016;1:1–16.

[148] Lin YC, Teng PY, Yeh CH, Koshino M, Chiu PW, Suenaga K. Structural and chemical dynamics of pyridinic-nitrogen defects in graphene. Nano Lett 2015;15:7408–13.

[149] Rybin M, Pereyaslavtsev A, Vasilieva T, Myasnikov V, Sokolov I, Pavlova A, Obraztsova E, Khomich A, Ralchenko V, Obraztsova E. Efficient nitrogen doping of graphene by plasma treatment. Carbon N. Y 2016;96:196–202.

[150] Lin YC, Lin CY, Chiu PW. Controllable graphene N-doping with ammonia plasma. Appl Phys Lett 2010;96.

[151] Ho JH, Lu CL, Hwang CC, Chang CP, Lin MF. Coulomb excitations in AA- and AB-stacked bilayer graphites. Phys Rev B Condens Matter 2006;74:85406.

[152] García-Vidal FJ, Pitarke JM. Optical absorption and energy-loss spectra of aligned carbon nanotubes. Eur Phys J B 2001;22:257–65.

[153] Lin MF, Shyu FL, Chang CP, Chen RB, Chiu CW. Magnetoelectronic and optical properties of carbon nanotubes. Phys Rev B Condens Matter 2003;67:45405.

[154] Miyake T, Saito S. Quasiparticle band structure of carbon nanotubes. AIP Conf Proc 2005;772:1037–8.

[155] Lin M, Huang C, Chuu D. Plasmons in graphite and stage-1 graphite intercalation compounds. Phys Rev B Condens Matter 1997;55:13961–71.

[156] Kociak M, Henrard L, Stéphan O, Suenaga K, Colliex C. Plasmons in layered nanospheres and nanotubes investigated by spatially resolved electron energy-loss spectroscopy. Phys Rev B Condens Matter 2000;61:13936–44.

[157] Rajasekaran S, Kaya S, Anniyev T, Ogasawara H, Nilsson A. Probing substrate effects in the carbon-projected band structure of graphene on Pt(111) through resonant inelastic x-ray scattering. Phys Rev B Condens Matter 2012;85:45419.

[158] Koch M, Ample F, Joachim C, Grill L. Voltage-dependent conductance of a single graphene nanoribbon. Nat Nanotechnol 2012;7:713–7.

[159] Lin MF, Shung KWK. Elementary excitations in cylindrical tubules. Phys Rev B 1993;47:6617–24.

[160] Liao W, Zhou G, Xi F. Optical properties for armchair-edge graphene nanoribbons. J Appl Phys 2008;104.

[161] Nelson FJ, Idrobo JC, Fite JD, Mišković ZL, Pennycook SJ, Pantelides ST, Lee JU, Diebold AC. Electronic excitations in graphene in the 1-50 eV range: the π and $\pi + \sigma$ peaks are not plasmons. Nano Lett 2014;14:3827–31.

[162] Lin MF, Shung KWK. Plasmons and optical properties of carbon nanotubes. Phys Rev B 1994;50:17744–7.

[163] Fujiwara H, Fujimoto S, Tamakoshi M, Kato M, Kadowaki H, Miyadera T, Tampo H, Chikamatsu M, Shibata H. Determination and interpretation of the optical constants for solar cell materials. Appl Surf Sci 2017;421:276–82.

[164] Naumis GG, Terrones M, Terrones H, Gaggero-Sager LM. Design of graphene electronic devices using nanoribbons of different widths. Appl Phys Lett 2009;95.

[165] Camblong HE, Levy PM. Novel results for quasiclassical linear transport in metallic multilayers. Phys Rev Lett 1992;69:2835–8.

[166] Low T, Rodin AS, Carvalho A, Jiang Y, Wang H, Xia F, Castro Neto AH. Tunable optical properties of multilayer black phosphorus thin films. Phys Rev B Condens Matter 2014;90.

[167] Bose SK. Electronic structure of liquid mercury. J Phys Condens Matter 1999;11:4597–615.

[168] Ho JH, Chang CP, Lin MF. Electronic excitations of the multilayered graphite. Phys Lett Sect A Gen at Solid State Phys 2006;352:446–50.

[169] Ho YH, Ho GW, Chen SC, Ho JH, Lin MF. Low-frequency excitation spectra in double-walled armchair carbon nanotubes. Phys Rev B Condens Matter 2007;76:115422.

[170] Huang YK, Chen SC, Ho YH, Lin CY, Lin MF. Feature-rich magnetic quantization in sliding bilayer graphenes. Sci Rep 2014;4:7509.

[171] Son YW, Choi SM, Hong YP, Woo S, Jhi SH. Electronic topological transition in sliding bilayer graphene. Phys Rev B Condens Matter 2011;84.

[172] Lee KW, Lee CE. Extreme sensitivity of the electric-field-induced band gap to the electronic topological transition in sliding bilayer graphene. Sci Rep 2015;5.
[173] Ho YH, Chiu YH, Lin DH, Chang CP, Lin MF. Magneto-optical selection rules in bilayer Bernal graphene. ACS Nano 2010;4:1465–72.
[174] Bhattacharyya S, Singh AK. Lifshitz transition and modulation of electronic and transport properties of bilayer graphene by sliding and applied normal compressive strain. Carbon N. Y 2016;99:432–8.
[175] Lin CY, Wu JY, Ou YJ, Chiu YH, Lin MF. Magneto-electronic properties of multilayer graphenes. Phys Chem Chem Phys 2015;17:26008–35.
[176] Lai YH, Ho JH, Chang CP, Lin MF. Magnetoelectronic properties of bilayer Bernal graphene. Phys Rev B Condens Matter 2008;77:85426.
[177] Lu CL, Chang CP, Huang YC, Chen RB, Lin ML. Influence of an electric field on the optical properties of few-layer graphene with AB stacking. Phys Rev B Condens Matter 2006;73:144427.
[178] Yin L-J, Shi L-J, Li S-Y, Zhang Y, Guo Z-H, He L. High-magnetic-field tunneling spectra of ABC -stacked trilayer graphene on graphite. Phys Rev Lett 2019;122:146802.
[179] Ho YH, Chiu YH, Su WP, Lin MF. Magneto-absorption spectra of Bernal graphite. Appl Phys Lett 2011;99.
[180] Orlita M, Faugeras C, Martinez G, Maude DK, Sadowski ML, Schneider JM, Potemski M. Magneto-transmission as a probe of Diracfermions in bulk graphite. J Phys Condens Matter 2008;20.
[181] Zaric S, Ostojic GN, Kono J, Shaver J, Moore VC, Strano MS, Hauge RH, Smalley RE, Wei X. Optical signatures of the Aharonov-Bohm phase in single-walled carbon nanotubes. Science 2004;304:1129–31.
[182] Suzuki H, Otsubo Y, Shen R, Kojima E, Murakami Y, Shiomi J, Maruyama S, Takeyama S. Magneto-absorption spectra from selected chirality of single-walled carbon nanotubes. J Low Temp Phys 2010;159:267–71.
[183] Shyu FL. Magneto-electronic and optical properties of zigzag silicene nanoribbons. Phys E Low-Dimensional Syst Nanostruct 2017;87:178–85.
[184] Chung HC, Yang PH, Li TS, Lin MF. Effects of transverse electric fields on Landau subbands in bilayer zigzag graphene nanoribbons. Philos Mag 2014;94:1859–72.
[185] Muoi D, Hieu NN, Nguyen CV, Hoi BD, Nguyen HV, Hien ND, et al. Magneto-optical absorption in silicene and germanene induced by electric and Zeeman fields. Phys Rev B 2020;101(20):205408.
[186] Do T-N, Shih P-H, Gumbs G, Lin M-F. Geometry- and field-diversified electronic and optical properties in bilayer silicene. arXiv 2018.
[187] Tabert CJ, Nicol EJ. Valley-spin polarization in the magneto-optical response of silicene and other similar 2D crystals. Phys Rev Lett 2013;110.
[188] Kumbhakar P, Chowde Gowda C, Mahapatra PL, Mukherjee M, Malviya KD, Chaker M, Chandra A, Lahiri B, Ajayan PM, Jariwala D, Singh A, Tiwary CS. Emerging 2D metal oxides and their applications. Mater Today 2021;45:142–68.
[189] Nguyen CV, Ngoc Hieu N, Duque CA, Quoc Khoa D, Van Hieu N, Van Tung L, Vinh Phuc H. Linear and nonlinear magneto-optical properties of monolayer phosphorene. J Appl Phys 2017;121.
[190] Wu JY, Chen SC, Do TN, Su WP, Gumbs G, Lin MF. The diverse magneto-optical selection rules in bilayer black phosphorus. Sci Rep 2018;8.
[191] Pfnür H, Tegenkamp C, Vattuone L. Plasmons in one and two dimensions. Springer Handbooks; 2020. p. 557–84.
[192] Umemoto Y, Sugawara K, Nakata Y, Takahashi T, Sato T. Pseudogap, Fermi arc, and Peierls-insulating phase induced by 3D–2D crossover in monolayer VSe2. Nano Res 2019;12:165–9.
[193] Pärnaste M, Marcellini M, Holmström E, Bock N, Fransson J, Eriksson O, Hjörvarsson B. Dimensionality crossover in the induced magnetization of Pd layers. J Phys Condens Matter 2007;19.
[194] Ho YH, Wang J, Chiu YH, Lin MF, Su WP. Characterization of Landau subbands in graphite: a tight-binding study. Phys Rev B 2011;83:121201.
[195] Uda T. Optical properties of graphite in the far-infrared region. J Phys Soc Japan 1970;28:946–54.
[196] Koshino M. Stacking-dependent optical absorption in multilayer graphene. New J Phys 2013;15:15010.
[197] Chen RB, Lee CH, Chang CP, Lin MF. Electronic and optical properties of finite carbon nanotubes in an electric field. Nanotechnology 2007;18:508–15.
[198] Chen RB, Chang CP, Shyu FL, Hwang JS, Lin MF. Optical excitations of finite carbon nanotubes. Carbon N. Y 2004;42:531–5.

[199] Wolf O, Dasog M, Yang Z, Balberg I, Veinot JGC, Millo O. Doping and quantum confinement effects in single Si nanocrystals observed by scanning tunneling spectroscopy. Nano Lett 2013;13:2516–21.
[200] Sofer Z, Bouša D, Luxa J, Mazanek V, Pumera M. Few-layer black phosphorus nanoparticles. Chem Commun 2016;52:1563–6.
[201] Shimizu A, Ogawa T, Sakaki H. Two-photon absorption spectra of quasi-low-dimensional exciton systems. Phys Rev B 1992;45(19):11338.
[202] Lin CH, Huang A, Pai WW, Chen WC, Chen TY, Chang TR, Yukawa R, Cheng CM, Mou CY, Matsuda I, Chiang TC, Jeng HT, Tang SJ. Single-layer dual germanene phases on Ag(111). Phys Rev Mater 2018;2:24003.
[203] Zhu FF, Chen WJ, Xu Y, Gao CL, Guan DD, Liu CH, Qian D, Zhang SC, Jia JF. Epitaxial growth of two-dimensional stanene. Nat Mater 2015;14:1020–5.
[204] Chiu YH, Ou YC, Lin MF. Optical properties of graphene in external fields. Graphene Sci Handb Electr Opt Prop 2016;18:469–88.
[205] Berdiyorov GR, Neek-Amal M, Peeters FM, Van Duin ACT. Stabilized silicene within bilayer graphene: a proposal based on molecular dynamics and density-functional tight-binding calculations. Phys Rev B Condens Matter 2014;89:24107.
[206] Du Y, Zhuang J, Wang J, Li Z, Liu H, Zhao J, Xu X, Feng H, Chen L, Wu K, Wang X, Dou SX. Quasi-freestanding epitaxial silicene on Ag(111) by oxygen intercalation. Sci Adv 2016;2:e1600067.
[207] Lian C, Ni J. Strain induced phase transitions in silicene bilayers: a first principles and tight-binding study. AIP Adv 2013;3:52102.
[208] Kane CL, Mele EJ. Quantum Spin hall effect in graphene. Phys Rev Lett 2005;95:226801.
[209] Zhang Y, Tan YW, Stormer HL, Kim P. Experimental observation of the quantum Hall effect and Berry's phase in graphene. Nature 2005;438:201–4.
[210] Ajiki H, Ando T. Energy bands of carbon nanotubes in magnetic fields. J Phys Soc Japan 1996;65:505–14.
[211] Wang Y, Zheng J, Ni Z, Fei R, Liu Q, Quhe R, Xu C, Zhou J, Gao Z, Lu J. Half-metallic silicene and germanene nanoribbons: towards high-performance spintronics device. Nano 2012;7:1250037.
[212] Xing Y, Sun QF, Wang J. Nernst and Seebeck effects in a graphene nanoribbon. Phys Rev B Condens Matter 2009;80:235411.
[213] Chen SC, Wang TS, Lee CH, Lin MF. Magneto-electronic properties of graphene nanoribbons in the spatially modulated electric field. Phys Lett Sect a Gen at Solid State Phys 2008;372:5999–6002.
[214] Guinea F, Castro Neto AH, Peres NMR. Electronic states and Landau levels in graphene stacks. Phys Rev B Condens Matter 2006;73:245426.
[215] Gusynin VP, Sharapov SG, Carbotte JP. Magneto-optical conductivity in graphene. J Phys Condens Matter 2007;19:26222.
[216] Gusynin VP, Sharapov SG, Carbotte JP. AC conductivity of graphene: from tight-binding model to 2 + 1-dimensional quantum electrodynamics. Int J Mod Phys B 2007;21:4611–58.

CHAPTER

3

Experimental measurements

Thi Dieu Hien Nguyen[1], Shih-Yang Lin[1], Vo Khuong Dien[1], Chi-Hsuan Lee[1], Hsin-Yi Liu[2], Hai Duong Pham[1], Thi My Duyen Huynh[1], Nguyen Thi Han[1], Ngoc Thanh Thuy Tran[3], Wei-Bang Li[1] and Ming-Fa Lin[1]

[1]Department of Physics, National Cheng Kung University, Tainan, Taiwan; [2]Center for Micro/Nano Science and Technology, National Cheng Kung University, Tainan, Taiwan; [3]Hierarchical Green-Energy Materials (Hi-GEM) Research Center, National Cheng Kung University, Tainan, Taiwan

Experimental measurements and theoretical calculations (details in Chapter 2) have been simultaneously illustrated in the previously published [1–3] and current books through a unified quasiparticle framework. Very interestingly, the former cover a plenty of physical [1,3], chemical [4–6] and material techniques [7,8] in terms of sample growths [9,10], crystal symmetries (e.g., stacking, layer number, bucking and edge structures in Refs. [3,11,12]), spatial charge densities [1,3], electronic energy spectra [12–14], van Hove singularities [12,13,15,16], net magnetic moments [17], ferromagnetic spin configurations [1,17], optical dielectric functions/reflectance spectra/absorption coefficients/electron energy loss functions [18], magneto-electronic quantization phenomena [19–21], magneto-optical selection rules [20], Coulomb/magneto-Coulomb electron-hole & collective excitations [22–27], quantum Hall conductivities [28,29], as well as chemical decorations/substitutions/adsorptions [13,17,30,30–34]. Obviously, the simultaneous examinations on certain important properties have been successfully done for low-dimensional materials, e.g., radii & chiralities/widths & edge structures of carbon nanotubes/graphene nanoribbons by STM & STS [35–39]. When the high-precision measurements are available up to now, the detailed comparisons with the predicted results would be fully expressed in Chapters 4–20 (discussed later). As for the absence of experimental data, the related facts might arise from the different intrinsic limits of equipment [40] and very complicated analysis processes. The close combinations

Fundamental Physicochemical Properties of Germanene-related Materials
https://doi.org/10.1016/B978-0-443-15801-8.00013-X

between experimental and theoretical viewpoints are rather useful in developing the diversified quasiparticle phenomena [1,3]. The frequently utilized observations are discussed in the following paragraphs.

3.1 STM, LEED/TEM, and X-ray

A successful synthesis on an emergent material needs to verify its crystal symmetry by observing surface or bulk morphology [41,42]. STM (scan tunneling microscopy in Fig. 3.1; [41,43,44]), LEED [low-energy electron diffraction in Fig. 3.2; [45];/TEM (tunneling electron microscopy in Fig. 3.3; [42,46]), and X-ray (Fig. 3.4; [47–49]); are available in examining

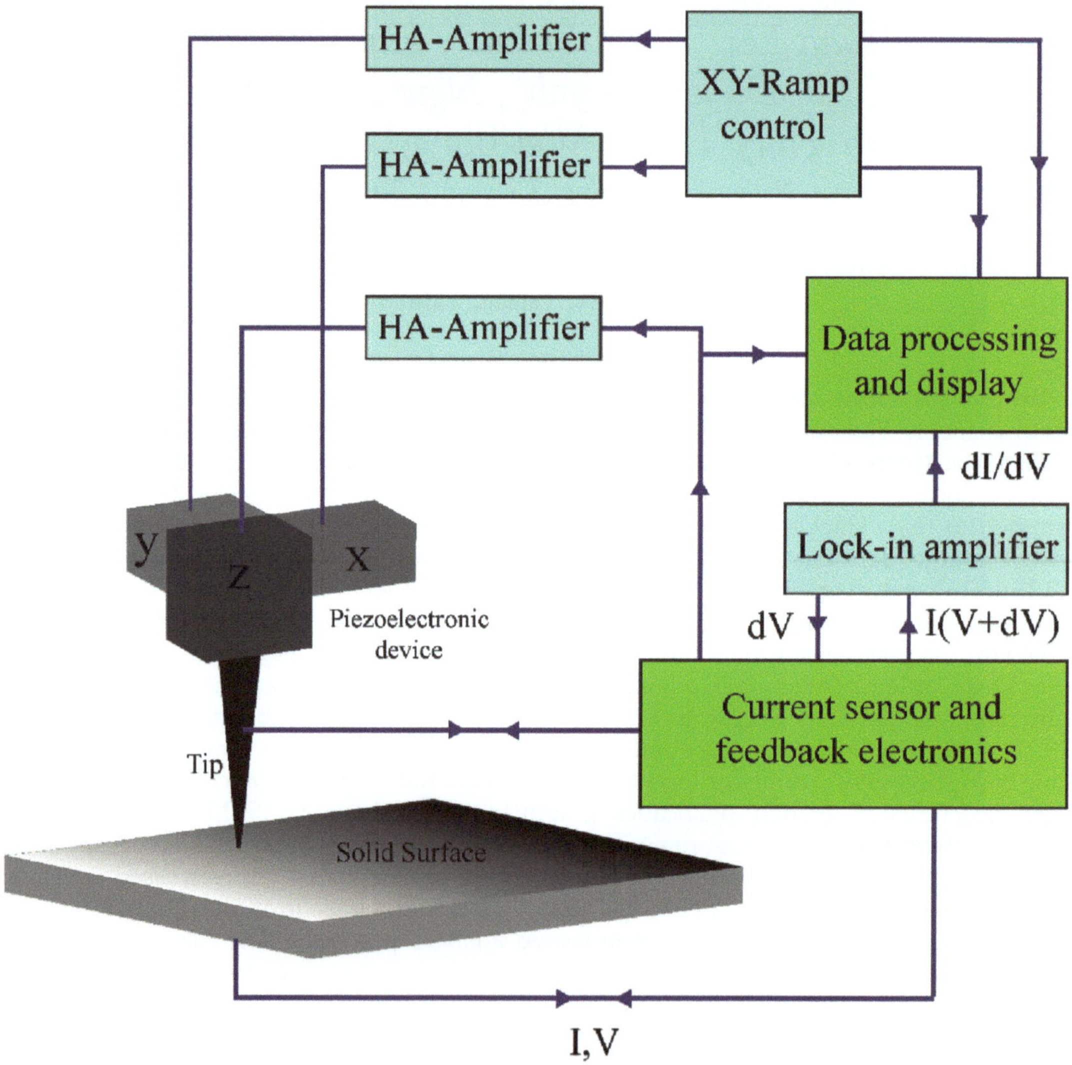

FIG. 3.1 The set-up of scanning tunneling microscopy.

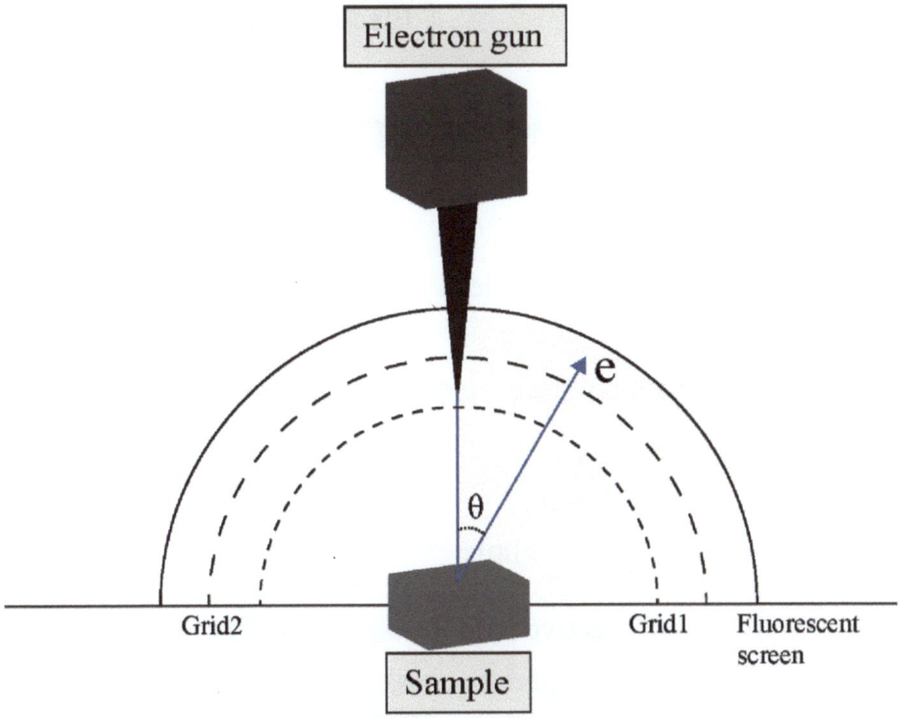

FIG. 3.2 Low-energy electron diffraction.

the atomic positions in a primitive unit cell. The former two and the last one are, respectively, suitable for the 2D/1D/0D [47,50–52] and 3D [49] condensed-matter systems, e.g., the honeycomb & buckled structure of germanene [53,54] and the 3D LiGeO compound (electrolyte of Li^+-based batteries; [55]) Such high-precision measurements are done, respectively, through the weak quantum tunneling current between a nanoscale probe and sample [41,56,57], the electron diffraction [49], and the elastic photon scatterings with a charge density distribution [58]. Whether these observations could provide very useful information about the estimated charge density distributions are worthy of detailed discussions. The close cooperation between the experimental and theoretical studies in geometric symmetries [55,59] and charge densities under a unified quasiparticle framework [55,59] should be one of studying focuses.

The high-precision STM measurements, as shown in Fig. 3.1, can clearly identify the various surface morphologies [41,60] since their first observation on a cylindrical carbon honeycomb crystal in 1998 [43], e.g., the planar [61], curved [62], folded [63,64], and scrolled [2,65] graphene nanoribbons. Such examinations are operated through a rather weak response of quantum tunneling current between a nanoscale probe [41,66] and a few-layer sample [67]. The gate voltage is able to greatly enhance the electronic transport abilities. Most important, the observable current at any positions is fixed through the modulation of distinct STM heights; therefore, a surface crystal structure is achieved under the delicate analyses. A lot of successful identifications have obviously indicated the STM merits (details in the following paragraph; [35,41,44]). Whether they could provide the information of spatial

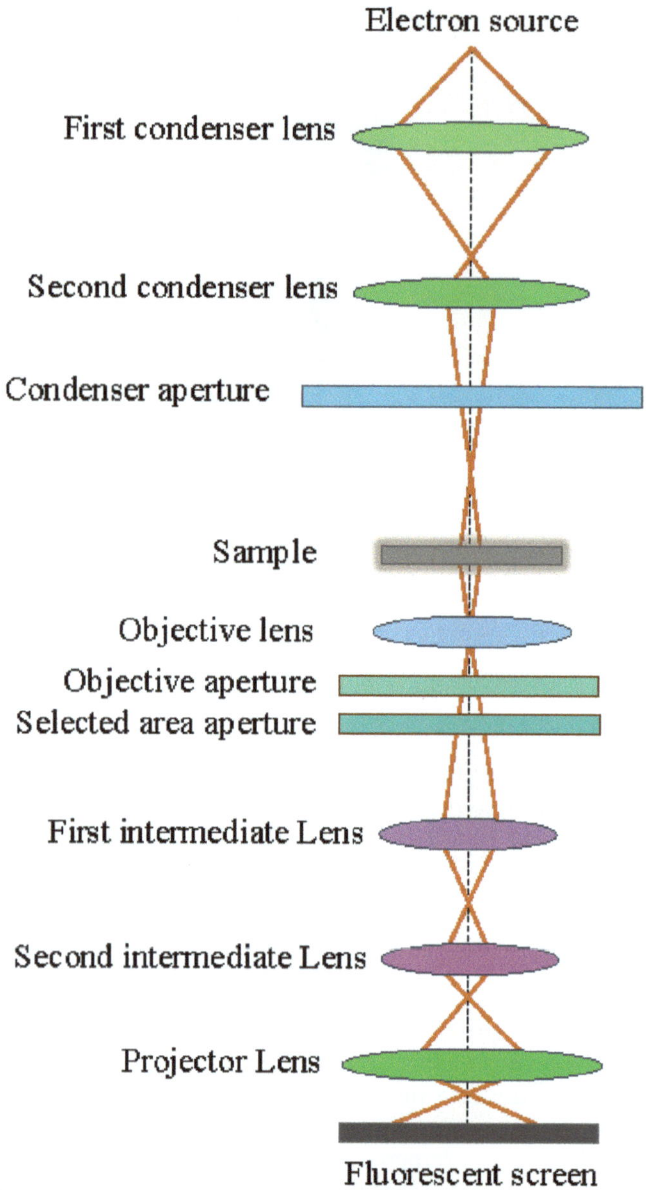

FIG. 3.3 Tunneling electron microscopy.

charge density distributions needs the strong cooperation of experimental and theoretical research under a unified quasiparticle framework. Concerning the quasiparticle diffractions of electrons and electrons/photons, the coupling od the identical/different quasiparticle are clearly presented in the high-precision measurements of LEED [45,68] and TEM [40,42,46]/X-ray [47–50]. According to the elastic scatterings associated with the composite

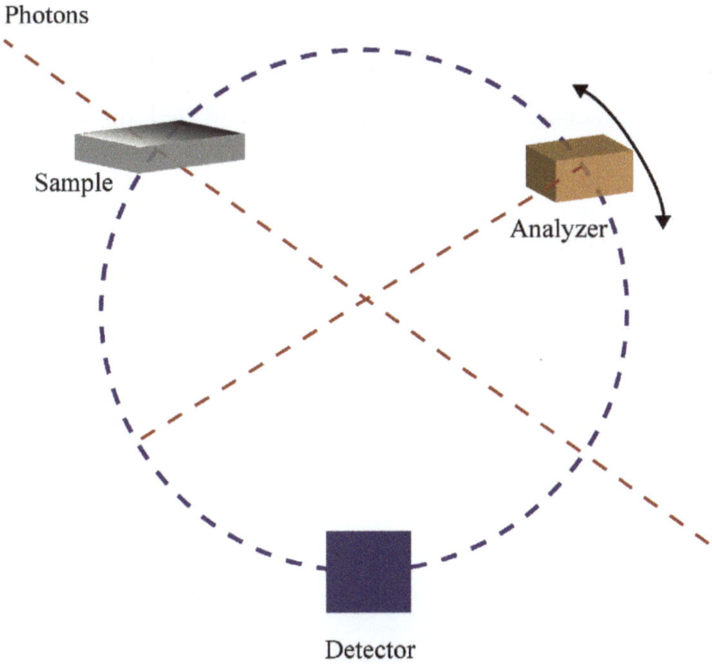

FIG. 3.4 X-ray.

quasiparticles [49], the prominent peaks come to exist under the specific conditions of the reciprocal lattices. The incident angle and wavelength, which are generated by the perturbed electron/photon beam, are responsible for the constructive scattering events [69]. One of the critical mechanisms is the periodical charge density, in which its distribution on a crystal structure could be estimated from the measured diffraction data (the following paragraph; [47,48,50]). This is very helpful in comparing with the VASP simulations about the spatial carrier density distributions and thus determine the significant orbital hybridizations in chemical bonds. Also noticed that how to utilize the optimal geometries of the VASP simulations in predicting the diffraction patterns should be a further studying focus.

The above-mentioned examinations have verified the rich and unique crystal symmetries. For example, the X-ray diffraction peaks, at least 50 years ago [70], clearly reveal an unusual structure about sodium chloride, where is consists of two faced cubic systems with a relative displacement along the $\hat{x}$, $\hat{y}$, or $\hat{z}$ direction. By using the ionic atomic configurations (their saturated orbitals [70]), the slight modifications on charge density distributions, corresponding to the weakly deformed isotropic form, could be achieved by means of the elastic diffraction model. Very interestingly, the high-resolution STM measurements are quite successful in verifying the diverse carbon honeycomb crystals since the first discovery about the achiral and chiral hexagonal arrangements on the cylindrical surfaces, such as, the 2D planar (graphenes [39,71])/cylindrical [35,43,72]/toroidal [73]/curved [74]/folded [75,76]/scrolled [63,65]/hatted/nondeformed (graphene nanoribbons; [38,60,61]/closed molecular [77,78]/ring-like [79]/nonstraight chain structures [80]). Such observations directly reflect the diversified chemical bondings of sp^2 [81,82] and sp [37], being attributed to the four half-occupied orbitals of carbon atoms in any condensed-matter systems [83]. Both delicate LEED and TEM detections can reveal the side-view morphologies, covering the normal ABA [84] & ABC [3,85] stackings in few-layer graphenes [67,86], the low/high rotation symmetry in bilayer graphenes [3,87] & doubled-wall carbon nanotubes [88], the stacked/curved/seriously deformed [2,24,89] graphene nanoribbons, and the interlayer distances of multi-layer graphene-related systems [3]. While the low-energy LEED or TEM is directly combined with the high-energy partner of electron energy loss spectroscopy (REELS or TEELS) ([90] or [91]), the Coulomb excitation spectrum is achieved from the in-situ well-characterized crystal structure. That is, the elastic and inelastic scattering events are delicately measured for the low-dimensional systems, such as, the discrete inter-π-band plasmon modes in a carbon nanotube bundle from the high-resolution measurements of TEM and EELS [46,92]. How to utilize STM and LEED/TEM in silicene-/germanene-/tinene-/plumbene-based materials in revealing the top and side views should overcome the nonnegligible substrate effects [41]. This requires a close cooperation of experimental and theoretical studies under a unified quasiparticle framework [2,3].

3.2 STS

Very importantly, the observations of scanning tunneling spectroscopy (Fig. 3.4) are able to directly reveal the density of states/van Hove singularities across the chemical potential under the absence of wave-vector dependence. STS is reliable for the low-dimensional materials because of the quite weak quantum tunneling currents between a needle probe and a sample surface. Its combination with STM has successfully in examining the geometric and electronic properties. For example, such tests can clearly identify the chiral/achiral hexagon arrangement on a cylindrical carbon nanotube surfaces [43] (two open edges of a planar graphene nanoribbon [93]) and simultaneously examine the featured band gaps & different van Hove singularities. Only each armchair carbon nanotube exhibits a finite free carrier density and thus belong to 1D metals [37,94]. The others are narrow- (about 1/3) or middle-gap semiconductors ($\sim$2/3). Furthermore, all graphene nanoribbons are small or middle-gap semiconductors mainly owing to the finite-width quantum confinement and the nonuniform chemical bond lengths [2]. The similar research could be generalized to group-IV and group-V nanoribbons [95–97] and nanotubes [96] without/with chemical modifications [37,98–100]. As for 3D materials, both STS and STM would become useless, e.g., the VASP simulations on van Hove singularities of LiGeO [55,59] in the absence of experimental observations. This might induce certain troubles for directly verifying the critical orbital hybridizations of chemical bonds within the specific energy ranges [55,59]. In addition, optical measurements are only related to the joint density of states between the initial and final ones discussed in Chapters 3 and 4.

STS is a standard extension of STM [41,66,101], in which quantum tunneling currents at any positions are delicately detected through the identical gate voltages. In general, the differential relation of dI/dV is taken in the detailed analyses, being roughly estimated to be proportional to the energy-dependent density of states [39,101]. Its magnitude is mainly determined from the position-dependent average value associated with the whole (x, y) plane

[41,72]. A lot of high-precision data are able to provide the form, number, energy and height of each van Hove singularity across the Fermi level [102]. These features, which are created by the band-edge states [the critical points in energy-wav-vector space; [103], are available in examining the predicted valence and conduction subbands through the VASP simulations/ the tight-binding model [19,20,104–107]. A semiconductor or metal is characterized by a vanishing or sufficiently high DOS near $E = 0$. Even under the absence of orbital-decomposed results, the unusual structures, coming to exist within the specific energy ranges, are very useful for the first-step comprehending of the concise orbital hybridizations [3,19]. The close relations between STS measurements and VASP calculations are expressed in Chapters 4–20. How to utilize both of strong cooperation in clearly presenting diverse phenomena should be the studying focuses [19].

The up-to-date STS measurements have successfully verified the diverse quasiparticle pictures close to the Fermi level [41,66,101], especially for the featured van Hove singularities in graphene-related sp^2-bonding materials [101,108]. Their close combinations with STS techniques can directly examine the specific relations between the geometric and electronic properties in the single-walled carbon nanotubes, such as, the metallic armchair (m, m) systems [109,110] and the middle-gap (m, n) ones of $2m + n = 3I$ (I an integer); band gap inversely proportional to radius due to the periodical boundary condition; [111,112]. As for the third kind of narrow-gap carbon nanotubes with $2m + n = 3I$, their small E_g s are created by the nonparallel orientation of carbon-$2p_z$ on a cylindrical surface, and the theoretical predictions of $1/r^2$ are also verified by the STS high-resolution observations [72]. The simultaneous examinations on the nanoscale width/open edge and band gap are thoroughly conducted on planar graphene nanoribbons [113]. A monolayer graphene is a zero-gap semiconductor since its Dirac cone shows a V-shaped structure with a zero DOS at the Fermi level in the STS observations [72,114]. Both AB- [115] and ABC-stacked graphenes [115] exhibit the semimetallic characteristics, being attributed to the weak, but significant band overlaps [115,116]. Also noticed that the layered ABC stackings are able to generate the partially flat bands across the chemical potential because of the localized surface states; that is, their DOSs are very high at $E = 0$. The trilayer AAB stacking, with rather complicated hopping integral, is examined to be a narrow-gap semiconductor, being consistent with the theoretical predictions from the first-principles method [3] and the tight-binding model [117,118]. The above-mentioned successful detections on the low-energy electronic properties are closely related to the well-behaved carbon honeycomb lattices with the concise π bondings. However, the similar STS measurements are almost absent in silicene-germanene-/tinene-/ plumbene-based materials, mainly owing to the nonnegligible chemical bondings of substrate [119–121]. How to greatly reduce the substrate effects is an open challenge of experimental and theoretical cooperation.

3.3 ARPES

The high-precision measurements of photoemission energy spectra are very successful in directly observing the wave-vector-dependent quasi-particle energies and their lifetimes for the occupied electron states [66,122]. The low-lying energy dispersions, being below the Fermi level, could be thoroughly examined from the delicate analyses [123], as well as the detailed comparisons with the theoretical predictions of the first-principles simulations/the tight-binding models [3,124,125]. The diverse behaviors are clearly revealed in a lot of emergent 2D materials, e.g., the stacking- and layer-number-enriched valence subbands in group-IV and group-V [1,3,15,126,127] (discussed later; details in books [1,3]). How to closely link the experimental observations and the theoretical models under a unified quasiparticle framework [1–3] is the studying focus. On the other hand, the ARPES examinations cannot be easily done for the intrinsic limits of physical and chemical environments. In general, there exist the nonconservation of (momentum, energy) transfers due to the surface morphology confinement [128], a lot of atoms in Moire superlattices (too many energy subbands to analyze simultaneously/very complicated multi-orbital hybridizations; [14,55,129]), and the undefined momentum direction for the deformed low-dimension systems (e.g., a very long, but curved carbon nanotubes). Overcoming these difficulties become open issues in the near-future research. For example, the ternary/quaternary lithium oxides play the critical roles in 3D core anode/electrolyte/cathode materials of lithium-ion-based batteries [14,55,130–134], while the predictions of electronic energy spectra need to be tested as soon as possible. The close cooperation of models and observations would achieve certain important conclusions.

The inelastic quasiparticle scatterings, many-particle photon- [135], electron- [135] and phonon-electron interactions at any temperatures, dominate the dynamic photoemission processes [136] because of the strong charge screening effects closely related to all electrons [135] and ions [135]. The electromagnetic perturbations are capable of generating the excited electron-hole pairs. Such intermediate states experience the various scattering evens, and then reach a surface boundary. While the excited quasiparticles have the sufficiently kinetic energies to overcome a specific work function [137], their momenta and energies could be detected within a small solid angle, as clearly indicated in Fig. 3.5. Apparently, too deep carriers, being below a planar surface beyond $\sim 10-20$ Å, can escape the quantum confinement only under an almost vanishing probability. This will lead to the heterojunction-enriched quasiparticle phenomena, e.g., the very strong substrate effects in the presence of dominating chemical bondings [137]. Very complicated decay behaviors of excited quasiparticles play the critical roles in the ARPES measurements [66,122,138,139], as previously predicted phenomena by the self-energy/modified self-energy methods [140,141]. Certain merits and drawbacks frequently come to exist under the intrinsic limits, so the delicate observations and analyses are strongly required for the identifications of quasiparticle energy spectra and their lifetimes. The former and the latter are, respectively, achieved from the strongest spectrum intensities and the effective width of strength distribution [139]. That is, the wave-vector-dependent photoemission spectra can evaluate energies and energy uncertainties of occupied electronic states simultaneously, with the various scatterings [66,142].

The up-to-date ARPES measurements clearly show that the low-dimensional systems clearly possess the diverse quasiparticle phenomena [122,138,139], especially for the full examinations about graphene-related 2D/1D/3D materials [122,123,143,144]/[145–147]. For example, monolayer [3], bilayer AB stacking [148], ABA- & ABC-stacked trilayer graphene [115,149] exhibit the different numbers of π-electronic valence subbands in the presence of stacking-enriched energy dispersion relations. Furthermore, the distribution width of ARPES intensities, being done for alkali-adsorbed monolayer graphene systems, show the strongly

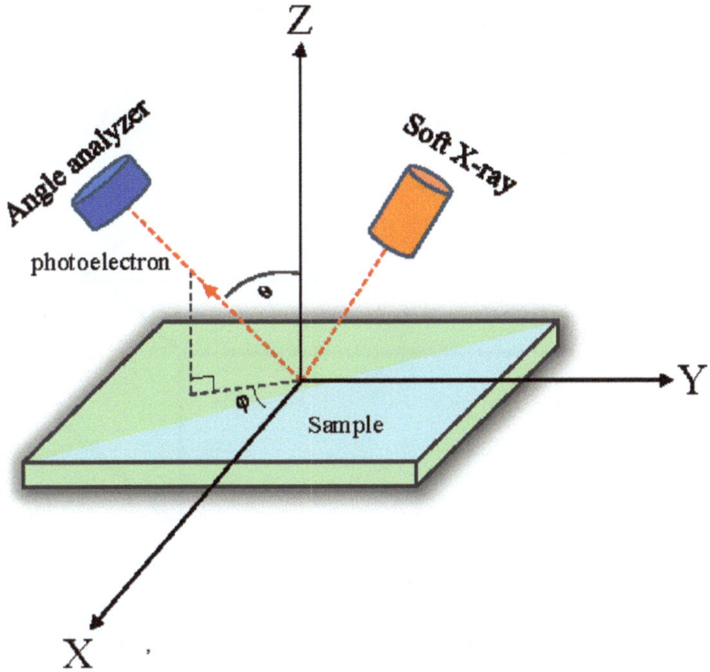

FIG. 3.5 Angle-resolved photoemission spectroscopy.

anisotropic dependence on wave vectors for the excited conduction electrons/holes and valence holes [150]. Only the latter below the valence Dirac point can present the Coulomb decays through the low-frequency 2D acoustic plasmon modes. This is consistent between experimental observations and theoretical calculations [3]. 1D graphene nanoribbons [2,151,152], with the finite-width quantum confinements, are verified to display the parabolic π-valence subbands with an observable band gap. However, the sensitive dependences of E_g's on widths and edge structures, which have been predicted by the first-principles simulations [2] and the tight-binding models [19,153,154], are not finished in the previous ARPES [66,138] and STS examinations [39,101]. The high-resolution ARPES examinations, which are conducted for few-layer silicene/germanene/tinene/plumbene grown on different substrates through the molecular beam epitaxy (MBE; [120,155]), have successful verified the heterojunction-enriched quasiparticle energy spectra. The very strong modifications on pristine results clearly illustrate that there exist, at least, three kinds of chemical bondings at two subsystems and heterojunction. The complicated orbital hybridizations are further identified from the reliable VASP simulations and analyses, in which four-/five-layer substrate widths are cover in the numerical calculations (meaningless for too-thick results [3]). As for a bulk graphite with a natural AB stacking [148], its featured π-electronic valence subbands are verified to survive along the high-symmetry paths of the 3D first Brillouin zone (Fig. 3.6). Most importantly, the z-direction surface confinement can be overcome by using the main features of electronic energy spectra. However, the opposite is true for 3D compounds with Moire supperlattices or complex orbital hybridizations, e.g., a lot of ternary/quaternary lithium

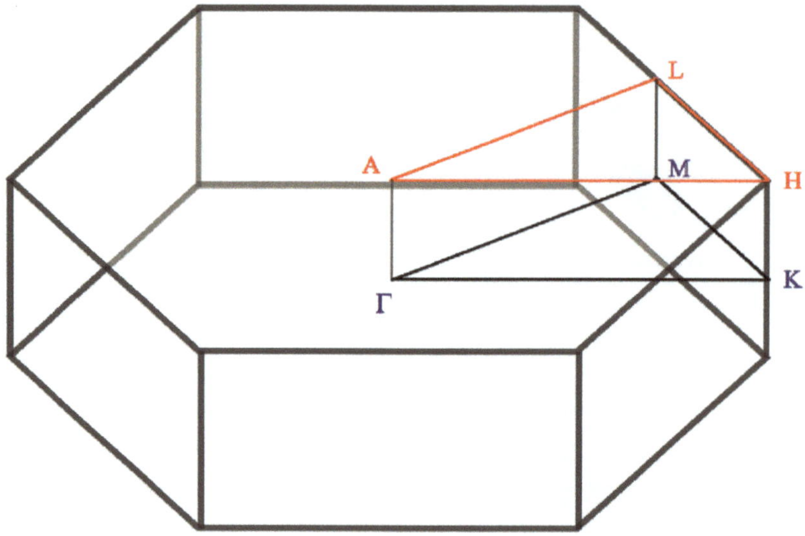

FIG. 3.6 The first brillouin zone of a bernal graphite with AB stacking.

oxides as the core cathode, electrolyte, and anode materials of Li^{+}-based batteries [130,134,156]. In this book, the ARPES observations are expected to be capable of providing the unusual quasiparticle energy spectra for the low-dimension germanene/germanene nanoribbons/germanene nanotubes [6,126,157,158] in the absence or presence of chemical modifications. This further indicates the consistent viewpoint of quasiparticle framework for the simultaneous progress of experimental and theoretical research.

3.4 Optical reflectance/absorption/transmission/photoluminescence spectroscopies

Experimental examinations on optical properties are very important in verifying the theoretical models and provide the highly potential applications by the featured transition behaviors. In general, the high-resolution experiments are conducted on the optical reflectance (Fig. 3.7A); [159,160]/absorption [161]/transmission [18,162]/photoluminescence spectra (Fig. 3.7B; [163,164]), in which the last ones are mainly focused on the excitonic effects. Whether the stable/meta-stable excitons come to exist is estimated to depend on the semiconducting properties, but not the metallic/semi-metallic ones (details in Chapter 7). Free carriers of a metal exhibit the strong screening effects so that the attractive Coulomb potential between the excited electron and hole will decline very rapidly as their distance grows. This is responsible for the absence of many-body optical effects, while one frequency-dependent optical property is achieved under the delicate observations and analyses, the Kramers-Kronig relations due to the physical events are available for getting the real and imaginary parts of the transverse dielectric function at long wavelength limit. However, the principle-value integrations require all optical data in an infinite frequency

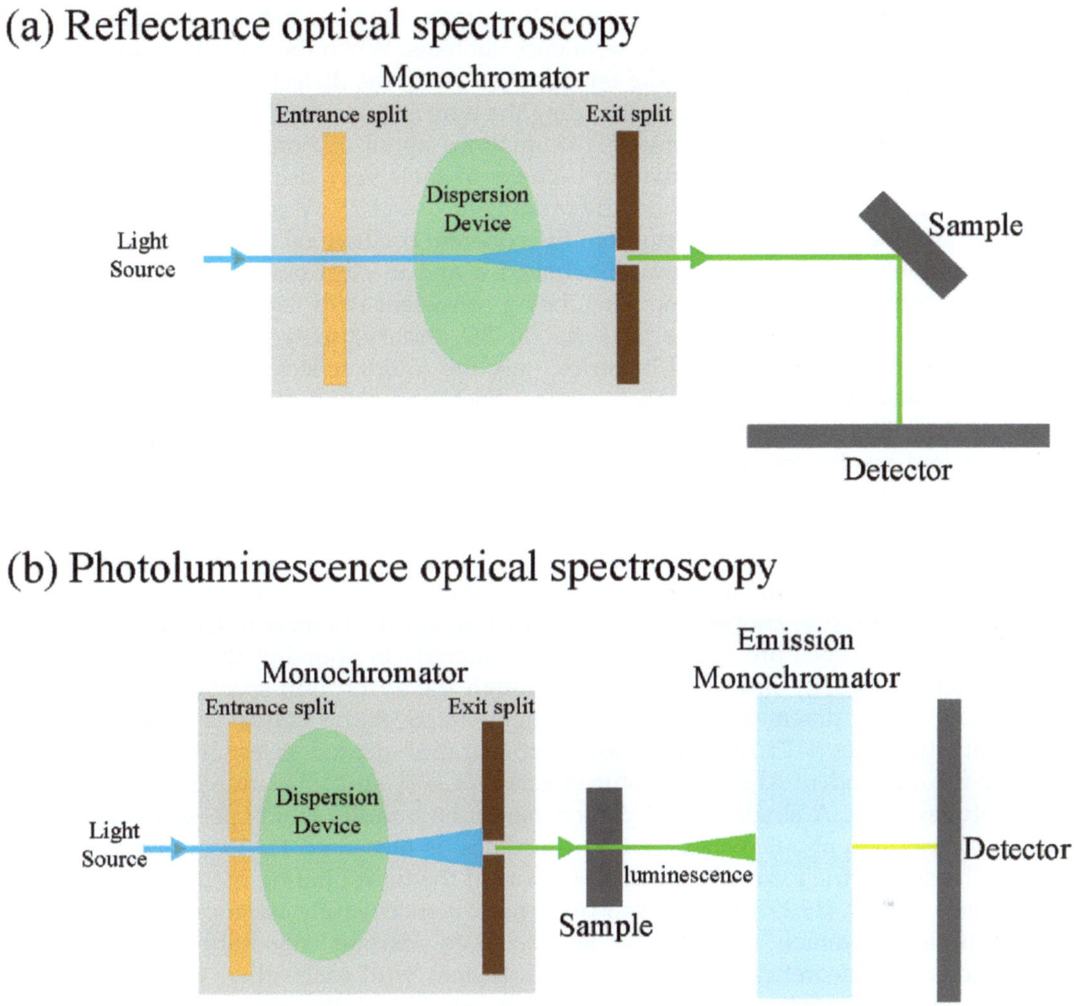

FIG. 3.7 Optical spectroscopies for (A) reflectance and (B) photoluminescence spectra.

range. The insufficiently experimental data need to be overcome by the different model approximations for the super-high-frequency vertical excitations. This could be partially supported through the specific phenomenological models [19,20]. Most important, how to directly link each prominent absorption structure with the critical orbital hybridizations of the initial and final states will be developed for emergent materials under the current quasiparticle framework [1–3].

The optical reflectance spectrum, which is done for the Bernal graphite by Taft and Phillip in 1965 within a sufficiently wide frequency range of $\omega < 26$ eV [165], clearly illustrates the concise vertical transitions of the almost perpendicular π- and σ-electronic states. The specific excitation structures, respectively, appearing at 0.8 eV, 5.3 eV, and 15 eV, are attributed to two pairs of band-overlap valence and conduction states, the inter-π-band excitations associated with the critical points, and three σ-valence subbans. When the measured data are reliably supported by the Kramers-Kronig relations [137,166], the dielectric function and energy loss function are very suitable in understanding the bare and screened response abilities of all orbital-dependent charges. The numerical estimations about carbon-1s excitation contributions could make a good approximation because of their very deep energy under a very strong binding. $Im(-1/\varepsilon)$ further show the prominent 7- and 25-eV plasmon peaks, in which the former and the latter mainly arise from the coherent oscillations of all π and $\pi + \sigma$ electrons, respectively. Apparently, the well-behaved π and σ -electronic bondings are responsible for the rich quasiparticle phenomena, being consistent with the developed framework in the previous and current books [1,3]. As for 3D ternary/quaternary lithium oxides of core materials in lithium-ion-based batteries, only the rough reflectance spectra are shown within the relatively narrow frequency ranges. Each absorption structure cannot identify the initial and final band-edge states and thus determine the critical chemical bondings. Such measured results are very difficult to further analyze the real and imaginary parts of the dielectric function since there might exist the optical transitions of the complicated multi-orbital hybridizations at lower and higher frequencies. That too many atoms in a unit cell create the highly nonuniform chemical environment and a lot of unusual energy dispersions is the main reason. How to greatly reduce the measured and analyzed uncertainties is to strengthen the close cooperation of experimental and theoretical research.

1D carbon nanotubes are very outstanding materials in showing the single [167] and many-particle [168–170] optical excitations through the absorption spectra. The successful observations on the diverse quasiparticle phenomena cover the radius- & chirality-created prominent structures [171,172], the electric-polarization-related selection rules [24], the magnetic-flux-dependent Aharonov-Bohm effects [173,174], and 1D excitonic in semiconducting nanotubes [170]. A single-walled carbon nanotube bundle, which possesses an almost uniform distribution about their radii and chiralities, is measured for its strong absorption peaks with asymmetric forms. By the full support of the tight-binding model and Kubo formula (details in Refs. [19,175]), each absorption peak is successfully assigned to a specific optical excitation channel of $(m, n)\%$ single-walled carbon nanotubes. Experimental examinations and theoretical predictions are consistent with each other. When the electric field of an electromagnetic wave is parallel/perpendicular to the nanotube axis, the initial and final states should have the same angular momentum/the difference of ± 1 according to the model calculations [23,24] and the VASP simulations [2,170,176]. This is verified under the measured absorption spectra [177,178]. The magneto-optical detections present the oscillatory absorption structures with the sensitive magnetic-flux dependences [179,180]. More importantly, all carbon nanotubes, metallic armchair carbon nanotubes excepted, could exhibit the sufficiently stable excitons [170,181] and thus display the reduced threshold absorption frequency and enhance the lower-frequency intensity. The prominent excitonic effects are directly reflected in the photoluminescence spectrum and thus the optical display applications [163,182]. The similar excitons might come to exist in ternary/quaternary lithium oxides, while the experimental analyses are very difficult because of the complicated band structures due to the quasi-Moire superlattices. The many-body VASP simulations [59] can provide the full support in overcoming the data-analysis barriers.

References

[1] Shih-Yang Lin H-YL, Nguyen TDH, Tran NTT, Pham HD, Chang S-L, Lin C-Y, et al. Silicene-based layered materials: essential properties. IOP Publishing; 2020.

[2] Lin SYT, N.T.T., Chang SL, Su WP, Lin MF. Structure- and adatom-enriched essential properties of graphene nanoribbons. CRC press; 2018.

[3] Tran NTT, Lin S-Y, Lin C-Y, Lin M-F. Geometric and electronic properties of graphene-related systems: chemical bonding schemes. CRC Press; 2017.

[4] Pan B, Xing B. Environ Sci Technol 2008;42:9005–13.

[5] James DK, Tour JM. Macromol Chem Phys 2012;213:1033–50.

[6] Pang Q, Zhang Y, Zhang J-M, Ji V, Xu K-W. Nanoscale 2011;3:4330–8.

[7] Singh V, Joung D, Zhai L, Das S, Khondaker SI, Seal S. Prog Mater Sci 2011;56:1178–271.

[8] Compton OC, Nguyen ST. Small 2010;6:711–23.

[9] Park HJ, Meyer J, Roth S, Skákalová V. Carbon 2010;48:1088–94.

[10] Lee J-K, Lee S-C, Ahn J-P, Kim S-C, Wilson JI, John P. J Chem Phys 2008;129:234709.

[11] Li W-B, Lin S-Y, Tran NTT, Lin M-F, Lin K-I. RSC Adv 2020;10:23573–81.

[12] Lin S-Y, Chang S-L, Tran NTT, Yang P-H, Lin M-F. Phys Chem Chem Phys 2015;17:26443–50.

[13] Pham HD, Su W-P, Nguyen TDH, Tran NTT, Lin M-F. R Soc Open Sci 2020;7:200723.

[14] Nguyen TDH, Pham HD, Lin SY, Lin MF. RSC Adv 2020;10:14071–9.

[15] Fu H, Zhang J, Ding Z, Li H, Meng S. Appl Phys Lett 2014;104:131904.

[16] Nguyen DK, Tran NTT, Chiu Y-H, Lin M-F. Sci Rep 2019;9:1–15.

[17] Pham HD, Gumbs G, Su W-P, Tran NTT, Lin M-F. RSC Adv 2020;10:32193–201.

[18] Fox M. Optical properties of solids. Oxford University Press; 2001.

[19] Lin C-Y, Ho C-H, Wu J-Y, Do T-N, Shih P-H, Lin S-Y, Lin M-F. Diverse quantization phenomena in layered materials. CRC Press; 2019.

[20] Chiun-Yan Lin J-YW, Chiu C-W, Lin M-F. Coulomb excitations and decays in graphene-related systems, in: Coulomb excitations and decays in graphene-related systems. CRC Press; 2019.

[21] Ho J-H, Lu C, Hwang C, Chang C, Lin M-F. Phys Rev B 2006;74:085406.

[22] Shyu F-L, Chang C-P, Chen R-B, Chiu C-W, Lin M-F. Phys Rev B 2003;67:045405.

[23] Chen S, Wang T, Lee C, Lin M-F. Phys Lett A 2008;372:5999–6002.

[24] Lin C-Y, Wu J-Y, Chang C-P, Lin M-F. Carbon 2014;69:151–61.

[25] Wang X-F, Chakraborty T. Phys Rev B 2007;75:041404.

[26] Wu F, Das Sarma S. Phys Rev Lett 2020;124:046403.

[27] Wang X-F, Chakraborty T. Phys Rev B 2007;75:033408.

[28] Jiang Z, Zhang Y, Tan Y-W, Stormer H, Kim P. Solid State Commun 2007;143:14–9.

[29] Barlas Y, Yang K, MacDonald A. Nanotechnology 2012;23:052001.

[30] Pham HD, Nguyen TDH, Vo KD, Huynh TMD, Lin M-F. APEX 2020;13:085502.

[31] Sivek J, Sahin H, Partoens B, Peeters FM. Phys Rev B 2013;87:085444.

[32] Chang S-L, Lin S-Y, Lin S-K, Lee C-H, Lin M-F. Sci Rep 2014;4:1–8.

[33] Fasel R, Osterwalder J. Surf Rev Lett 1995;2:359–86.

[34] Praveen C, Piccinin S, Fabris S. Phys Rev B 2015;92:075403.

[35] Venema L, Wildöer J, Dekker C, Rinzler G, Smalley R. Appl Phys A 1998;66:S153–5.

[36] Benguediab S, Tounsi A, Zidour M, Semmah A. Compos B Eng 2014;57:21–4.

[37] Ho Y, Chang C, Shyu F, Chen R, Chen S, Lin M-F. Carbon 2004;42:3159–67.

[38] Celis A, Nair MN, Taleb-Ibrahimi A, Conrad E, Berger C, De Heer W, Tejeda A. J Phys Appl Phys 2016;49:143001.

[39] Ridene M, Girard J, Travers L, David C, Ouerghi A. Surf Sci 2012;606:1289–92.

[40] Egerton R, Li P, Malac M. Micron 2004;35:399–409.

[41] Kano S, Tada T, Majima Y. Chem Soc Rev 2015;44:970–87.

[42] Wang Z, Poncharal P, De Heer W. J Phys Chem Solid 2000;61:1025–30.

[43] Biró L, Gyulai J, Lambin P, Nagy JB, Lazarescu S, Márk G, Fonseca A, Surján P, Szekeres Z, Thiry P. Carbon 1998;36:689–96.

[44] Nguyen H, Cutler P, Feuchtwang TE, Huang Z-H, Kuk Y, Silverman P, Lucas A, Sullivan TE. IEEE Trans Electron Dev 1989;36:2671–8.

[45] Metois J, Le Lay G. Surf Sci 1983;133:422–42.
[46] Egerton RF. Rep Prog Phys 2008;72:016502.
[47] Kerber S, Barr T, Mann G, Brantley W, Papazoglou E, Mitchell J. J Mater Eng Perform 1998;7:329–33.
[48] Baltes H, Yacoby Y, Pindak R, Clarke R, Pfeiffer L, Berman L. Phys Rev Lett 1997;79:1285–8.
[49] Li J, Sun JL. Acc Chem Res 2017;50:2737–45.
[50] Speakman SA. Center for materials science and engineering at MIT. 2013.
[51] Reznik D, Olk C, Neumann D, Copley J. Phys Rev B 1995;52:116.
[52] Tomboulian D, Bedo D. Phys Rev 1956;104:590.
[53] Matthes L, Pulci O, Bechstedt F. New J Phys 2014;16.
[54] Shih P-H, Chiu Y-H, Wu J-Y, Shyu F-L, Lin M-F. Sci Rep 2017;7:40600.
[55] Khuong Dien V, Thi Han N, Nguyen TDH, Huynh TMD, Pham HD, Lin M-F. Front Mater 2020;7.
[56] Trukhin AN, Rogulis U, Spingis M. J Lumin 1997;72:890–2.
[57] Schwarz A, Bode M, Wiesendanger R. Handbook of magnetism and advanced magnetic materials. 2007.
[58] Deshpande A, Bao W, Zhao Z, Lau C, LeRoy BJ. Phys Rev B 2011;83:155409.
[59] Dien VK, Pham HD, Tran NTT, Han NT, Huynh TMD, Nguyen TDH, Fa-Lin M. Sci Rep 2021;11:1–10.
[60] Tapasztó L, Dobrik G, Lambin P, Biro LP. Nat Nanotechnol 2008;3:397–401.
[61] Cai J, Ruffieux P, Jaafar R, Bieri M, Braun T, Blankenburg S, Muoth M, Seitsonen AP, Saleh M, Feng X. Nature 2010;466:470–3.
[62] Zhang X, Yazyev OV, Feng J, Xie L, Tao C, Chen Y-C, Jiao L, Pedramrazi Z, Zettl A, Louie SG. ACS Nano 2013;7:198–202.
[63] Persichetti L, Tombolini F, Casciardi S, Diociaiuti M, Fanfoni M, Palleschi G, Sgarlata A, Valentini F, Balzarotti A. Appl Phys Lett 2011;99:041904.
[64] Tao C, Jiao L, Yazyev OV, Chen Y-C, Feng J, Zhang X, Capaz RB, Tour JM, Zettl A, Louie SG. Nat Phys 2011;7:616–20.
[65] Schaper AK, Hou H, Wang M, Bando Y, Golberg D. Carbon 2011;49:1821–8.
[66] Bussolotti F, Chi D, Goh KJ, Huang YL, Wee AT. STM/STS and ARPES characterization—structure and electronic properties. In: 2D semiconductor materials and devices. Elsevier; 2020. p. 199–220.
[67] Hibino H, Mizuno S, Kageshima H, Nagase M, Yamaguchi H. Phys Rev B 2009;80:085406.
[68] VanHove MA, Weinberg WH, Chan C-M. Low-energy electron diffraction: experiment, theory and surface structure determination. Springer Science & Business Media; 2012.
[69] Jackson JD. American association of physics teachers. 1999.
[70] Nickels J, Fineman M, Wallace W. J Phys Chem 1949;53:625–8.
[71] Stolyarova E, Rim KT, Ryu S, Maultzsch J, Kim P, Brus LE, Heinz TF, Hybertsen MS, Flynn GW. Proc Natl Acad Sci U S A 2007;104:9209–12.
[72] Osvath Z, Fulcheri L, Mark GI, Tapaszto L, Gyulai J, Biro LP. Nanotechnology. International Society for Optics and Photonics; 2003. p. 587–93.
[73] Grushevskaya H, Krylov G, Kruchinin S, Vlahovic B, Bellucci S. Phys Rev B 2021;103:235102.
[74] Kholmanov I, Cavaliere E, Fanetti M, Cepek C, Gavioli L. Phys Rev B 2009;79:233403.
[75] Shi L-J, Yang L-Z, Deng J-Q, Tong L-H, Wu Q, Zhang L, Zhang L, Yin L-J, Qin Z. Carbon 2020;165:169–74.
[76] Pandey DK, Chung TF, Prakash G, Piner R, Chen YP, Reifenberger R. Surf Sci 2011;605:1669–75.
[77] Gangopadhyay S. Scanning tunneling microscopy (STM) imaging of carbon nanotropes: C 60, CNT and graphene. In: Carbon nanomaterial electronics: devices and applications. Springer; 2021. p. 47–75.
[78] Xu J, Liu W, Geng Y, Deng K, Zhan C, Zeng Q. Nanoscale 2017;9:2579–84.
[79] Balog R, Jørgensen B, Nilsson L, Andersen M, Rienks E, Bianchi M, Fanetti M, Lægsgaard E, Baraldi A, Lizzit S. Nat Mater 2010;9:315–9.
[80] Kimouche A, Ervasti MM, Drost R, Halonen S, Harju A, Joensuu PM, Sainio J, Liljeroth P. Nat Commun 2015;6:1–6.
[81] Campos-Delgado J, Romo-Herrera JM, Jia X, Cullen DA, Muramatsu H, Kim YA, Hayashi T, Ren Z, Smith DJ, Okuno Y. Nano Lett 2008;8:2773–8.
[82] Unwin PR, Guell AG, Zhang G. Acc Chem Res 2016;49:2041–8.
[83] Huang H-C, Lin S-Y, Wu C-L, Lin M-F. Carbon 2016;103:84–93.
[84] Charlier A, McRae E, Heyd R, Charlier M, Moretti D. Carbon 1999;37:1779–83.
[85] Brown L, Hovden R, Huang P, Wojcik M, Muller DA, Park J. Nano Lett 2012;12:1609–15.

[86] Warner JH, Rummeli MH, Gemming T, Buchner B, Briggs GAD. Nano Lett 2009;9:102–6.
[87] Liu Z, Suenaga K, Harris PJ, Iijima S. Phys Rev Lett 2009;102:015501.
[88] Shen C, Brozena AH, Wang Y. Nanoscale 2011;3:503–18.
[89] Sun L, Li Q, Ren H, Su H, Shi Q, Yang J. J Chem Phys 2008;129:074704.
[90] Vicanek M. Surf Sci 1999;440:1–40.
[91] Froitzheim H, Schulze M. Surf Sci 1989;211:837–43.
[92] Egerton R, Malac M. J Electron Spectrosc Relat Phenom 2005;143:43–50.
[93] Xu Z, Buehler MJ. Nanotechnology 2009;20:375704.
[94] Meunier V, Nardelli MB, Roland C, Bernholc J. Phys Rev B 2001;64:195419.
[95] Tsai H-S, Wang S-W, Hsiao C-H, Chen C-W, Ouyang H, Chueh Y-L, Kuo H-C, Liang J-H. Chem Mater 2016;28:425–9.
[96] Nagarajan V, Chandiramouli R. Phys E Low-dimens Syst Nanostruct 2018;97:98–104.
[97] Fadaie M, Dideban D, Gülseren Ou. Appl Phys A 2020;126:1–8.
[98] Shyu FL, Lin MF. J Phys Soc Jpn 2002;71:1820–3.
[99] Qin C, Luo X, Xie Q, Nie T, Guo X, Xu J. Mater Sci Semicond Process 2021;133:105962.
[100] Yang Y-Y, Gong P, Ma W-D, Li Y-L, Fang X-Y, Jia Y-H, Cao M-S. Phys Lett A 2020;384:126602.
[101] Hipps K. Scanning tunneling spectroscopy (STS). In: Handbook of applied solid state spectroscopy. Springer; 2006. p. 305–50.
[102] Li G, Luican A, Dos Santos JL, Neto AC, Reina A, Kong J, Andrei E. Nat Phys 2010;6:109–13.
[103] Yan W, Liu M, Dou R-F, Meng L, Feng L, Chu Z-D, Zhang Y, Liu Z, Nie J-C, He L. Phys Rev Lett 2012;109:126801.
[104] Hafner J. J Comput Chem 2008;29:2044–78.
[105] Lozhnikov VE, Mamonov AV, Borzilov VO, Mamonova MV, Prudnikov PV, Sorokin AA, Baksheev GG. In: CEUR workshop proceedings; 2019. p. 24–9.
[106] Foulkes WMC, Haydock R. Phys Rev B 1989;39:12520.
[107] Konschuh S, Gmitra M, Fabian J. Phys Rev B 2010;82:245412.
[108] Ma C, Sun H, Zhao Y, Li B, Li Q, Zhao A, Wang X, Luo Y, Yang J, Wang B. Phys Rev Lett 2014;112:226802.
[109] Odom TW, Huang J-L, Kim P, Lieber CM. Nature 1998;391:62–4.
[110] Ouyang M, Huang J-L, Cheung CL, Lieber CM. Science 2001;292:702–5.
[111] Arghavan S, Singh A. J Sound Vib 2011;330:3102–22.
[112] Cao J, Yan X, Xiao Y, Tang Y, Ding J. Phys Rev B 2003;67:045413.
[113] Huang H, Wei D, Sun J, Wong SL, Feng YP, Neto AC, Wee ATS. Sci Rep 2012;2:1–7.
[114] Gao L. Graphene 2D Mater 2014;1.
[115] Lu CL, Chang CP, Huang YC, Ho JH, Hwang CC, Lin MF. J Phys Soc Jpn 2007;76:024701.
[116] Yelgel C. J Phys Conf 2016:012022.
[117] Do T-N, Lin C-Y, Lin Y-P, Shih P-H, Lin M-F. Carbon 2015;94:619–32.
[118] Lin C-Y, Huang B-L, Ho C-H, Gumbs G, Lin M-F. Phys Rev B 2018;98:195442.
[119] Roome NJ, Carey JD. ACS Appl Mater Interfac 2014;6:7743–50.
[120] Yuhara J, Le Lay G. J Appl Phys 2020;59:SN0801.
[121] Galbiati M, Motta N, De Crescenzi M, Camilli L. Appl Phys Rev 2019;6:041310.
[122] Kordyuk A. Low Temp Phys 2014;40:286–96.
[123] Pletikosić I, Kralj M, Pervan P, Brako R, Coraux J, N'diaye A, Busse C, Michely T. Phys Rev Lett 2009;102:056808.
[124] Rohlfing M, Louie SG. Phys Rev B 2000;62:4927.
[125] Wu J-Y, Chen L-H, Li T-S, Lin M-F. Appl Phys Lett 2010;97:031114.
[126] Vogt P. Beilstein-Institute. 2018.
[127] Cardona M, Greenaway DL. Phys Rev 1964;133:A1685.
[128] Lin S, Shih C-J, Strano MS, Blankschtein D. J Am Chem Soc 2011;133:12810–23.
[129] Orellana W, Miwa R, Fazzio A. Phys Rev Lett 2003;91:166802.
[130] Lu J, Chen Z, Pan F, Cui Y, Amine K. Electrochem Energy Rev 2018;1:35–53.
[131] Zhang SS. J Power Sources 2006;162:1379–94.
[132] Balbuena PB, Wang Y. Lithium-ion batteries: solid-electrolyte interphase. Imperial college press; 2004.
[133] Wang L, He X, Sun W, Wang J, Li Y, Fan S. Nano Lett 2012;12:5632–6.

[134] Li C, Zhang H, Fu L, Liu H, Wu Y, Rahm E, Holze R, Wu H. Electrochim Acta 2006;51:3872–83.
[135] Mahan GD. Many-particle physics. Springer Science & Business Media; 2013.
[136] Wei Z, Schumann F, Li C, Behnke L, Di Filippo G, Stefani G, Kirschner J. Phys Rev Lett 2014;113:267603.
[137] Kittel C. Introduction to Solid State Physics. New York: Wiley; 1996.
[138] Palczewski AD. Ames Laboratory (AMES), ames, IA (United States). 2010.
[139] Damascelli A. Phys Scripta 2004:61. 2004.
[140] Rice T, Yang K-Y, Zhang F-C. Physica C: Superconductivity and its applications 460; 2007. p. 252–5.
[141] Hwang E, Sarma SD. Phys Rev B 2008;77:081412.
[142] Kulić M, Dolgov O. Phys Stat Solidi (B) 2005;242:151–78.
[143] Sprinkle M, Siegel D, Hu Y, Hicks J, Tejeda A, Taleb-Ibrahimi A, Le Fèvre P, Bertran F, Vizzini S, Enriquez H. Phys Rev Lett 2009;103:226803.
[144] Zhou S, Siegel D, Fedorov A, El Gabaly F, Schmid A, Neto AC, Lee D-H, Lanzara A. Nat Mater 2008;7:259–60.
[145] Fink JH, Lambin P. Carbon nanotubes. 2001. p. 247–72.
[146] Nicholas RJ, Chuang K-C. Int J Mod Phys B 2009;23:2655–64.
[147] García N, Esquinazi P, Barzola-Quiquia J, Dusari S. New J Phys 2012;14:053015.
[148] Lin C-Y, Do T-N, Huang Y-K, Lin M-F. Optical properties of graphene in magnetic and electric fields. IOP Publishing; 2017.
[149] Yuan S, Roldán R, Katsnelson MI. Phys Rev B 2011;84:125455.
[150] Ludbrook B, Levy G, Nigge P, Zonno M, Schneider M, Dvorak D, Veenstra C, Zhdanovich S, Wong D, Dosanjh P. Proc Natl Acad Sci U S A 2015;112:11795–9.
[151] Yin W-J, Xie Y-E, Liu L-M, Chen Y-P, Wang R-Z, Wei X-L, Zhong J-X, Lau L. J Appl Phys 2013;113:173506.
[152] Yang X, Dou X, Rouhanipour A, Zhi L, Räder HJ, Müllen K. J Am Chem Soc 2008;130:4216–7.
[153] Zhang XG, Krstić P, Butler W. Int J Quant Chem 2003;95:394–403.
[154] Chen S, Wu J, Lin C, Lin M. New J Phys 2016;18:103024.
[155] Vogt P, De Padova P, Quaresima C, Avila J, Frantzeskakis E, Asensio MC, Resta A, Ealet B, Le Lay G. Phys Rev Lett 2012;108:155501.
[156] Goriparti S, Miele E, De Angelis F, Di Fabrizio E, Zaccaria RP, Capiglia C. J Power Sour 2014;257:421–43.
[157] Shiraz AK, Goharrizi AY, Hamidi SM. Phys E Low-dimens Syst Nanostruct 2019;107:150–3.
[158] Snehha P, Nagarajan V, Chandiramouli R. J Comput Electron 2019;18:308–18.
[159] Kortüm G. Reflectance spectroscopy: principles, methods, applications. Springer Science & Business Media; 2012.
[160] Dowell FE, Pearson TC, Maghirang EB, Xie F, Wicklow DT. Cereal Chem 2002;79:222–6.
[161] Dawlaty JM, Shivaraman S, Strait J, George P, Chandrashekhar M, Rana F, Spencer MG, Veksler D, Chen Y. Appl Phys Lett 2008;93:131905.
[162] Sadowski M, Martinez G, Potemski M, Berger C, de Heer WA. Solid State Commun 2007;143:123–5.
[163] Högele A, Galland C, Winger M, Imamoğlu A. Phys Rev Lett 2008;100:217401.
[164] Lorchat E, López LEP, Robert C, Lagarde D, Froehlicher G, Taniguchi T, Watanabe K, Marie X, Berciaud S. Nat Nanotechnol 2020;15:283–8.
[165] Taft E, Philipp H. Phys Rev 1965;138:A197.
[166] Jackson JD. Classical electrodynamics. John Wiley & Sons; 2007.
[167] Dumont G, Boulanger P, Côté M, Ernzerhof M. Phys Rev B 2010;82:035419.
[168] Marty L, Adam E, Albert L, Doyon R, Menard D, Martel R. Phys Rev Lett 2006;96:136803.
[169] Ichida M, Mizuno S, Tani Y, Saito Y, Nakamura A. J Phys Soc 1999;68:3131–3.
[170] Spataru CD, Ismail-Beigi S, Benedict LX, Louie SG. Phys Rev Lett 2004;92:077402.
[171] Sanchez-Valencia JR, Dienel T, Gröning O, Shorubalko I, Mueller A, Jansen M, Amsharov K, Ruffieux P, Fasel R. Nature 2014;512:61–4.
[172] Pierce N, Chen G, Rajukumar LP, Chou NH, Koh AL, Sinclair R, Maruyama S, Terrones M, Harutyunyan AR. ACS Nano 2017;11:9941–9.
[173] Peshkin M. The aharonov-bohm effect part one: theory. In: The aharonov-bohm effect. Springer; 1989. p. 1–34.
[174] Ajiki H, Ando T. Phys B Condens Matter 1994;201:349–52.
[175] Nakayama T, Shima H. Phys Rev 1998;58:3984.
[176] Dresselhaus M, Dresselhaus G, Saito R. Carbon 1995;33:883–91.
[177] Kataura H, Kumazawa Y, Maniwa Y, Umezu I, Suzuki S, Ohtsuka Y, Achiba Y. Synth Met 1999;103:2555–8.

[178] Malić E, Hirtschulz M, Milde F, Knorr A, Reich S. Phys Rev B 2006;74:195431.
[179] Booshehri LG, Mielke CH, Rickel DG, Crooker SA, Zhang Q, Ren L, et al. Phys Rev B 2012;85:205407.
[180] Weisman RB, Kono J. Handbook of carbon nanomaterials. 2019. p. 1−43.
[181] Matsuda K, Inoue T, Murakami Y, Maruyama S, Kanemitsu Y. Phys Rev B 2008;77:033406.
[182] Lefebvre J, Fraser J, Homma Y, Finnie P. Appl Phys A 2004;78:1107−10.

CHAPTER

4

Monolayer and bilayer germanene systems

Hsin-Yi Liu[1], *Shih-Yang Lin*[1], *Vo Khuong Dien*[1], *Chi-Hsuan Lee*[1], *Hai Duong Pham*[1], *Thi My Duyen Huynh*[1], *Nguyen Thi Han*[1], *Ngoc Thanh Thuy Tran*[2], *Thi Dieu Hien Nguyen*[1], *Wei-Bang Li*[1] *and Ming-Fa Lin*[1]

[1]Department of Physics, National Cheng Kung University, Tainan, Taiwan; [2]Hierarchical Green-Energy Materials (Hi-GEM) Research Center, National Cheng Kung University, Tainan, Taiwan

Since 2D materials of graphene with a hexagonal lattice structure has be found by a method of the mechanical exfoliation in 2004 [1], plenty of interests arouse everyone's attention due to its novel electronic properties [2]. It is not only for theoretical research, but also for applications of electronic devices. For instance, one of the best material candidates for flexible nanoelectronics due to its excellent flexibility, which have been produced experimentally [3,4]. A relative high thermal conductivity at room temperature, the value reaches to 5000 $Wm^{-1}K^{-1}$ and is suitable for membrane technology [5–7]. Massless Dirac fermion, high carrier mobility of 20,000 cm^2/vs and half-integer Hall conductance, which show many advantages in applied nanoelectronics [8]. Besides, its special ambient stability benefit sustainable exploitation in nanoelectronics [9,10]. Graphene can be also used for a topological insulator (TI) with an opening band gap at Dirac point caused by strain or other modification of functional group with spin-orbit coupling (SOC) [11,12]. However, the band gap is too small to be investigated. Germanene counterparts of graphene with a honeycomb monolayer shows a large spin-orbit coupling are suitable for applications of topological insulators (TI) and spintronic devices [13]. Production of germanene requires a metal substrate and the interaction between germanene and substrate is very strong. A relative weak interaction of germanene on a semiconducting GaAs (0001) substrate is reduced by intercalation of H at interface

Fundamental Physicochemical Properties of Germanene-related Materials
https://doi.org/10.1016/B978-0-443-15801-8.00002-5

[14] to formation of the quasi freestanding germanene. After that, buckled germanene has been successfully synthesized on a Pt (111) substrate [15], where the structure is characterized as a $\sqrt{19}X\sqrt{19}$ superstructure about the substrate lattice analyzed by low-energy electron diffraction (LEED) and scanning tunneling microscopy (STM) [16]. Due to the nontrivial topological properties the Quantum Spin Hall effect (QSHE) of buckled and H-doped germanene can be realized [16]. Quantum anomalous Hall Effect can be hosted by V-adsorbed germanene [17] and high-T_c superconductivity is influenced by doped germanene [18]. Layered materials play an important role in highly diversifying the fundamental properties and enhancing the potential applications. There are a lot of experimental [1,19–21] and theoretical [22–27] researches on the graphene-related materials so far. Such studies clearly illustrate that the rich and unique phenomena are dominated by the π bonding of carbon-$2p_z$ orbitals. For example, few-layer graphenes have been identified to present the AAA [28], ABA [29], ABC [28,29], AAB [30], twisted [31], sliding [32], modulated [33] and turbostratic stackings [34]. Recently few layered germanene has been grown on an Au (111) template by a dry epitaxy method and the reduction of few layers germanene possesses better Dirac cones. In order to formation of bilayer germanene with AB-stacking configuration, germanene flakes torn off on an Al (111) substrate and are redeposited on monolayer germanene [35]. The mechanical properties of bilayer germanene show higher mechanical strength and exhibit higher fracture strength and strain. Therefore, germanene exhibit some advantages compared to graphene: (1) tunable band gap, which are influenced by physical and chemical conditions, (2) controllable electronic properties because of diverse substrate, (3) compatibility with electronic devices, (4) large spin orbital coupling lead to observe the Quantum Spin Hall effect (QSHE). Buckled monolayer germanene shows a band gap opening around 22.9–23.9 meV at K point by DFT calculations [13,36,37].

4.1 Monolayer germanene

Monolayer germanene systems shown in Fig. 4.1A, presents a stable honeycomb lattice in the isolated case, where a is the lattice constant. Such matcrials can provide full information about the π, sp^2 and sp^3 bonding due to, respectively, the 4pz, (4s, $4p_x$, $4p_y$) and (4s, $4p_x$, $4p_y$,

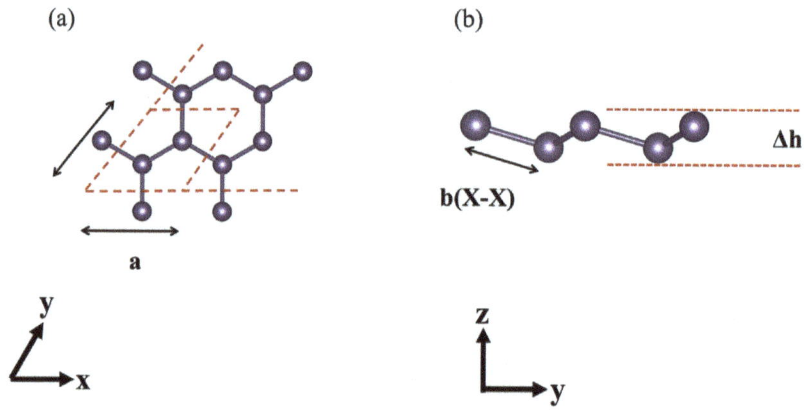

FIGURE 4.1 Geometric structure of monolayer germanene in x-y (A) and y-z (B) projection; Δh: Height differences.

$4p_z$) orbital hybridization. The main features, planar and buckled structures, ground state energies, lattice constants, and heights of the A and B sublattices, are consistent with the significant chemical bonding of the outermost 4s, $4p_x$, $4p_y$ and $4p_z$ orbitals. Only graphene consists of normal hexagons, while germanene possesses different heights in the A and B sublattices and the height differences is 0.69 Å shown in Fig. 4.1B. The two sublattices A and B are equivalent on the (x,y)-plane for any fundamental properties of all the group-IV materials. This result illustrates clearly that the dominance of sp^2 bonding is reduced or enhanced gradually with the increase of atomic number, while the opposite is true for the importance of sp^3 bonding. Further, the sp^2-bonding strength is much stronger than that of π bonding. The ground state energy E_0 of monolayer germanene is −4.01 eV/atom and its lattice constant is 4.06 Å [37]. Comparing with graphene, the E_0 of germanene is much higher due to the buckling structure and not dominated sp^2 hybridization. b is the Ge−Ge bond length and the bond length of germanene is 2.44 Å. The spin-orbital coupling configurations have been included in the VASP calculations of the total ground state energies, in which the spin-orbit interactions play a significant role in the geometric and electronic properties [38−41]. In short, multi-orbital hybridization dominates the optimal geometric structures fully and thus enrich the fundamental properties greatly. Monolayer germanene, illustrated in Fig. 4.2, presents various valence and conduction bands. The electronic states of germanene are consist of the spin-up- and spin-downcombined states with double degeneracy, which can be classified into three groups according to the effects of sp^3 bonding and spin-orbit coupling. The low-lying valence and conduction bands of germanene are dominated by

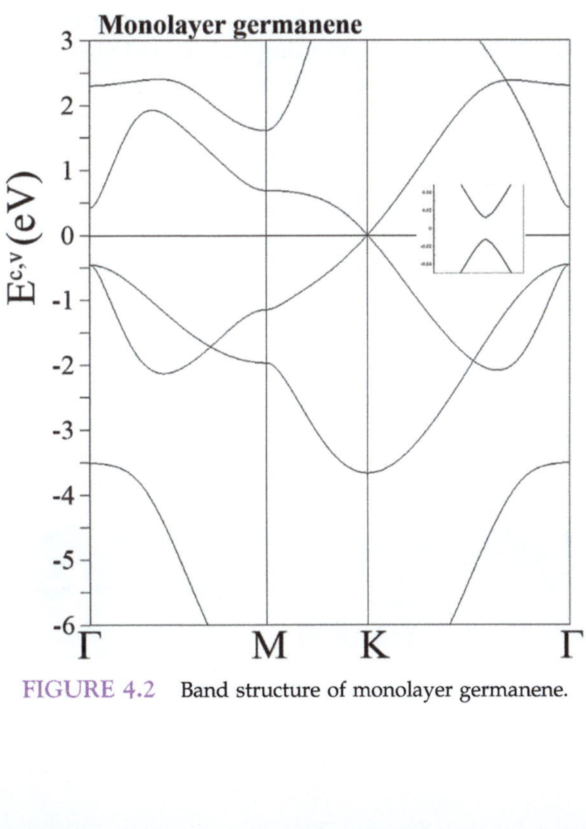

FIGURE 4.2 Band structure of monolayer germanene.

the π bonding of $4p_z$ orbital. The low-energy ranges of $E^{c,v}$ without σ bands are around 0.5 eV and indicates reduced π bonding under larger lattice constants and wider orbitals. Silicene exists some dissimilarities compared with germanene and its The low-energy ranges of $E^{c,v}$ without σ bands are located around 1.0 eV [37]. Germanene is attributed to enhanced spin-orbital coupling in the buckled honeycomb lattice. The Dirac-cone structure presents a narrow band gaps of E_g 24.0 meV shown in Fig. 4.2. The narrow band gaps of E_g are observable by silicene, which is 1.55 meV [37]. The results are also consistent with the tight-binding model with nearest-neighbor hopping integrals and next-nearest-neighbor spin-orbit interactions [42]. Graphene is a zero-gap semiconductor, as verified directly from STS measurements [43—49]. The low-energy fundamental properties are dominated by the π bonding of $2p_z$ orbitals. It can be deduced that single-p_z-orbital hybridization is roughly suitable for the low-energy physics of monolayer graphene, silicene and germanene. Alternatively, the energy spacing between the initial σ bands and the Fermi level at the Γ valley becomes small with the increase of atomic number, owing mainly to the enhanced sp^3 bonding and spin-orbital coupling. Furthermore, anticrossing behavior of σ and π bands might occur frequently along any direction related to the Γ-M and K-Γ point.

Spatial charge distributions are available for identifying and examining multi-and single orbital hybridization due to various chemical bonds [50—53]. Both the total charge density distribution and its variation, ρ and $\Delta\rho$ in Fig. 4.3A and B with (x,z)-plane projection by methods of first-principles calculations. The variation of the total charge density distribution $\rho\Delta$ is obtained by subtracting the superposition of that in an isolated atom. Monolayer germanene presents σ and π chemical bonding and it becomes nonorthogonal to each other, as observed clearly in Fig. 4.3A. The bonding strengths decline as the atomic number of the group-IV materials grows. This leads to longer bond lengths. However, the opposite is true for the observable or significant sp^3 bonding in buckled systems. Specifically, graphene has the largest valence electron density with the shortest C—C bond length, compared to the germanene and other group-IV systems. The height difference on the A- and B-sublattice can be also observed by the total charge density, which is enhanced with an increase in atomic number. Large buckling indicates that it is relatively difficult to synthesize few-layer group-IV materials with heavy masses. To comprehend the fundamental properties fully, nonnegligible sp^3 multi-orbital hybridization and spin-orbital coupling might have to be included in the tight-binding model simultaneously [36,54—56].

The orbital-projected density of states (PDOS) is a beneficial tool to understanding the strong effects of multi-orbital hybridization on fundamental properties. The various band edge states of electronic energy properties, shown in Fig. 4.4, can create diverse 2D van Hove singularities. Generally, the critical points in energy-wave-vector space, which appear

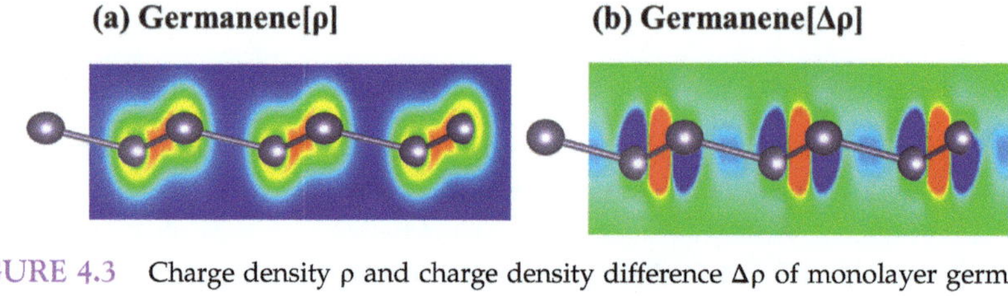

FIGURE 4.3 Charge density ρ and charge density difference Δρ of monolayer germanene.

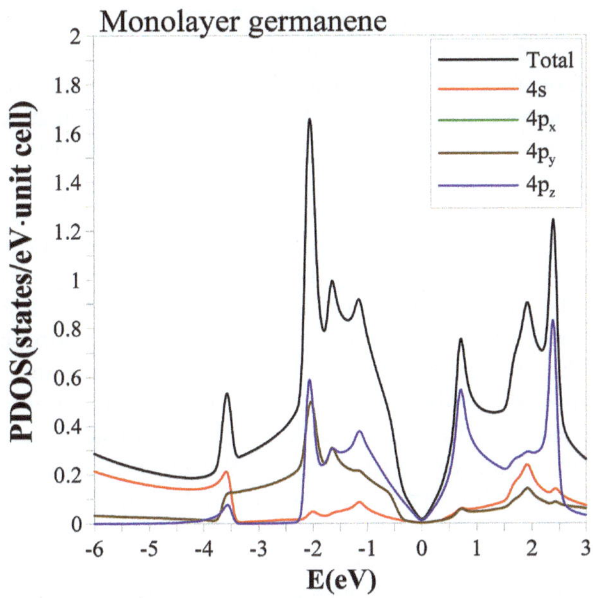

FIGURE 4.4 Orbital-projected density of states of monolayer germanene.

in the energy dispersions of Fig. 4.2, include the Dirac-cone bottoms, the extreme states of parabolic dispersions, the saddle points and the constant-energy loops. The electronic states exhibit V-shaped structures, shoulders, logarithmically symmetric peaks and square-root form asymmetric structures, respectively. For pristine monolayer germanene, its particular orbital-projected density of states (PDOSs) presents a linear energy dependence that is asymmetric around the Fermi level, the -1.0 eV π peak and the 0.8 eV π^* peak, and the -0.5 eV shoulder arising from the $4p_x$ and $4p_y$ orbitals. The energies of the symmetric π^* peak and the σ shoulder are decreased compared with monolayer graphene and silicene. The critical mechanisms are the reduced π bonding and the enhanced dominance of sp^3 bonding and spin-orbital coupling. There exists an asymmetric peak after the formation of the σ bands, indicating clearly the mixing of orbital dominated energy bands, such as the -1.6 eV peak structures. This indicates the existence of significant sp^3 bonding. Studies using STS measurements on monolayer germanene materials have been able to identify the main features for the π, sp^2 and sp^3 bonding, as well as the spin-orbit interaction.

4.2 AA-bt and AA-bb

Graphene systems present planar geometries because of their dominant σ bonding [57]. In addition to layer numbers, stacking configurations will also play a critical role in the various fundamental properties. To date, AA [58], AB [59], ABC [30,60], AAB [20] and twisted [31] stacking has been identified in 2D graphenes. Recently, few-layer of group-IV elements have been synthesized successfully on various substrate surfaces. Such as silicene on surfaces of Ag (111), Ir (111) and ZrB_2 [61–67], germanene on surfaces of Pt (111), Au (111) and Al

(111) [68–70], stanene (tinene) on Bi_2Te_3 surfaces [71,72] and plumbene on Si (111) surfaces [73,74]. This indicates significant mixing of sp^2 and sp^3 orbital hybridization, since the relatively weak chemical bonding between larger atoms cannot maintain a planar structure. Bilayer systems of germanene are expected to have more complicated stacking symmetry structures when using experimental growth methods due to the buckling effects. Both the AB-bt and AB-bb configurations of bilayer silicene systems are analyzed in details by use of high-angle annular dark-field scanning transmission electron microscopy [19,75].

Bilayer germanene systems are completely different to monolayer systems, as illustrated clearly by the detailed comparison in Fig. 4.5A–F. The significant differences are mainly from the buckling heights, thus leading to giant interlayer hopping integrals, more prominent sp^3 bonding and layer- and sublattice-dependent spin-orbital coupling in configurations. The differences of buckling heights cause the AA and AB systems separated to two types individually. Here can be names as AA-bb, AA-bt, AB-bb and AB-bt for the AA and AB system, respectively, and be illustrated in Fig. 4.5A–D in the y-z projection. The symbol b means at the bottom site of the sublattice and t means at the top site. The AA-bb configuration of bilayer germanene present its bottom sublattices are right opposite the other layer of bottom sublattices and conversely, the AA-bt configuration present its bottom sublattices are right opposite the other layer of top sublattices.

Most importantly, the strongly modified electronic structures include the stacking enriched valence and conduction bands, semimetallic behaviors, the obvious electron-hole asymmetric energy spectra around the Fermi level, more oscillatory (nonuniform) valence and conduction bands even close to E_F, frequent band crossing and anticrossing, and many band-edge states and valley structures. As a result, it is very difficult to characterize the stable valleys due to the initial and final σ and π bands; that is, the separation of π and σ bonding would become meaningless. Dominated π bonding, which comes from the

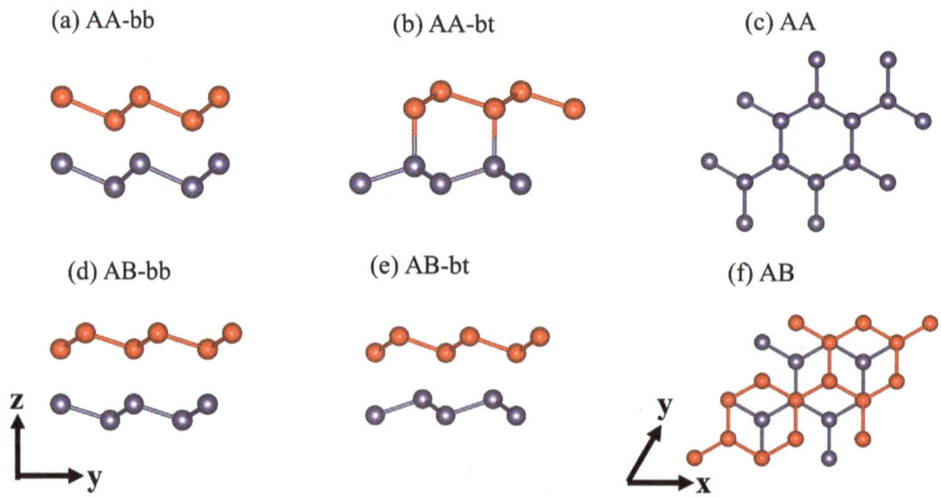

FIGURE 4.5 Geometric structure of bilayer germanene AA-bb (A), AA-bt (B), AB-bb (D) and AB-bt (E) in y-z and in x-y (C) and (F) projection.

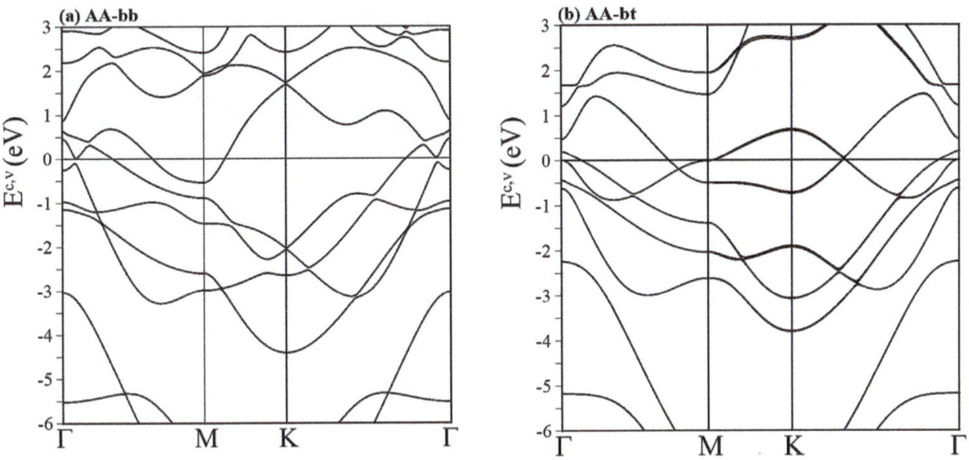

FIGURE 4.6 (A) and (B): band structure of bilayer germanene AA-bb and AA-bt. (C) and (D): band structure of bilayer germanene AB-bb and AB-bt.

$3p_z$ orbital, are not appropriate for the low-energy fundamental properties of the phenomenological models.

The AA-bb ($\delta_a = 0$) system shown in Fig. 4.6A, presents six valence and four conduction bands, which is double those of monolayer germanene (Fig. 4.2). However, the AA-bt system (Fig. 4.6B) of bilayer germanene presents only three subbands in conduction region. The overlap of the valence and conduction bands at M point indicates semimetallic behavior. There are some small overlaps near the Γ points at fermi level and exhibits anticrossings with valence and conduction bands, respectively. The valence and conduction bands of AA-bt bilayer germanene around the E_F level at Γ points are not overlapping. The highest occupied valence band and the lowest unoccupied conduction band form two crossing bands located between Γ and M point and K and Γ point. The energy dispersion relations of AA-bt germanene present partial flat in both valence and conduction regions. In the conduction region of AA-bt germanene more anticrossing subbands are observed between M K points and KΓ points. The AA-bt stacking of silicene exists.

Dirac-cone structures along the M K and M Γ directions based on π-band evidence [31,76]. However, one of the Dirac-cone structures from germanene AA-bt stacking is deviate from the fermi level. The band structure of AB-bb exhibits two crossing dispersion relations around K point and one crossing dispersion relation located at K point for the AB-bt system. The nondegenerate bands are observable in both AA-bt and AB-bt systems, when the spin-orbital coupling are taken into consideration by first-principle calculations. The four types of different bilayer stacking have semimetallic characteristics and similar features of energy dispersions, for example, parabolic valleys at K points, saddle M point structures and their valence and conduction bands, which are near E_F, exhibit oscillatory energy dispersions. Their valence states above the Fermi level possess free holes, meanwhile the conduction states below the Fermi level possess electrons. The electronic energy spectra of these four bilayer stacking are doubly degenerate in the absence of spin configurations. The spin-dependent

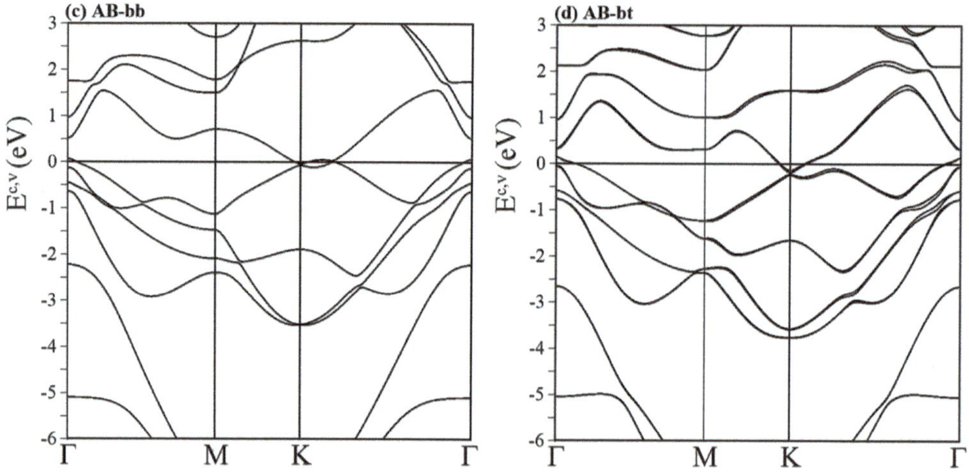

FIGURE 4.6 cont'd.

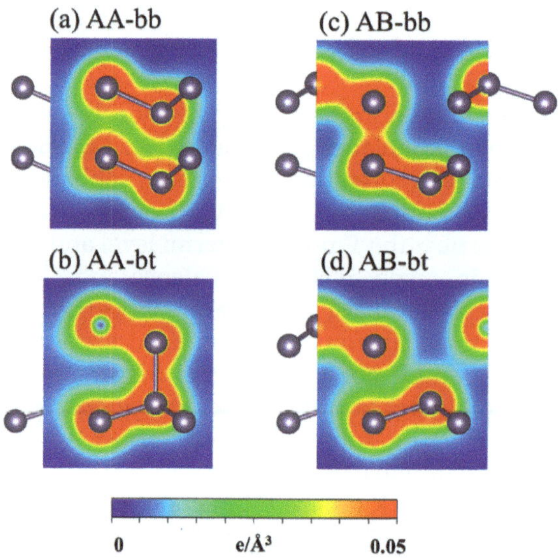

FIGURE 4.7 Charge density of bilayer germanene AA and AB.

interactions are fully negligible in bilayer AA-bb stacking, the reason could be due to the highest-symmetry stacking and the highly strong interlayer interactions.

The spatial charge density distributions including of interlayer atomic interactions present impressive transformations. The bilayer germanene systems extremely different from monolayer systems (Fig. 4.3A) shown in Fig. 4.7A–D. Compared to isolated germanium, the charge density of the two neighboring germanium atoms distributed on the same layer (the red regions), is reduced or transferred greatly and so does the π bonding (the green regions). This means more charge densities of interlayer atoms come to exist in the interlayer

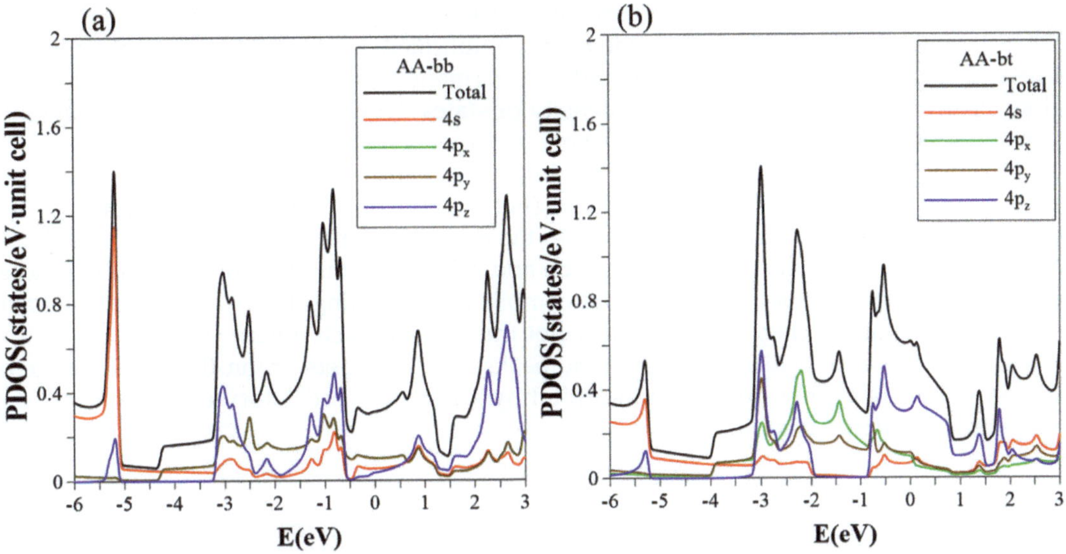

FIGURE 4.8 (A) and (B): orbital-projected density of states of bilayer germanene AA-bb and AA-bt. (C) and (D): orbital-projected density of states of bilayer germanene AB-bb and AB-bt.

spacing (the orange or green regions). Electron carriers are clearly illustrated that transferred from the first and second layers to the interlayer spacing and exists important four-orbital hybridization (the (4s, $4p_x$, $4p_y$, $4p_z$)-(4s, $4p_x$, $4p_y$, $4p_z$) bonding). Thus, it is difficult to observe those phenomena via the sp^3-based tight-binding model [77].

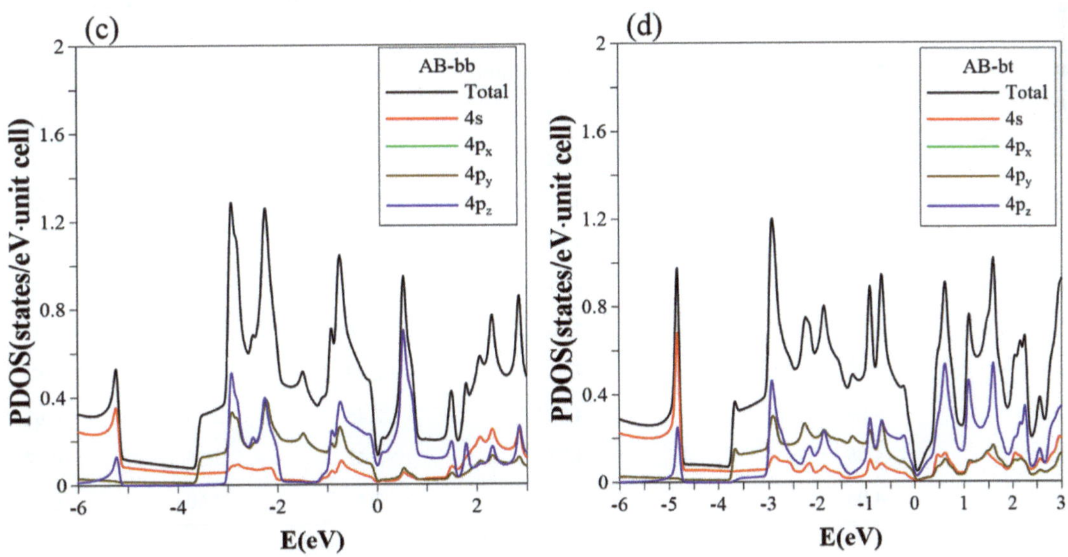

FIGURE 4.8 cont'd.

The main features of the bilayer germanene systems in orbital-projected density of states (PDOSs) shown in Fig. 4.8A–D, become more complicated when interlayer coupling is concerned. The semimetallic bilayer AA and AB germanene systems are characterized with values near the Fermi level. The finite density of states (DOSs) are accompanied frequently by strong special structures and the weak energy dispersions are across fermi level. However, the vanishing density-of-state ranges of the bilayer AA and AB stacking, which are close to zero, are not observable. The AB-bt system (Fig. 4.8D) of bilayer germanene presents the asymmetric V-shapes at $E_F = 0$ correspond to the modified Dirac-cone structures. The dominated $4p_z$ orbitals are observed in both AB-bb and AB-bt systems. The initial, middle saddle-point structure and final π and σ bands are complicated to distinguish. Therefore, bilayer and monolayer germanene (Fig. 4.2) are totally not the same as each other.

A lot of symmetric peaks, antisymmetric peaks and shoulders are in the whole range of electron volt, it is mainly due to multi-orbital hybridization, oscillatory energy dispersions and frequent anticrossing.

4.3 Sliding bilayer germanene systems

Sliding bilayer germanene materials are created through a relative shift between two subsystems, where δ is in units of the lattice constant ($\sqrt{3}b\cos\theta = 4.16$ Å and b = 2.44 Å). The typical sliding types are divided two parts and illustrated in Fig. 4.1A $\delta_a = 0$ (AAbb), (b) $\delta_a = 1/8$, (c) $\delta_a = 4/8$, (d) $\delta_a = 8/8$ (AB-bt), (e) $\delta_a = 11/8$ and (f) $\delta_a = 12/8$ along the armchair direction, and (g) $\delta_z = 1/8$ and (j) $\delta_z = 3/8$ along the zigzag direction. Four germanium atoms are in the unit cell for each stacking configuration. The ground state energies E_0 are

TABLE 4.1 Lattice constant, ground state energies E_0, Ge–Ge bond length, interlayer distance Δd, buckling height Δh.

Bilayer Ge	Lattice constant (Å)	E_0 (eV/atom)	x-x (Å)	Δd (Å)	Δh (Å)
AA-bb	4.06	−4.32	2.56	2.94	1.03
δ a = 1/8	4.06	−4.32	2.56	2.94	1.03
δ a = 4/8	4.06	−4.34	2.50	2.56	0.85
AB-bt	4.06	−4.29	2.49	3.15	0.66
δ a = 11/8	4.06	−4.31	2.48	2.62	0.79
δ a = 12/8	4.06	−4.30	2.48	2.65	0.81
δ z = 1/8	4.06	−4.31	2.49	2.63	0.83
δ z = 3/8	4.06	−4.30	2.48	2.65	0.81
AA-bt	4.06	−4.27	2.47	2.56	0.77
AB-bb	4.06	−4.32	2.46	2.60	0.75

from −4.29 to −4.34 eV per atom and are summarized in Table 4.1. Compared to monolayer germanene, its ground state energy E_0 is −4.04 eV per unit atom [37]. This indicates significant attractive forces between the two monolayer subsystems.

Among the different stacking configurations, the AA $\delta_a = 4/8$ configuration shows the lowest total energy, suggesting the most stable geometric structure [78]. The height difference Δh between the intralayer A and B sublattices is enhanced greatly, from 0.69 Å in the monolayer [37] to 0.66−1.03 Å in the bilayer systems. Further, the interlayer distance Δd of 3.15−2.56 Å is shorter than the 3.52 and 3.26 Å for the AA- and AB-stacked bilayer graphenes, respectively [79]. All of the bilayer buckled honeycomb lattices including of the AA-bt and AB-bb stacking configurations, the AA-bb and AB-bt present the highest and lowest buckling distance, respectively. Briefly, the increased buckling and the reduced interlayer distance in bilayer germanene systems are expected to induce more prominent multi-orbital hybridization of Ge-4s, -4p_x, -4p_y and -4p_z.

The electronic energy band structure of sliding bilayer germanene are sensitive from armchair δ_a to zigzag direction δ_z. For $\delta_a = 1/8$ in Fig. 4.7A, the bilayer system retains the semimetallic property near the Fermi level, while the main features of the electronic structures present diverse behaviors. The anticrossing of the valence and conduction states occurs at different wave vectors, not only located at the Γ point. The saddle M point structures are capable of accumulating a lot of electronic states shown in the results of density of states in Fig. 4.4A. Therefore, they are expected to exhibit very prominent optical absorption peaks. This phenomenon can also be found in carbon-related sp^2-bonding systems, such as layered graphenes [80−85], carbon nanotubes [86−88] and graphite [89−94]. Some energy bands of

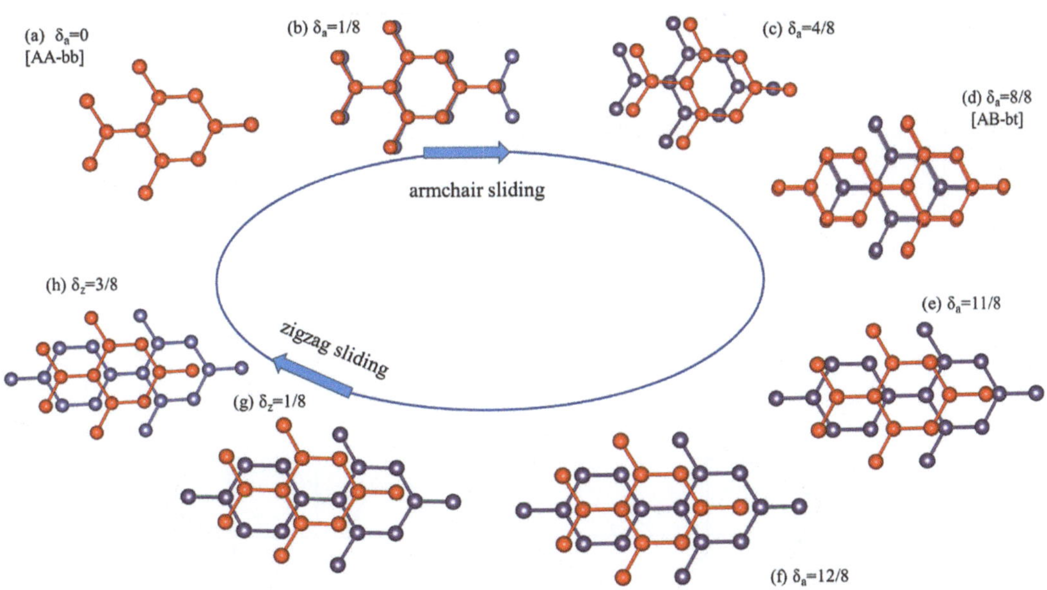

FIGURE 4.9 Geometric structure of sliding bilayer germanene in armchair and zigzag direction; $\delta_a = 0$ (AA-bb) (A), $\delta_a = 1/8$ (B), $\delta_a = 4/8$ (C), $\delta_a = 8/8$ (AB-bt) (D), $\delta_a = 11/8$ (E), $\delta_a = 12/8$ (F), $\delta_z = 1/8$ (G), $\delta_z = 3/8$ (H).

the electronic states create parabolic valleys near the M points. All the band-edge states appear at the Γ, M, and K points and between them. Similar semimetallic behavior is revealed as a weak band overlap under $\delta_a = 11/8$, $\delta_a = 12/8$, $\delta_z = 1/8$ and $\delta_z = 3/8$ shwon in Fig. 4.9C–F, respectively. The other stacking configuration such as $\delta_a = 4/8$ belongs to the indirec gap (143 meV) semiconductors. Some stacking configurations such as $\delta_a = 11/8$, $\delta_a = 12/8$ and $\delta_z = 3/8$, present modified Dirac-cone structures. stacking configurations.

The spatial charge density distributions present dramatic transformations after including interlayer atomic interactions. The bilayer sliding germanene systems are completely different from monolayer systems, as shown in Fig. 4.10A–F. The charge density, which is distributed between two adjacent germanium atoms on the same layer (the red regions), is reduced or transferred greatly, compared to isolated germanene. The lower part of the π and σ bonding in the first layer and the upper part in the second layer show an obvious and dramatic charge transformation, when sliding to $\delta_a = 4/8$. After keep sliding to AB-bb (Fig. 4.7C), $\delta_a = 11/8$, 12/8 and zigzag direction, the formation of the σ and π bonding in red and in green-yellow region, respectively, from intralayer disappeared. The charge densities of Electron carriers are transferred from the first and second layers to the interlayer spacing and exist important four-orbital hybridization (the (4s, $4p_x$, $4p_y$, $4p_z$)-(4s, $4p_x$, $4p_y$, $4p_z$) bonding). Furthermore, the orbital-created intralayer and interlayer hopping integrals are very sensitive to the stacking configurations. As a result, diverse electronic properties are revealed directly in sliding bilayer germanene such as their band structures (Fig. 4.10A–F) and van Hove singularities. It is almost impossible to simulate the stacking-dependent electronic energy spectra through the sp^3-based tight-binding model [77]. For example, the theoretical simulation has only been performed for the first pair of valence and conduction bands nearest to the Fermi level in AB-bt bilayer silicene [77]. Similar chemical bonding and

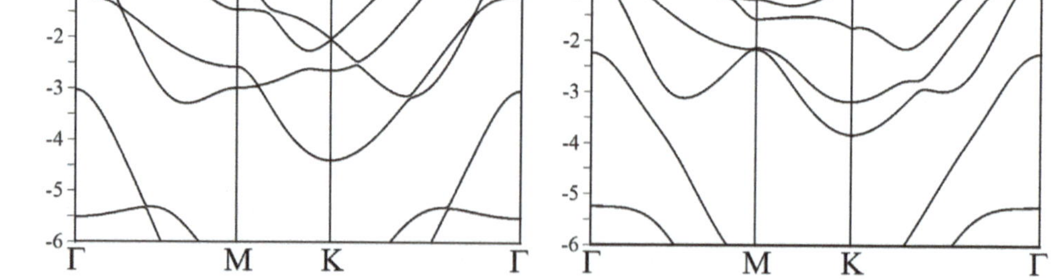

FIGURE 4.10 (A) and (B): band structure of sliding bilayer germanene AA $\delta_a = 1/8$ and $\delta_a = 4/8$. (C) and (D): band structure of sliding bilayer germanene AA $\delta_a = 11/8$ and $\delta_a = 12/8$. (E) and (F): band structure of sliding bilayer germanene AA $\delta_z = 1/8$ and $\delta_z = 3/8$.

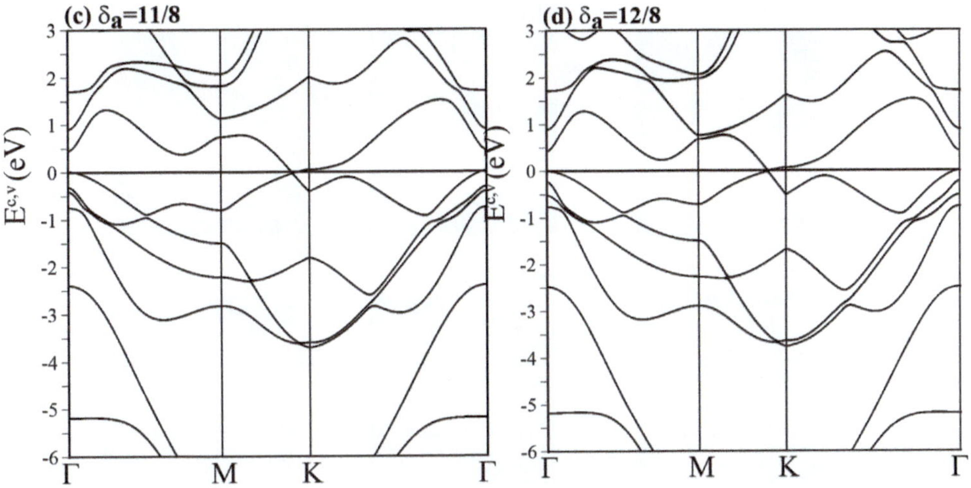

FIGURE 4.10 cont'd.

difficulties in phenomenological models are expected to be revealed in more complicated few-layer silicene systems, e.g., tri and four-layer systems [31,53,75,95,96].

The main features of sliding bilayer systems in the densities of states (DOSs) are illustrated in Figs. 4.11 and 4.12A–F. It becomes very complicated under any interlayer coupling in bilayer germanene systems. The semimetallic and semiconducting bilayer germanene systems are characterized by their values near the Fermi level. The weak energy dispersions across the fermi level $E_F = 0$ lead the finite densities of states accompanied frequently by strong special structures shown in Fig. 4.12A, C and F. However, the vanishing density of-

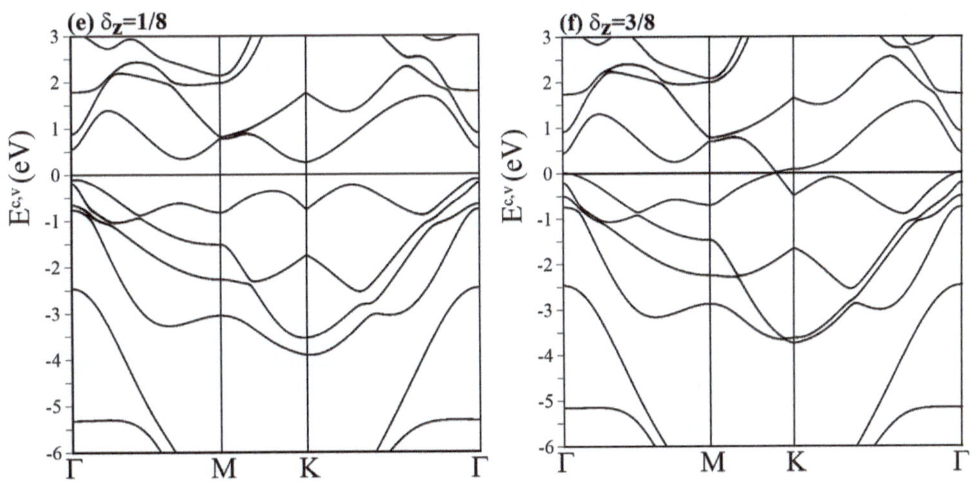

FIGURE 4.10 cont'd.

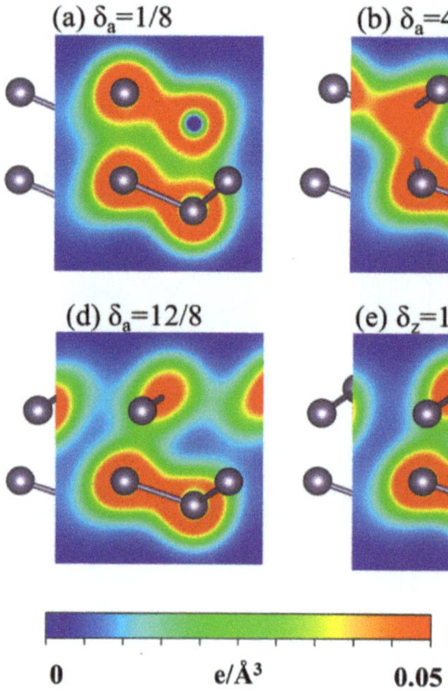

FIGURE 4.11 Charge density of sliding bilayer germanene.

state ranges of $\delta_a = 4/8$ are close to zero and are observable the asymmetric V-shapes in

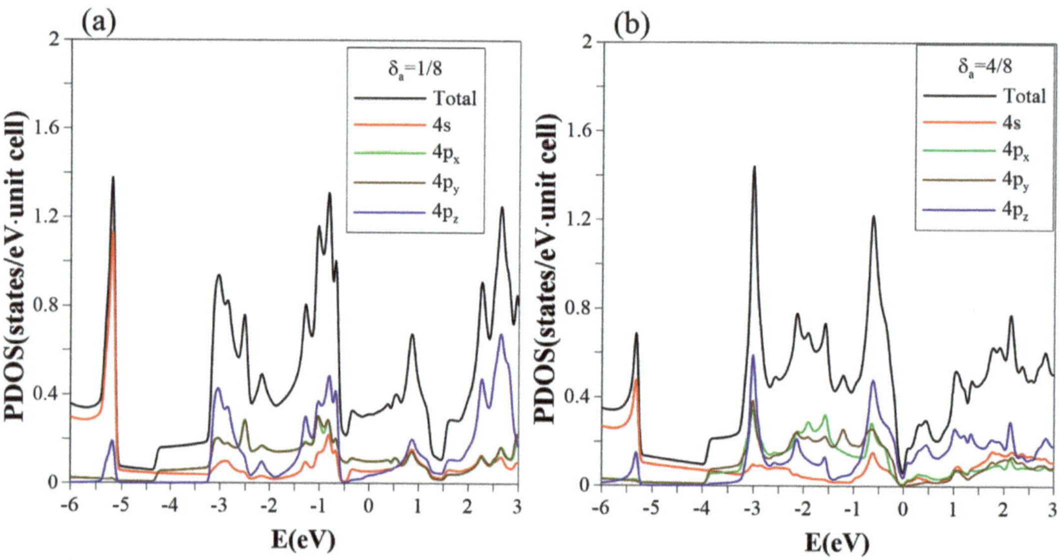

FIGURE 4.12 (A) and(B): orbital-projected density of states of sliding bilayer germanene AA $\delta_a = 1/8$ and $\delta_a = 4/8$. (C) and (D): orbital-projected density of states of sliding bilayer germanene AA $\delta_a = 11/8$ and $\delta_a = 12/8$. (E) and (F): orbital-projected density of states of sliding bilayer germanene AA $\delta_z = 1/8$ and $\delta_z = 4/8$.

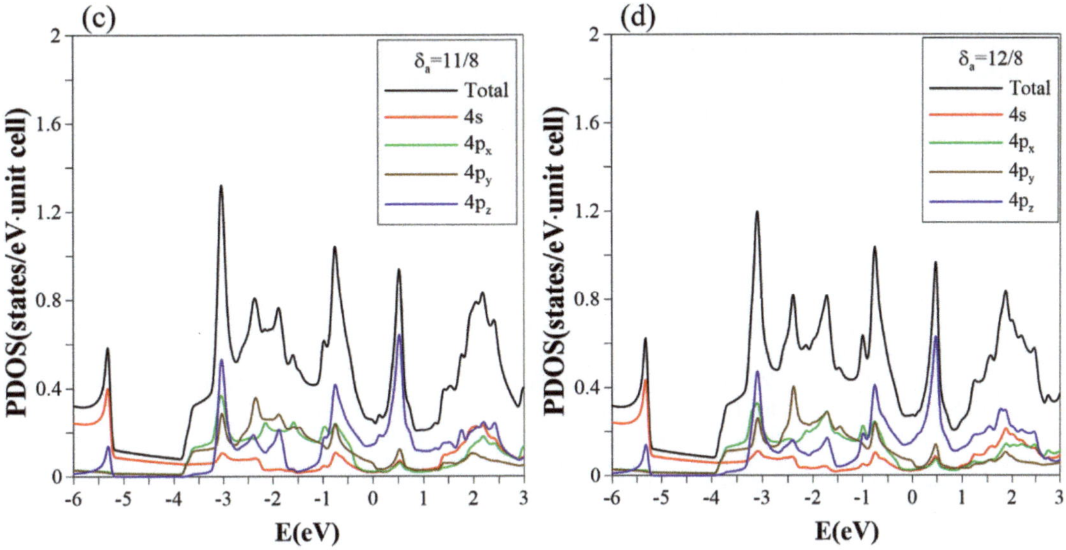

FIGURE 4.12 cont'd.

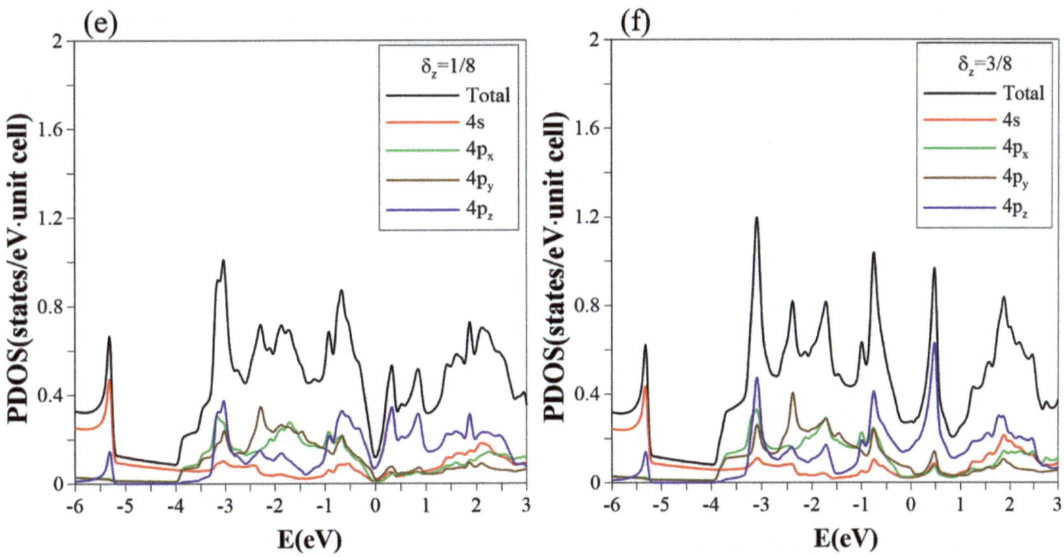

FIGURE 4.12 cont'd.

Fig. 4.12B. In the low-energy range of $|E|<.5$ eV, some stacking configurations are dominated by the $4p_z$ orbitals such as $\delta_a = 4/8$, $11/8$, $12/8$ and $\delta_z = 1/8$, $3/8$. In generally, it is very difficult to distinguish the initial, middle saddle-point structure and final π and σ bands are complicated to tell. There are a lot of symmetric peaks, antisymmetric peaks and shoulders in the whole energy range, owing mainly to multi-orbital hybridization, oscillatory energy dispersions and frequent anticrossing.

References

[1] Novoselov KS, Geim AK, Morozov SV. Electric field effect in atomically thin carbon films. Science 2004;306:666.
[2] Castro Neto AH, Guinea F, Peres NMR, Novoselov KS, Geim AK. The electronic properties of graphene. Rev Mod Phys 2009;81:109.
[3] Ho KI, Huang CH, Liao JH, Zhang W, Li LJ, Lai CS, Su CY. Fluorinated graphene as high performance dielectric materials and the applications for graphene nanoelectronics. Sci Rep 2014;45893.
[4] Sato S. Graphene for nanoelectronics. Jpn J Appl Phys 2014;54:040102.
[5] Balandin AA, Ghosh S, Bao W, Calizo I, Teweldebrhan D, Miao F, Lau CN. Superior thermal conductivity of single-layer graphene. Nano Lett 2008;8:902.
[6] Stoeberl U, Wurstbauer U, Wegscheider W, Weiss D, Eroms J. Morphology and flexibility of graphene and few-layer graphene on various substrates. Appl Phys Lett 2008;93:051906.
[7] Novoselov KS, Falko VI, Colombo L, Gellert PR, Schwab MG, Kim K. A roadmap for graphene. Nature 2012;490:192.
[8] Wu X, Hu Y, Ruan M, Madiomanana NK, Hankinson JH, Sprinkle M, Berger C, de Heer WA. Half integer quantum Hall effect in high mobility single layer epitaxial graphene. Appl Phys Lett 2009;95:223108.
[9] Akinwande D, Petrone N, Hone J. Two-dimensional flexible nanoelectronics. Nat Commun 2014;5:5678.
[10] Ryan K, Neumayer SM, Maraka HVR, Buchete NV, Kholkin AL, Rice JH, Rodriguez BJ. Thermal and aqueous stability improvement of graphene oxide enhanced diphenylalanine nanocomposites. Sci Technol Adv Mater 2017;18:172.
[11] Khokhriakov D, Hoque MA, Karpiak B, Dash SP. Gate-tunable spingalvanic effect in graphene-topological insulator van der Waals heterostructures at room temperature. Nat Commun 2020;11:3657.
[12] Kane CL, Mele EJ. Quantum spin Hall effect in graphene. Phys Rev Lett 2005;95:226801.
[13] Liu CC, Feng W, Yao Y. Quantum spin Hall effect in silicene and TwoDimensional germanium. Phys Rev Lett 2011;107:076802.
[14] Kaloni TP, Schwingenschloegl U. Weak interaction between germanene and GaAs(0001) by H intercalation: a route to exfoliation. J Appl Phys 2013;114:184307.
[15] Li L, Lu S, Pan J, Qin Z, Wang Y, Wang Y, et al. Buckled germanene formation on Pt(111). Adv Mater 2014;26:4820.
[16] Seixas L, Padilha JE, Fazzio A. Quantum spin Hall effect on germanene nanorod embedded in completely hydrogenated germanene. Phys Rev B 2014;89:195403.
[17] Kaloni TP. Tuning the structural, electronic, and magnetic properties of germanene by the adsorption of 3d transition metal atoms. J Phys Chem C 2014;118:25200.
[18] Sante DD, Wu X, Fink M, Hanke W, Thomale R. Triplet superconductivity in the Dirac semimetal germanene on a substrate. Phys Rev B 2019;99:201106.
[19] Yaokawa R, Ohsuna T, Morishita T, Hayasaka Y, Spencer MJS, Nakano H. Monolayer-to-bilayer transformation of silicenes and their structural analysis. Nat Commun 2016;7:10657.
[20] Li H, Fu HX, Meng S. Silicene: from monolayer to multilayer - a concise review. Chin Phys B 2015;24:086102.
[21] Lee JK, Lee SC, Ahn JP, Kim SC, Wilson JI, John P. The growth of AA graphite on (111) diamond. J Chem Phys 2008;129:234709.
[22] D'avila ME, Lay GL. Few layer epitaxial germanene: a novel two-dimensional Dirac material. Sci Rep 2016;6:20714.
[23] Zhu FF, Chen WJ, Xu Y, Gao CI, Guan DD, Jia JF. Epitaxial growth of two-dimensional stanene. Nat Mater 2015;14:2020.

[24] Liu H, Neal AT, Zhu Z, Luo Z, Xu XF, Ye PD, et al. Phosphorene: an unexplored 2D semiconductor with a high hole mobility. ACS Nano 2014;8:4033.
[25] Yang QQ, Liu RT, Huang C, Huang YF, Gao LF, Zhang HL, et al. Phos-phorene: 2D bismuthene fabricated via acid-intercalated exfoliation showing strong nonlinear near-infrared responses for mode-locking lasers. Nanoscale 2018;10:21106.
[26] Shao Y, Liu ZL, Cheng C, Wu X, Liu H, Gao HJ, et al. Epitaxial growth of flat antimonene monolayer: a new honeycomb analogue of graphene. Nano Lett 2018;18:2133.
[27] Pumera M, Sofer Z. 2D monoelemental arsenene, antimonene, and bismuthene: beyond black phosphorus. Adv Mater 2017;29:1605299.
[28] Zhou L, Fox L, Wlodek M, Islas L, Slastanova A, Briscoe WH, et al. Surface structure of few layer graphene. Carbon 2018;136:255.
[29] Ping J, Fuhrer MS. Layer number and stacking sequence imaging of few-layer graphene by transmission electron microscopy. Nano Lett 2012;12:4635.
[30] Lin CY, Huang BL, Ho CH, Gumbs G, Lin MF. Geometry-diversified Coulomb excitations in trilayer AAB stacking graphene. Phys Rev B 2018;98:195442.
[31] Sharpe AL, Fox EJ, Barnard AW, Finney J, Watanabe K, Kastner MA. Emergent ferromagnetism near three-quarters filling in twisted bilayer graphene. Science 2019;10:1126.
[32] Huang YK, Chen SC, Ho YH, Lin CY, Lin MF. Feature-rich magnetic quantization in sliding bilayer graphenes. Sci Rep 2014;4:7509.
[33] Chiu YH, Lai YH, Ho JH, Chuu S, Lin MF. Electronic structure of a two-dimensional graphene monolayer in a spatially modulated magnetic field: peierls tight-binding model. Phys Rev B 2008;77:045407.
[34] Uemura K, Ikuta T, Maehashi K. Turbostratic stacked CVD graphene for high-performance devices. Jpn J Appl Phys 2018;57:030311.
[35] Bampoulis P, Zhang L, Safaei A, van Gastel R, Poelsema B, Zandvliet HJW. Germanene termination of Ge2Pt crystals on Ge(110). J Phys Condens Matter 2014;26:442001.
[36] Kaloni TP, Modarresi M, Tahir M, Roknabadi MR. Electrically engineered band gap in two-dimensional Ge, Sn, and Pb: a first-principles and tight-binding approach. J Phys Chem C 2015;119:11896.
[37] Liu HY, Lin SY, Wu JY. Stacking-configuration-enriched essential properties of bilayer graphenes and silicenes. J Chem Phys 2020;153:154707.
[38] Shih PH, Chiu YH, Wu JY, Shyu FL, Lin MF. Coulomb excitations of monolayer germanene. Sci Rep 2017;7:40600.
[39] Singh R. Spin-orbit coupling in graphene, silicene and germanene: dependence on the configuration of full hydrogenation and fluorination. Bull Mater Sci 2018;41:158.
[40] Acun A, Zhang L, Bampoulis P, Farmanbar M, van Houselt A, Zandvliet HJW, et al. Germanene: the germanium analogue of graphene. J Phys Condens Matter 2015;27:443002.
[41] Xu Y, Yan B, Zhang HJ, Wang J, Xu G, Zhang SC, et al. Large-gap quantum spin Hall insulators in tin films. Phys Rev Lett 2013;111:136804.
[42] Liu CC, Jiang H, Yao Y. Low-energy effective Hamiltonian involving spinorbit coupling in silicene and two-dimensional germanium and tin. Phys Rev B 2011;84:195430.
[43] Que Y, Xiao W, Chen H, Wang D, Du S, Gao HJ. Stacking-dependent electronic property of trilayer graphene epitaxially grown on Ru (0001). Appl Phys Lett 2015;107:263101.
[44] Luican A, Li G, Reina A, Kong J, Nair R, Geim AK, Andrei E, et al. Singlelayer behavior and its breakdown in twisted graphene layers. Phys Rev Lett 2011;106:126802.
[45] Li G, Luican A, Dos Santos JL, Neto AC, Reina A, Andrei E, et al. Observation of Van Hove singularities in twisted graphene layers. Nat Phys 2010;6:109.
[46] Cherkez V, de Laissardiere GT, Mallet P, Veuillen JY. Van Hove singularities in doped twisted graphene bilayers studied by scanning tunneling spectroscopy. Phys Rev B 2015;95:155428.
[47] Lauffer P, Emtsev K, Graupner R, Seyller T, Ley L, Weber H, et al. Atomic and electronic structure of few-layer graphene on SiC (0001) studied with scanning tunneling microscopy and spectroscopy. Phys Rev B 2008;77:155426.
[48] Yankowitz M, Wang F, Lau CN, LeRoy BJ. Local spectroscopy of the electrically tunable band gap in trilayer graphene. Phys Rev B 2013;87:165102.
[49] Pierucci D, Sediri H, Hajlaoui M, Girard JC, Brumme T, Ouerghi A, et al. Evidence for flat bands near the fermi level in epitaxial rhombohedral multilayer graphene. ACS Nano 2015;9:5432.

[50] Bena C, Montambaux G. Remarks on the tight-binding model of graphene. New J Phys 2009;11:095003.

[51] Crepaldi A, Pons S, Frantzeskakis E, Calleja F, Etzkorn M, Grioni M, et al. Combined ARPES and STM study of Pb/Au(111) Moir'estructure: one overlayer, two symmetries. Phys Rev B 2013;87:115138.

[52] Nguyen DK, Lin YT, Lin SY, Chiu YH, Thuy Trana NT, Lin MF. Fluorination-enriched electronic and magnetic properties in graphene nanoribbons. Phys Chem Chem Phys 2017;19:20667.

[53] Lin SY, Lin YT, Thuy Tran NT, Su WP, Lin MF. Feature-rich electronic properties of aluminum-adsorbed graphenes. Carbon 2017;120:209.

[54] Do TN, Gumbs G, Shih PH, Huang D, Chiu CW, Lin MF, et al. Pecu-liar optical properties of bilayer silicene under the influence of external electric and magnetic fields. Sci Rep 2019;9:36547.

[55] Fan Y, Liu X, Wang J, Ai H, Zhao M. Silicene and germanene on InSe substrates: structures and tunable electronic properties. Phys Chem Chem Phys 2018;20:11369.

[56] Jiang L, Marconcini P, Hossian MS, Qiu W, Evans R, Skafidas E, et al. A tight binding and $--\rightarrow k\cdot --\rightarrow p$ study of monolayer stanene. Sci Rep 2017;7:12069.

[57] Song MK, do Hong S, No KT. The structure of lithium intercalated graphite using an effective atomic charge of lithium. J Electrochem Soc 2001;148:1159.

[58] McCann E, Abergel DS, Fal'ko VI. The low energy electronic band structure of bilayer graphene. Eur Phys J Spec Top 2007;148:91.

[59] Warner JH, Mukai M, Kirkland AI. Atomic structure of ABC rhombohedral stacked trilayer graphene. ACS Nano 2012;6:5680.

[60] Zhang F, Sahu B, Min H, MacDonald AH. Band structure of ABC-stacked graphene trilayers. Phys Rev B 2010;82:035409.

[61] Osborne DA, Morishita T, Tawfik SA, Yayamab T, Spencer MJS. Adsorption of toxic gases on silicene/Ag(111). Phys Chem Chem Phys 2019;10:1039.

[62] Shu H, Cao D, Liang P, Wang X, Chen X, Lu W. Two-dimensional silicene nucleation on a Ag(111) surface: structural evolution and the role of surface diffusion. Phys Chem Chem Phys 2014;16:304.

[63] Arafune R, Lin CL, Kawahara K, Tsukahara N, Minamitani E, Kawai M, et al. Structural transition of silicene on Ag(111). Surf Sci 2013;608:297.

[64] Meng L, Wang Y, Zhang L, Du S, Wu R, Gao HJ, et al. Buckled silicene formation on Ir(111). Nano Lett 2013;13:685.

[65] Gill TG, Fleurence A, Warner B, Prüser H, Friedlein R, Yamada-Takamura Y, et al. Metallic atomically-thin layered silicon epitaxially grown on silicene/ZrB_2 *2D*. Mater 2017;4:021015.

[66] Wiggers FB, Bui HV, Friedlein R, Yamada-Takamura Y, Schmitz J, Jong MP. A nitride-based epitaxial surface layer formed by ammonia treatment of siliceneterminated ZrB_2. J Chem Phys 2016;144:134703.

[67] Fleurence A, Yamada-Takamura Y. Insights into the spontaneous formation of silicene sheet on diboride thin films. Appl Phys Lett 2017;110:041601.

[68] Li L, Lu SZ, Pan J, Qin Z, Wang YQ, Gao HJ, et al. Buckled germanene formation on Pt(111). Adv Mater 2014;26:4820.

[69] Cantero ED, Solis LM, Tong Y, Fuhr JD, Martiarena ML, Sánchez EA, et al. Growth of germanium on Au(111): formation of germanene or intermixing of Au and Ge atoms? Phys Chem Chem Phys 2017;19:18580.

[70] Martínez EA, Fuhr JD, Grizzi O, Sánchez EA, Cantero ED. Growth of germanene on Al(111) hindered by surface alloy formation. J Phys Chem C 2019;123:12910.

[71] Chen SW, Wu CY, Wu HJ, Chiu WT. Interfacial reactions in Sn/Bi_2Te_3, Sn/Bi_2Se_3 and Sn/$Bi_2(Te_1-xSe_x)_3$ couples. J Alloys Compd 2014;611:313.

[72] Kim KC, Kwon B, Kim HJ, Baek SH, Hyun DB, Kim JS, et al. Sn doping in thermoelectric Bi_2Te_3 films by metal-organic chemical vapor deposition. Appl Surf Sci 2015;353:232.

[73] Ganz E, Hwang IS, Xiong F, Theiss SK, Golovchenko J. Growth and morphology of Pb on Si(111). Surf Sci 1991;257:259.

[74] Yakes M, Tringides MC. Probing the buried Pb/Si(111) interface with SPA LEED and STM on Si(111)-Pb $\propto \sqrt{3} \times \sqrt{3}$. J Phys Chem A 2011;115:7096.

[75] Johns JE, Hersam MC. Atomic covalent functionalization of graphene. Acc Chem Res 2013;46:77.

[76] Yuan S, Roldn R, Katsnelson MI. Landau level spectrum of ABA-and ABCstacked trilayer graphene. Phys Rev B 2011;84:125455.

[77] Zhou XY, Zhang R, Sun JP, Zou L, Zhang D, Chang K, et al. Landau levels and magneto-transport property of monolayer phosphorene. Sci Rep 2015;5:12295.
[78] Liu Y, Shu H, Liang P, Cao D, Chen X, Lu W. Structural, electronic, and optical properties of hydrogenated few-layer silicene: size and stacking effects. J Appl Phys 2013;114:094308.
[79] Thuy Tran NT, Lin SY, Glukhova OE, Lin M. Configuration-induced rich electronic properties of bilayer graphene. J Phys Chem C 2015;119:10623.
[80] Do TN, Shih PH, Chang CP, Lin CY, Lin MF. Rich magneto-absorption spectra in AAB-stacked trilayer graphene. Phys Chem Chem Phys 2016;18:17597.
[81] Koshino M. Stacking-dependent optical absorption in multilayer graphene. New J Phys 2013;15:015010.
[82] Chiu CW, Chen RB. Influence of electric fields on absorption spectra of AAB-stacked trilayer graphene. Appl Phys Express 2016;9:065103.
[83] Chen SC, Chiu CW, Wu CL, Lin MF. Shift- enriched optical properties in bilayer graphene. RSC Adv 2014;4:63779.
[84] Chiu CW, Huang YC, Shyu FL, Lin MF. Excitation spectra of ABC-stacked graphene superlattice. Appl Phys Lett 2011;98:2631920.
[85] Shyu FL, Lin MF. Electronic and optical properties of narrow-gap carbon nanotubes. J Phys Soc Jpn 2002;71:1820.
[86] Lin MF, Shung KWK. Plasmons and optical properties of carbon nanotubes. Phys Rev B 1994;50:17744.
[87] Ruzicka B, Degiorgi L, Gaal R, Thien-Nga L, Bacsa R, Forro L, et al. Optical and dc conductivity study of potassium-doped single-walled carbon nanotube films. Phys Rev B 2000;61:2468.
[88] Djurisic AB, Li EH. Optical properties of graphite. J Appl Phys 1999;85:7404.
[89] Chiu CW, Lee SH, Chen SC, Shyu FL, Lin MF. Absorption spectra of AA-stacked graphite. New J Phys 2010;12:083060.
[90] Taft EA, Philipp HR. Optical properties of graphite. Phys Rev 1965;138:197.
[91] Ichikawa YH, Kobayashi K. Optical properties of graphite in the infrared region. Carbon 1996;3:401.
[92] Dovbeshko GI, Romanyuk VR, Pidgirnyi DV, Cherepanov VV, Andreev EO, Svirko YP, et al. Optical properties of pyrolytic carbon films versus graphite and graphene. Nanoscale Res Lett 2015;10:234.
[93] Ergun S, Yasinsky JB, Townsend JR. Transverse and longitudinal optical properties of graphite. Carbon 1967;5:403.
[94] Padova PD, Generosi A, Paci B, Ottaviani C, Quaresima C, Lay GL, et al. Multilayer silicene: clear evidence. 2D Mater 2016;3:049501.
[95] Peters JL, Altantzis T, Lobato I, Jazi MA, Overbeek CV, Sinai SB, et al. Mono- and multilayer silicene-type honeycomb lattices by oriented attachment of PbSe nanocrystals: synthesis, structural characterization, and analysis of the disorder. Chem Mater 2018;30:4831.
[96] Coletti C, Forti S, Principi A, Emtsev KV, Zakharov AA, Starke1 U, et al. Revealing the electronic band structure of trilayer graphene on SiC: an angle-resolved photoemission study. Phys Rev B 2013;88:155439.

CHAPTER

5

BCN-adsorbed germanenes

Wei-Bang Li[1], Shih-Yang Lin[1], Vo Khuong Dien[1], Chi-Hsuan Lee[1], Hsin-Yi Liu[1], Hai Duong Pham[1], Thi My Duyen Huynh[1], Nguyen Thi Han[1], Ngoc Thanh Thuy Tran[2], Thi Dieu Hien Nguyen[1] and Ming-Fa Lin[1]

[1]**Department of Physics, National Cheng Kung University, Tainan, Taiwan;** [2]**Hierarchical Green-Energy Materials (Hi-GEM) Research Center, National Cheng Kung University, Tainan, Taiwan**

5.1 Introduction

Group-III, group-IV and group-V elements [1–7], as clearly indicated in a periodic table [8,9], can exhibit the rich and unique quasiparticle phenomena under the various crystal structures [10]. The unusual behaviors could be thoroughly explored from the theoretical (details in Chapter 2) and experimental viewpoints (Chapter 3), mainly owing the concise pictures of the atomic configurations. For example, 3D bulk graphite [11,12] and diamond/silicon/germanium [13–15], respectively, have the perpendicular sp^2 & bondings and sp^3 ones. The similar chemical bonding forms of the former further generate layered graphene systems, graphene nanoribbons, single-/multi-walled carbon nanotubes, C_{60}-associated fullerenes, carbon onions, carbon tori, carbon rings, and carbon chains [16–23]. Very interesting, the binary/ternary/quaternary compounds [24–26] have been successfully generated through the delicate synthesis methods. The chemical modifications will enrich or create the different chemical bondings, leading to the diversified quasiparticle behaviors. In this chapter, boron-, carbon- and nitrogen-adatom chemisorptions on geremanene honeycomb surface are capable of fully illustrating the dramatic transformations of chemical bondings and their prominent effects on all essential properties. The critical mechanisms would be fully achieved from the first-principles simulations [27], being partially supported by the phenomenological models [28].

Fundamental Physicochemical Properties of Germanene-related Materials
https://doi.org/10.1016/B978-0-443-15801-8.00023-2

Any condensed-matter systems, which are successfully synthesized in experimental laboratories, reveal open edges associate with finite-size quantum confinements [29]. The creations of dangling bonds are capable of providing the very active chemical environments. The chemical boundary modulations are available in obviously strengthening the structural stabilities [30], drastically changing all essential properties [31–33], greatly enhancing the special functionalities [34], and largely promoting the potential applications [35]. Such unusual phenomena frequently come to exist in 2D/1D/0D-dimensional systems [36–38], such as, those in few-layer graphenes with (x,y)-plane confinements [39], finite-long carbon nanotubes [23], planar/curved/scr hydrogen-adatom olled/folded graphene nanoribbons [40–43], and quantum dots of carbon disks [44,45], In general, the hydrogen-adatom decorations are relatively easily to passivate the large edge deformations fluctuations [46–49], e.g., the successful chemical modifications on the open boundaries of the graphene-related materials [50–52]. These edge-decorated emergent materials are expected/predicted to reveal the diverse quasi-particle phenomena, being very useful in greatly promoting the development of theoretical framework [53,54].

In addition to edge passivation, honeycomb surfaces of graphene/silicene/germanene/tinene/plumbene [1–7,55] are very suitable for adatom chemisorptions in observing the unusual quasiparticle behaviors [56–58] and creating the functional nano-devices [59–65]. According to the previously theoretical predictions and experimental examinations [66,67], on the planar/buckled structures, there exist the perpendicular C-$2p_z$ orbitals/the non-perpendicular (s, p_x, p_y, p_z) ones in the graphene-related/other materials [28,68]. Such active dangling orbitals are able to very friendly environments in realizing the various chemisorptions. The very successful syntheses have been systematically conducted on the graphene-based compounds, covering the double- and single-side hydrogenations [69–71], alkalizations [72], oxidizations [73], and halogenations [74]. Very inter esting, the VASP calculations are clearly identified to be available for completely understanding the concise mechanisms of orbital hybridizations in distinct chemical bonds and spin-dependent configurations [75].

For example, these chemisorption cases, respectively, induce the semiconducting [76–79], metallic (n-type doping with free conduction electrons [80–85]), as well as, semiconducting/metallic (p-type doping with free valence holes [86–88]). Apparently, all quasiparticle properties are greatly diversified by the intrinsic interactions due to charges and spins, be consistent with the current viewpoints.

The first-principles simulations are utilized to thoroughly investigate the essential properties of boron-, carbon- and nitrogen-chemisorbed germanene systems. The critical mechanisms, the single-/multi-orbital hybridizations [89]; the spin-orbital couplings [90] and/or the on-site Coulomb interactions [91,92], are closely related to the Ge–Ge, Ge–B/Ge–C/Ge–N, and B–B/C–C/N–N bonds, as well as, the magnetic configurations of host and guest atoms. The delicate identifications are examined from the full consistence among the relevant physical and chemical quantities. That is, the consistent pictures will be revealed in the highly non-uniform chemical environment in a Moire superlattice [93,94], the featured band structures & atom-dominated Bloch wave functions [95], the spatial charge/spin density distributions & their variations after the significant chemisorptions [96], and the atom-, orbital- and spin-decomposed density of states [97]. The merged structures of orbital-dependent van Hove singularities, which would survive at the specific energy ranges, could serve as the basic evaluation rules. Furthermore, they are fully assisted by the other characteristics. Certain important similarities and differences, arising from the germanene-, silicene- and graphene-related substitution compounds, will be discussed in details. Most important, the concise pictures of chemical bondings are available in evaluating whether the phenomenological models are suitable and reliable in the near-future studies of quasiparticle framework [98]. The high-resolution measurements are required to systematically examine the theoretical predictions (details in Chapter 3).

5.2 Theoretical calculations and framework

The diverse physical/chemical/material properties, which are frequently revealed in emergent materials [99–101], provide outstanding phenomena in developing the theoretical frameworks [102–104]. This can unify the various viewpoints inside the numerical simulations and phenomenological models.

Computer modeling and simulation have emerged as indespensable methods for scientific materials research, in addition to experiments and theories. Quantum mechanic wave functions provides all information about a specific system. For a many-particle condensed-matter system, it is very hard to solve the Schrodinger equation exactly. Some approximations could be made to obtain the reliable solution. Up to now, the first-principles calculation are the dominant method in the quantum mechanical simulation of periodic systems. They are efficient in solving the very complicated electron-electron and electron-ion interactions through a series of approximations and simplifications to get eigenvalues and eigenfunctions, and thus geometric, electronic, and magnetic properties. For recently decades, Vienna ab initio simulation package (VASP) [27] computes an approximate solution to the many-body Schrodinger equation, either within the density functional theory (DFT), solving the Kohn-Sham equations, or within the Hartree-Fock approximation, solving the Roothaan equations.

In this work, the essential properties are investigated by the first-principles DFT using VASP. The exchange-correlation energy due to the electron-electron interactions is calculated from the Perdew-Burke-Ernzerhof functional under the generalized gradient approximation [105]. The projector-augmented wave pseudopotentials are employed to evaluate the electron-ion interactions [106]. The wave functions are built from the plane waves with a maximum energy cutoff of 500 eV, in which they are test to achieve a very high accuracy. All the atoms in the relaxation process are subjected the adjustments of their positions in order to have an optimized structure with the lowest total ground state energy. And then, the most stable configuration is used as an input in for the further calculations on 16th fundamental physical properties. The vacuum distance along the z-axis is set to be 10 Å for avoiding the interaction between two neighboring cells. To fully understand the adsorption effects on the magnetism, the spin configurations are taken into account for the adatom-adsorbed graphene systems. Furthermore, the van der Waals force is employed in the calculations using the semi-emprical DFT-D2 correction of Grimme [107] to correctly describe the significant atomic interactions between two graphene layers. The first Brillouin zone is sampled in a Gamma scheme along the two-dimensional periodic direction by $12 \times 12 \times 2$ k points for structure relaxations, and by $100 \times 100 \times 5$ for accurate evaluations on electronic properties of the 3×3 supercell. An equivalent k-point mesh is set for other supercells based on their sizes. The convergence criterion for one full relaxation is mainly determined by setting the Hellmann-Feynman forces smaller than 0.01 eV/ Å and the total energy difference of $\Delta E_{total} < 10^{-5}$ eV.

The atomic orbitals of Ge and adatoms are included in the calculations simultaneously, so that the dramatic changes of chemical bondings in honeycomb structure and the critical hybridization of distinct atoms are investigated in detail. For the pristine structure-dependent germanene and the adatom-germanene, the calculated results include the ground state energy, binding energy, bond length, position and height of adatom, band structure, adatom-induced energy gap and free carriers, spatial charge distribution, charge variation; total and orbital-projected DOS.

5.3 Results and discussion

5.3.1 Geometric structures

The well-known silicene and germanene exhibit the similar electronic properties as graphene and honeycomb structures as graphene, shown in Fig. 5.1. However, graphene possesses pure sp2 orbitals between carbon atoms and forms planar sheet, the silicene and germanene is conjectured to have a buckled honeycomb structure due to the hybridizations of sp2 and sp3 orbitals between Si—Si atom and between Ge—Ge atoms. Furthermore, for the Si and Ge, the buckling occurs at various points in the unit cell, resulting in different types of lattice arrangements such as chair-, boat-, and washboard-like structures [108]. The density functional theory reveals that the boat- and washboard-like arrangements are unstable and will tend to converge into the flat structure, while the chair-like remains more stable than planar structure [108]. According to the previous studies, the buckling height of silicene and germanene are 0.46 Å [109,110] and 0.69 Å [110], respectively, and the buckling is responsible for the tunable band gap [111—113].

The main characteristics of geometric structures of single-side/double-side B-, C-, and N-adsorbed germanene, including the bond lengths and bucklings, are calculated in the delicate first-principles calculations. The unit cell of germanene possesses two atoms, and we define that the top adsorbed one is situated upon the valley of Ge atom and bottom one is located bellowed the top of Ge atom, shown in Fig. 5.2. The results reveal that all cases still display uniform stable honeycomb with bucklings.

For single-side B-, C-, and N-adsorbed germanene, shown in Fig. 5.3A, C and E, the bond lengths of Ge—Ge are enlarged from 2.340 to 3.145, 3.257, and 2.960 Å, respectively. The heights of buckling are increased from 0.690 to 2.210, 2.367, 1.938 Å. The bond lengths of Ge-adatoms (top) are 2.095, 1.935, and 1.956 Å. It reveals that the adatom will provide a chemical modification. Compare with the pristine germanene, all Ge—Ge bond length get larger implying the weaker covalent bonds. The heights of buckling are all much larger

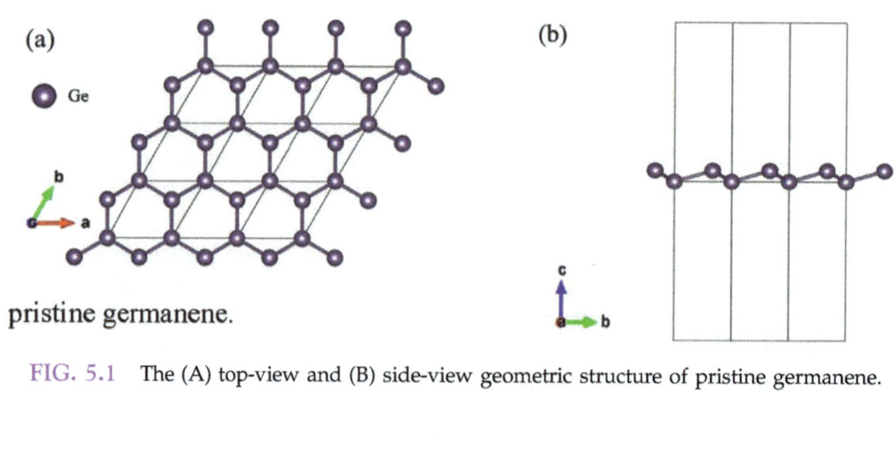

FIG. 5.1 The (A) top-view and (B) side-view geometric structure of pristine germanene.

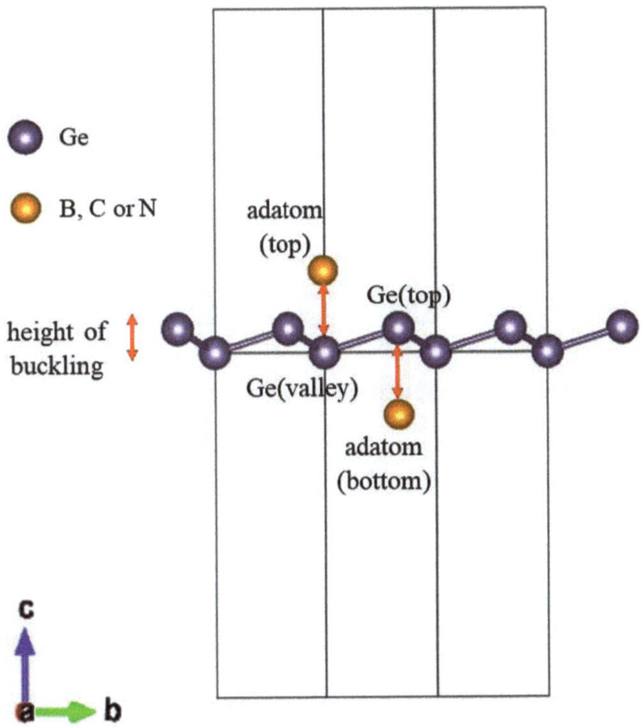

FIG. 5.2 The side-view boron-, carbon-, nitrogen-adsorbed germanene system.

even three times than pristine case, which shows that the adatoms provide the very non-uniform chemical environment. The C-adatom one exhibit the largest buckling and Ge–Ge bond length, but the shortest Ge-adatom bond length. This might reveal that the carbon atoms experience the strong guest-host interaction among the three cases.

For the double-side B-, C-, and N-adsorbed germanene, shown in Fig. 5.3B, D and F, the bond lengths of Ge–Ge, which are different from the single-side adsorbed cases, are 3.321, 3.075, and 3.090 Å, respectively. The B-adsorbed germanene has the larger bond length of Ge–Ge, namely, the Ge–Ge bond in B-adsorbed germanene is weaker than C- and N-adsorbed ones. The bond lengths of Ge-adatoms (top)/(bottom) are 2.453/2.457, 2.082/2.081, and 1.984/1.982 Å, respectively, for B-, C-, and N-adsorbed germanene. Moreover, the heights of buckling for B-, C-, and N-adsorbed germanene are 2.455, 2.109, and 2.132 Å, respectively, indicating the double-side B-adsorption create the much non-uniform chemical environment. The graphene only exhibits the pure sp2 orbitals bringing about the planar structure, but the germanene possesses hybridization of sp2 and sp3 orbitals causing the buckling. For B-, C-, and N-adsorbed germanene, the drastic changes of heights of buckling indicate that there still exist the hybridizations of sp2 and sp3, and the sp3 get more significant than in pristine germanene, shown in Table 5.1. Interestingly, our results of magnetic moment (μ_B) for all single-side and double-side adsorbed germanene reveal that only B-single-side case possesses weak magnetic moment of $0.057\mu_B$, but other cases do not contain magnetic configuration.

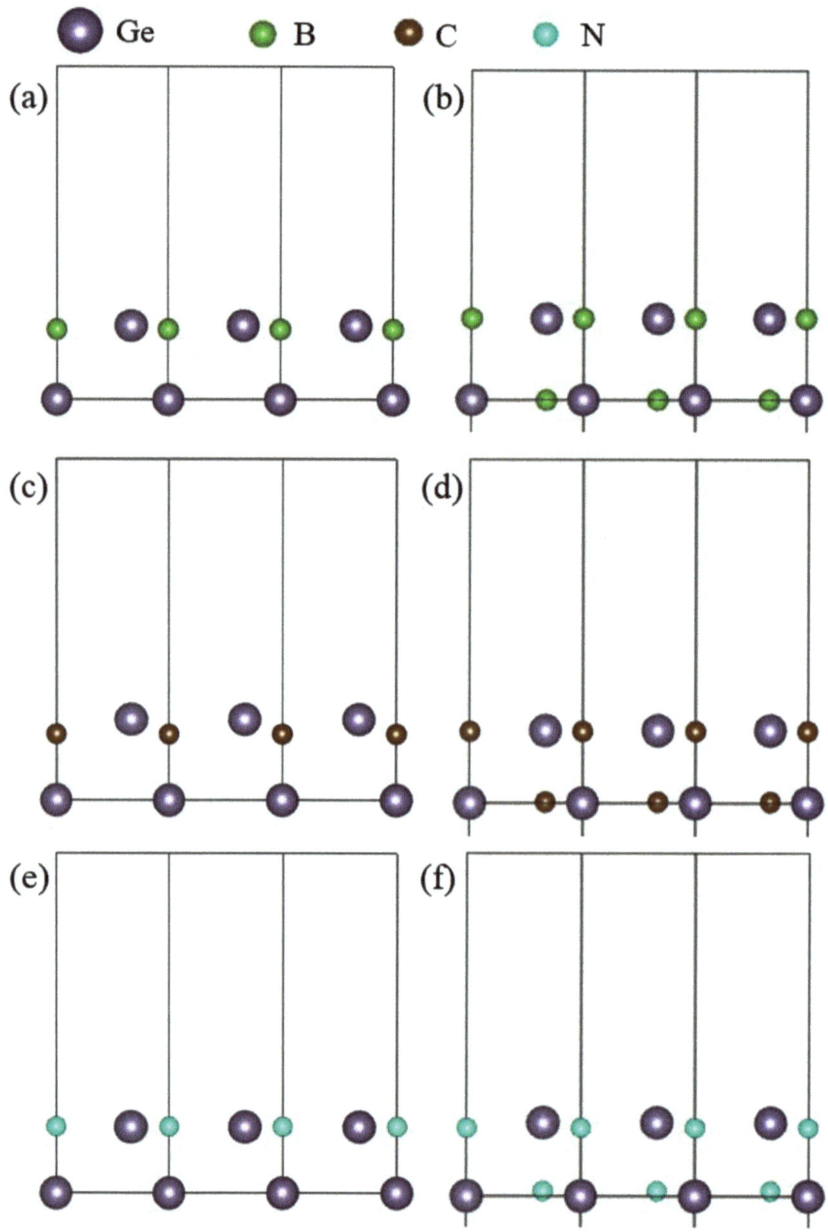

FIG. 5.3 The optimal crystal structures for the single-side/double-side (A)/(B) boron-, (C)/(D) carbon-, (E)/(F) nitrogen-adsorbed germanene systems.

TABLE 5.1 The calculated geometric structures, magnetic moment, and ground-state energy.

	height of buckling (Å)	Ge-Ge bond length (Å)	Ge-adatom (top) bond length (Å)	Ge-adatom (bottom) bond length (Å)	Ground state energy (eV)	Magnetic moment (μ_B)
Germanene	**0.690**	**2.340**				**0**
single-side GeB	**2.210**	**3.145**	**2.095**		**-13.129**	**0.057**
single-side GeC	**2.367**	**3.257**	**1.935**		**-14.460**	**0**
single-side GeN	**1.938**	**2.960**	**1.956**		**-14.796**	**0**
double-side GeB	**2.455**	**3.321**	**2.453**	**2.457**	**-18.768**	**0**
double-side GeC	**2.109**	**3.075**	**2.082**	**2.081**	**-21.983**	**0**
double-side GeN	**2.132**	**3.090**	**1.984**	**1.982**	**-22.506**	**0**

5.3.2 Band structures

The specific relations between the π, σ, and sp3 bondings in pristine monolayer germanene describe the main features of the band structure. The anti-crossing behavior between σ and π valence bands along the ΓK and ΓM directions implies there exists a weak, but significant sp3 bonding. Generally, in monolayer germanene, all π-bandwidth could be well-defined. However, it is very difficult to measure the σ-bandwidth.

It is known that boron, carbon, and nitrogen possess three, four, and five valence electrons in the 2s, 2px, 2py, and 2pz orbitals, which can create unique electronic band structures and totally different from the pristine one. All the electronic states in the Brillouin zone scheme, which are plotted along Γ-K-M-Γ-A-H-L-A paths, are presented in Fig. 5.5. For the pristine germanene, the anti-crossing of π band and σ band at roughly $-1.9 \sim -2.2$ eV reveals the existence of buckling in Fig. 5.4. For the single-side adsorbed germanene, the first σ bands could be defined along the K-M-Γ path according to the concave-downward parabolic dispersions, and they initiate from −0.5eV, −2.1eV and −3.1eV, respectively, for B-, C-, and N-adsorbed germanene. The dominated atoms in low-lying energy range are Ge in Fig. 5.4 (blue circle), and the almost perfect Dirac cone from pristine germanene are destroyed in these three single-side cases. The symmetries of band structures vanish as well after adsorptions.

For the double-side adsorbed germanene, the B-, C-, N-atoms dominate much more in the low-lying energy range, very different from the single-side ones. The band structures get more because there are one more atom in the unit cell. According to the geometric structures after adsorptions in Fig. 5.3, the double-side adsorbed germanene display like bilayer

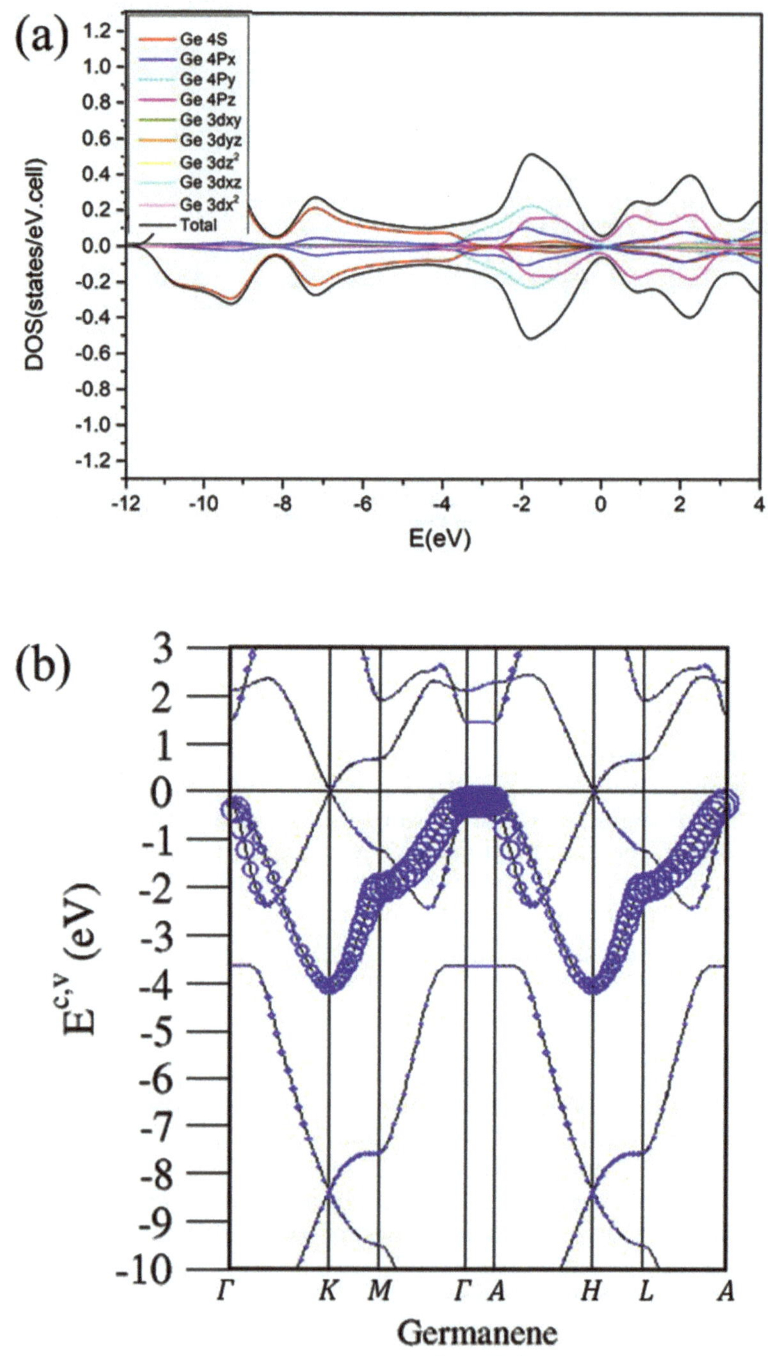

FIG. 5.4 The (A) energy-dependent density of states, as well as the atom/orbital decompositions, and (B) energy band structures for the pristine germanene.

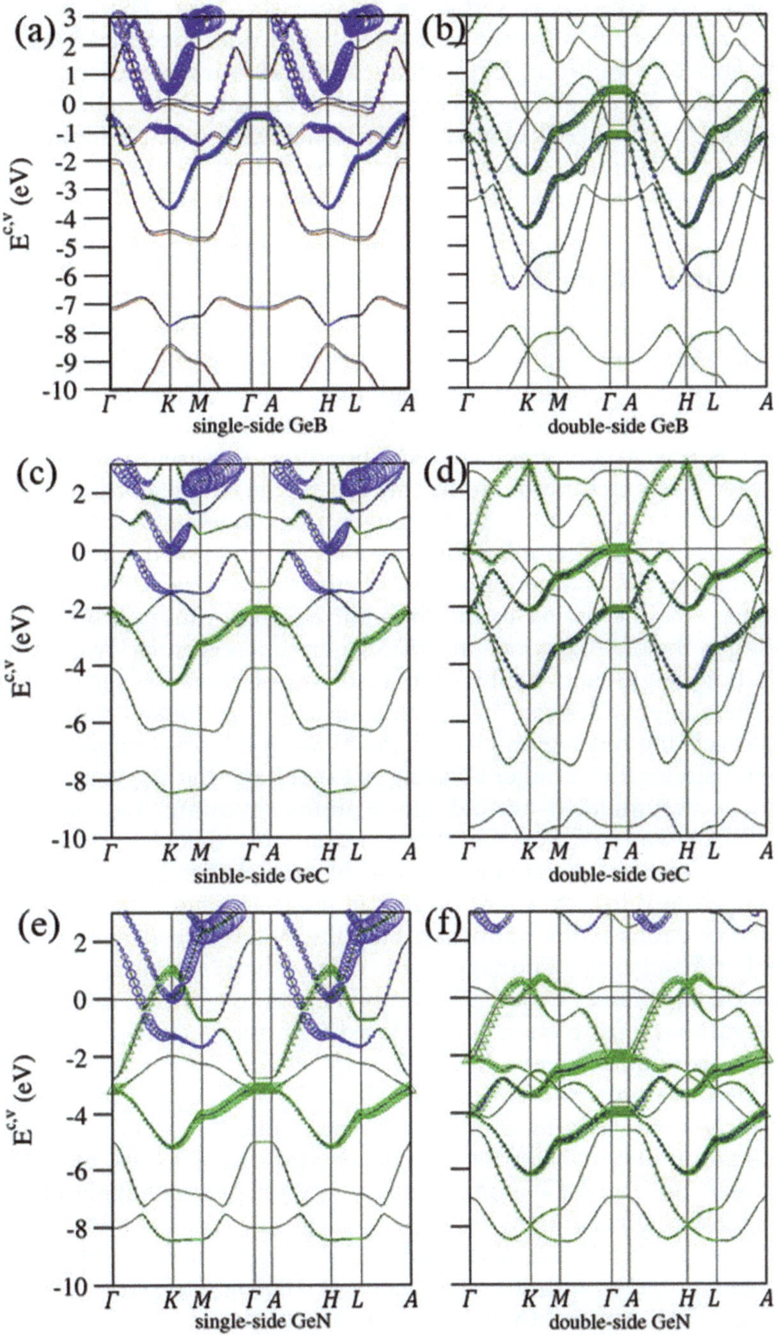

FIG. 5.5 Band structures for the single-side/double-side (A)/(B) boron-, (C)/(D) carbon-, (E)/(F) nitrogen-adsorbed germanene systems.

materials, namely, the Ge—B/Ge—C/Ge—N bilayer structure. The contributions of the adsorbed atoms are also quite different because of the distinct concentrations of adsorptions.

The buckling structure and four/three, four/four, and four/five valence electrons in Ge/B, Ge/C, and Ge/N, respectively, atoms cause the complicated orbital-hybridization of (2s, 2px, 2py)-(3s, 3px, 3py) in N—Si bonds, the ferromagnetic spin configuration, and create an unusual Ge-B-, Ge-C-, and Ge-N-based electronic states. Obviously, the disappearance of Dirac cone along with the unwell-behavior of π and σ bands can be realized.

5.3.3 Spatial charge distributions and variations

The characterization of the impure/pure π and σ chemical bondings, as well as their non-orthogonality or orthogonality, are able to scrutinize in the spatial charge density distributions. Both π and σ chemical bondings that survive in pristine germanene become non-orthogonal to each other, as clearly observed in Fig. 5.6A. This feature implies that the pristine system has a weak but significant sp3 bonding. Apparently, most of the charge is assembled between two germanium atoms (the red region) in proportion to the very strong σ bonding of three orbitals (4s, 4px, 4py). Moreover, perpendicular to the (x; y) plane, the parallel 4pz orbitals which create the π bonding are also observed by the light green in the outer region. For adsorption, the charge density is highly asymmetric with respect to the B/C/N—Ge bond center, mainly owing to the different electron affinities of the guest and host atoms. Interestingly, the Ge—Ge bonds in pristine germanene get very weak after all adsorption cases including single- and double-side adsorbed germanene, shown in Fig. 5.6B—M. That is to say, the adatoms destroy the pristine structures seriously. Also, comparing these three adsorbed cases, the red region around the N atom occupies more than B but slightly than C case demonstrates the stronger bonding of Ge—N. For the single-side cases, the higher charge density are accumulated around guest atoms revealing the charge transfer from germanium to B/C/N-atoms. For the charge variations, the blue region around Ge in Ge—B case is the largest but the smallest in Ge—N case, that indicates the charge transfers much more from Ge to B than to C aor N. It is in good agreement with the fact that boron just possess three electrons in the outmost orbital but N has five electrons.

For the double-side adsorbed germanene, the charge distributions and variations are very unusual. They look like bilayer structures with two kinds of atoms in each plane. The charge distributions reveal that most carriers are gathered with respect to B-/C-/N-atoms. According to the charge distribution shown in Fig. 5.6I—M, it is apparent that the Ge—Ge bond is much weaker than adatoms. Moreover, the vertical red regions around guest atoms are higher in C- and N-atom cases than in B-atom one, while the differences between C- and N-atom are difficult to identify. But for both single and double-cases, the B-atom has great red region between horizonal Ge-atom, namely, B-atom gets relatively strong interaction with horizon-neighboring Ge-atoms.

Interestingly, compared with C-, and N-adsorbed cases, only the B-adsorbed system possesses the magnetic property with the weak value of $0.057\mu_B$. Th spin up and spin down are illustrated in (Fig. 5.5A). Apparently in this case, the valence electronic states determine the net magnetic moment.

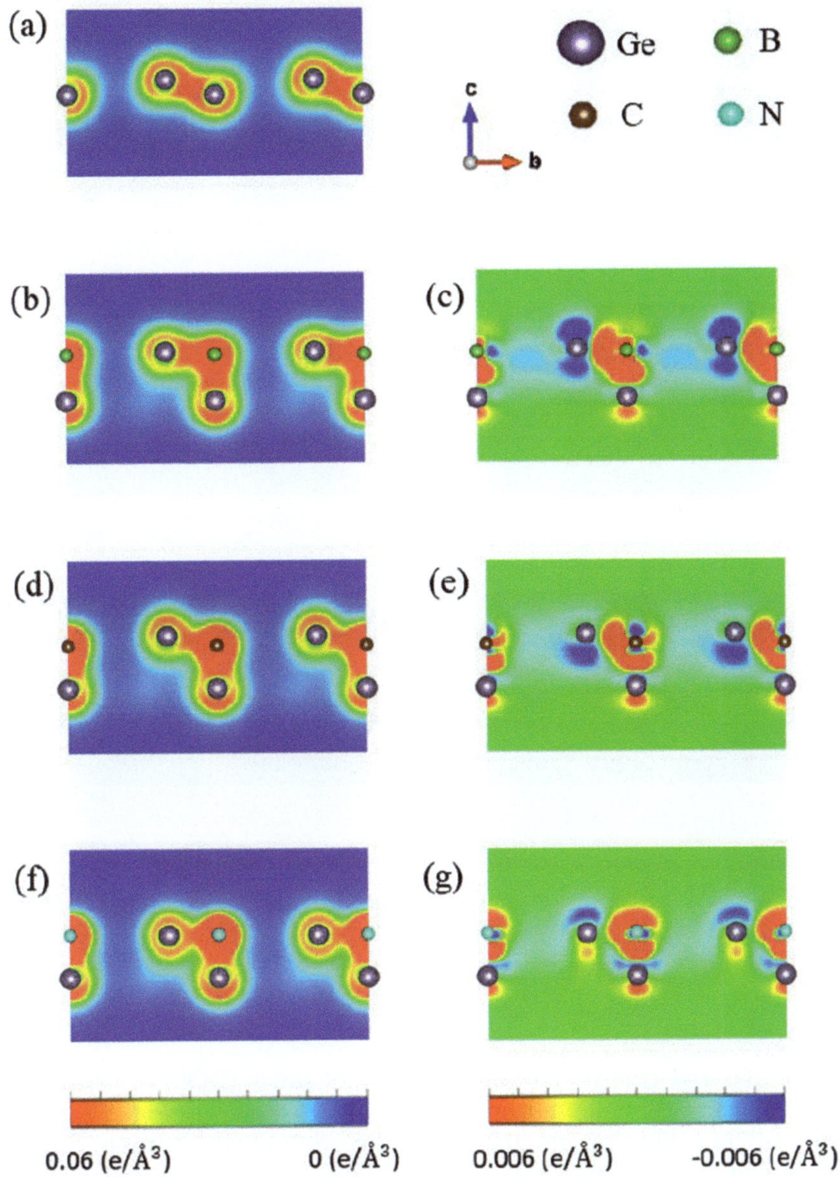

FIG. 5.6 The charge density distributions for (A) the pristine germanene, the single-side/double-side (B)/(H) boron-, (D)/(J)carbon-, (F)/(L) nitrogen-adsorbed germanene; and the charge variations after adsorptions for the single-side/double-side (C)/(I) boron-, (E)/(K) carbon-, (G)/(M) nitrogen-adsorbed germanene.

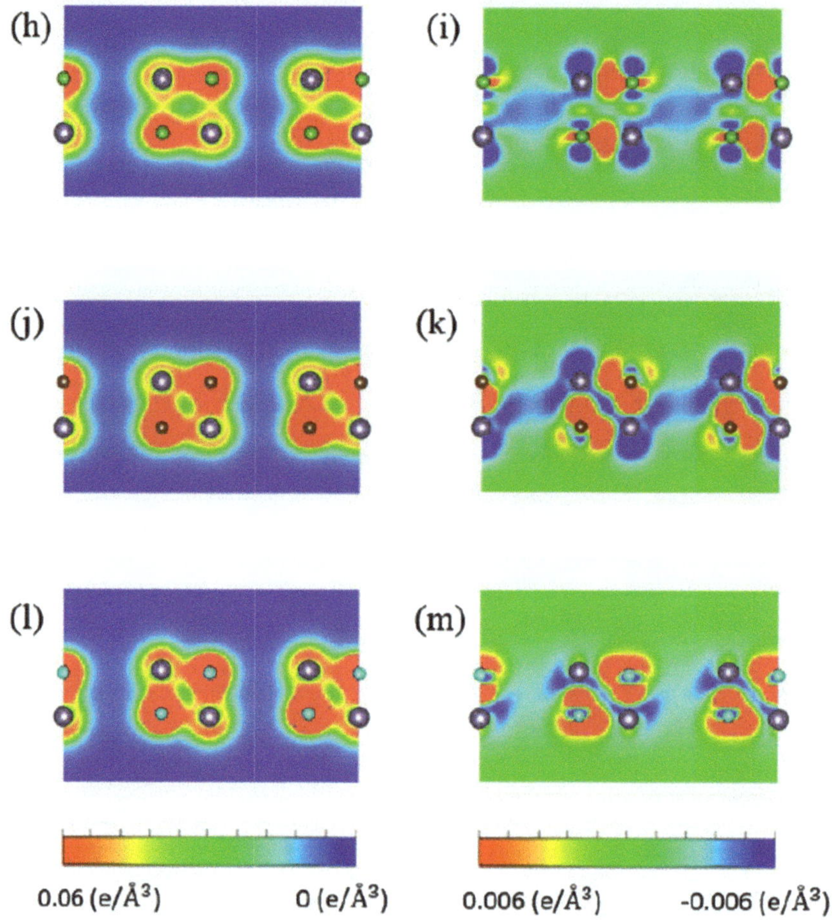

FIG. 5.6 cont'd

5.3.4 Density of states

The atom-, orbital and spin-decomposed density of states (DOS) of pristine germanene, shown in Fig. 5.4A, and B-, C-, N-adsorbed germanene, shown in Fig. 5.7, as clearly illustrated in, can provide further information about the orbital-hybridization, doping phenomena. The band-edge states of parabolic and oscillatory energy dispersions mainly create many special structures in the density of states. For example, the Dirac cone structure, local minima or maxima of parabolic electronic band structure, saddle points, constant energy loops, the flat or partial flat band are close related to the various kinds of van Hove singularities like the V-shape form, discontinuous shoulders, logarithmically divergent peaks, asymmetric peaks in the square root divergence, and delta function-like peaks, respectively. Obviously, Ge-(4s, 4px, 4py, 4pz) orbitals make larger contribution in every region. The combination of density of states and band structures exhibit the three-orbital of (4s, 4px, 4py)-(2s,

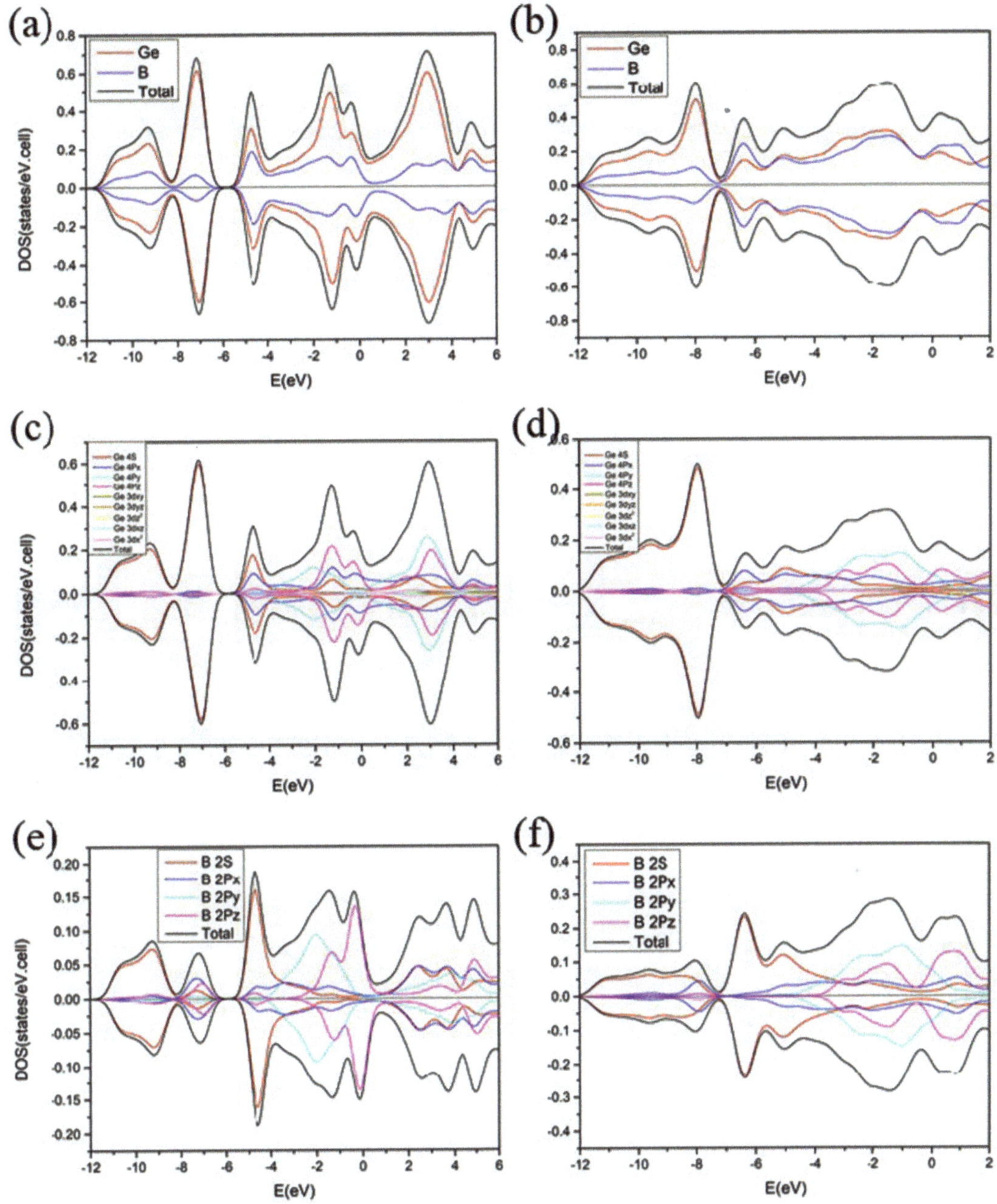

FIG. 5.7 The energy-dependent density of states, as well as the atom/orbital/orbital decompositions, for the single-side adsorbed germanene systems: (A)/(C)/(E) boron-, (G)/(I)/(K) carbon- (M)/(O)/(Q) nitrogen-adsorptions; for the double-side adsorbed germanene systems: (B)/(D)/(F) boron-, (H)/(J)/(L) carbon- (N)/(P)/(R) nitrogen-adsorptions.

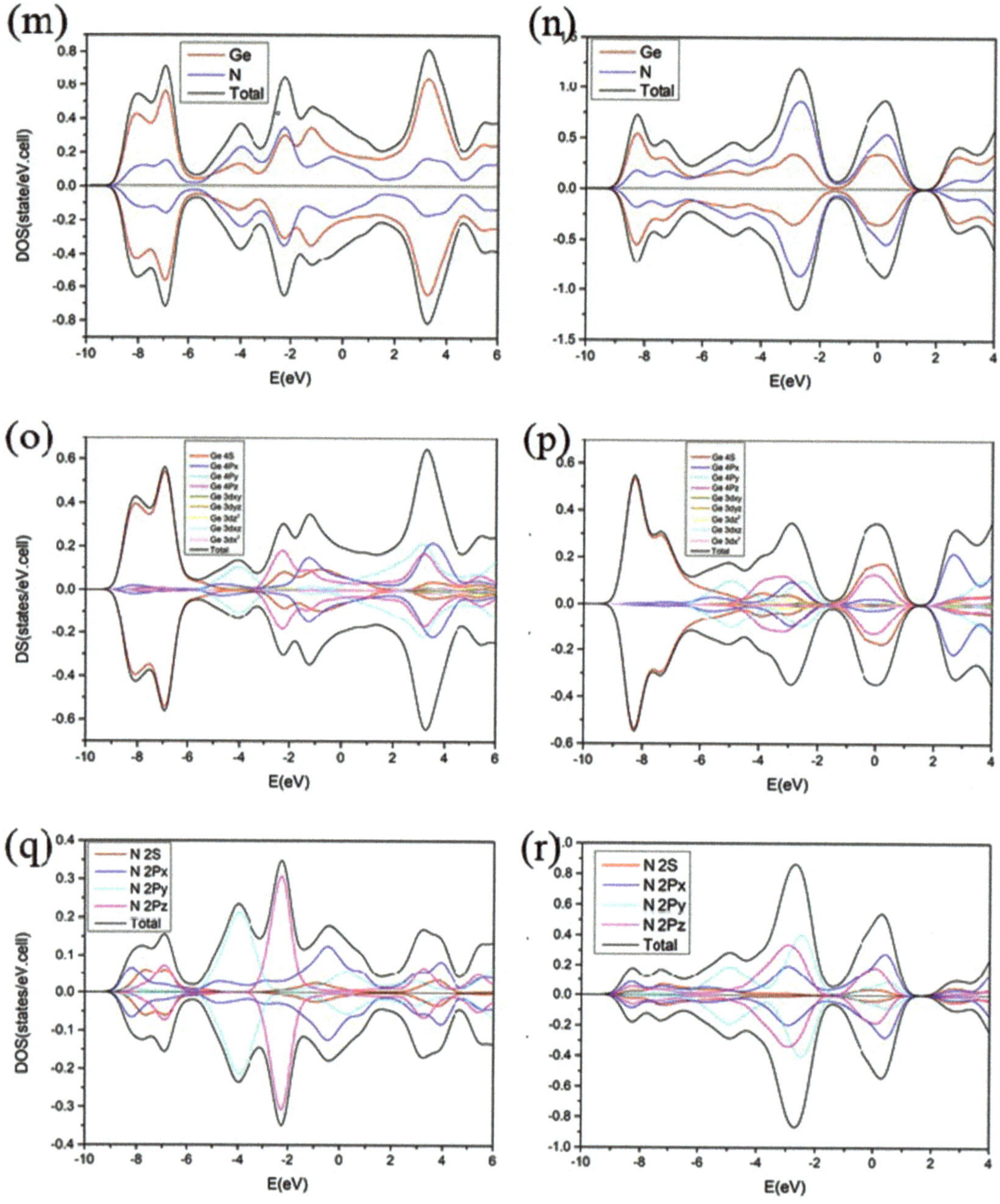
(m)
Ge
N
Total
DOS(state/eV.cell)
E(eV)
(n)
Ge
N
Total
E(eV)
(o)
DS(states/eV.cell)
E(eV)
(p)
E(eV)
(q)
N 2S
N 2Px
N 2Py
N 2Pz
Total
DOS(states/eV.cell)
E(eV)
(r)
N 2S
N 2Px
N 2Py
N 2Pz
Total
E(eV)

FIG. 5.7 cont'd

2px, 2py), and the four-orbital of (4s, 4px, 4py, 4pz)-(2s, 2px, 2py, 2pz). That is to say, the anti-crossings of band structures and the features of van Hove singularities reveal the co-existence of hybridization of sp2 and sp3. The buckling honeycomb lattice might create the complicated (2s, 2px, 2py, 2pz)-(4s, 4px, 4py, 4pz) sp3-orbital hybridizations and thus lead to difficult to trace the path as well as to measure the π and σ bandwidths. In other word, the complicated orbital hybridization makes the identification of the π and σ bands is impossible. We cannot verify the bandwidths, the initial and final state of π and σ bands. The presence of mixed four-orbital hybridizations implies that it would be difficult to simulate the VASP band structures by employing a tight-binding model with orbital-induced hopping integrals, site energies, and the spin-dependent interactions.

5.4 Conclusion

In conclusion, on the basis of the first-principles calculations, we display the optimal structures, spatial charge distributions, charge variations, band structures, and density of states of boron-, carbon-, and nitrogen-adsorbed germanene. The heights of buckling are drastically changed to be 2.210/2.367/1.938 and 2.455/2.109/2.132 Å, respectively, for the single-side adsorbed and double-side adsorbed germanene. And the host-guest bond length of Ge−B/Ge−C/Ge−N for single-cases are 2.095/1.935/1.956 Å, respectively; and for the double-cases of the top adatom (the bottom adatom) are 2.453/2.082/1.984 (2.457/2.081/1.982 Å), respectively; and the ground state energy are −13.129 eV/-14.460 eV/-14.796 eV (−18.768 eV/-21.983 eV/-22.506 eV). The boron atoms get stronger interaction with the horizon-neighboring germanium atoms, but the nitrogen has the stronger interactions with the germanium atoms blow. Whether single-ones or double-side ones, the nitrogen-adsorbed germanene shows the most stability among all. The calculated results all reveal that there exist very strong but complicated (2s, 2px, 2py, 2pz)-(4s, 4px, 4py, 4pz) sp3-orbital hybridizations. Even the double-side adsorbed cases exhibit like bilayer structures. When the first-principles electronic energy spectra along the high-symmetry paths are successfully simulated by the tight-binding model with the uniform & multi-/single-orbital hopping integrals, the diversified essential properties could be fully explored in the near-future studies, e.g., the rich and unique magnetic quantization phenomena, the quantum spin Hall effect, Coulomb excitation, and optical properties.

References

[1] An XT, Zhang YY, Liu JJ, Li SS. Quantum spin Hall effect induced by electric field in silicene. Appl Phys Lett 2013;102.

[2] Zhao FL, Wang Y, Zhang X, Liang XJ, Zhang F, Wang L, Li Y, Feng YY, Feng W. Few-layer methyl-terminated germanene-graphene nanocomposite with high capacity for stable lithium storage. Carbon 2020;161:287−98.

[3] Chen RB, Chen SC, Chiu CW, Lin MF. Optical properties of monolayer tinene in electric fields. Sci Rep 2017;7.

[4] Mahmud S, Alam MK. Large bandgap quantum spin Hall insulator in methyl decorated plumbene monolayer: a first-principles study. RSC Adv 2019;9:42194−203.

[5] Suragtkhuu S, Bat-Erdene M, Bati ASR, Shapter JG, Davaasambuu S, Batmunkh M. Few-layer black phosphorus and boron-doped graphene based hetero electrocatalyst for enhanced hydrogen evolution. J Mater Chem 2020;8:20446−52.

[6] Zhong W, Zhao Y, Zhu BB, Sha JJ, Walker ES, Bank S, Chen YF, Akinwande D, Tao L. Anisotropic thermoelectric effect and field-effect devices in epitaxial bismuthene on Si (111). Nanotechnology 2020;31.
[7] Zou H, Zhang H, Yang ZX, Zhang ZH. Magneto-electronic and spin transport properties of transition metal doped antimonene nanoribbons. Physica E 2021;126.
[8] Waber JT, Cromer DT. Orbital radii of atom and ions. J Phys Chem 1965;42:4116.
[9] Clementi C, Raimondi DL. Atomic screening constants of from SCF functions. J Phys Chem 1963;38:2686.
[10] Bostwick A, et al. Quasiparticle dynamics in graphene. Nat Phys 2007;3:36–40.
[11] Tuinstra F, Koenig JL. Raman spectrum of graphite. J Chem Phys 1970;53:11261130.
[12] Huang CS, Lin MF, Chuu DS. Collective excitations in graphite layers. Solid State Commun 1997;103:603606.
[13] Savin A, et al. Electron localization in solid-state structures of the elements: the diamond structure. Angew Chem Int Ed 1992;31:187–8. in English.
[14] LeVier R, Harrison M, Cook R, Lane T. Plast Reconstr Surg 1993;92:12–167.
[15] Bundy FP, Kasper JS. A new dense form of solid germanium. Science 1963;139:340–1.
[16] Erohin SV, Ruan QY, Sorokin PB, Yakobson BI. Nano-thermodynamics of chemically induced graphene-diamond transformation. Small 2020;16.
[17] Meyer RA, Dirling RB, Kratsch KM. Influence of microstructure on sublimation characteristics of bulk graphites. Carbon 1972;10:338. -&.
[18] Weibel A, Flaureau A, Pham A, Chevallier G, Esvan J, Estournes C, Laurent C. One-step synthesis of few-layered-graphene/alumina powders for strong and tough composites with high electrical conductivity. J Eur Ceram Soc 2020;40:5779–89.
[19] Stefan-van Staden RI, Comnea-Stancu IR. Chiral single-walled carbon nanotubes as chiral selectors in multimode enantioselective sensors. Chirality 2021;33:51–8.
[20] Pawlak R, Liu XS, Ninova S, D'Astolfo P, Drechsel C, Sangtarash S, Haner R, Decurtins S, Sadeghi H, Lambert CJ, Aschauer U, Liu SX, Meyer E. Bottom-up synthesis of nitrogen-doped porous graphene nanoribbons. J Am Chem Soc 2020;142:12568–73.
[21] Artyukh AA, Chernozatonskii LA. Simulation of the formation and mechanical properties of layered structures with polymerized fullerene-graphene components. Jetp Lett 2020;111:109–15.
[22] Liu YY, Kim DY. Ultraviolet and blue emitting graphene quantum dots synthesized from carbon nano-onions and their comparison for metal ion sensing. Chem Commun 2015;51:4176–9.
[23] Gao EL, Li RS, Baughman RH. Predicted confinement-enhanced stability and extraordinary mechanical properties for carbon nanotube wrapped chains of linear carbon. ACS Nano 2020;14:17071–9.
[24] Qiu Y, Gao L. Chemical synthesis of turbostratic carbon nitride, containing CN crystallites, at atmospheric pressure. Chem Commun 2003;18:2378–9.
[25] Deng X, Kousaka H, Tokoroyama T, Umehara N. Deposition and tribological behaviors of ternary BCN coatings at elevated temperatures. Surf Coat Technol 2014;259:2–6.
[26] Cizek J, et al. Mechanical and optical properties of quaternary Si-B-C-N films prepared by reactive magnetron aputerring. Thin Solid Films 2007;516:7286–93.
[27] Kresse G, Furthmüller J. Efficient iterative schemes for ab initio total-energy calculations using a plane-wave basis set. Phys Rev B 1996;54:11169.
[28] Grüneis A, et al. Tight-binding description of the quasiparticle dispersion of graphite and few-layer graphene. Phys Rev B 2008;78:205425.
[29] Shang YR. Electronic state in crystals of finite size: quantum confinement of Bloch waves. First edition. Springer; 2006.
[30] Zhang H, Yu L, Chen T, Zhou W, Lou XW. Surface modulation of hierarchical MoS2 nanosheets by Ni single atoms for enhanced electrocatalytic hydrogen evolution. Adv Func Mater 2018;28:1807086.
[31] Gupta S, Sil S, Bhattachayya B. Ground state of a chemically modulated Hubbard chain at half fillin Ground state of a chemically modulated Hubbard chain at half filling. Phys Rev B 2001;63:125113.
[32] Li H, et al. Activating and optimizing MoS2 basal planes for hydrogen evolution through the formation of strained sulphur vacancies. Nat Mater 2016;15:48.
[33] Ye R, et al. High-Performance hydrogen evolution from MoS2(1-x) P(x) solid solution. Adv Mater 2016;28:1427.
[34] Mishra MK, Kadambi SB, Ramamurty U, Ghosh S. Elastic flexibility tuning via interaction factor modulation in molecular crystals. Chem Commun 2018;54:9047–50.
[35] Desiraju GR. Crystal engineering: from molecule to crystal. J Am Chem Soc 2013;135:9952–67.

[36] Zhang C, Liu TX. A review on hybridization modification of graphene and its polymer nanocomposites. Chin Sci Bull 2012;57:3010–21.
[37] Wu ZS, Ren WC, Gao LB, Liu BL, Zhao JP, Cheng HM. Efficient synthesis of graphene nanoribbons sonochemically cut from graphene sheets. Nano Res 2010;3:16–22.
[38] Tang C, Oppenheim T, Tung VC, Martini A. Structure-stability relationships for graphene-wrapped fullerene-coated carbon nanotubes. Carbon 2013;61:458–66.
[39] Villamagua L, Carini M, Stashans A, Gomez CV. Band gap engineering of graphene through quantum confinement and edge distortions. Ricerche Matemat 2016;65:579–84.
[40] Song JJ, Zhang HJ, Cai YL, Zhang YX, Bao SN, He PM. Bottom-up fabrication of graphene nanostructures on Ru(1010). Nanotechnology 2016;27:055602.
[41] Kim K, Sussman A, Zettl A. Graphene nanoribbons obtained by electrically unwrapping carbon nanotubes. ACS Nano 2010;4:1362.
[42] Enoki T, Fujii S, Takai K. Zigzag and armchair edges in graphene. Carbon 2012;50:3141.
[43] Patra N, Wang B, Kral P. Nanodroplet activated and guided folding of graphene nanostructures. Nano Lett 2009;9:3766.
[44] Ghosh S, Schwingenschlog. Hexagonal graphene quantum dots. RRL 2016;11:1600226.
[45] Antonova IV, Neboratikova NA, Rinz VY. J Appl Phys 2016;11922:4302.
[46] Balog R, et al. Bandgap opening in graphene induced by patterned hydrogen adsorption. Nat Mater 2010;9:315–9.
[47] Grassi R, Low T, Lundstrom M. Scaling of the energy gap in pattern-hydrogenated graphene. Nano Lett 2011;11:4574–8.
[48] Guisinger NP, Rutter GM, Crain JN, First PN, Stroscio JA. Exposure of epitaxial graphene on SiC (0001) to atomic hydrogen. Nano Lett 2009;9:1462–6.
[49] Castellanos-Gomez A, Wojtaszek M, Tombros N, vanWees BJ. Reversible hydrogenation and bandgap opening of graphene and graphite surfaces probed by scanning tunneling spectroscopy. Small 2012;8:1607–13.
[50] Chang SL, Lin SY, Lin SK, Lee CH, Lin MF. Geometric and electronic properties of edge-decorated graphene nanoribbons. Sci Rep 2014;4.
[51] Lin YT, Chung HC, Yang PH, Lin SY, Lin MF. Adatom bond-induced geometric and electronic properties of passivated armchair graphene nanoribbons. Phys Chem Chem Phys 2015;17:16545–52.
[52] Mao Y, Hao W, Wei X, Yuan J, Zhong J. Edge-adsorption of potassium adatoms on graphene nanoribbon: a first principle study. Appl Surf Sci 2013;280:698704.
[53] Kher-Elden M, et al. Metallic bands in chevron-type polyacenes. RSC Adv 2020;10:33844–50.
[54] Muley S, Ravindra NM. Graphene: properties, synthesis, and applications. Springer semiconductors; 2019. p. 219–332.
[55] Zhu YW, et al. Graphene and graphene oxide: synthesis, properties, and applications. Adv Mater 2010;22:3906–24.
[56] Wu JY, Chen SC, Do TN, Su WP, Gumbs G, Lin MF. The diverse magneto-optical selection rules in bilayer black phosphorus. Sci Rep 2018;8.
[57] Chiu CW, Shyu FL, Chang CP, Chen RB, Lin MF. Optical spectra of AB- and AA-stacked nanographite ribbons. J Phys Soc Jpn 2003;72:170–7.
[58] Zhang H, Duley W. First-principles study of edge chemical modifications in graphene nanodots. *Phys Rev* B 2008;78:045421.
[59] Singh D, Gupta SK, Sonvane Y, Lukacevic I. Antimonene: a monolayer material for ultraviolet optical nanodevices. J Mater Chem C 2016;4:6386–90.
[60] Ao Z, Yang J, Li S, Jiang Q. Enhancement of CO detection in Al doped graphene. Chem Phys Lett 2008;461:276279.
[61] Yan J, Liu J, Fan Z, Wei T, Zhang L. High-performance supercapacitor electrodes based on highly corrugated graphene sheets. Carbon 2012;50:2179–88.
[62] Chi M, Zhao YP. Adsorption of formaldehyde molecule on the intrinsic and Al-doped graphene: a first principle study. Comput Mater Sci 2009;46:10851090.
[63] Tozzini V, Pellegrini V. Reversible hydrogen storage by controlled buckling of graphene layers. J Phys Chem C 2011;115:25523–8.
[64] Ning G, Xu C, Mu L, Chen G, Wang G, Gao J, Fan Z, Qian W, Wei F. High capacity gas storage in corrugated porous graphene with a specific surface area-lossless tightly stacking manner. Chem Commun 2012;48:6815–7.

[65] Tozzini V, Pellegrini V. Prospects for hydrogen storage in graphene. Phys Chem Chem Phys 2013;15:8089.
[66] Bianchi M, et al. Electron−phonon coupling in potassium-doped graphene: angle-resolved photoemission spectroscopy. Phys Rev B 2010;81:041403.
[67] Avila J, et al. Presence of gapped silicene-derived band in the prototypical (3 × 3) silicene phase on silver (111) surfaces. J Phys Condens Matter 2013;25:262001.
[68] Singh R. Spin-orbit coupling in graphene, silicene and germanene: dependence on the configuration of full hydrogenation and fluorination. Bull Mater Sci 2018;41:158.
[69] Eng AYS, Sofer Z, Simek P, Kosina J, Pumera M. Highly hydrogenated graphene through microwave exfoliation of graphite oxide in hydrogen plasma: towards electrochemical applications. Chem Eur J 2013;19:15583−92.
[70] Luo Z, Yu T, Kim K j, Ni Z, You Y, Lim S, Shen Z, Wang S, Lin J. Thickness-dependent reversible hydrogenation of graphene layers. ACS Nano 2009;3:1781−8.
[71] Wojtaszek M, Tombros N, Caretta A, Van Loosdrecht P, Van Wees B. A road to hydrogenating graphene by a reactive ion etching plasma. J Appl Phys 2011;110:063715.
[72] Lin YT, Lin SY, Chiu YH, Lin MF. Alkali-created rich properties in graphene nanoribbons: chemical bondings. Sci Rep 2017;7:1−14.
[73] Tran NTT, Lin S-Y, Glukhova OE, Lin M-F. Bonding-dominated energy gaps in graphene oxide. RSC Adv 2016 π-;6:24458−63.
[74] Tran N-T-T, Nguyen D-K, Glukhova O-E, Lin M-F. Coverage-dependent essential properties of halogenated graphene: a DFT study. Sci Rep 2017;7:17858.
[75] Giese TJ, York DM. Improvement of semiempirical response properties with charge-dependent response density. J Chem Phys 2005;123:164108.
[76] Nair RR, et al. Fluorographene: a two-dimensional counterpart of teflon. Small 2010;6:2877−84.
[77] Jeon KJ, et al. Fluorographene: a wide bandgap semiconductor with ultraviolet luminescence. ACS Nano 2011;5:1042−104.
[78] Medeiros PV, Mascarenhas AJ, de Brito Mota F, de Castilho CMCA. DFT study of halogen atoms adsorbed on graphene layers. Nanotech 2010;21:485701.
[79] Sahin H, Ciraci S. Chlorine adsorption on graphene: Chlorographene. J Phys Chem C 2012;116:24075−83.
[80] Hannay N, Geballe T, Matthias B, Andres K, Schmidt P, MacNair D. Superconductivity in graphitic compounds. Phys Rev Lett 1965;14:225.
[81] Dresselhaus M, Dresselhaus G. Intercalation compounds of graphite. Adv Phys 1981;30:139−326.
[82] Lee R, Kim H, Fischer J, Thess A, Smalley RE. Conductivity enhancement in single-walled carbon nanotube bundles doped with K and Br. Nature 1997;388:255−7. 1997.
[83] Rao AM, Eklund P, Bandow S, Thess, Smalley RE. Evidence for charge transfer in doped carbon nanotube bundles from Raman scattering. Nature 1997;388:257−9.
[84] Hebard A, Rosseinky M, Haddon R, Murphy D, Glarum S, Palstra T, Ramirez A, Karton A. Potassium-doped C60. Nature 1991;350:600601.
[85] Haddon R. Electronic structure, conductivity and superconductivity of alkali metal doped (C60). Acc Chem Res 1992;25:127−33.
[86] Wu J, Xie L, Li Y, Wang H, Ouyang Y, Guo J, Dai H. Controlled chlorine plasma reaction for noninvasive graphene doping. J Am Chem Soc 2011;133:19668−71.
[87] Chu SW, Baek SJ, Kim DC, Seo S, Kim JS, Park YW. Charge transport in graphene doped with diatomic halogen molecules (I2, Br2) near Dirac point. Synthetic Met 2012;162:1689−93.
[88] Subrahmanyam K, Voggu R, Govindaraj A, Rao C. A comparative Raman study of the interaction of electron donor and acceptor molecules with graphene prepared by different methods. Chem Phys Lett 2009;472:9698.
[89] Wu JY, Chen SC, Gumbs G, Lin MF. Field-induced diverse quantizations in monolayer and bilayer black phosphorus. Phys Rev B 2017;95.
[90] Chen SC, Wu JY, Lin MF. Feature-rich magneto-electronic properties of bismuthene. New J Phys 2018;20.
[91] Tran NTT, Lin S-Y, Lin C-Y, Lin M-F. Geometric and electronic properties of graphene-related systems: chemical bonding schemes. CRC Press; 2017.
[92] Lin S-Y, Tran NTT, Chang S-L, Su W-P, Lin M-F. Structure-and adatom-enriched essential properties of graphene nanoribbons. CRC press; 2018.
[93] Nguyen TDH, Pham HD, Lin SY, Lin MF. Featured properties of Li+-based battery anode: $Li_4Ti_5O_{12}$. RSC Adv 2020;10:14071−9.

[94] Sivek J, Sahin H, Partoens B, Peeters FM. Adsorption and absorption of boron, nitrogen, aluminum, and phosphorus on silicene: stability and electronic and phonon properties. Phys Rev B 2013;87:085444.
[95] Pamungkas MA, Dobirin K, Abdurrouf. Effects of doping Na and Cl atom on electronic structure of silicene: density functional theory calculation. In: AIP conference proceedings, vol. 1948; 2018. p. 020001.
[96] Monshi MM, Aghaei SM, Calizo I. Doping and defect-induced germanene: a superior media for sensing H_2S, SO_2, and CO_2 gas molecules. Surf Sci 2017;665:96–102.
[97] Bechstedt F, Gori P, Pulci O. Beyond graphene: clean, hydrogenated and halogenated silicene, germanene, stanene, and plumbene. Progr Surface Sci J Progsurf 2021;2021:100615.
[98] Lin SY, et al. Silicene-based layered materials. IOP Publishing); 2020.
[99] Li JY, Zhang C, Li DD, Qiao L, Long R, Yang LY, Fang XH, Lin YY, Chen YG, Yin M. Substrate-free flexible thin film solar cells by graphene-mediated peel-off technology. J Mater Sci Mater Electron 2020;31:10279–87.
[100] Luan HX, Zhang CW, Zheng FB, Wang PJ. First-principles study of the electronic properties of B/N atom doped silicene nanoribbons. J Phys Chem C 2013;117:13620–6.
[101] Liu CC, Feng WX, Yao YG. Quantum spin Hall effect in silicene and two-dimensional germanium. Phys Rev Lett 2011;107.
[102] Popov VN, Henrard L. Optical properties of single-walled carbon nanotubes within a nonorthogonal tight-binding model. Fullerenes, Nanotub Carbon Nanostruct 2005;13:45–52.
[103] Bates SP, Kresse G, Gillan MJ. A systematic study of the surface energetics and structure of TiO_2(110) by first-principles calculations. Surf Sci 1997;385:386–94.
[104] Hummel F, Gruneis A, Kresse G, Ziesche P. Screened exchange corrections to the random phase approximation from many-body perturbation theory. J Chem Theor Comput 2019;15:3223–36.
[105] Perdew JP, Burke K, Ernzerhof M. Generalized gradient approximation made simple. Phys Rev Lett 1996;77:3865.
[106] Blochl PE. Projector augmented-wave method. Phys Rev B 1994;50:17953.
[107] Grimme S. Semiempirical GGA-type density functional constructed with a long-range dispersion correction. J Comput Chem 2006;27:1787–99.
[108] Ding Y, Wang YL. Density functional theory study of the silicene-like SiX and XSi3 (X = B, C, N, Al, P) honeycomb lattices: the various buckled structures and versatile electronic properties. J Phys Chem C 2013;117:18266.
[109] Matusalem F, Marques F, Teles LK, Bechstedt F. Stability and electronic structure of two-dimensional allotropes of group-IV materials. Phys Rev B 2015;92:45436.
[110] Matthes L, Pulci O, Bechstedt F. Massive Dirac quasiparticles in the optical absorbance of graphene, silicene, germanene, and tinene. J Phys Condens Matter 2013;25:395305.
[111] Ni Z, Liu Q, Tang K, Zheng J, Zhou J, Qin R, Gao Z, Yu D, Lu J. Tunable bandgap in silicene and germanene. Nano Lett 2012;12:113–8.
[112] Kaloni TP, Tahir M, Schwingenschlogl U. Quasi free-standing silicene in a superlattice with hexagonal boron nitride. Sci Rep 2013;3:3192.
[113] Kaloni TP, Schreckenbach G, Freund MS. Large enhancement and tunable band gap in silicene by small organic molecule adsorption. J Phys Chem C 2014. https://doi.org/10.1021/jp505814v.

CHAPTER

6

Carbon-/boron-/nitrogen-substituted germaneness

Yu-Ming Wang[1], *Jheng-Hong Shih*[1], *Wei-Bang Li*[1] *and Ming-Fa Lin*[1,2]

[1]Department of Physics, National Cheng Kung University, Tainan, Taiwan; [2]Center for Micro/Nano Science and Technology, National Cheng Kung University, Tainan, Taiwan

6.1 Introduction

Group-IV elements in a periodic table have generated a lot of dimension-dependent crystal structures [1], with the various lattice symmetries [2]. Such main-stream materials are outstanding candidates in thoroughly exploring the basic [3] and applied sciences [4]. For example, the three-dimensional non-layered crystal systems of carbon [5], silicon/germanium [6,7], as well as, tin/lead atoms [8], respectively, create the insulating, semiconducting and metallic properties near the Fermi level. Furthermore, the dilute impurities could be added into the diamond structures associated C, Si and Ge elements [9], greatly enriching the featured band structures [10]. This could be regarded as the hydrogen-like topics [11], in which the bound states of charged impurities are strongly affected by the effective mass and the static dielectric constant of a diamond crystal. Very interesting, group-III B/Al/Ga/In/Tl and group-V N/P/As/Sb/Bi, belong to the acceptor- [12] and donor-type impurities [13], respectively. Such open issues might come to exist in 2D graphene, silicene, germanene, tinene and plumbene. The similar topics, the chemical substitutions at high guest-atom concentrations, are worthy of systematic investigations. The binary compounds, being based on germanene, are modulated by the various B-. C- and N-guest atoms. According to their unusual atomic configurations [14], the various orbital hybridizations will be created under the different chemical modifications, thus leading to the diverse quasiparticle phenomena in the essential properties.

Very interesting, the creation of all condensed-matter systems has been clearly illustrated in the periodic table, with the normal 102 elements (details in Ref. [15]). The various atomic

Fundamental Physicochemical Properties of Germanene-related Materials
https://doi.org/10.1016/B978-0-443-15801-8.00016-5

configurations can generate the diverse chemical bondings and frequently form the dimensionality-related crystals. Each crystal structure presents a plenty of merits and drawbacks in terms of basic [3] and applied sciences [4,16–19]. For example, few-layer graphene systems, with honeycomb lattices [20], are rather suitable for studying the diversities of physical/chemical/material phenomena, e.g., the unusual behaviors of geometric [21], electronic [22], optical [23], transport [24] and magnetic properties [25]. That they possess the concise orbital hybridizations is the main reason. The different elements could be further synthesized by the various approaches in experimental laboratories. The chemical modifications are quite powerful and efficient for generating the specific materials with the pronounced functionalities, especial for graphene-based semimetals [26], metals [27] and semiconductors [16]. On the theoretical side, the fundamental properties, with/without the many-body effects [29–31], need to be thoroughly explored for many diversified phenomena. A quasiparticle framework, being based on the first-principles simulations [20] and phenomenological models [33], will be proposed to account for them. Under the current investigations, the former are very suitable in thoroughly exploring the germanene-related chemical substitutions. Their close relations are clearly illustrated through the diverse phenomena of chemical modification effects [34–36].

Very interesting, group-III boron, group-IV carbon, and group-V nitrogen have been successfully synthesized together to form single-element [37], binary [38] and ternary crystals [39]. $B_xC_yN_z$ represents a series of condensed-matter systems, being greatly diversified by the different dimensionalities [40], the planar/curved/folded/cylindrical structures [40–43], and the various rations [44], such as, 3D graphite/diamond [5,45], binary BC/CN/BN materials [46–48], few-layer graphenes [49], different compounds, 1D nanotubes [50], and nanoribbons [51]. Their diverse quasiparticle properties are worthy of systematic investigations, as done for the carbon-related systems (details in books of [52]). When boron, carbon or nitrogen are chosen as guest atoms within a honeycomb germanene, the mainstream binary compounds are formed through the various substitution cases. Obviously, the distinct atomic configurations of group-III, group-IV and group-V can generate the diverse orbital hybridizations with the [3s, $3p_x$, $3p_y$, $3p_z$] orbitals of germanium atoms. Under the various chemical bondings due to the different substitution cases, the physical/chemical/material phenomena will be largely modified under their sensitive modulations. Similarly, the diversified quasiparticle phenomena could be achieved by using the chemisorption ways [53]. The chemical substitutions and absorptions might be quite different from each other.

It is well known that 2D layered materials have stirred a lot of theoretical [54] and experimental researches [55] since the discovery of few-layer graphene systems in 2004 by the mechanical exfoliation [56]. The main-stream systems of group-IV elements, graphene/silicene/germanene/tinene/plumbene [56–60], are one of the studying focuses in observing the rich and unique quasiparticle phenomena, e.g., the featured behaviors of geometric [51], electronic [62], optical [63], magnetic [64], Coulomb-excitation [29] and transport properties [65]. The main reason is that there exist the unusual chemical bondings and spin configurations, i.e., the perpendicular or non-perpendicular π & σ bondings (the sp^2 and sp^3 ones; [66]), as well as, the absence or existence of spin-orbital coupling, respectively, correspond to the honeycomb lattices of graphene [33] and other systems [67]. Specifically, germanium and silicon atoms could generate the functional semiconductors, being frequently utilized in the commercial applications and even leading to the industrial revolutions [68]. The up-to-now results only show few viewpoints about the chemical modifications of germanene and silicene materials. From the theoretical thoughts, the VASP simulations, but not the phenomenological models [69], are quite suitable in thoroughly exploring the essential properties [70], revealing their close relations [71], and promoting the potential applications [72].

In this chapter, the delicate calculations and analyses of VASP simulations are available in clearly identifying the boron-, carbon- and nitrogen-substitution-diversified germanene systems [14]. The critical quasiparticle mechanisms, the charge- and spin-dependent ones (details in Chapter 2; [73]), will be thoroughly examined through the full consistence of the relevant physical and chemical quantities. The close relations between the first-principles method and the phenomenological models are discussed in detail. The theoretical predictions require the necessary measurement examinations, such as, the high-precision tests on surface morphologies, occupied energy spectra and van Hove singularities, respectively, by STM/TEM (details in Chapter 3; [74]), ARPES [75], and STS [76,77]. After the best relaxations, the optimal crystal symmetries, the atom- and spin-dominated band structure & wave functions, the spatial charge/spin density distributions & their variations for the substitutions, and the atom-, orbital-, and spin-projected density of states are very sensitive the changes of guest-atom concentrations and arrangements. The highly non-uniform chemical environments are expected to be responsible for the other fundamental quasiparticle properties. A lot of diverse behaviors are sufficient in obtaining the concise physical and chemical pictures. The important differences of the chemical substitutions and chemisorptions are worthy a detailed comparison. Also, the significant similarities and differences among germanene-, silicene- and graphene-related systems are also investigated in detail. The similar first-principles studies could be generalized to other more complicated chemical modifications, such as, the chemical substitutions of transition metals [78,79] and rare earth ones [80] in observing the stable ferromagnetic spin configurations.

6.2 Diverse crystal symmetries of binary germanene compounds

The unusual features of nitrogen substitutions in silicene are investigated in detail by using the VASP code [81,82]. The exchange and correlation energies due to many-particle electron–electron interactions are calculated with the use of the Perdew–Burke– Ernzerhof (PBE) functional under the generalized gradient approximation [83]. Furthermore, the electron–ion interactions can be characterized by the projector augmented wave (PAW) pseudopotentials [84]. A space of 15 A width is inserted between the silicene planes to avoid their interaction. In general, a planewave basis set with a maximum kinetic energy cutoff of 500 eV is chosen to expand the wave function. The first Brillouin zone is sampled by $9\times 9\times 1$ and $100 \times 100 \times 1$ k-point meshes within the Monkhorst–Pack scheme for further calculations of the electronic and magnetic properties. The energy convergence is set to be 10^{-5} eV between two consecutive steps and the maximum Hellmann–Feynman force acting on each atom is less than 0.01 eV Å^{-1} during the ionic relaxations.

Very interesting, the diversities of the primitive unit cells are easily generated by changing the concentrations and distributions of guest atoms in monolayer germanene, as clearly indicated in Figs. 6.1A–6.6L, respectively, for boron, carbon and nitrogen substitutions. The dilute approaches obviously illustrate enlarged Moire superlattices with more guest and host atoms (or more active orbitals and thus chemical bonds; [85–87]). According to the

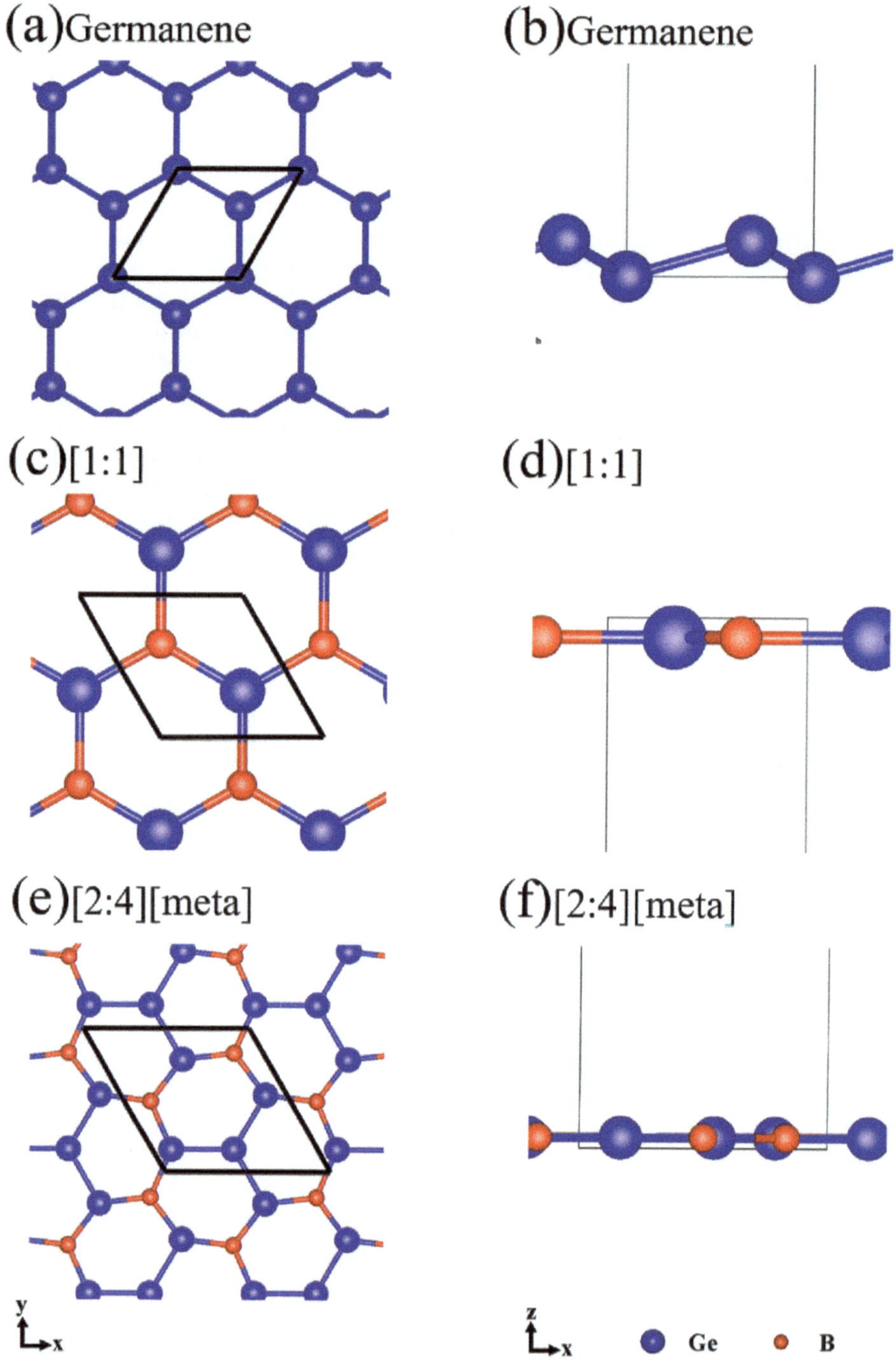

FIG. 6.1 (A–F) The optimal crystal structures, with the top/side views (the left/right plots) for the boron-substituted germanene systems under the guest-atom substitution cases.

edge structures, there exist three kinds of well-behaved atom arrangements, namely, meta [88], ortho [14], and para [89]. The optimal crystal structures mainly consist of the position- and atom-related chemical bondings. The different bonds of Ge–Ge, B-/C-/N–Ge, and B–B/C–C/N–N pairs could survive under the specific substitution cases. Such highly non-uniform chemical environments are responsible for the rich quasiparticle properties. The theoretical predictions on surface morphologies could be examined by the STM and TEM measurements for the top and side views, respectively (details in Chapter 3).

After the reliable VASP simulations, the featured honeycomb crystals of germanene-based binary compounds are obviously revealed through the various substitution cases, as clearly shown in Figs. 6.1–6.8 and Tables 6.1–6.3. Most of them have stable structures with/without the deformation of hexagonal symmetries, and only very few systems belong to unstable bondings, e.g., the unstable structure of the [C, Ge]-[2:6]-[ortho] substitution (not shown). The main features in a primitive unit cell cover the positions of host and guest atoms (the anisotropic distributions), the distinct Ge–Ge, Ge–B/Ge–C/Ge–N, B–B/C–C/N–N bonds (the highly non-uniform chemical environments), the existence or absence of buckled structures (side views in Figs. 6.1–6.7; Tables 6.1–6.3), and the observable fluctuations of bond lengths (Fig. 6.8). The prominent characteristics should be sufficient in evaluating the form, strength, and composition/separation of chemical bondings. This will be fully supported by the energy spectra/wave functions (Figs. 6.9–6.14), spatial charge densities (Figs. 6.15–6.21), and van Hove singularities (discussed later in Figs. 6.22–6.31).

Very interesting, the saturated binary compounds (Figs. 6.1C,D, 6.3A,B, and 6.5A,B) exhibit the planar honeycomb structures, compared with a pristine germanene/graphene (Fig. 6.1A and B; [63]). These clearly indicate the almost perpendicular π and σ bondings, in which the former and the latter, respectively, arise from $4p_z$-$2p_z$ and [4s, $4p_x$, $4p_y$]-[2s, $2p_x$, $2p_y$] orbital hybridizations under the uniform Ge–B/Ge–C/Ge–N bonds. Such atomic interactions should be weaker than those of graphene because of longer bond lengths [90]. Furthermore, the spin-orbital couplings might be greatly suppressed by the prominent orbital mixings [91]. There exist observable differences among three kinds of B-/C-/N–Ge bond lengths, respectively, corresponding to 1.997, 2.543 and 2.387 Å (2.750 Å of Ge–Ge bonds). The strengths of orbital hybridizations (the hopping integrals; [14]) are roughly inversely proportional to bond length according to the Harrison's rule [92]. The planar crystals, with the non-uniform bonds, are revealed in certain non-saturated compounds or under very dilute guest-atom substitutions, e.g., the non-buckled geometries of [B, Ge]-[1:1] (Fig. 6.1C and D), [B, Ge]-[2:4]-[meta] (Fig. 6.1E and F), [B, Ge]-[2: 4]-[ortho] (Fig. 6.2G and H), [C, Ge]-[1:1] (Fig. 6.3A and B), [N, Ge]-[2:6]-(para) (Fig. 6.4K and L) As to buckled honeycomb crystals, the high differences of A and B two sublattices are non-homogeneous through most of guest-atom substitutions (the details in Tables 6.1–6.3 and Figs. 6.1–6.7). Under the present substitution concentrations, their buckling values are roughly within the observable ranges of 0.163–2.484, 0.6512.909, 0.705–2.783 Ås, respectively for the B-, C-, N-cases. The maximum buckling for each case is indicated in the right-hand side Figure. These deformed structures along the [x, y, z]-direction suggest the rich and complicated multi-orbital hybridizations, covering [4s, $4p_x$, $4p_y$, $4p_z$]-[4s, $4p_x$, $4p_y$, $4p_z$], [4s, $4p_x$, $4p_y$, $4p_z$]-[2s, $2p_x$, $2p_y$, $2p_z$], and [2s, $2p_x$, $2p_y$, $2p_z$]-[2s, $2p_x$, $2p_y$, $2p_z$] ones, respectively, in Ge–Ge, Ge–B/Ge–C/Ge–N and B–B/C–C/N–N bonds. The above-mentioned geometric properties are closely related to the spatial charge density distributions (Figs. 6.15–6.21).

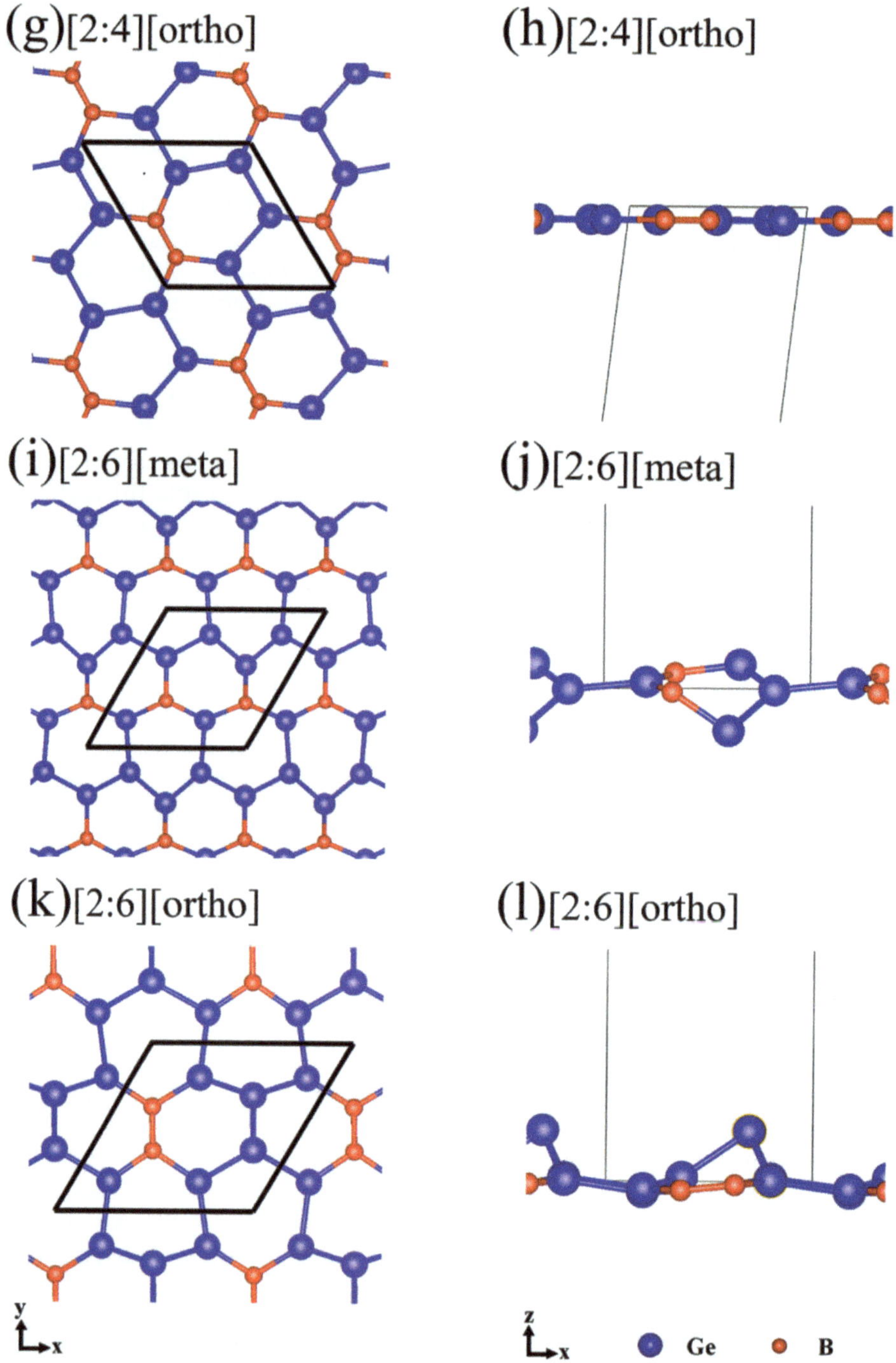

FIG. 6.2 (G–L) The optimal crystal structures, with the top/side views (the left/right plots) for the boron-substituted germanene systems under the guest-atom substitution cases.

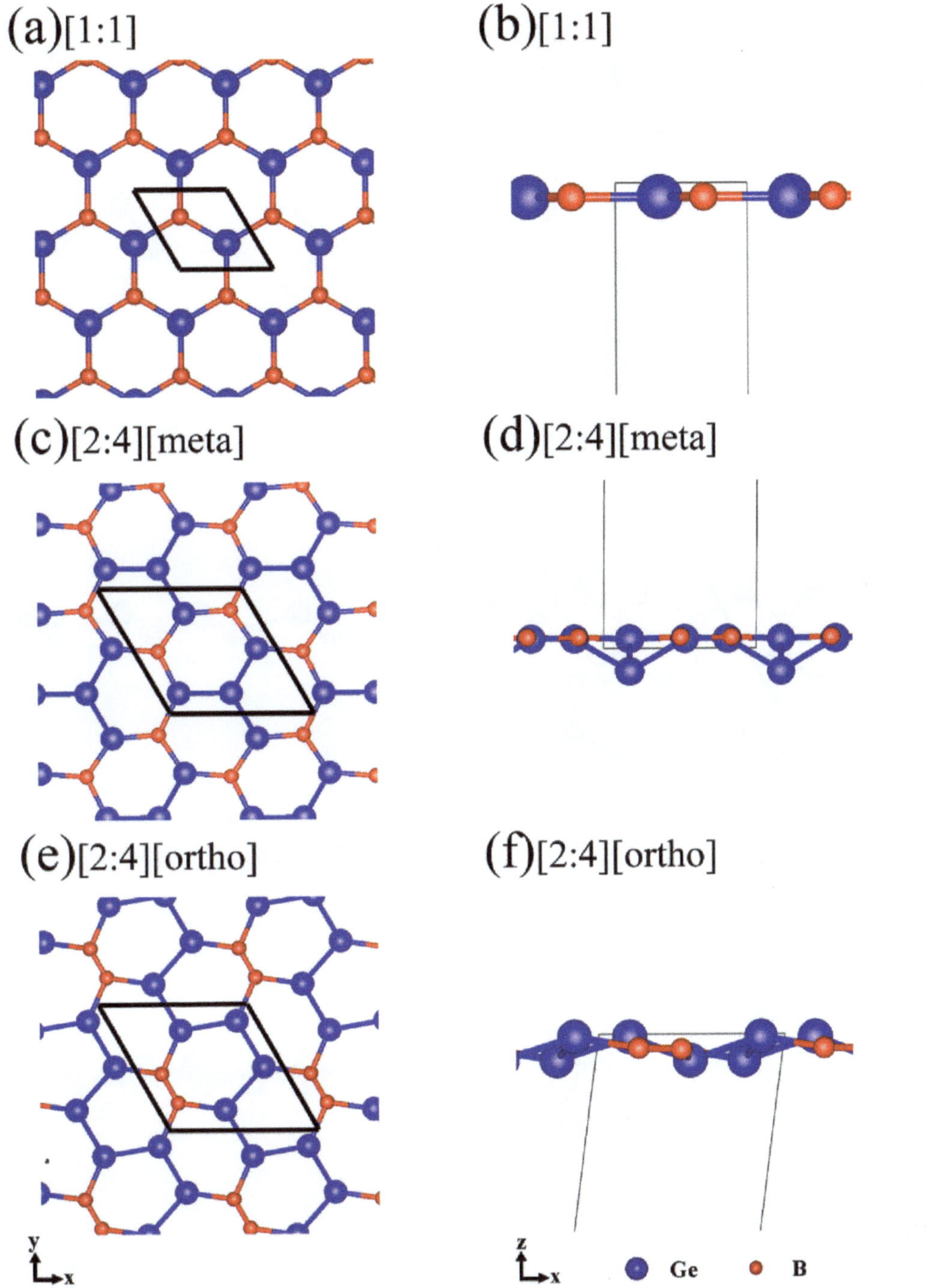

FIG. 6.3 (A–F) The optimal crystal structures, with the top/side views (the left/right plots) for the carbon-substituted germanene systems under the guest-atom substitution cases.

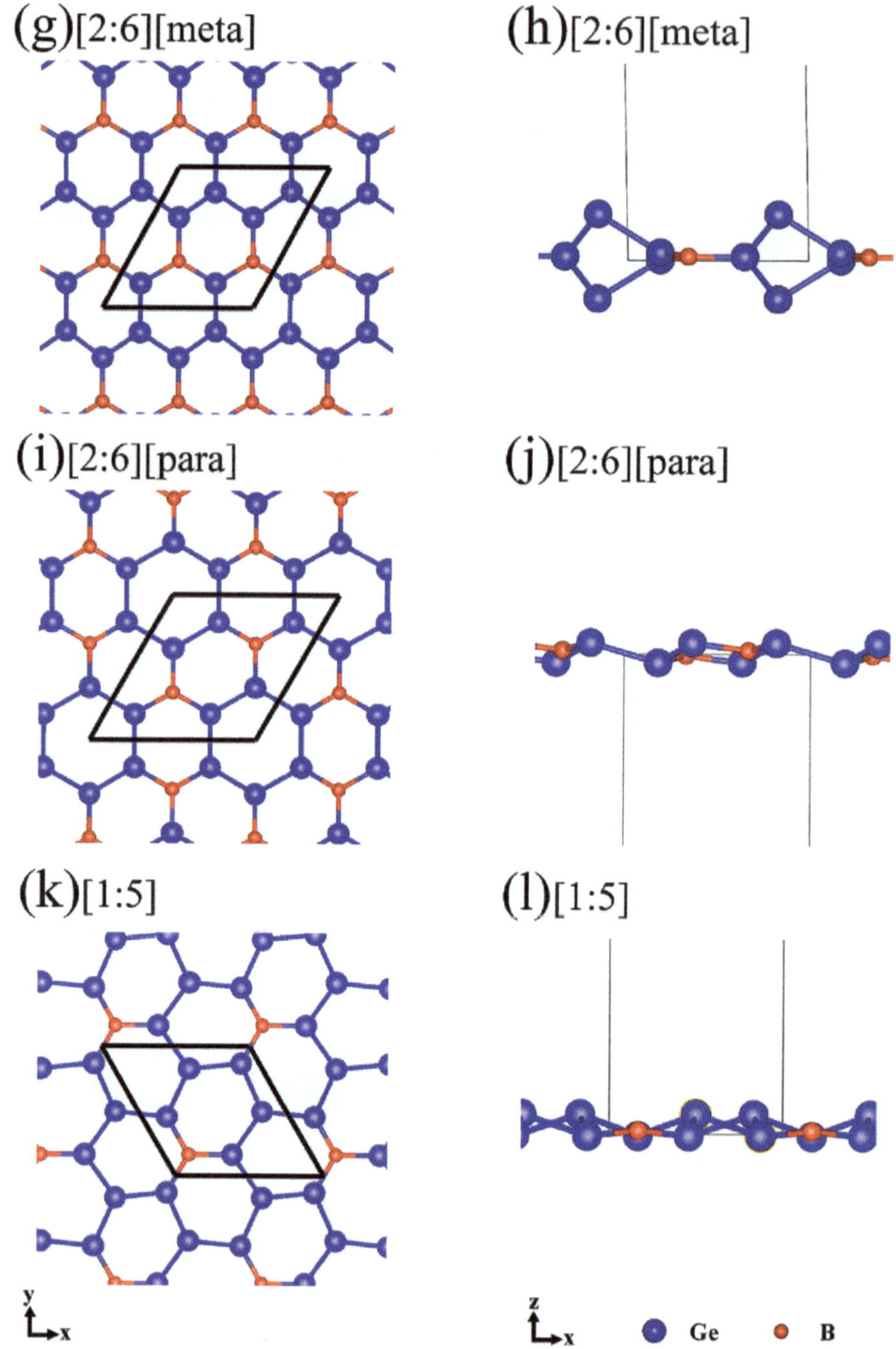

FIG. 6.4 (G–L) The optimal crystal structures, with the top/side views (the left/right plots) for the carbon-substituted germanene systems under the guest-atom substitution cases.

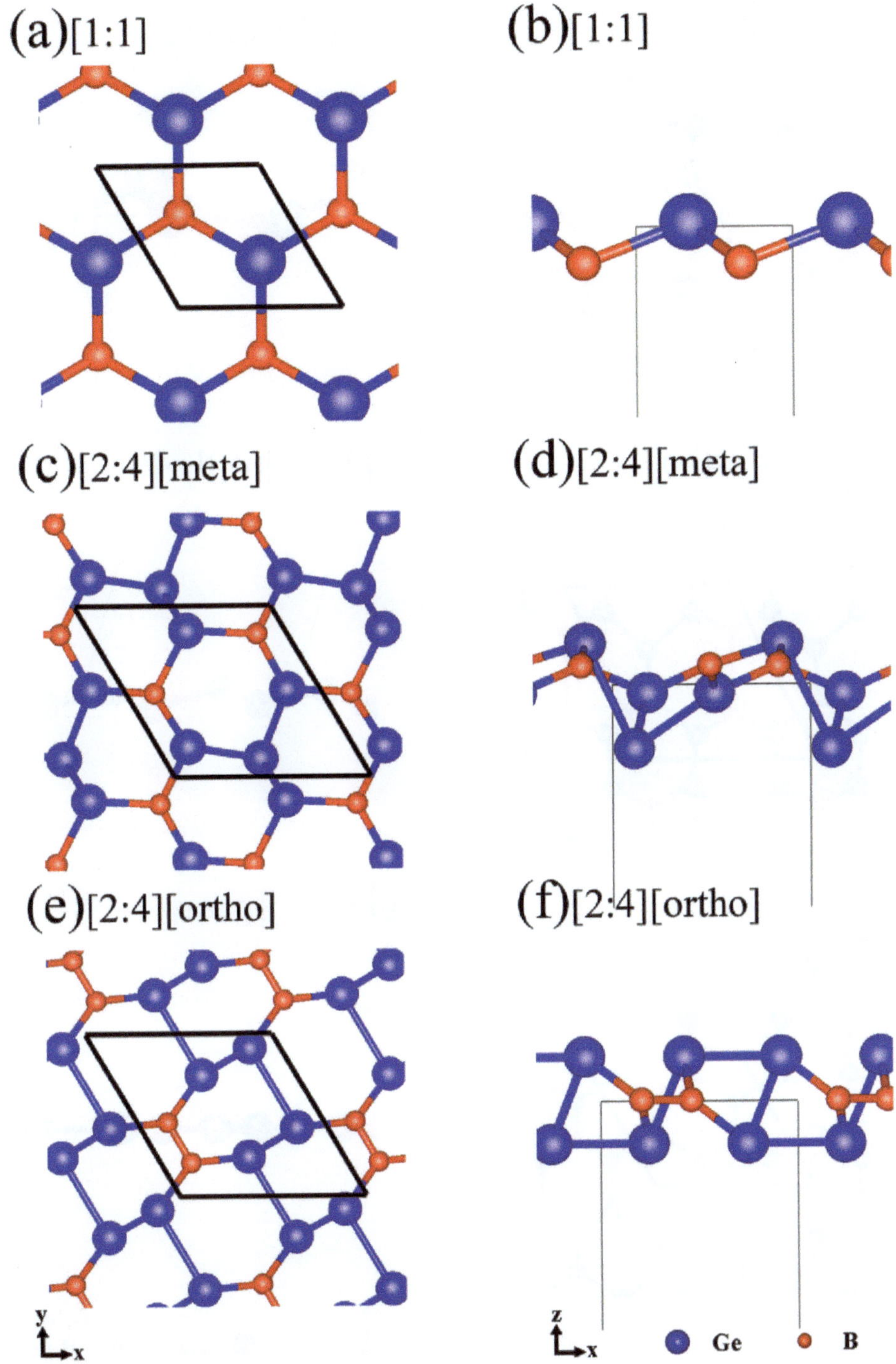

FIG. 6.5 (A–F) The optimal crystal structures, with the top/side views (the left/right plots) for the nitrogen-substituted germanene systems under the guest-atom substitution cases.

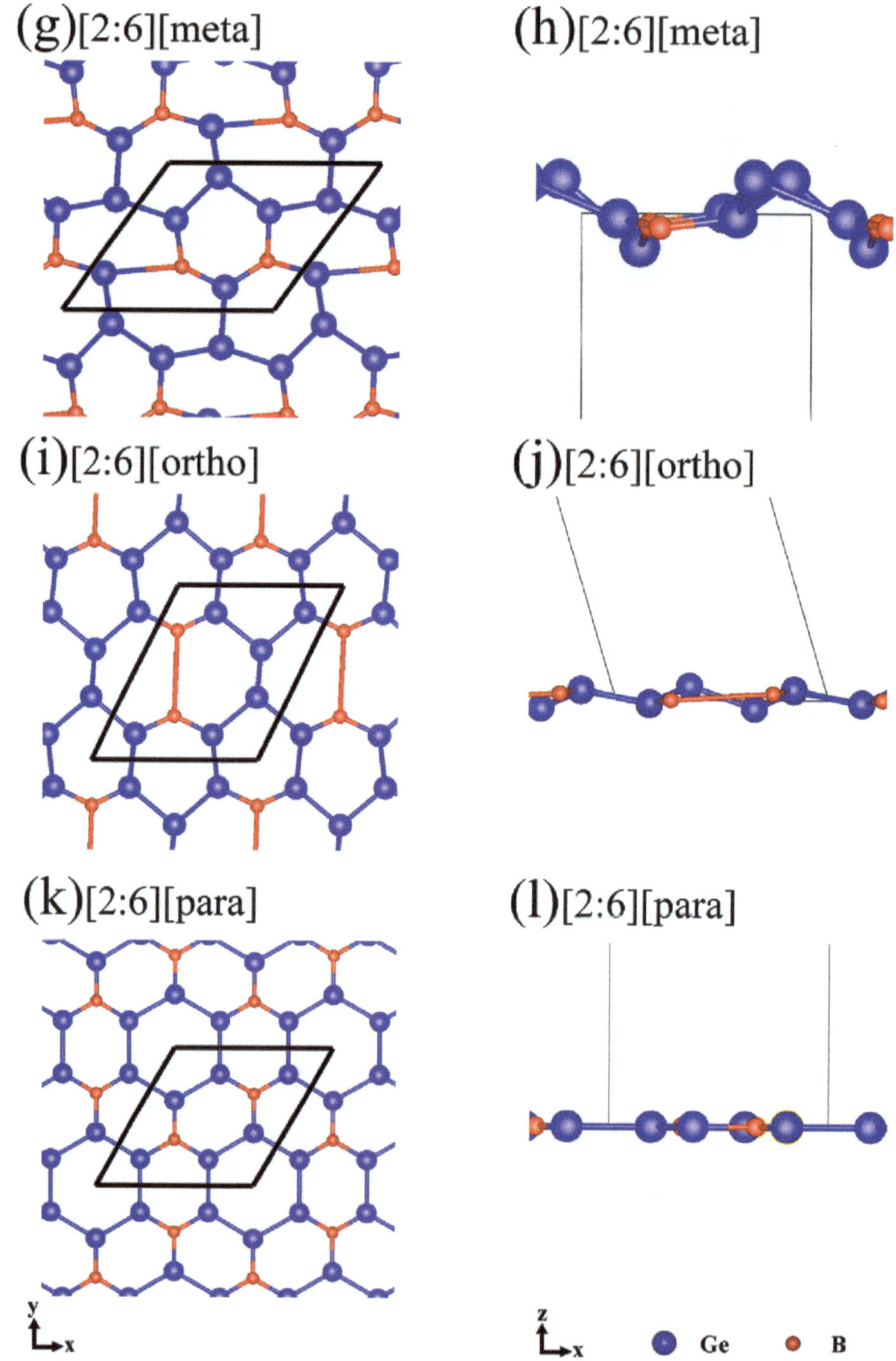

FIG. 6.6 (G–L) The optimal crystal structures, with the top/side views (the left/right plots) for the nitrogen-substituted germanene systems under the guest-atom substitution cases.

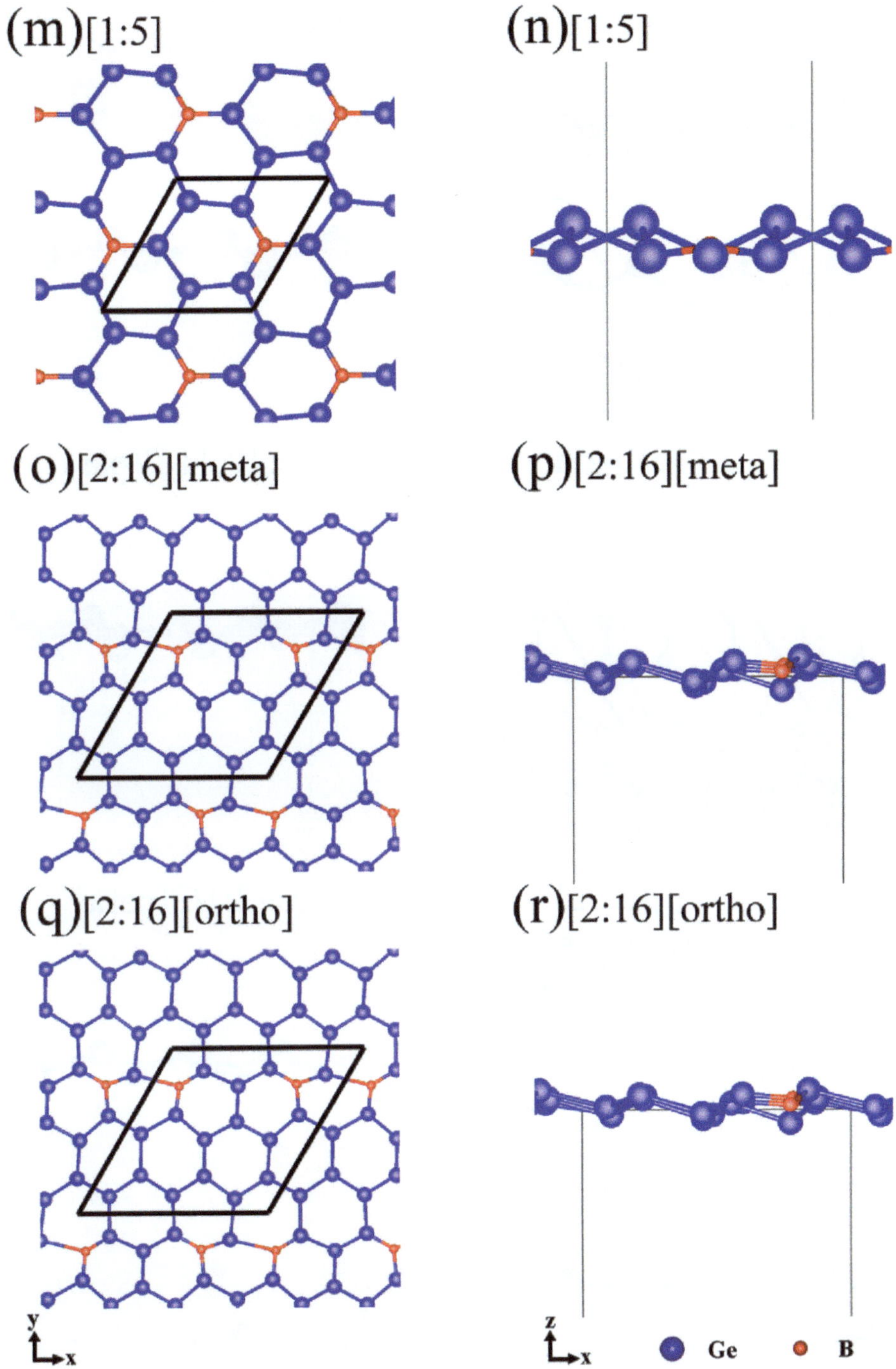

FIG. 6.7 (M–R) The optimal crystal structures, with the top/side views (the left/right plots) for the nitrogen-substituted germanene systems under the guest-atom substitution cases.

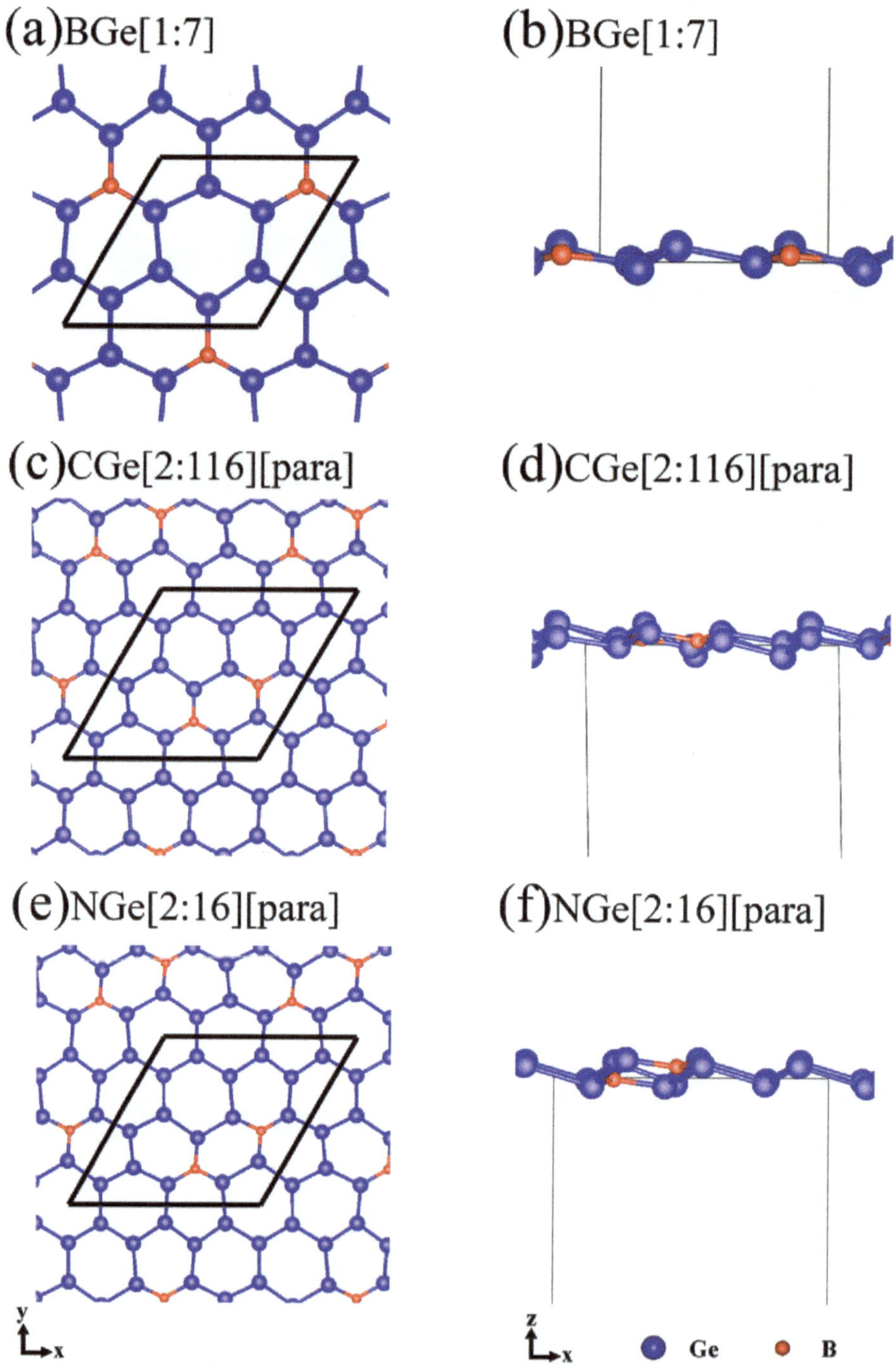

FIG. 6.8 (A–F) The observable bond-length fluctuations under the specific substitution cases of [2:16] (meta) for (A) [B: Ge], (B) [C: Ge], as well as, (C) [N, Ge].

TABLE 6.1 The geometric and electronic properties for the boron-substituted germanene systems.

The optimal geometric structures of boron-substituted gemanene systems at various concentrations and configurations with the B–B, B–Ge, and Ge–Ge bond lengths, the band gaps, and the ground state energies per unit cell, together with the height differences between the A and B sublattices.

Number of atoms/unit cell	Percentage	Ratio of B/Ge	B–B bond length (Å)	B–Ge bond length (Å)	Nearest Ge–Ge distance (Å)	Second nearest Ge–Ge distance (Å)	Magnetic moment (μ_B)	E_g^d (eV)/ metal	Optimized E_0 (eV)	Δ (Å)
2	–	Pristine	–	–	2.443	–	–	0.031	−8.014	0.675
	100%	[1:1]		2.023	–	–	0	0.360	−9.255	0
6	50%	[2:4] meta		2.012	2.413	–	0	0.017	−26.686	0
	50%	[2:4] ortho/ para	1.618	2.020	2.409	2.429	0	0	−26.947	0
	20%	[1:5]	–	2.017	2.406	–	0	0.242	−25.268	0.461
8	33.3%	[2:6] meta	–	1.997 −2.090	2.396	2.461	0	0.252	−34.179	2.162
	33.3%	[2:6] ortho	1.635	2.008 −2.047	2.384 −2.453	2.405 −2.543	0	0.080	−34.764	1.844
	33.3%	[2:6] para		–	1.997	2.406	0	0	−34.617	0.163
	14.3%	[1:7]	–	1.978	2.378 −2.385	2.457 −2.463	0	0.080	−33.027	0.844
18	12.5%	[2:16] meta	–	2.020 −2.037	2.403 −2.429	2.403 −2.441	0	0.095	−74.275	2.028
	12.5%	[2:16] ortho	1.650	2.054	2.401 −2.429	2.425 −2.460	0	0.198	−74.391	2.484
	12.5%	[2:16] para	–	2.020	2.398 −2.431	2.422 −2.449	0	0.064	−74.408	1.041

In addition to the planar or buckled crystal structures, the significant fluctuations of bound lengths are very easy to be observed in most of substitution cases except for the uniform binary compounds. For example, Fig. 6.8A–F under the [2:16]-[meta] substitution configurations clearly illustrate the highly non-uniform chemical environments within the significant bond-length ranges of 2.020–2.475, 2.035–2.750, and 2.413–2.724 Å, respectively, for boron, carbon and nitrogen substitutions. This will generate the quite complicated orbital-hybridization strengths, i.e., the position- and multi-orbital hopping integrals/the on-site Coulomb potentials need to be considered within the phenomenological models

TABLE 6.2 The geometric and electronic properties for the carbon-substituted germanene systems.

The optimal geometric structures of carbon-substituted gemanene systems at various concentrations and configurations with the C–C, C–Ge, and Ge–Ge bond lengths, the band gaps, and the ground state energies per unit cell, together with the height differences between the A and B sublattices.

Number of atoms/unit cell	Percentage	Ratio of C/Ge	C–C bond length (Å)	C–Ge bond length (Å)	Nearest Ge–Ge distance (Å)	Second nearest Ge–Ge distance (Å)	Magnetic moment (μ_B)	E_g^d (eV)/ metal	Optimized E_0 (eV)	Δ (Å)
2	100%	[1:1]	–	1.883	–	–	0	2.060	−11.835	0
6	50%	[2:4] meta		1.916	2.413	–	0	1.493	−31.133	1.110
	50%	[2:4] ortho/ para	1.385	1.968	2.411 −2.424	–	0	0.390	−31.591	0.672
	20%	[1:5]	–	1.927	2.433	–	0	0.081	−27.399	0.707
8	33.3%	[2:6] meta	–	1.910	2.470	2.507	0	0.928	−39.138	2.909
	33.3%	[2:6] para	–	1.930	2.425	–	0	Semi-metal	−38.762	0.680
	14.3%	[1:7]	–	1.977	2.378 −2.385	2.459 −2.463	0	Semi-metal	−34.709	1.019
18	12.5%	[2:16] meta	–	2.036	2.429 −2.437	2.511 −2.750	0	0.381	−76.741	0.825
	12.5%	[2:16] ortho	1.378	2.154	2.425 −2.509	2.440 −2.674	0	Semi-metal	−77.036	0.651
	12.5%	[2:16] para	–	1.946	2.387 −2.439	2.428 −2.483	0	0.208	−78.392	1.263

simultaneously, (the tight-/generalized tight-binding models; [93]). The Harrison rule would be helpful in solving this open issue. Obviously, band structures/wave functions, spatial charge densities and density of states will become more complex, e.g., the great enhancement of band-edges states and thus special structures of van Hove singularities. The VASP predictions on the non-homogeneous honeycomb crystals could be verified by the STM measurements, while these unusual quasiparticle behaviors will strongly suppress the high-precision examinations through ARPES [94] and STS [95,96]. In short, the diversified honeycomb crystals, which are formed by the various chemical bondings, are responsible for all essential quasiparticle properties [97].

The direct combination of STM and STS are very powerful in the simultaneous identifications of nano-scaled surface morphologies and their energy-dependent electronic properties

TABLE 6.3 The geometric and electronic properties for the nitrogen-substituted germanene systems.

The optimal geometric structures of nitrogen-substituted gemanene systems at various concentrations and configurations with the N–N, N–Ge, and Ge–Ge bond lengths, the band gaps, and the ground state energies per unit cell, together with the height differences between the A and B sublattices.

Number of atoms/ unit cell	Percentage	Ratio of N/Ge	N–N bond length (Å)	N–Ge bond length (Å)	Nearest Ge–Ge distance (Å)	Second nearest Ge –Ge distance (Å)	Magnetic moment (μ_B)	$E_g{}^d$ (eV)/ metal	Optimized E_0 (eV)	Δ (Å)
2	100%	[1:1]	–	1.937	–	–	0.496	0.212	−11.519	0.705
6	50%	[2:4] meta	–	1.851 −2.018	2.423 −2.839	–	0	0.825	−31.686	2.783
	50%	[2:4] ortho/ para	1.341	1.975	2.623 −2.673	–	0	Semi-metal	−31.049	2.297
	20%	[1:5]	–	1.918	2.480	–	0.486	0.182	−10.867	0.909
8	33.3%	[2:6] meta	–	1.817 −3.019	2.443 −2.567	2.484–2.500	0	0.712	−38.881	1.660
	33.3%	[2:6] ortho	3.694	1.799 −1.809	2.450 −2.482	2.451–2.535	0	0.591	−37.949	0.730
	33.3%	[2:6] para	–	1.912	2.570	2.570	0	Semi-metal	−37.791	0
	14.3%	[1:7]	–	1.976	2.371	2.499	0	0.281	−34.774	0.859
13	9.1%	[1:11]	–	2.927 −1.940	2.415 −2.439	2.428–2.482	0	Semi-metal	−50.984	2.447
18	12.5%	[2:16] meta	–	1.844 −3.044	2.413 −2.497	2.426–2.606	0	0.146	−77.142	1.295
	12.5%	[2:16] ortho	3.517	1.813	2.399 −2.444	2.423–2.486	0	0.061	−77.822	1.608
	12.5%	[2:16] para	–	1.970	2.400 −2.524	2.425–2.710	0	0.047	−77.833	1.033

since the first high-precision measurements on single-walled carbon nanotubes in 1998 [98]. The close relations between the radius/chirality of carbon hexagons and the semiconducting or metallic properties near the Fermi level are clearly verified from the measurements of van Hove singularities, being consistent with the VASP simulations [54] and the tight-binding model calculations [99]. For example, a semiconducting/metallic carbon nanotube exhibits a pair of asymmetric valence and conduction peaks in the square root form within a band-gap region/a plateau structure including the chemical potential [100]. The similar identifications have been successfully generalized to graphene nanoribbons, in which there exist the

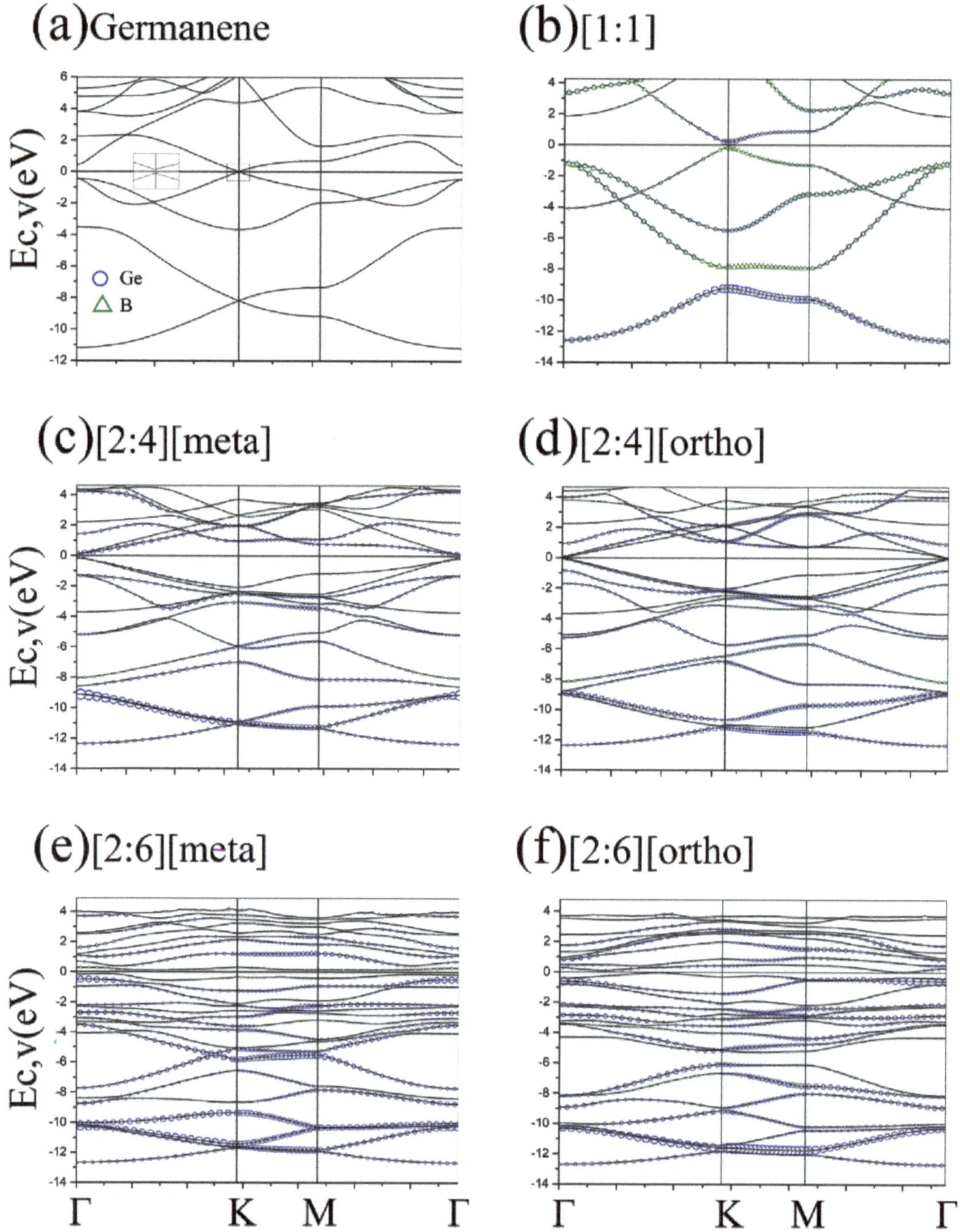

FIG. 6.9 (A–F) Electronic energy spectra and their atom dominances of the boron-substituted germanene systems, with the sufficiently wide valence subbands, under the different substitutions.

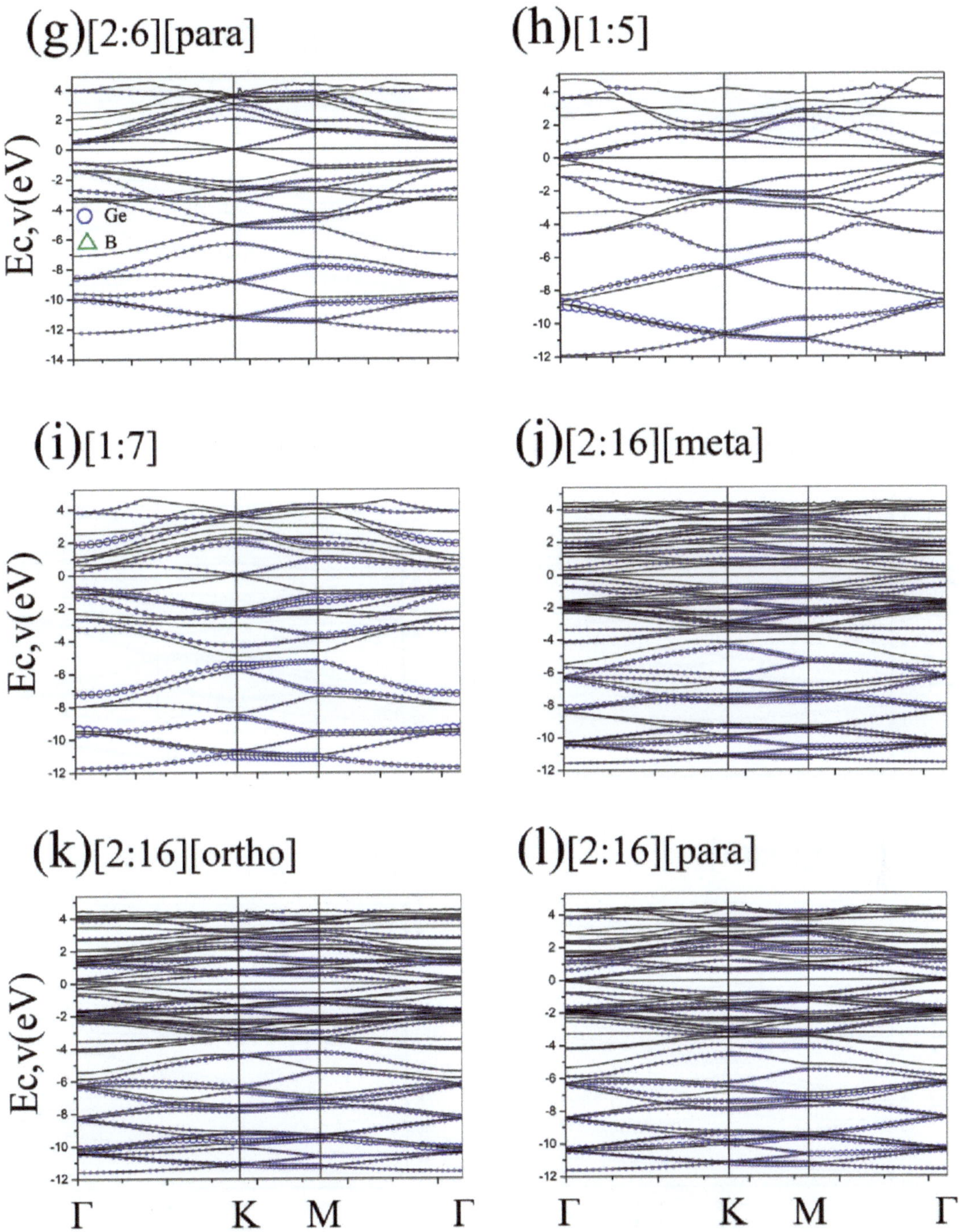

FIG. 6.10 (G–L) Electronic energy spectra and their atom dominances of the boron-substituted germanene systems, with the sufficiently wide valence subbands, under the different substitutions.

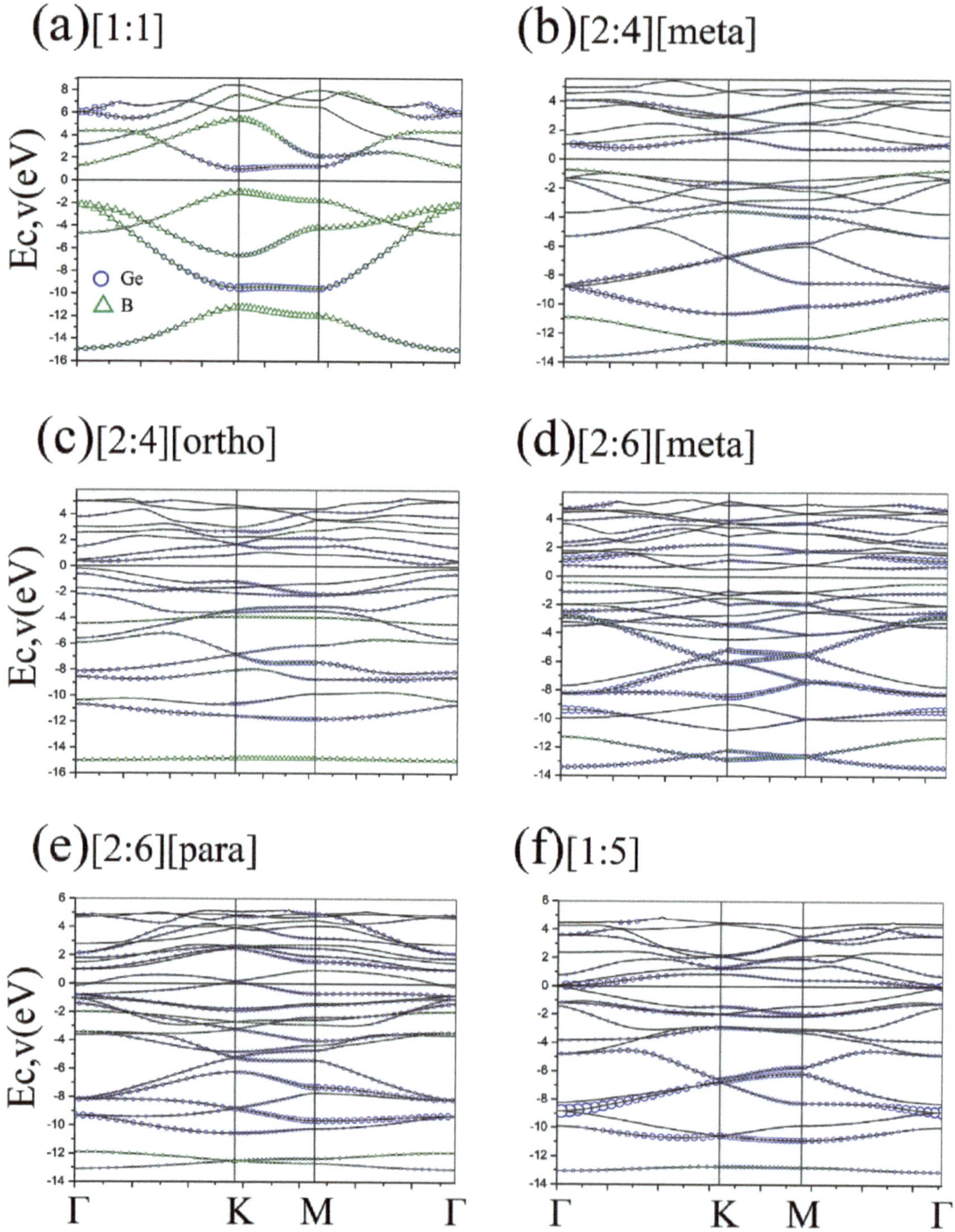

FIG. 6.11 (A–F) Electronic energy spectra and their atom dominances of the carbon-substituted germanene systems, with the sufficiently wide valence subbands, under the different substitutions.

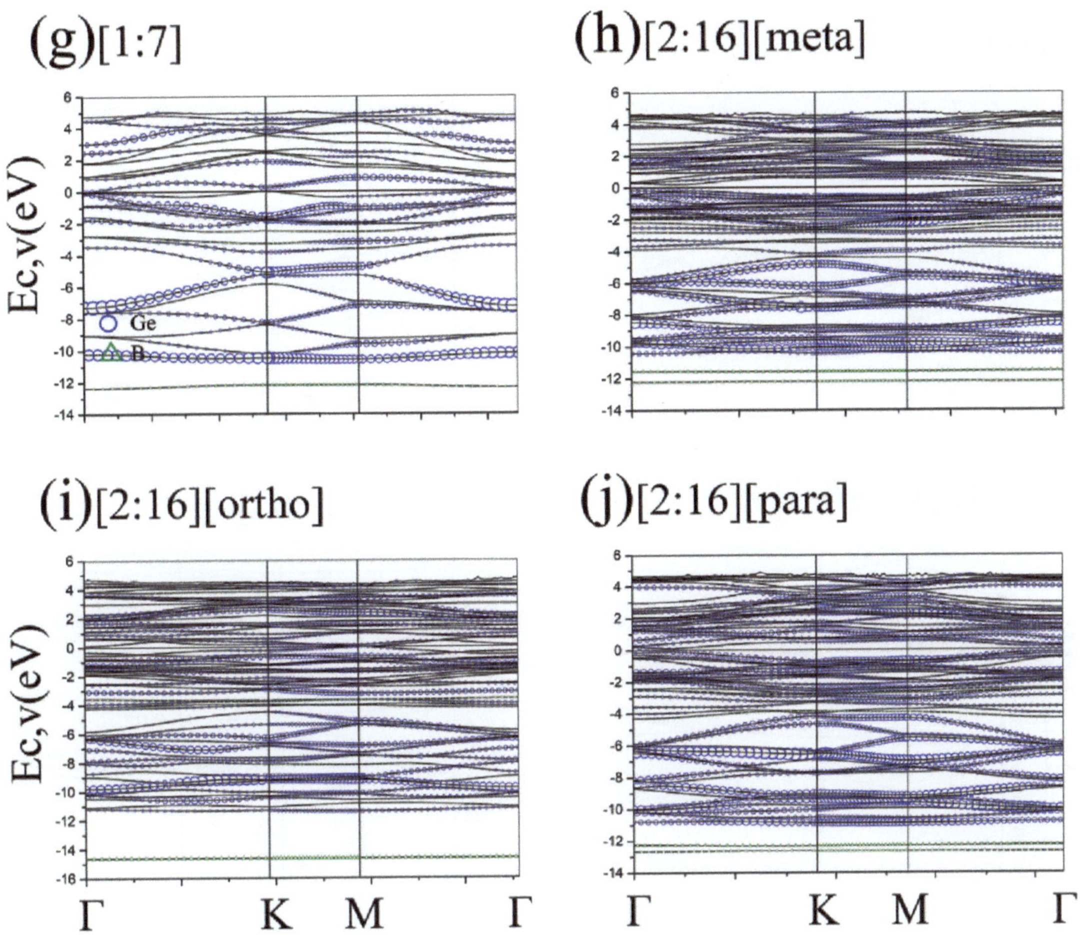

FIG. 6.12 (G–J) Electronic energy spectra and their atom dominances of the carbon-substituted germanene systems, with the sufficiently wide valence subbands, under the different substitutions.

sensitive dependences of band gaps on their widths and open edge structures [101]. For example, with the increase of nanoribbon width, the quick decline of energy gaps shows three approaches/one monotonous form for a chiral armchair/zigzag systems [102]. As for the association of STM and TEM, their examinations are quite successful in verifying the full crystal symmetries of top- and side-view morphologies, e.g., the stacking configuration and layer number [interlayer distance in coaxial carbon nanotubes [104], few-layer graphene systems [105], and graphene nanoribbons [106], as well as, the non-planar honeycomb structures in curve/folded/scrolled/hatted nanoribbon systems [106–109]. That is, such combined techniques can reveal the various deformations [110] and composite structures (significant bond-length fluctuations; [88,111]). They are also available in fully exploring the intrinsic symmetries of substitution-enriched germanene systems. The position-dependent chemical bonds are responsible for all other essential properties [112].

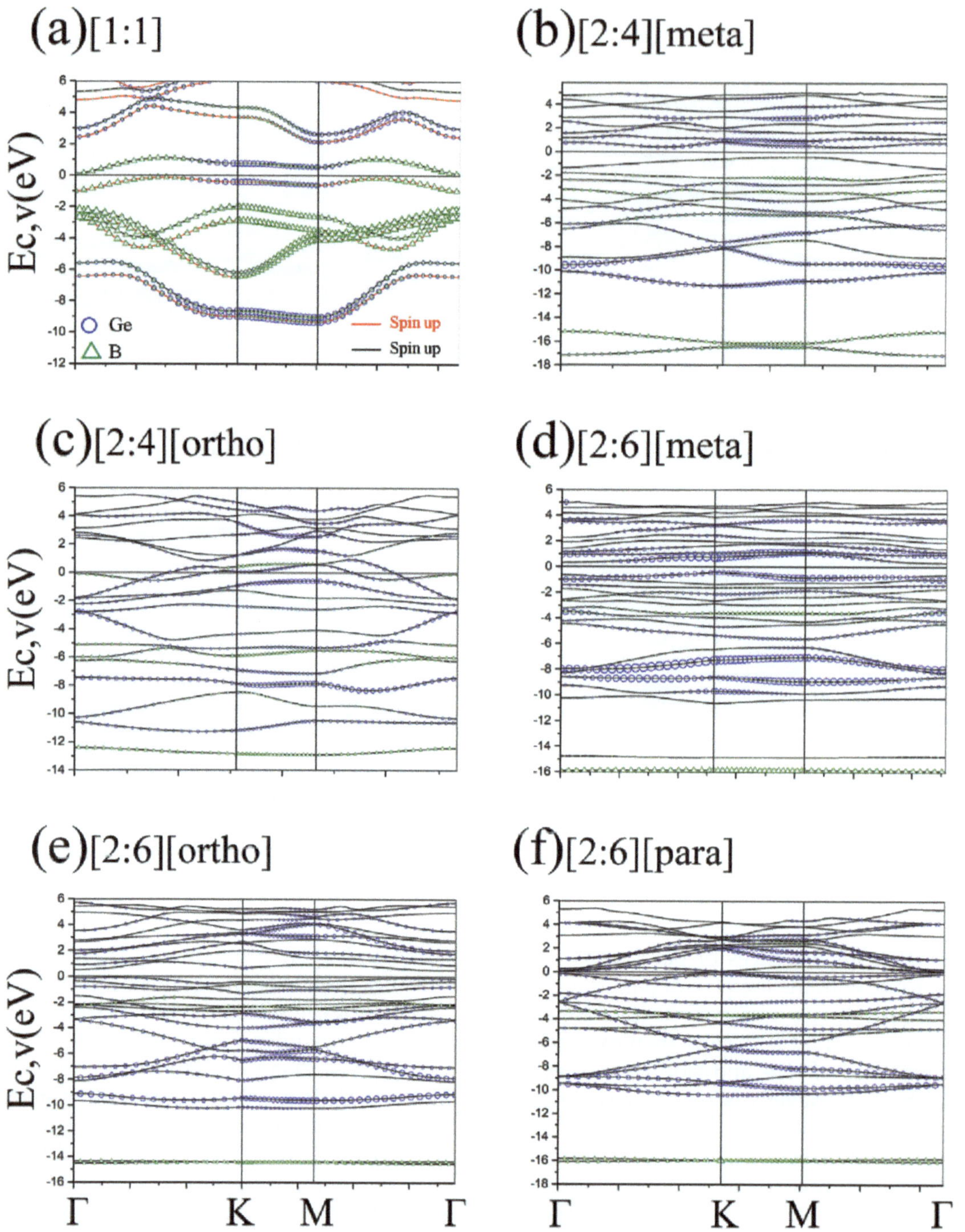

FIG. 6.13 (A–F) Electronic energy spectra and their atom dominances of the nitrogen-substituted germanene systems, with the sufficiently wide valence subbands, under the different substitutions.

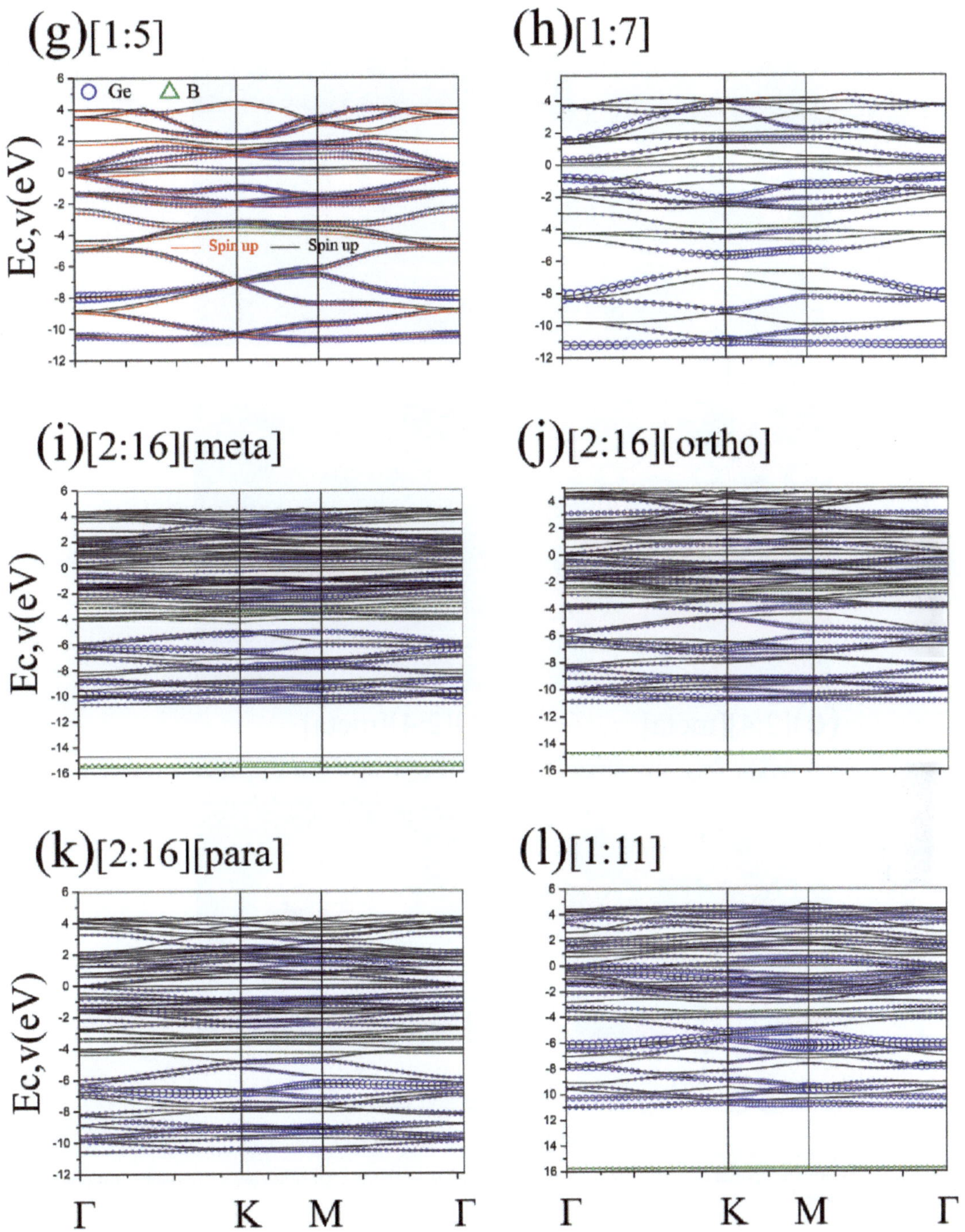

FIG. 6.14 (G–L) Electronic energy spectra and their atom dominances of the nitrogen-substituted germanene systems, with the sufficiently wide valence subbands, under the different substitutions.

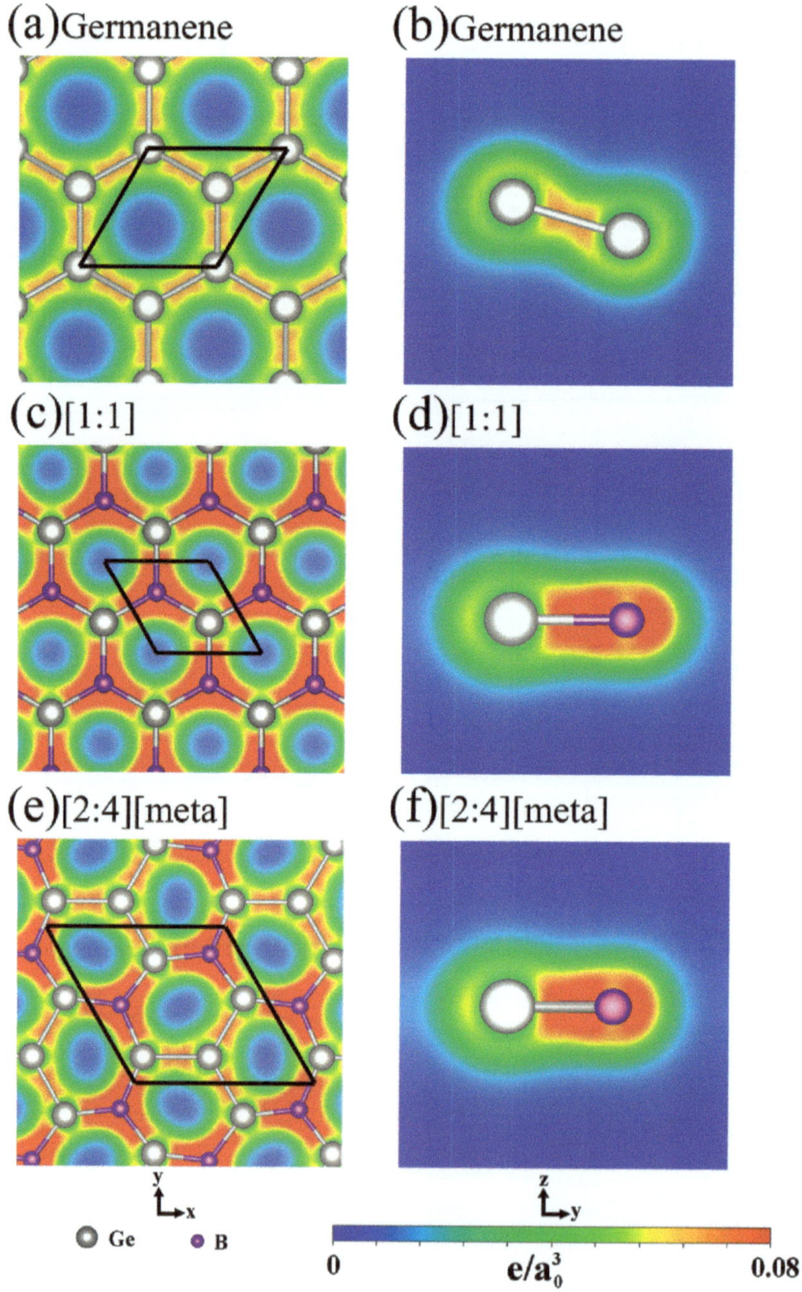

FIG. 6.15 (A–F) The spatial charge density distributions [ρ(**r**)], with the top/side views, after substitutions for the various boron cases.

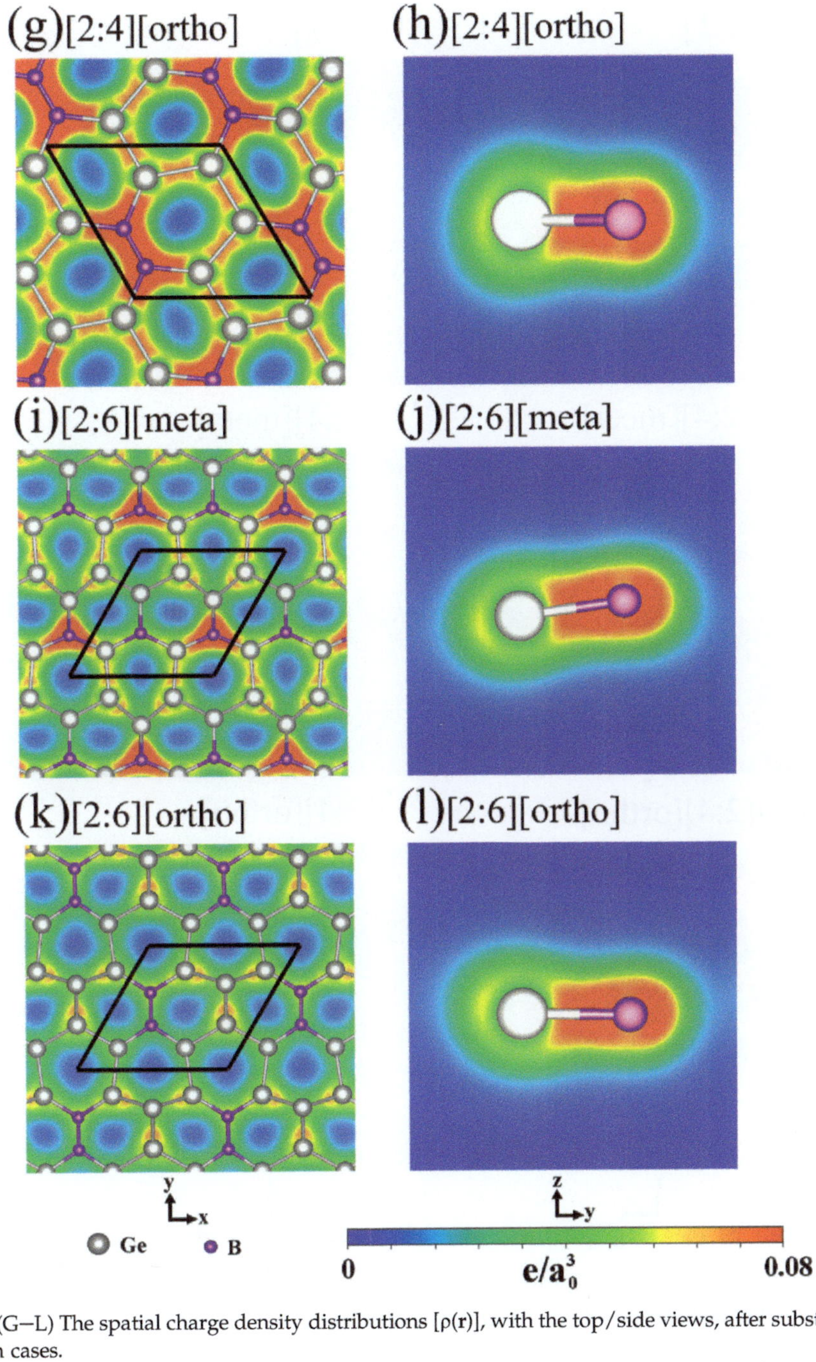

FIG. 6.16 (G–L) The spatial charge density distributions [ρ(**r**)], with the top/side views, after substitutions for the various boron cases.

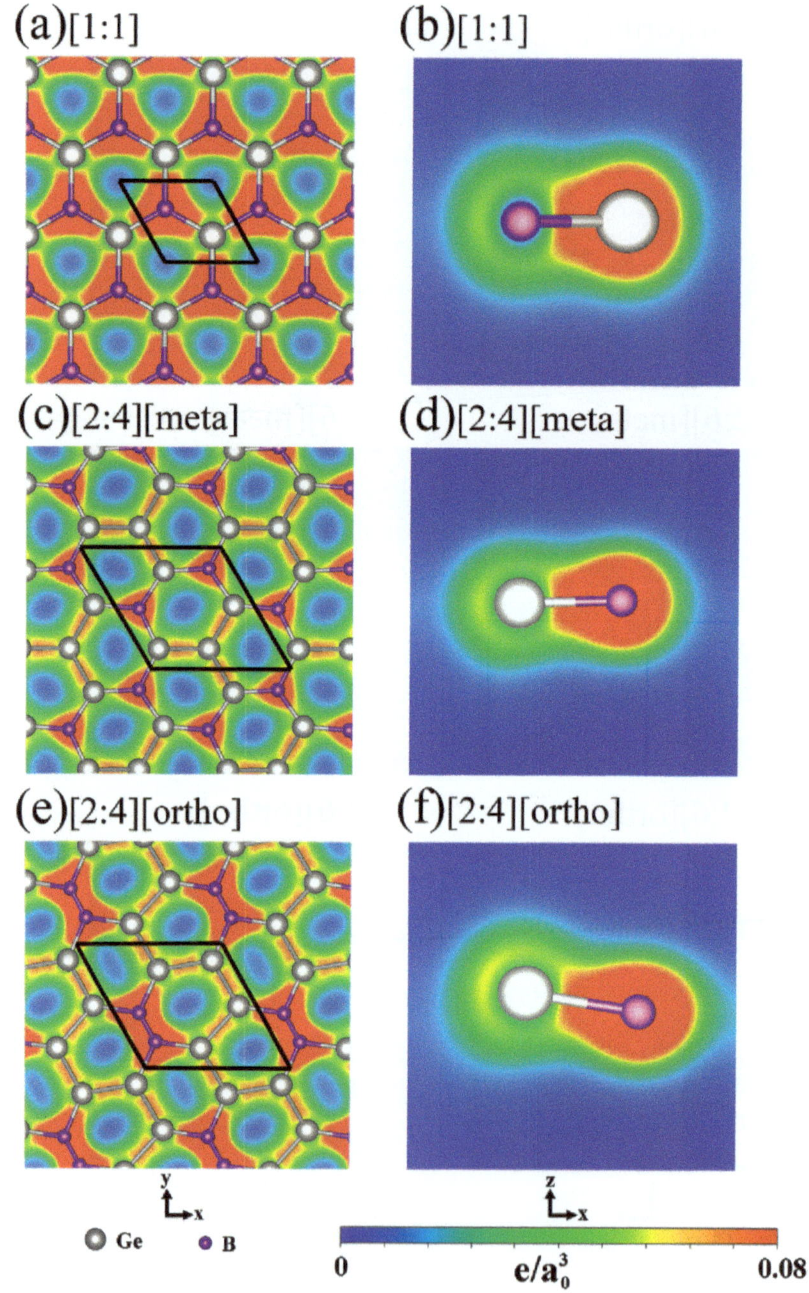

FIG. 6.17 (A–F) The spatial charge density distributions [ρ(**r**)], with the top/side views, after substitutions for the various carbon cases.

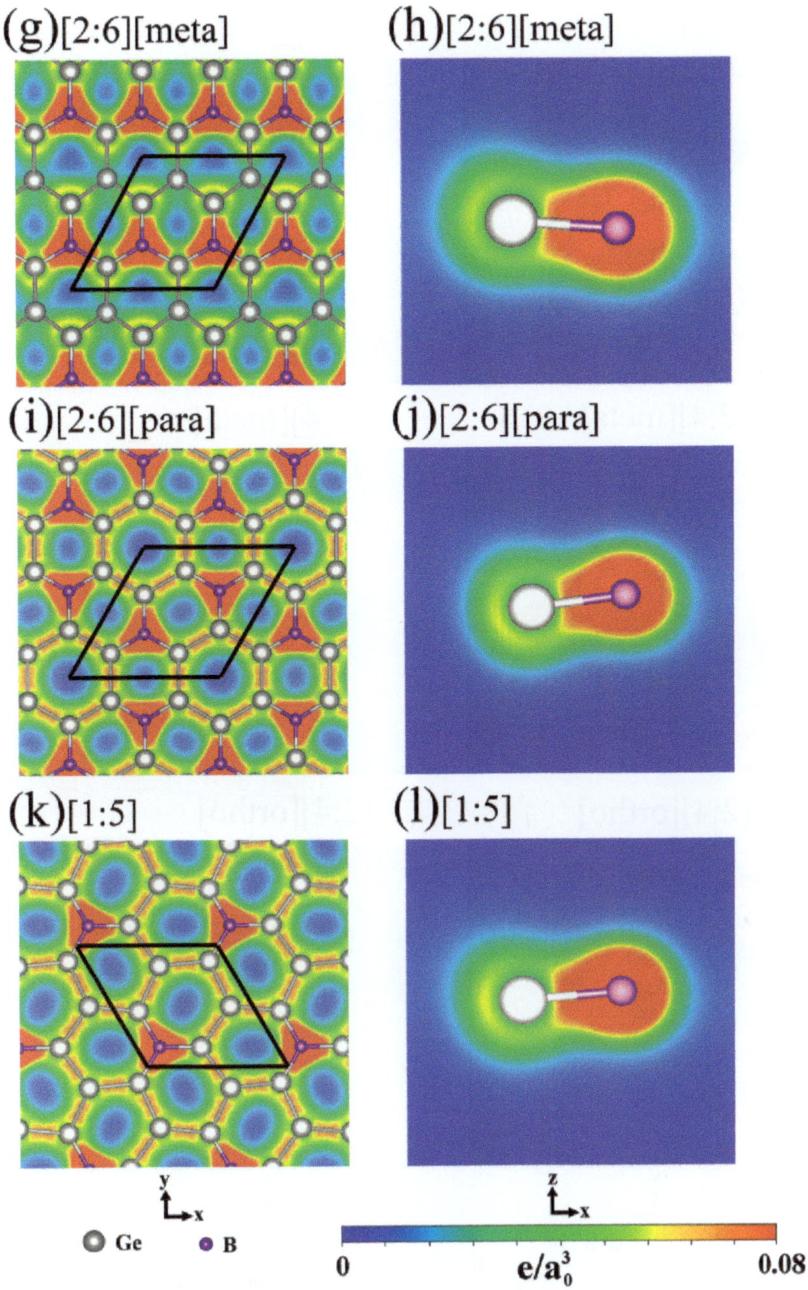

FIG. 6.18 (G–L) The spatial charge density distributions [ρ(**r**)], with the top/side views, after substitutions for the various carbon cases.

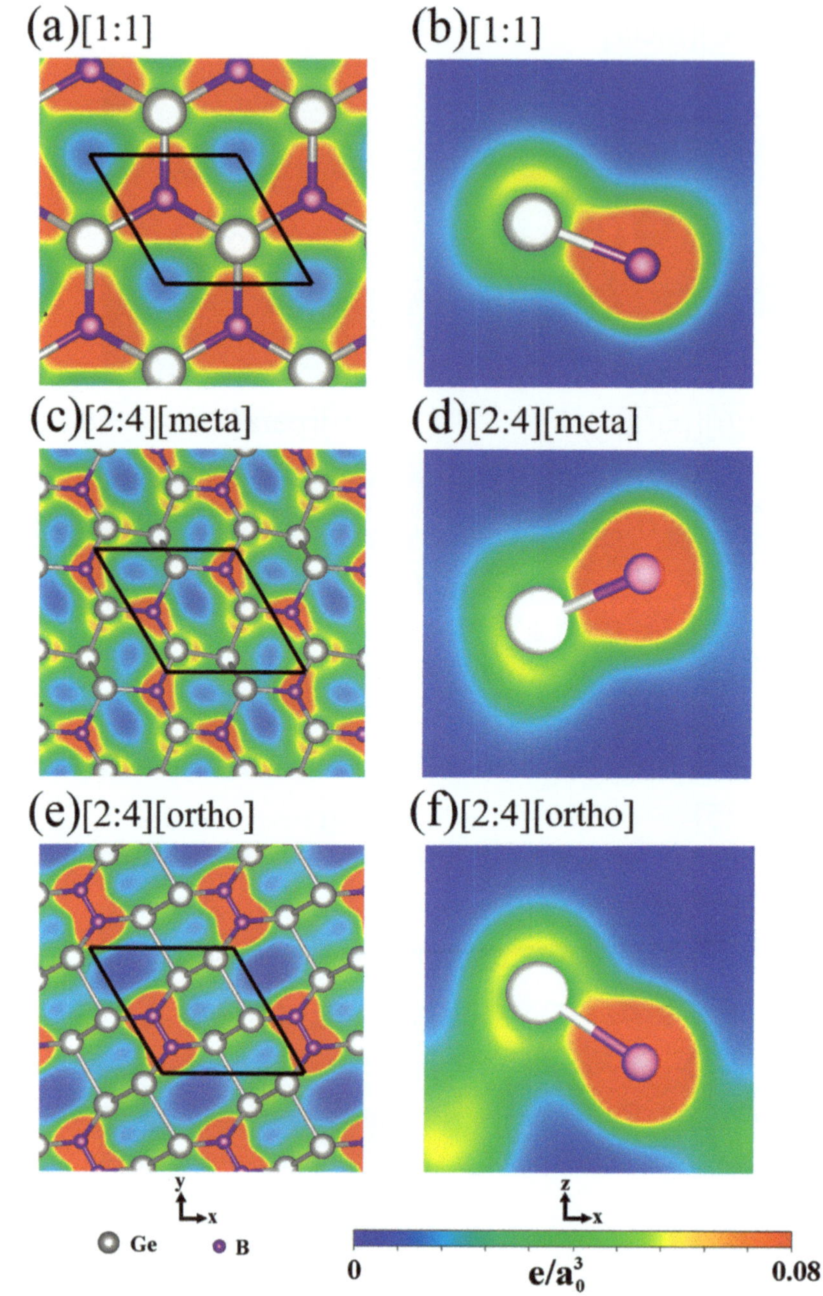

FIG. 6.19 (A–F) The spatial charge density distributions [ρ(**r**)], with the top/side views, after substitutions for the various nitrogen cases.

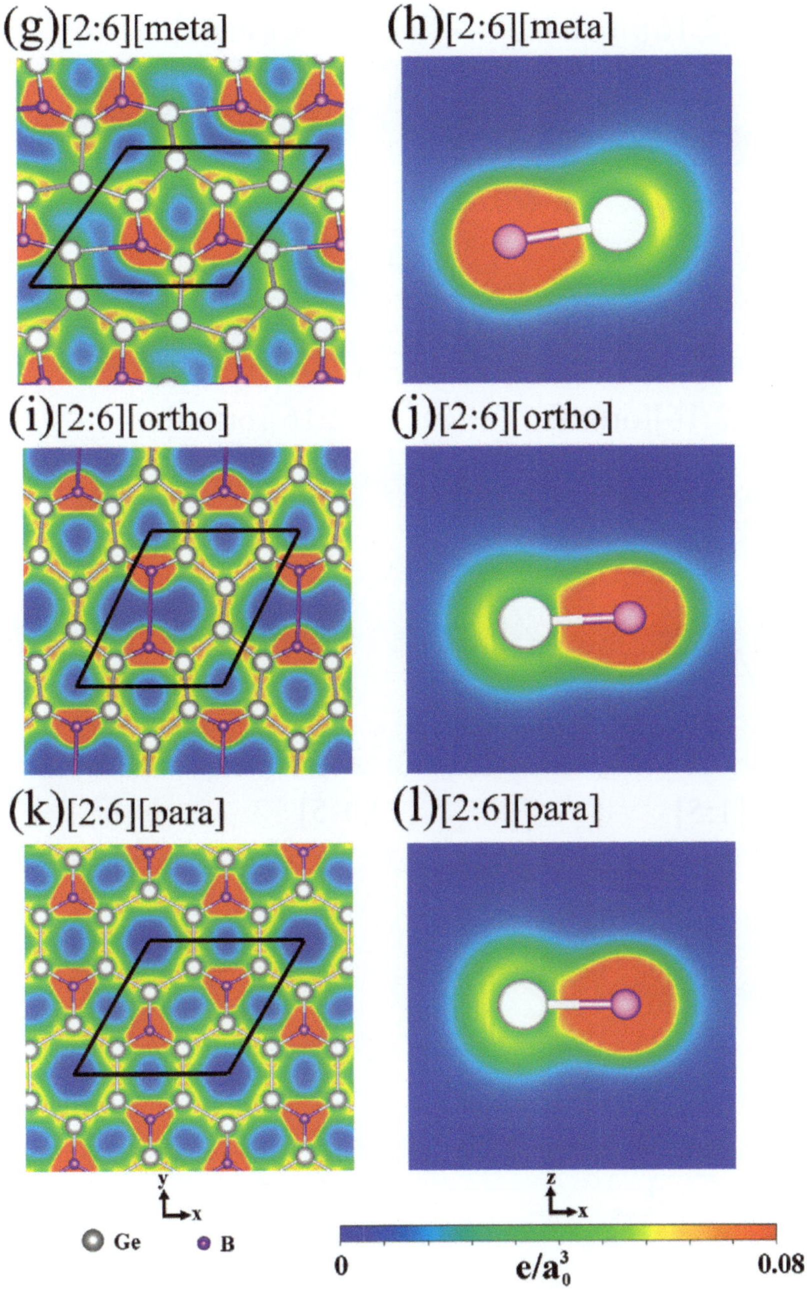

FIG. 6.20 (G–L) The spatial charge density distributions [ρ(r)], with the top/side views, after substitutions for the various nitrogen cases.

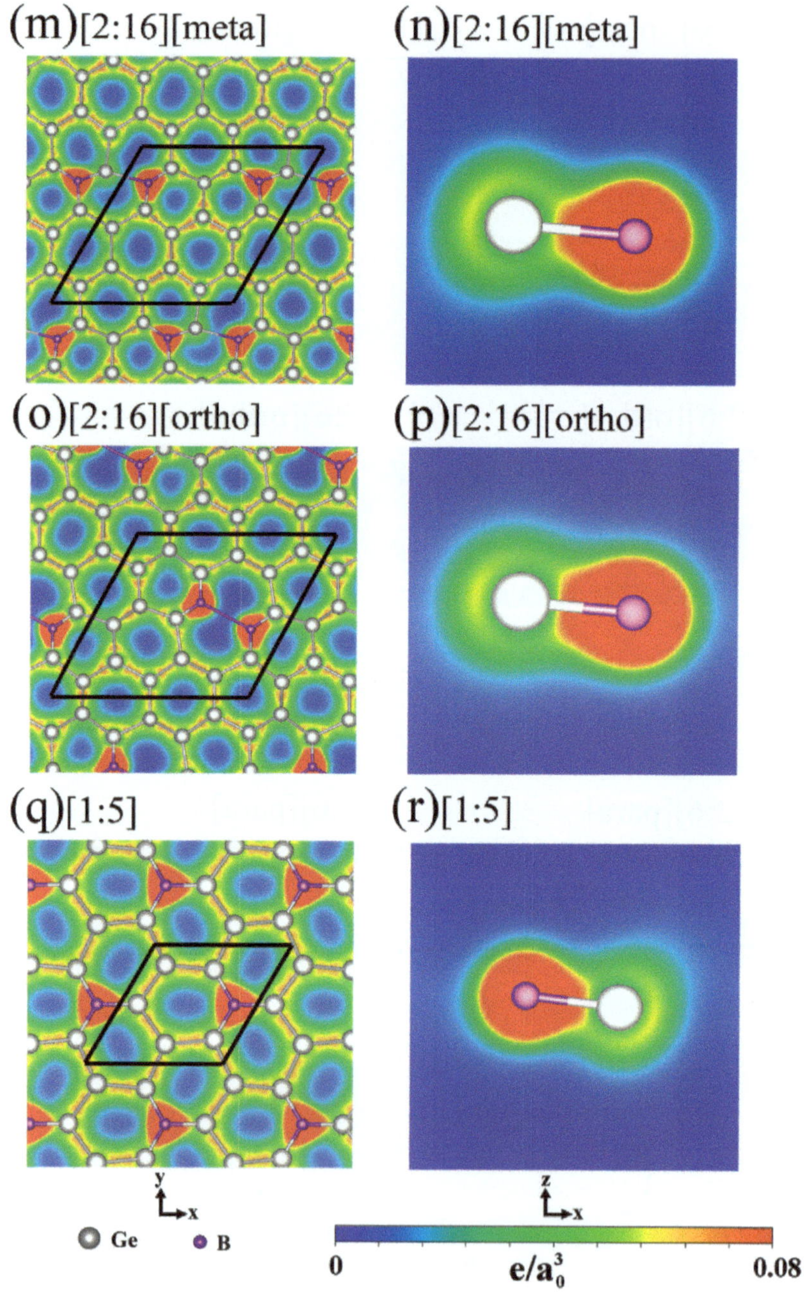

FIG. 6.21 (M–R) The spatial charge density distributions [ρ(**r**)], with the top/side views, after substitutions for the various nitrogen cases.

6.3 Featured band structures and atom-dominated wave functions in binary germanene compounds

Electronic structures and atom dominances, which are conducted along the high-symmetry paths of the first Brillouin zone (Figs. 6.9A–F, 6.10G–L, 6.11A–F, 6.12G–J, 6.13A–F and 6.14G–L for the boron-carbon and nitrogen substitution cases, respectively) clearly show the diverse behaviors in terms of their characteristics. The main features cover a lot of valence and conduction subbands (many orbitals in a unit cell; [14]), the semiconducting/semi-metallic/metallic properties near the chemical potential (more strong covalent bondings or charge transfers; [113]), an obvious asymmetry of valence hole and conduction electron spectra about E_F (the multi-orbital hybridizations; [114]), the weak perturbations/major modifications/thorough destructions for the perpendicular π and σ bodings, the various energy dispersions initiated from the stable valleys (the linear/parabolic/oscillatory/partially flat ones; [115]), the band-edge states with zero group velocities (the critical points in energy-wave-vector space; [88]), the non-crossing/crossing/anti-crossing behaviors of energy subbands, and the spin-split ferromagnetic band structures. The close relations with the chemical bondings are further established by the atom dominances within the specific energy ranges. Such predictions, being associated with the occupied electronic states, could be thoroughly examined by the high-precision ARPES measurements. It is worthy of developing the reliable parameters in the tight-binding model by comparing with the low-energy of the VASP results.

A pristine germanene, as clearly shown in Fig. 6.9A, exhibits an unusual band structure. This is attributed to the non-orthogonal π and σ bondings, respectively, due to the significant hybridizations of the neighboring $4p_z$ and [4s, $4p_x$, $4p_z$] orbitals (consistent with a buckled honeycomb crystal; [89]). A direct band gap of $E^d{}_g$ ~31 meV is revealed by a slightly separated and modified Dirac-cone structure centered at the K/K′ points. However, graphene is a zero-gap semiconductor. The former is created by another critical mechanism, the observable spin-orbital coupling [67]. The π band is well defined along the KMΓ path, in which M and Γ points, respectively, the saddle and extreme points. Its width is about 3.62 eV smaller than the half of that in graphene (larger than 7 eV; [14]), mainly owing the weaker π bonding (the longer bond length). Two initial σ valence bands of parabolic dispersions, being dominated by $4p_x$ and $4p_y$ orbitals [14], initiate from the stable Γ valley at E^v ~ −0.26 eV, split, and then forward along the ΓMK path with the deepest energies of E^v ~ −3.51 and −8.18 eV. Their band widths are quite different from each other. It should be noticed that the π band has a weak/strong anti-crossing phenomenon with the second σ band. Such behavior can create an one-dimensional constant-energy loop, i.e., 2D energy dispersions are dramatically changed into 1D parabolic ones (1D van Hove singularities; [88]). The obviously anti-crossing behaviors indicate the significant sp^3. Finally, the third σ band, which is dominated by the 4s orbitals [14], is along the KMΓ path with the initial and final valence-state energies of −8.18 and −11.17 eV, respectively.

Boron, carbon and nitrogen substitutions on germanene can induce the slightly, strongly and thoroughly modified band structures and wave functions, as clearly indicated under the 1:1 cases, respectively, in Figs. 6.9B, 6.11A and 6.13A. Obviously, the first system behaves like a pristine material (Fig. 6.9A and B) in terms of π and σ bands. That is, its electronic energy spectrum is well characterized by the π and σ bondings in the absence of buckling (Fig. 6.1D). However, the energy spacing of a pair of Dirac points at the K point is greatly enhanced from 0.18 meV to 0.58 eV. The subband anti-crossings, which occur between the π band and the first/second σ ones, become very weak as a result of the almost perpendicular features about two kinds of chemical bindings. The π, the first-, second- and third-σ valence bands, respectively, along the KMΓ, ΓMK, ΓMK, and KMΓ, possess the band widths of ~3.91, 4.28, 6.74, and 3.34 eVs. Most important, the Fermi level of $E_F = 1.76$ eV, is situated at the valence Dirac-cone structure. There exist free valence holes due to the *p*-type doping, or the substitution atoms are able to attract electrons from the neighboring host ones. Boron, being relative to germanium, can serve as the accepter-type guest atoms. As for the carbon-substituted germanene binary compound, a middle-gap semiconductor of $E^d{}_g$ ~2.06 eV is clearly revealed at the K valley. It is mainly determined by the prominently modified Dirac-cone structure. This is totally different from the boron-substitution case (Fig. 6.9B). Such system should exhibit a more strong covalent bond, but mot a metallic one. All σ valence bands remain the parabolic energy dispersions, and their widths are, respectively, ~3.65, 4.47, 7.40, and 3.70 eVs. Very interesting, the Dirac-cone structure completely vanishes through the saturated nitrogen substitution (Fig. 6.13A). A metallic behavior is identified from the blue shift of the Fermi level, since E_F is above the band edge of the first conduction band at the Γ point roughly 0.102 eV. Many free conduction electrons (an n-type doping; [116]) could be generated under the nitrogen substitution [89]. It is mission impossible in defining the well-behaved π-band width along the high-symmetry points. It should be noticed that the first, second and third σ, respectively, have 0.88, 2.00, and 1.78 eVs. Based on the above-mentioned features, the B-, C- and N-substitutions, respectively, display the normal, unusual and irregular π bondings due to the $4p_zz$-$2p_z$ hybridizations. Furthermore, they are almost perpendicular to the σ bondings associated with the [4s, $4p_x$, $4p_y$]-[4s, $4p_x$, $4p_y$] mixings. The concise pictures would be very useful in the successful establishment of the tight-binding model/the generalized tight-binding model [26,117].

When the concentration of guest atoms is gradually reduced under the chemical substation, band structure become more complicated, as clearly shown in Figs. 6.9C, 6.10L, 6.11B, 6.12J, 6.13B and 6.14L. The rich and unique energy spectra, corresponding to the formation of Moire superlattices [85], are induced by the cooperation of the different chemical bondings, the spin-dependent interactions and the zone-folding effect. Also noticed that the non-saturated/lower-concentration systems have Ge–Ge and B–B/C–C/N–N bonds (the distinct multi-orbital hybridizations; details in Figs. 6.9–6.14). It might be very difficult to precisely analyze the existences of π and σ valence bands through their important characteristics near the high-symmetry points, mainly owing to the close energies of band-edge states. Such issues could be solved by the full support from the atom- and orbital-decomposed van Hove singularities (details in Figs. 6.22–6.31). Each significant critical points could be assigned the specific atom- and orbital-dominances. The thorough examinations would be very helpful in exploring the electronic excitations due to Coulomb [118] or electromagnetic fields [119].

The hole-doped behaviors remain similar during the decline of boron-centration (Figs. 6.9C and 6.10L), being attributed to the prominent affinity of guest atoms. Whether free hole densities are linearly proportional to boron concentrations is worthy of a further study. This should check the enclosed [k_x, k_y]-area below the Fermi-momentum states. The close relations between chemical bondings and the orbital hybridizations and the π/σ valence bands will be clarified from the featured van Hove singularities. Concerning carbon-substitution cases, most of binary compounds exhibit the semiconducting behaviors, while certain of them show the semi-metallic/zero-gap behaviors with the weak/toughing valence and conduction band overlaps, e.g., the non-zero-gap features under the substitution of [1:1], [2:4]-[meta], [2:4-[ortho], [2:6]-[meta], [1:5], [2:16]-[meta], & [2:16]-[para] (Figs. 6.11A–F and 6.12G–J). Specifically, the concentrations lower than 2% will show a recovered Dirac-cone structure (not shown). The highly non-uniform chemical bonds and anisotropic orbitals (the various hopping integrals and on-site Coulomb potentials; [120]) have strong effects on the low-lying band properties. The similar mechanisms might dominate the metallic or semi-conducting behaviors for the nitrogen-germanium binary compounds. For example, [1:1], [2:4]-[meta], [1:5], [2:6]-[meta], [2:6]-[ortho], [1:7], [2:16]-[meta], and [2:16]-[ortho], [2:16]-[para] N-substitutions, respectively, display indirect/direct band gaps of $E_g{}^i = 0.121$ eV, $E_g{}^i = 0.825$ eV, $E_g{}^d = 0.182$ eV $E_g{}^d = 0.712$ eV $E_g{}^d = 0.591$ eV $E_g{}^d = 0.281$ eV $E_g{}^d = 0.643$ eV $E_g{}^d = 0.712$ eV $E_g{}^d = 0.061$ eV $E_g{}^d = 0.047$ eV. In addition, the previous VASP simulations show such quasiparticle phenomena for B- and N-substituted graphene compounds [121], i.e., the former and the latter, respectively, belongs to the hole-doped metals and the donor-type metals/semiconductors.

Angle-resolved photoemission spectroscopy is very powerful in the detailed verifications about the predicted 2D occupied band structures, especially for the high substitution cases with concise energy spectra [116]. ARPES is suitable for examining the planar/buckled crystal structures under the conservation of the 2D transferred momenta and energies [122]. The high-precision measurements have been successfully utilized to explore graphene-related materials with the well-defined π and σ valence bands, e.g., the low-lying modified Dirac-cone structures, with weak, but signify-cant band overlaps, in the ABA- [123], ABC- [124] & AAB-stacked trilayer graphene systems, the monolayer- & bilayer-like electronic structures for the 3D Bernal graphite [125], as well as, the 1D parabolic subbands for the semiconducting graphene nanoribbons [108]. These unusual 2D and 3D results clearly illustrate the critical roles of the interlayer C-2p_z-orbital interactions in the dramatic transformation of energy dispersions and band overlaps (free carrier densities of holes and electrons; [126]). Furthermore, the rich 1D observations directly reflect the quantum-confinement effects [127].

As to silicene/germanene/tinene/plumbene grown on the different substrates [57,128–133], [134–136], [60,137–139], the unique chemical bondings are verified to dominate the electronic structures near the Fermi level (details in Chapter 3). Also noticed that the ARPES examinations on 1D carbon nanotubes cannot be finished up to now because of the undefined momentum transfers along a deformed tubule axis [140]. Such tests are expected to be reliable for the boron-, carbon- and nitrogen-substituted germanene compounds, covering, the semiconducting [14], semi-metallic [14] or metallic properties [63], the hole or electron doping the red or blue shift of the Fermi level; [14], the moderately/strongly/thoroughly destroyed Dirac-cone structure, and two initial σ valence subbbands due to the dominating [4p_x, 4p_y] orbitals from the Γ point [14]. The successful identifications are able to clarify the basic chemical bondings, being consistent with the theoretical development of quasiparticle framework [141].

6.4 Critical orbital hybridizations in distinct chemical bonds

How to construct the direct comparisons between the theoretical predictions and the experimental evaluations becomes an open issue. The up-to-date STM and TEM data (details in Chapter 3) are able to reveal each atom position on the [x, y] and [x, z]/[y, z] planes, respectively. While they are fully assisted by the atomic orbitals through the neighboring orbital hoppings, the spatial charge densities would come to exist in the highly deformed ways. The critical factors should be the different strengths of various orbital hybridizations [the distinct hopping integrals in the tight-binding model; [142]. On the theoretical side, the reliable establishments of phenomenological models might be available in determining the prominent diffraction peaks and the corresponding carrier densities. That is, the direct combination of the tight-binding model [142] and the elastic diffraction theory [143] could achieve the optimal crystal structures and their charge density distributions.

The spatial charge densities [$\rho(\mathbf{r})$] after guest-atom substitutions are capable of directly revealing the π, σ and sp^3 bondings, being useful in fully supporting more complicated orbital hybridizations of the merge/non-merged van Hove singularities (discussed later in Figs. 6.22–6.31). The diverse quasiparticle behaviors are able to be observed in a pristine germanene (Fig. 6.15A and B), the non-buckled binary compounds (Figs. 6.15A–F, 6.16G–L, 6.17A,B and 6.18I,J), and the bucked ones (the others). A uniform Ge–Ge bonds can create the well-characterized σ bondings, as clearly indicated by the red region between two neighboring germanium atoms on the [x, y] plane of Fig. 6.15A. Their non-orthogonal relations with the π bondings are obviously shown on the [x, z]/[y, z] plane of Fig. 6.15B, in which the latter correspond to the wave-like distributions of two neighboring $4p_z$ orbitals in the green regions. As a result, there exist the dominating σ bondings of [4s, $4p_x$, $4p_y$]-[4s, $4p_x$, $4p_y$], as well as, the weak, but significant sp^3 bndings ones of [4s, $4p_x$, $4p_y$, $4p_z$]-[4s, $4p_x$, $4p_y$, $4p_z$]. As for the planar binary compounds, the σ and π bondings almost perpendicular to each other. Furthermore, the charge density distributions between two neighboring atoms become highly non-uniform, e.g., the obvious deformations of the σ bondings for the B-, C- and N-saturated compounds in Figs. 6.15C, 6.17A and 6.19A, respectively. Such cases have clearly shown the complicated orbital hybridizations of [4s, $4p_x$, $4p_y$]-[2s, $2p_x$, $2p_y$] & $4p_z$-$2p_z$. Apparently, the chemical environments are getting complex for the buckled binary crystals at the lower-concentration substitutions, in which the sp^3 bondings will play the critical roles in determining all low-energy essential properties, such as, [4s, $4p_x$, $4p_y$, $4p_z$]-[4s, $4p_x$, $4p_y$, $4p_z$] [Ge–Ge bonds], [4s, $4p_x$, $4p_y$, $4p_z$]-[2s, $2p_x$, $2p_y$, $2p_z$] [Ge–B, Ge–C and Ge–N bonds], as well as, [2s, $2p_x$, $2p_y$, $2p_z$]-[2s, $2p_x$, $2p_y$, $2p_z$] [B–B/C–C/N–N bonds]. The above-mentioned multi-/single-orbital hybridization in distinct chemical bonds will be fully supported by the merged van Hove singularities (discussed later in Figs. 6.22–6.31).

The energy-pendent densities of states, corresponding to the 2D pristine materials and binary compounds, are characterized by the integration about the inverse of group velocity on a/the constant-energy loop/loops [144]. It is well known that their prominent structures are directly induced by the zero/discontinuous group velocities. The previous studies clearly show the V-shape, plateau, logarithmic divergence, delta-function-like structure and asymmetrically divergent peak in the square-root form, respectively, arising from the linear Dirac cones [145], extreme points (local maxima or minima; [145]), saddle points [146], partially flat bands [147], and constant-energy loops [148]. The merged van Hove singularities, which are strongly associated with the different atoms and orbitals, are very useful in achieving the critical single-/multi-orbital hybridizations of the distinct chemical bonds. The concise physical and chemical pictures are thoroughly consistent with the main features in band structures (Figs. 6.9–6.14) and spatial charge density distributions (Figs. 6.15–6.21). All VASP predictions on valence and conduction density of states could be further verified by the high-precision STS measurements (details in Chapter 3).

A pristine germanene, as clearly shown in Fig. 6.22A, exhibits the unusual density of states across the Fermi level. Furthermore, the similar van Hove singularities due to the specific band-edge states are also revealed in any substitution compounds. A dip-like structure is centered about the Fermi level with a very low height within a broadening factor ([~0.02 eV), corresponding to a narrow gap of 0031 eV (Fig. 6.9A). It is characteristic of a slightly separated Dirac-cone structure closely related to the C-$2p_z$ π bonding (the pink curve in Figs. 6.9A and 6.22A; [62]). Along the KM/ΓM path of band structure, the saddle point clearly reveals a logarithmically divergent peak at ~−1.15 eV. On the other side, the σ bondings of the dominant [$4p_x$, $4p_y$] orbitals the green and blue curves display two merged shoulders structures near −0.36 eV. Both contributions are different each other because of the buckling-induced destructions of the $z = 0$-plane mirror symmetry [149]. In addition, the opposite is true for a pure graphene; [62]. And then, an obvious anti-crossing of the former and the second σ band is well characterized by a pair of anti-symmetric peaks (the total black curve associated with a 1D constant-energy loop; [14]), as well as, the vanishing contributions of the Ge-$4p_z$ orbitals within the specific energy range of −2.08 to −3.66 eV. Also noticed that another anti-crossing phenomenon is too weak to be observed through a finite broadening. The π- and first-σ-band widths are evaluated to be 3.308 and 2.08 eV, respectively. Very interesting, the 4s-orbital-related van Hove singularities (the orange curve) come to exits at −1.10, −2.10, − 3.70, −7.35, −9.30, and −11.10 eV, in which the former/latter three structures combine with those of [$4p_x$, $4p_y$, $4p_z$]/[$4p_x$, $4p_y$] orbitals. All main characteristics of van Hove singularities have clearly indicated the significant sp^3 bondings and the dominant σ ones in Ge–Ge bonds. This is very useful in establishing the phenomenological models and thus fully comprehending the quasiparticle phenomena [141].

The featured van Hove singularities are greatly diversified through the guest-atom substitutions, as clearly illustrated in Figs. 6.22B, 6.24P, 6.25A, 6.27R, 6.28A and 6.31X, respectively, associated with B-, C- and N-substitution cases. This directly reflects more complicated multi-orbital hybridizations under the highly non-uniform chemical environments [150], as well as, the zone-folding effects [151]. For example, the saturated binary compounds (Figs. 6.22B,C, 6.25A,C and 6.28A,C), obviously display the strongly modified density of states. There exists a large blue shift of the Fermi level (~4.80 eV), an enlarged band gap $E_g{}^d = 2.06$ eV, and a giant red-shift E_F (~10.03 eV). Furthermore, a certain van Hove singularity is very close to the Fermi level, indicating a rather high carrier density of free valence holes/conduction electrons. The merged van Hove singularities are very sensitive to the distinct guest atoms. For example, the $4p_z$-/$2p_z$-, [$4p_x$, $4p_y$]-/[$2p_x$, $2p_y$]-, 4s-/2s-p projected densities of states, which are thoroughly analyzed for Ge/B components, show the dip/logarithmic peak/shoulder structures at 1.00 eV/0.00 eV/−1.36 eV, the shoulder/logarithmic peak/shoulder/asymmetric peak/asymmetric peak structures at 0.90 eV/−1.20 eV/−4.01 eV/−6.01 eV/−8.21 eV, and the asymmetric peak/asymmetric peak/shoulder structures at −6.01 eV/−8.21 eV/−11.02 eV. These unusual results can be attributed to the almost perpendicular π and σ bondings, since the simultaneous mergences of four-orbital are absent. Consequently, the multi-orbital hybridizations of the uniform Ge−B bonds consist of $4p_z$-$2p_z$ and [4s, $4p_x$, $4p_y$]-[4s, $4p_x$, $4p_y$]. Similarly, the merged/non-merged structures are observed in the van Hove singularities of full carbon and nitrogen substitutions. Three kinds of saturated binary compounds should have the similar hopping integrals under the different strengths and densities of occupied electron densities. Such critical factors need to be included in the establishment of phenomenological models [152]. Such bonding behaviors are relatively easily observed in the non-buckled substitution cases, covering the diverse van Hove singularities under [B, Ge]-[1:1] (Table 6.1; Fig. 6.22B,D, [B. Ge]-[2: 4]-[meta] (Figs. 6.22E and 6.23G), [B: Ge]-[2: 4]-[ortho] (Fig. 6.23H,J), [C, Ge]-[1: 1] (Table 6.2; Fig. 6.25A,B), [N, Ge]-[2: 6]-[para] (Table 6.3; Fig. 6.30P,R). Also noticed that certain merged strictures are induced by the zone-folding effects (more band-edge states).

On the other side, the buckled crystal structures, which survive under most substitution cases (Tables 6.1−6.3), clearly reveal another type of merged van Hove singularities. For example, [B, Ge]-[2: 6]-[meta] (Table 6.1; Figs. 6.23K and 6.24M), [B. Ge]-[2:6]-[ortho] (Fig. 6.24N,P), [C, Ge]-[23K: 6]-[meta] (Table 6.2: Fig. 6.26J,L), and [N, Ge]-[2: 6]-[ortho] (Table 6.3; Fig. 6.30M,O), exhibit four- and three orbital mergences of [4s, $4p_x$, $4p_y$, $4p_z$]/[2s, $2p_x$, $2p_y$, $2p_z$] and [4s, $4p_x$, $4p_y$]/[2s, $2p_x$, $2p_y$], respectively, within the energy ranges of higher and deeper than −5.00 eV. This clearly illustrates three kinds of sp^3 bondings, namely, [4s, $4p_x$, $4p_y$, $4p_z$]-[2s, $2p_x$, $2p_y$, $2p_z$], [4s, $4p_x$, $4p_y$, $4p_z$]-[2s, $2p_x$, $2p_y$, $2p_z$], and [4s, $4p_x$, $4p_y$, $4p_z$]-[2s, $2p_x$, $2p_y$, $2p_z$] in Ge−Ge, Ge−B/Ge−C/Ge−N, and B−B/C−C/N−N bonds. A strong correlation between the geometric symmetries and the single-/multi-orbital hybridizations are responsible for all the other fundamental properties. This viewpoint has been successfully utilized in the theoretical development of quasi-particle framework.

While the clear identifications of orbital hybridizations/spin configurations are finished for the different chemical bonds/atoms [14], the phenomenological models could be further established to explore the other quasiparticle properties, e.g., the diversified quantization phenomenon due to a uniform perpendicular magnetic field [97]. As to the tight-binding model, the reliable parameters might cover the hopping integrals [153], the ionization energies [154], the spin-orbital couplings [155]. and the spin-dependent on-site Coulomb interactions [156] under the boron-/carbon-/nitrogen-substitutions on germanene [14]. They are achieved from the well fitting with the low-lying valence and conduction subbands of VASP simulations. A unit cell, which is associated with the periodic vector potential (the vmagnetic Peierls phase; [63]), will become a giant Moire superlattice. Its area is inversely proportional to the strength of magnetic field. The generalized tight-binding model, being based on the layer-dependent sublattices, has been successfully developed for certain main-stream materials, such as, the diverse magneto-electronic properties in AA-/AB-/ABC-stacked bulk graphites [157−159], layered graphene systems [158], carbon nanotubes [160], and graphene nanoribbons [161]. Very interesting, its direct combinations with the modified random phase approximation, the static Kubo formula, and the dynamic

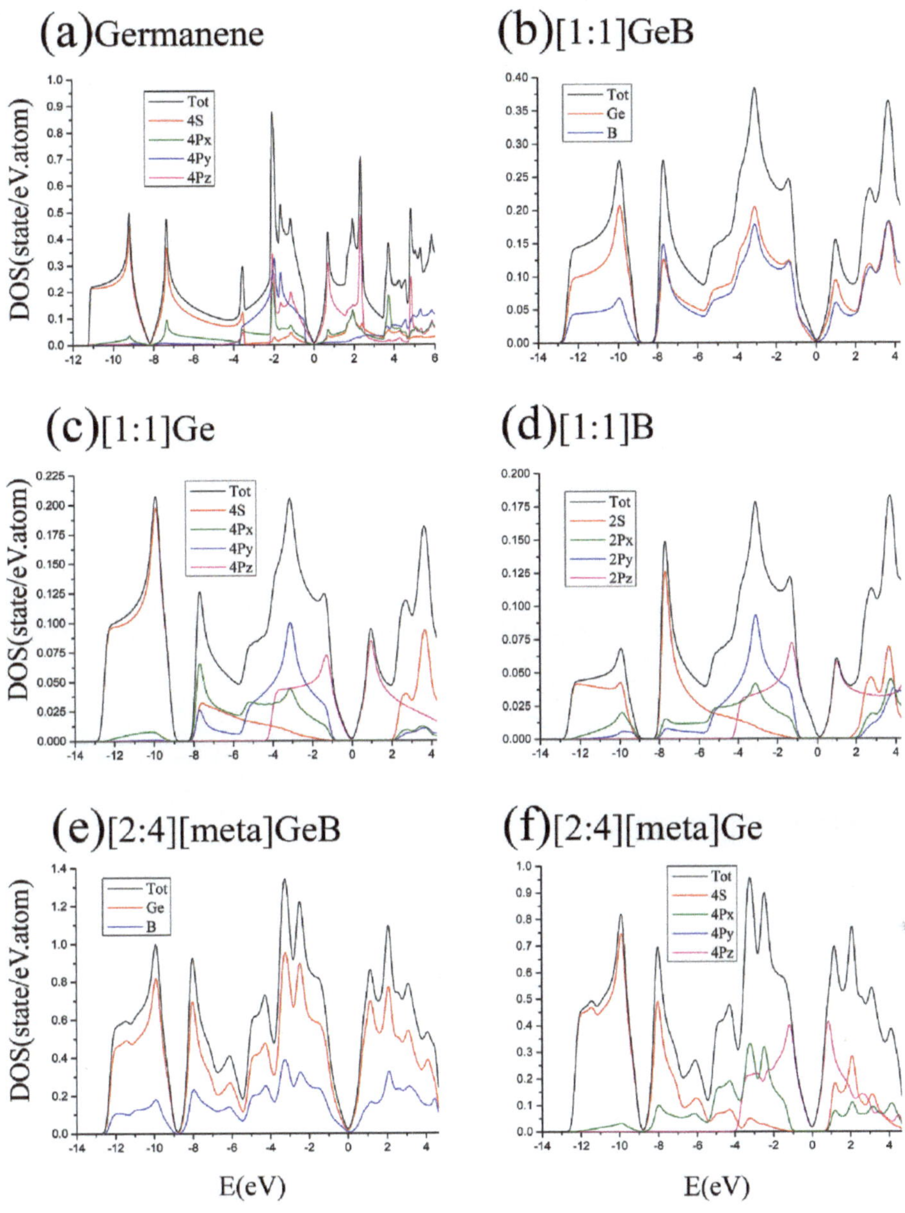

FIG. 6.22 (A–F) The energy-dependent density of states, as well as the atom/orbital/orbital decompositions, for the boron-substituted germanene materials.

one are, respectively, very useful in revealing the inter-Landau-level electron-hole excitations & magnetoplasmon modes [162], the unusual Hall effects [24], and the magneto-optical selection rules [23]. In short, the chemical substitutions are expected to greatly diversify the magnetic configurations.

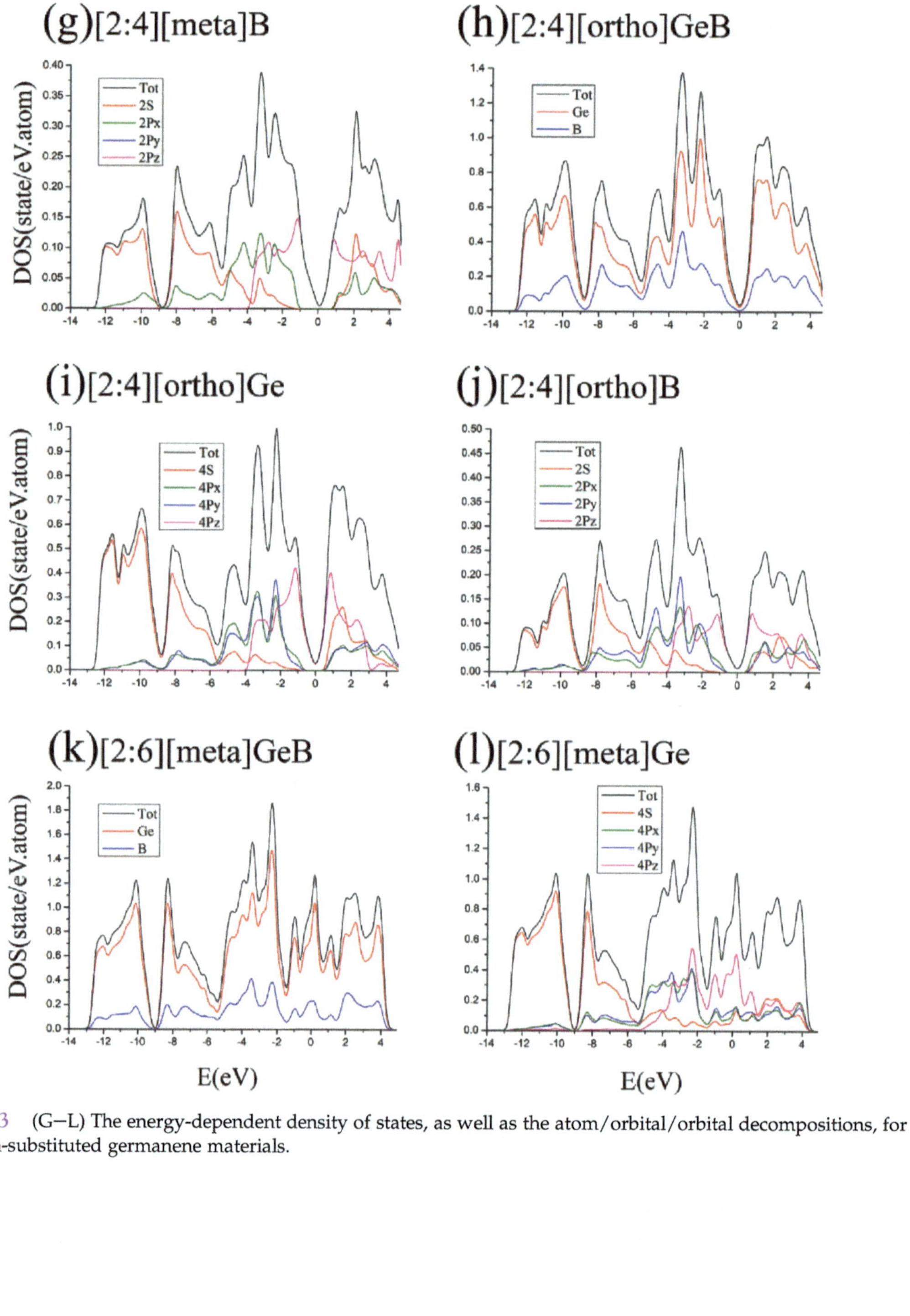

FIG. 6.23 (G–L) The energy-dependent density of states, as well as the atom/orbital/orbital decompositions, for the boron-substituted germanene materials.

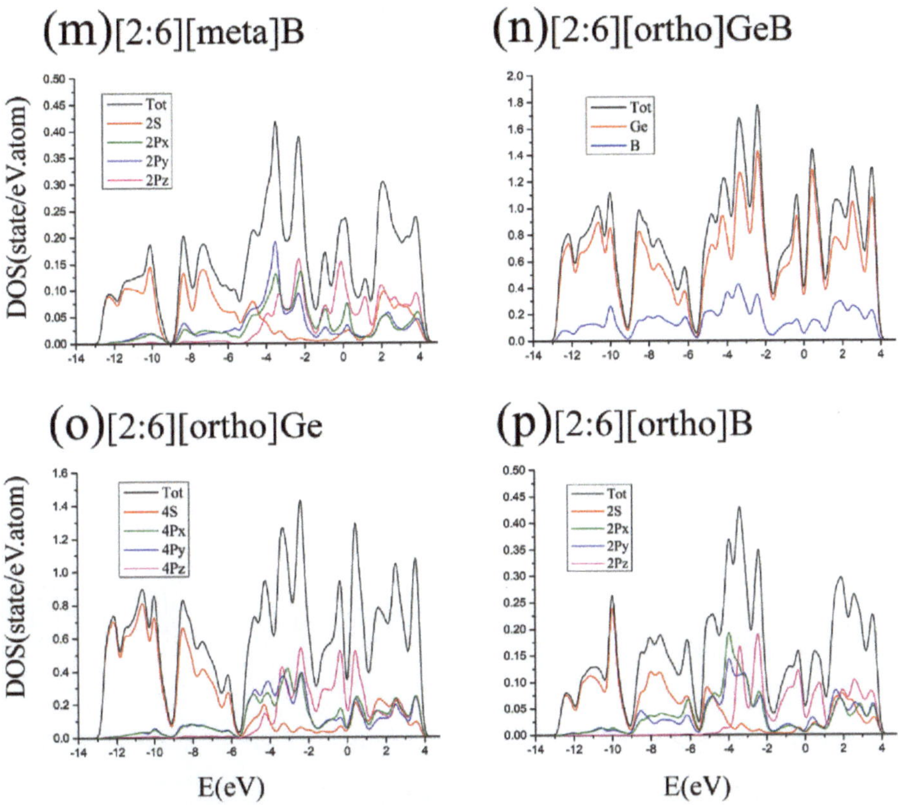

FIG. 6.24 (M–P) The energy-dependent density of states, as well as the atom/orbital/orbital decompositions, for the boron-substituted germanene materials.

6.5 Spin-featured magnetic properties

Certain nitrogen-substituted germanene systems exhibit the unusual ferromagnetic configurations, e.g., the [1:1] and [1:5] chemisorption cases in Figs. 6.32 and 6.33, respectively. However, a pristine system [14] and boron-/carbon-guest-atom substitutions [63] only present the non-magnetic quasiparticle behaviors. The specific nitrogen atoms might induce the spin density distribution, when the spin-dependent on-site Coulomb interactions (but not the spin-orbital couplings; [119]) could cooperate with the multi-orbital hybridizations of chemical bonds in greatly reducing the total ground state energy (details in Chapter 2 about the VASP simulations on magnetic many-particle interactions; [163]). Concerning the featured band structures and wave functions (Fig. 6.13 of [1: 1]/Fig. 6.14 of [1: 5]), both spin-up (the red curve) and spin-down (the black curve) energy subbands are split under the different occupation/non-occupation forms across the Fermi level ($E_F = 0$).

That is, the former and the latter, being nearest to E_F, are the occupied valence and unoccupied conduction states, respectively. Such systems belong to indirect small-gap semiconductors, in which the highest occupied spin-up state and the lowest unoccupied one possess the different wave vectors (the colored arrows). This is obviously responsible for

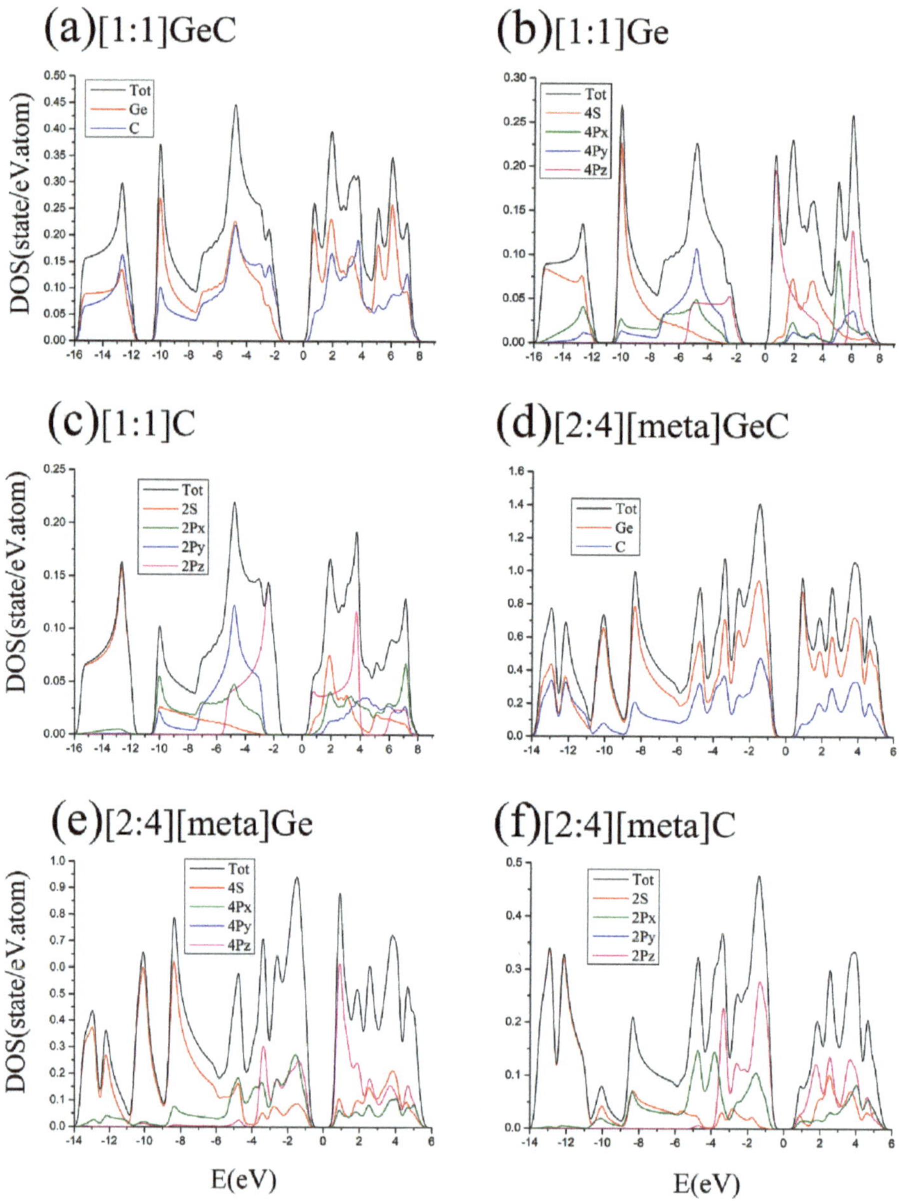

FIG. 6.25 (A–F) The energy-dependent density of states, as well as the atom/orbital/orbital decompositions, for the carbon-substituted germanene materials.

the net magnetic moments (Tables 6.4, 6.5A and 6.5B), the spin density distributions (Figs. 6.32 and 6.33), and spin-projected van Hove singularities (Figs. 6.28A and 6.31X). Very interesting, the spin-split energy spectra could be directly verified by the spin-polarized angle-resolved photoemission spectroscopy [164,165], in which ferromagnetic

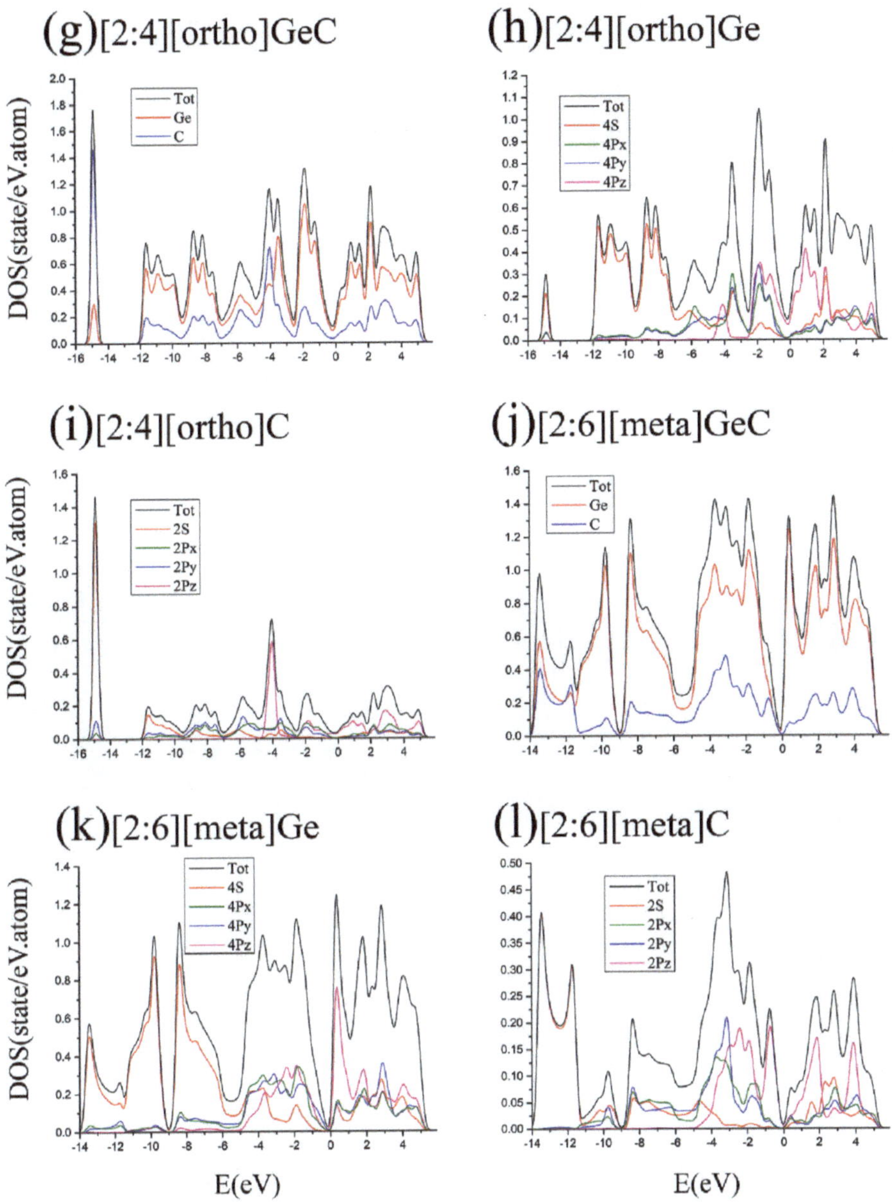

FIG. 6.26 (G–L) The energy-dependent density of states, as well as the atom/orbital/orbital decompositions, for the carbon-substituted germanene materials.

nanoscale probe is able to select the same-spin scattered electrons. Such observations are expected to be reachable under the concise orbital hybridizations and valence subbands below the Fermi level. The close cooperation between the delicate VASP predictions and SP-ARPES measurements can overcome this open issue and reveal the unusual magnetic phenomena.

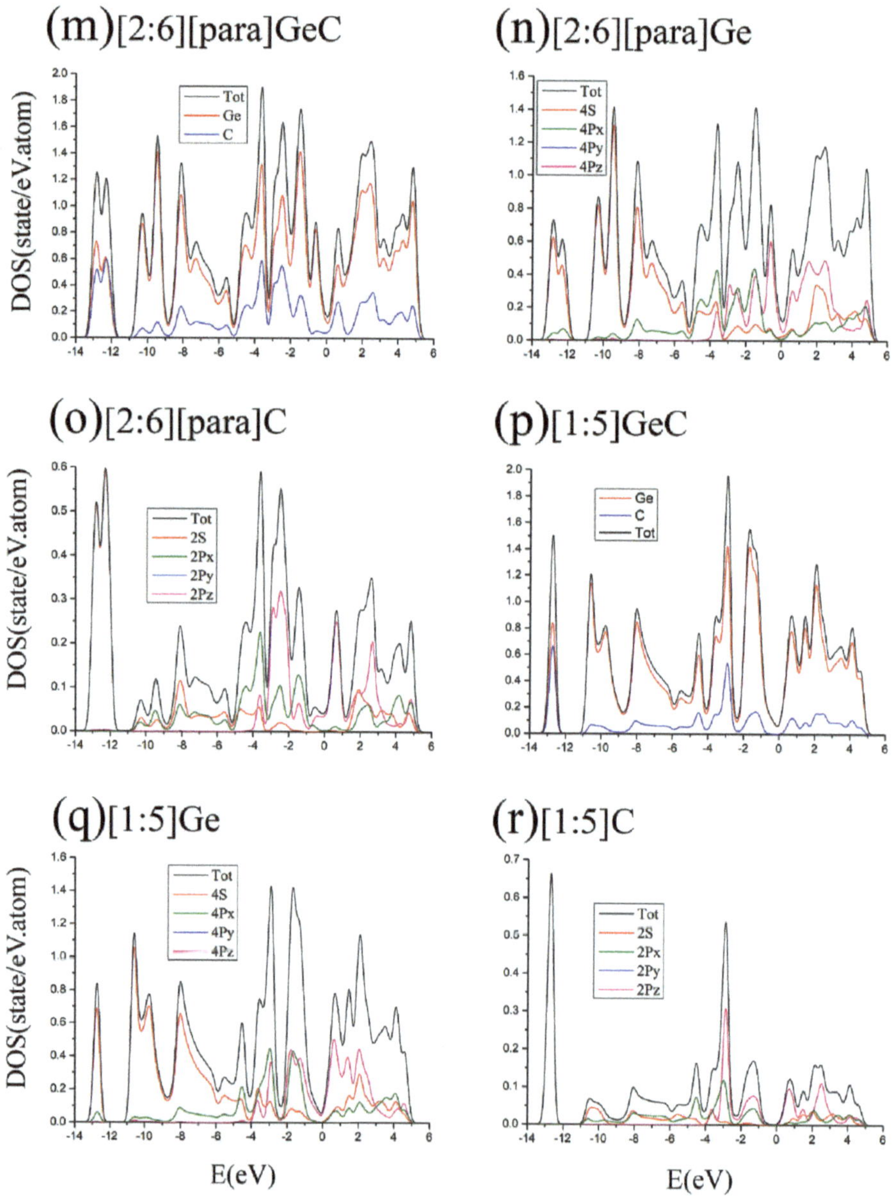

FIG. 6.27 (M−R) The energy-dependent density of states, as well as the atom/orbital/orbital decompositions, for the carbon-substituted germanene materials.

The net magnetic momenta, as clearly illustrated in Tables 6.4, 6.5A and 6.5B, are very sensitive to the guest and host atoms, as well as their active four orbitals of [2s, $2p_x$, $2p_y$, $2p_z$] & [4s, $4p_x$, $4p_y$, $4p_z$]. Their magnitudes are closely associated with the wave-vector-, atom- and orbital-decomposed Bloch wave functions. Apparently, the saturated binary compounds

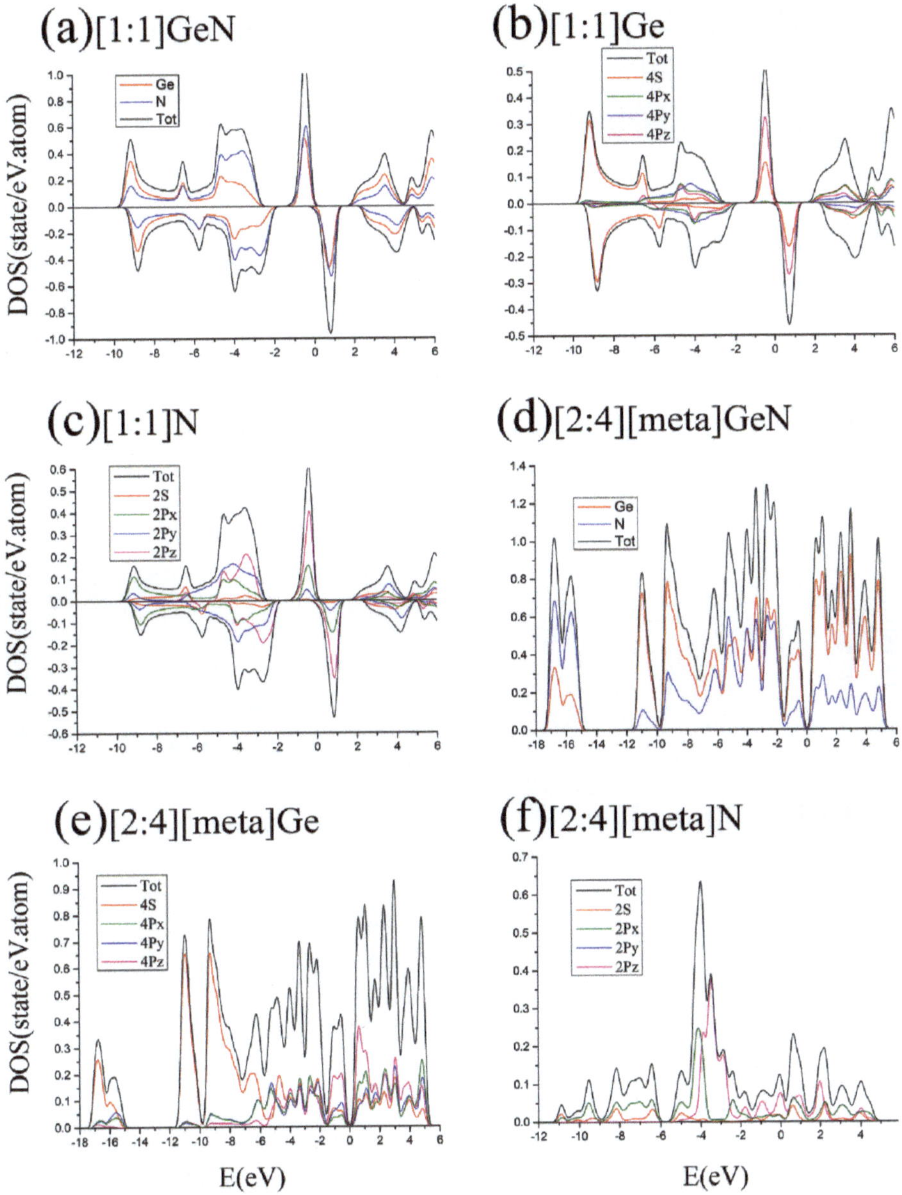

FIG. 6.28 (A–F) The energy-dependent density of states, as well as the atom/orbital/orbital decompositions, for the nitrogen-substituted germanene materials.

present the higher- and lower for the guest (green triangles for Ns) and host atoms (blue circles for Ges). The former can drive the ferromagnetic spin arrangement; furthermore, the relative ratio is about 1.2 (Table 6.4). The total magnetic moment is ~0.490 μ_B per unitcell. This value is greatly reduced to ~0.108 μ_B under the [1: 5] case, being roughly consistent with to

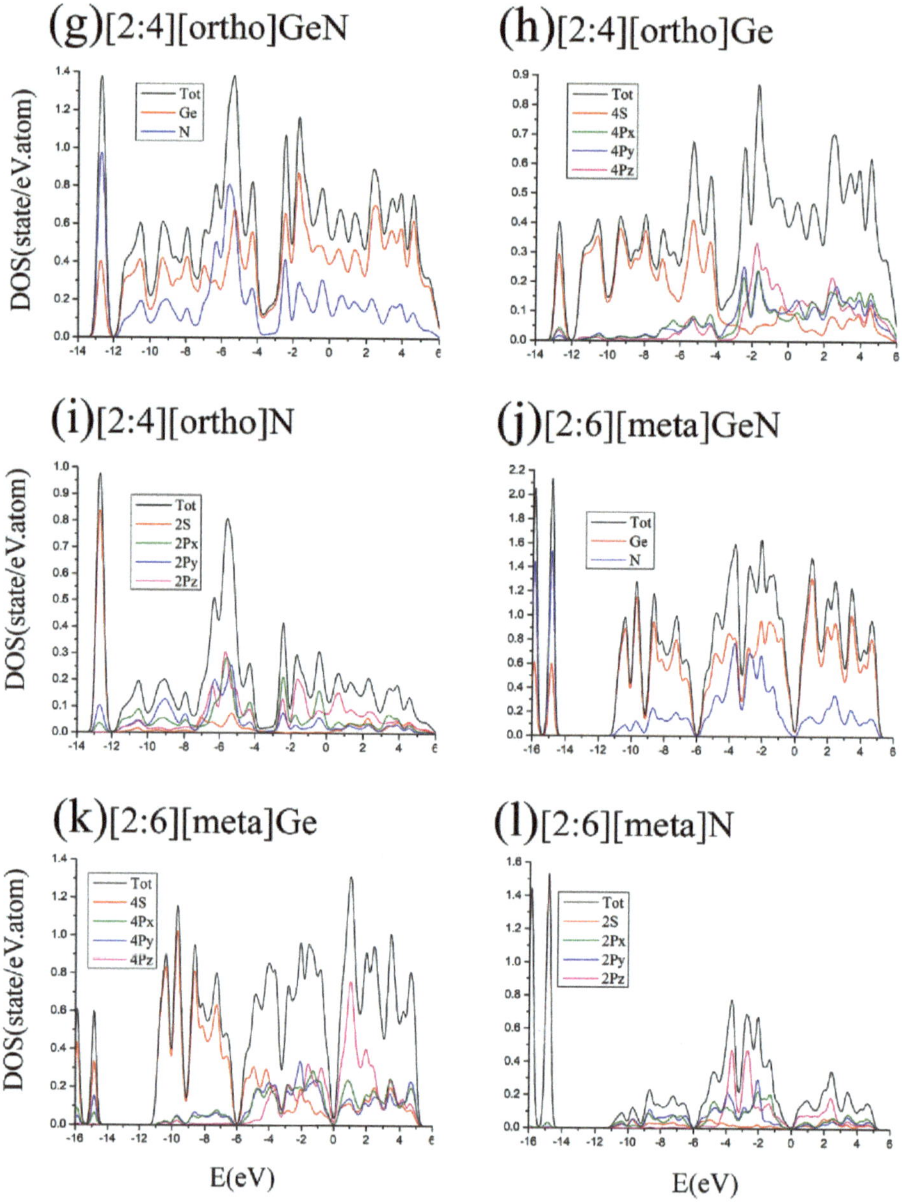

FIG. 6.29 (G–L) The energy-dependent density of states, as well as the atom/orbital/orbital decompositions, for the nitrogen-substituted germanene materials.

an obvious decrease of guest-atom concentration (Tables 6.5A and 6.5B). The predicted results should be sufficiently large in the high-resolution magnetic examinations, e.g., the magnetic detections by the NMR observations [166,167]. Each atom contribution magnetism mainly determined by the different orbitals. Apparently, the orbital-dependences are highly

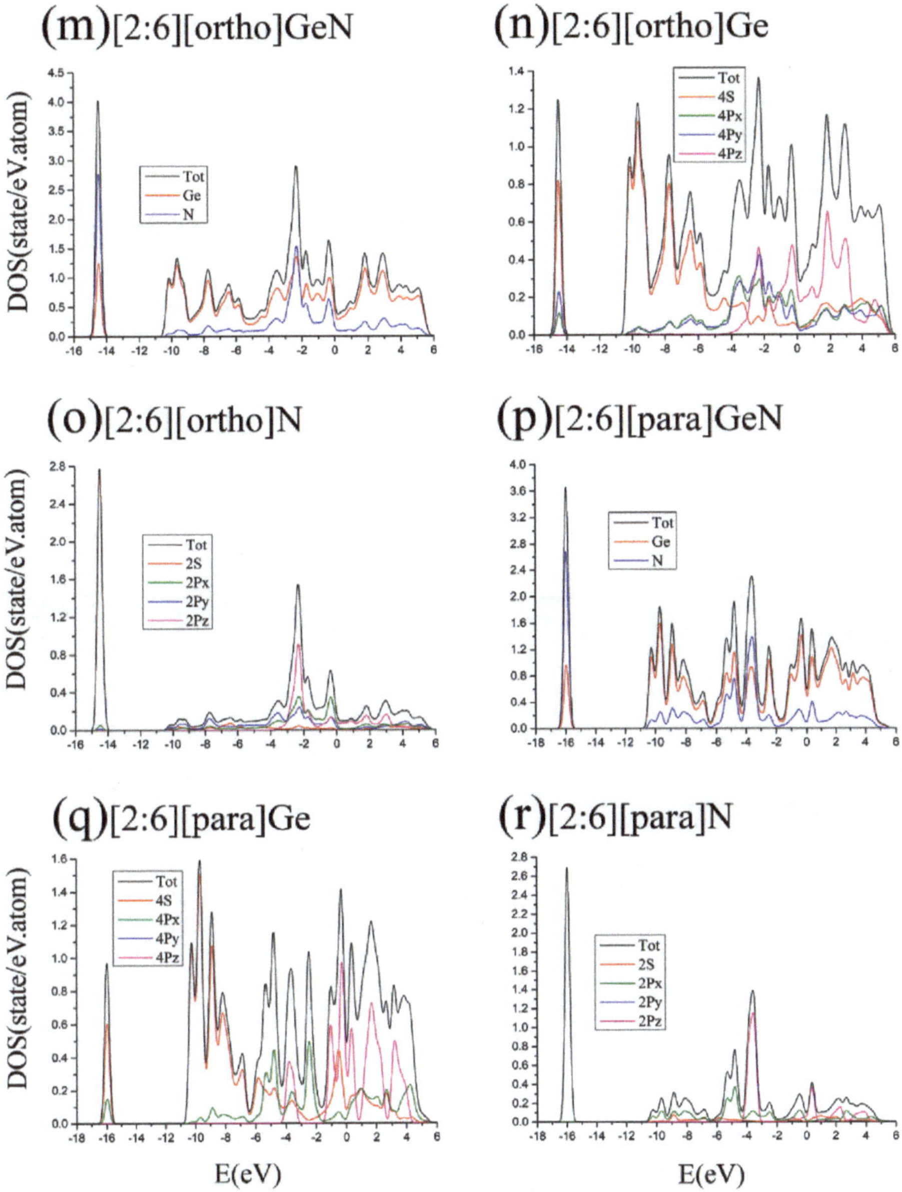

FIG. 6.30 (M–R) The energy-dependent density of states, as well as the atom/orbital/orbital decompositions, for the nitrogen-substituted germanene materials.

non-homogeneous for 4s, $4p_x+4p_y$ and $4p_z$ orbital 2s, $2p_x+2p_y$ and $2p_z$. This directly reflects the rather strong cooperation or competition between the spin-induced Coulomb interactions with the multi-orbital orbital bondings, leadings to the lower ground state energies. In general, the third and second ones have the highest and lowest (even negligible) contributions to

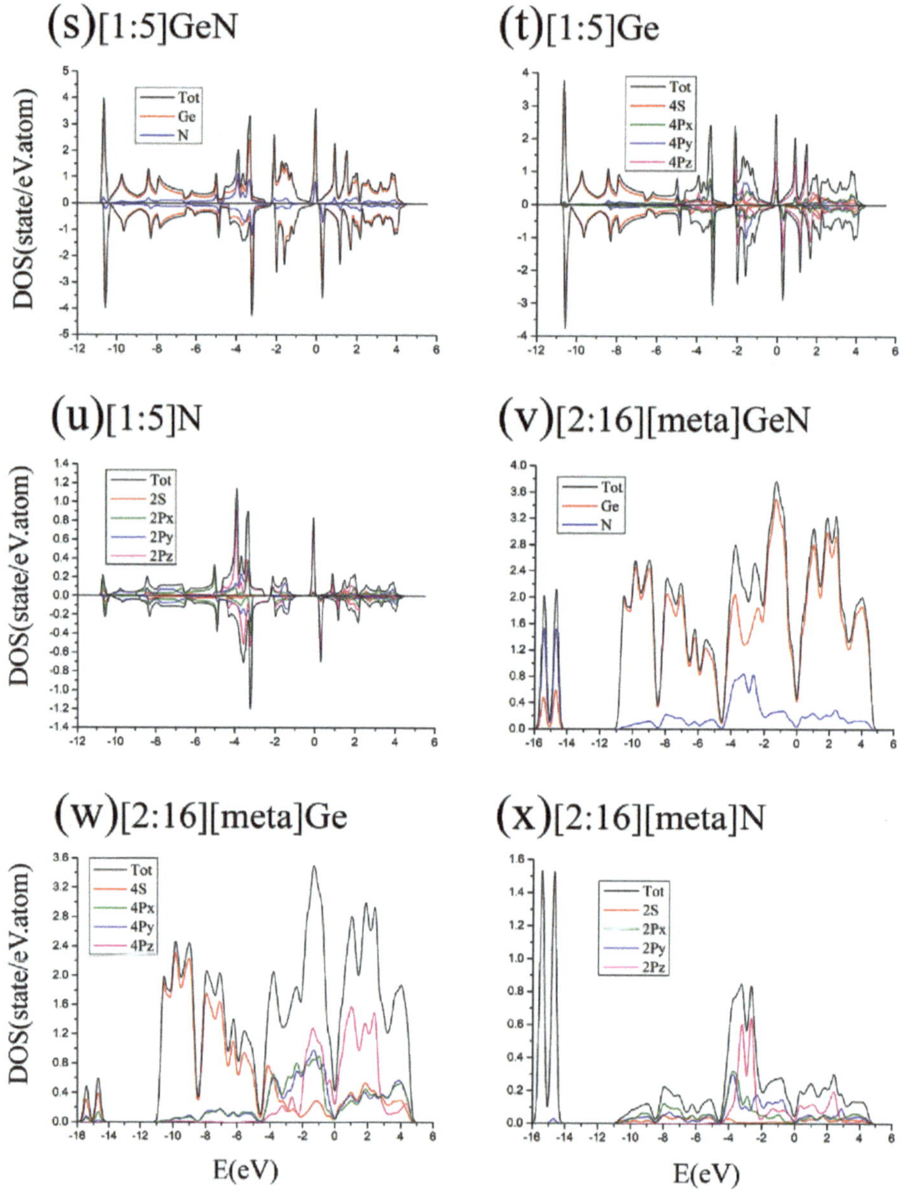

FIG. 6.31 (S–X) The energy-dependent density of states, as well as the atom/orbital/orbital decompositions, for the nitrogen-substituted germanene materials.

ferromagnetic configurations. It should be noticed that the five germanium host atoms in a unit cell (Fig. 6.14G) make the distinct contributions. This clearly indicate a highly anisotropic physical environment. The spin-induced bi-particle interactions will exhibit the complicated form through the Hubbard-like model [163]. This becomes a challenge in the current development of phenomenological models.

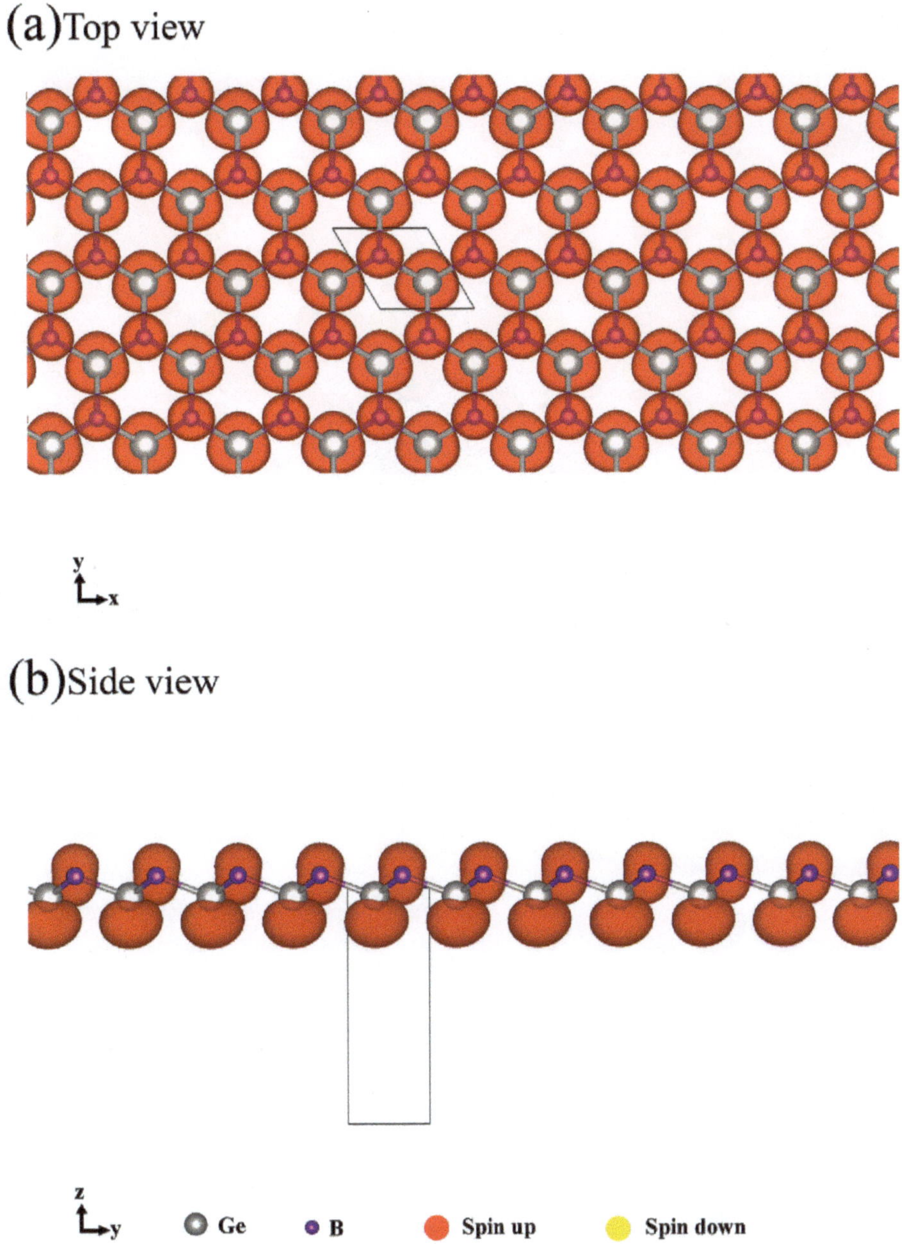

FIG. 6.32 (A and B) The spin-charge distributions in nitrogen-substituted silicene systems for [1:1] cases: Showing top and side views. Red and yellow correspond to spinup and spin-down.

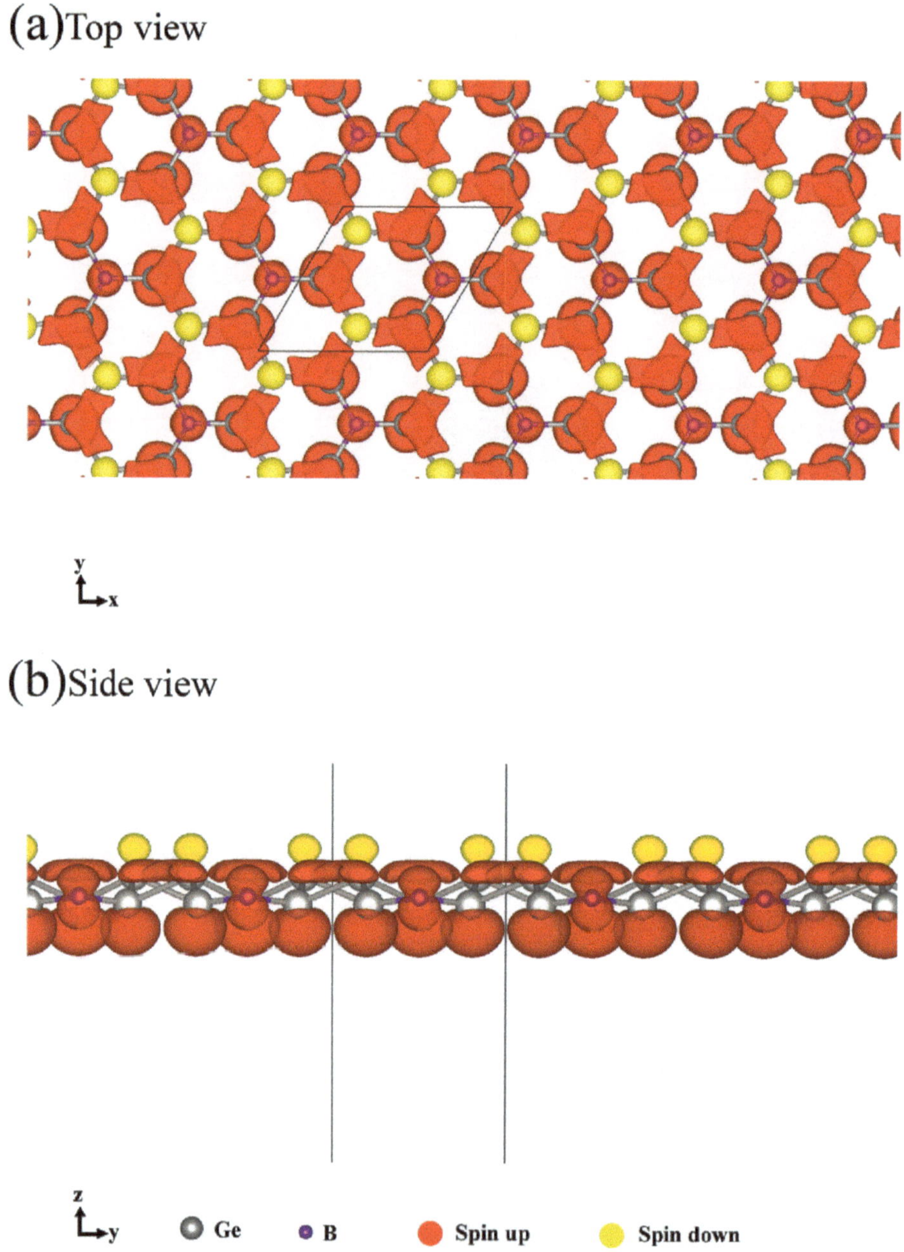

FIG. 6.33 (A and B) The spin-charge distributions in nitrogen-substituted silicene systems for [1:5] cases: Showing top and side views. Red and yellow correspond to spinup and spin-down.

TABLE 6.4 The great fluctuations of bond lengths under the specific substitution cases.

Substituted atom X	Geometric structure	X–X Bond length (Å)	X–Ge Bond length (Å)	Ge–Ge distance (Å)	Δ (Å)
X = B	[2:16] meta	1.650	2.054	2.401–2.471	2.484
X = C	[2:6]meta	–	1.910	2.470–2.507	2.909
X = N	[2:4]meta	–	1.851–2.018	2.423–2.839	2.783

TABLE 6.5A The spin-charge distributions in nitrogen-substituted gemanene systems for [1:1].

Atom	Orbital	x + y direction	z direction	Total
Ge	S	0.001	0.064	0.065
	P	0	0.156	0.156
	Total(Ge)	0.001	0.220	0.221
N	S	0	0.006	0.006
	P	0.002	0.261	0.263
	Total(N)	0.002	0.267	0.269
Total(Ge)+Total(N)		0.003	0.487	0.490

The highly non-uniform/anisotropic spin density distributions, as clearly in Fig. 6.32A and B (the top and side view) and Fig. 6.33A and B, respectively, for the [1: 1] and [1: 5] substitution cases. Such unusual ferromagnetic configurations are full supported by the spatial charge distributions (Figs. 6.19A,B and 6.21Q,R). The main features cover the atom-, orbital-, position- and concen-tration-dependent. A whole crystal only shows the spin-up periodical distribution, while that of the spin-down one (anti-ferromagnetism; [168]) is thoroughly absent. Guest-nitrogen and host-germanium atoms make the neighboring ferromagnetism [79] for the N–Ge and Ge–Ge bonds. The 3D volume, being covered by each atom, have the strong weight along the z-direction, compared with the transverse (x, y) plane. This is consistent with the orbital-projected net magnetic moments (Tables 6.5A and 6.5B). The magnetic behaviors are very sensitive to the relative positions between guest and host atoms. For example, five germanium atoms in a unit of the [1:5] Moire superlattice (Figs. 6.32A and B) exhibit the different structures and volumes within the closed surfaces. The magnetic moments become smaller for the longer distances (Tables 6.5A and 6.5B). Apparently, the reduced guest-atom concentrations are able to create more dilute and complicated spin density distributions. This will induce much trouble in developing the phenomenological models, e.g., the establishment of the Hubbard model [169]. Whether

TABLE 6.5B The spin-charge distributions in nitrogen-substituted gemanene systems for [1:5].

Atom	Orbital	x + y direction	z direction	Total
Ge1	S	0	−0.004	−0.004
	P	−0.001	0.035	0.034
	Total(Ge1)	−0.001	0.031	0.030
Ge2	S	0	0.017	0.017
	P	−0.002	0.083	0.081
	Total(Ge2)	−0.002	0.100	0.098
Ge3	S	0	0.026	0.026
	P	−0.002	0.084	0.082
	Total(Ge3)	−0.002	0.110	0.108
Ge4	S	−0.001	0.024	0.023
	P	0.002	0.074	0.076
	Total(Ge4)	0.001	0.098	0.099
Ge5	S	0	−0.004	−0.004
	P	0.002	0.035	0.037
	Total(Ge5)	0.002	0.031	0.033
N	S	0	0.010	0.010
	P	0	0.108	0.108
	Total(N)	0	0.118	0.118
Total(Ge1+Ge2+Ge3+Ge4+Ge5)		−0.003	0.371	0.368
Total(Ge1+Ge2+Ge3+Ge4+Ge5+N)		−0.003	0.489	0.486

the VASP predictions of ferromagnetic spin configurations, with enough large magnetic moments could be verified by the spin-polarized scanning tunneling microscopy (SP-STM; [170−172]) requires the close cooperation of the theoretical [79] and experimental researches [173,174] under a unified quasiparticle framework [141].

The spin-decomposed density of states can identify the energy dependence of the ferromagnetic/anti-ferromagnetic/non-magnetic properties. In general, the low-lying valence and conduction subbands exhibit the prominent splitting of spin-up and spin-down states, e.g., those of the [1: 1] (Fig. 6.13A) and [1: 5] cases (Fig. 6.14G). Their critical points (the band-edge states) possess the vanishing group velocities and thus the singular van Hove singularities. Very interesting, the prominent structures below and above the Fermi level are quite different from each other. This clearly means an unequal probability in making the effective and zero moments, or it accounts for a ferromagnetic moment. Only a narrow band gap is widened by a specific broadening factor of ~20 meV in the VASP calculations. The spin-split prominent structures across the Fermi level could be verified by the spin-polarized scanning tunneling spectroscopy (SP-STS; [170–172]), in which a ferromagnetic nanoscale probe could effectively select the spin-up- or spin-down-related quantum current. It should notice that the multi-orbital hybridizations are similar in the presence and absence of ferromagnetic configurations (details in Chapter 6; [14,88]). How to build a reliable Hubbard-like tight-binding model is one of the near-future research focuses. In short, there exist three kinds of Hamiltonians, revealing the zero spin-dependent interactions [175], the significant spin-orbital couplings [119], and the prominent on-site Coulomb interactions [176]. This is worthy of systematic investigations under a unified quasiparticle framework, as done in the previous [141] a current book.

6.6 Similarities and differences among boron-, carbon- and nitrogen-substitution germanene, silicene and graphene compounds

There are important similarities and differences for the germanene-, silicene- and graphene-related binary compounds. Boron-, carbon- and nitrogen-substitutions are outstanding candidates in illustrating the diverse quasiparticle properties. The chemical modifications have clearly shown the position- and atom-dependent chemical environment and thus the different orbital hybridizations with the various strengths. Three types of group-IV compounds might exhibit the planar or buckled honeycomb crystals, the creation of valence holes, enlarged band gaps & conduction electrons, the almost or perpendicular π & σ bondings or sp^3 ones, and the merged van Hove singularities of [s, p_x, p_y] or [s, p_x, p_y, p_z] orbitals. The similar quasiparticle phenomena are very helping in establishing the reliable phenomenological models, e.g., the tight-binding model [177] and the generalized tight-binding model [178], especially for the magnetic quantization greatly enriched by the chemical substitutions. This is under the current investigations.

It is well known that the pristine germanene [14], silicene [88], and graphene [4] quite differ from one another in lattice constants and side-view crystal structures. The first system possesses the longest bond lengths and the largest buckling, while the third one has the shortest ones in the absence of any deformation. These clearly indicate the weakest covalent strengths of Ge–Ge bonds, compared with those of Si–Si and C–C ones. The obvious/vanishing height differences of A and B two sublattices illustrate the non-orthogonal/perpendicular relations between the sp^2 and sp^3 bondings, corresponding to germanene & silicene/graphene. The four-orbital hybridizations should play the most important roles for the germanene-related buckled compounds/pristine materials [14]. The geometric symmetries are responsible for the other quasiparticle properties, since they are closely related to the various orbital mixings and the spin-dependent. Specifically, the spin-orbital coupling has created a small band gap of E_g^d ~45 meV/5 meV in germanene/silicene at the K valley [179,180]. The main features of electronic properties cover the narrowest/middle/widest π- & σ-band widths, and the obvious/observable/vanishing π-σ subband anti-crossings in a pure germanene/silicene/graphene. Very interesting, the *p*-type metals, the gap-enlarged semiconductor or semimetals, and the *n*-type dopings or the semiconducting behaviors, are respectively, revealed in B, C, and N-substitution cases. Apparently, free carrier densities of valence holes/conduction electrons and band gaps are very sensitive to host atom. The critical mechanisms lie in the various orbital hybridizations of the distinct chemical bonds. For example, the buckled binary germanene/silicene/graphene compounds exhibit the sp^3 bondigs of [4s, $4p_x$, $4p_y$, $4p_z$]-[2s, $2p_x$, $2p_y$, $2p_z$]/[3s, $3p_x$, $3p_y$, $3p_z$]-[2s, $2p_x$, $2p_y$, $2p_z$]/[2s, $2p_x$, $2p_y$, $2p_z$]-[2s, $2p_x$, $2p_y$, $2p_z$] through the boron substitutions.

Both orbital hybridizations and spin configurations are greatly diversified through the distribution and concentration of host/quest atoms, the uniform/non-uniform chemical environments, as well as, the buckled/non-buckled & deformed/uniform honeycomb crystals. All pristine and binary substitution compounds are worthy of the systematic investigations from the first-principles simulations and the phenomenological models, being fully consistent with the quasiparticle frameworks in the previously published books [181], this book, and near-future ones [141]. The various hopping integrals, on-site Coulomb potentials and spin-dependent interactions (the single-particle spin-orbital couplings and the on-site Coulomb interactions of opposite spins; [119]), which are very useful parameters in the tight-binding model [182], need to be achieved from the rather detailed fittings with the VASP calculations about the low-lying valence and conduction bands. Very interesting, the first ones mainly arise from the pure and composite orbital hybridizations in the chemical bonds of the same and different neighboring atoms, respectively. Furthermore, they are further classified the non-orthogonal or orthogonal π & σ bondings for germanene/silicene or graphene, the non-hybridized and/or hybridized π/σ bondings in Ge—Ge/B—B/C—C/N—N bonds and/or Be—B/Ge—C/Ge—N ones for the non-buckled binary germanium compounds, and the impure sp^3 hybridizations for the buckled systems. Also, such critical mechanisms strongly depend on the occupied orbital numbers of distinct an host and guest atoms. The featured orbital mixings are expected to create the diverse quasiparticle phenomena. For example, the high substitution cases in this chapter are very suitable in fully exploring the magnetic quantization by developing the generalized tight-binding model (details in Ref. [141]). The main features, which are under a perpendicular uniform magnetic field (B**z**, might include the various Moire superlattices, the large crystal periods due to the vector potential), the well definitions of magnetic quantum numbers ($n^{c,v}$; the zero-point numbers of localized and oscillatory probability functions on distinct sublattices), the different Landau-level subgroups, the sensitive dependences of magnetic energy spectra on B and $n^{c,v}$, and the non-crossing/crossing/anti-crossing behaviors during the variation of the magnetic field strength. The magneto-electronic properties are available in the further investigations of magneto-plasmon modes, magneto-optical selection rules [23], and quantum Hall transports [183]. The direct combinations of the geometric symmetries, the intrinsic interactions, the chemical modifications, and the external magnetic/electric fields are able to provide very rich platforms in basic and applied science researches [184,185].

When the concentrations of guest atoms are very dilute, the semiconducting substitutions would become very interesting charged impurities [186]. It is well known that 3D diamond crystals of carbon/silicon/germanium atoms [6,7,9], with large band gaps, clearly show the acceptor- and donor-type charged impurity levels under the B/Al/Ga/In/Tl and N/P/As/Sb/Bi substitution perturbations. The negative and positive ion cores, respectively, create the attractive and repulsive Coulomb fields for an electronic state. This could be regarded as a hydrogen-like topic, while the effective mass (m^*; [187,188]) due to the crystal potential and the static dielectric functions (ε_0; [187,188]) associated with all valence carriers need to be taken into account for the kinetic energy and the effective Coulomb potential, respectively. Bohr radius and 1s bound-state energy are, respectively, proportional to $\varepsilon_0 m^*$ and $\varepsilon_0 m^{*2}$ [187]. It should be noticed that the latter is measured from the top and bottom of valence and conduction subbands [89]. The similar studies could be generalized to the semiconducting materials of germanene and silicene with observable direct band gaps. That is, the reasonable m^* and ε_0 in 2D Schrodinger equation play the critical roles in determining the featured impurity behavior. Such topic is under the current studies.

6.7 Concise conclusions

Boron-, carbon- and nitrogen-substituted germanene systems show the rich and unique essential properties, being very helpful in promoting the theoretical development of quasiparticle framework [189] and the potential applications [190]. The various orbital hybridizations and spin-dependent magnetic distributions, which are driven by the distinct and/or same atomic configurations, as well as, the concentration and arrangements, are responsible for all quasiparticle phenomena [123,191]. The critical mechanisms are clearly identified from the position-dependent chemical bonds, the atom-dominated energy spectra & wave functions, the spatial charge densities & their variations after the chemical substitutions, and the atom-, orbital- and spin-decomposed density of states. Among of them. the merged structures of van Hove singularities within the specific energy ranges could serve as the most important criteria, being fully supported by the other relevant physical/chemical quantities. Such concise pictures are further available in developing the phenomenological models (discussed earlier), such as, the tight-binding model/the generalized tight-binding model [120,192], the modified random-phase approximation/the corrected self-energy method [29,193], and the static/dynamic Kubo formulas [69,188]. The similar viewpoints are suitable for silicene- [194] and graphene-related [195] chemical substitutions, mainly owing to certain significant differences.

The main features of quasiparticle phenomena include the position-dependent non-uniform chemical bonds through the different substitutions [196], the diverse energy subbands & wave functions (semiconductors or metals, direct or indirect band gaps, asymmetric hole and electron bands about the Fermi level, linear/parabolic/oscillatory/partially flat dispersion relations, critical points, existence or absence of the π & σ subbands, non-crossing/crossing/anti-crossing behaviors, and atom dominances at different energy ranges; [14]), the diversified orbital hybridizations (the sp^2 or sp^3 bondings from top and side views; [88]), and various van Hove singularities arising from different band-edge states [14,89]. Through the delicate analyses of all consistent quantities, the specific orbital mixings are deduced to be the dominating π & σ bondings/the minor sp^3 ones in a pristine germanene, the [4s, $4p_x$, $4p_y$, $4p_z$]-[4s, $4p_x$, $4p_y$, $4p_z$] & [4s, $4p_x$, $4p_y$, $4p_z$]-[4s, $4p_x$, $4p_y$, $4p_z$] orbital hybridizations, respectively, in Ge–Ge & Ge–B/Ge–C/Ge–N bonds for buckled binary compounds, and the almost perpendicular to each other of π & σ bondings in non-buckled ones. Also noticed that the orbital-dependent on-site Coulomb potentials and hopping integrals [120] are quite different from one another among these complicated orbital mixings through the distinct atomic configurations. Such clear evidences are very useful in the reliable establishment of the tight-binding model/the generalized tight-binding model [63,197], as well as, the other phenomenological models [198]. The high-precision measurements under STM/TEM, ARPES and STS (details in Chapter 3), are required to verify the first-principles simulations on the crystal structures, occupied electronic states, and density of states across E_F, respectively.

Both chemical substitutions and absorptions, which are able to dramatically transform all quasiparticle behaviors, have strong effects even at very dilute concentrations of guest atoms. They will exhibit the stable/quasi-stable bound states in a semiconducting material [199], or make a significant contribution to the residual resistivity of a metallic system [200]. Pristine germanene and silicene systems are direrct-gap semiconductors, while graphene and tinene/plubmene, respectively have a zero gap [179] and the multi-orbital-dominated band properties near the Fermi level [14]. Only the former exhibit the hydrogen-like charged impurities after few substitutions/adsorptions of acceptor-type B/Al/Ga/In/Tl [201–205] or donor-type N/P/As/Sb/Bi [206–210], respectively, leadings to the discrete states above the highest valence-band top or below the lowest conduction-band bottom [11]. The 2D Schrodinger equation, being modified by the effective mass and the static dielectric screening, is available in exploring the featured impurity behaviors. As for transport properties due to charged impurities, the Boltzmann equation [211], being assisted by the Born scattering approximation [212], is suitable in evaluating the residual resistivity. For example, this is successful in comprehending the rich transport behaviors of free conduction electrons for graphite-alkali intercalation compounds [213,214]. The similar researches could be generalized to 2D metallic systems, e.g., the impurity-concentration-enriched residual transports for metallic group-IV materials.

Very interesting, boron-, carbon- and nitrogen-substituted germanene systems exhibit the diverse orbital hybridizations and spin configurations, as thoroughly examined from the first-principles simulations [14] for the highly non-uniform chemical bonds, the atom- & spin-dominated band structures within the specific energy ranges, the spatial charge/spin densities & their variations after the chemical substitutions, and the atom-, orbital- & spin-projected density of states. The featured quasiparticle properties cover the position-dependent chemical bonds under the various substitution cases [14], the diversified energy spectra & wave functions (in terms of semiconductors/metals, direct/indirect band gap, asymmetry of hole and electron bands, distinct energy dispersions, more band-edge states, zone-folding effect, existence/absence of π & σ bands, non-crossing/crossing/anti-crossing behaviors; [89]), the diversified orbital hybri-dizations (the sp^2 or sp^3 ones from top and side views; [14]), and various van Hove singularities due to different critical points [88]. Through the detailed analyses of all relevant quantities, the specific orbital hybridizations are the dominant π & σ bondings and the minor sp^3 ones in a pristine monolayer germanene, the [4s, $4p_x$, $4p_y$, $4p_z$]-[4s, $4p_x$, $4p_y$, $4p_z$] and [4s, $4p_x$, $4p_y$, $4p_z$]-[2s, $2p_x$, $2p_y$, $2p_z$] mixings, respectively, in Ge–Ge, Ge–B/Ge–C/Ge–N and B–B/C–C/N–N bonds of buckled binary compounds, as well as, the almost perpendicular π & σ bondings for non-buckled systems. Also noticed that the hopping integrals and the one-site Coulomb potentials very differ from one another among these complex orbital mixings under the distinct atomic configurations. The high-precision measurements of STM/TEM, ARPES and STS (details n Chapter 3), are necessary in, respectively, examining the theoretical predictions about the optimal geometries, occupied energy bands, and density of states across the Fermi level.

The concise pictures are achieved by the merged van Hove singularities [14], being consistent with the relevant quantities. Such viewpoints are rather helpful in the reliable establishment of the phenomenological models, e.g., the suitable parameters in the tight-binding model/the generalized tight-binding model for the featured electronic/magneto-electronic properties [118,215]. These interesting topics are under the systematic investigations. The chemical substitution cases are reduced to charged impurities under very dilute guest-atom concentrations [216]. This corresponds to a 2D hydrogen-like Schrodinger equation [216,217], with the significant effects of an effective mass and a static dielectric screening [187,188].

References

[1] Sahin H, Cahangirov S, Topsakal M, Bekaroglu E, Akturk E, Senger RT, Ciraci S. Monolayer honeycomb structures of group IV elements and III-V binary compounds. Phys Rev B 2018;80:155453.

[2] Özdamar B, et al. Structural, vibrational, and electronic properties of single-layer hexagonal crystals of group IV and V elements. Phys Rev B 2018;98:045431.

[3] Novoselov KS, Geim AK, Morozov S, Jiang D, Katsnelson MI, Grigorieva I, Dubonos SV, Firsov AA. Two-dimensional gas of massless Dirac fermions in graphene. Nature 2005;438:197.

[4] Zhu M, Diao G. Review on the progress in synthesis and application of magnetic carbon nanocomposites. Nanoscale 2011;3:2748.

[5] Williams OA. Nanocrystalline diamond. Diam Relat Mater 2011;20:621–40.

[6] Moon JH, Yang S, Dong W, Perry JW, Adibi A, Yang SM. Core–shell diamond-like silicon photonic crystals from 3D polymer templates created by holographic lithography. Opt Express 2006;14:6297–302.

[7] Newman R, Tyler WW. Effect of impurities on free-hole infrared absorption in p-type germanium. Phys Rev 1957;105:885.

[8] Vunk F, DeMonte M, Smith RW. The composition dependence of the grey tin → white tin transition in dilute tin-germanium alloys. Mat Lett 1983;2:67–70.

[9] Shiryaev AA, Frost DJ, Langenhorst F. Impurity diffusion and microstructure in diamonds deformed at high pressures and temperatures. Diam Relat Mater 2007;16:503–11.

[10] Bassani F, Yoshimine M. Electronic band structure of group IV elements and of III-V compounds. Phys Rev 1963;130:20.

[11] Eagles D. Optical absorption and recombination radiation in semiconductors due to transitions between hydrogen-like acceptor impurity levels and the conduction band. J Phys Chem Solid 1960;16:76–83.

[12] Jones RL, Fisher P. Excitation spectra of group III impurities in germanium. J Phys Chem Solid 1965;26:1125–31.

[13] Spitzer WG, Trumbore FA, Logan RA. Properties of heavily doped n-type germanium. J Appl Phys 1961;32:1822.

[14] Pham HD, et al. Rich essential properties of boron, carbon, and nitrogen substituted germanenes. Appl Phys Express 2020;13:085502.

[15] Semenova AA, et al. 2019 Periodic table of elements and nanotechnology. Mendeleev Commun 2019;29:479–85.

[16] Chen L-F, et al. Synthesis of nitrogen-doped porous carbon nanofibers as an efficient electrode material for supercapacitors. ACS Nano 2012;6:7092–102.

[17] Zhang H-F, Dohnalkova AC, Wang C-M, Young JS, Buck EC, Wang L-S. Lithium-assisted self-assembly of aluminum carbide nanowires and nanoribbons. Am Chem Soc 2001;2:105–8.

[18] Ho KI, et al. Fluorinated graphene as high performance dielectric materials and the applications for graphene nanoelectronics. Sci Rep 2014;4:5893.

[19] Zhang X, Rajaraman BRS, Liu H, Ramakrishna S. Graphene's potential in materials science and engineering. RSC Adv 2014;4:28987.

[20] Vishwakarma R, et al. Transfer free graphene growth on SiO_2 substrate at 250°C. Sci Rep 2017;7:43756.

[21] Pham KD, Nguyen CV. First principles calculations of geometric structures and electronic properties of van der Waals heterostructure based on graphene, hexagonal boron nitride and molybdenum diselenide. Diam Relat Mater 2018;88:151.
[22] Xu Y, et al. Large-gap quantum spin Hall insulators in tin films. Phys Rev Lett 2013;111:136804.
[23] Wu JY, Chen SC, Do TN, Su WP, Gumbs G, Lin MF. The diverse magneto-optical selection rules in bilayer Bernal graphene. ACS Nano 2018;4:1465.
[24] Zhang Y, Tan YW, Stormer HL, Kim P. Experimental observation of the quantum Hall effect and Berry's phase in graphene. Nature 2005;438:201.
[25] Wu J-Y, Chen S-C, Gumbs G, Lin M-F. Field-induced diverse quantizations in monolayer and bilayer black phosphorus. Phys Rev B 2017;95:115411.
[26] Sharpe AL, Fox EJ, Barnard AW, Finney J, Watanabe K, Kastner MA. Emergent ferromagnetism near three-quarters filling in twisted bilayer graphene. Science 2019;10:1126.
[27] Thuy Tran NT, Lin SY, Glukhova OE, Lin MF. Configuration-induced rich electronic properties of bilayer graphene. J Phys Chem C 2015;119:10623.
[28] Deleted in review.
[29] Ho JH, Lu CL, Hwang CC, Chang CP, Lin MF. Coulomb excitations in AA- and AB-stacked bilayer graphites. Phys Rev B 2006;74:085406.
[30] Shyu FL, Lin MF. Loss spectra of graphite-related systems: a multiwall carbon nanotube, a single-wall carbon nanotube bundle, and graphite layers. Phys Rev B 2000;62:8508.
[31] Wei W, Dai Y, Huang B, Jacob T. Many-body effects in silicene, silicane, germanene and germanane. Phys Chem Chem Phys 2013;15:8789–94.
[32] Deleted in review.
[33] Grüneis A, et al. Tight-binding description of the quasiparticle dispersion of graphite and few-layer graphene. Phys Rev B 2008;78:205425.
[34] Nguyen DK, Lin YT, Lin SY, Chiu YH, Tran NTT, Lin MF. Fluorination-enriched electronic and magnetic properties in graphene nanoribbons. Phys Chem Chem Phys 2017;19:20667–76.
[35] Nguyen DK, Tran NTT, Nguyen TT, Lin MF. Diverse electronic and magnetic properties of chlorination-related graphene nanoribbons. Sci Rep 2018;8:17859.
[36] Wang X, et al. High-yield production of highly fluorinated graphene by direct heating fluorination of graphene-oxide. ACS Appl Mater Interfac 2013;5:8294–9.
[37] Andrew JM, et al. Synthesis of borophenes: anisotropic, two-dimensional boron polymorphs. Science 2015;350:1513–6.
[38] Qiu Y, Gao L. Chemical synthesis of turbostratic carbon nitride, containing CN crystallites, at atmospheric pressure. Chem Commun 2003;18:2378–9.
[39] Deng X, Kousaka H, Tokoroyama T, Umehara N. Deposition and tribological behaviors of ternary BCN coatings at elevated temperatures. Surf Coat Technol 2014;259:2–6.
[40] Terrones M, et al. Pyrolytically grown BxCyNz nanomaterials: nanofibres and nanotubes. Chem Phy Lett 1996;257:576–82.
[41] Prakash A, Sundaram KB, Campiglia AD. Photoluminescence studies on BCN thin flims synthesized by RF magnetron sputtering. Mater Lett 2016;183:355–8.
[42] Kosynkin DV, et al. Longitudinal unzipping of carbon nanotubes to form graphene nanoribbons. Nature 2009;458:872–6.
[43] Liu Z, Suenaga K, Harris PJF, Iijima S. Open and closed edges of graphene layers. Phys Rev Lett 2009;102:015501.
[44] Wang J, Ma F, Sun M. Graphene, hexagonal boron nitride, and their heterostructures: properties and applications. RSC Adv 2017;7:16801.
[45] Bernal JD. The structure of graphite. Proc R Soc Lond A 1924;106:749.
[46] Wang Y, Zhang J, Wang X, Antonietti M, Li H. Boron- and fluorine-containing mesoporous carbon nitride polymers: metal-free catalysts for cyclohexane oxidation. Angew Chem 2010;49:3356–9.
[47] Bojdys MJ, Muller J-O, Antonietti M, Thomas A. Ionothermal synthesis of crystalline, condensed, graphitic carbon nitride. Chem—Eur J 2008;4:8177–82.
[48] Mirkarimi PB, McCarty KF, Medlin DL. Review of advances in cubic boron nitride film synthesis. Mater Sci Eng 1997;21:47–100.

[49] Faugeras C, et al. Few-layer graphene on SiC, pyrolitic graphite, and graphene: a Raman scattering study. Appl Phys Lett 2008;92:011914.
[50] Han W, Bando Y, Kurashima K, Sato T. Synthesis of boron nitride nanotubes from carbon nanotubes by a substitution reaction. Appl Phys Lett 1998;73:3085.
[51] Erickson KJ, Gibb AL, Sinitskii A, Rousseas M, Alem N, Tour JM, Zettl AK. Longitudinal splitting of boron nitride nanotubes for the facile synthesis of high quality boron nitride nanoribbons. Nano Lett 2011;11:3221−6.
[52] Tran NTT, Lin SY, Lin CY, Lin MF. Geometric and electronic properties of graphene-related systems: chemical bonding schemes. Boca Raton, FL: CRC Press; 2017.
[53] Zhang H, Duley W. First-principles study of edge chemical modifications in graphene nanodots. Phy Rev B 2008;78:045421.
[54] [a] Wang YP, Li XG, Fry JN, Cheng HP. First-principles studies of electric field effects on the electronic structure of trilayer graphene. Phys Rev B 2016;94:165428.
[b] Bostwick A, Ohta T, Seyller T, Horn K, Rotenberg E. Quasiparticle dynamics in graphene. Nat Phys 2007;3:36−40.
[55] Novoselov KS, et al. Electric field effect in atomically thin carbon film. Science 2004;306:666.
[56] Shao Y, et al. Epitaxial growth of flat antimonene monolayer: a new honeycomb analouge of graphene. Nano Lett 2018;18:2133.
[57] Shu H, Cao D, Liang P, Wang X, Chen X, Lu W. Two-dimensional silicene nucleation on a Ag(111) surface: structural evolution and the role of surface diffusion. Phys Chem Chem Phys 2014;16:304.
[58] Dávila ME, Lay GL. Few layer epitaxial germanene: a novel two-dimensional Dirac material. Sci Rep 2016;6:20714.
[59] Zhu FF, et al. Epitaxial growth of two-dimensional stanene. Nat Mater 2015;14:1020.
[60] Ganz E, Hwang IS, Xiong F, Theiss SK, Golovchenko J. Growth and morphology of Pb on Si(111). Surf Sci 1991;257:259.
[61] Deleted in review.
[62] Castro Neto AH, Guinea F, Peres NMR, Novoselov KS, Geim AK. The electronic properties of graphene. Rev Mod Phys 2009;81:109.
[63] Chiu YH, Lai YH, Ho JH, Chuu S, Lin MF. Electronic structure of a two-dimensional graphene monolayer in a spatially modulated magnetic field: Peierls tight-binding model. Phys Rev B 2008;82:245412.
[64] Lin CY, Wu JY, Chiu YH, Lin MF. Magneto-electronic properties of multilayer graphenes. Phys Chem 2015;17:26008.
[65] Gusynin VP, Sharapov SG. Unconventional integer quantum Hall effect in graphene. Phys Rev Lett 2005;95:146801.
[66] Johns JE, Hersam MC. Atomic covalent functionalization of graphene. Acc Chem Res 2013;46:77.
[67] Singh R. Spin-orbit coupling in graphene, silicene and germanene: dependence on the configuration of full hydrogenation and fluorination. Bull Mater Sci 2018;41:158.
[68] Molle A, Grazianetti C, Tao L, Taneja D, Alam MH, Akinwande D. Silicene, silicene derivatives, and their device applications. Chem Soc Rev 2018;47:6370.
[69] Ho YH, Chiu CW, Su WP, Lin MF. Magneto-optical spectra of transition metal dichalcogenides: a comparative study. Appl Phys Lett 2014;105:222411.
[70] Xia W, Hu W, Li Z, Yang J. A first-principles study of gas adsorption on germanene. Phys Chem 2014;16:22495.
[71] Shih PH, Do TN, Gumbs G, Huang D, Pham HD, Lin MF. Rich magnetic quantization phenomena in AA bilayer silicene. Sci Rep 2019;9:14799.
[72] Monshi MM, Aghaei SM, Calizo I. Doping and defect-induced germanene: a superior media for sensing H_2S, SO_2, and CO_2 gas molecules. Surf Sci 2017;665:96.
[73] Giese TJ, York DM. Improvement of semiempirical response properties with charge-dependent response density. J Chem Phys 2005;123:164108.
[74] Ruska E. The development of electron microscope and of electron microscopy. Rev Mod Phys 1987;59:627.
[75] Acun A, et al. Germanene: the germanium analogue of graphene. J Phys Condens Matter 2015;27:443002.
[76] Hou Z, Terakura K. Effect of nitrogen doping on the migration of the carbon adatom and monovacancy in graphene. J Phys Chem C 2015;119:4922.
[77] Varns R, Strange P. Stability of gold atoms and dimers adsorbed on graphene. J Phys Condens Matter 2008;20:22.

[78] Sun X, Wang L, Lin H, Hou T, Li Y. Induce magnetism into silicene by embedding transition-metal atoms. Appl Phys Lett 2015;106:222401.
[79] Tran NTT, Nguyen DK, Lin SY, Gumbs G, Fa-Lin M. Fundamental properties of transition-metals-adsorbed graphene. Chem Phys Chem 2019;10:1905.
[80] Sun M, Ren Q, Wang S, Zhang Y, Du Y, Yu J, Tang W. Magnetism in transition-metal-doped germanene: a first-principles study. Comput Mater Sci 2016;118:112.
[81] Kresse G, Furthmuller J. Phys Rev B: Condens Matter Mater Phys 1996;54:11169.
[82] Kresse G, Joubert D. Phys Rev B: Condens Matter Phys 1999;59:1758.
[83] Perdew JP, Burke K, Ernzerhof M. Phys Rev Lett 1996;77:3865.
[84] Blochl PE. Phys Rev B: Condens Matter Mater Phys 1994;50:17953.
[85] Feng B, et al. Evidence of silicene in honeycomb structures of silicon on Ag (111). Nano Lett 2012;12:3507–11.
[86] Fleurence A, et al. Experimental evidence for epitaxial silicene on diboride thin films. Phys Rev Lett 2012;108:245501.
[87] Lin CL, et al. Structure of silicene grown on Ag (111). Appl Phys Express 2012;5:045802.
[88] Pham HD, Su WP, Nguyen TDH, Tran NTT, Lin MF. Rich p-type-doping phenomena in boron-substituted silicene systems. R Soc Open Sci 2020;7:200723.
[89] Pham HD, Gumbs G, Su WP, Tran NTT, Lin MF. Unusual features of nitrogen substitutions in silicene. RSC Adv 2020;10:32193.
[90] Yu D, Lupton EM, Gao HJ, Zhang C, Liu F. A unified geometric rule for designing nanomagnetism in graphene. Nano Res 2008;1:497.
[91] Guinea F, Katsnelson M, Geim A. Energy gaps and a zero-field quantum Hall effect in graphene by strain engineering. Nat Phys 2010;6:30.
[92] Harrison WA, Tersoff J. Tight binding theory of heterojunction band lineups and interface dipoles. J Vac Sci Technol B 1986;4:1068.
[93] Bena C, Montambaux G. Remarks on the tight-binding model of graphene. New J Phys 2009;11:095003.
[94] Chen L, et al. Evidence for dirac fermions in a honeycomb lattice based on silicon. Phys Rev Lett 2012;109:056804.
[95] Li G, Luican A, Andrei EY. Scanning tunneling spectroscopy of graphene on graphite. Phys Rev Lett 2009;102:176804.
[96] Brar VW, et al. Scanning tunneling spectroscopy of inhomogeneous electronic structure in monolayer and bilayer graphene on SiC. Appl Phys Lett 2007;91:122102.
[97] Lin CY, Do TN, Wu JY, Shih PH, Lin SY, Ho CH, Lin MF. Diverse quantization phenomena in Emergent layered materials. Boca Raton, FL: CRC Press; 2019.
[98] Wilder JW, Venema LC, Rinzler AG, Smalley RE, Dekker C. Electronic structure of atomically resolved carbon nanotubes. Nature 1998;391:59.
[99] Uryu S, Ando T. Electronic intertube transfer in double-wall carbon nanotubes with impurities: tight-binding calculations. Phys Rev B 2007;76:155434.
[100] Iijima S. Microtubules of graphitic carbon. Nature 1991 Helical;354:56–8.
[101] Söde H, et al. Electronic band dispersion of graphene nanoribbons via Fourier-transformed scanning tunneling spectroscopy. Phys Rev B 2015;91:045429.
[102] Lin SY, Tran NTT, Chang SL, Su WP, Lin MF. Structure- and adatom-enriched essential properties of graphene nanoribbons. Boca Raton, FL: CRC Press; 2019.
[103] Deleted in review.
[104] Lee J, Novoselov KS, Shin HS. Interaction between metal and graphene: dependence on the layer number of graphene. ACS Nano 2010;5:608–12.
[105] Song JJ, Zhang HJ, Cai YL, Zhang YX, Bao SN, He PM. Bottom-up fabrication of graphene nanostructures on Ru(1010). Nanotechnology 2016;27:055602.
[106] Kim K, Sussman A, Zettl A. Graphene nanoribbons obtained by electrically unwrapping carbon nanotubes. ACS Nano 2010;4:1362.
[107] Patra N, Wang B, Kral P. Nanodroplet activated and guided folding of graphene nanostructures. Nano Lett 2009;9:3766.
[108] Enoki T, Fujii S, Takai K. Zigzag and armchair edges in graphene. Carbon 2012;50:3141.
[109] Tao C, et al. Spatially resolving edge states of chiral graphene nanoribbons. Nat Phys 2011;7:616.

[110] Min K, Aluru NR. Mechanical properties of graphene under shear deformation. Appl Phys Lett 2011;98:013113.
[111] Tsai W, Huang C, Chang T, Lin H, Jeng H, Bansil A. Gated silicene as a tunable source of nearly 100% spin-polarized electrons. Nat Commun 2013;4:1500.
[112] Chiu YH, Ho JH, Chang CP, Chuu DS, Lin MF. Low-frequency magneto-optical excitations of a graphene monolayer: Peierls tight-binding model and gradient approximation calculation. Phys Rev B 2018;78:245411.
[113] Joucken F, et al. Charge transfer and electronic doping in nitrogen-doped graphene. Sci Rep 2015;5:14564.
[114] Yan W, et al. Angle-dependent van Hove singularities in a slightly twisted graphene bilayer. Phys Rev Lett 2012;109:126801.
[115] Lin SY, Chang SL, Shyu FL, Lu JM, Lin MF. Feature-rich electronic properties in graphene ripples. Carbon 2015;86:207–16.
[116] Bianchi M, et al. Electron–phonon coupling in potassium-doped graphene: angle-resolved photoemission spectroscopy. Phys Rev B 2010;81:041403.
[117] Berdiyorov GR, Neek-Amal M, Peeters FM, van Duin AC. Stabilized silicene within bilayer graphene: a proposal based on molecular dynamics and density-functional tight-binding calculations. Phys Rev B 2014;89:024107.
[118] Shih PH, Chiu YH, Wu JY, Shyu FL, Lin MF. Coulomb excitations of monolayer germanene. Sci Rep 2017;7:40600.
[119] Wu J-Y, Lin C-Y, Gumbsb G, Lin M-F. The effect of perpendicular electric field on temperature-induced plasmon excitations for intrinsic silicene. RSC Adv 2015;5:51912.
[120] Shih P-H, Do T-N, Gumbs G, Pham HD, Lin M-F. Electric-field-diversified optical properties of bilayer silicene. Opt Lett 2019;44:4721–4.
[121] Muhich CL, et al. The effect of N and B doping on graphene and the adsorption and migration behavior of Pt atoms. J Phys Chem C 2013;117:10523–35.
[122] Avila J, et al. Presence of gapped silicene-derived band in the prototypical (3 × 3) silicene phase on silver (111) surfaces. J Phys Condens Matter 2013;25:262001.
[123] Kamal C, Chakrabarti A, Banerjee A, Deb SK. Silicene beyond mono-layers: different stacking configurations and their properties. J Phys Condens Matter 2013;25:085508.
[124] Zhang F, Sahu B, Min H, MacDonald AH. Band structure of ABC-stacked graphene trilayers. Phys Rev B 2010;82:035409.
[125] Rong ZY, Kuiper P. Electronic effects in scanning tunneling microscopy: moiré pattern on a graphite surface. Phys Rev B 1993;48:17427.
[126] Johannsen JC, et al. Direct view of hot carrier dynamics in graphene. Phys Rev Lett 2013;111:027403.
[127] Fang DQ, Zhang SL, Xu H. Tuning the electronic and magnetic properties of zigzag silicene nanoribbons by edge hydrogenation and doping. RSC Adv 2013;3:24075–80.
[128] Osborne DA, Morishita T, Tawfik SA, Yayamab T, Spencer MJS. Adsorption of toxic gases on silicene/Ag(111). Phys Chem Chem Phys 2019;10:1039.
[129] Arafune R, et al. Structural transition of silicene on Ag(111). Surf Sci 2013;608:297.
[130] Meng L, et al. Buckled silicene formation on Ir(111). Nano Lett 2013;13:685.
[131] Gill TG, et al. Metallic atomically-thin layered silicon epitaxially grown on silicene/ZrB2. 2D Mater 2017;4:021015.
[132] Wiggers FB, Bui HV, Friedlein R, Yamada-Takamura Y, Schmitz J, Jong MP. A nitride-based epitaxial surface layer formed by ammonia treatment of silicene-terminated ZrB2. J Chem Phys 2016;144:134703.
[133] Fleurence A, Yamada-Takamura Y. Insights into the spontaneous formation of silicene sheet on diboride thin films. Appl Phys Lett 2017;110:041601.
[134] Li L, et al. Buckled germanene formation on Pt(111). Adv Mater 2014;26:4820.
[135] Cantero ED, et al. Growth of germanium on Au(111): formation of germanene or intermixing of Au and Ge atoms? Phys Chem Chem Phys 2017;19:18580.
[136] Martinez EA, Fuhr JD, Grizzi O, Sánchez EA, Cantero ED. Growth of germanene on Al(111) hindered by surface alloy formation. J Phys Chem C 2019;123:12910.
[137] Chen SW, Wu CY, Wu HJ, Chiu WT. Interfacial reactions in Sn/Bi_2Te_3, Sn/Bi_2Se_3 and Sn/Bi_2 (Te1-xSe x)3 couples. J Alloy Compd 2014;611:313.

[138] Kim KC, et al. Sn doping in thermoelectric Bi_2Te_3 films by metal—organic chemical vapor deposition. Appl Surf Sci 2015;353:232.

[139] Yakes M, Tringides MC. Probing the buried Pb/Si(111) interface with SPA LEED and STM on Si(111)-Pb ∝ 3×3. J Phys Chem A 2011;115:7096.

[140] Yang L, Han J. Electronic structure of deformed carbon nanotubes. Phys Rev Lett 2000;85:1.

[141] Lin CY, Ho CH, Wu JY, Do TN, Shih PH, Lin SY, Lin MF. Diverse quantization phenomena in layered materials. Boca Raton, FL: CRC Press; 2020.

[142] Kahl K, Kintscher N. Geometric multigrid for the tight-binding Hamiltonian of graphene. SIAM J Numer Anal 2018;56:499.

[143] Scheibner EJ, Germer LH, Hartman CD. Apparatus for direct observation of low-energy electron diffraction patterns. Rev Sci Instrum 1960;31:112.

[144] Gnani E, Reggiani S, Rudan M. Density of states and group velocity of electrons in SiO_2 calculated from a full band structure. Phy Rev B 2002;66:195205.

[145] Ohta T, et al. Morphology of graphene thin film growth on SiC (0001). New J Phys 2008;10:023034.

[146] Chiu CW, Huang YC, Shyu FL, Lin MF. Excitation spectra of ABC-stacked graphene superlattice. Appl Phys Lett 2011;98:2631920.

[147] Pierucci D, et al. Evidence for flat bands near the fermi level in epitaxial rhombohedral multilayer graphene. ACS Nano 2015;9:5432.

[148] Lin CY, Do TN, Huang YK, Lin MF. Electronic and optical properties of graphene in magnetic and electric fields. San Rafael, CA: Morgan and Claypool; 2017.

[149] Peng X, Wei Q, Copple A. Strain-engineered direct-indirect band gap transition and its mechanism in two-dimensional phosphorene. Phys Rev B 2014;90:085402.

[150] Chiu Y-H, Ho J-H, Ho Y-H, Chuu D-S, Lin M-F. Effects of a modulated electric field on the optical absorption spectra in a single-layer graphene. J Nanosci Nanotechnol 2009;9:6579—86.

[151] Reich S, Maultzsch J, Thomsen C. Tight-binding description of graphene. Phys Rev B 2002;66:035412.

[152] Ou YC, Sheu JK, Chiu YH, Chen RB, Lin MF. Influence of modulated fields on the Landau level properties of graphene. Phys Rev B 2011;83:195405.

[153] Chen S-C, Wu J-Y, Lin M-F. Feature-rich magneto-electronic properties of bismuthene. New J Phys 2018;20:062001.

[154] Esrafili MD, Behzadi H. A comparative study on carbon, boron-nitride, boron-phosphide and silicon-carbide nanotubes based on surface electrostatic potentials and average local ionization energies. J Mol Model 2013;19:2375—82.

[155] Türeci HE, Alhassid Y. Spin—orbit interaction in quantum dots in the presence of exchange correlations: an approach based on a good-spin basis of the universal Hamiltonian. Phys Rev B 2006;74:165333.

[156] Rakhmanov AL, Rozhkov AV, Sboychakov AO, Franco N. Instabilities of the AA-stacked graphene bilayer. Phys Rev Lett 2012;109:206801.

[157] Chiu CW, Lee SH, Chen SC, Shyu FL, Lin MF. Absorption spectra of AA-stacked graphite. New J Phys 2010;12:083060.

[158] Lai YH, Ho JH, Chang CP, Lin MF. Magnetoelectronic properties of bilayer bernal graphene. Phys Rev B 2008;77:085426.

[159] Lu CL, Chang CP, Huang YC, Ho JH, Hwang CC, Lin MF. Electronic properties of AA- and ABC-stacked few-layer graphites. J Phys Soc Jpn 2007;76:024701.

[160] Shyu FL, Chang CP, Chen RB, Chiu CW, Lin MF. Magnetoelectronic and optical properties of carbon nanotubes. Phys Rev B 2003;67:045405.

[161] Lin CY, Chen SC, Wu JY, Lin MF. Curvature effects on magnetoelectronic properties of nanographene ribbons. J Phys Soc Jpn 2012;1:064719.

[162] Lin CY, Wu JY, Ou YJ, Chiu YH, Lin MF. Magneto-electronic properties of multilayer graphenes. Phys Chem Chem Phys 2015;17:26008—35.

[163] Wei P, Qi ZQ. Insulating gap in the transition-metal oxides: a calculation using the local-spin-density approximation with the on-site Coulomb U correlation correction. Phys Rev B 1994;49:10864.

[164] Qi XL, Zhang SC. Topological insulators and superconductors. Rev Mod Phys 2011;83:1057.

[165] de Jong S, et al. Droplet-like Fermi surfaces in the anti-ferromagnetic phase of EuFe2As2, an Fe-pnictide superconductor parent compound. EPL 2010;89:27007.

[166] Lauterbur PC. Image Formation by induced local interactions: examples employing nuclear magnetic resonance. Nature 1973;242:190–1.

[167] Fry CG. The nobel prize in medicine for magnetic resonance imaging. J Chem Educ 2004;81:922.

[168] Salvador JR, Gour JR, Bilc D, Mahanti SD, Kanatzidis MG. Stabilization of new forms of the intermetallic phases β-RENiGe2(RE = Dy, Ho, Er, Tm, Yb, Lu) in liquid indium. Inorg Chem 2004;43:1403–10.

[169] Dong YJ, Wang XF, Vasilopoulos P, Zhai MX, Wu XM. Half-metallicity in aluminum-doped zigzag silicene nanoribbons. J Phys D Appl Phys 2014;47:105304.

[170] Ravlić R, Bode M, Kubetzka A, Wiesendanger R. Correlation of dislocation and domain structure of Cr(001) investigated by spin-polarized scanning tunneling microscopy. Phys Rev B 2003;67:174411.

[171] Pietzsch O, Kubetzka A, Bode M, Wiesendanger R. Real-space observation of dipolar antiferromagnetism in magnetic nanowires by spin-polarized scanning tunneling spectroscopy. Phys Rev Lett 2000;84:5212.

[172] Heinze S, Bode M, Kubetzka1 A, Pietzsch O, Nie X, Blüge S, Wiesendanger R. Real-space imaging of two-dimensional antiferromagnetism on the atomic scale Science, vol. 288; 2000. p. 5472.

[173] Yayon Y, Brar VW, Senapati L, Erwin SC, Crommie MF. Observing spin polarization of individual magnetic adatoms. Phys Rev B 2007;99:067202.

[174] Brune H, Gambardella P. Magnetism of individual atoms adsorbed on surfaces. Surf Sci 2009;603:1812.

[175] Sivek J, Sahin H, Partoens B, Peeters FM. Adsorption and absorption of boron, nitrogen, aluminum, and phosphorus on silicene: stability and electronic and phonon properties. Phys Rev B 2013;87:085444.

[176] Jin H, Jeong H, Ozaki T, Yu J. Anisotropic exchange interactions of spin–orbit-integrated states in Sr_2IrO_4. Phys Rev B 2009;80:075112.

[177] Shih PH, Chiu CW, Wu JY, Do TN, Lin MF. Coulomb scattering rates of excited states in monolayer electron-doped germanene. Phys Rev B 2018;97:195302.

[178] C Y, Do TN, Huang YK, Lin MF. Optical properties of graphene in magnetic and electric fields. Bristol: IOP Publishing); 2017.

[179] Kaloni TP, Modarresi M, Tahir M, Roknabadi MR. Electrically engineered band gap in two-dimensional Ge, Sn, and Pb: a first-principles and tight-binding approach. J Phys Chem C 2015;119:11896.

[180] Zaminpayma E, Nayebi P. Band gap engineering in silicene: a theoretical study of density functional tight-binding theory. Physica E 2016;84:555–63.

[181] Lin SY, Liu HY, Nguyen DK, Tran NTT, Pham HD, Chang SL, Lin CY, Lin MF. Silicene-based layered materials. Bristol: IOP Publishing); 2020.

[182] Do TN, Shih PH, Gumbs G, Huang D, Chiu CW, Lin MF. Diverse magnetic quantization in bilayer silicene. Phys Rev B 2018;97:125416.

[183] Kiryu T, Koshino M. Quantum Hall effect in three-dimensional graphene. Phy Rev B 2019;99:085443.

[184] Sahin H, Topsakal M, Ciraci S. Structures of fluorinated graphene and their signatures. Phys Rev B 2011;83:115432.

[185] Peelaers H, Hernandez-Nieves AD, Leenaerts O, Partoens B, Peeters FM. Vibrational properties of graphene fluoride and graphane. Appl Phys Lett 2011;98:051914.

[186] Wang D, Shi J. Effect of charged impurities on the thermoelectric power ofgraphene near the Dirac point. Phys Rev B 2011;83:113403.

[187] Ando T. Screening effect and impurity scattering in monolayer graphene. J Phys Soc Jpn 2006;75:074716.

[188] Hwang EH, Das Sarma S. Dielectric function, screening, and plasmons in two-dimensional graphene. Phys Rev B 2007;75:205418.

[189] Popov VV, Polischuk OV, Shur MS. Resonant excitation of plasma oscillations in a partially gated two-dimensional electron layer. J Appl Phys 2005;98:033510.

[190] Chen Y, Gao B, Zhao JX, Cai QH, Fu HG. Si-doped graphene: an ideal sensor for NO or NO_2-detection and metal-free catalyst for N_2O-reduction. J Mol Model 2012;18:2043–54.

[191] Zhang WB, Song ZB, Dou LM. The tunable electronic structure and mechanical properties of halogenated silicene: a firstprinciples study. J Mater Chem C 2015;3:3087–94.

[192] Das Sarma S, Hwang EH, Tse WK. Many-body interaction effects in doped and undoped graphene: fermi liquid versus non-fermi liquid. Phys Rev B 2007;75:121406.

[193] Lin CY, Chen RB, Ho YH, Lin MF. Electronic and optical properties of graphite-related systems. Boca Raton, FL: CRC Press; 2018.

[194] Do TN, et al. Peculiar optical properties of bilayer silicene under the influence of external electric and magnetic fields. Sci Rep 2019;9:36547.
[195] Koshino M. Interlayer screening effect in graphene multilayers with ABA and ABC stacking. Phys Rev B 2010;81:125304.
[196] Barraza SJ, Denmark SE. Synthesis, reactivity, functionalization, and ADMET properties of silicon-containing nitrogen heterocycles. J Am Chem Soc 2018;140:6668–84.
[197] Fan Y, Liu X, Wang J, Ai H, Zhao M. Silicene and germanene on InSe substrates: structures and tunable electronic properties. Phys Chem Chem Phys 2018;20:11369.
[198] Lin CY, Huang BL, Ho CH, Gumbs G, Lin MF. Geometry-diversified coulomb excitations in trilayer AAB stacking graphene. Phys Rev B 2018;98:195442.
[199] Podlesny I, et al. Metastable bound states of the quasi–bimagnetoexcitons in the lowest landau levels approximation. Phys E Low-dimens Syst Nanostruct 2020;115:113638.
[200] Chen JH, et al. Charged-impurity scattering in graphene. Nat Phys 2008;4:377–81.
[201] Gebhardt J, et al. Growth and electronic structure of boron-doped graphene. Phys Rev B 2013;87:155437.
[202] Ao ZM, et al. Al doped graphene: a promising material for hydrogen storage at room temperature. J Appl Phys 2009;105:074307.
[203] Lv YA, et al. Enhanced role of Al or Ga-doped graphene on the adsorption and dissociation of N_2O under electric field. Phys Chem Chem Phys 2011;13:12472–7.
[204] Chandni U, Henriksen EA, Eisenstein JP. Transport in indium-decorated graphene. Phys Rev B 2015;91:245402.
[205] Yang Y, et al. High-temperature quantum spin Hall insulator in compensated n-p codoped graphene. J Phys D Appl Phys 2016;49:075004.
[206] Wei D, et al. Synthesis of N- doped graphene by chemical vapor deposition and its electrical properties. Nano Lett 2009;9:5 1752–1758.
[207] Dai J, Yuan J. Modulating the electronic and magnetic structures of P-doped graphene by molecule doping. J Phys Condens Matter 2010;22:225501.
[208] Zhao C, Wu H. Density functional investigation of mercury and arsenic adsorption on nitrogen doped graphene decorated with palladium clusters: a promising heavy metal sensing material in farmland. Appl Surf Sci 2017;399:55–66.
[209] Jeon IY, et al. Metalloid tellurium-doped graphene nanoplatelets as ultimately stable electrocatalysts for cobalt reduction reaction in dye-sensitized solar cells. Nano Energy 2016;30:867–76.
[210] Aktürk OÜ, Tomak M. Bismuth doping of graphene. Appl Phys Lett 2010;96:081914.
[211] Hwang EH, Adam S, Das Sarma S. Carrier transport in two-dimensional graphene layers. Phys Rev Lett 2007;98:186806.
[212] Koshino M, Ando T. Transport in bilayer graphene: calculations within a self-consistent born approximation. Phys Rev B 2006;73:245403.
[213] Okamoto Y. Density functional theory calculations of alkali metal (Li, Na, and K) graphite intercalation compounds. J Phys Chem C 2013;118:16–9.
[214] Thinius S, Islam MM, Heitjans P, Bredow T. Theoretical study of Li migration in lithium–graphite intercalation compounds with dispersion-corrected DFT methods. J Phys Chem C 2014;118:2273–80.
[215] Wu JY, Chen SC, Lin MF. Temperature-dependent coulomb excitations in silicene. New J Phys 2014;16:125002.
[216] Yan XZ, Romiah Y, Ting CS. Electric transport theory of dirac fermions in graphene. Phys Rev B 2008;77:125409.
[217] Nomura K, MacDonald AH. Quantum transport of massless dirac fermions. Phys Rev Lett 2007;98:076602.

CHAPTER

7

Hydrogen-adsorbed group-IV materials

Vo Khuong Dien[1], Shih-Yang Lin[1], Chi-Hsuan Lee[1], Hsin-Yi Liu[1], Thi My Duyen Huynh[1], Nguyen Thi Han[1], Ngoc Thanh Thuy Tran[2], Thi Dieu Hien Nguyen[1], Wei-Bang Li[1] and Ming-Fa Lin[1]

[1]Department of Physics, National Cheng Kung University, Tainan, Taiwan; [2]Hierarchical Green-Energy Materials (Hi-GEM) Research Center, National Cheng Kung University, Tainan, Taiwan

7.1 Introduction

Elements in a periodic table creates this beautiful universe by providing the various pure crystals and the multi-component compounds. For example, the first element, hydrogen, frequently come to exist in the diverse forms of single-atom/molecular gases [1—3], liquids [4], and solid-state molecule crystals [5]. However, a metallic hydrogen crystal has not been successfully synthesized through a super-high pressure [6,7]. As a result of its rather active half-occupied 1s orbital, hydrogen atoms play one of the critical roles in the basic and applied sciences [8—11], such as, the outstanding candidates for chemical modifications [12,13], energy storages [14,15], and theoretical models under the single- and many-particle pictures. The active chemical environments are very convenient in observing the rich and unique physical/chemical/material phenomena [16,17]. One the group-IV side, there exist four significant [s, p_x, p_y, p_z]-orbitals, clearly indicating the rather strong covalent bonding. The single- and multi-orbital hybridizations are frequent revealed in the main-stream materials, the sp^3 bonding in the diamond-like crystals [18,19], the well-separated p_z-π and [s, p_x, p_y]-σ bondings for the graphene-related systems, and the non-negligible mixing of π and σ bonding due to the buckled silicene/germanene/tinene/plumbene crystals [21]. Very interesting, the 2D emergent systems display a lot of unusual quasiparticle properties, as done

Fundamental Physicochemical Properties of Germanene-related Materials
https://doi.org/10.1016/B978-0-443-15801-8.00001-3

in a series of books [16,17,22]. The development of quasiparticle framework is based on the phenomenological models [22,23] and/or the first-principles simulations [16,17,24], in which their close relations need to be established by the relevant physical properties.

The dramatic transformation of electronic properties could be modulated by the chemical modifications [25,26], Coulomb-field excitations of plasma beam [27], electromagnetic radiations [28], mechanical strains [29], and thermal excitations [30]. Specifically, the first ones are very powerful in creating the various multi-component compounds [24], diversifying the quasiparticle properties, greatly enhancing the material functions, and enlarging the potential applications. According the significant characteristics of chemical environments, the up-to-date experimental syntheses could be classified into four types: (I) intercalations [31,32], (II) decorations [33,34], (III) substitutions [35], and adsorptions [36]. For example, the neighboring carbon-honeycomb lattices, being attracted by the weak, but important van der Waals intercalations, are very easily intercalated by the various guest atoms since the discovery of acceptor- and donor-type graphite intercalation compounds 40 years ago [37]. The semimetallic behavior is changed into the metallic one [38], and the traditional superconductivity appears at very low temperature [39,40]. Most important, the lithium- [41] and sodium-related [41,42] graphite interaction compounds play the critical roles in ion-based batteries, being under the current science researches [43,44]. The similar intercalations could be successfully synthesized for 3D carbon nanotube bundles [41,45], 1D coaxial carbon nanotubes [46], planar/curved/folded graphene nanoribbons [47,48], and C_{60}-related 0D and 3D fullerenes [49,50]. Hydrogen atoms are rather efficient in creating the obvious passivation of dangling bonds at the open boundaries, as done for the achiral or chiral edges of finite-width graphene nanoribbons/finite-length carbon nanotubes, and the vacancies on carbon-honeycomb lattices [41,51,52]. Both of essential properties sharply contrast to each other in the presence and absence of hydrogen decorations, e.g., the fluctuation of C–C bond lengths [53,54] and the magnitude of band gap [17]. Chemical substitutions are frequently utilized to generate the binary, ternary, and quaternary compounds, such as, 3D P_4S_x isomers [$x \leq 10$], diamond-like and few-layer SiC [55–57], graphene/silicene/germanene/stanene/plumbene bilayer composites [58–63], $B_xC_yN_z$ nanotubes/layers [64,65], Li/Ti–/Si–/Ge–/Fe–/Co–/Ni–O (the core materials in ion-based batteries [66–71]), and Li-, P-, S- and Cl-related materials (very efficient solid-state electrolytes in lithium sulfur batters [72]). This could modulate the metal-semiconductor and semimetal-semiconductor transitions.

Specifically, chemical adsorptions are able to totally change the pristine fundamental properties [17,24], largely strengthen the material functions [24], and greatly enhancing the potential applications [73,74]. Among all the adsorption-induced compounds, carbon honeycomb lattices, which possess the strong σ bondings of the neighboring [2s, $2p_x$, $2p_y$] orbitals [75,76], can provide very active surface structures closely related to the perpendicular $2p_z$ orbital [77]. A plenty of local carbon-dangling bonds are easy to experience chemical reactions with guest atoms [78,79], molecules [79] and functional root [80]. Such experimental syntheses have generated the various n- or p-type metals [81], semimetals [82,83], and narrow- or wide-gap semiconductors [53]. For example, the potassium-doped graphene is identified to display a large blue shift by using the high-precision the angle-resolved photoemission spectroscopy (ARPES [84]). On the theoretical side, the systematic investigations have been done for few-layer graphene systems [17,22], graphene nanoribbons [24], and silicene-based materials [16]. Very interesting, each alkali adatom is predicted to totally transfer the outmost s-orbital onto carbon atom by using the 1D characteristics of graphene nanoribbons [47,48]. Furthermore, the concentration and distribution of adatoms strongly dominate the magnetic configurations in zigzag nanoribbon [85,86]; that, there exist the anti-ferromagnetic [87], ferro-magnetic [88] and non-magnetic [89] spin arrangements by controlling the adatom adsorptions. Most important, the chemical modifications are covered under the framework developments of quasiparticle viewpoints [17,22–24,27,90,91].

It is well known that monolayer graphene is a zero-gap semiconductor with a pair of linear valence and conduction bands intersecting at the Fermi level [75]. This system is potentially an ideal candidate for next-generation electronic devices due to the performance of a high electron mobility [92,93]. However, the potential applications of graphene is limited because of the absence of an energy gap. In order to further open up a sufficiently large energy gap, the chemisorption of hydrogen atoms is a very efficient way [94–96]. Elias et al. have successfully synthesized graphane in 2009, where they hydrogenated graphene via exposing it to a cold hydrogen plasma, and the predicted gap-opening phenomenon is delicately examined by the experimental measurements of the electric conductance. And then, Balog et al. can modulate the adatom adsorption by exposing graphene to hydrogen for different lengths of time, clearly illustrating that hydrogen concentration and energy gap grow with increasing exposure time. Apparently, the electronic properties are drastically changed by the density of adsorbed hydrogen atoms. The single-side hydrogenated plumbene systems are successfully synthesized at low concentrations, without the specific distribution configurations [97]. However, plumbane and plumbone are not discovered in experimental laboratories.

On the theoretical side, the optimal crystal symmetries and featured energy spectra have been studied for all the hydrogenated group-IV materials [98,99]. A new hydrocarbon material, graphene (100% percentage hydrogenation in Ref. [100]), is predicted by Sofo et al. in 2007 [100] through the first-principles density functional theory. Only few works [17,101] are conducted on the double- and single-side hydrogenations under various concentrations and arrangements. The top-side H-adatoms can create or enhance the buckled structure of a honeycomb lattice because of the strong four-orbital s-sp^3 bondings, e.g., a dramatic transformation from the planar into deformed crystal for the hydrogen-adsorbed graphene systems at higher concentrations [17]. Very interesting, certain adsorption cases even present the ferromagnetic spin configurations. The highly non-uniform chemical environments, atom-dominated band structures, spatial charge densities before and after hydrogenation, atom- and orbital-decomposed magnetic moments, as well as, atom-, orbital- and spin-projected density of states, are utilized to identify the critical charge and spin mechanisms of the quasiparticle framework [17]. As for the phenomenological methods, they are only developed for graphane up to now [102], since it is rather difficult to obtain the reliable parameters for the complicated s-sp^3 on buckled honeycomb lattices, such as, the distinct hopping integrals of the neighboring four orbitals and on-site Coulomb potentials in the tight-binding model [103], Furthermore, the gradient approximation is used to evaluate the electric diploe matrix elements [104], and then the optical absorption spectrum is analyzed using the orbital dependence. There exist a very large threshold frequency of $\sim$3.5 eV and some prominent absorptions closely related to the significant orbital hybridizations, being in sharp contrast with the optical properties of monolayer graphene [105,106], This clearly indicates the total change of chemical bonds after hydrogenation. However, the theoretical predictions on optical transitions cannot include the many-body effects of excitons [107]. The systematic investigations on the unusual hydrogenations and orbital-dependent optical excitations will be finished in this work simultaneously.

The hydrogenated group-IV layered systems [22,91,108,109] are outstanding candidates in revealing the diverse physical/chemical/material phenomena, as well as, verifying the theoretical framework of quasiparticle viewpoints [24,91]. The first-principles simulations [17,24] are very suitable for providing the full information, in which the critical mechanisms and the fundamental properties are combined together simultaneously. The significant Hamiltonian can account for all the intrinsic interactions related to the single- and/or multi-orbital hybridizations [24] and the spin density distributions [24]. It covers kinetic energy, crystal potential, exchange and correlation energies, on-site spin-dependent electron-electron interactions (the Hubbard-like ones), and the non-negligible spin-orbital couplings (SOC). To clarify the active orbital hybridizations in C−/Si−/Ge−/Sn−/Pb−H bonds and C−C/Si−Si/Ge−Ge/Sn−Sn/Pb−Pb bonds, as well as the spin distributions of adatoms and host atoms, the delicate analyses are thoroughly conducted on the optimal crystal symmetries before/after hydrogenation [24], the featured electronic properties with/without chemisorptions (band structures, spatial charge densities and van Hove singularities in Refs. [22,24]), and the prominent optical transitions in the presence/absence of excitonic effects (dielectric functions, reflectance spectra, absorption coefficients; electron energy loss spectra in Refs. [110,111]). Furthermore, the various hydrogen-adsorption cases are included in the numerical calculations, e.g., the single- and double-side chemical adsorptions under the distinct concentrations and distributions [17]. Very interesting, the different orbital hybridizations/magnetic configurations, which correspond to the specific energy ranges, could be clearly identified from the merged van Hove singularities. Such examinations are partially assisted by the atom-dominated by the atom-dominated energy spectra and the drastic variation of charge density after hydrogenations [17]. This work is closely relate to the binding and release of hydrogen atoms, being very helpful in energy storage [73,112]. How to develop the phenomenological models and their direct associations with the numerical simulations is under the current investigations [17,22,27].

7.2 Computational details

In this work, the density functional theory (DFT) calculations via the Vienna Ab-initio Simulation Package (VASP) [113] was utilized to estimate the ground-state features of the lithium ferrite compound. To approximate the exchange and correlation potential we used the Perdew-Burke-Ernzerhof (PBE) method [114] in generalized gradient approximation. We employed the projected augmented wave (PAW) pseudopotentials to describe the electronic wave functions in the core region [115]. The cutoff energy for the expansion of the plane wave basis was set to 600 eV. The Brillouin zone was integrated with a special k-point mesh of $50 \times 50 \times 1$ in the Monkhorst-Pack sampling technique for geometric optimization. The convergence condition of the ground-state was set to 10^{-8} eV between two consecutive simulation steps, and all atoms were allowed to fully relax during the geometric optimization until the Hellmann-Feynman force acting on each atom was smaller than 0.01 eV/Å.

7.2.1 Quasiparticle calculations

Based on the ground-state Kohn-Sham wave functions and the corresponding eigenvalues (E_{KS}), the quasiparticle energy spectrum was obtained by using GW approximation on the exchange-correlation self-energy. We performed the single shoot quasi-particle Green's function (G_0W_0). The screening effects are described by using the plasmon mode model of Hybertsen and Louie [116]. These approaches using an energy cutoff of 600 eV for expansion the plan wave, and energy cutoff of 300 eV for the response functions. The Brillouin zone was integrated with a special k-points mesh of 80 × 80 × 1 in the Γ sampling technique. The electronic quasi-particle band structure of $LiFeO_2$ was achieved under WANNIER90 codes [117].

7.2.2 Optical excitations

Under the perturbation of an electromagnetic wave, the electrons are vertically excited from the occupied states to the unoccupied ones in the quasi-particle energy spectrum. The interactions between photons and the charge carriers of the systems can be well characterized by the macroscopic dielectric functions $\varepsilon(\omega)$. This frequency-dependent function is very useful in understanding the main features of the energy loss functions, reflections and the absorption coefficients.

According to the Fermi's golden rule [118], the picture for single-particle excitations can be expressed by the imaginary part of the dielectric function:

$$\varepsilon_2^{\uparrow(\downarrow)}(\omega) = \frac{8\pi^2 e^2}{\omega^2}\sum_{vck}\left|e.\langle v\boldsymbol{k}^{\uparrow(\downarrow)}\left|\mathrm{v}\right|c\boldsymbol{k}^{\uparrow(\downarrow)}\rangle\right|^2\delta\left(\omega - E_{ck}^{\uparrow(\downarrow)} - E_{vk}^{\uparrow(\downarrow)}\right),$$

where the transition energy is associated with a prominence peak in the optical spectrum, and the oscillation strength of each excitation peak is directly related to the joined density of state $\delta\left(\omega - E_{ck}^{\uparrow(\downarrow)} - E_{vk}^{\uparrow(\downarrow)}\right)$, and the velocity matrix element, $\left|e.\langle v\boldsymbol{k}^{\uparrow(\downarrow)}\left|v\right|c\boldsymbol{k}^{\uparrow(\downarrow)}\rangle\right|^2$, respectively.

In general, the electron-hole pairs would be effectively bound together through Coulomb interactions under suitable conditions, e.g., low screening effects with the important Coulomb electron-electron interactions. The strongly correlated electron-hole pairs, called excitons, dominate the absorption spectra of the materials. Using the same k-point sampling, energy cutoff, and number of bands as in the GW calculations, these electron-hole states can be determined by solving the standard Bethe-Salpeter equation (BSE) [119]. The imaginary part of the dielectric function, $\varepsilon_2(\omega)$, is modified to:

$$\varepsilon_2^{\uparrow(\downarrow)}(\omega) = \frac{8\pi^2 e^2}{\omega^2}\sum_{vck}\left|e.\langle 0^{\uparrow(\downarrow)}\left|\mathrm{v}\right|S^{\uparrow(\downarrow)}\rangle\right|^2\delta\left(\omega - \Omega_S^{\uparrow(\downarrow)}\right).$$

The presence of electron-hole pairs may significantly affect the main features of the optical spectra. The close connection of the quasi-particle charges, spin-polarizations, orbital hybridizations and the effects of the coupled quasi-particles on optical excitations will be discussed in detail in the current work.

The hydrogenation of monolayer plumbene systems exhibit the rich crystal symmetries, as clearly illustrated in Fig. 7.1A–F through the top and side views and Table 7.1. The Pb–Pb bong lengths, Pb–H bond lengths (H-top-site heights [120,121]), honeycomb lattice on the [x, y]-plane (Fig. 7.1A, C and E), as well as, the height difference between A and B two sublattices (Δ in Fig. 7.1A, C and E), present the featured crystal symmetries. As for a pristine

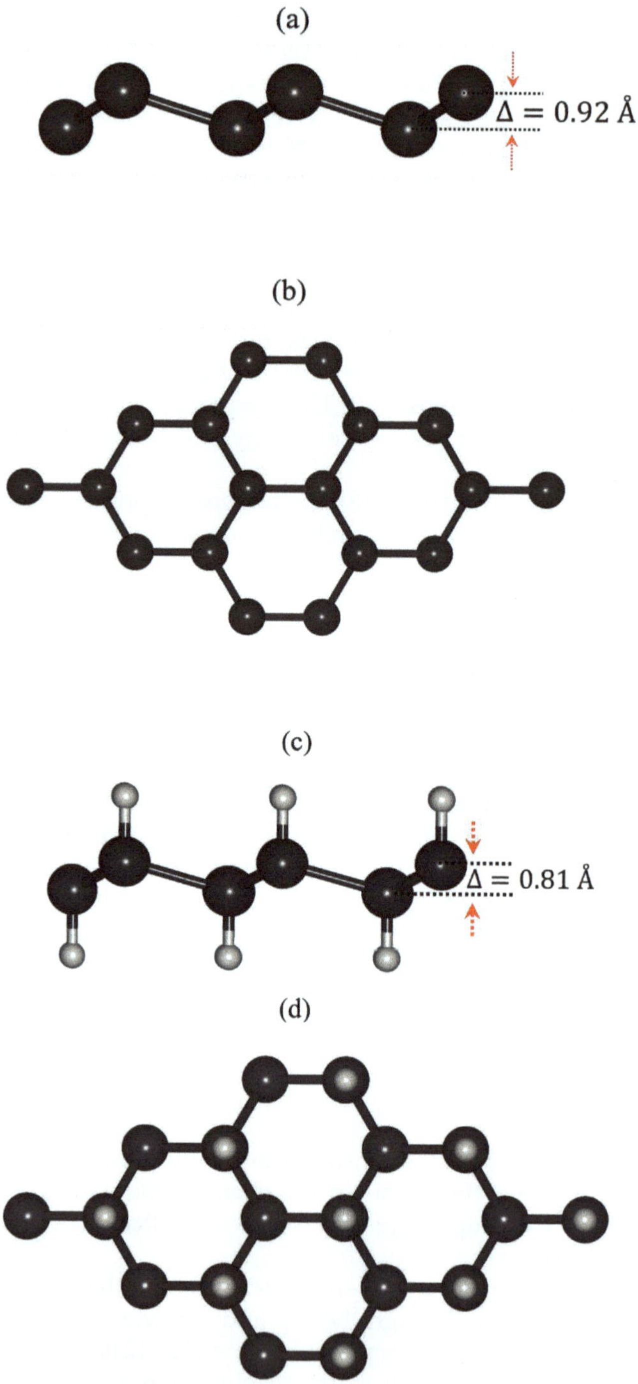

FIG. 7.1 Optimal crystal structures for monolayer (A)/(B) plumbene, and (C)/(D) plumbane under the side/top views, clearly illustrating the buckled geometries characterized by the height difference of A and B sublattice.

TABLE 7.1 The geometric parameters and electronic properties for monolayer pristine graphene, silicene, germanene, stanine, plumbene, as well as graphene, silicane, germanane, stanine and plumbane.

Material	Buckling height	X–X bonding	X–H bonding	Eg	
				W/O SOC	SOC
Graphene	0	1.42	–	0	0
Graphane	0.23	1.53	1.12	3.57	3.60
Silicene	0.45	2.23	–	1.9 (meV)	
Silicane	0.72	2.35	1.50	3.7	3.71
Germanene	0.68	2.43	–	0.2 (meV)	0.3(meV)
Germanane	0.73	2.46	1.55	2.1	2.105
Stanene	0.85	2.82	–	0.07	0.1
Stanane	0.82	2.84	1.74	0.4	0.9
Plumbene	0.92	3.0	–	Semimetal	0.6
Plumbane	0.81	3.01	1.81	0	1.3

system (Fig. 7.1A and B), the strong competitions of the sp^2 and sp^3 bondings (the non-orthogonal $6p_z$-π and [6s, $6p_x$, $6p_y$]-σ bondings) lead to the shorter Pb–Pb bond length about 3.00 Å (3.50 Å in a 3D crystal [122,123]), and the significant buckling of $\Delta = 0.92$ Å. The highly non-uniform chemical environment will be responsible for other essential properties, e.g., electronic and optical properties (discussed later). After the hydrogenation reactions, the optimal position of H-adatoms is shown at the top site for any concentrations and distributions, where there exist the strongest multi-orbital hybridizations in Pb–H bonds. Plumbane and plumbone, respectively, possess, 3.01 and 3.08 Å of Pb–Pb bond lengths; 0.81 and 1.95 Å of Δ. The former only a minor modification on Pb–Pb bond length, but an observable variation about the buckled structure (Fig. 7.1C). However, the drastic changes on them (the greatly longagated Pb–Pb bong length and buckling) are clearly revealed in the single-side absorption case (Fig. 7.1C). Such important differences lie in the symmetric or asymmetric H-distributions about the $z = 0$ plane. Furthermore, the main features indicate the dramatic charge redistributions, being very sensitive to the H-absorption environments. It should be noticed that the non-uniform host-guest bond lengths would be relatively easily observed in the non-saturated single-side chemisorptions of group-IV materials [17].

The first-principles predictions on the top- and side-view crystal structures could be verified by the high-precision measurements of scanning tunneling microscopy (STM in Refs. [124,125]; details in Chapter 3.1) and tunneling electron microscopy (TEM in Ref. [126]; discussed earlier in Chapter 3.1), respectively. In general, the former/the latter achieves the nanoscale surface morphologies/stacking configurations through a very weak, but observable quantum current/a sufficiently strong electron beam pattern. Both are very successful in examining/identifying/verifying the low-dimensional nanostructures, such as, armchair, zigzag and chiral single-walled carbon nanotubes (Fig. 7.8A–C in Refs. [127–131]), chiral and achiral open edges of graphene nanoribbons (Fig. 7.9A–C in Refs. [132,133]]), AB, ABC and AAB stackings of few-layer graphene systems (Fig. 7.11A–C in Refs. [134–136]). Very interesting, the well consistent between the experimental and theoretical studies [137,138] clearly illustrate the well separation/the significant hybridization between $2p_z$-π and [2s, $2p_x$, $2p_y$]-σ bonding on the planar/deformed honeycomb lattice. The sp^3 bonding in monolayer plumbene becomes strong under an enhanced buckling. Whether the π and σ bondings are meaningful before and after hydrogenations will be thoroughly test by their main features.

7.3 Featured energy spectra and wave functions

The theoretical predictions on energy spectra and wave functions within the hexagonal first Brilloun zone, as clearly illustrated in Figs. 7.2A–D and 7.3A–D, are greatly diversified by the coupling effects due to the distinct crystal symmetries, prominent sp^3 bondings, spin-orbital interactions. The main features cover spin-degenerate band structures [87,139], spin-dependent wave functions on A and B sublattices (Fig. 7.1A–F [91]), the number of valence subbands (related to single- or double side hydrogenations [17,24]), a direct/an indirect/vanishing band gap [17,24], the weak/prominent/zero subband overlaps [17,24], a large asymmetry of valence and conduction subbands (asymmetric hole and electron subbands [24]), energy dispersions (linear, parabolic, oscillatory and partially flat ones [24]), critical points in energy-wave-vector space (zero group velocities; van Hove singularities in Ref. [17]), non-crossing/crossing/anti-crossing behaviors [17], specific constant-energy loops (in anti-crossing subands [101]), the well-defined or undefined $\pi/\sigma/sp^3/s\text{-}sp^3$ bonding and their band widths along the hexagonal high-symmetry points [140,141], and significant contributions hydrogen adatoms and lead host atoms within a whole energy range [121,142].

A monolayer plumbene in the absence (Fig. 7.2A) and presence (Fig. 7.2B) spin-orbital couplings sharply contrast to each other in terms of featured band structures. Two cases show four valence subbands associated with the significant hybridizations of [6s, $6p_x$, $6p_y$, $6p_z$] orbitals, being doubly degenerate at the high-symmetry points. Their whole valence widths are about ~10.20 eV regardless of the strong SOC's effects, or the valence states of $E^v < -5.o\ eV$ are independent of SOC. Furthermore, energy spectrum is spin-degenerate, while each electronic states, with SOCs, is the linear superposition of the spin-up and spin-down configurations on two A and B sublattice [91]. That is, there exist four-component wave functions based on the sublattices of $[A_\uparrow,\ B_\downarrow,\ A_\downarrow,\ B_\uparrow]$ (four spin-dependent tight-binding functions in Ref. [91]). On the other hand, the low-energy electronic properties, which come to exist near the stable K and Γ valleys (very strong sp^3 bondings in Ref. [17]), are drastically changed the significant SOCs. The former is a semimetal with an almost vanishing subband overlap. A separated and slightly modified Dirac-cone structure appear at the K point, where the lowest conduction state touches the Fermi level. However, a pair of parabolic valence and conduction subbands, with a small energy spacing, occur at the Γ point. They have the highest occupied valence states intersects with $E_F = 0$. This separation is greatly enhanced through the cooperation of SOC and sp^3 bonding; therefore, a direct gap of ~ 0.5 eV is generated at the K/K′ valleys. It should be noticed that the shallow two valence subbands present the

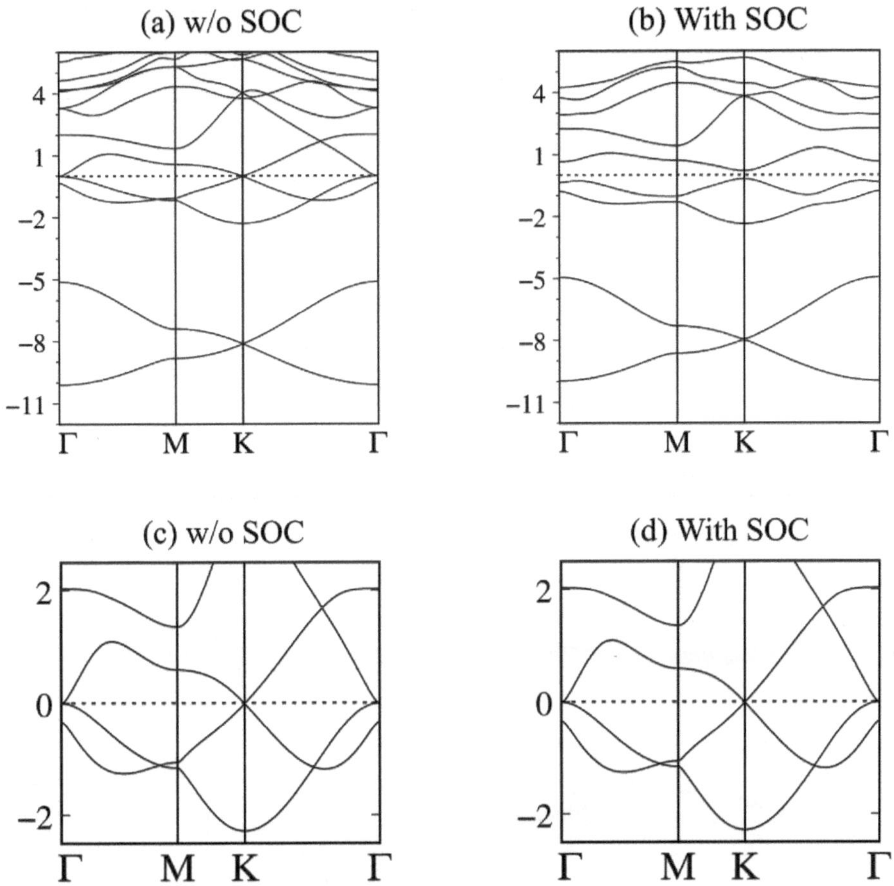

FIG. 7.2 Band structure of monolayer pristine plumbene in the (A)/(B) absence/presence spin-orbital coupling along the high-symmetry points of the hexagonal first Brillouin zone. The valence energy subbands are wide enough to cover those associated with the four-orbital sp^3 bondings. The inset in (A) is the hexagonal first Brillouin zone. (C)/(D) Correspond to those of the enlarged low-lying valence and conduction energy subbbands across the Fermi level [$E_F = 0$].

observable/very obviously anti-crossing behaviors along the ΓK and ΓM paths under four-orbital hybridizations/the composite effects [17]. The rich and unique energy dispersions (linear/parabolic/partially flat/oscillatory dispersions and constant-energy loops) will induce the unusual van Hove singularities and thus the other essential properties [17].

Hydrogenations are able to dramatically modify the low- and middle-energy states (Fig. 7.2C–F), as well as, the deep-energy ones under the asymmetric H-adatom adsorption (the former case). Plumbone, as shown in Fig. 7.3A, display, at least, five valence energy subbands. This illustrates an obvious evidence of s-sp^3 chemical bondings (the multi-orbital hybridizations of 1s-(6s, $6p_x$, $6p_y$, $6p_z$) in H–Pb bonds). Very interesting, the first and second ones can cross the Fermi level to become partially unoccupied near the Γ point, leading to the creation of free hole carriers. Electrons are transferred from the lead host atoms to the

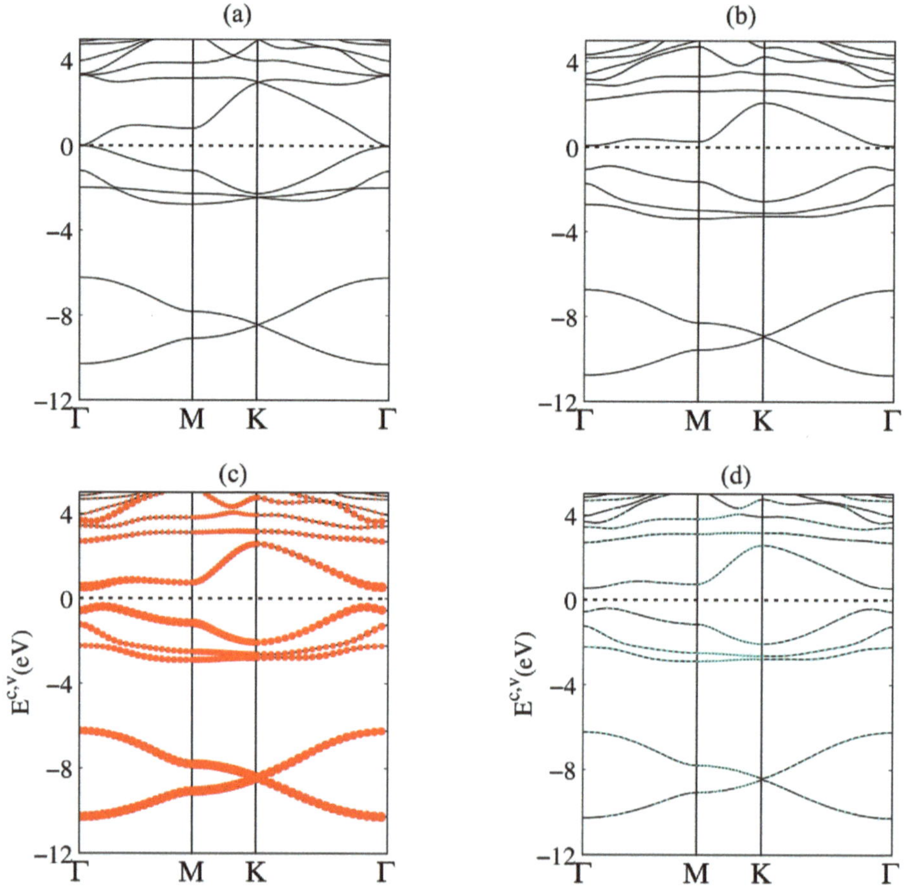

FIG. 7.3 Electronic energy spectra and wave functions for plumbane under the various cases: (A) without the spin-orbital coupling, as well as, (B) such interactions-induced band structures, being decomposed into the significant contributions of (C) lead and (D) hydrogen atoms (red and light blue balls, respectively).

hydrogen adatoms. Such chemical reaction processes, which mainly originate from the prominent coupling effects of s-sp^3 bonding and low-symmetry hydrogen adsorption, belong to the p-type doping. Plumbone is deduced to possess a composite feature of covalent and metallic bonds. Whether this 2D metallic systems will exhibit a good carrier mobility is worthy of near-future transport examinations. The Dirac-cone structure near the K/K* valley is totally vanishing after hydrogenation; therefore, the π bonding due to 6p$_z$ orbital cannot formed in plumbone. The similar phenomena are revealed in the high-concentration hydrogen adsorptions of group-IV materials (the diversified hydrogenations in distinct systems [53,101]). As for the third and fourth valence subbands, their band widths are only below ~1.00 eV, and the partially flat energy dispersions, with sufficient wide wave-vector ranges, come to exist to be centered at the Γ point. This unusual behavior will be directly reflected in van Hove singularities (discussed later in Fig. 7.6) and thus is very helpful in identifying the specific multi-orbital hybridization. Moreover, the fifth valence band, with the deepest electronic states, become more wide. It is attributed to the dominating 6s orbitals because of the largest ionization energy among sp^3 four orbitals.

While the significant SOC, s-sp^3 bonding, and asymmetric H-adatom adsorption configuration are combined together in plumbone, band structure and wave function become more complicated within the energy range of $E^{c,v} < 2.0$ eV. Both valence and conduction energy subbands, being initiated from the Γ valley, clearly illustrate the state splitting. Each electronic state is characterized the 20 tight-binding functions (five orbitals of sp^3, A and B sublattices, and spin-up and spin-down configurations [91]) according to the phenomenological model [91]. It is very sensitive the change of wave vector, as revealed in a pristine tinene systems. The third valence subbands are only slightly split through the spin-orbital coupling; furthermore, the fourth and fifth ones are independent of SOC. The identical results should be found under any adsorption cases and other Hydrogenated group-IV systems. Concerning the atom-decomposed weights, lead host atoms dominate within a whole energy spectrum covering all valence and conduction subbands, as obviously indicated in Fig. 7.3C. On the other side, hydrogen adatoms make the minor, but contributions in every energy subband. Such result directly reflects the number of occupied 1s and [6s, $6p_x$, $6p_y$, $6p_z$] orbitals in Pb—H bonds.

The theoretical predictions, which are conducted on the occupied energy spectra along the high-symmetry points of 2D emergent materials [143,144], could be verified by the high-resolution angle-resolved photoemission spectroscopy. ARPES is available for the layered honeycomb lattices, like few-layer graphene systems with various stacking configurations [145], as well as, 1D graphene nanoribbons [146]. However, it is not reliable for carbon nanotubes [147], and curved/folded graphene nanoribbons [146,148], mainly owing to the undefined momentum transfers along the longitudinal direction under obvious deformations. There are no published works up to now. Most important, the very successful ARPES measurements have identified the (k_x, k_y, k_z)-dependent 3D band structures of AB-stacked graphite with a weak, but significant semi-metallic band overlap [149,150], 1D parabolic energy subbands in semiconducting zigzag/armchair graphene nanoribbons (the strong evidences of quantum-confinement effects [151]), the linearly isotropic Dirac cones intersecting at the K/K* valleys for monolayer graphene [145] and few-layer AA-stacked graphene [152], two parabolic valence energy dispersions in AB-stacked bilayer graphene [153,154], the coexistence of linear and parabolic dispersion relations for symmetry-broken bilayer graphene, the linear and parabolic bands in trilayer ABA stacking [155], the linear, partially flat and sombrero-shaped bands for trilayer ABC stacking [156], as well as, and the semi-metal-semiconductor and semiconductor-metal transitions (the tunable low-lying valence and conduction bands) after the molecule/adatom absorptions on graphene surfaces [17]. However, the π- and σ-band widths need to be thoroughly examined in the further experimental examinations, providing whether both of chemical bondings are well defined under their almost orthogonal relations [17]. The similar observations are urgently required to test the diversified quasiparticle phenomena in hydrogenated plumbene systems through the important couplings of sp^3 bonding and SOC. This will very helpful in establishing the reliable phenomenological models, e.g., the accurate hopping integrals and strengths of SOC for the tight-binding model. In addition, these parameters become more complicated in the increment of group-IV atomic number [91]. In addition to the quasiparticle energy spectra, the ARPES measurement are available in observing the wave-vector-dependent lifetime from the inverse of the photoemission scattering width [143,144].

7.4 Active orbital hybridizations

Density of states in a 2D condensed-matter system is well characterized by the integration of group velocity on a/the constant-energy loop/loops [91]. The van Hove singularities, being due to the zero wave-packet velocities, correspond to the critical points in the energy-wave-vector space. Such quasi-particle behaviors are closely related to the distinct energy dispersions. The linear, parabolic, partially flat, saddle-point dispersion relations, respectively, generate the V-shape [157,158], shoulder [159], delta-function-like [160] and logarithmically divergent structures [161]. It should be noticed that the special structures in DOS are very sensitive to dimensionalities, such as, the similar energy dispersion relations in 3D graphite [162], 2D layered graphene systems [163], 1D carbon nanotubes/graphene nanoribbons [164–167], and 0D fullerenes [167], but the van Hove singularities quite different from one another. For example, as to a monolayer graphene/an armchair carbon nanotube, the Dirac-cone structure, which both valence and conduction bands linearly intersect at the Fermi level, displays a V-shape with a zero density of states at E_F/a finite-height plateau [168,169]. This have verified by the high-precision experimental measurements (STS [170]). In addition to the direct responses in STS, the joint van Hove singularities, being associated the initial and final states, frequently appear in the various scattering processes, e.g., the vertical transitions of optical absorptions [22], and the Coulomb-field excitations with the transferred momenta and frequencies simultaneously [27]. In short, the featured of van Hove singularities can dominate the prominent responses of the essential physical properties [27,91].

When the buckled/planar honeycomb lattices, the s-sp^3 bondigs, and the strengths of SOCs are clearly identified from the VASP simulations about the geometric and electronic properties, the tight-binding models could be explored by fitting their low-lying valence and conduction bands, The reliable parameters, which need to be conducted for this phenomenological methods, the ob-site Coulomb potentials due to the distinct orbitals (the different ionization energies in Ref. [91]), the various hopping integrals associated with s-sp^3 and sp^3-sp^3 in H–Pb and Pb–Pb bonds, respectively, and the significant SOCs of host lead atom [171]. Obviously, the numerical fittings become more difficult in the increment of the atomic number, since the competition of the critical factors is getting strong. That is, it is relatively difficult/easy to understand the rich quasiparticle phenomena of plumbone/graphane, according to the intrinsic interactions and lattice symmetries of hydrogenated group-IV materials. The tight-binding model could be generalized to study the magnetic quantization [54]; furthermore, its direct combinations with the static and dynamic Kubo formulas [172–174], the modified random-phase approximation [174], and the modified self-energy method [27] are useful in understanding Hall quantum transports [175], optical absorption spectra [176,177], Coulomb-field excitations [27], and inelastic scattering rates (quasiparticle lifetimes [178,179]).

7.5 Single-particle optical transitions

When the 2D pristine/hydrogenated group-IV materials are perturbed by an electromagnetic wave, the occupied electronic states are vertically excited to the unoccupied ones. All the charge carriers are capable of dynamically screening this external field and thus generate the induced current density [104]. Within the linear response, the screening ability of charges (the dynamic process in Ref. [180]) is well characterized by the transverse dielectric function at the long wavelength limit [$\varepsilon(q \rightarrow 0, \omega)$]. According to the basic electrodynamics [181], this frequency-dependent function is very useful in understanding the main features of reflectance spectrum, absorption coefficient, refractive index, and energy loss function. The relevant physical quantities are consistent with one another using the same theoretical framework. Furthermore, the single-particle electronic excitations (electron-hole pairs) and collective excitations (plasmon modes) are expected to dominate the couplings of charges and electromagnetic waves. The delicate analyses are done by the direct linking among the orbital-dominated band-edge states (band structures), the joint Hove singularities (the large density of states), and the specific absorption frequency (optical spectra). Each prominent absorption structure will be identified from the corresponding orbital hybridizations, thus, illustrating the quasiparticle viewpoints [17,22,27].

The calculated optical properties cover Re $[\varepsilon(\omega)]$ and Im $[\varepsilon(\omega)]$ (the real- and imaginary part of the bare response function), Im $[-1/\varepsilon(\omega)]$ (the screened response function), $R(\omega) = \left|\frac{\sqrt{\varepsilon}-1}{\sqrt{\varepsilon}+1}\right|^2$ (the normal reflectance), and $\alpha(\omega) = \frac{\sqrt{2}}{c}\omega\sqrt{\varepsilon(\omega) - Re[\varepsilon(\omega)]}$ (the effective decay length of electromagnetic wave). Im $[\varepsilon(\omega)]$, which characterizes the creations electron-holes pairs, is mainly determined by the available excitation channels and the electric dipole moments, In general, it is first evaluated using the linear Kubo formula [118] and the Fermi golden rule [182], and then Re $[\varepsilon(\omega)]$ is obtained from the Krammers-Kronig relations under a physical event [183]. While electromagnetic waves, with the suitable energy ranges, are normally incident on a semi-infinite sample, part of them are directly reflected on the vacuum side; the others are transmitted into a material and rapidly declines in the exponential way through a lot of strong interband transitions. Such decay feature is well described by the absorption coefficient. The significant couplings between the propagating electromagnetic waves and carrier charges will create the unusual quasiparticle optical phenomena, e.g., the plasmon peaks/edges in the energy loss/reflectance spectra, being closely related to the coherent charge oscillations after the screened responses. The prominent optical structures are very useful in identifying the important relations of vertically optical transitions and orbital hybridizations.

A pristine graphene clearly exhibits the unusual optical excitation phenomena in the absence/presence of excitonic effects, as indicated in Fig. 7.7A and B and Table 7.2A. Its frequency-dependent bare response function, which is due to the vertically interband electron-hole excitations, show four prominent absorption structures, respectively at 0.21, 5.510, 12.60 eV, and 16.2 eV/0.35 eV, 4.55 eV, 12.20 eV, and 15.80 eV. The featured results only reflects the different orbital hybridizations of 2D carbon honeycomb crystal, respectively, arising from the inter-π-band transitions inside a Dirac-cone structure (the red arrow in Fig. 7.4A), the higher-energy ones associated with the saddle M-point (the blue arrow), the composite excitations of the σ band and π band from the K and M points (the purple arrow), and the deepest σ band and π band near the M point (the yellow arrow). Such identifications are delicately achieved from the orbital-decomposed van Hove singularities (details in Fig. 7.5A); that is, the initial and/or the final states must possess the singular densities of states. The third and fourth strong responses are closely related to [2s, $2p_x$, $2p_y$]-[$2p_z$], but dominated by [$2p_x$, $2p_y$]-[$2p_z$] and 2s, respectively, these very active excitation channels are

TABLE 7.2 (A)–(O): The following 10 tables, covering the specific relations between the prominent absorption structures and the dominated orbital hybridizations in the absence and presence of excitonic effects for (A)–(E): (graphene, silicene, germanene, stanene, plumbene) and (F)–(J) (graphane, silicane, germanane, stanane, plumbane).

(A) Graphene

	Energy (eV)			
Colors	**w/o excitonic effects**	**With excitonic effects**	**Specific orbital hybridizations**	
Red	0.2	0.35	C-(2pz) → C-(2pz)	*K*
Blue	5.1	4.55	C-(2pz) → C-(2pz)	*M*
Purple	13	12.2	C-(2s, 2px, 2py) → C-(2pz)	*M*
Yellow	16.2	15.8	C-(2s, 2px, 2py) → C-(2pz)	*M*

(B) Silicene

	Energy (eV)			
Colors	**w/o excitonic effects**	**With excitonic effects**	**Specific orbital hybridizations**	
Red	0.28	0.27	Si-(3pz) → Si-(3pz)	K
Blue	1.90	1.72	Si-(3pz) → Si-(3pz)	M
Purple	5.25	4.74	Si-(3px, 3py) → Si-(3pz)	M

(C) Germanene

	Energy (eV)			
Colors	**w/o excitonic effects**	**With excitonic effects**	**Specific orbital hybridizations**	
Red	0.25	0.26	Ge-(4pz) → Ge-(4pz)	K
Blue	2.21	2.05	Ge-(4pz) → Ge-(4pz)	M
Purple	4.23	3.95	Ge-(4px, 4py) → Ge-(4pz)	M
Yellow	4.80	4.45	Ge-(4px, 4py, 4pz) → Ge-(4pz)	Γ*M*

(D) Stanene

	Energy (eV)			
Colors	**w/o excitonic effects**	**With excitonic effects**	**Specific orbital hybridizations**	
Red	0.29	0.3	Sn-(5pz) → Sn-(5pz)	K
Purple	0.69	0.67	Sn-(5px, 5py) → Sn-(5px, 5py)	Γ
Blue	1.54	1.51	Sn-(5pz) → Sn-(5pz)	M
Green	1.90	1.85	Sn-(5px, 5py) → Sn-(5pz)	M
Black	2.79	2.62	Sn-(5pz) → Sn-(5pz)	M

TABLE 7.2 (A)–(O): The following 10 tables, covering the specific relations between the prominent absorption structures and the dominated orbital hybridizations in the absence and presence of excitonic effects for (A)–(E): (graphene, silicene, germanene, stanene, plumbene) and (F)–(J) (graphane, silicane, germanane, stanane, plumbane).—cont'd

(D) Stanene

	Energy (eV)			
Colors	**w/o excitonic effects**	**With excitonic effects**	**Specific orbital hybridizations**	
Pink	3.50	3.41	Sn-(5px, 5py, 5pz) → Sn-(5pz)	ΓM
Yellow	5.60	5.52	Sn-(5s) → Sn-(5pz)	Γ
Cyan	7.20	6.90	Sn-(5s) → Sn-(5pz)	M

(E) Plumbene

	Energy (eV)			
Colors	**w/o excitonic effects**	**With excitonic effects**	**Specific orbital hybridizations**	
Black	–	0.49	–	
Red	0.6	0.55	Pb-(6pz) → Pb-(6pz)	K
Blue	1.73	1.56	Pb-(6pz) → Pb-(6pz)	M
Light green	2.20	1.88	Pb-(6px, 6py, 6pz) → Pb-(6pz)	M
Green	2.95	2.74	Pb-(6px, 6py, 6pz) → Pb-(6px, 6py)	KΓ
Cyan	3.55	3.50	Pb-(6px, 6py, 6pz) → Pb-(6px, 6py, 6pz)	ΓM
Brown	5.31	4.75	Pb-(6s) → Pb-(6px, 6py)	Γ
Pink	6.15	5.90	Pb-(6s) → Pb-(6pz)	Γ
Yellow	8.14	7.90	Pb-(6s) → Pb-(6pz)	M

(F) Graphane

	Energy (eV)			
Colors	**w/o excitonic effects**	**With excitonic effects**	**Specific orbital hybridizations**	
Black	–	3.3	–	
Black	–	4.2	–	
Black	4.7	4.8	C-(2px + 2py) → C-(2pz)	Γ
Blue	8.0	6.6	C-(2px + 2py) → C-(2s, 2px, 2py, 2pz)	Γ
Light-green	9.7	7.9	C-(2pz) + H-1s → C-(2pz)	Γ
Green	10.2	8.2	C-(2px, 2py) → C-(2s, 2px, 2py, 2pz) + H-1s	c
Yellow	11.4	9.0	C-(2px, 2py) → C-(2s, 2px, 2py, 2pz)	M
Light-blue	13.1	11.0	C-(2px, 2py) → C-(2s, 2px, 2py, 2pz) + H-1s	KΓ

(*Continued*)

TABLE 7.2 (A)–(O): The following 10 tables, covering the specific relations between the prominent absorption structures and the dominated orbital hybridizations in the absence and presence of excitonic effects for (A)–(E): (graphene, silicene, germanene, stanene, plumbene) and (F)–(J) (graphane, silicane, germanane, stanane, plumbane).—cont'd

(F) Graphane

	Energy (eV)			
Colors	**w/o excitonic effects**	**With excitonic effects**	**Specific orbital hybridizations**	
Purple	14.1	10.5	C-(2px, 2py) → C-(2s, 2px, 2py, 2pz) + H-1s	*MK*
Pink	15.3	12.0	C-(2s, 2px, 2py, 2pz) + H-1s → C-(2s, 2px, 2py, 2pz) + H-1s	*MK*
Gray	16.9	13.2	C-(2s, 2px, 2py, 2pz) + H-1s → C-(2s, 2px, 2py, 2pz) + H-1s	Γ*M*
White-red	18.3	14.7	C-(2s, 2px, 2py, 2pz) + H-1s → C-(2s, 2px, 2py, 2pz) + H-1s	Γ*M*
White-black	20.1	17.8	C-(2s, 2px, 2py) + H-1s → C-(2pz)	M

(G) Silicane

	Energy (eV)			
Colors	**w/o excitonic effects**	**With excitonic effects**	**Specific orbital hybridizations**	
Black	–	3.3	–	
Red	3.8	3.8	Si-(3px, 3py) → Si-(3s, 3px, 3py, 3pz)	Γ
Blue	5.2	4.2	Si-(3px, 3py) → Si-(3s, 3px, 3py, 3pz)	M
Green	6.8	5.0	Si-(3s, 3px, 3py, 3pz) + H-1s → Si-(3s, 3px, 3py, 3pz) + H-1s	ΓM
Yellow	9.4	7.7	Si-(3s, 3px, 3py, 3pz) + H-1s → Si-(3s, 3px, 3py, 3pz) + H-1s	M
Pink	12.3	11.2	Si-(3s, 3px, 3py) + H-1s → Si-(3s, 3px, 3py, 3pz) + H-1s	M

(H) Germanane

	Energy (eV)			
Colors	**w/o excitonic effects**	**With excitonic effects**	**Specific orbital hybridizations**	
Black	–	2.05	–	
Red	2.15	2.16	Ge-(4px, 4py) → Ge-(4s, 4px, 4py, 4pz)	Γ
Blue	4.7	3.8	Ge-(4px, 4py) → Ge-(4s, 4px, 4py, 4pz)	M
Green	6.6	4.8	Ge-(4s, 4px, 4py, 4pz) + H-1s → Ge-(4s, 4px, 4py, 4pz)	ΓM

TABLE 7.2 (A)–(O): The following 10 tables, covering the specific relations between the prominent absorption structures and the dominated orbital hybridizations in the absence and presence of excitonic effects for (A)–(E): (graphene, silicene, germanene, stanene, plumbene) and (F)–(J) (graphane, silicane, germanane, stanane, plumbane).—cont'd

(H) Germanane

Colors	Energy (eV) w/o excitonic effects	Energy (eV) With excitonic effects	Specific orbital hybridizations	
Yellow	8.7	6.4	Ge-(4s, 4px, 4py, 4pz) + H-1s → Ge-(4s, 4px, 4py, 4pz) + H-1s	Γ M
Pink	9.4	7.4	Ge-(4s, 4px, 4py, 4pz) + H-1s → Ge-(4s, 4px, 4py, 4pz) + H-1s	MK
Brown	11.0	9.2	Ge-(4s, 4px, 4py) + H-1s → Ge-(4s, 4px, 4py, 4pz)	M
Cyan	13.1	11.9	Ge-(4s, 4px, 4py) + H-1s → Ge-(4s, 4px, 4py, 4pz)	M

(I) Stanane

Colors	Energy (eV) w/o excitonic effects	Energy (eV) With excitonic effects	Specific orbital hybridizations	
Black	–	0.75	–	
Red	0.96	0.97	Sn-(5px, 5py) → Sn-(5px, 5py, 5pz)	Γ
Blue	1.4	1.3	Sn-(5px, 5py) → Sn-(5px, 5py, 5pz)	Γ
Green	3.7	3.1	Sn-(4px, 4py) → Sn-(5px, 5py, 5pz)	M
Yellow	4.4	4.0	Sn-(5s, 5px, 5py, 5pz) + H-1s → Sn-(5px, 5py, 5pz) + H-1s	M
Purple	7.7	6.6	Sn-(5s, 5pz) + H-1s → Sn-(5px, 5py, 5pz) + H-1s	ΓM
Pink	8.4	8.1	Sn-(5s, 5px, 5py) + H-1s → Sn-(5px, 5py, 5pz)	M
Cyan	11.0	10.3	Sn-(5s, 5px, 5py) + H-1s → Sn-(5px, 5py, 5pz)	M

(J) Plumbane

Colors	Energy (eV) w/o excitonic effects	Energy (eV) With excitonic effects	Specific orbital hybridizations	
Black	–	1.1	–	
Red	1.6	1.6	Pb-(6px, 6py) → Pb-(6s, 6px, 6py)	ΓM/ ΓK
Blue	2.3	1.9	Pb-(6px, 6py) → Pb-(6s, 6px, 6py)	M
	2.9	2.7	Pb-(6s, 6px, 6py, 6pz) → Pb-(6s, 6px, 6py)	Γ

(*Continued*)

TABLE 7.2 (A)–(O): The following 10 tables, covering the specific relations between the prominent absorption structures and the dominated orbital hybridizations in the absence and presence of excitonic effects for (A)–(E): (graphene, silicene, germanene, stanene, plumbene) and (F)–(J) (graphane, silicane, germanane, stanane, plumbane).—cont'd

(J) Plumbane			
	Energy (eV)		
Colors	**w/o excitonic effects**	**With excitonic effects**	**Specific orbital hybridizations**
Light green			
Yellow	3.7	3.5	Pb-(6s, 6px, 6py, 6pz) + H-1s → Pb-(6s, 6px, 6py) Γ M
Pink	4.6	4.1	Pb-(6s, 6px, 6py, 6pz) + H-1s → Pb-(6s, 6px, 6py) M
Brown	7	5.9	Pb-(6s, 6px, 6py) + H-1s → Pb-(6s, 6px, 6py) Γ
Cyan	8.0	7.3	Pb-(6s, 6px, 6py) + H-1s → Pb-(6s, 6px, 6py) M
Green	11.7	10.8	Pb-(6s, 6px, 6py) + H-1s → Pb-(6s, 6px, 6py) M

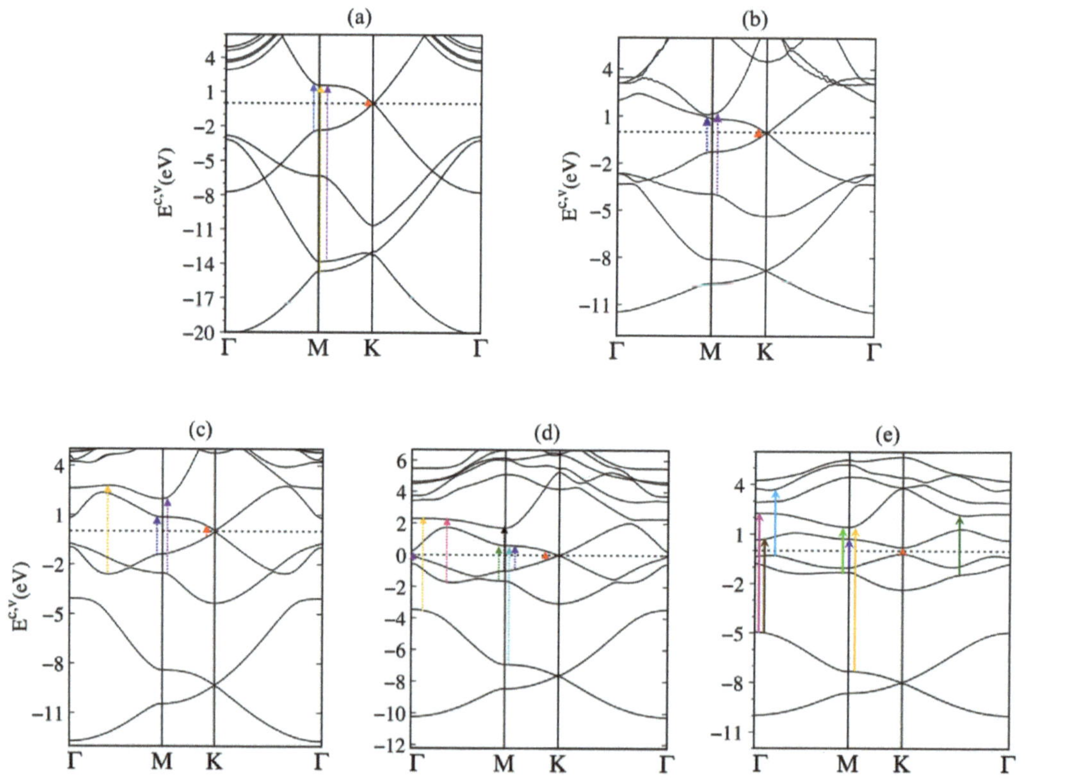

FIG. 7.4 Band structure of monolayer pristine (A)/(B)/(C)/(D)/(E) graphene/silicene/germanene/stanine/plumbene and (F)/(G)/(H)/(I)/(J) graphene/silicane/germanane/stanine/plumbane in the presence spin-orbital coupling along the high-symmetry points of the hexagonal first Brillouin zone. The color arrows corresponding to the electron vertical excitation under the photon absorption (Table 7.2).

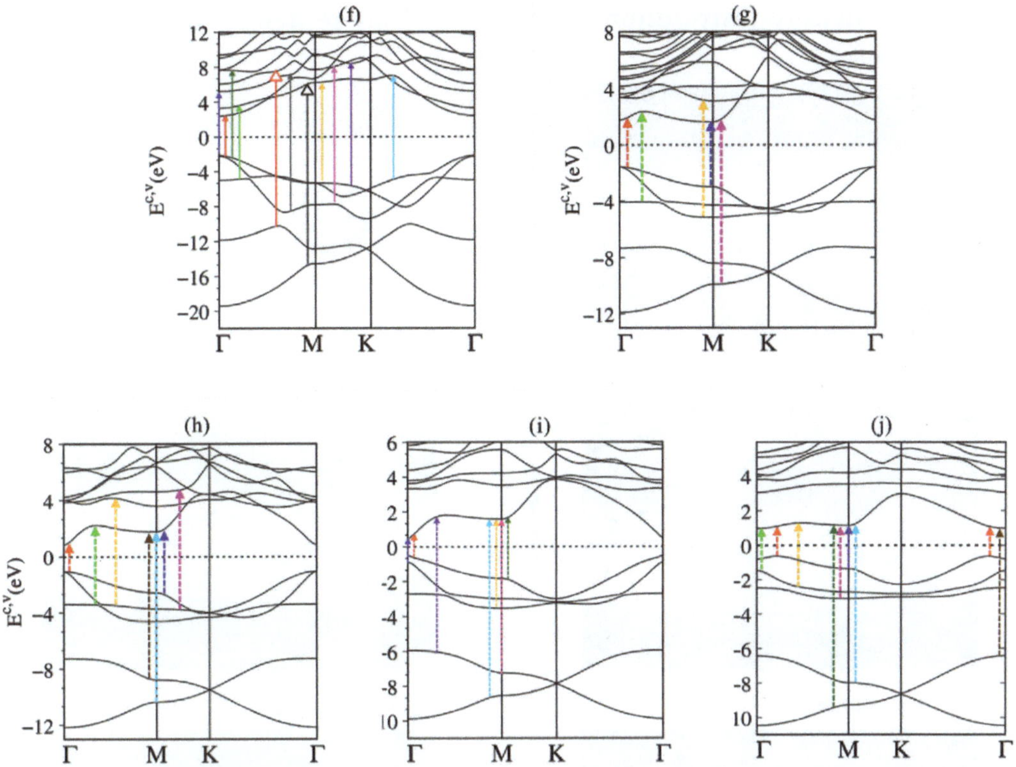

FIG. 7.4 cont'd

fully supported by orbital-decomposed density of states. Also noticed that the π and σ bondings are almost perpendicular to each, while their electric dipole moments might survive under the quantum averages of the gradient operator. Their values, being associated with the first σ band at the M point, could be vanishing because of the absence of a special absorption structure. Concerning the excitonic effects, they only slightly induce a blue shift of the first peak and red shifts of the others (details in Table 7.2A). Any stable or meta-stable excitonic peaks do not come to exist and thus cannot strongly modify the low-frequency optical behaviors. This unusual quasiparticle phenomenon could be attributed to the zero-gap semiconducting property [75]. The excitonic states are expected to be absent in semi-metallic few-layer graphene systems and bulk graphites [184]. However, the finite-gap semiconductors of 1D carbon nanotubes [185] and graphene nanoribbons [186] are outstanding candidates in observing the strongly many-body effects. In addition, the spin-orbital couplings are negligible in the pure carbon-related materials, i.e., they hardly affect electronic properties (Fig. 7.4A) and optical excitations (Fig. 7.7A and B).

After the charge screenings of all π and σ valence electrons, the energy loss functions (the red/blue curves), with/without excitonic effects, absorption coefficients, and reflectance spectra (Fig. 7.8A), clearly reveal the featured plasmon modes, respectively, corresponding

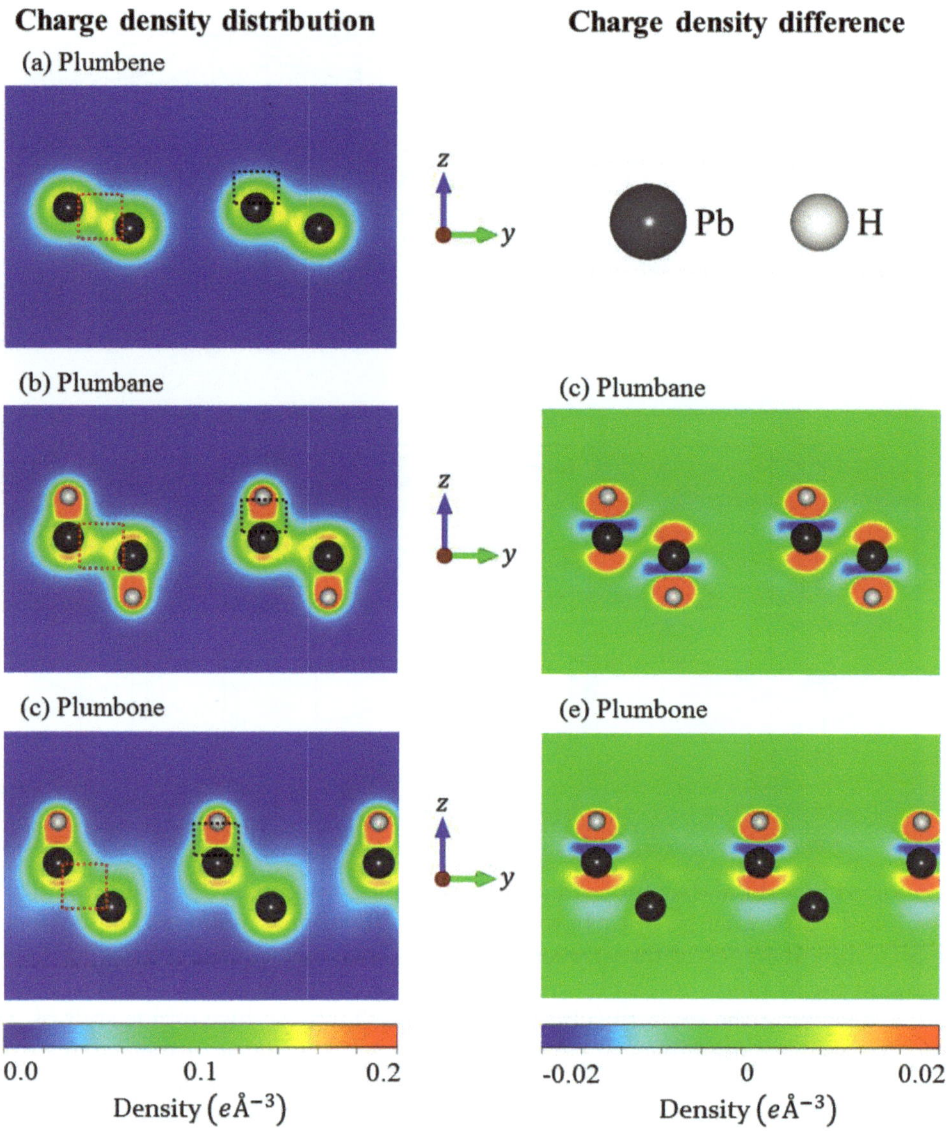

FIG. 7.5 To present the active multi-orbital hybridizations of Pb—Pb and Pb—H bonds in hydrogenated monolayer plumbene system, the charge density distributions [ρ] and their variations [$\Delta\rho$] after hydrogenations are clearly displayed through the distinct conditions: (A) a pristine system (B)/(C) plumbane and (D)/(E) plumbone.

to the prominent peaks, the similar ones, and the drastic plasmon-edges. From the calculated results of the former two, the π plasmon, the second σ-electronic collective excitations, and the third σ plasmon, respectively, survive 5.810, 13.10, and 16.80 eV/5.15, 12.70, and 16.20 eV in the absence/presence of excitonic effects (details in Table 7.1). Apparently, such energy-dependent plasmons belong to the quantization modes of the coherent charge

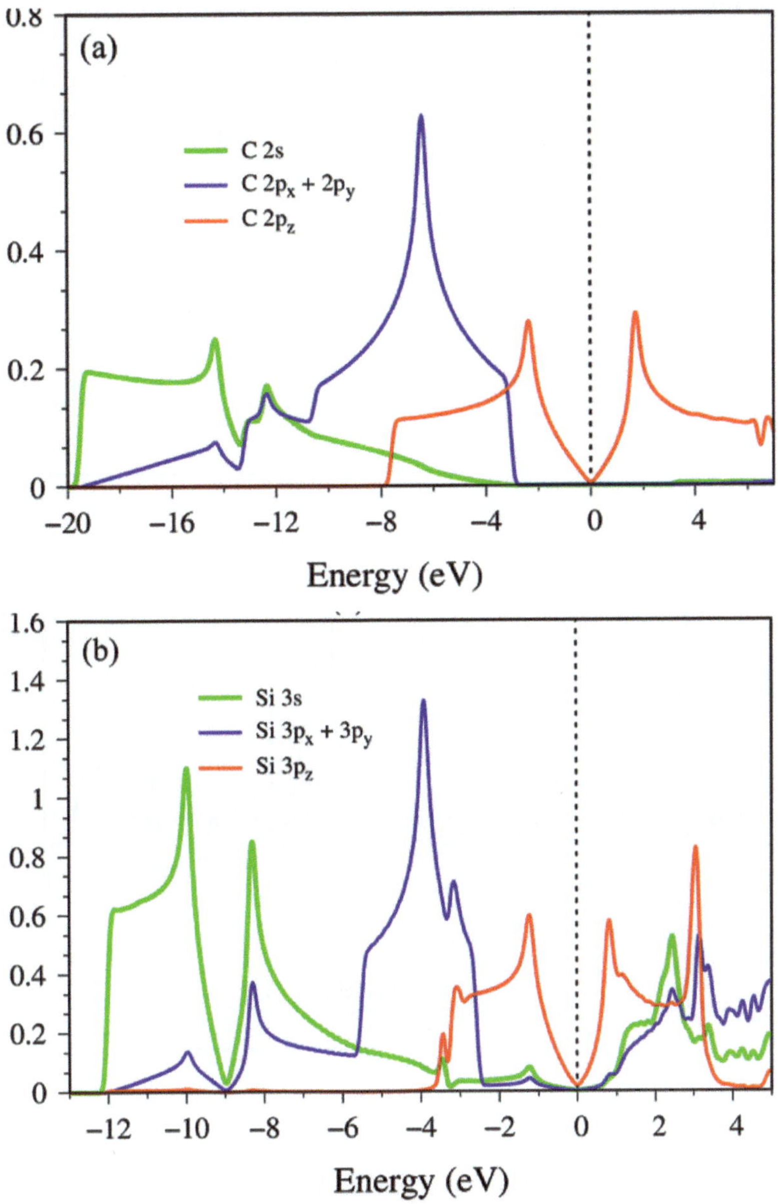

FIG. 7.6 To identify the significant multi-orbital hybridizations of X—X and/or X—H bonds (X=C/Si/Ge/Sn/Pb), the van Hove singularities within a sufficiently wide energy range are decomposed into the atom- and orbital-dependent parts in the presence of spin-orbital for (A)/(B)/(C)/(D)/(E) pristine graphene/silicene/germanene/stanine/plumbene and for (F)/(G)/(H)/(I)/(J) graphene/silicane/germanane/stanine/plumbane.

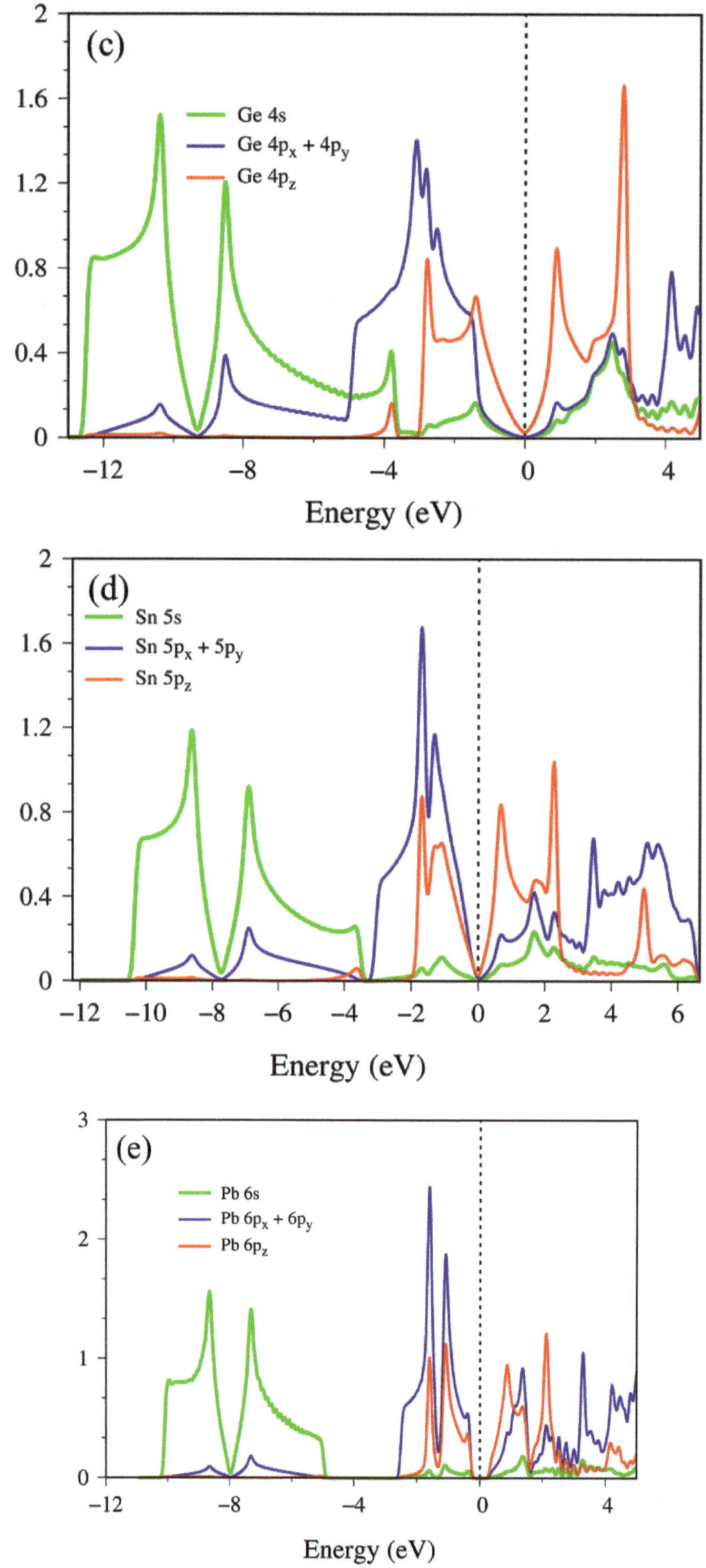
(c)
Ge 4s
Ge $4p_x + 4p_y$
Ge $4p_z$
Energy (eV)
(d)
Sn 5s
Sn $5p_x + 5p_y$
Sn $5p_z$
Energy (eV)
(e)
Pb 6s
Pb $6p_x + 6p_y$
Pb $6p_z$
Energy (eV)

FIG. 7.6 cont'd

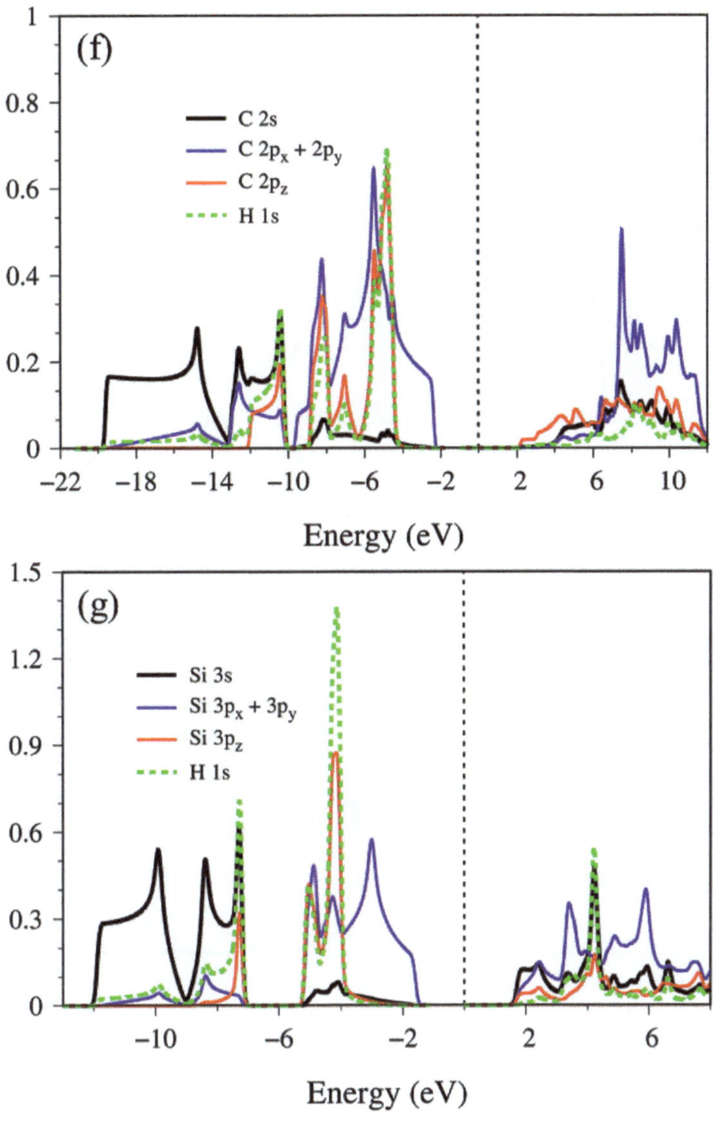

FIG. 7.6 cont'd

oscillations under the long wavelength limit, corresponding to the different orbital-dominated valence electrons. Compared with those in the Bernal bulk graphite, the later only exhibits the π and $\pi + \sigma$ plasmon modes, with frequencies higher than 7 and 20 eV [187], respectively. The interlayer atomic interactions should play the critical roles in greatly enhancing the frequency and strength of collective excitations. Of course, the dimensionality crossover behaviors of plasmon modes should be an interesting issue. There also exist the drastic decline near the plasmon frequencies for the frequency-related reflectance spectra

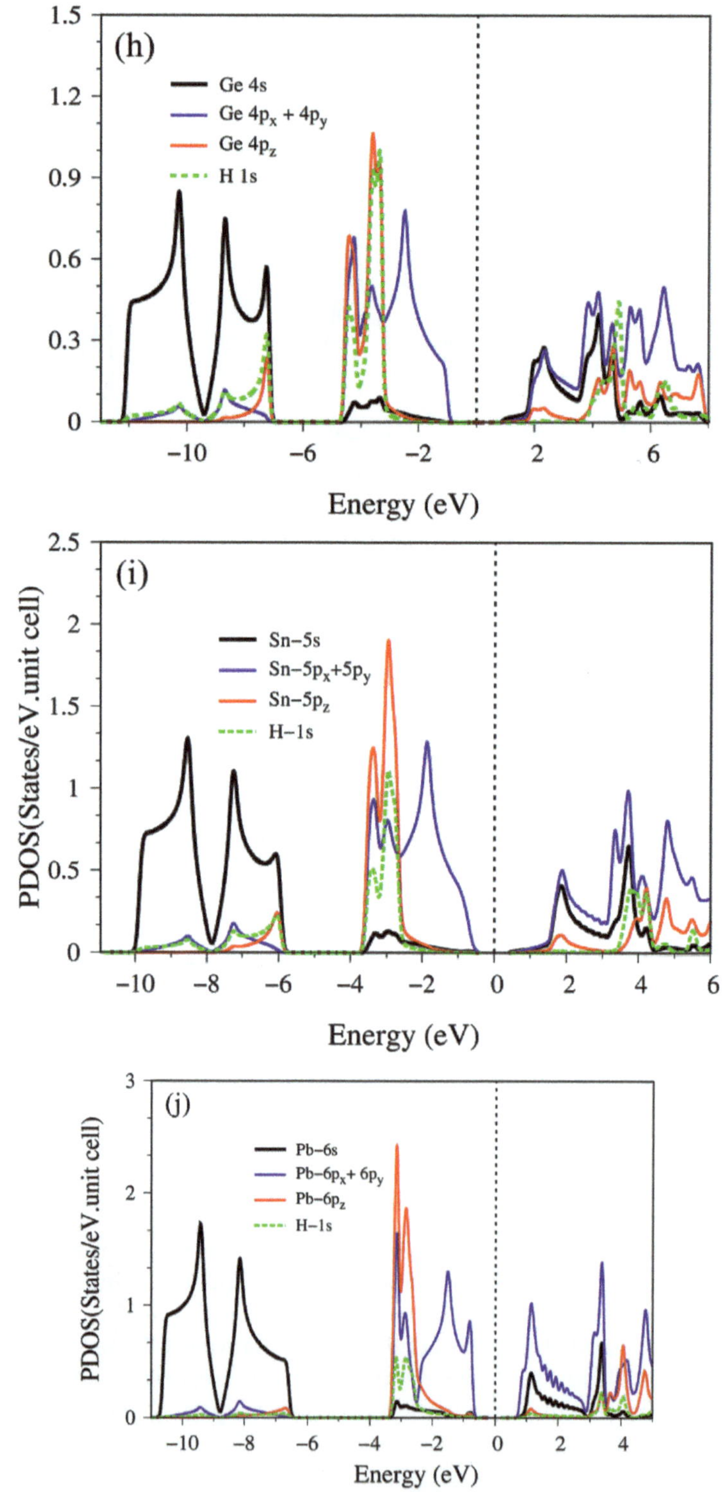
(h)
Ge 4s
Ge 4px + 4py
Ge 4pz
H 1s
Energy (eV)
(i)
Sn−5s
Sn−5px+5py
Sn−5pz
H−1s
PDOS(States/eV.unit cell)
Energy (eV)
(j)
Pb−6s
Pb−6px+ 6py
Pb−6pz
H−1s
PDOS(States/eV.unit cell)
Energy (eV)

FIG. 7.6 cont'd

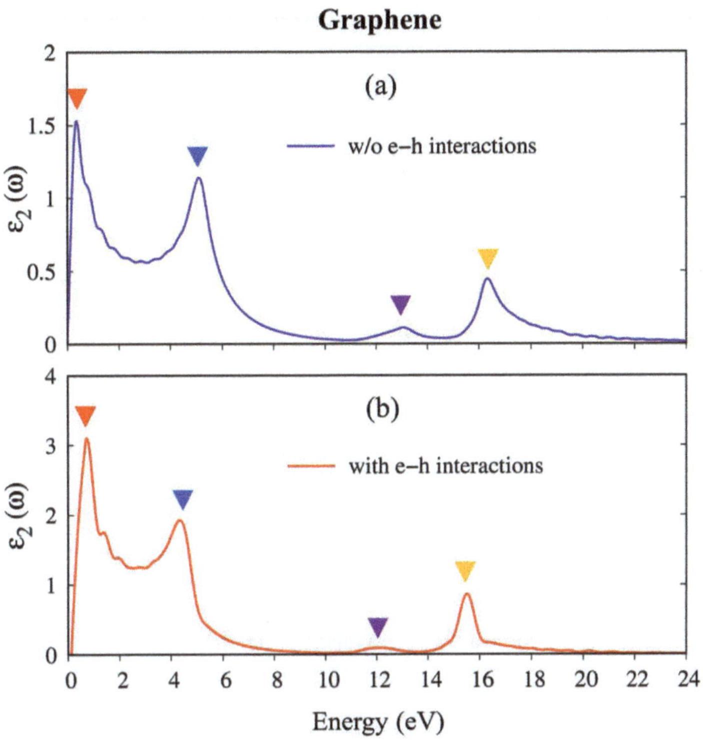

FIG. 7.7 The imaginary parts of dielectric function, without/with the inelastic electron-hole Coulomb scatterings (the red and blue curves) for (A)/(B) graphene, (C)/(D) silicene, (E)/(F) germanene, (G)/(H) stanene, (I)/(J) plumbene as well as, (K)/(L) graphene, (M)/(N) silicane, (O)/(P) germanane, (Q)/(R) stanine and (S)/(T) plumbane. Each prominent absorption structure mainly arises from the specific orbital hybridization using the joint van Hove singularity.

(Fig. 7.8A). Furthermore, the π and the σ-2s-dominated plasmons exhibit such features. Very interesting, $R(\omega)$ totally vanishes within the specific range 8 eV $< \omega <$ 15 eV. Most of the incident light are transmitted through a single-layer graphene where only a partial perturbation is absorbed by material (transferred to heat reservoir). The sensitive dependences of reflectance and transmittance on the layer number and stacking configuration is worthy of the systematic investigations (Fig. 7.10).

The orbital- and spin-dependent intrinsic interactions can greatly diversify the main features of optical excitations in group-IV pristine systems. Graphene (Fig. 7.7A and B), silicene (Fig. 7.7C and D), germanene (Fig. 7.7E and F), tinene (Fig. 7.7G and H) and plumbene (Fig. 7.7I and J) are quite different from each other in all relevant physical quantities. Monolayer silicene, which has the weakest sp^3 bondings and spin-orbital couplings (the smallest buckling crystal [188]), exhibit the featured quasiparticle transitions in terms of the effective frequency range, the enhanced electron-hole excitation intensities, and few prominent interband absorption channels. The silicon-[3s, $3p_x$, $3p_y$, $3p_z$] orbital are able to generate a total occupied bandwidth of ~12.0 eV (Fig. 7.4B), being much smaller than that of graphene (~19.60 eV in Fig. 7.4A). This directly reflects the larger Si—Si bond lengths and the relatively

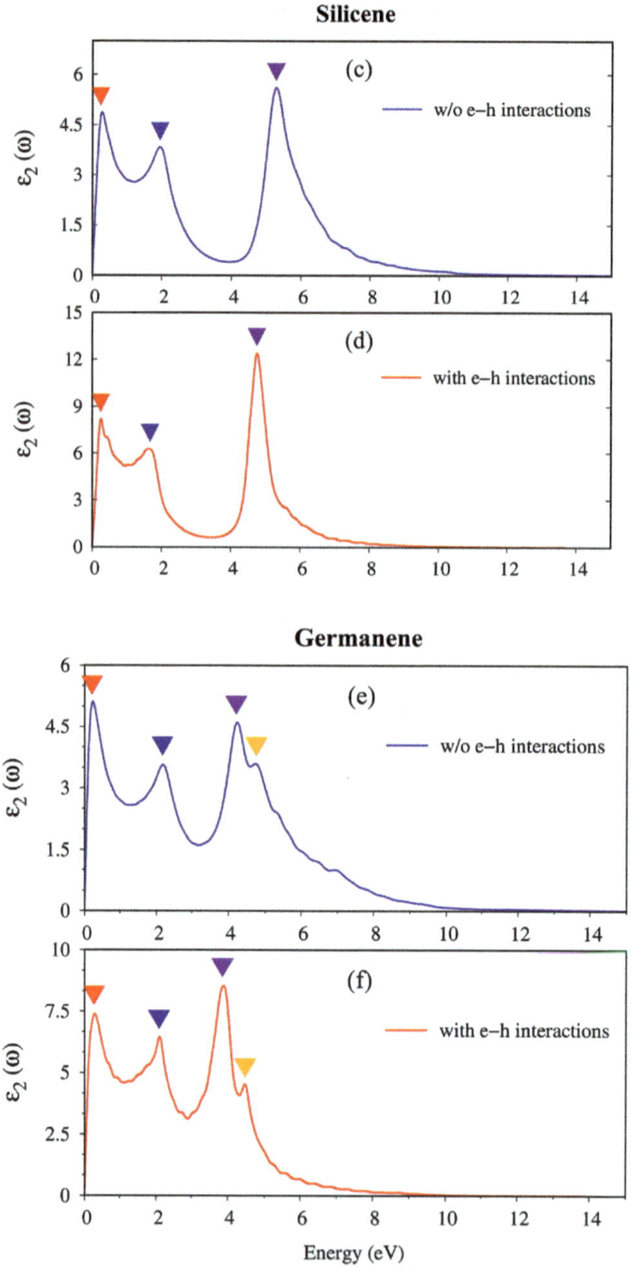
Silicene
(c)
w/o e–h interactions
ε2 (ω)
(d)
with e–h interactions
ε2 (ω)
Germanene
(e)
w/o e–h interactions
ε2 (ω)
(f)
with e–h interactions
ε2 (ω)
Energy (eV)

FIG. 7.7 cont'd

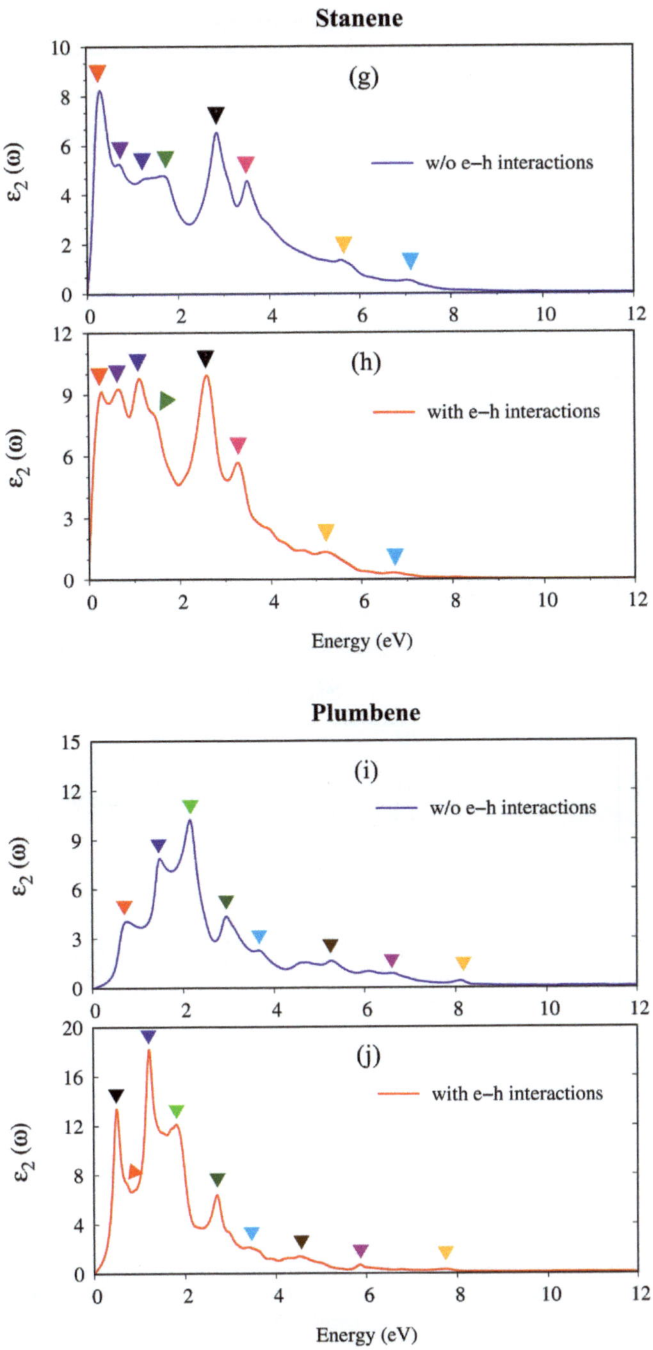
Stanene
(g)
w/o e–h interactions
(h)
with e–h interactions
ε2 (ω)
Energy (eV)
Plumbene
(i)
w/o e–h interactions
(j)
with e–h interactions
ε2 (ω)
Energy (eV)

FIG. 7.7 cont'd

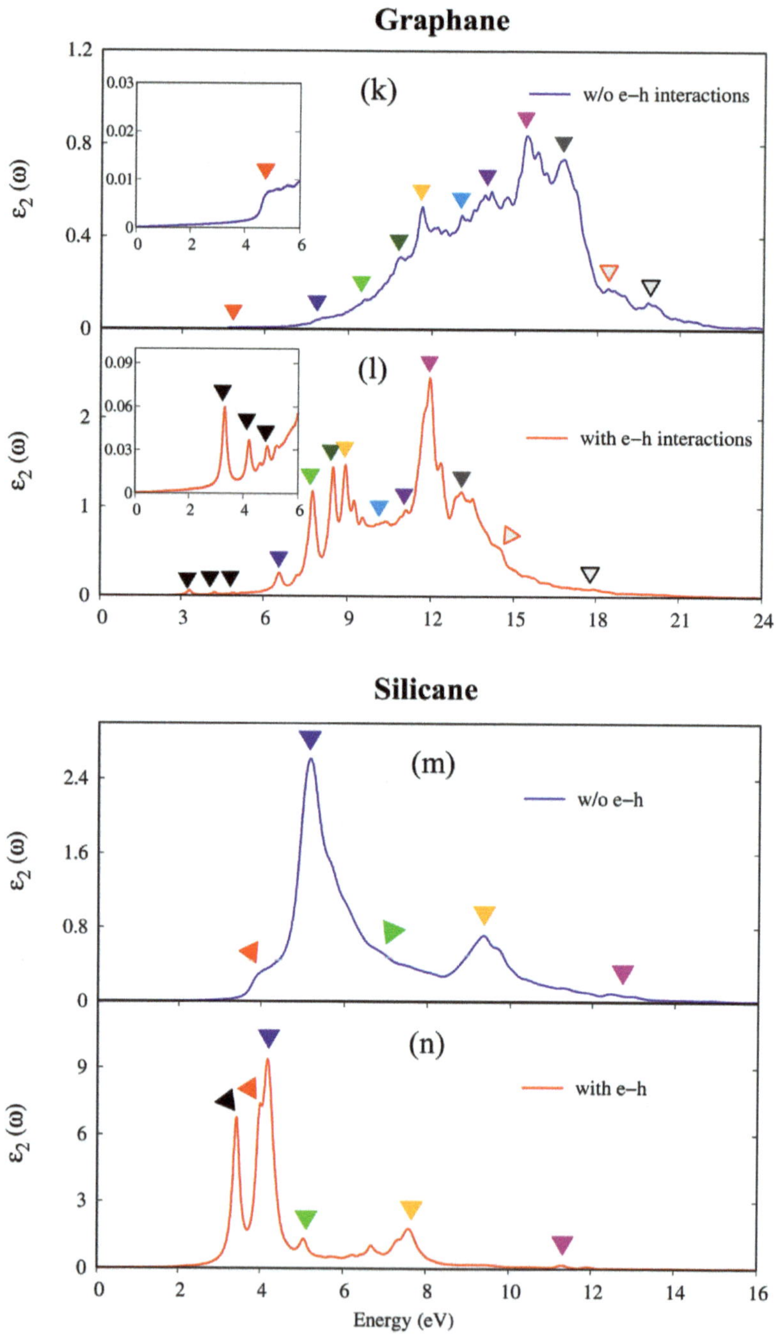
Graphane
(k)
w/o e–h interactions
(l)
with e–h interactions
ε2 (ω)
Silicane
(m)
w/o e–h
(n)
with e–h
Energy (eV)

FIG. 7.7 cont'd

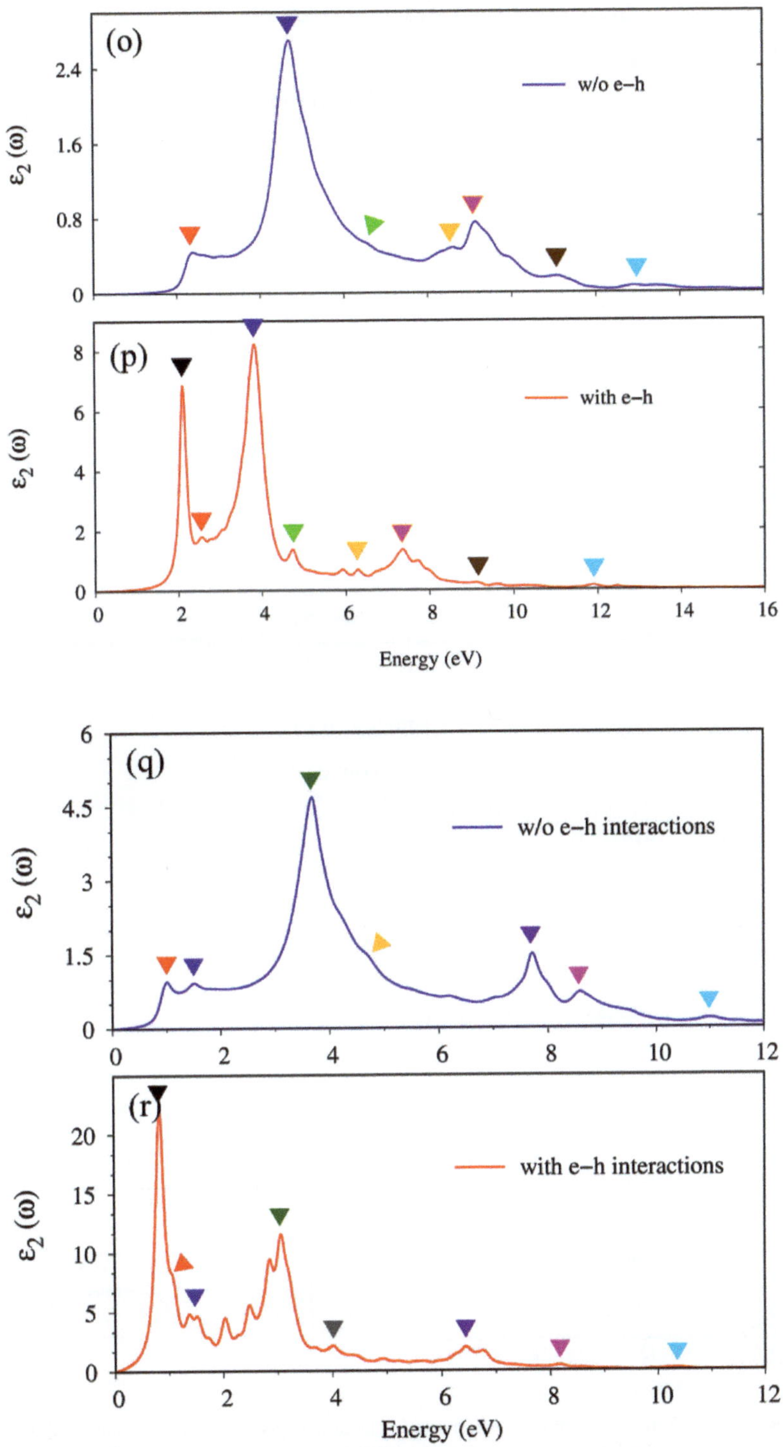
(o)
w/o e–h
(p)
with e–h
ε2 (ω)
Energy (eV)
(q)
w/o e–h interactions
(r)
with e–h interactions
ε2 (ω)
Energy (eV)

FIG. 7.7 cont'd

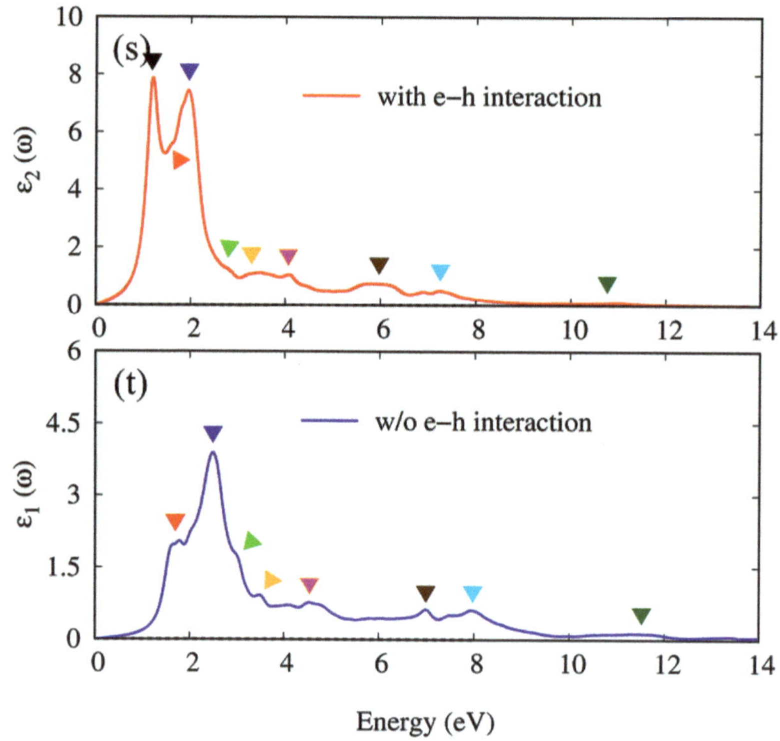

FIG. 7.7 cont'd

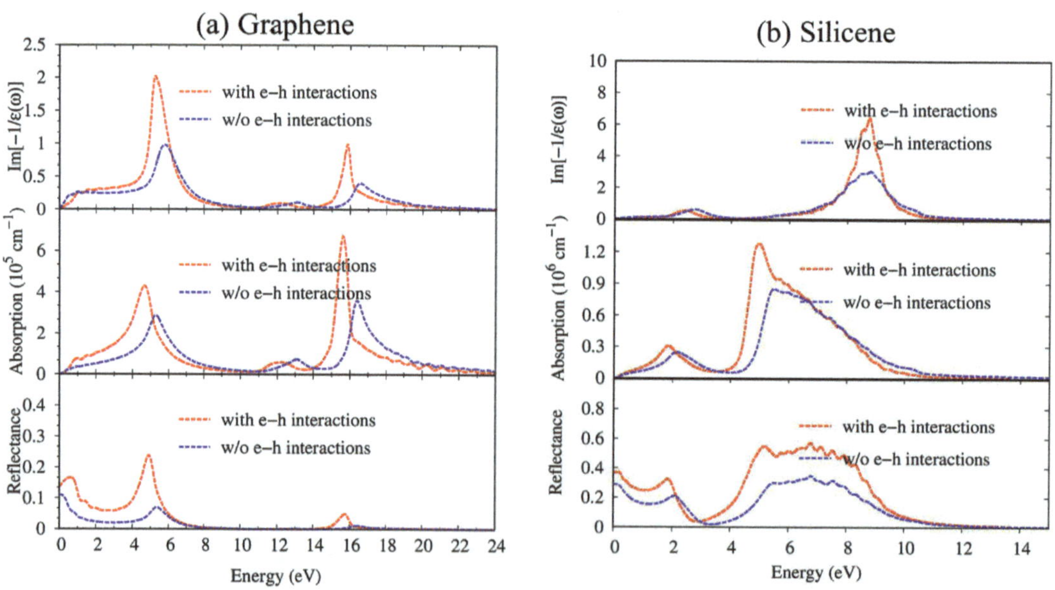

FIG. 7.8 Energy loss functions/absorption coefficients/reflectance spectra in the presence and absence of excitonic effects (the solid red and dashed blue curves) for (A)/(B)/(C)/(D)/(E) graphene/silicene/germanene/stanine/plumbene and (F)/(G)/(H)/(I)/(J) graphene/silicane/germanane/stanine/plumbane. Such screened response functions cover all the significant effects of band structure, charge screening, and Coulomb scattering.

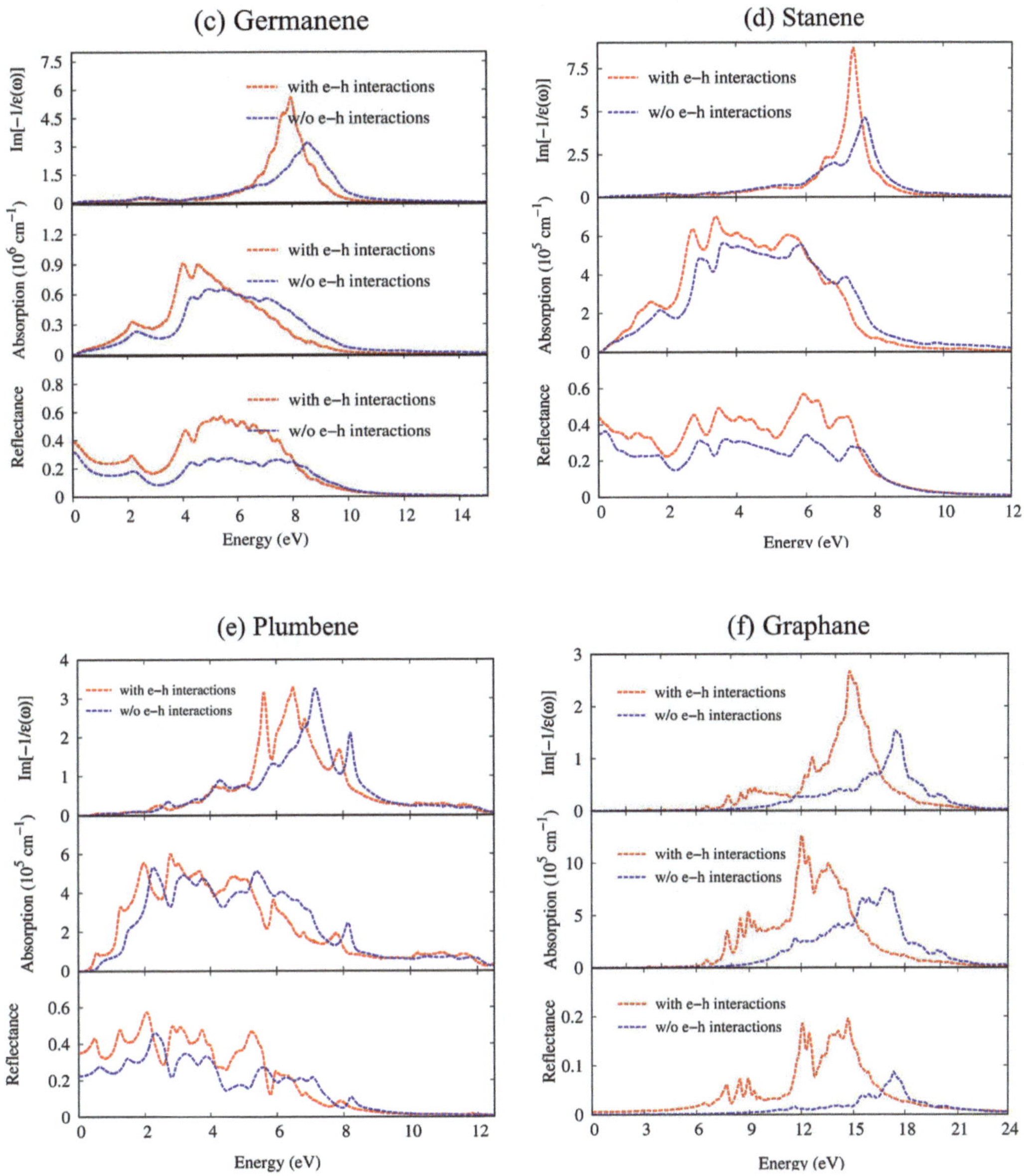

FIG. 7.8 cont'd

weak σ bondings. Furthermore, the 3s-orbitals do not induce the strong optical responses by the delicate calculations and analyses, as clearly illustrated in Table 7.2B. The critical factor should be the negligible dipole matrix elements associated their band-edge states. The efficient optical transitions are only revealed within a narrow window below 8 eV. The magnitude of dielectric function, the bare response function, become stronger, e.g., the maxima of

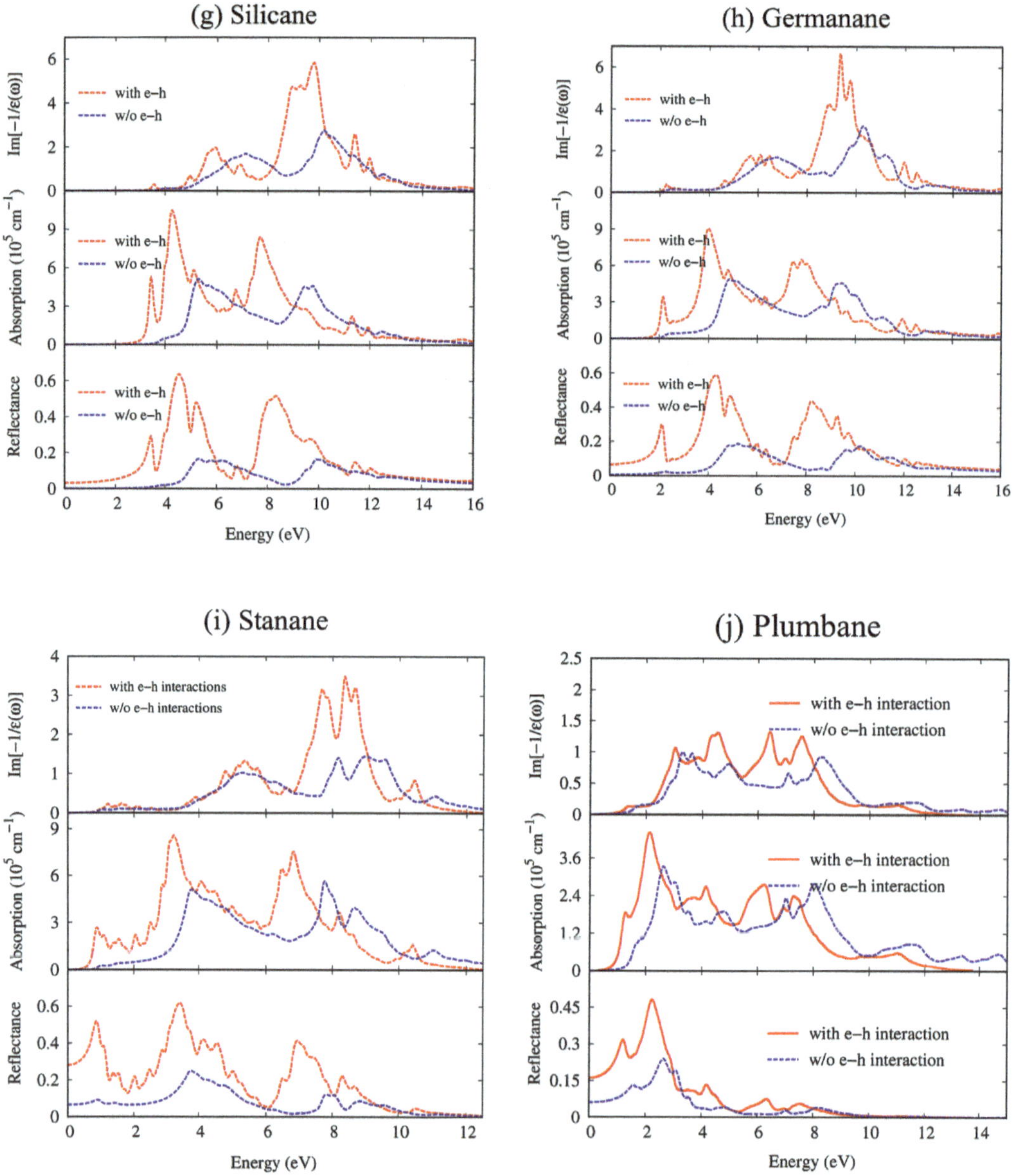

FIG. 7.8 cont'd

$\varepsilon_2(\omega)$, respectively, corresponding to ~6 and ~3 eV for silicene (Fig. 7.7C and D) and graphene (Fig. 7.7A and B). Such behavior is closely related to the sum rule of the total single-particle excitations (the total charge density). Apparently, silicene shows three single-particle absorption structures at 0.28 eV/0.27 eV (red arrow), 1.90 eV/1.72 eV (blue arrow), and 5.25 eV/4.74 eV (purple), in which they are, respectively, examined to mainly arise

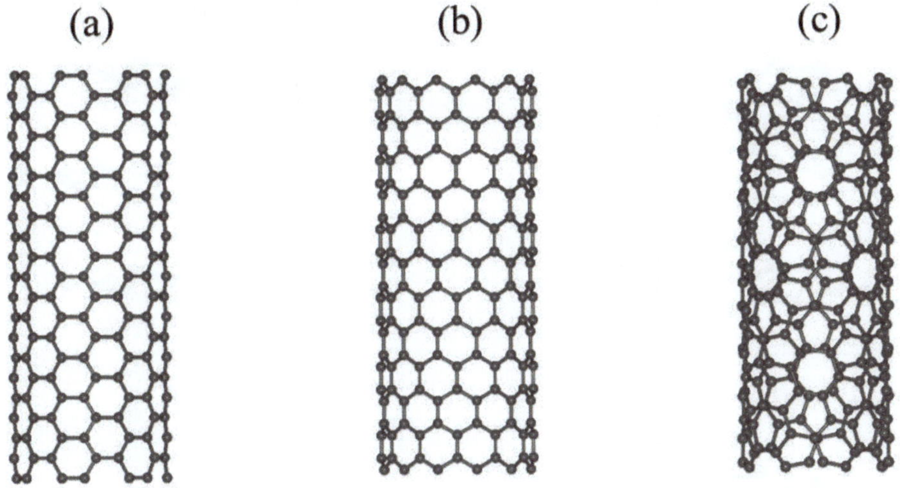

FIG. 7.9 Geometric structures of cylindrical single-walled carbon nanotubes for (A) armchair, (B) zigzag and (C) chiral types.

from the valence and conduction Dirac cones (the vertical arrow in Fig. 7.4B), the inter-π-band transitions of Si-3p_z orbitals near the M saddle, and the σ- and π-electronic excitation channel for through [3p_x, 3p_y] → 3p_z (the second valence and conduction energy subbands). The excitonic effects only lead to the observable red shifts in the second and third absorption

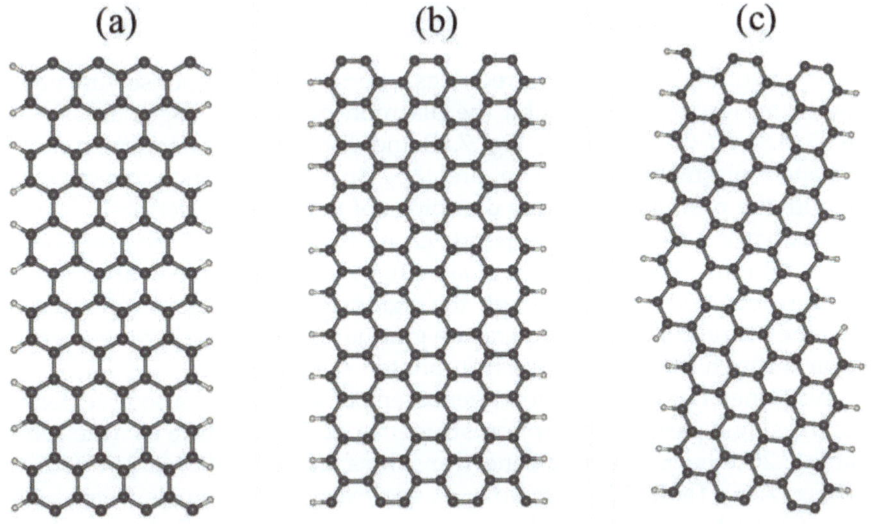

FIG. 7.10 Crystal structures for (A) armchair, (B) zigzag and (C) chiral graphene nanoribbons with hydrogen passivation on two open edges.

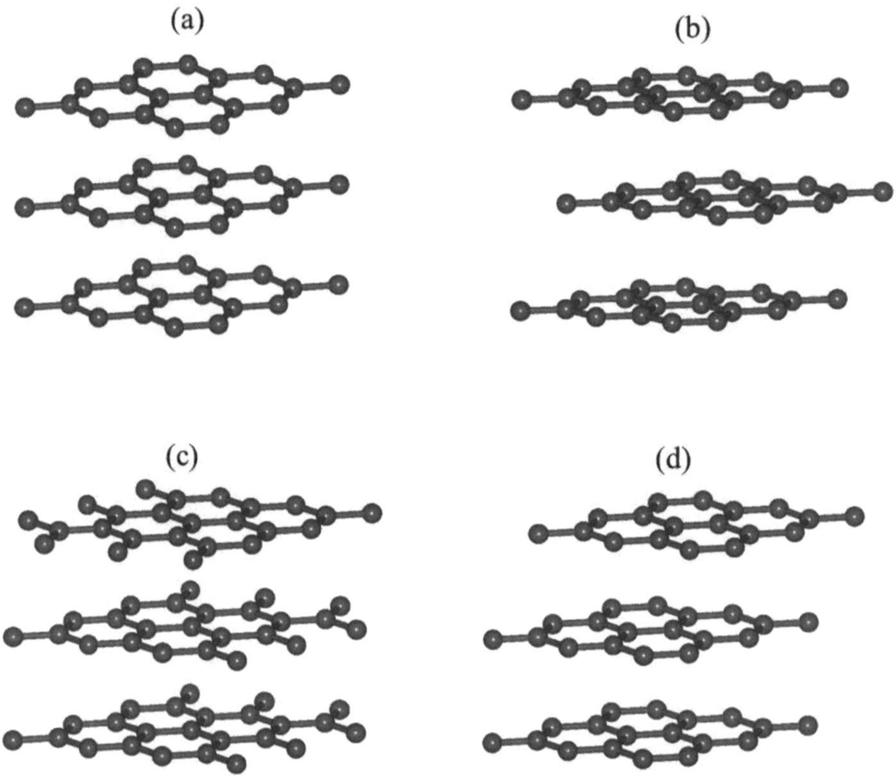

FIG. 7.11 Trilayer graphene systems with the stacking configurations of (A) AAA, (B) ABA, (C) AAB.

structures. After including all charge screenings, the screened response function and the reflectance spectrum, respectively, exhibit the plasmon peaks and edges at $\omega_P \sim 2.80$ eV/ 2.37 eV and 8.85/8.83 eV (the details in Fig. 7.8B and Table 7.4B). The former and the latter belong to the π and σ plasmon modes of $3p_z$ and [$3p_x$, $3p_y$] orbitals. Such collective excitations could co-exist with the prominent Landau dampings of the excited electron-hole pairs. These two kinds of quasiparticle excitations cooperate each other to present the prominent absorption coefficient within the frequency range of $\omega < 10$ eV. They are able to absorb the incident energy of an electromagnetic wave through the significant couplings of phonons and electrons, and then the average sample temperature gradually grows during the energy transfer to heat reservoir through the inelastic electron-photon scatterings.

The above-mentioned optical excitation phenomena of silicene are also revealed in germanene, such as, the single-particle absorption structures at 0.25 eV/0.26 eV (the red arrow in Fig. 7.4C; Table 7.2C; the inter-Dirac-cone transitions in Fig. 7.4C), 2.21 eV/2.05 eV (the blue arrow in Fig. 7.4C; Table 7.2C; the 4pz $\rightarrow$ 4pz excitation channels from the M saddle point in Fig. 7.4C), and 4.23 eV/3.95 eV (the purple arrow in Fig. 7.4C; the ($4p_x$, $4p_y$) $\rightarrow$ $4p_z$ transitions associated with it). However, there exists an additionally prominent peak at 4.80 eV/4.45 due to the dominating orbitals of ($4p_x$, $4p_y$, $4p_z$) $\rightarrow$ $4p_z$. Such structure is purely induced by the band-edge states under the π and σ anti-crossing behaviors (the vertically yellow arrow in Fig. 7.4C), since the joint density of states, corresponding the valence and conduction energy subbands, is sufficiently large for the hole and electron energy spectra. That is to say, this is an obvious evidence of the non-perpendicular π and σ in a buckled Ge-honeycomb crystal. As a result, the π- and σ-electronic collective excitations, respectively, come to exist 2.60 eV/2.57 eV ($4p_z$) and 8.60 eV/7.90 eV ($4p_x$, $4p_y$, $4p_z$).

Very interesting, both pristine tinene and plumbene systems exhibit more rich and complicated absorption structures in the imaginary-/real-part dielectric functions, screened response functions, absorption coefficients, and reflectance spectra functions without/with the excitonic effects (the dashed blue/red curves). Apparently, the featured optical transitions directly reflected the unusual band structures and wave functions. The previously delicate works, which are done by the VASP simulations [17] and the tight-binding models [22], have shown a close cooperation between the significant sp^3 chemical bondings and spin-orbital couplings. This further leads to the strong modifications on the band properties near the Fermi level (Fig. 7.4D and E), the energy dispersions of valence and conduction energy subbands near the K, M and Γ points, their crossing and anti-crossing phenomena, and the greatly enhanced [5s, $5p_x$, $5p_y$, $5p_z$]-/[6s, $6p_x$, $6p_y$, $6p_z$]-dependences for all electronic states. As a result, there are more band-edge states, with zero group velocities, being responsible for many joint van Hove singularities. Furthermore, this is fully assisted by the finite photon-electron scattering probabilities through the four-orbital significances. By the direct combination of the orbital-dominated energy spectrum/critical points (Fig. 7.4D and E) and density of states (Fig. 7.7D and E), the rich single-particle absorption structures of monolayer tinene (Fig. 7.7D) appear at 0.29 eV/0.3 eV (the red arrow due to inter-Dirac-cone transitions of Sn-$5p_z$ borbitals near the K valleys), 0.69 eV/0.67 eV (the purple arrow associated with the vertical inter-σ-band excitations of ($5p_x$, $5p_y$) $\rightarrow$ ($5p_x$, $5p_y$) from the Γ point), 1.54 eV/1.51 eV (the blue arrow due to the inter-π-band of $5p_z$ $\rightarrow$ $5p_z$ from the saddle M point), 1.90 eV/1.85 eV (the green arrow arising from the hybridized channels of ($5p_x$, $\rightarrow$ $5p_y$) $5p_z$ at the M point), 2.79 eV/2.62 eV (the black arrow corresponding to $5p_z$ $\rightarrow$ $5p_z$ of the first valence and second conduction band from the M point), 3.50 eV/3.41 eV (the pink arrow indicating the ($5p_x$, $5p_y$, $5p_z$) $\rightarrow$ $5p_z$ on the ΓM path), 5.60 eV/5.52 eV (the yellow arrow illustrating the available 5s $\rightarrow$ $5p_z$ at the Γ valley), and 7.20 eV/6.90 eV (the cyan arrow showing the dominating 5s $\rightarrow$ $5p_z$ channel under the M point), The very complicated electron-hole excitations can largely suppress the coherent charge oscillations of [5s, $5p_x$, $5p_y$, $5p_z$] orbitals within the lower-frequency range; that is, they experience rather strong Landau dampings for $\omega < 5$ eV. The frequency-dependent energy loss function, absorption coefficient and reflectance spectrum (Fig. 7.7D) only two featured plasmon modes at 6.00 eV/5.80 eV and 7.40 eV/7.2 eV. Apparently, the prominent π-electronic collective excitations of graphene/silicene/germanene systems disappears in tinene/plumbene ones.

The unusual optical behaviors, directly reflects the featured band structure (Fig. 7.4E), are also revealed in a pristine monolayer plumbene (Figs. 7.6E and 7.7E; Tables 7.2E and 7.4E), such as, the prominent electron-hole excitation structures at 0.60 eV/0.55 eV (the red arrow from the vertical inter-Dirac-cone excitations dominated by Pb-$6p_z$ orbitals near the K valleys), 1.73 eV/1.56 eV (the blue arrow related to the inter-π-band transitions of the saddle M point), 2.20 eV/1.88 eV (the purple arrow due to the ($6p_x$, $6p_y$, $6p_z$) $\rightarrow$ $6p_z$ from the second valence and first conduction subbands at the M point), 2.95 eV/2.74 eV (the green arrow arising from the hybridized channels of ($6p_x$, $6p_y$, $6p_z$) → ($6p_x$, $6p_y$) in the second pair of valence and conduction subbands along the $K\Gamma$ path), 3.55 eV/3.50 eV (the cyan arrow corresponding to ($6p_x$, $6p_y$, $6p_z$) → ($6p_x$, $6p_y$, $6p_z$) through the first valence and third conduction subbands under the ΓM path), 5.31 eV/4.75 eV (the brown arrow indicating the dominant 6s → ($6p_x$, $6p_y$ channel of the third valence and first conduction subbands near the Γ point), 6.15 eV/5.90 eV (the pink arrow illustrating the available 6s → $6p_z$ transition of the third valence and second conduction subbands at the Γ valley), and 8.14 eV/7.90 eV (the yellow arrow showing the dominating 6s → $6p_z$ channel through the third valence and second conduction subbbands at the M point). In general, each kind of prominent single-particle excitation channel is accompanied by a specific plasmon mode. However, the various Landau dampings frequently overlap one another, leading to the strong suppression of the lower-frequency plasmon modes. As a result, the screened response function, absorption coefficient and reflectance spectrum (Fig. 7.8E) display four plasmon modes at 6.0 eV/5.6 eV, 7.0 eV/6.60 eV, 7.60 eV/7.20 eV and 8.30 eV/8.00 eV. It should be noticed that the well-known π plasmon modes in graphene/silicene/germanene systems characteristic of the dominating sp^2 bondings are absent in tinene/plumbene ones because of too strong sp^3 hybridizations. Very interesting, among monolayer group-IV materials, only plumbene shows a stable exciton peak at the specific threshold frequency of 0.49 eV (the black arrow), being stronger than the band-gap structure initiated from the K valley (the red arrow). This middle-gap system of E_g^d, which has the weakest charge screening ability, cannot quickly reduce the long-range attractive Coulomb potential and thus built the stable/quasi-stable excitonic bound states. The exciton-induced optical excitations could be further verified by the photoluminescence spectroscopy [189,190].

7.6 Coupling effects of hydrogenations and excitons

When the excited valence holes and conduction electrons are simultaneously created by the external photons with almost vanishing momenta, such pairs would be effective bound together through the significant charge screenings of all carriers. The non-linear many-body effects (the excited electron-hole pairs with the important Coulomb electron-electron interactions), which dominate the featured absorption spectra, could be included in the first-principles simulations by solving the bi-particle Bethe Salpeter equation (the theoretical details in Chapter 2). The stable/metastable excitonic states are determined by the Coulomb-field screening abilities so that their existences strongly depend on the intrinsic material properties (the longitudinal static dielectric function [118]). They can drastically change the pronounced characteristics of optical properties, such as, the frequency, number, intensity and form of non-negligible absorption structures, especially for threshold frequency (an optical gap in Refs. [110,111]). The observable decays of excitonic states, being due to the inelastic scatterings, are associated with the photoluminescence spectra, being frequently utilized in laser applications [191,192].

Very interesting, the strong couplings of hydrogenations and excitons lead to the diversified optical excitation phenomena in group-IV systems, as clearly illustrated in Fig. 7.7F–J under the double-side saturated hydrogen adsorption. The top-side hydrogen-adatom chemisorption cases have generated the different crystal structures for graphene/silicane/germanane/tinane/plumbane through the rather strong chemical bondings of ssp^3 bondings, the destruction of π-electronic hybridizations and the observation modifications of three σ-orbital ones. As a result, there are more complicated valence and conduction energy subbands with the unusual energy dispersions (Fig. 7.4F–J) and band-edge states. The drastic changes in the wave-vector-dependent energy spectra and wave functions further create the dramatic variations about the whole vertical transition spectra, mainly owing to more available joint van Hove singularities. For example, the number, form, intensity, and frequency of prominent absorption structures are quite different from among five hydrogenated systems, e.g., the distinct excitonic states below or near the single-particle threshold frequency. The full VASP simulations, which are conducted on the rich and unique optical properties associated with chemical adsorptions and many-body effects, could be verified by using the high-precision measurements about reflectance/absorption/transmission/photoluminescence spectroscopies (details in Chapter 3.4). How to build the 2D exciton theory through the phenomenological manners is an another research topic.

Graphane, with an unusual electronic structures (Fig. 7.4F), clearly reveals the rich optical transitions in the imaginary-and real-part bare dielectric functions (Fig. 7.7L and K), screened energy loss spectra, absorption coefficients and reflectance spectra (Fig. 7.7F) in the absence/presence of excitonic effects. Compared with those of a pristine graphene, there exist more prominent absorption structures, directly reflecting the featured band structure after hydrogen adatom chemisorptions. The concise orbital hybridizations of the initial valence and final conduction states, being responsible for all significant transitions, are thoroughly examined from the consistent physical quantities, in which the atom-dominated energy spectrum, the atom- and orbital-decomposed van Hove singularities (Fig. 7.6F), and the strong absorption peaks (Fig. 7.7L and K) are unified together under a quasiparticle framework (books). The evaluated results are clearly shown in Table 7.2F, being assisted by the various vertical arrows of the featured band structure (Fig. 7.4F). The distinct excitation channels cover the pronounced absorption peaks in the absence/presence of excitonic effects: (I) the single-particle threshold frequency of 4.70 eV/4.80 eV due to the first pair of valence and conduction energy sub-bands of C-($2p_x + 2p_y$) $\rightarrow$ C-$2p_z$ from the Γ valley (the red arrow in Fig. 7.4F), (II) the 8.00 eV/6.60 eV absorption peak from the first valence and third conduction energy subbands through C-($2p_x + 2p_y$) $\rightarrow$ C-(2s, $2p_x$, $2p_y$, $2p_z$) near the Γ point (the blue arrow), (III) the prominent 9.70-/7.90-eV excitations related to the third valence and second conduction subbands under the available H-1s and C-$2p_z$ $\rightarrow$ C-$2p_z$ channel from the Γ valley (the light green arrow), (IV) the 10.20-/8.20-eV absorption structure due to the first and second valence and the fifth and sixth conduction sub-bands through the C-($2p_x$, $2p_y$) $\rightarrow$ H-1s and C-(2s, $2p_x$, $2p_y$, $2p_z$) close to the Γ valley (the green arrow), (V) the 11.40 eV-/9.00 eV-strong responses associated with the first pair of valence and conduction sub-bands using the C-($2p_x$, $2p_y$) $\rightarrow$ C-(2s, $2p_x$, $2p_y$, $2p_z$) at the M valley (the yellow arrow), (VI) the 13.10 eV-/11.00 eV-absorption peak from the valence and conduction sub-bands nearest to E_F by the C-($2p_x$, $2p_y$) $\rightarrow$ H-1s and C-(2s, $2p_x$, $2p_y$, $2p_z$) along the K Γ path (the light blue arrow), (VII) the observable 14.10-/10.50-eV excitation peak from the first valence and fourth conduction sub-bands along the MK path through the channel of C-($2p_x$, $2p_y$) $\rightarrow$ H-1s and C-(2s, $2p_x$, $2p_y$, $2p_z$) (the purple arrow), (VIII) the significant 15.30-/12.00-eV transition between the third valence and second conduction band on the MK path under the multi-orbital hybridizations of H-1s and C-(2s, $2p_x$, $2p_y$, $2p_z$) $\rightarrow$ H-1s and C-(2s, $2p_x$, $2p_y$) (the pink arrow), (IX) the pronounced 16.90-/13.20-eV excitation peak dominated by the third pair of valence and conduction sub-bands about the Γ M path using the whole ssp^3 mixings in C−H bonds (the gray arrow), (X) the strong 18.30-/14.70-eV absorption peak arising from the fourth pair of valence and conduction sub-bands with the complete ssp^3 bondings (the white-red arrow), as well as, (XI) the highest 20.10-/17.80-eV absorption structure determined by the fifth valence and the first conduction subbands in the presence of H-1s and C-(2s, $2p_x$, $2p_y$) → C-$2p_z$ from the M valley (the white-black arrow). As to the featured plasmon modes, their characteristic frequencies are obtained from rather strong structures in energy loss functions, absorption coefficients and reflectance spectra (Fig. 7.8F), Roughly speaking, there exist three well-characterized collective excitations; that is, the whole ssp^3 valence carriers can coherently oscillate within three different energy ranges of 11.60 eV/8.80 eV, 15.90 eV/12.30 and 17.50 eV/15.00 eV.

Very interesting, graphane exhibits the significant excitonic effects in terms of the number, frequency, intensity and form of prominent absorption structures. Specifically, there exist here lower-frequency observable absorption peaks in the inset of Fig. 7.7L. Their characteristic frequencies come to exist at 3.30 eV, 4.20 and 4.75 eV, where they are measured from the lowest unoccupied conduction band-edge states. Whether the stable/quasi-stable bound states of the excited electron-hole pairs could survive is mainly determined by their effective masses (the inverse of band curvatures) and all charge screening abilities (the great reduce of the attractive bare Coulomb potentials). The latter might be too strong to form them in the metallic/semi-metallic/zero-gap semiconducting systems, being obviously indicated for the pristine group-IV materials. The VASP simulations, which predict three kinds of excitonic states in a wide-gap graphene, could be verified from the high-resolution photoluminescence measurements [193]. The similar VASP studies have been successfully done for the other hydrogenations of silicene/germanene/tinene/plumbene (discussed later). According to the diversified transition behaviors, the reliable 2D exciton theory, corresponding to the many-particle effects, is urgently required in the near-future science researches because of a plenty of emergent layered materials [194]. The strong cooperation among the first-principles simulations, the phenomenological models, and the high-precision optical examinations are capable of creating much progress in the basic and applied sciences.

Silicane is quite different from graphene in optical excitations, as clearly illustrated in the dielectric functions (Fig. 7.7M and N), screened energy loss functions, absorption coefficients, and reflectance spectra (Fig. 7.8G). Through the delicate VASP calculations and analyses among all consistent quantities, the well-characterized vertical transitions (Table 7.3G), corresponding to the specific orbital hybridizations, could be classified into five typical channels: (I) the single-particle threshold 3.80-/3.80-eV absorption structure arising from the first pair of valence and conduction sub-bands through the Si-($3p_x$, $3p_y$) → Si-(3s, $3p_x$, $3p_y$, $3p_z$) multi-orbital mixings at the stable Γ valley (the red arrow in Fig. 7.4G; the details in the atom- and orbital-projected van Hove singularities of Fig. 7.6G), (II) a rather strong 5.20-/4.20-eV absorption peak from the saddle M point of the former sub-bands in (I) using the identical orbital hybridizations (the blue arrow), (III) the observable 6.80-/5.00-eV absorption peak associated with the third valence and first conduction sub-bands on the Γ M path under the whole ssp^3 bondings for the initial and final states (the green arrow), (VI) the prominent 9.40-/7.70-eV absorption structure initiated from the third valence and second conduction sub-bands at the M point by previous bondings (the yellow arrow), and (V) the highest-frequency 12.30/11.20-eV absorption peak between the fifth valence and first conduction

TABLE 7.3 (A)–(O): The following 10 tables, covering the specific relations between the prominent plasmons and the dominated orbital hybridizations in the absence and presence of excitonic effects for (A)–(E): (graphene, silicene, germanene, stanene, plumbene) and (F)–(J) (graphane, silicane, germanane, stanane, plumbane).

(A) Graphene

Energy (eV)		
w/o excitonic effects	**With excitonic effects**	**Specific orbital hybridizations**
5.81	5.15	π [C-(2pz)]
13.1	12.7	σ [C-(2s, 2px,2py)]
16.8	16.20	σ [C-(2s, 2px, 2py)]

(B) Silicene

	Energy (eV)			
Colors	**w/o excitonic effects**	**With excitonic effects**	**Specific orbital hybridizations**	
Red	0.28	0.27	Si-(3pz) → Si-(3pz)	K
Blue	1.90	1.72	Si-(3pz) → Si-(3pz)	M
Purple	5.25	4.74	Si-(3px, 3py) → Si-(3pz)	M

(C) Germanene

Energy (eV)		
w/o excitonic effects	**With excitonic effects**	**Specific orbital hybridizations**
2.6	2.57	π [Ge –(4*pz*)]
8.6	7.9	σ [Ge-(4s, 4px, 4py)]

(D) Stanene

Energy (eV)		
w/o excitonic effects	**With excitonic effects**	**Specific orbital hybridizations**
7.4	7.2	σ [Sn-(5s, 5px, 5py)]

(E) Plumbene

Energy (eV)		
w/o excitonic effects	**With excitonic effects**	**Specific orbital hybridizations**
5.9	5.7	σ [Pb-(6s, 6px, 6py)]
7.2	6.5	σ [Pb-(6s, 6px, 6py)]
8.3	7.9	σ [Pb-(6s, 6px, 6py)]

(Continued)

TABLE 7.3 (A)−(O): The following 10 tables, covering the specific relations between the prominent plasmons and the dominated orbital hybridizations in the absence and presence of excitonic effects for (A)−(E): (graphene, silicene, germanene, stanene, plumbene) and (F)−(J) (graphane, silicane, germanane, stanane, plumbane).—cont'd

(F) Graphane

Energy (eV)		
w/o excitonic effects	**With excitonic effects**	**Specific orbital hybridizations**
11.6	8.8	σ [C-(2s, 2px, 2py) & H-1s] + π [C-(2pz)]
15.9	12.3	σ [C-(2s, 2px, 2py) & H-1s] + π [C-(2pz)]
17.5	15.0	σ [C-(2s, 2px, 2py) & H-1s]

(G) Silicane

Energy (eV)		
w/o excitonic effects	**With excitonic effects**	**Specific orbital hybridizations**
7.0	5.9	σ [Si-(3s, 3px, 3py) & H-1s] + π [Si-(3pz)]
10.1	9.6	σ [Si-(3s, 3px, 3py) & H-1s]

(H) Germanane

Energy (eV)		
w/o excitonic effects	**With excitonic effects**	**Specific orbital hybridizations**
6.6	6.0	σ [Ge-(4s, 4px, 4py) & H-1s] + π [Ge-(4pz)]
10.3	9.65	σ [Ge-(4s, 4px, 4py) & H-1s]

(I) Stanane

Energy (eV)		
w/o excitonic effects	**With excitonic effects**	**Specific orbital hybridizations**
5.4	5.3	σ [Sn-(5s, 5px, 5py) & H-1s] + π [Sn-(5pz)]
8.1	7.7	σ [Sn-(5s, 5px, 5py) & H-1s] + π [Sn-(5pz)]
9.1	8.2	σ [Sn-(5s, 5px, 5py) & H-1s]

(J) Plumbane

Energy (eV)		
w/o excitonic effects	**With excitonic effects**	**Specific orbital hybridizations**
3.3	3.0	σ [Pb-(6s, 6px, 6py) & H-1s]
5.0	4.5	σ [Pb-(6s, 6px, 6py) & H-1s] + π [Pb-6p_z]
7.0	6.5	σ [Pb-(6s, 6px, 6py) & H-1s]
8.5	7.5	σ [Pb-(6s, 6px, 6py) & H-1s]

subbbands with the similar ones (the pink ones). When the incident frequency is beyond the excitonic absorption one of ~3.30 eV (Table 7.2G), the Landau dampings would greatly suppress the coherent carrier oscillations of coupled orbitals. This leads to only two plasmon modes at 7.00 eV/5.90 and 10.10 eV/9.60 eV, being attributed to the collective excitations of ssp^3 multi-orbital hybridizations. Very interesting, both π- and σ-electronic plasmons cannot/can survive after/before hydrogenations.

Obviously, there are five and 11 prominent excitation channels, respectively in silicane and graphene (Table 7.2F and G). Such significant differences, which directly reflect the main features of electronic energy spectra and atom- and orbital-projected van Hove singularities (Fig. 7.6F and G), principally arise from the complicated orbital hybridizations in the highly non-uniform chemical environments of buckled honeycomb crystals (Table 7.1). The former and the latter, respectively, have utilized the complex orbital hybridizations of [$3p_x$, $3p_y$]/[3s, $3p_x$, $3p_y$]/1s-[3s, $3p_x$, $3p_y$/[3s, $3p_x$, $3p_y$, $3p_z$]/1s-[3s, $3p_x$, $3p_y$, $3p_z$], and $2p_z$/1s-$2p_z$/[$2p_x$, $2p_y$]/[2s, $2p_x$, $2p_y$]/1s-[2s, $2p_x$, $2p_y$]/[2s, $2p_x$, $2p_y$, $2p_z$]/1s-[2s, $2p_x$, $2p_y$, $2p_z$]. These rich orbital mixings, taking part in the initial and final excitation states clearly illustrate a sharp contrast between these two hydrogenated systems. The further explorations show the strong evidences of the featured band structures and density of states. The atom- and orbital-dominated energy dispersions and band-edge states strongly affect the number, frequency, intensity and structure of the well-behaved absorption peaks, since they determine the strengths of available transition channels through the joint density if states and electric diploe moments. Although both systems possess the similar ssp^3 bonding after the double-side hydrogenation adsorption, the very weak, but coexistent π, σ, and sp^3-electronic orbital mixings (the strongly modified ones) are able to induce the diversified quasiparticle phenomena. These intermediate chemical bondings, which could survive during the vertical transitions, would play the critical roles in creating the diversified phenomena of electron-photon/orbital-photon couplings. Such viewpoints, being covered in the current book, are further developed for other group-IV hydrogenated materials.

Germanane is relatively close to silicane in terms of the fundamental properties. Its position-dependent chemical bonds of Ge—Ge and Ge—H generate the high-non-uniform physical/chemical/material environments, thus, accounting for the complicated multi-orbital hybridizations about (4s, $4p_x$, $4p_y$, $4p_z$)-(4s, $4p_x$, $4p_y$, $4p_z$) and 1s-(4s, $4p_x$, $4p_y$, $4p_z$). The concise chemical picture is clearly identified from the top-site optimal geometry, the atom-dominated energy spectrum, the spatial charge density with the deformed orbitals, and the orbital-merged van Hove singularities within the specific energy ranges. All consistent quantities obviously reveal that the hopping integrals among the various orbitals are very sensitive to the nearest-neighbor atoms in a primitive unit cell. The orbital-determined bonding strengths are strongly affected by the fact whether the original π-, σ- and sp^3-electronic bondings could survive after the prominent double-side hydrogenation. The detailed mechanisms are thoroughly explored from the whole optical behaviors. The main features of optical excitation spectra are classified into seven types: (I) the threshold 2.15—2.16-eV absorption peak from the first pair of valence and conduction subbands under the Ge-($4p_x$, $4p_y$) $\rightarrow$ Ge-(4s, $4p_x$, $4p_y$, $4p_z$) multi-orbital hybridizations at the Γ valley (the red arrow), (II) the very pronounced 4.70-/3.80-eV absorption structure due to the similar mechanism, but initiated from the saddle M point (the blue arrow), (III) the observable 6.60-/4.80-eV response associated with the third valence and first conduction subbands along the Γ M path through the initial and final states of 1s-(4s, $4p_x$, $4p_y$, $4p_z$) five-orbital mixings (the green arrow), (IV) the detectable 8.70-/6.40-eV excitation structure related to the third pair of valence and conduction subbands on the Γ M direction using the whole five-orbital hybridizations in the initial and final states (the yellow arrow), (V) the prominent 9.40-/7.40-eV excitation structure arising from the second valence and third conduction subbands along the MK path under the maximum orbital mixings (the pink arrow), (VI) the weak, but significant 11.00-/9.20-eV absorption structure coming from the fourth valence and first conduction subbands by the 1s-(4s, $4p_x$, $4p_y$) → (4s, $4p_x$, $4p_y$, $4p_z$) multi-orbital hybridizations at the saddle M point (the brown arrow), and (VII) the highest-frequency 13.10-/11.90-eV absorption peak corresponding to the fifth valence and first conduction subbands, but the similar orbital-photon couplings (the cyan arrow). There also exists a very strong exciton peak at $\omega = 2.05$ eV below the single-particle threshold one (the black arrow in Fig. 7.4H), where it is very stable because of the resonant response. After the full charge screenings, the coherent orbital oscillations are revealed at 6.60 eV/6.00 eV and 10.30 eV/9.65 eV, in which these plasmon modes are identified from the four-orbital hybridizations of 1s-(4s, $4p_x$, $4p_y$), but with the negligible $4p_z$ contribution. The above-mentioned optical excitation behaviors have successfully illustrated the diverse quasiparticle phenomena using the developed viewpoints of the stale and intermediate chemical bondings and the orbital-photon/electron-photon couplings. As a result, the theoretical framework, being clearly indicated in the previous books [195], is greatly enlarged by using the dynamic processes of optical excitations. The extended viewpoints could be verified through a close cooperation among the phenomenological models, the VASP simulations, and the high-resolution measurements of optical spectra.

When the atomic number quickly grow, both Sn-(5s, $5p_x$, $5p_y$, $5p_z$) and Pb-(6s, $6p_x$, $6p_y$, $6p_z$) will exhibit more complicated and sensitive multi-orbital hybridizations after the double-side hydrogenations. This is responsible for the diversified quasiparticle phenomena of the single-particle and collective excitations. The stable chemical bondings of sp^3 and ssp^3 between Sn—Sn/Pb—Pb and Sn—H/Pb—H neighboring atoms are thoroughly examined from all consistent geometric and electronic properties. Also, the coexistence of intermediate orbital hybridizations of the initial and final states (Tables 7.2H and 7.4H) can generate the diverse optical excitation behaviors. Stanane can exhibit, at least, seven types of vertical optical transitions: (I) the threshold 0.96-eV/0.97-eV absorption peak due to the first pair of valence and conduction energy subbands at the Γ valley under the multi-orbital hybridizations of Sn-($5p_x$, $5p_y$) → Sn-($5p_x$, $5p_y$, $5p_z$) (the red arrow of Fig. 7.4H), (II) the prominent 1.40-eV/1.30-eV absorption structure arising from the second valence and first conduction subbands at the Γ point using the similar chemical bondings in (I) (the blue arrow), (III) the very strong 3.70-/3.10-eV absorption structure associated with the identical subbands and mechanisms of (I), but initiated from the Γ saddle point (the green arrow), (IV) the weak, but important 4.40-/4.10-eV excitation structure corresponding to the third valence and the first conduction subbbands near the Γ point through the complicated orbital hybridizations of 1s-(5s, $5p_x$, $5p_y$, $5p_z$) → 1s-($5p_x$, $5p_y$, $5p_z$) (the yellow arrow), (V) the detectable 7.70 eV-/6.60 eV response structure between the fourth valence and first conduction subbands on the Γ M path through the 1s-(5s, $5p_z$) → 1s-($5p_x$, $5p_y$, $5p_z$) chemical bondings (the purple arrow), (VI) the significant 8.40 eV-/8.10 eV-absorption peak from the fourth valence and first conduction subbands at the Γ saddle point under the intermediate multi-orbital mixings of 1s-(5s, $5p_x$, $5p_y$) →

($5p_x$, $5p_y$, $5p_z$) chemical bondings (the pink arrow), (VII) the highest-frequency 11.00-/10.30-eV excitation peak the fifth valence and first conduction subbands initiated from the M saddle point under the unusual 1s-(5s, $5p_x$, $5p_y$) $\rightarrow$ ($5p_x$, $5p_y$, $5p_z$) hybridization mechanism (the cyan arrow). The stable and intermediate chemical bondings are identified to be, respectively, sp^3/ssp^3 and (5s, $5p_z$)/($5p_x$, $5p_y$)/($5p_x$, $5p_y$, $5p_z$)/(5s, $5p_x$, $5p_y$). The strong Coulomb attractions of the excited holes and electrons are capable of generating more than one featured absorption structure, e.g., the most obvious one at 0.75 eV and the other minor results. Within the whole charge screenings, the collective excitations of all active orbitals are examined to appear at 5.40 eV/5.30 eV, 8.10 eV/7.70 eV and 9.10 eV/8.20 eV. Three kinds of coherent charge oscillations, with the significant Landau dampings, are further attributed to 1s-(5s, $5p_x$, $5p_y$) in regardless of $5p_z$ orbitals. through the delicate analyses.

The significant optical excitations in plumbane, as clearly achieved from the dielectric function (Fig. 7.7S and T), energy loss functions, absorption coefficient and reflectance (Fig. 7.8J), could be classified eight types: (I) the single-particle 1.60 eV-/1.60 eV threshold frequency initiated from the first pair of valence and conduction energy subbands at the Γ valley through the multi-orbital mixings about Pb-($6p_x$, $6p_y$) $\rightarrow$ Pb-(6s, $6p_x$, $6p_y$) (the red arrow in Fig. 7.4J), (II) the strongest 2.30-eV/1.90-eV absorption peak using the similar subbands and orbital hybridizations, but related to the saddle M point (the blue arrow), (III) the observable 2.90-/2.70-eV excitation structure associated with the second valence and first conduction subbands at the Γ valley through (6s, $6p_x$, $6p_y$, $6p_z$) $\rightarrow$ (6s, $6p_x$,$6p_y$) (the light green arrow), (IV) the weak, but significant 3.70-/3.50-eV absorption structure related to the second valence and the first conduction subbbands along the ΓM path by the multi-orbital hybridizations of 1s-(6s, $6p_x$, $6p_y$, $6p_z$) $\rightarrow$ (6s, $6p_x$, $6p_y$) (the yellow arrow), (V) the important 4.60 eV-/4.10 eV responses closely related to the third valence and first conduction subbands from the M valley using the 1s-(6s, $6p_x$, $6p_y$, $6p_z$) $\rightarrow$ (6s, $6p_x$, $6p_y$) chemical bondings (the pink arrow), (VI) the non-negligible 7.00 eV-/5.90 eV-absorption peak from the fourth valence and first conduction subbands at the stable Γ valley by the similar chemical bondings in (V), (VII) the observable 8.00-/7.30-eV excitation peak the fourth valence and first conduction subbands from the M point using the previous ones (the cyan arrow), (VIII) the highest-frequency 11.70-/10.80-eV absorption peak between the fifth valence and first conduction band under the similar mechanisms of (V) (the green arrow). After including the full charge screenings, there exist four prominent plasmon modes. Their characteristics frequencies come to exist at 3.30 eV/3.00 eV, 5.00 eV/4.50 eV, 7.00 eV/6.59 and 8.50 eV/7.50 eV. Furthermore, the coherent charge oscillations are identified to mainly come from the 1s-(6s, $6p_x$, $6p_y$) multi-orbital hybridizations in Pb—Pb and Pb—H bonds. Apparently, the stable and quasi-stable excitons have strongly affected the main features of optical absorption, e.g., a very prominent exciton peak at 1.19 eV (Fig. 7.7S and T).

There are a lot of important differences among hydrogenated group-IV materials in terms of single-particle and many-body optical excitations, as clearly indicated in Fig. 7.7S and T. The main features in various optical spectra, the number, frequency, intensity and form of the prominent response structures, are greatly diversified by the different elements and hydrogenation effects. The rich and unique quasiparticle behaviors can be explained by the intrinsic characteristics, namely, the multi-/single-orbital hybridizations ion chemical bonds [91] and the spin-induced magnetic configurations (the spin-orbital couplings and/or the on-site electron-electron Coulomb interactions [91]). Each interband absorption structure has been clearly identified to the specific orbitals of the initial and/or the final band-edge states. This is consistent with the featured energy dispersions and van Hove singularities. Its characteristics directly reflect the perpendicular π and σ/sp^3 and ssp^3 chemical bondings of the graphene/(silicene, germanene, tinene, plumbene) hydrogenated group-IV systems, respectively. The plasmon modes, corresponding to the prominent peaks of the energy loss function at long wavelength, are very different from one another (the 15 cases). The coherent charge oscillations strongly depend on electronic energy spectra and wave functions, as clearly illustrate from the random-phase approximation [27]. Their existences are mainly determined by the Landau damping [27], e.g., the different collective excitations due to the specific valence orbitals. Of course, the unusual plasmon edges in reflectance spectra present the various height variations near ω_P. Moreover, the inverse of $\alpha(\omega)$ obviously show the frequency-dependent transmission depth. Such diverse optical properties would be very useful in developing the phenomenological models [27], as well as, the modulations of electromagnetic waves, further leading to the highly potential device applications [196].

Through the current investigations, both pristine and hydrogenated group-IV systems are outstanding candidates in greatly enhancing the quasiparticle framework based on the first-principles simulations [17] and the phenomenological models [22]. The rich and unique mechanisms cover the strong π and σ, sp^3 and ssp^3 bondings, as well as, the significant spin-orbital couplings and the on-site Coulomb interactions of spin-up and spin-down states. The systematic investigations, being conducted on the geometric symmetries, electronic energy spectra/wave functions, van Hove singularities, and the single-particle and many-body optical transitions, need to be constructed from the unified phenomenological models. First, the reliable tight-binding model is achieved from an excellent fitting with the VASP calculations, in which the hopping integrals of the different orbitals, the on-site Coulomb potentials; the spin-dependent single-particle and bi-particle interactions are the critical mechanisms. The generalized tight-binding mode could be further developed for the diverse magnetic quantization phenomena (details in Refs. [23,27]). While such models directly combine with the modified random-phase approximation, the corrected self-energy method, dynamic Kubo formula, and static one, their combinations are very powerful in revealing the unusual quasi-particle properties of electronic/magneto-electronic excitations [22], Coulomb decay rates [27], optical/magneto-optical selection rules [197], and quantum Hall conductivities [198]. Specifically, the 2D excitonic theories are worthy of the further development to fully understand the Coulomb-field-created optical scatterings and charge screening effects in different layered systems.

7.7 Similarities and differences among group-IV materials in basic and applied sciences

The layered group-IV-related materials have attracted a lot of theoretical and experimental studies [17,22,24], as done for the systematic investigations in terms of review works and books [17,24] covering the basic and applied sciences [199,200]. Pristine and chemically modified systems clearly exhibit the diverse physical/chemical/material phenomena, according to the fundamental quasiparticle properties (geometric, electronic, Coulomb-excitation optical, transport, thermal, mechanical and phonon properties), as well as, potential applications (electronic, optical, and biological devices in Refs. [201–204]).

7.7.1 Experimental syntheses and measurements

Many multi-component compounds are successfully synthesized on the planar/buckled/curved/folded honeycomb crystals [75,205–207] through the efficient adatom/molecule chemisorptions [208–210] and the guest-atom substitutions, such as, the various hydrogenations, oxidizations, halogenations, and alkalizations of few-layer binary graphene material [211–213], the similar chemical reactions for coaxial carbon nanotubes/carbon nanotube bundles [214], and ternary $B_xC_yN_z$ layered/nanotube systems [64,65]. The up-to-date experimental syntheses are mainly focused on graphene-related materials, in which they could be generated by the various growth methods, e.g., the mechanical exfoliation (the first discovery in 2004 in Ref. [215]), the chemical vapor deposition (CVD [216]), the electrostatic manipulation of scanning tunneling microscopy [217], chemically and electro-chemically reduced from graphene oxides [218,219], and the arc discharge [220]. Among such synthesis methods, CVD is able to generate a large-scale honeycomb lattices of pure planar graphenes for the high-resolution experimental examinations [221,222]. However, the other monolayer group-IV systems are frequently grown on the specific substrate only using the molecular beam epitaxy (MBE; details in Refs. [223–226]). The very important chemical bondings between group-IV adlayers and substrate are predicted to play critical roles in the fundamental properties through the accurate VASP simulations and analyses, e.g., the significant reconstruction on surface and substrate morphologies, The MBE synthesis is hardly utilized in layered graphene materials, while the mechanical technique cannot obtain few-layer silicene/germanene/tinene/plumbene systems because of the absence of weak van der Waals in thin film systems. The latter should be totally dominated by the 3D chemical bondings (the rapid dimension crossover in the increase of layer number). Their further hydrogenations have not been successfully synthesized in the experimental laboratories up to now. This will induce giant barriers for the basic and applied science researches. Of course, the high-resolution measurements on the surface morphologies, wave-vector-dependent occupied bands, van Hove singularities, and prominent absorption structures, which need to be done by STM/TEM, ARPES, STS and optical reflectance/absorption/transmission spectroscopies (details in Chapter 3), are completely absent. Such examinations are expected to clearly show certain important differences among hydrogenated group-IV materials (discussed later in Section 7.7.2 for theoretical predictions). These will lead to very high barriers for the basic and applied science researches.

7.7.2 Theoretical predictions on quasiparticle properties

On the theoretical side, the first-principles simulations and phenomenological models are frequently used in fully explored the rich and unique quasiparticle phenomena of group-IV materials. The latter is relatively suitable under the high-symmetry chemical environments with the concise interaction pictures [108,227]. A series of systematic investigations, as clearly illustrated for the critical mechanisms (the active multi-/single-orbital hybridizations and the ferromagnetic/anti-ferromagnetic spin configurations in Refs. [54,228]), have been successfully conducted by choosing the graphene- and silicene-related pristine systems [17,24] and various compounds [110,111], such as, 3D AA- [229], AB- [230] and ABC-stacked bulk graphite systems [106,231], few-layer graphene with the distinct stackings and layer numbers [108,232], armchair/zigzag/chiral carbon nanotubes [233–235], achiral/chiral graphene nanoribbons [236], the different adatoms chemisorptions/guest-atom substitutions [17], and the similar partners in silicene/germanene/tinene/plumbene [91,226]. The quasiparticle framework is greatly developed under the well separation/cooperation of two different approaches; furthermore, the similar strategy would be generalized to cover more unusual quasiparticle phenomena due to the other and/or coupled quasiparticles, e.g., phonons [104], photons [104], and their significant couplings with electrons [22,27] under the current investigations.

Monolayer group-IV materials are quite different from one another in terms of the geometric [75,188,237], electronic/magneto-electronic [177,236,238,239], optical/magneto-optical [106,176,227,236], Coulomb-field excitations/magneto-electronic excitations/decay rates [27,240], quantum transport properties [241,242], chemical modifications (absorptions in Ref. [17]; substitutions in Refs. [243,244]), and magnetic spin configurations [54], according to the previous predictions. Such essential quasiparticle phenomena, which are thoroughly examined by the relevant physical and chemical quantities [17,27], strongly depend on the crystal symmetries, the dominances of π, sp^2 or sp^3 chemical bondings, the strength of spin-orbital coupling, spin density distributions, and concentrations and distributions of adatoms/guest atoms.

Only monolayer graphene exhibits the planar honeycomb crystal, while single-layer silicene/germanene/tinene/plumbene show the buckled structures. This important differences mainly come from the well-separated and non-orthogonal features of π and σ bondings, respectively [91]. Compared with the dominance of sp^2, that of sp^3 is getting strong in the increase for atomic number because of the reduced σ-covalent bonding. Also, the intrinsic spin-orbital couplings also play one of the critical roles in determining the lowest ground state energy. The host-host and host-hydrogen bonds possess the shortest lengths in graphene systems, clearly indicating the most active and thus strongest orbital hybridizations in C—H bonds [101]. The double-side hydrogenation can create an observable buckling in graphene [101], while the deformation might be slightly reduced for other fully hydrogenated group-IV materials [53]. Obviously, the asymmetric H-adatom distributions of single-side chemisorption display the enhanced bucklings for all the systems, where the largest effects on the essential properties are revealed in plumbone (Fig. 7.1C, and Table 7.1). The position-dependent chemical bonds are closely related to the whole intrinsic interactions through the Bloch phases; therefore, the anisotropic and non-uniform chemical environments in establishing the reliable phenomenological models, e.g., the various hopping integrals, onside-Coulomb potentials and strengths of SOCs in the tight-binding model [91].

Very interesting, band structures, wave functions, spatial charge density distributions and density of states in group-IV monolayer systems clearly illustrate the diverse quasiparticle phenomena [91,245]. It is well known that graphene has a linear and isotropic Dirac-cone structure being initiated from the stable K/K′ valleys [75], A gapless behavior just appears at the Dirac point because of the hexagonal symmetry, leading to a zero-gap semiconductor without free carriers at the Fermi level. The $2p_z$ orbital—π valence bands are extended along the $KM\Gamma$ path with a whole bandwidth of ~7.10 eV, where K and Γ are extreme critical points of linear and parabolic dispersions, respectively, and M belongs to a saddle point. Such band-edge states are able to generate three kinds of 2D van Hove singularities (V-shape, shoulder and logarithmic divergence). Concerning the orthogonal σ bondings, more strong covalent strengths generate the deeper valence states at the Γ point from about −3.10 eV with the specific double degeneracy of $2p_x$ and $2p_y$ orbital (the same ionization energy). When such σ bands are explored along the ΓMK path, they only crossing the π bands without any anticrossings. That is, these kinds of valence subbands are deduced to be orthogonal to each other (the absence of comparable normal modes). The fourth valence, which mainly arises from 2s orbital, appear at the deepest energy range of -19.20 eV $< E^v <$ -12.80 eV along the ΓMK path (Fig. 7.4A). All the valence and conduction subbands can induce the unusual behaviors under the prominent hydrogenation effects [101].

With the increment of atomic number, four valence subbands and two conduction ones nearest to the Fermi level present the dramatic transformation, as clearly illustrated for silicene/germanene/stanene/plumbene in Fig. 7.2B–D and F. The rich energy spectra and orbital dominances lie in the strong competitions of sp^2 and sp^3 bondings and the greatly enhanced spin-orbital interactions. Band gaps of group-IV materials are mainly determined by the separated and slightly modified Dirac-cone structure at the K/K* points. They belong to direct energy gasp, and their value are, respectively, 1.30, 23.90, 76.50 meVs and 0.65 eV for silicene [246], germanene [247], stanine [247], and plumbene [248]. Both σ valence subbands, being initiated from the Γ point, become shallow and splitting through the quick approach to the Fermi level, e.g., their initial split energies of $E^v = -0.40/-0.60$ eV, $-1.20/-1.40$ eV, $-0.05/-0.50$ eV, and $-0,40/-0.75$ eV for silicene, germanene, stanene and plumbene. This obviously indicates that the sp^3 four orbitals need to be taken into account simultaneously for the phenomenological models (the multi-orbital hybridizations done for stanene in Refs. [249,250]). The non-orthogonal relation of π and σ bondings leads to the serious anti-crossing behaviors between the first and second/third valence subbands, in which the former and the latter, are, respectively, along the $KM\Gamma$ and ΓMK paths, Such phenomena further create the specific 1D constant-energy loops, being directly reflected in the unusual van Hove singularities. Finally, the strong coupling of sp^3 bonding and SOC can largely reduce the initial energy of the fourth valence band (~ -9.00 to -11.30 eV) and its bandwidth ($\sim$ 2–3 eV). The featured valence subbands could be verified by using the high-resolution measurements of the angle-resolved photoemission spectroscopy [144] and scanning tunneling spectroscopy [251]. In addition, a buckled honeycomb lattice is very suitable for studying gate-voltage-enriched electronic properties, such as, the obvious spin splittings and semiconductor-semimetal transitions during the variation of a perpendicular electric field [252,253].

A 2D hexagonal crystal is outstanding in creating the diverse magnetic quantization phenomena, being successfully conducted on group-IV- and group-V-related materials through the systematic studies [17,22,24,27]. All the intrinsic interactions, the geometric symmetries, the magnetic field could be taken into account simultaneously through the generalized tight-binding model (all the calculation details in Refs. [22,27]). Furthermore, the developed strategy is based on the distinct sublattices inside a Moire superlattice arising from a periodic vector potential. The physical pictures of dispersionless Landau levels (LLs [229,254]), which present the highly localized distributions with the symmetric or anti-symmetric oscillations, are available in overcoming the super-high barriers of numerical calculations. The main features of magnetic configurations cover the well characterized quantum numbers from normal LL's oscillational and several LL subgroups ($n_i^{c,v}$ i for subgroup), the coexistence of normal, perturbed and undefined LLs, the sensitive dependences of LL energies on the magnetic-field strength (due to energy dispersions before quantization), the non-crossing/crossing/anti-crossing behaviors during the variation of magnetic/electric fields, the composite effects of ¯magnetic and electric effects (e.g., the splitting of K- and K*-valley LLs in buckled or few-layer crystals [229]), and the spin-split LL subgroups under the significant spin-orbital interactions [227]. The diverse quasiparticle phenomena are directly reflected in magneto-optical selection rules [227], quantum Hall conductivities [198], and magneto-plasmon modes [238].

Monolayer graphene exhibits the unusual magnetic quantization behaviors because of the low-lying isotropic Dirac-cone structure. A pair of valence and conduction LL groups, which possess a zero energy spacing at the neutral point [255], are well characterized by the square-root dependences on the quantum number [$n^{c,v}$] and magnetic-field strength [B_z]. The similar LL energy spectra are revealed in few-layer AA-stacked graphene systems [229]. It should be noticed that the normal LL wave functions are similar to the well-behaved ones of a simple harmonic oscillator, i.e., their discrete probability distributions on the lattice sites could be regarded as continuous presentations under the enough long magneto-length (even for a very weak field below $B_z = 0.01$ T [229]). In addition, the linear relations with $n^{c,v}/B_z$ are delicately identified from the parabolic energy dispersions. e.g., those in bilayer graphene of AB stacking [256], As the oscillatory energy subbands, their magnetically quantized LLs show the irregular LL_anti-crossing phenomena, such as, trilayer ABC [106] and AAB stackings [257]. All the low-lying LLs are dominated by the π bondings of $2p_z$ orbitals. The single-orbital approach is suitable for most of layered graphene systems and even 3D bulk graphite ones, according to the previous predictions [258,259]. That is, this model can explore the dimensionality crossover behaviors in the increment of layer number. However, it cannot be utilized in the other group-IV materials because of too strong interlayer atomic interactions [91].

The magnetic quantization of four-orbital hybridizations need to be taken into consideration for silicene/germanene/tinene/plumbene. Very interesting, the diverse quantization phenomena are induced by the low-lying electronic states near the K/K* and Γ valleys with the coexistent constant-energy loops. The four-orbital- and spin-decomposed LLs, which are driven by the significant couplings of the multi-orbital hybridizations, SOCs and magnetic field, are clearly expressed through the 16 tight-biding functions [91]. As a result, the numerical calculations become more complicate. Each LL state is well characterized by [s, p_x, p_y, p_z]-, spin-up- and spin-down-, and A- and B-related sublattices. Such complicated LLs are able to present the unusual quasiparticle behaviors, covering the 16 subgroups of four-fold degenerate LLs (half of that in monolayer graphene [260]), the distinct localizations for different subgroups, a lot of crossing and anti-crossing behaviors during the variation of B_z/V_z [261], the irregular magnetic-field and quantum-number dependences [262], and the further LL splitting of K-/K*-valley LLS under the application of gate voltage (V_z on a buckled honeycomb crystal [245]). Also noticed that electronic states close to the K/K′ Γ points are quantized into the specific probability distributions centered at [1/6, 4/6]/[2/6, 5/6]/[0, 1/2] (a rectangular armchair Moire superlattices; details in Refs. [23,91]), The other abnormal quasiparticle behaviors are closely related to the original energy dispersions, directly reflecting the strong couplings of the intrinsic interactions and the external actions [236,263]. These rich properties clearly illustrate that group-IV materials are very outstanding candidates in fully exploring valleytronics [264] and spintronics simultaneously [265].

Optical properties, which arise from the perturbation of an electromagnetic wave (details in Chapter 2.1.4), are greatly diversified by all the intrinsic and extrinsic interactions in terms of the frequency, number, intensity and form of the special absorption structures, as well as, plasmon modes (coherent charge oscillations). They are thoroughly analyzed from the transverse dielectric functions, reflectance spectra, absorption coefficients and energy loss functions [110,111] at long wavelength limit. An open band gap, being defined by the threshold absorption frequency, just corresponds to a direct energy gap of any group-IV material (Table 7.1). This is due to the SOC-induced separation of a modified Dirac-cone structures (Fig. 7.2). More prominent excitation structures come to exist in the low- and middle-frequency ranges, since there exist certain anti-crossing behaviors through the sp^3 multi-orbital hybridizations. Each asymmetric or symmetric absorption peak is identified to mainly come from the specific orbitals in the presence/absence of excitonic effects (Table 7.2). The application of gate voltage can generate pairs of absorption peaks under the spin splitting, where they are relatively easily observed for the larger SOCs (e.g., tinene and plumbene in Refs. [248,266]). Furthermore, its critical value might achieve the non-monotonous variation for the threshold absorption frequency, i.e., an optical gap would become zero under a certain Vz [245].

As for the vertical magneto-optical excitations, the spectral intensities are mainly determined by the spatial symmetries of the initial and final LLs. Each well-defined/perturbed LL has an independent oscillation mode/the linear superposition of certain ones, in which the former is just a composite form of the exponential decay and Hermite polynomial functions (wave function of an oscillator). Every prominent absorption peak appears in the delta-function-like structure, mainly owing to an infinite joint density of states between the occupied and unoccupied LLs. Whether its height is finite or vanishing relies on the intrinsic symmetries of LLs. As a result, the well-behaved magneto-optical selection rule of $\Delta n = \pm 1$ is frequently revealed in most magneto-absorption peaks, e.g., that in monolayer graphene However, the extra magneto-optical ones of $\Delta n = \pm 0, 2, 3, \ldots$, etc, could also be shown in certain important magneto-absorption peaks, Such unusual magneto-optical phenomena directly reflect the irregular LL anti-crossing behaviors (non-monotonous Bz-dependent LL energy spectra) [267]. These results are predicted to be clearly shown in monolayer tinene and plumbene [250], or silicene, germanene, tinene and plumbene under a composite Lorentz field [268]. In addition, all the magneto-optical rules have been thoroughly examined for layered graphene-related systems and bulk graphites [27], such as, their coexistence in bilayer sliding graphene systems [269], as well as, trilayer/few-layer ABC [231,256], AAB [136], and ABA stackings [136]. Also noticed that the similar magneto-scattering rules are very useful in comprehending quantum Hall conductivities (discussed later [270]).

The Coulomb-field electronic excitations are well characterized by a longitudinal dielectric function under the various momentum and frequency transfers $[q, \omega]$. Generally speaking, the perturbed plasma waves of charge oscillations propagate along the direction of $\mathbf{q}$ through a vibrational wavelength of $2\pi/q$. $\varepsilon(q, \omega)$ and $Im[-1/\varepsilon(q, \omega)]$, respectively, represent the bare and screened response functions of valence and conduction charge screenings [271]. Both functions could be calculated by the modified random-phase approximation and/or the time-dependent first-principle simulations, where the latter requires much computer time because of an infinite intermediate excitation configuration. The former has been successfully utilized for the systematic investigations about the diverse $[q, \omega]$-phase excitation diagrams, magneto-plasmon modes with/without Landau dampings, and inelastic Coulomb decay rates in graphene-related 2D systems, such as, those in AA- [229], AB- [230] and ABC [106] -stacked bulk graphite systems, monolayer graphene [255], silicene [272], germanene [273], bi-layer graphenes with AA an AB stackings [273], tri-layer graphenes in the presence of AAA, ABA [274], ABC and AAB stackings [275], as well as, single-/multi-walled carbon nanotubes, a 2D aligned nanotube superlattice [276,277], and a 3D carbon nanotube bundle [278]. As for the low-energy electronic excitations, the significant couplings, which originate from the π bonding, temperature, spin-orbital interaction, gate voltage, perpendicular magnetic field, and carrier doping, are deduced to be reasonable for the above-mentioned emergent materials. This consistence is based on the suitable predictions of phenomenological models, compared with certain high-precision measurements through energy loss spectroscopy (EELS [279]).

An intrinsic monolayer graphene only exhibits the purely temperature-induced plasmon modes, as observed in a 2D electron gas (the square-root momentum dependence of a 2D plasmon frequency). The thermal excitations can create the low-density free electrons and holes in the conduction and valence Dirac-cone structure, respectively (an analytic T^2 relation). With the increasing temperature, the original inter-π-band electron-hole excitations are quickly replaced by intraband-π-band ones. The collective excitations of high-mobility carriers will come to exist under a sufficiently strong energy loss spectrum, in which they are accompanied with the relatively weak Landau dampings of inter-π-band transitions. A (momentum, frequency)-phase diagram covers two different e-h regions, other vacant regions, and a T-created plasmon dispersion relation. The critical Very interesting, 2D plasmon modes could survive above/below the critical temperature/momentum of T_c/q_c. When the free carrier density is greatly enhanced by the n-type or p-type doping, the single-particle and collective excitations are drastically changed the Fermi surface of the π-electronic structure. The undamped plasmon modes, with very strong screened response function, appear at small momentum transfers through the well-behaved $\sqrt{q}$ relation. And then, they could coexist with the inter-π-band e-h excitations at larger q and would almost disappear beyond the critical momentum. On silicene and germanene sides [99], the prominent coupling effects, which cover the significant low-energy π bonding [280], SOC [281], temperature [282], doping [283], and gate voltage [284], are responsible for the diversified Coulomb excitations/deexcitations. As a result of the modified energy gap, carrier density, Dirac-cone structure, and state degeneracy, there exist the diverse phase diagrams being characterized by the featured single-particle excitations and plasmon modes. The former possess the distinct e-h boundaries (associated with the Fermi level and band-edge state energies of valence and conduction subbands [285]); furthermore, the latter present in the specific momentum/temperature ranges with the normal or irregular dispersion relations [286].

The magneto-Coulomb excitations, which are induced by a uniform perpendicular magnetic field [287], show the rich and unique charge screenings. The concise picture is a very strong competition between the longitudinal Coulomb forces and the transverse magnetic ones [288], being fully driven by the featured LLs (the dispersionless energy spectra and normal/perturbed modes). The coherent carrier vibrations/the propagating plasmon waves are strongly combined with the cyclotron motions. After a normal magnetic quantization, the continuous intraband and interband single-particle excitations dramatically transform into the discrete inter-LL ones with the delta-function-like dielectric/energy loss functions, directly reflecting their van Hove singularities. The bare and screened response functions, respectively, display the prominent single-particle and collective excitations (magnetoplasmon modes [289]) under sufficiently high temperatures [290]. The magnetoplasmon frequencies grow (positive group velocities [291]), decline (negative ones) and terminate in the increments of momentum transfer from zero. The critical momentum transfer is very sensitive to the changes of temperature and magnetic-field strength. In addition to interband magnetoplasmons, there also exist the intraband modes for an extrinsic graphene with n- or p-type dopings. Furthermore, such magnetoplasmons behave as plasmons of 2D electron gas in terms of their main features, i.e., they mainly come from the coherent oscillations of conduction electrons or valence holes using the magneto-plasma waves with positive group velocities. As to magnet-electronic excitations of monolayer silicene [292]. As for magneto-electronic excitations of silicene [109], quasiparticle behaviors are greatly diversified by the significant SOC [293] and the application gate voltage [294]. The close relations among the Coulomb interactions, the intrinsic interactions, the external electric and magnetic fields, as well as, the carrier dopings, are responsible for the featured excitation spectra. Both of low-frequency interband and intraband magnetoplasmon modes could co-survive at the low magnetic-field strength and/or doping. Moreover, the Bz-dependent magneto-plasmon spectra are dominated by the wavelength of collective carrier oscillations, magneto-length and LL degeneracy. With the increasing of Bz, intraband magnetoplasmons are considerably transformed into interband modes, as clearly illustrated by an abrupt change of the spectral width, frequency, and intensity. An electric field is able to create spin- and valley-relate polarizations and induce more localized magnetoplasmon modes. In other words, a direct combination of spin valleytronics and magnetoplasmonics could be revealed/examined in monolayer silicene. The similar magneto-quasiparticle excitation phenomena are deduced to be relatively easily observed in germanene with a larger SOC and buckling.

In addition to quasiparticle energy spectra (band structures with elastic and inelastic scatterings [295]), electron-electron Coulomb [296] and electron-phonon interactions [297] dominate their lifetimes in all the condensed-matter systems. The high-resolution ARPES measurements could directly examine both wave-vector-dependent properties for the occupied states through the spectral frequencies and widths [110,111,298–300]. Specifically, Coulomb deexcitations have been successfully explored for layered graphene-related systems under the modified random-phase approximation and self-energy method, such as, those in a 3D graphite, 2D few-layer graphenes with the different stacking configurations, and 1D coaxial carbon nanotubes [27]. By the development of quasiparticle framework, the intralayer/interlayer orbital hybridizations, as well as, the intralayer/interlayer Coulomb interactions are included in the theoretical predictions simultaneously. A lot of unusual quasiparticle behaviors might be in great contrast with the well-behaved ones of 3D/2D/1D electron gases in terms of energy, wave-vector and temperature dependences [301]. For example, a n-type monolayer graphene, which possesses the excited conduction electrons (above the Fermi level [293]), conduction holes (below E_F) and valence holes (lower than the Dirac point), display the diversified quasiparticle phenomena in three energy ranges. The Coulomb decay rates are mainly determined by the available deexcitation channels of electron-hole pairs and plasmon modes, being confined by the conservation of energy and momentum in the inelastic scatterings [302], Only the excited holes in valence bands, show a prominent peak structure, since they have the enough decay energy through collective excitations. As result this leads to very strong broadening effects on the measured energy spectra, e.g., an obvious and unusual width in ARPES measurements, and the widened absorption structures. As for an extrinsic germanene/silicene, the Coulomb decay rates are drastically changed by the separated and modified Dirac-cone structure, such as, the great enhancement of non-monotonous energy/wave-vector dependences and anisotropic behaviors along the different wave-vector directions. Certain important differences among graphene, silicene and germanene lie in the nearest-neighbor hopping integral of π bonds, the strength of spin-orbital coupling, and the buckled honeycomb crystal [245]. In addition, bilayer silicene/germanene are totally different from bilayer graphene, since the former possess very strong interlayer atomic interactions under the complicated stacking and buckling configurations [16]. The perturbation viewpoint, e.g., the effective-mass approximation about a certain high-symmetry valley [303], is not suitable for such emergent materials for all the essential properties [304].

The featured energy spectra and wave functions of LLs are able to create quantum Hall conductivities in emergent 2D materials at very low temperatures [305]. The magnetic transport properties clearly present the diversified quasiparticle behaviors in terms of the unique dependences on the magnetic-field strength and Fermi level/free carrier density. Very interesting, the up-to-date experimental measurements and theoretical predictions, which have been systematically conducted on few-layer graphene systems with distinct stacking configurations (AB and ABC ones) and layer numbers, are consistent with each other [91]. The latter are achieved by the direct combination of the generalized-tight model (discussed earlier [306]) and the static Kubo formula (the dynamic Kubo formula through the reduced case of $\omega = 0$ in Refs.). Under T = 0 K, the stationary magneto-current is driven by the vertical transitions without any momentum transfer from the occupied to unoccupied Land-level states. Basically, the significant inter-LL scatterings need to satisfy the normal and/or extra magneto-selection rules (the similar ones in magneto-optical absorption spectra [249]). In general, there are a lot of valence and conduction LLs in making the non-negligible contributions, in which both positive and negative conductivities might cancel seriously and then present the distinct quantum structures (integral, rational, and irrational times of $G = \frac{4e^2}{h}$). For example, the sliding bilayer graphene systems are predicted to exhibit the diverse magneto-transport phenomena, covering the specific non-integer conductivities, the integer conductivities with the distinct steps, the splitting-induced reduction and complexity of quantum conductivity, a vanishing or non-zero conductivity at the Dirac point, and the quasi-well, staircase, composite, and irregular plateau structures during the variation of the magnetic-field strength. Such model calculations will become very complicated and difficult for other bilayer group-IV materials, since the single-p_z-orbital hybridization is not reliable in fully comprehending the low-energy magnetic quantization, e.g., very abnormal LLs in bilayer silicene systems [109]. The high-precision magnetic transport measurements on monolayer graphene have clearly revealed the unconventional half-integer Hall conductivity [307] of $\sigma_{xy} = (m+1/2)\frac{4e^2}{h}$, where m is an integer and the coefficient of 4 represents the spin and sublattice degeneracy. This unusual quantization arises from the quantum anomaly of the $n^{c,v} = 0$ LLs from the gapless Dirac points. The diversified quantum Hall effects would be clearly revealed in monolayer and bilayer silicene systems [91] through the strong couplings of orbital hybridizations, spin-orbital interaction, gate voltage, and magnetic field, since there exist the valley- and spin-featured LLs. The rich and unique QHEs of buckled few-layer silicene/germanene/tinene/plumbene materials are worthy of the systematic investigations for the complete development of quasiparticle framework.

Very interesting, three kinds of magnetic configurations, ferromagnetism [308], anti-ferromagnetism [309] and non-magnetism [310], frequently come to exist in low-dimensional group-IV materials through the substrate effects, the specific edge structures [311], the unusual local defects [312], as well as the chemical modification of transition metal/rare earth atoms (discussed later [313]). For example, 1D zigzag graphene nanoribbons/0D armchair carbon nanotubes are predicted to exhibit the ferromagnetic/anti-ferromagnetic spin distribution along two open boundaries/across the ribbon center [314,315]. The up-to-date experimental syntheses clearly show that few-layer silicene/germanene/tinene/plumbene systems could be grown on the specific metals [16], e.g., the clear evidences of various Moire superlattices for Si-layer/Ag(111) compounds: $3 \times 3/\ 4 \times 4$, 3×3, $2\sqrt{3} \times 2\sqrt{3}$, $\sqrt{3} \times \sqrt{3}$, $\sqrt{7} \times \sqrt{7}$, and $\sqrt{13} \times \sqrt{13}$. The rich stacking configurations of the Si–Ag heterojunctions represent the outstanding candidates in examining the quasiparticle framework of first-principles simulations [16]. Monolayer silicene and AB-bt bilayer system on Ag (111) are thoroughly explored by the first-principles simulations, where two Moire superlattices in subsystems and an enough thick substrate must be taken into account simultaneously under the VASP calculations. Both silicon and silver atoms, with atomic numbers of 14 and 47, respectively, possess the unusual orbital configuration. The different multi-orbital hybridizations, being active in Si–Si, Si–Ag and Ag–Ag bonds, should be closely combined together using the extremely non-uniform and anisotropic charge distribution. Furthermore, there exist the obvious fluctuations near the heterojunction. These lead to the mission impossible for the establishment of phenomenological models [316]. When their cooperative relations are fully assisted by the zone-folding effects of Moire superlattices/spin configurations of substrate, the essential properties are greatly diversified for these two model systems. Compared with the pristine cases, the dramatic transformations come to exist in electronic and magnetic properties. Monolayer and bilayer silicenes/Ag(111), respectively, belong to non-magnetic and ferromagnetic semimetals with valence and conduction band overlaps/without modified Dirac-cone structures. Concerning the latter, the spin-up electronic configurations principally arise from silver atoms on the first, second, third and fourth layers, and their contributions are comparable to one another. However, the relative contribution to spin density distribution is almost negligible for the double-layer silicon atoms. The similar quasiparticle phenomena will be revealed in heavier group-IV materials [16].

Experimental syntheses and theoretical predictions clearly illustrate that chemical adsorptions and substitutions are very powerful in creating various binary/ternary/quaternary/pentanary compounds [317–320]. As for group-IV systems, their honeycomb lattices are capable of providing very suitable/stable chemical environments in chemical modifications, since such structures possess the almost perpendicular dangle bonds/the strong σ bondings associated with the pz/[s px, py] orbitals. The significant chemical modifications, which greatly diversify the quasiparticle phenomena (orbital hybridizations and spin configurations), cover hydrogenations, oxidizations, halogenations, alkalizations, as well as, the direct combinations with transition-metal and rare-earth guest atoms. For example, hydrogen adatoms atoms can passivate the dangling bonds at two open edges of graphene/silicene/germanene nanoribbons, stabilize the crystal symmetry and thus drastically change the other fundament properties.

Specifically, the silicon-doped graphene systems [321], with various Si-concentrations and configurations, have been successfully explored by the generalized tight-binding model for the diverse magnetic quantization. The non-uniform bond lengths, site energies and hopping integrals, as well as, a uniform perpendicular magnetic field, are included in the calculations simultaneously. The quantized Landau levels could be classified into four different groups based on the probability distributions and oscillation modes. The main characteristics of the LLs are responsible for the magneto-optical selection rules which cover the dominant $\Delta n = |n^c - n^v| = 0$, the coexistence of $\Delta n = 0$ and $\Delta n = 1$, along with a specific $\Delta n = 1$. These rules for inter-LL excitations are attributed to the non-equivalence or equivalence of the A_i and B_i sublattices in a Moire supercell. The diversified quasi-particle behaviors are expected to reveal in carbon-/boron-/nitrogen-substituted silicene/germanene/tinene/plumbene compounds [16], or vice versa. Such binary compounds are outstanding candidates in examining the phenomenological models of quasiparticle framework (books), being under the current investigations.

7.8 In summary and future challenges

Hydrogenations of group-IV materials have clearly shown the diverse quasiparticle phenomena [250], covering crystal symmetries. electronic properties and many-body optical excitations in the presence/absence of active chemisorptions. Their main features are linked together in determining the charge- and spin-dependent-mechanisms and greatly promoting the theoretical development of quasiparticle framework based on the first-principles simulations [16].

Very interesting, π and σ bondings, sp^2, and sp^3 bondings are, respectively, identified for graphene [236], silicene/germanene/tinene/plumbene, and hydrogenated group-IV systems [16]. The spin-orbital couplings are significant for the non-graphene-based materials. Moreover, the spin-dependent electron-electron Coulomb interactions only come to exist in few ferromagnetic cases. By the delicate analyses, this should be the first work for establishing the close relations between the specific orbital hybridizations and absorption structures, being never predicted in other published papers. All the hydrogenation cases are never studied by the phenomenological model except for graphane. How to solve the complicated atom-, adatom-, orbital- and spin-related Hamiltonian become an interesting issue. As to the experimental syntheses and examinations, most of them are absent up to now. The successful synthesis is insufficient in thoroughly exploring the rich essential properties; therefore, the potential applications will be largely suppressed by the absence of basic science researches. The similar investigations could be generalized to chemically modified 2D materials, e.g., the oxidized, halogenated, and alkalized group-IV/group-V systems [16]. Part of them are under the current studies. Obviously, crystal structures, being under the optimal statuses of VASP relaxations, are largely diversified through the planar/buckled honeycomb lattices, the various orbital hybridizations, the H-adatom concentrations and distributions, and the spin-dependent magnetic configurations. The featured geometries include the IV—IV & IV—H bond lengths, the observable fluctuation of the former, the top-site H-heights, the IV—H bonding angles, and the buckling of A and B two sublattices. Furthermore, STM and TEM are available in examining the surface morphologies. These unusual behaviors, which are very sensitive to chemisorption cases, are closely related to the various chemical bondings. Hydrogenations have created the highly non-uniform chemical environments, being directly reflected in the main features of electronic properties and optical transitions. The orbital hybridizations and magnetic configurations become more complex; therefore, they are able to create the diverse quasiparticle behaviors in this current work. On the other hand, the phenomenological models of quasiparticle framework might be difficult for getting the reliable ones.

The diverse orbital hybridizations [54], being assisted by the spin-orbital couplings and/or the Hubbard-like Coulomb interactions are responsible for the rich and unique electronic energy spectra and wave functions. The perpendicular π and σ bondings, the significant sp^2 hybridizations and sp^3 ones, respectively, arise from graphene, silicene/germanene/tinene/plumbene, and hydrogenated group-IV materials [16]. A lot of significant characteristics are clearly revealed in the VASP calculations, such as, a zero-gap semiconductor due to the normal Dirac-cone structure, the small-, middle- and large-gap semiconductors from the spin-dependent interactions and/or the hydrogenation effects, the perturbed/modified/destroyed π and σ valence subbands. the strong asymmetry of hole and electron spectra, linear/parabolic/oscillatory/partially flat energy dispersions, non-crossing/crossing/anti-crossing subband behaviors, as well as, more band-edge states (van Hove singularities). These diverse phenomena clearly indicate that a tight-binding model could be well established only through each chemical environment. On the ARPES/STS sides (details in Chapter 3; in Ref. [16]), the high-precision measurements on the occupied valence subbands/the strong van Hove singularities are absent. Their examinations are required in the near-future studies.

Very interesting, the frequency-dependent optical properties, the transverse dielectric functions [322], energy loss functions [323], reflectance spectra [324], and absorption coefficients [325], present the unusual single-particle excitations and different plasmon modes.

The rich quasiparticle phenomena, corresponding to optical transitions at the long wavelengths, are delicately explored in the presence and absence of many-body effects, such as, the vertical transitions due to band-edge states [252] and the coherent charge oscillations for the different energy-range valence carriers (the distinct plasmons [326]). The prominent absorption structures and the excitonic effects have been successfully licked with the joint van Hove singularities (details in Tables 7.2 and 7.3). Each pronounced peak is clearly identified to a specific set of orbital hybridizations from the initial to final band-edge states, being fully supported by the featured band structures. The lowest/lower excitonic peaks are due to the stable or quasi-stable electron-hole bound states, depending on the dielectric screening of the total valence charges. The red shifts of threshold absorption frequency clearly illustrate that optical gaps are smaller than bad gaps, and their differences are sensitive to the intrinsic interactions. Reflectance, absorption, transmission and photoluminescence spectroscopies could be utilized to verify the first-principles predictions. In addition, the excitonic theory of emergent 2D materials could be developed in the further studies, while their numerical calculations might present a giant barrier associated with the Coulomb-field screening effects (the evaluation of the longitudinal dielectric function [327]).

References

[1] Snelgrove DW, Lusztyk J, Banks J, Mulder P, Ingold K. J Am Chem Soc 2001;123:469–77.
[2] Field GB, Somerville W, Dressler K. Annu Rev Astron Astrophys 1966;4:207–44.
[3] Matta CF, Hernández-Trujillo J, Tang TH, Bader RF. Chem–Eur J 2003;9:1940–51.
[4] Roohnikan M, Toader V, Rey A, Reven L. Langmuir 2016;32:8442–50.
[5] Steiner T. Angew Chem Int Ed 2002;41:48–76.
[6] Ramaker DE, Kumar L, Harris FE. Phys Rev Lett 1975;34:812.
[7] Caron LG. Comments Solid State Phys 1975;6:103–15.
[8] Graetz J. Chem Soc Rev 2009;38:73–82.
[9] Cheng H-M, Yang Q-H, Liu C. Carbon 2001;39:1447–54.
[10] Geusic M, Morse M, Smalley R. J Chem Phys 1985;82:590–1.
[11] Newns D. Phys Rev 1969;178:1123.
[12] Hamza AV, Kubiak GD, Stulen RH. Surf Sci 1990;237:35–52.
[13] Chabal Y, Higashi G, Christman S. Phys Rev B 1983;28:4472.
[14] Almahdi M, Dincer I, Rosen M. Renew Energy 2016;91:302–14.
[15] Wolf E. Large-scale hydrogen energy storage. In: Electrochemical energy storage for renewable sources and grid balancing. Elsevier; 2015. p. 129–42.
[16] Lin S-Y, Liu H-Y, Nguyen DK, Tran NTT, Pham HD, Chang S-L, et al. Silicene-based layered materials. IOP Publishing Limited; 2020.
[17] Tran NTT, Lin S-Y, Lin C-Y, Lin M-F. Geometric and electronic properties of graphene-related systems: chemical bonding schemes. CRC Press; 2017.
[18] Gibaja C, Rodríguez-San-Miguel D, Paz WS, Torres I, Salagre E, Segovia P, et al. Adv Mater 2021;33:2006826.
[19] Cahangirov S, Sahin H, Le Lay G, Rubio A. Germanene, stanene and other 2D materials. In: Introduction to the physics of silicene and other 2D materials. Springer; 2017. p. 63–85.
[20] Deleted in review
[21] Barraza-Lopez S, Fregoso BM, Villanova JW, Parkin SS, Chang K. Rev Mod Phys 2021;93:011001.
[22] Lin C-Y, Chen R-B, Ho Y-H, Lin M-F. Electronic and optical properties of graphite-related systems. CRC Press; 2017.
[23] Chen S-C, Wu J-Y, Lin C-Y, Lin M-F. Theory of magnetoelectric properties of 2d systems. IOP Publishing; 2017.
[24] Lin S-Y, Tran NTT, Chang S-L, Su W-P, Lin M-F. Structure-and adatom-enriched essential properties of graphene nanoribbons. CRC press; 2018.
[25] Lonkar SP, Deshmukh YS, Abdala AA. Nano Res 2015;8:1039–74.

[26] Batzill M. Surf Sci Rep 2012;67:83–115.
[27] Lin C-Y, Wu J-Y, Chiu C-W, Lin M-F. Coulomb excitations and decays in graphene-related systems. CRC Press; 2019.
[28] Arief I, Biswas S, Bose S. Nano-Struct Nano-Obj 2017;11:94–101.
[29] Kariuki VM, Fasih-Ahmad SA, Osonga FJ, Sadik OA. Analyst 2016;141(7):2259–69.
[30] Fugallo G, Cepellotti A, Paulatto L, Lazzeri M, Marzari N, Mauri F. Nano Lett 2014;14:6109–14.
[31] Petrović M, Rakić IŠ, Runte S, Busse C, Sadowski J, Lazić P, et al. Nat Commun 2013;4:1–8.
[32] Sutter P, Sadowski JT, Sutter EA. J Am Chem Soc 2010;132:8175–9.
[33] Li J, Zhang L, Zhang L, Hao W, Wang H, Qu Q, et al. J Power Sources 2014;249:311–9.
[34] Singh R, Bester G. Phys Rev B 2011;84:155427.
[35] Wang H, Wang Q, Cheng Y, Li K, Yao Y, Zhang Q, et al. Nano Lett 2012;12:141–4.
[36] Casolo S, Løvvik OM, Martinazzo R, Tantardini GF. J Chem Phys 2009;130:054704.
[37] Enoki T, Suzuki M, Endo M. Graphite intercalation compounds and applications. Oxford University Press; 2003.
[38] Liu J, Li X, Wang Q, Kawazoe Y, Jena P. J Mater Chem 2018;6:13816–24.
[39] Sadovskii MV. Phys Rep 1997;282:225–348.
[40] Monthoux P, Pines D, Lonzarich G. Nature 2007;450:1177–83.
[41] Fauster T, Himpsel F, Fischer J, Plummer E. Phys Rev Lett 1983;51:430.
[42] Bordet F, Ahlbrecht K, Tübke J, Ufheil J, Hoes T, Oetken M, et al. Electrochim Acta 2015;174:1317–23.
[43] Pruvost S, Hérold C, Hérold A, Lagrange P. Carbon 2004;42:1825–31.
[44] Kim H, Hong J, Yoon G, Kim H, Park K-Y, Park M-S, et al. Energy Environ Sci 2015;8:2963–9.
[45] Chew SY, Ng SH, Wang J, Novák P, Krumeich F, Chou SL, et al. Carbon 2009;47:2976–83.
[46] Dresselhaus M, Dresselhaus G, Saito R. Carbon 1995;33:883–91.
[47] Son Y-W, Cohen ML, Louie SG. Phys Rev Lett 2006;97:216803.
[48] Son Y-W, Cohen ML, Louie SG. Nature 2006;444:347–9.
[49] Kim K, Lee TH, Santos EJ, Jo PS, Salleo A, Nishi Y, et al. ACS Nano 2015;9:5922–8.
[50] Cerón MR, Zhan C, Campbell PG, Freyman MC, Santoyo C, Echegoyen L, et al. ACS Appl Mater Interfaces 2019;11:28818–22.
[51] Ori O, Cataldo F, Putz MV, Kaatz F, Bultheel A. Fullerenes, Nanotub Carbon Nanostruct 2016;24:353–62.
[52] Nakanishi T, Ando T. J Phys Soc Jpn 2009;78:114708.
[53] Lin S-Y, Chang S-L, Tran NTT, Yang P-H, Lin M-F. Phys Chem Chem Phys 2015;17:26443–50.
[54] Nguyen DK, Tran NTT, Nguyen TT, Lin M-F. Sci Rep 2018;8:1–12.
[55] Mizokawa Y, Miyasato T, Nakamura S, Geib KM, Wilmsen CW. Surf Sci 1987;182:431–8.
[56] Kimoto T, Itoh A, Matsunami H. Physica Status Solidi (B) 1997;202:247–62.
[57] Ferralis N, Maboudian R, Carraro C. Phys Rev Lett 2008;101:156801.
[58] McCann E, Koshino M. Rep Prog Phys 2013;76:056503.
[59] Liu F, Liu C-C, Wu K, Yang F, Yao Y. Phys Rev Lett 2013;111:066804.
[60] Fu H, Zhang J, Ding Z, Li H, Meng S. Appl Phys Lett 2014;104:131904.
[61] Cai Y, Chuu C-P, Wei C, Chou M. Phys Rev B 2013;88:245408.
[62] Hong Y, Zhu C, Ju M, Zhang J, Zeng XC. Phys Chem Chem Phys 2017;19:6554–62.
[63] Zhang L, Zhao H, Ji W-x, Zhang C-w, Li P, Wang P-j. Chem Phys Lett 2018;712:78–82.
[64] Blase X, Charlier J-C, De Vita A, Car R. Appl Phys A 1999;68:293–300.
[65] Mazzoni MS, Nunes R, Azevedo S, Chacham H. Phys Rev B 2006;73:073108.
[66] Liebert B, Huggins R. Mater Res Bull 1976;11:533–8.
[67] Bai J, Gong Z, Lv D, Li Y, Zou H, Yang Y. J Mater Chem 2012;22:12128–32.
[68] Sakurai Y, Arai H, Okada S, Yamaki J-i. J Power Sources 1997;68:711–5.
[69] Kanno R, Shirane T, Kawamoto Y, Takeda Y, Takano M, Ohashi M, et al. J Electrochem Soc 1996;143:2435.
[70] Laubach S, Laubach S, Schmidt PC, Ensling D, Schmid S, Jaegermann W, et al. Phys Chem Chem Phys 2009;11:3278–89.
[71] Van Elp J, Wieland J, Eskes H, Kuiper P, Sawatzky G, De Groot F, et al. Phys Rev B 1991;44:6090.
[72] Musaev D, Charkin O. J Struct Chem 1992;33:1–7.
[73] Tozzini V, Pellegrini V. Phys Chem Chem Phys 2013;15:80–9.
[74] Wang L, Lee K, Sun Y-Y, Lucking M, Chen Z, Zhao JJ, et al. ACS Nano 2009;3:2995–3000.

[75] Geim AK, Novoselov KS. The rise of graphene. In: Nanoscience and technology: a collection of reviews from nature journals. World Scientific; 2010. p. 11–9.
[76] Alvial-Palavicino C, Konrad K. Futures 2019;109:192–202.
[77] Radadiya TM. Eur J Mater Sci 2015;2:6–18.
[78] Ma Y, Lu Y, Yi J, Feng Y, Herng T, Liu X, et al. Nat Commun 2012;3:1–8.
[79] Müllen K. ACS Nano 2014;8:6531–41.
[80] Weng Y, You Y, Lu Q, Zhong A, Liu S, Liu H, et al. Environ Pollut 2020;262:114224.
[81] Pi K, McCreary K, Bao W, Han W, Chiang Y, Li Y, et al. Phys Rev B 2009;80:075406.
[82] Zhong C, Chen Y, Xie Y, Yang SA, Cohen ML, Zhang S. Nanoscale 2016;8:7232–9.
[83] Miransky VA, Shovkovy IA. Phys Rep 2015;576:1–209.
[84] Bianchi M, Rienks E, Lizzit S, Baraldi A, Balog R, Hornekær L, et al. Phys Rev B 2010;81:041403.
[85] Kan E-j, Li Z, Yang J, Hou J. J Am Chem Soc 2008;130:4224–5.
[86] Hod O, Barone V, Peralta JE, Scuseria GE. Nano Lett 2007;7:2295–9.
[87] Nguyen DK, Lin Y-T, Lin S-Y, Chiu Y-H, Tran NTT, Fa-Lin M. Phys Chem Chem Phys 2017;19:20667–76.
[88] Kan M, Zhou J, Sun Q, Wang Q, Kawazoe Y, Jena P. Phys Rev B 2012;85:155450.
[89] Sahin H, Ataca C, Ciraci S. Phys Rev B 2010;81:205417.
[90] Lin S-Y, Chang S-L, Shyu F-L, Lu J-M, Lin M-F. Carbon 2015;86:207–16.
[91] Lin C-Y, Ho C-H, Wu J-Y, Do T-N, Shih P-H, Lin S-Y, et al. Diverse quantization phenomena in layered materials. CRC Press; 2019.
[92] Pallecchi E, Lafont F, Cavaliere V, Schopfer F, Mailly D, Poirier W, et al. Sci Rep 2014;4:1–7.
[93] Hajlaoui M, Sediri H, Pierucci D, Henck H, Phuphachong T, Silly MG, et al. Sci Rep 2016;6:1–8.
[94] Balog R, Jørgensen B, Nilsson L, Andersen M, Rienks E, Bianchi M, et al. Nat Mater 2010;9:315–9.
[95] Srinivas G, Zhu Y, Piner R, Skipper N, Ellerby M, Ruoff R. Carbon 2010;48:630–5.
[96] Wang L, Stuckert NR, Yang RT. AIChE J 2011;57:2902–8.
[97] Zhao H, Zhang C-w, Ji W-x, Zhang R-w, Li S-s, Yan S-s, et al. Sci Rep 2016;6:1–8.
[98] Pumera M, Wong CHA. Chem Soc Rev 2013;42:5987–95.
[99] Houssa M, Scalise E, Sankaran K, Pourtois G, Afanas' Ev V, Stesmans A. Appl Phys Lett 2011;98:223107.
[100] Sofo JO, Chaudhari AS, Barber GD. Phys Rev B 2007;75:153401.
[101] Huang H-C, Lin S-Y, Wu C-L, Lin M-F. Carbon 2016;103:84–93.
[102] Lee M, Chung H, Lu J, Chang C, Lin M. Phil Mag 2015;95:2717–30.
[103] Bena C, Montambaux G. New J Phys 2009;11:095003.
[104] Mahan G. Many-body physics. New York: Kluwer Academic/Plenum Publishers; 2000.
[105] Huang B-L, Chuu C-P, Lin M-F. Sci Rep 2019;9:1–12.
[106] Lin Y-P, Lin C-Y, Ho Y-H, Do T-N, Lin M-F. Phys Chem Chem Phys 2015;17:15921–7.
[107] Yang L, Deslippe J, Park C-H, Cohen ML, Louie SG. Phys Rev Lett 2009;103:186802.
[108] Wong J-H, Wu B-R, Lin M-F. J Phys Chem C 2012;116:8271–7.
[109] Do T-N, Shih P-H, Gumbs G, Huang D, Chiu C-W, Lin M-F. Phys Rev B 2018;97:125416.
[110] Dien VK, Pham HD, Tran NTT, Han NT, Huynh TMD, Nguyen TDH, et al. Sci Rep 2021;11:1–10.
[111] Han NT, Dien VK, Lin M-F. Sci Rep 2021;11:1–10.
[112] Ataca C, Aktürk E, Ciraci S, Ustunel H. Appl Phys Lett 2008;93:043123.
[113] Hafner J. J Comput Chem 2008;29:2044–78.
[114] Peng H, Perdew JP. Phys Rev B 2017;95:081105.
[115] Blöchl PE. Phys Rev B 1994;50:17953.
[116] Hybertsen MS, Louie SG. Phys Rev B 1986;34:5390–413.
[117] Pizzi G, Vitale V, Arita R, Blügel S, Freimuth F, Géranton G, et al. J Phys Condens Matter 2020;32:165902.
[118] Mahan GD. Many-particle physics. Springer Science & Business Media; 2013.
[119] Nakanishi N. Prog Theor Phys Suppl 1969;43:1–81.
[120] Mahmood A, Rahman G. J Phys Condens Matter 2020;32:205501.
[121] Gao G, Huan H, Bao H, Xue Y, Zhao B, Yu R, et al. Phys Rev B 2020;102:205114.
[122] Zubizarreta X, Silkin VM, Chulkov EV. Phys Rev B 2011;84:115144.
[123] Zubizarreta X, Silkin VM, Chulkov EV. Phys Rev B 2013;87:115112.
[124] Binnig G, Rohrer H. Surf Sci 1983;126:236–44.
[125] Hansma PK, Tersoff J. J Appl Phys 1987;61:R1–24.

[126] Garnaes J, Kragh F, Mo/rch K, Thölén A. J Vac Sci Technol: Vacuum Surfaces Films 1990;8:441–4.
[127] Cui S, Zhuang G, Lu D, Huang Q, Jia H, Wang Y, et al. Angew Chem 2018;130:9474–9.
[128] Xia Z, Pun SH, Chen H, Miao Q. Angew Chem 2021;133:10399–406.
[129] Hároz EH, Duque JG, Tu X, Zheng M, Walker ARH, Hauge RH, et al. Nanoscale 2013;5:1411–39.
[130] Liu B, Wu F, Gui H, Zheng M, Zhou C. ACS Nano 2017;11:31–53.
[131] Zhang Y, Zheng L. Nanoscale 2010;2:1919–29.
[132] Charlier J-C, Lambin P. Phys Rev B 1998;57:R15037.
[133] Saroka V, Pushkarchuk A, Kuten S, Portnoi M. J Saudi Chem Soc 2018;22:985–92.
[134] Lu C, Chang C-P, Huang Y-C, Chen R-B, Lin M. Phys Rev B 2006;73:144427.
[135] Koshino M. Phys Rev B 2010;81:125304.
[136] Lin C-Y, Huang B-L, Ho C-H, Gumbs G, Lin M-F. Phys Rev B 2018;98:195442.
[137] Mak KF, Shan J, Heinz TF. Phys Rev Lett 2010;104:176404.
[138] Xu Y, Li X, Dong J. Nanotechnology 2010;21:065711.
[139] Chung H-C, Chiu C-W, Lin M-F. Sci Rep 2019;9:1–12.
[140] Tran NTT, Nguyen DK, Glukhova OE, Lin M-F. Sci Rep 2017;7:1–13.
[141] Nguyen DK, Tran NTT, Chiu Y-H, Lin M-F. Sci Rep 2019;9:1–15.
[142] Ren M, Yuan M, Chen X, Ji W, Li P, Li F. Phys Chem Chem Phys 2018;20:9610–5.
[143] Hermanson J. Solid State Commun 1977;22:9–11.
[144] Lv B, Qian T, Ding H. Nat Rev Phys 2019;1:609–26.
[145] Ohta T, Bostwick A, McChesney JL, Seyller T, Horn K, Rotenberg E. Phys Rev Lett 2007;98:206802.
[146] Senkovskiy BV, Usachov DY, Fedorov AV, Haberer D, Ehlen N, Fischer FR, et al. 2D Mater 2018;5:035007.
[147] Choi J, Lee SM, Choi YC, Lee YH, Jiang J. Chem Phys Lett 2001;349:185–90.
[148] Kajiwara T, Nakamori Y, Visikovskiy A, Iimori T, Komori F, Nakatsuji K, et al. Phys Rev B 2013;87:121407.
[149] Lu C, Chang C-P, Huang Y-C, Lu J, Hwang C-C, Lin M-F. J Phys Condens Matter 2006;18:5849.
[150] Rong ZY, Kuiper P. Phys Rev B 1993;48:17427.
[151] Ahmadi MT, Johari Z, Amin NA, Fallahpour AH, Ismail R. J Nanomater 2010.
[152] Moreau E, Godey S, Wallart X, Razado-Colambo I, Avila J, Asensio M-C, et al. Phys Rev B 2013;88:075406.
[153] Bao C, Yao W, Wang E, Chen C, Avila J, Asensio MC, et al. Nano Lett 2017;17:1564–8.
[154] Joucken F, Avila J, Ge Z, Quezada-Lopez EA, Yi H, Le Goff R, et al. Nano Lett 2019;19:2682–7.
[155] Cong C, Yu T, Sato K, Shang J, Saito R, Dresselhaus GF, et al. ACS Nano 2011;5:8760–8.
[156] Lin Y-P, Wang J, Lu J-M, Lin C-Y, Lin M-F. RSC Adv 2014;4:56552–60.
[157] McChesney JL, Bostwick A, Ohta T, Seyller T, Horn K, González J, et al. Phys Rev Lett 2010;104:136803.
[158] Li G, Luican A, Dos Santos JL, Neto AC, Reina A, Kong J, et al. Nat Phys 2010;6:109–13.
[159] Chumakov A, Monaco G, Monaco A, Crichton W, Bosak A, Rüffer R, et al. Phys Rev Lett 2011;106:225501.
[160] Yuan NF, Isobe H, Fu L. Nat Commun 2019;10:1–7.
[161] Ziletti A, Huang SM, Coker DF, Lin H. Phys Rev B 2015;92:085423.
[162] Zhang X, Luo H. Appl Phys Lett 2013;103:231602.
[163] Rosenzweig P, Karakachian H, Marchenko D, Küster K, Starke U. Phys Rev Lett 2020;125:176403.
[164] Jorio A, Souza Filho A, Dresselhaus G, Dresselhaus M, Saito R, Hafner J, et al. Phys Rev B 2001;63:245416.
[165] Ohno Y, Kishimoto S, Mizutani T, Okazaki T, Shinohara H. Appl Phys Lett 2004;84:1368–70.
[166] Hartmann R, Saroka V, Portnoi M. J Appl Phys 2019;125:151607.
[167] Zhang G, Fang X, Yao Y-X, Wang C, Ding Z, Ho K-M. J Phys Condens Matter 2010;23:025302.
[168] Van Bentum P, Tsuei C. Tunneling into high Tc superconductors—the effect of a two-dimensional van Hove singularity. In: AIP conference proceedings; 1992. p. 277–91.
[169] Sala G, Stone M, Rai BK, May A, Laurell P, Garlea V, et al. Nat Commun 2021;12:1–8.
[170] Marchini S, Günther S, Wintterlin J. Phys Rev B 2007;76:075429.
[171] Liu X-J, Borunda MF, Liu X, Sinova J. Phys Rev Lett 2009;102:046402.
[172] Crépieux A, Bruno P. Phys Rev B 2001;64:014416.
[173] Nakayama T, Shima H. Phys Rev 1998;58:3984.
[174] Anderson PW. Phys Rev 1958;112:1900.
[175] Kane C, Fisher MP. Phys Rev B 1997;55:15832.
[176] Lin M-F, Shyu F-L. J Phys Soc Jpn 2000;69:3529–32.
[177] Shih P-H, Do T-N, Gumbs G, Lin M-F. Phys E Low-dimens Syst Nanostruct 2020;118:113894.

[178] Altshuler BL, Gefen Y, Kamenev A, Levitov LS. Phys Rev Lett 1997;78:2803.
[179] Abrahams E, Anderson P, Lee P, Ramakrishnan T. Phys Rev B 1981;24:6783.
[180] Dukelsky J, Röpke G, Schuck P. Nucl Phys 1998;628:17–40.
[181] Jackson JD. American association of physics teachers. 1999.
[182] Sakurai JJ, Commins ED. American association of physics teachers. 1995.
[183] Kittel C. Solid state physics. Shell Development Company; 1955.
[184] Trevisanutto PE, Holzmann M, Côté M, Olevano V. Phys Rev B 2010;81:121405.
[185] Perebeinos V, Tersoff J, Avouris P. Phys Rev Lett 2004;92:257402.
[186] Yang L, Cohen ML, Louie SG. Nano Lett 2007;7:3112–5.
[187] Lin M, Huang C, Chuu D. Phys Rev B 1997;55:13961.
[188] Lalmi B, Oughaddou H, Enriquez H, Kara A, Vizzini S, Ealet B, et al. Appl Phys Lett 2010;97:223109.
[189] Ekimov AI, Hache F, Schanne-Klein M, Ricard al, D, Flytzanis C, et al. JOSA B 1993;10:100–7.
[190] Pophristic M, Long F, Tran C, Ferguson I, Karlicek Jr R. Appl Phys Lett 1998;73:3550–2.
[191] Osche GR. Optical detection theory for laser applications. 2002.
[192] Duarte FJ. Tunable laser applications. CRC Press; 2008.
[193] Kornitzer K, Ebner T, Grehl M, Thonke K, Sauer R, Kirchner C, et al. Physica Status Solidi (B) 1999;216:5–9.
[194] Liu Z-L, Wang M-X, Xu J-P, Ge J-F, Le Lay G, Vogt P, et al. New J Phys 2014;16:075006.
[195] Gou J, Zhong Q, Sheng S, Li W, Cheng P, Li H, et al. 2D Mater 2016;3:045005.
[196] Guo B, Xiao QL, Wang SH, Zhang H. Laser Photon Rev 2019;13:1800327.
[197] Shih P-H, Do T-N, Huang B-L, Gumbs G, Huang D, Lin M-F. Carbon 2019;144:608–14.
[198] Do T-N, Gumbs G, Shih P-H, Huang D, Lin M-F. Phys Rev B 2019;100:155403.
[199] Berry V. Carbon 2013;62:1–10.
[200] Zhu J, Schwingenschlögl U. 2D Mater 2016;3:035012.
[201] Qi X, Tan C, Wei J, Zhang H. Nanoscale 2013;5:1440–51.
[202] Chen X, Yang Q, Meng R, Jiang J, Liang Q, Tan C, et al. J Mater Chem C 2016;4:5434–41.
[203] Liu Y, Dong X, Chen P. Chem Soc Rev 2012;41:2283–307.
[204] Castagnola V, Zhao W, Boselli L, Giudice ML, Meder F, Polo E, et al. Nat Commun 2018;9:1–9.
[205] Derivaz M, Dentel D, Stephan R, Hanf M-C, Mehdaoui A, Sonnet P, et al. Nano Lett 2015;15:2510–6.
[206] Cortijo A, Vozmediano MA. EPL 2007;77:47002.
[207] Kim K, Lee Z, Malone BD, Chan KT, Alemán B, Regan W, et al. Phys Rev B 2011;83:245433.
[208] Chan KT, Neaton J, Cohen ML. Phys Rev B 2008;77:235430.
[209] Kong L, Enders A, Rahman TS, Dowben PA. J Phys Condens Matter 2014;26:443001.
[210] Dien VK, Le OK, Chihaia V, Pham-Ho M-P. J Comput Electron 2021;20:135–50.
[211] Poh HL, Šaněk F, Sofer Z, Pumera M. Nanoscale 2012;4:7006–11.
[212] Nováček M, Jankovský O, Luxa J, Sedmidubský D, Pumera M, Fila V, et al. J Mater Chem 2017;5:2739–48.
[213] Yang Y, Jiang X, Li Y, Jin W, Huang X. Compos Commun 2021;25:100771.
[214] Zhao J, Han J, Lu JP. Phys Rev B 2002;65:193401.
[215] Hashimoto A, Suenaga K, Gloter A, Urita K, Iijima S. Nature 2004;430:870–3.
[216] Muñoz R, Gómez-Aleixandre C. Chem Vap Depos 2013;19:297–322.
[217] Bao W, Myhro K, Zhao Z, Chen Z, Jang W, Jing L, et al. Nano Lett 2012;12:5470–4.
[218] Ju H-M, Huh SH, Choi S-H, Lee H-L. Mater Lett 2010;64:357–60.
[219] Zhou M, Zhai Y, Dong S. Anal Chem 2009;81:5603–13.
[220] Subrahmanyam K, Panchakarla L, Govindaraj A, Rao C. J Phys Chem C 2009;113:4257–9.
[221] Suk JW, Kitt A, Magnuson CW, Hao Y, Ahmed S, An J, et al. ACS Nano 2011;5:6916–24.
[222] Huang Y, Dong X, Shi Y, Li CM, Li L-J, Chen P. Nanoscale 2010;2:1485–8.
[223] Chen H-D, Lin D-S. ACS Omega 2016;1:357–62.
[224] Yuhara J, Shimazu H, Ito K, Ohta A, Araidai M, Kurosawa M, et al. ACS Nano 2018;12:11632–7.
[225] Zhu F-F, Chen W-J, Xu Y, Gao C-L, Guan D-D, Liu C-H, et al. Nat Mater 2015;14:1020–5.
[226] Yuhara J, He B, Matsunami N, Nakatake M, Le Lay G. Adv Mater 2019;31:1901017.
[227] Ho Y-H, Chiu Y-H, Lin D-H, Chang C-P, Lin M-F. ACS Nano 2010;4:1465–72.
[228] Chang S-L, Lin S-Y, Lin S-K, Lee C-H, Lin M-F. Sci Rep 2014;4:1–8.
[229] Ho Y-H, Wu J-Y, Chen R-B, Chiu Y-H, Lin M-F. Appl Phys Lett 2010;97:101905.
[230] Chuang Y-C, Wu J-Y, Lin M-F. Sci Rep 2013;3:1–5.

[231] Lin C-Y, Lee M-H, Lin M-F. Phys Rev B 2018;98:041408.
[232] Lin M-F, Chuang Y-C, Wu J-Y. Phys Rev B 2012;86:125434.
[233] Gharbavi K, Badehian H. Comput Mater Sci 2014;82:159–64.
[234] Kupka T, Stachow M, Stobiński L, Kaminský J. J Mol Graph Model 2016;67:14–9.
[235] Farhana S, Alam AZ, Khan S, Motakabber S. Modeling of optimum chiral carbon nanotube using DFT. In: 2013 13th IEEE international conference on nanotechnology (IEEE-NANO 2013); 2013. p. 853–7.
[236] Chung H-C, Chang C-P, Lin C-Y, Lin M-F. Phys Chem Chem Phys 2016;18:7573–616.
[237] Li L, Lu SZ, Pan J, Qin Z, Wang YQ, Wang Y, et al. Adv Mater 2014;26:4820–4.
[238] Lin C-Y, Wu J-Y, Ou Y-J, Chiu Y-H, Lin M-F. Phys Chem Chem Phys 2015;17:26008–35.
[239] Chen S-C, Wu J-Y, Lin M-F. New J Phys 2018;20:062001.
[240] Lin S-Y, Ho Y-H, Huang Y-C, Lin M-F. J Phys Soc Jpn 2012;81:084602.
[241] Do T-N, Chang C-P, Shih P-H, Wu J-Y, Lin M-F. Phys Chem Chem Phys 2017;19:29525–33.
[242] Shih P-H, Do T-N, Gumbs G, Huang D, Pham TP, Lin M-F. Carbon 2020;160:211–8.
[243] Pham HD, Gumbs G, Su W-P, Tran NTT, Lin M-F. RSC Adv 2020;10:32193–201.
[244] Pham HD, Su W-P, Nguyen TDH, Tran NTT, Lin Royal Society Open M-F. Science 2020;7:200723.
[245] Chen R-B, Chen S-C, Chiu C-W, Lin M-F. Sci Rep 2017;7:1–9.
[246] Drummond N, Zolyomi V, Fal'Ko V. Phys Rev B 2012;85:075423.
[247] Ye M, Quhe R, Zheng J, Ni Z, Wang Y, Yuan Y, et al. Phys E Low-dimens Syst Nanostruct 2014;59:60–5.
[248] Mahdavifar S, Tagani MB. Phys E Low-dimens Syst Nanostruct 2021:114837.
[249] Lin SY, Liu HY, Nguyen DK, Tran NT, Pham HD, Chang SL, et al. Silicene-Based Layered Materials. IOP Publishing Limited; 2020.
[250] Lin CY, Ho CH, Wu JY, Do TN, Shih PH, Lin SY, et al. Diverse quantization phenomena in layered materials. CRC Press; 2019.
[251] Feenstra RM. Surf Sci 1994;299:965–79.
[252] Shih P-H, Do T-N, Gumbs G, Pham HD, Lin M-F. Opt Lett 2019;44:4721–4.
[253] Chen R-B, Jang D-J, Lin M-C, Lin M-F. Opt Lett 2018;43:6089–92.
[254] Ho Y-H, Tsai S-J, Lin M-F, Su W-P. Phys Rev B 2013;87:075417.
[255] Ho J-H, Lai Y-H, Chiu Y-H, Lin M-F. arXiv preprint arXiv:07080216. 2007.
[256] Lin C-Y, Wu J-Y, Chiu Y-H, Chang C-P, Lin M-F. Phys Rev B 2014;90:205434.
[257] Do T-N, Lin C-Y, Lin Y-P, Shih P-H, Lin M-F. Carbon 2015;94:619–32.
[258] Kondo T, Iwasaki Y, Honma Y, Takagi Y, Okada S, Nakamura J. Phys Rev B 2009;80:233408.
[259] Li W-B, Lin S-Y, Tran NTT, Lin M-F, Lin K-I. RSC Adv 2020;10:23573–81.
[260] Wu J-Y, Chen S-C, Gumbs G, Lin M-F. Phys Rev B 2017;95:115411.
[261] Do T-N, Gumbs G, Shih P-H, Huang D, Chiu C-W, Chen C-Y, et al. Sci Rep 2019;9:1–15.
[262] Shih P-H, Do T-N, Gumbs G, Huang D, Pham HD, Lin M-F. Sci Rep 2019;9:1–8.
[263] Li T-S, Lin M-F. Phys Rev B 2006;73:075432.
[264] Schaibley JR, Yu H, Clark G, Rivera P, Ross JS, Seyler KL, et al. Nat Rev Mater 2016;1:1–15.
[265] Bader S, Parkin S. Annu Rev Condens Matter Phys 2010;1:71–88.
[266] Matthes L, Pulci O, Bechstedt F. J Phys Condens Matter 2013;25:395305.
[267] Chen SC, Wu CL, Wu JY, Lin MF. Phys Rev B 2016;94(4):045410.
[268] Gadomsky O, Altunin K, Stepin S, Katnov V, Rusin A, Pereskokov E. Opt Commun 2014;315:286–94.
[269] Kim KS, Walter AL, Moreschini L, Seyller T, Horn K, Rotenberg E, et al. Nat Mater 2013;12:887–92.
[270] Gusynin V, Sharapov S. Phys Rev Lett 2005;95:146801.
[271] Li L, Partoens B, Peeters F. Phys Rev B 2018;97:155424.
[272] Richter CA, Hacker CA, Richter LJ. J Phys Chem B 2005;109:21836–41.
[273] Shih P-H, Chiu Y-H, Wu J-Y, Shyu F-L, Lin M-F. Sci Rep 2017;7:40600.
[274] Wang J, Wang Z, Zhang R, Zheng Y, Chen L, Wang S, et al. Phys Chem Chem Phys 2018;20:18110–6.
[275] Do T-N, Shih P-H, Chang C-P, Lin C-Y, Lin M-F. Phys Chem Chem Phys 2016;18:17597–605.
[276] Bachtold A, Strunk C, Salvetat J-P, Bonard J-M, Forró L, Nussbaumer T, et al. Nature 1999;397:673–5.
[277] Guzmán de Villoria R, Hart AJ, Wardle BL. ACS Nano 2011;5:4850–7.
[278] Lin M-F. Phys Rev B 2000;62:13153.
[279] Brydson R. Electron energy loss spectroscopy. Garland Science; 2020.
[280] Barrios R, Skurski P, Simons J. J Phys Chem B 2002;106:7991–4.

[281] Winkler R, Papadakis S, De Poortere E, Shayegan M. Spin-orbit coupling in two-dimensional electron and hole systems. Springer; 2003.
[282] Brock TD. Science 1985;230:132–8.
[283] Schubert EF. Doping in III-V semiconductors. E. Fred Schubert; 2015.
[284] Horowitz G, Hajlaoui R, Fichou D, El Kassmi A. J Appl Phys 1999;85:3202–6.
[285] Jasieniak J, Califano M, Watkins SE. ACS Nano 2011;5:5888–902.
[286] Aliabadi AA, Krayenhoff ES, Nazarian N, Chew LW, Armstrong PR, Afshari A, et al. Boundary-Layer Meteorol 2017;164:249–79.
[287] Wu J-Y, Gumbs G, Lin M-F. Phys Rev B 2014;89:165407.
[288] Matsushita T. Jpn J Appl Phys 2011;51:010111.
[289] Wu J-Y, Lin C-Y, Do T-N, Shih P-H, Lin S-Y, Ho C-H, et al. Diverse Quantization Phenomena in Layered Materials 2019;249.
[290] Nakamura T. Sol Energy 1977;19:467–75.
[291] Park HW, Oh JH. Sci Rep 2019;9:1–13.
[292] Wu J-Y, Chen S-C, Gumbs G, Lin M-F. Phys Rev B 2016;94:205427.
[293] Shih P-H, Chiu C-W, Wu J-Y, Do T-N, Lin M-F. Phys Rev B 2018;97:195302.
[294] Kahng D, Sze SM. Bell Syst Tech J 1967;46:1288–95.
[295] Bengtsson R, Frauendorf S. Nucl Phys 1979;327:139–71.
[296] Blanter YM, Mirlin A, Muzykantskii B. Phys Rev Lett 1997;78:2449.
[297] Giustino F. Rev Mod Phys 2017;89:015003.
[298] Han NT, Dien VK, Tran NTT, Nguyen DK, Su W-P, Lin M-F. RSC Adv 2020;10:24721–9.
[299] Thi HN, Tran NTT, Dien VK, Nguyen DK, Lin M-F. Diversified properties in 3D ternary oxide compound: Li_2SiO_3. In: Lithium-ion batteries and solar cells. CRC Press; 2021. p. 79–101.
[300] Han NT, Dien VK, Lin M-F. arXiv preprint arXiv:201011621. 2020.
[301] Yang C-H, Lin C-Y, Chiu C-W, Liu C-T, Lin M-F. arXiv preprint arXiv:200408209. 2020.
[302] Rocca M. Surf Sci Rep 1995;22:1–71.
[303] Jin Z, Li X, Mullen JT, Kim KW. Phys Rev B 2014;90:045422.
[304] Murtola T, Bunker A, Vattulainen I, Deserno M, Karttunen M. Phys Chem Chem Phys 2009;11:1869–92.
[305] Bultinck N, Chatterjee S, Zaletel MP. Phys Rev Lett 2020;124:166601.
[306] Manes JL, de Juan F, Sturla M, Vozmediano MA. Phys Rev B 2013;88:155405.
[307] Wu J-Y, Su W-P, Gumbs G. Sci Rep 2020;10:1–10.
[308] Sandratskii L, Kübler J. Phys Rev Lett 1996;76:4963.
[309] Tao H-S, Chen Y-H, Lin H-F, Liu H-D, Liu W-M. Sci Rep 2014;4:1–7.
[310] Ishii T, Inuishi Y. J Phys Soc Jpn 1968;25:1406–13.
[311] Gomes LC, Carvalho A. J Appl Phys 2020;128:121101.
[312] Chen G, Pei L-J, Huang J, Song X-M, Lin L-m, Gu X, et al. Biomed Environ Sci 2009;22:340–4.
[313] Zhao J-W, Li Y-Z, Chen L-J, Yang G-Y. Chem Commun 2016;52:4418–45.
[314] Dutta S, Manna AK, Pati SK. Phys Rev Lett 2009;102:096601.
[315] Yildirim T, Gülseren O, Ciraci S. Phys Rev B 2001;64:075404.
[316] Oden JT, Prudencio EE, Hawkins-Daarud A. Math Model Methods Appl Sci 2013;23:1309–38.
[317] Bruck RH. A survey of binary systems. Springer; 1971.
[318] Chou KC, Austin Chang Y. Ber Bunsen Ges Phys Chem 1989;93:735–41.
[319] Lowe JJ, Walker M. Reconstructing quaternary environments. Routledge; 2014.
[320] Goushi Y, Hakuma H, Tabuchi K, Kijima S, Kushiya K. Sol Energy Mater Sol Cell 2009;93:1318–20.
[321] Shahrokhi M, Leonard C. J Alloys Compd 2017;693:1185–96.
[322] Kliewer K, Fuchs R. Phys Rev 1969;181:552.
[323] Nagatomi T, Shimizu R, Ritchie R. Surf Sci 1999;419:158–73.
[324] Clark RN. Rock physics and phase relations: a handbook of physical constants. 1995. p. 178–88.
[325] Hubbell JH. Int J Appl Radiat Isot 1982;33:1269–90.
[326] Shao L, Fang C, Chen H, Man YC, Wang J, Lin H-Q. Nano Lett 2012;12:1424–30.
[327] Chen S, Hseih W, Lin M-F. Phys Rev B 2005;72:193412.

tunneling microscopy/transmission electron microscopy (STM/TEM) [17,18], angle-resolved photoemission spectroscopy (ARPES) [19], scanning tunneling spectroscopy (STS) [20], and superconductor quantum interference device (SQUID) measurements [21], respectively.

8.2 Computational methods

We studied the structure, electronic and magnetic properties of transition metals-adsorbed germanene systems on the first principles of density functional theory (DFT). This process is implemented by the Vienna Ab initio Simulation Package (VASP) [22,23]. The generalized gradient approximation (GGA) to exchange-correlation functional of the Perdew–Burke–Ernzerhof (PBE) form [24] is used to calculate the electron exchange and correlation energies whereas the projector augmented wave (PAW) pseudopotentials method is utilized to deal with the electron-ion interactions [25]. The 2D periodic boundary condition is along the x- and y-axis. To avoid the interaction between adjacent unit cells, a vacuum distance along the z-axis is set to be 15 Å. The cut-off energies of the wave function expanded by plane waves were chosen to be 500 eV for all calculations. The pristine first Brillouin zone was sampled by $9 \times 9 \times 1$ and $100 \times 100 \times 1$ K-points, respectively, for the geometry optimization and further evaluations of electronic, magnetic properties within the Gamma scheme. During the relaxations process, the Hellmann–Feynman force convergence on each atom was set to be 0.01 eV/Å and the difference of eigenvalues between two simulation steps was less than 10^{-5} eV.

8.3 Results and discussions

8.3.1 Geometric structures

To get a systematic study, four possible adsorption sites for a single TM atom on germanene were considered including the top side (on the top of the upper Ge atom), valley-side (on the top of the lower Ge atom), hollow-side (above of the center of Ge honeycomb lattices), and bridge-side (on the top of the midway of Ge–Ge bond), as illustrated in Fig. 8.1A. To find the optimal structure, we calculated the absorbed energy using formula $E_{ads} = E_{TM} + E_{Ge} - E_{(Ge + TM)}$, where E_{TM}, E_{Ge}, and $E_{(Ge + TM)}$ are the total energy of an isolated TM atom, the total energies for germanene supercell, and the total energy of the transition metal adsorbed-germanene systems [26]. The larger E_{ads} represents the favorer site of adatom TM to germanene. Table 8.1 demonstrates the adsorption energies (E_{ads}) for TM atoms at four top, hollow, valley, and bridge sites on the germanene, including the corresponding geometrical parameters such as the adatom height over the germanene surface (h), the bond lengths between adatom and its nearest Ge atoms, and between two Ge atoms, the bucklings of adsorption systems.

As result, all transition metals Fe, Co, and Ni favor bonding on the hollow site of the germanene systems and the valley site on the low-lying germanene system is the second preferable site. Fig. 8.1B demonstrates the comparative energy plot of all four adsorption sites. Comparing with the transition metal atoms adsorbed-silicene systems [26], we found that the adsorption energy is larger in the case of germanene systems. That result demonstrates

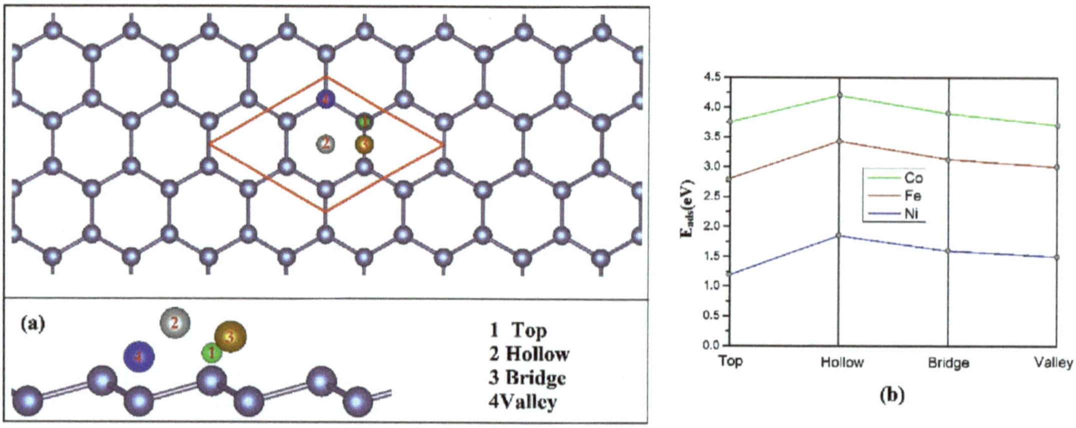

FIG. 8.1 (A) Four possible adsorption sites are displayed by top (1), hollow (2), bridge (3), and valley (4); (B) the plot shows the absorption energies of Co, Fe, and Ni for green, red and blue, respectively.

TABLE 8.1 The optimal geometric structures of iron-, cobalt-, and nickel-adsorbed germanene systems with the adsorption energies, the distance of adatom to germanene layer, X–Ge bond lengths (X = Fe, Co, Ni), the first and second nearest Ge–Ge bond lengths, magnetic moments together with the buckling height.

Adatoms	Sites	E_{ads} (eV)	X-high (h)	X–Ge (Å)	1st Ge–Ge (Å)	2nd Ge–Ge (Å)	Magnetic moment (μ_B)	Buckling (Å)
Iron	Valley	3.0	0	2.381	2.752	2.452	2.74	1.57
	Bridge	3.13	0	2.454	2.655	2.462	2.96	
	Hollow	3.43	1.02	2.459–2.710	2.651	2.535	1.77	1.13
	Top	2.8	2.332	2.382	2.754	2.454	2.73	
Cobalt	Valley	3.6	0	2.261	2.459	2.459	0	0.51
	Bridge	3.8	0	2.326	2.463	2.463		0.55
	Hollow	4.20	1.05	2.437–2.718	2.595	2.490	1.00	0.95
	Top	3.7	2.370	2.401	2.466	2.466	1.75	1.44
Nickel	Valley	1.19	0	2.423	2.577	2.468	0	1.36
	Bridge	1.60	0	2.421	2.579	2.467	0	1.36
	Hollow	1.85	0.95	2436–2649	2.552	2.438	0	1.35
	Top	1.5	2.411	2.420	2.579	2.467	0	1.37

that germanene might be the best candidate to accommodate TM atoms among all three 2D materials. Very interestingly, in the bridge site case, when the transition atoms are put on the top of the midway of the Ge–Ge bond, they always move to the valley position after the relaxation processes are completed along with create a non-magnetic system in these two sites. That is to say, bridge site adsorption of transition metal is not achievable on a germanene system. Because the hollow site is the most energetically favorable of all the cases under study, in this chapter, our discussions only focus on the results of geometric structures, the electronic and magnetic properties of the TM adsorbed-germanene transition metal in this side (Fig. 8.2).

It should be recalled that in our previous studies, pristine germanene has a lattice constant of 4.08 Å, a Ge–Ge bond length of 2.42 Å, and a buckling height of 0.66 Å [27]. After the TM atoms adsorbed-germanene, the structure was completely changed compared with the pristine case. In general, the buckling height became more serious and bond lengths between Ge atoms significantly changed. Since there is more than one buckling value in germanene after adsorption, we only show the maximum value in each structure in Table 8.1. The maximum value of buckling height in the two cases of Fe and Co is equal to 1.12 Å, while it is 1.35 Å in the case of Ni. The height of TM atom to G sheet (h) will be independent of the initial position (h-initial) of TM atoms before the relaxation process if h-initial is in the range 1 Å $< d <$ 3.5 Å. Instead, h depends on the type of TM where it has values of 1.02, 1.05, and 0.95 Å for Fe, Co, and Ni, respectively.

Due to the high buckling, the nearest TM-Ge bond lengths have a huge fluctuation around 2.459–2.710 Å for Fe, 2.437–2.718 Å for Co, and 2.436–2.649 Å for Ni. The smaller and longer

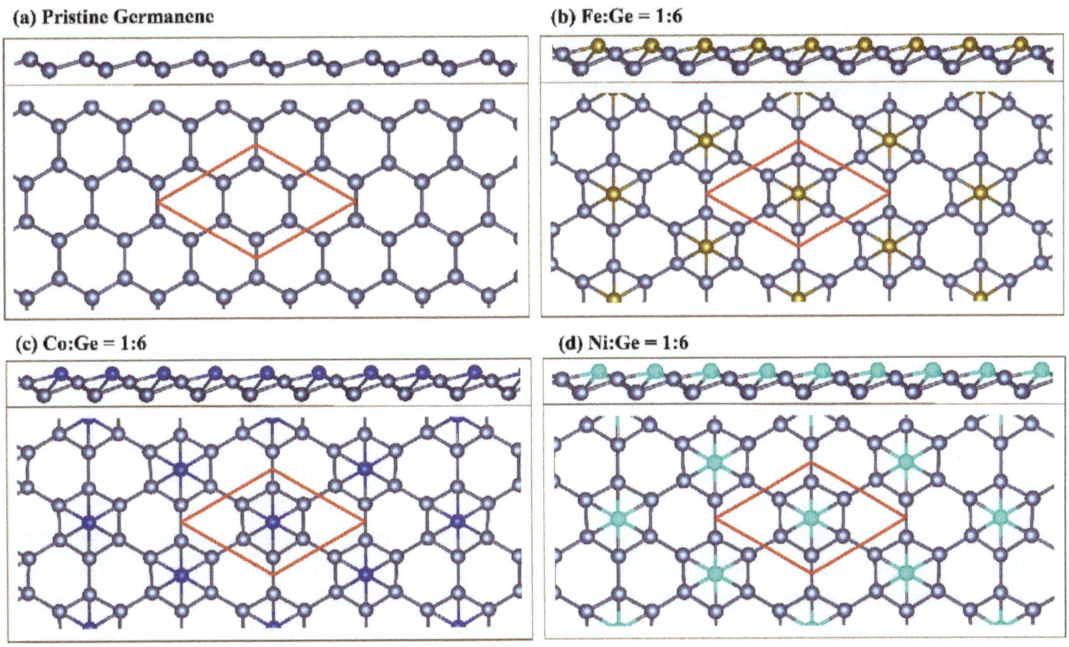

FIG. 8.2 The side and top views of optimal geometric structures for (A) pristine germanene, (B) iron-, (C) cobalt-, and (D) nickel-adsorption.

values of such bond lengths correspond to the distance between TM and the upper/lower Ge atoms. Very interestingly, the Ge–Ge bond length is greatly changed as compared to pristine germanene with a value of 2.44 Å, especially for the first nearest Ge–Ge bonds, as shown in Table 8.1. This result demonstrates that there is a charge transfer from the Ge atoms to the TM atoms. Obviously, the presence of TM atoms has totally changed the structure of the system and created the highly asymmetric chemical environments, which are described as the noticeable fluctuations ranges of TM-Ge bond lengths and buckled structures, which are expected to generate complicated charge distributions and thus diversifying the other essential properties.

8.3.2 Band structure tailoring of transition metal adsorbed-germanene

In this section, we present the atom-dominated and spin-polarized electronic band structure for Co-, Fe-, and Ni-adsorbed on the most favorable site (hollow site) on the germanene surface. It should be noted that transition metal adsorption significantly disturbs the hexagonal lattice symmetry, so that the electron band dispersion along the high symmetry points (Γ-M-K-Γ) of perfect germanene may not represent the exact electronic properties of the entire structure. In order to get reliable results, the band structure dispersion must be consistent with the density of state results in a large number of k-points in the Brillouin zone.

Fig. 8.3 demonstrates the band structure of pristine germanene and three adsorption cases. In pristine monolayer germanene, as shown in Fig. 8.3A, the main features of the band structure are determined by the relation of π, σ, and sp^3 bondings. The system exhibits a modified-Dirac cone structure at K valley. Near the Fermi level, the valence and conduction bands are symmetric with a narrow gap of about 0.02 eV that causes by spin-orbit coupling. The valence/conduction π/π* band, which are generated by $4p_z$ orbital, have an initial point at Γ valley and show the saddle M-point at −1.15/0.7 eV. On the other hand, the four-fold degenerate σ bands are mainly created by (4px,4py) exhibit parabolic energy dispersion with the origin at −0.6 eV in Γ valley. An observable mixing of π and σ bands in both the ΓM and ΓK paths demonstrates a weak but significant sp^3 bonding (nonorthogonality of π and σ bondings).

Compare with the pristine germanene, the iron-adsorbed system presents an unusual electronic band dispersion, as indicated in Fig. 8.3B. The low-lying energy bands are mostly provided by the greatly distorted π bonding on the germanene surface. The almost perfect Dirac cone is completely destroyed along with the highly asymmetry between the occupied electron energy spectrum and the unoccupied hole spectrum about the Fermi level. Apparently, these novel features mainly due to the complicated multi-orbital hybridizations of the non-uniform chemical bonds. As a result, the system becomes an indirect gap semiconductor of 0.22 eV. The lowest unoccupied conduction-band state locates at the K point whereas the highest occupied valence state locates in between the Γ and M points. Very interestingly, the iron-adsorbed germanene exhibits the ferromagnetic configuration that is characterized by two spin-split valence and conduction subbands near the Fermi level. The non-monotonous wave-vector dependencies of such subbands might closely be related to the different contributions of the guest adatom and host atoms. Specifically, according to atom dominance, both iron and germanene atoms make significant contributions everywhere in

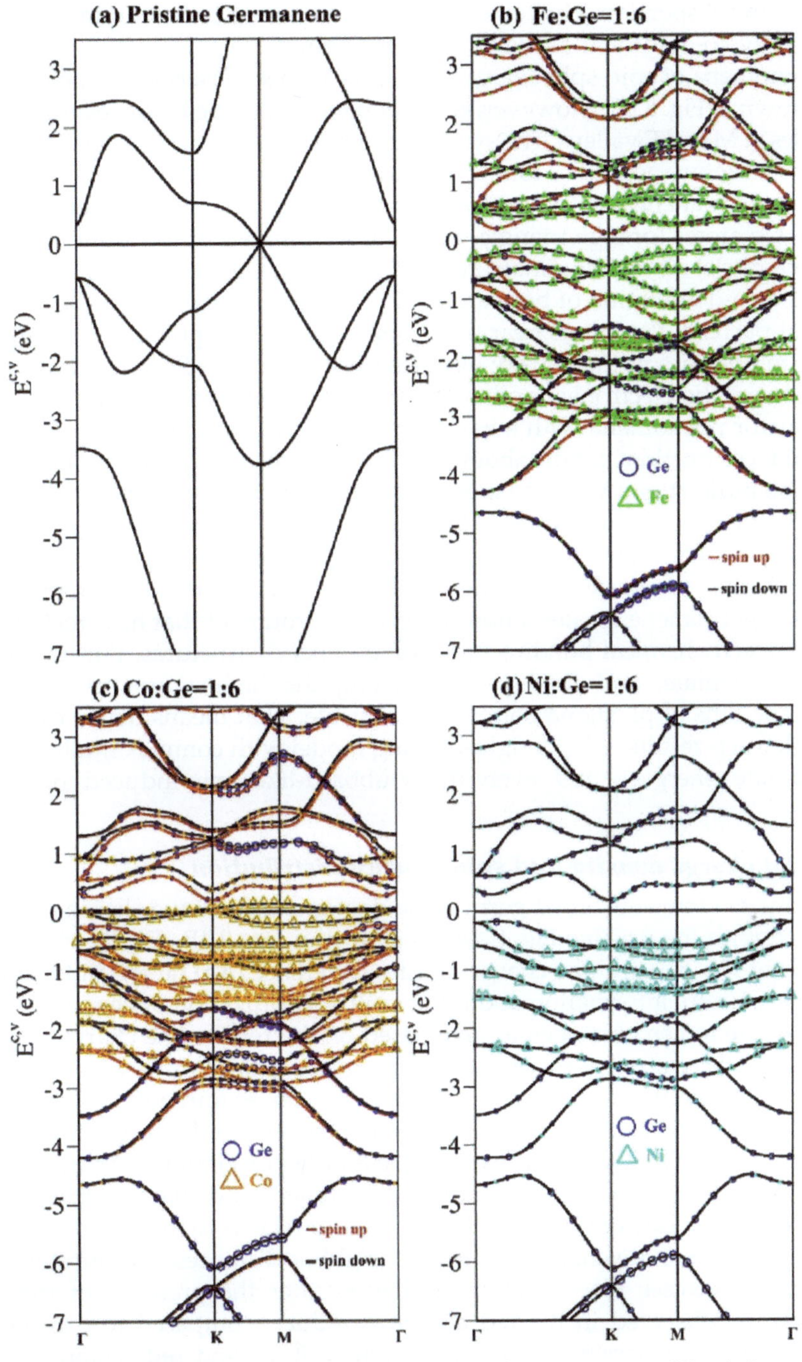

FIG. 8.3 Band structures together with the atom-dominances under various cases: (A) pristine germanene (blue), (B) cobalt (green)-, (C) iron (orange)-, and (D) nickel (cyan)-adsorption.

the electronic band dispersion. However, the iron atom seems to be dominant at low energies and they are only co-dominant at higher/lower energies.

Similar phenomena of spin-split energy subbands are also observed in cobalt-adsorbed germanene, as shown in Fig. 8.3C. However, such subbands cross the Fermi level and create plenty of valence holes at M and Γ valleys. The system becomes metal behavior together with the modified Dirac cone structure near EF is totally absent. These results also indicate that the original π bond in pristine germanene has undergone a drastic transformation. Based on the results, in the valence band, Co atoms (orange triangle) completely dominate at the low-lying energy, while Ge atoms (blue circle) almost dominate at lower energies. On the other hand, the conduction band presents a co-domination of both atoms. There are several energy subbands, with weak oscillatory and partially flat bands near the Γ point/the K & M points. This result is reflected in van Hove singularities in the density of states (discuss later in Section 3.2.2).

Fig. 8.3D shows the electronic band structure of Ni-adsorbed germanene that belongs to the semiconductor phenomena with a bandgap of 0.36 eV. The asymmetric energy spectrum of valence and conduction bands about the Fermi level is extremely enhanced after Co adsorption. The indirect gap with the lowest unoccupied conduction band stable at K valley while the highest valence band is located between K and G valleys. After detailed and careful calculation, our results show that the Ni-adsorbed germanene system is classified as a non-magnetic material regardless of the position of the adsorbed adatom. In short, TM atoms-absorbed germanene creates a non-uniform environment that has created complex hybridization between chemical bondings and unique band structures. Furthermore, defining the initial, intermediate, and final p-bands along the high-symmetry paths as well as measuring the widths of pi, sigma bands is impossible. That means that it can be very difficult to simulate such results using a tight-binding model with complex multi-orbital hopping integrals and site energies, and even the Hubbard-like spin-induced on-site Coulomb interactions.

8.3.2.1 Spatial charge density and spin density distribution

Transition metal atoms-absorbed germanene demonstrate unique charge density distributions compared with the pristine Fig. 8.4A, as shown in Fig. 8.4B–D. The charge density (ρ) (at the left side of Fig. 8.4) and the charge density difference ($\Delta\rho$) (at the right side of Fig. 8.4) can provide very useful information about the multi-hybridization of chemical bondings and charge transfer phenomena between atoms. Based on the results, the charge density appears to be similar in all three absorption cases. The charges are concentrated around the transition metal, shown in the red area, being responsible for an asymmetry charge accumulation around the TMs-Ge bond center. The optimal geometry, with a deformed honeycomb lattice, including TMs-Ge and Ge–Ge bonds, explain the highly non-uniform chemical environment. Their distinct bond lengths specify chemical bond strength and charge density distributions. It is easy to see that the carrier density, which is closely related to the electron orbitals of each atom, is very sensitive to the change of nearest neighbor distances. Specifically, the chemical bonding strength between Ge–Ge atom decreased after the presence of transition metal atoms. This can be observed in the red area was reduce, compared to pristine case. Most importantly, it could be classified into light-green, yellow, and red regions, which mainly correspond to $4p_z$ ($4s$, $4p_x$, $4p_y$), and ($3d_{x^2-y^2}$, $3d_{xy}$, $3d_{yx}$, $3d_{xz}$, $3d_{z^2}$), respectively.

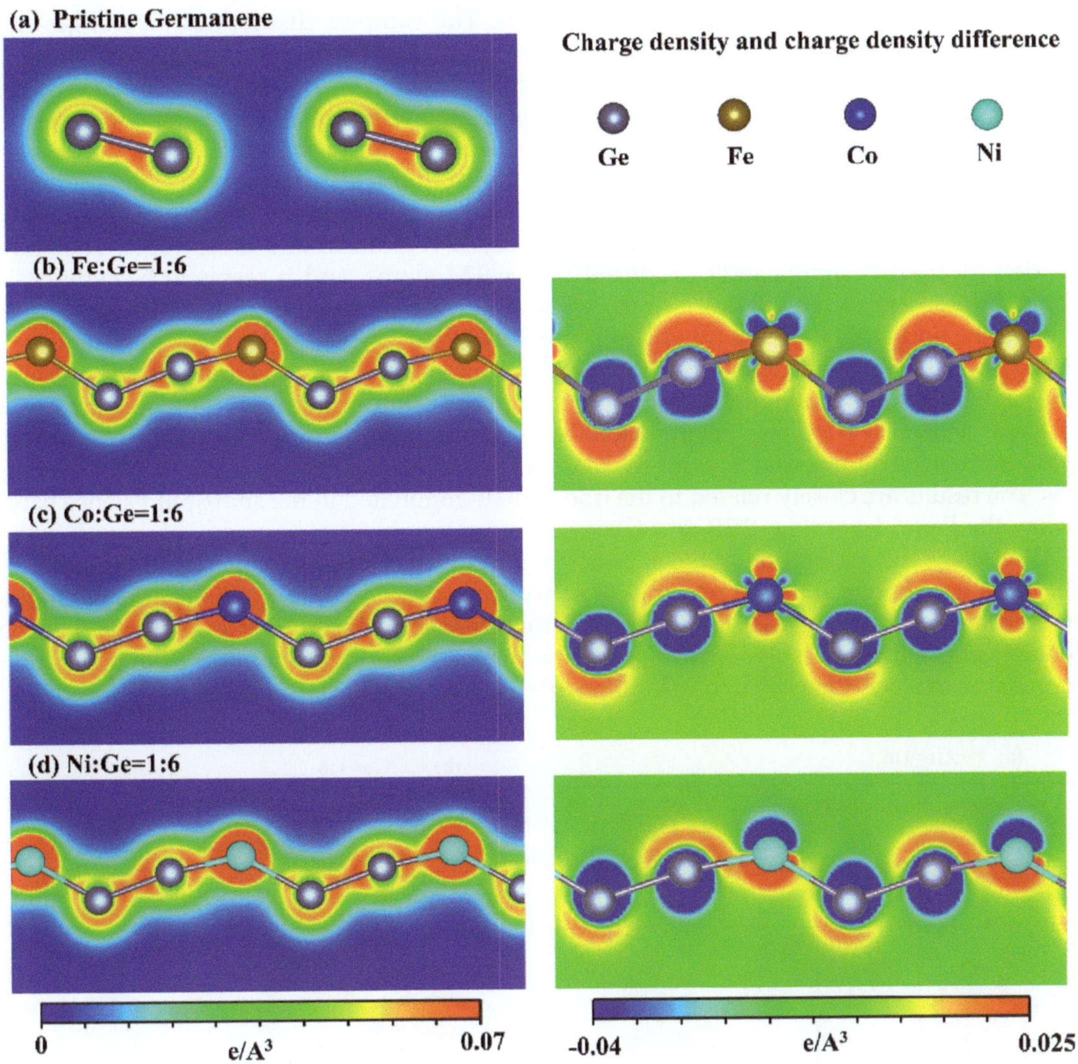

FIG. 8.4 Spatial charge density (left side) and charge density difference (right side) of (A) without adsorption, (B) iron-, (C) cobalt-, and (D) nickel-adsorbed germanene.

The charge density difference, which is calculated by the charge of adatom-adsorbed germanene systems minus the total charges of pristine germanene and isolated adatoms, is very useful in understanding the electron transfer and charge distribution. The charge redistribution represented by the loss or gain of some/more electron carriers is illustrated in blue, yellow, and red colors, respectively. There is clearly charge transfer occurring in the Ge atoms. These electrons are transferred from the bottom (blue) to the top of the Ge atom as well as transferring to TM atoms (red), leading to a decrease in the bonding strength between the Ge—Ge atoms and create a non-uniform chemical environment. The complex distribution of charges has created unique band structures that also give rise to interesting magnetic properties.

The configuration for the spin density distribution can provide more information on their magnetic properties, as illustrated in Fig. 8.5A and B. The ferromagnetic behavior comes to exist in all doping cases of Fe on germanene, however, in stark contrast to Fe, Ni-doped germanene exhibits a non-magnetic nature in all absorption sites. Interestingly, the magnetic phenomenon only occurs in the hollow and topsides configuration of the Co-Ge systems. In general, the distribution of spin density is very non-uniform and anisotropic. The spin-up density (red) is significantly produced by the TM atoms, and it seems to be symmetric about the centers of these guest atoms, while it is only partially expressed in the upper/around the host atoms. On the other hand, spin down density (yellow) is mostly concentrated in the lower half-plane of $z < 0$ for the host atoms and a small part is around the TM atoms. The highly asymmetrical spin distribution may play an important role in the generation of ferromagnetism. The red regions produced by iron/cobalt are comparable to those associated with germanene atoms, that is, both make important contributions to magnetic field properties. The results are closely related to the iron/cobalt dominated in the spin-split valence band below the Fermi level (Fig. 8.2B and C).

Obviously, the magnitude of the spin-down is quite small when compared to the spin-up. The difference between these two types of spin creates a rather large magnetic moment in the two absorption cases of 1.77 for Fe and 1.0 for Co-adsorbed germanene, respectively. The theoretical prediction results of the spatial ferromagnetic configuration and spin-generating moments can be verified by high-resolution SP-STM and SQUID measurements, respectively.

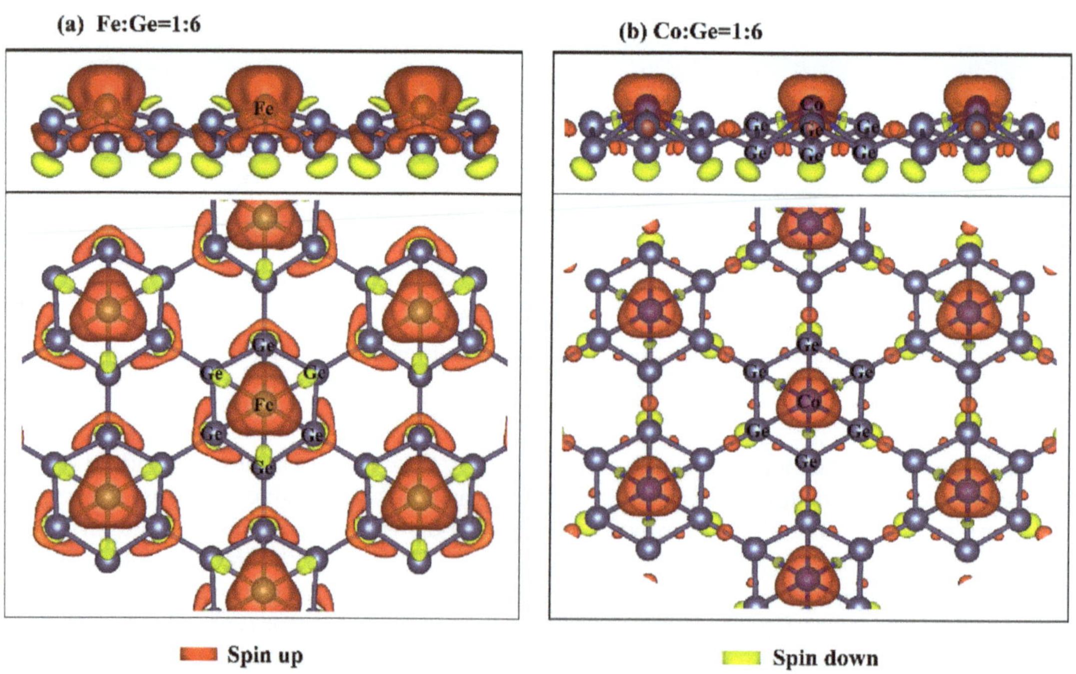

FIG. 8.5 The spin-charge distributions of (A) iron-, and (B) cobalt-adsorbed germanene, showing side and top views. Red and yellow correspond to spin up and spin-down, respectively.

8.3.2.2 *Density of state*

The main features of the band structure are directly reflected in the atom-, orbital- and spin-decomposed density of states, which are able to explain the orbital contributions and hybridization in chemical bonds, as clearly illustrated in Fig. 8.6A–D for the pristine and adsorption cases. These special critical points are closely related to the extreme states in parabolic subbands, flat or partially flat bands in a certain direction. Such extreme points can

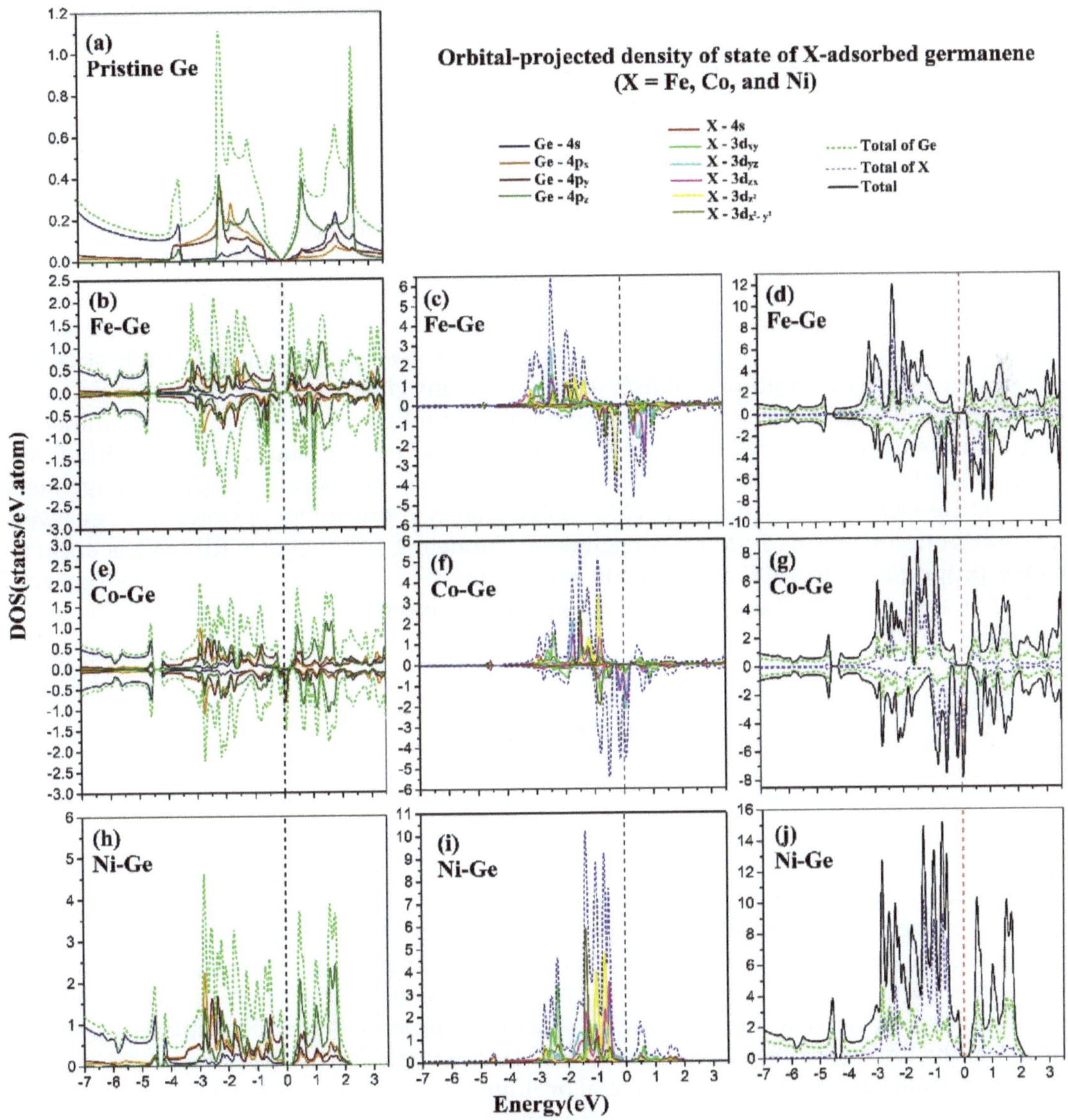

FIG. 8.6 The atom-, orbital-, and spin-projected density of states for (A) pristine germanene, (by iron), cobalt-, and (D) nickel-absorbed germanene.

create many forms of van Hove singularities such as shoulders, asymmetric V-shapes, delta-function-like peaks, divergent peaks in the square-root form, or logarithmic peaks. For example, the V-shaped peak, due to the Ge-4p_z orbitals occurs at -0.5 eV $< E < 0.5$ eV, is created by the modified Dirac cone structure. However, these V-shaped structures completely disappeared when TM atoms-doped germanene.

According to the density of states near the Fermi level, we can distinguish the system as belonging to metal or semiconductor behavior. Only the Co-Ge case is clearly metal, while the other ones are semiconductors. In particular, cobalt-adsorbed germanene has a sufficiently high number of free carriers, as well illustrated by the large density of states at E_F, due to the intersection of partially flat subbands with the Fermi level. In contrast, the absence of density of states near the Fermi level demonstrates the semiconducting nature of Fe-/Ni-doped germanene. Obviously, the orbital and spin-projected DOSs reveal highly complex van Hove singularities and thus identify important multi-orbital hybrids in the TM-Ge bondings.

There exist a significant number of anomalous dominated van Hove singularities near Fermi level, mainly arising from the multi-hybridizations between the (4s, 4p_x, 4p_y, 4p_z) orbitals of Ge and (3$d_{x^2-y^2}$, 3d_{xy}, 3d_{yx}, 3d_{xz}, 3d_{z^2}) orbitals of TM atoms as a result of the TM-Ge bonds. As a result, the p and s subbands do not exist in any of the adsorbed germanene systems. A series of extended van Hove singularities are difficult to identify among the distinctive structures. The six/four orbitals of the guest and host atoms might overlap at certain energies illustrates the multi-orbital hybridizations in any chemical bond. Interestingly, based on the spin-split density of states, the ferromagnetic configuration is mainly determined by the subbands near the Fermi level. The spin-up and spin-down states shown at $E < 0$ and $E > 0$, respectively. However, at lower/higher energies, two kinds of spin arrangements coexist at Ge-DOSs and disappear at TMs-DOSs. That is to say at higher/lower energy levels, the contribution of the TM atoms is negligible. Tunneling conductance (dI/dV) is approximately proportional to DOS and directly reflects the main features in DOS. This powerful method has been successfully used to explore diverse electronic properties in graphene-like systems.

8.4 Conclusion

The geometric and electronic properties of Fe-/Co-/Ni-adsorbed germanene are investigated using the first-principles calculations. They are shown to be dominated by multi-orbital hybridization between 4p_z orbitals of the Ge atoms and the (3$d_{x^2-y^2}$, 3d_{xy}, 3d_{yx}, 3d_{xz}, 3d_{z^2}) orbitals of Fe/Co/Ni. In general, adsorbed systems exhibit optimal positions on the hollow side through the highest adsorption energies. After adsorption, the hexagonal honeycomb structure is partially deformed and enhanced buckling degrees, thus produce a non-uniform environment and creating novel properties. Most importantly, adsorption-enriched electronic structures are expected to present diverse essential properties. First-principles calculations show that the low-energy electronic states could be dominated by the TM atoms. The systems present unusual band structures with metal or finite gap semiconductor behavior. Obviously, the spin-split sub-bands, as well as the net moments are found to be important for creating diverse magnetic configurations. The spin-up densities are mostly produced from TM atoms where the spin-down component is only very small near the host atoms. Most importantly, the adsorption-enriched electronic structures are expected to exhibit diverse essential properties. First-principle calculations show that the low-energy electronic and magnetic properties are dominated by the TM atoms. This work is the first step toward further studies of other essential properties of the transition metal adsorbed to germanene. Furthermore, besides adsorption, there are several events that can occur during experimental growth, such as guest atom substitutions, defects, and heterojunctions that can be exploited with a similar method.

References

[1] Pham HD, Gumbs G, Su WP, Tran NTT, Lin MF. Unusual features of nitrogen substitutions in silicene. RSC Adv 2020;10:32193–201.

[2] Pflugradt P, Matthes L, Bechstedt F. Unexpected symmetry and AA stacking of bilayer silicene on Ag(111). Phys Rev B 2014;89.

[3] Shen M, Zhang YY, An XT, Liu JJ, Li SS. Valley polarization in graphene-silicene-graphene heterojunction in zigzag nanoribbon. J Appl Phys 2014;115.

[4] Ye M, Quhe R, Zheng JX, Ni ZY, Wang YY, Yuan YK, Tse G, Shi JJ, Gao ZX, Lu J. Tunable band gap in germanene by surface adsorption. Physica E 2014;59:60–5.

[5] Lee KW, Lee CE. Spin-orbit coupling-induced band inversion and spin Chern insulator phase in plumbene and stanene. Curr Appl Phys 2020;20:413–8.

[6] Kuzmany H, Shi L, Martinati M, Cambre S, Wenseleers W, Kurti J, Koltai J, Kukucska G, Cao KC, Kaiser U, Saito T, Pichler T. Well-defined sub-nanometer graphene ribbons synthesized inside carbon nanotubes. Carbon 2021;171:221–9.

[7] Usachov DY, Fedorov AV, Vilkov OY, Senkovskiy BV, Adamchuk VK, Andryushechkin BV, Vyalikh DV. Synthesis and electronic structure of nitrogen-doped graphene. Phys Solid State 2013;55:1325–32.

[8] Ileri N, Goldman N. Graphene and nano-diamond synthesis in expansions of molten liquid carbon. J Chem Phys 2014;141.

[9] Du M, Xu CH, Sun J, Gao L. One step synthesis of Fe_2O_3/nitrogen-doped graphene composite as anode materials for lithium ion batteries. Electrochim Acta 2012;80:302–7.

[10] Shukla A, Kumar R, Mazher J, Balan A. Graphene made easy: high quality, large-area samples. Solid State Commun 2009;149:718–21.

[11] Zhang L, Liang JJ, Huang Y, Ma YF, Wang Y, Chen YS. Size-controlled synthesis of graphene oxide sheets on a large scale using chemical exfoliation. Carbon 2009;47:3365–8.

[12] Chen L, Hernandez Y, Feng XL, Mullen K. From nanographene and graphene nanoribbons to graphene sheets: chemical synthesis. Angew Chem, Int Ed 2012;51:7640–54.

[13] Zeng J, Ji XX, Ma YH, Zhang ZX, Wang SG, Ren ZH, Zhi CY, Yu J. 3D graphene fibers grown by thermal chemical vapor deposition. Adv Mater 2018;30.

[14] Hassan HMA, Abdelsayed V, Khder AERS, AbouZeid KM, Terner J, El-Shall MS, Al-Resayes SI, El-Azhary AA. Microwave synthesis of graphene sheets supporting metal nanocrystals in aqueous and organic media. J Mater Chem 2009;19:3832–7.

[15] Pang Q, Li L, Zhang CL, Wei XM, Song YL. Structural, electronic and magnetic properties of 3d transition metal atom adsorbed germanene: a first-principles study. Mater Chem Phys 2015;160:96–104.

[16] Sun ML, Ren QQ, Wang S, Zhang YJ, Du YH, Yu J, Tang WC. Magnetism in transition-metal-doped germanene: a first-principles study. Comput Mater Sci 2016;118:112–6.

[17] Nemoto R, Kruger P, Hartini ANP, Hosokai T, Horie M, Kera S, Yamada TK. Well-ordered monolayer growth of crown-ether ring molecules on Cu(111) in ultra-high vacuum: an STM, UPS, and DFT study. J Phys Chem C 2019;123:18939–50.

[18] Heidarzadeh A. Tensile behavior, microstructure, and substructure of the friction stir welded 70/30 brass joints: RSM, EBSD, and TEM study. Arch Civ Mech Eng 2019;19:137–46.

[19] Cheng ZW, Zhang ZY, Sun HG, Li SJ, Yuan H, Wang ZJ, Cao Y, Shao ZB, Bian Q, Zhang X, Li FS, Feng JG, Ding SA, Mao ZQ, Pan MH. Visualizing Dirac nodal-line band structure of topological semimetal ZrGeSe by ARPES. Apl Mater 2019;7.
[20] Ridene M, Girard JC, Travers L, David C, Ouerghi A. STM/STS investigation of edge structure in epitaxial graphene. Surf Sci 2012;606:1289–92.
[21] Wang YL, Zhang GF, Zhang SL, Rong LL, Wang Y, Xie XM. Two-terminal wide range SQUID amplifier with hysteretic quasi linear transfer characteristics. Physica C 2019;567.
[22] Kresse G, Furthmuller J. Efficient iterative schemes for ab initio total-energy calculations using a plane-wave basis set. Phys Rev B 1996;54:11169–86.
[23] Kresse G, Joubert D. From ultrasoft pseudopotentials to the projector augmented-wave method. Phys Rev B 1999;59:1758–75.
[24] Perdew JP, Burke K, Ernzerhof M. Generalized gradient approximation made simple. Phys Rev Lett 1996;77:3865–8.
[25] Blanc X, Cances E, Dupuy MS. Variational projector augmented-wave method. Cr Math 2017;355:665–70.
[26] Kaloni TP, Singh N, Schwingenschlogl U. Prediction of a quantum anomalous Hall state in Co-decorated silicene. Phys Rev B 2014;89.
[27] Pham HD, Nguyen TDH, Vo KD, Huynh TMD, Lin MF. Rich essential properties of boron, carbon, and nitrogen substituted germanenes. Appl Phys Express 2020;13.

CHAPTER

9

Geometric structures, electronic, and excitonic optical properties of germanium dioxide α − quartz GeO_2

Thi Dieu Hien Nguyen, Shih-Yang Lin, Vo Khuong Dien, Chi-Hsuan Lee and Ming-Fa Lin

Department of Physics, National Cheng Kung University, Tainan, Taiwan

9.1 Introduction

Germanium dioxide (GeO_2) is known as germania, or salt of germanium, is an inorganic compound. In general, GeO_2 comes into existence in many phases, such as $\alpha-$ quartz, $\beta-$ quartz, or rutile structure. In this book chapter, $\alpha-$ quartz is our main research, which belongs to trigonal crystal system. Recently, GeO_2 based nanomaterial has attracted many interests due to its various applications including optoelectronic devices [1−3], photosensors, vacuum technology.

The binary compound GeO_2 presents a crystal symmetry structure and classifies into semiconductor. In previous studies, there are many methods to synthesize this material such as low-temperature chemical synthesis [4], chemical precipitation method [5]. GeO_2 belongs to Moire super lattice unit cell, which shows a highly non-uniform chemical bondings [6]. This result might have the challenge in using the first-principles calculations on some rich properties of condense matter materials such as optimal geometric structures, electronic and optical characteristic.

There are many theoretical studies of this three-dimensional binary compound, in which the developed framework in terms of hybridization might not be mentioned in detail. In

Fundamental Physicochemical Properties of Germanene-related Materials
https://doi.org/10.1016/B978-0-443-15801-8.00011-6

this chapter, the focused studies cover the significant charge density around germanium and oxygen atoms for isolated case and chemical interaction configuration, the separated van Hove singularities for atom and orbitals, the complex dielectric functions with real and imaginary parts, absorbance, energy loss spectrum, and reflectance. We will combine these properties with the unified theoretical framework based on single or multi hybridizations. Furthermore, the creation of electron and hole in the conduction and valence bands and their interaction can be called an exciton with the interesting phenomenon in condensed matter systems. Especially, the optical properties will provide in the numerical analysis which can directly connect with orbital hybridizations. The measurement for optical properties, particularly on absorbance, the threshold frequency will attract much attention rather than other optical ranges. Additionally, the diverse prominent peaks are worthy to investigate in theoretical aspect, which directly link with electronic phenomena. Furthermore, other optical characteristics such as energy loss function and reflectance should be examined which lead to a completed picture of the binary compound GeO_2. Different from literature research, we also focus on the excitonic effect, originated from the electron-hole pair for the semiconductor material. It is well-known that the excitonic characteristic is unable to apply for metallic system because the electrons will be screening heavily, leading to the difficulty to form electron-hole phenomenon. Moreover, the current work can combine all the essential properties of germanium dioxide using first-principles analysis. In identified the final result, these calculations need to provide further high-resolution measurement.

9.2 Computational method

Three-dimensional compound is thoroughly examined by using the Vienna Ab initio Simulation Package (VASP) [7], under the density functional theory (DFT; [8]). By using the Perdew–Burke–Ernzerhof (PBE; [9]) method, the interaction between electron and electron, the exchange and correlation terms can be implemented in the simulation with the generalized gradient approximation (GGA) [10]. Additionally, the projector-augmented wave (PAW; [11,12]) pseudopotentials are available to identify the electron-ion interactions. It is widely known that the Hamiltonian for the single and many particles for the aforementioned intrinsic interactions cannot express precise formulas so a certain approximation needs to be included in the theoretical analysis [13]. The energy cutoff is set at 520 eV, which is important for controlling the number of plane-wave basis sets [7]. The convergence ground state energy is 10^{-6} eV for continuous relaxation steps. During the ion relaxation process, the Hellmann–Feynman force needs to consider a value smaller than 0.01 eV/ Å during the ionic relaxation. The k-meshes are $6 \times 6 \times 4$ and $48 \times 48 \times 32$ for running stability structure and electronic properties. The excitonic characteristics require a long time of simulation steps compared to other properties. It is necessary to run with four steps, including (1) standard relaxation, (2) DFT calculation, (3) GW approximation, and (4) Bethe-Salpeter equation.

9.3 Rich geometric structures

The optimal relaxation of this structure exhibits an unusual crystal symmetry, shown in Table 9.1. It is noted that $\alpha -$ quartz GeO_2 compound belongs to the trigonal system, P 3_221 Hermann Mauguin, P 32 2″ Hall, and 32-point group. The lattice parameters are

TABLE 9.1 The geometric properties of GeO_2 compound.

Lattice constant	
This work	**Experiment**
a = b = 5.08460 Å, c = 5.76210 Å $\alpha = \beta = 90^o, \gamma = 120^o$	a = b = 4.987(4) Å, c = 5.652(5) Å [5]
Structure symmetry	
Hermann mauguin	P $3_2$21 [154]
Hall	P 32 2″
Point group	32
Crystal system	Trigonal
12 Ge—O chemical bonds (6 Ge—O bonds of 1.7699 Å and 6 Ge—O bonds of 1.7644 Å)	
Unit cell: 9 atoms (3 Ge - 6 O)	

a = b = 5.08460 Å, c = 5.76210 Å, the title angle along the x, y, and z directions $\alpha = \beta = 90^o$, $\gamma = 120^o$ (Figs. 9.1 and 9.2). Compared with other experimental value of a = b = 4.987(4) Å, c = 5.652(5) Å [5], the current value is quite consistent.

In this structure, the diverse chemical bonds [14] in the unit cell play a crucial role in comprehending the feature properties of GeO_2. After obtaining the optimal structures, the bond lengths appear with 1.7966 and 1.7644 Å, mainly for Ge—O bonds. The various chemical interaction bonds of germanene and oxygen present a nonuniform structure, associated with the anisotropic phenomenon. In further investigation, the distribution of charge density will help to understand the essential chemical aspects of the binary structure (shown later in Fig. 9.4). In terms of multi-orbital hybridizations, significant bonds in the primitive cell can be explained in detail (presented by orbital-dependent van Hove singularities in Fig. 9.5). On the other hand, the phenomenological model can be useful to support first-principles calculations by calculating the hopping integrals. However, it might have difficulty with this calculation due to the complicated Hamiltonian matrix.

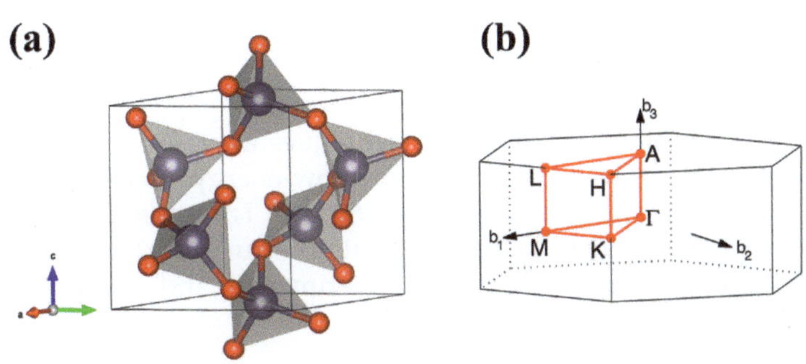

FIG. 9.1 (A) the optimal structure of GeO_2 and (B) the first brillouin zone.

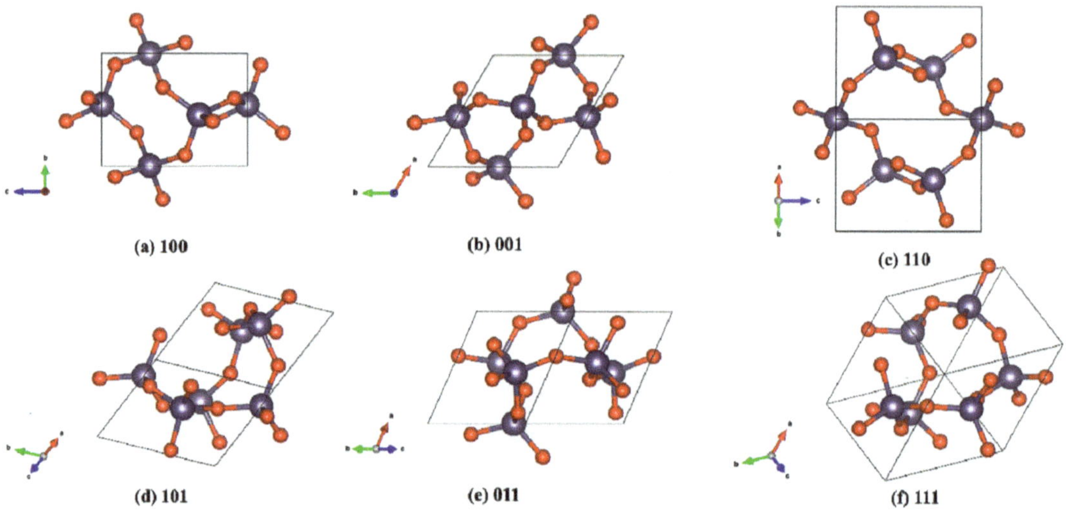

FIG. 9.2 The optimal structure of GeO_2 in (A) 100, (B) 001, (C) 110, (D) 101, (E) 011, (F) 111.

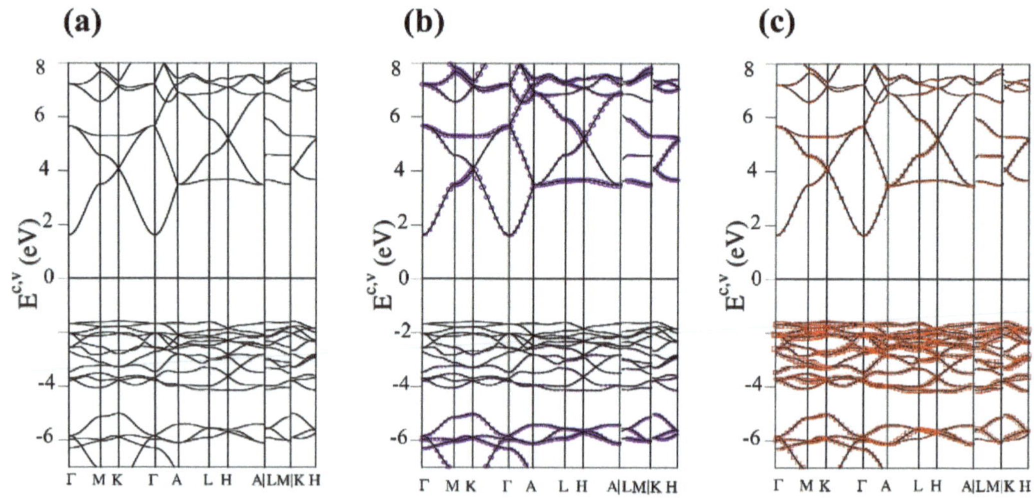

FIG. 9.3 Band structure along high-symmetric points.

In the experimental measurements, many techniques are used for identifying the lattice constants such as X-ray diffraction (XRD) or Low-energy electron diffraction (LEED) [15,16]. Scanning Electron Microscope (SEM) [16] and Transmission electron microscopy (TEM) are powerful tools to achieve high-resolution structure details as well as the size and morphology. The comparison between experimental and theoretical calculations is important for basic and applied science.

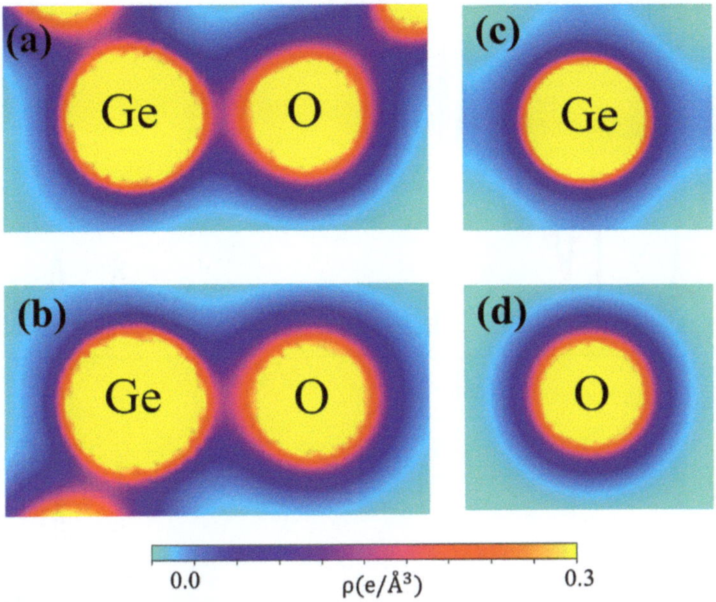

FIG. 9.4 Charge density for (A)1.7699 Å Ge—O bonds (B)1.7644 Å Ge—O bonds, (C) isolated Ge atom (D) isolated O atom.

9.4 Featured electronic properties

The unique electronic structures of GeO_2 mainly focus on the wave vector-dependent band structure, the atom dominances in the specific energy ranges, the spatial charge densities, and the atom- and orbital-decomposed van Hove singularities. Fig. 9.3A depicts the band characteristics of the binary compound in the specific energy range of −8 to 8 eV. The appearance of many subbands along with the high symmetric points for trigonal first Brillouin zone present feature-rich properties of GeO_2. The asymmetric behaviors are available between the occupied electron states and unoccupied states about the Fermi level. It is generally known that the Fermi level is located in the middle of the lowest conduction and the highest valence bands. This phenomenon is resulted from the contribution of various activate multi-orbital hybridizations of Ge (4s, $4p_x$, $4p_y$, $4p_z$) and O (2s, $2p_x$, $2p_y$, $2p_z$) in the non-uniform chemical bonds. Interestingly, the indirect gap of $E^i_g \sim 3.2193$ eV is measured from Γ to K points, which shows the semiconductor behavior of germanium dioxide. The band gap is an important factor, which is measured between the highest occupied hole states and the lowest unoccupied electron spectrum. Different from the direct gap, the indirect gap result from the creation and absorption of phonons. It is important to know that the indirect gap is smaller than the threshold energy. Due to the large bandgap semiconductor, the spin-split phenomenon is difficult to form, leading to a vanishing magnetic moment in the system. Therefore, spin-dependent interactions will not consider in this system. The number of subbands that

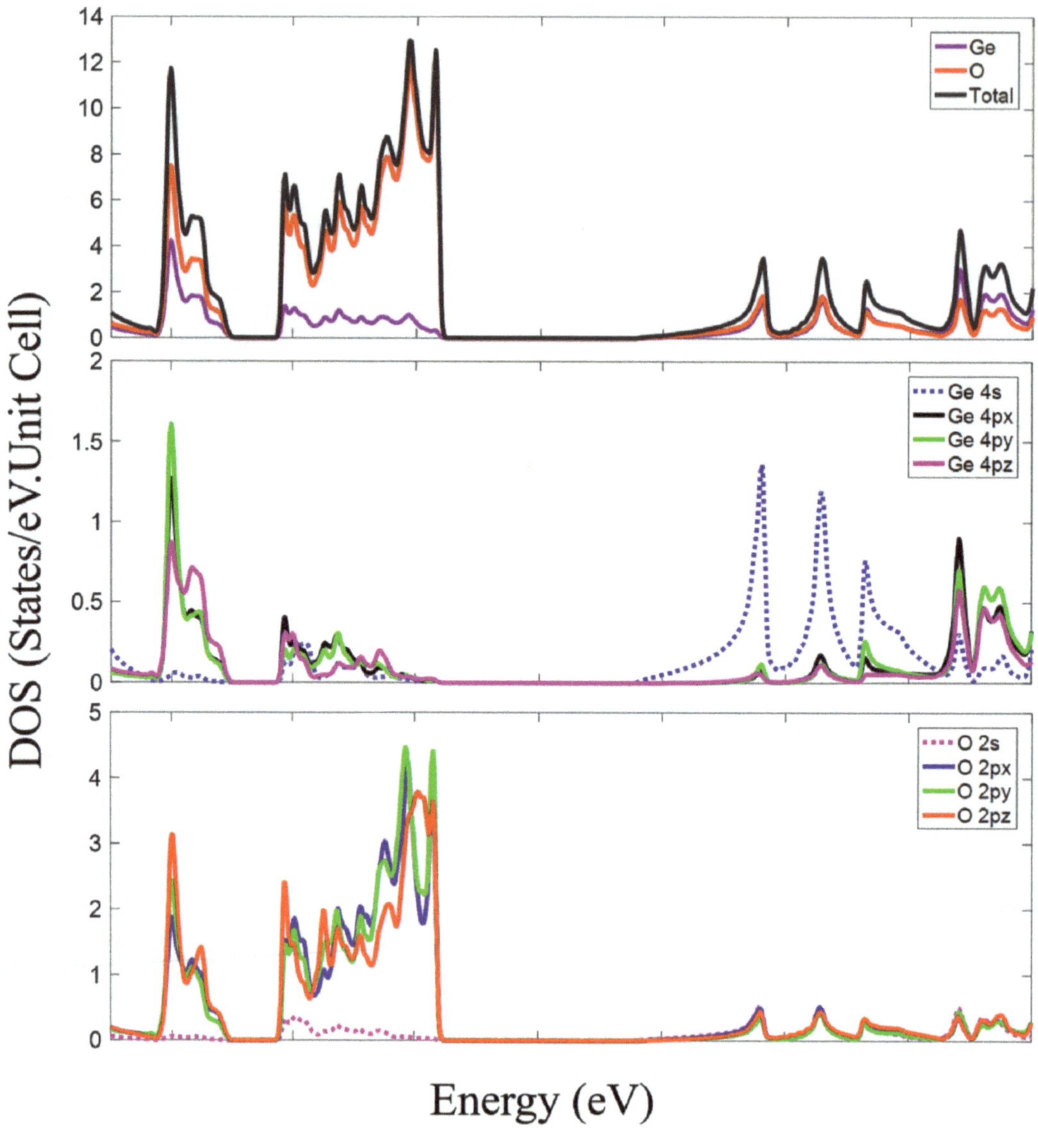

FIG. 9.5 Density of states (A) total, (B) Ge atom, and (C) O atom.

come into existence of GeO_2 will depend on two main factors, the number of atoms in the unit cell and the orbitals of the outer shell in each atom. Various dispersions can be observed such as parabolic, oscillatory, linear, and partially flat bands. Remarkably, the band-edge states show the local minimum or maximum, presenting zero group velocities. The energy width stays between 0.303 and 0.786 eV. Additionally, there is a specific energy space from −5.10 to −4.15 eV. Furthermore, non-crossing, crossing, and anti-crossing behaviors, mainly come from the orthogonal or nonorthogonal characteristics of π and σ bands in the diverse energy subbands, thus leading to complicated electronic behaviors. The band-edge states, which correspond to zero group velocity come into existence frequently along with the high-symmetric points. Consequently, band structures show many rich behaviors, which can be supported by the density of states with specific van Hove singularities.

The energy range also gives the atom dominance properties, which is important to understand the rich chemical bonds between Ge and O. In Fig. 9.3B and C, Germanium and oxygen are presented in purple and red color, they almost come into existence in the energy range from −8.00 to 8.00 eV. Observably, Ge atoms dominate in the range of 1.609−8.00 eV and the deep range of −8.00 to −5.52 eV, while oxygen atoms present strong dominance in the range of −4.2−1.609 eV. The coexistence of Ge and O atoms in the band structure will determine the chemical bonds between them, which is based on the energy of (2s, $2p_x$, $2p_y$, $2p_z$) and (4s, $4p_x$, $4p_y$, $4p_z$) orbitals. This may partly explain under the critical single and multi-orbital hybridization theory.

It is important to compare with the phenomenological model, however, several physical properties such as free charge density, crossing, and anti-crossing behaviors might be difficult to achieve. How to generate the hopping integral to build a Hamiltonian matrix will become a challenge for basic science because it provides us with concise chemical and physical understanding. The optimal structure shows non-uniform features, which result from the critical orbital hybridizations of germanium and oxygen.

The numerical simulation results, which are conducted on the wave-vector-dependent energy spectra below the Fermi level, especially for the occupied electronic states, can be tested with Angle-Resolved Photoemission Spectroscopy (ARPES) [17,18]. This technique is very successful in observing the diversified band structures such as the diverse energy dispersions in the sp^2-bonding carbon systems, 3D transitions metal oxide compounds, and group-IV and group-V layered materials. In detail, it has been utilized to test in the linear Dirac-cone structure in twisted bilayer graphene (Moire superlattices with very large unit cells), monolayer graphene, the parabolic/parabolic and linear dispersions in bilayer/tri-layer AB stackings, the linear, partially flat, and Sombrero-shaped bands in tri-layer ABC stacking, and the semi-metallic property of bulk Bernal graphite. Among them, monolayer graphene is a zero-gap semiconductor, while the others are semimetals with weak valence and conduction band overlaps. Similar experimental measurements are available in examining the main features of the GeO_2 compound, such as the wave vector dependence, the large bandgap, the highly asymmetric unoccupied and occupied states, the parabolic or partial flat or weak energy dispersions, and the frequent non-crossing/crossing/anti-crossing properties. The experimental analysis plays a substantial role in comprehending the critical orbital hybridizations of chemical bonds. However, for this 3D material, the conservation of momentum and energy in the z-direction can face a struggle to determine due to the inexact measurement of momentum.

Fig. 9.4A−D shows the spatial charge density of different chemical bonds and the isolated atoms. The partial comprehending of orbital hybridizations for late investigation can be systematically combined with other electronic behaviors such as atom- and orbital-density of states with specific van Hove singularities (further presented in Fig. 9.5) and the atom dominance in the energy range from −8 to 8 eV (Fig. 9.3B and C). The trigonal structure depicts the GeO_4 chemical bonds, where one germanium interacts with four oxygen by the covalent bonds. This creates a high non-uniform chemical environment due to the modulation of chemical bonds. As a result, the various charge density needs to be considered in diverse cases of bond lengths. The bond length comes with the shortest value of 1.7644 Å, presented in Fig. 9.4B. Obviously, the overlapping range between germanene and oxygen is getting heavier, while this range will be extended if the bond length is longer. The overlap range is approximately measured by 0.592 – 0.888 Å (from oxygen core atom). It is noted that the isolated atoms of germanene and oxygen appear with spherical-like shapes with yellow (4s), pink and blue colors ($4p_x$, $4p_y$, $4p_z$ or $2p_x$, $2p_y$, $2p_z$). The charge distributions of Ge–O will strongly depend on the atomic number, e.g., germanene will effectively take part in the chemical interaction more than oxygen. The dense charge density will create around the germanium atom, presenting ($4s, 4p_x, 4p_y, 4p_z$)-orbitals. From the center of the germanium ion to the enlarged range of ($4s, 4p_x, 4p_y, 4p_z$), the effective range is about 0.71 Å. The oxygen atom has contributed four orbitals, including 2s and ($2p_x$, $2p_y$, $2p_z$), however, 2s is located in the deep core and almost fulfilled by electron configurations. Therefore, the 2s shell will be absent in the chemical interactions with germanium. The deformation of Ge and O charge density exhibits the essential charge transfer from the former to the latter because of the important multi-orbital hybridizations. The high non-uniform environment will induce an enormous challenge to calculate reliable and precise hopping integrals for the tight-binding models. From this state, we can predict that the chemical bond between germanium and oxygen is caused by (2s, $2p_x$, $2p_y$, $2p_z$) - (4s, $4p_x$, $4p_y$, $4p_z$) orbital hybridizations. A similar analysis can be applied for the increasing cases of bond lengths with the diminishing overlap phenomenon.

The density of states are characterized as the number of states considering in a small energy from E to E + dE using the integrals in the close loop, which show energy-dependent features in the specific valence and conduction bands. For diverse 0D, 1D, 2D, and 3D materials, DOSs are inversely proportional to the integration of group velocity on the identical constant energy space domain, e.g., discrete point/segment/loop or surface. This electronic property creates unusual van Hove singularities under the derivative of energy depending on the wave vector. It is noted that van Hove singularities originate from the band-edge states (local minimum or local maximum) of the energy subbands, in which the group velocity tends to zero (Fig. 9.3). Fig. 9.5A–C give a full picture of comprehending the multi orbital hybridization theory by presenting the atom and partial orbital van Hove singularities. In the bandgap range, the intensity of DOS will reach vanishing, which is in good agreement with the band structure of $E^i_g \sim 3.219$ eV. Additionally, the graph expresses the dominance of germanium and oxygen in a certain energy range, particularly -5.785 eV $\leq E^v \leq -1.609$ and 1.609 eV $\leq E^c \leq$ 5.058 eV, respectively with purple and red colors in Fig. 9.5A. On the experimental side, scanning tunneling spectroscopy (STS) [13] is an efficient method to investigate DOS features.

The diverse orbitals for Ge (4s, $4p_x$, $4p_y$, $4p_z$) and O-(2s, $2p_x$, $2p_y$, $2p_z$) will help to complete the whole essential properties of GeO_2. They reveal green, blue, red, magenta, black, and cyan curves in Fig. 9.5B, and dashed magenta, blue, green, and red curves in Fig. 9.5C) for the former and latter, respectively. Both Ge and O significantly determine the rich property of Ge–O chemical bonds. Interestingly, DOSs in Fig. 9.5B–C come into existence with diverse van Hove singularities, corresponding to the orbital decomposed significant forms, intensities, energies, and numbers of prominent peaks. The orbitals $4p_x$, $4p_y$, $4p_z$ of Ge (Fig. 9.5B) take part in the important covalent bonds of Ge—O, while 4s presents small DOS, this it will not participate in the essential interaction. Similarly, oxygen also reveal $2p_x$, $2p_y$, $2p_z$ contributions and 2s shell just make a minor value in the energy range of 10 eV. All complicated phenomena of Ge—O result from (4s, $4p_x$, $4p_y$, $4p_z$)-($2p_x$, $2p_y$, $2p_z$) multi-orbital hybridizations (sp^3-sp^3) chemical bonding and the various bond lengths of 1.7644—1.7966 Å. The significant multi-orbital hybridizations also partly support the prominent absorption peaks in the imaginary part of dielectric functions (discussed later in the optical properties section).

9.5 Excitonic optical properties

Electromagnetic (EM) waves incident the surface of condensed matter systems will cause the excitation of electrons from occupied states to unoccupied states [19]. At the same time, the charge carriers gradually screen and create an important coupling of charges and wave vectors. In general, the transverse component of EM will join this process, but not the longitudinal one. The significant mechanism reveals the creation of electron-hole pair, which is known as single-particle electronic excitations. Besides, collective excitations with plasmon modes can be observed in the energy loss function. Other optical properties such as reflectance, and absorptance can be carried out under the long wave-length limit, where the wave vector tends to zero. The excitation absorption spectra present several prominent peaks, and they have a close combination with some physical qualities such as the band-edge states of band structure transitions and the initial and final van Hove singularities in density of states.

The complex dielectric functions, i.e., $Re[\varepsilon(\omega)]$ and $Im[\varepsilon(\omega)]$ play a vital role in calculating several significant properties such as the energy loss function $Im\left[-\frac{1}{\varepsilon(\omega)}\right]$, the normal reflectance $R(\omega) = \left|\frac{\sqrt{\varepsilon(\omega)}-1}{\sqrt{\varepsilon(\omega)}+1}\right|^2$, and absorption coefficient $\alpha(\omega) = \sqrt{2}\omega\left[\sqrt{\varepsilon_1^2(\omega)+\varepsilon_2^2(\omega)} - \varepsilon_1(\omega)\right]^2$. The optical excitation phenomena reveal in the cases with and without excitonic effects, specific for dielectric functions are shown in Fig. 9.6. There are 17 absorption peaks can be observed, covering 5.563, 6.042, 6.402, 6.762, 7.002, 7.362, 8.003, 8.643, 9.603, 10.36, 11.2,

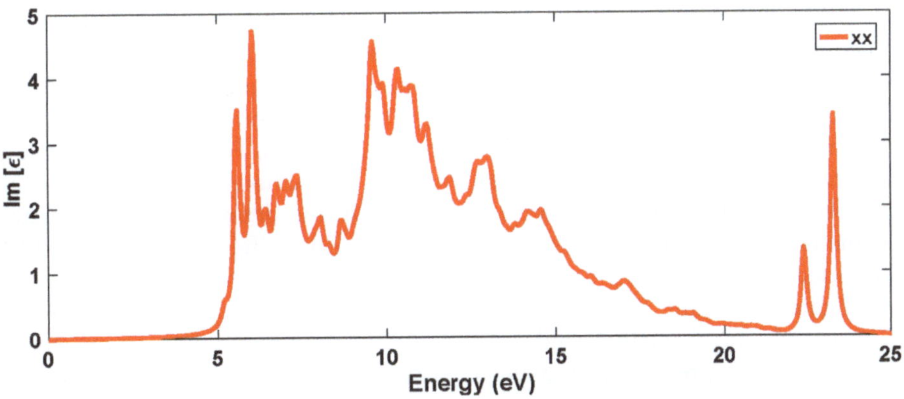

FIG. 9.6 Imaginary part of dielectric function in x-direction.

11.84, 12.96, 14.52, 17.05, 22.41, and 23.25 eV. Remarkably, there are three extra absorptions 5.563, 6.042, and 6.402 eV come into existence due to the excitonic effects. The specific results present the different orbital hybridization, which creates by interband transitions from the band edge states in band structures. Also, these peaks might be resulted from the orbital transitions of (4s, $4p_x$, $4p_y$, $4p_z$) and (2s, $2p_x$, $2p_y$, $2p_z$).

The energy loss function (ELF) ($Im[1/\varepsilon]$), is known as the screened response ability of the important valence carrier charge of germanium and oxygen under the Coulomb field perturbation. As a result, the creation of the coherent charge oscillations at the long-wavelength limit can be useful for comprehending the certain plasmon mode or small peaks in ELF. Fig. 9.7B shows the strongest peaks at 23.73 eV in x- or y-directions and 24.14 eV in the z-direction, which results from Coulomb excitation phenomena. The contributed orbitals of O- ($2p_x$, $2p_y$, $2p_z$) and Ge- (4s, $4p_x$, $4p_y$, $4p_z$) determine all the feature of GeO_2 structure. The plasmon peaks come into existence where the real part $\varepsilon_1(\omega)$ is vanishing. In this case, the O-2s contribution for the most prominent plasma wave is negligible because of its minor dominating valence spectrum in comparison with ($2p_x$, $2p_y$, $2p_z$) (the dash pink curve in Fig. 9.5C).

The reflectance spectra and absorption characteristics are shown in Fig. 9.5A and C in three polarized directions (xx/yy/zz). These behaviors have close relation with the complex dielectric functions, e.g., imaginary and real parts. The absorption appears when the light propagates in the medium, where the frequency of light is in resonance with the transition frequencies materials. Consequently, the absorption spectrum presents rich properties of diverse absorption channels (Fig. 9.5A). It has a remarkable combination of orbital-projected van Hove singularities and the transition band structure, which can determine the physical, chemical, and material properties of the 3D GeO_2. The reflectance spectrum is clearly depicted in Fig. 9.5C, resulting from an electromagnetic wave propagating at a smaller velocity than in the free space. The formula to identify the reflectance from real and imaginary components is given by $\left|\frac{\sqrt{\varepsilon(\omega)}+1}{\sqrt{\varepsilon(\omega)}-1}\right|^2$. The reflectance $R(\omega)$ and the absorption coefficient $\alpha(\omega)$ could be useful to understand a single particle and the collective excitations. Typically, these properties are inversely proportional to the propagation decay rate [20]. In the low-frequency reflection, the electron-hole excitation is negligible, so it can create a strong transmission due to the full electromagnetic wave propagation. When the frequency is increasing, reflectance sensitively changes and creates a large fluctuation under the various inter-band excitation channels. The drastic change of plasmon edge is available at ω_p, about 20% variation in the case of the electric field in the x-direction (the red line).

The current experimental analysis shows that the optimal geometric, electronic, and optical properties can be investigated by many developed techniques. For example, lattice parameters of significant structures can be examined by X-ray diffractions [21,22] or low-energy electron diffractions (LEED, [23]) with precise resolution. In the detail of 3D GeO_2 material, the mentioned methods are available to totally identify the diversified chemical bond lengths of Ge–O. Consequently, the complicated multi-orbital hybridizations in all the chemical bonds can be investigated in detail. It is well-known that the wave-vector dependence of occupied valence states could be directly tested by the angle-resolved photoemission spectroscopy (ARPES) [17,24]. There are many unusual energy subbands (Fig. 9.3A) due to a unit cell

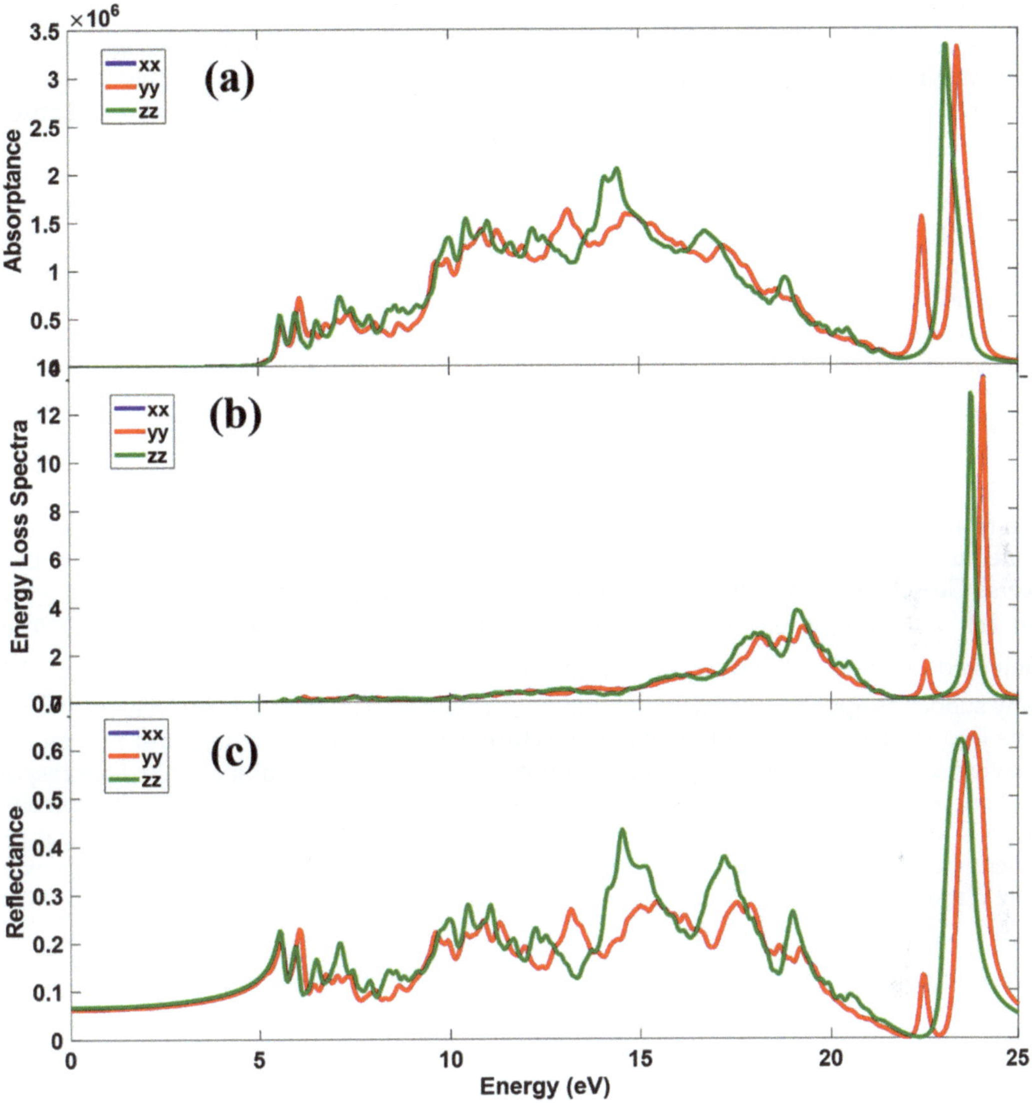

FIG. 9.7 Optical properties include (A) absorptance, (B) energy loss spectra, and (C) reflectance.

and diverse orbitals contributions, which would induce a high challenge in the ARPES measurements. Ultraviolet-visible (UV-Vis) absorption spectroscopy or photoluminescence spectra (PL) are popular techniques to evaluate the band gap of condensed matter systems. Experimental measurement of dielectric properties directly refers on the analysis of the complex relative permittivity (ε_r) and complex relative permeability (μ_r) of the materials. The real part of the complex permittivity is useful to propose the dielectric constant, which identifies by the amount of energy from an external electrical field screening in the material. The developed methods measure the complex permittivity, which depend on constraints in specific frequencies, kinds of materials, and their significant practical applications such as transmission or reflection line method [25], open-ended coaxial probe measurement [26,27], free space method [28], frequency resonant analysis [29]. On the other hand, reflectance, absorptance, and energy loss function could be identified by optical spectroscopies, for example, analysis of the specular reflection arising from the surface of materials can evaluate the complex index of refraction using the Kramers-Kronig relation [30]. The experimental measurement could fully examine the initial excitonic peaks, the variation of threshold excitation frequency, the prominent absorption channels, and the strongest plasmon mode at $\omega_P > 20$ eV.

The research on diverse stability structure, electronic and optical characteristics of low dimensional materials requires further examination in experimental aspect.

9.6 Conclusion

In this chapter, the geometric, electronic, and optical properties of GeO_2 could be comprehended under the orbital hybridization framework. The chemical bond lengths exist with the diverse Ge–O bonds, which create nonuniform environment. The system possesses a large unit cell and unique electronic properties such as large band gap of $E_g^i = 3.2193$ eV. Compared with GW band structure, this band gap is much smaller, $E_g^O = 6.2193$ eV. Many subbands appear asymmetric about the Fermi level. Flat/oscillatory/weak dispersions come into existence in the whole band structure. Crossing, anti-crossing, and non-crossing behaviors is created in the occupied and unoccupied bands. The combination of bare respond function with different prominent absorption peaks could identified under the transition channels in different band edge states and orbital density of states. Rich and unusual phenomena in this binary compound are worthy to deeply investigate and compare with experimental measurements.

References

[1] Terakado N, Tanaka K. Photo-induced phenomena in GeO_2 glass. J Non-Cryst Solids 2006;352:3815–22.
[2] Terakado N, Tanaka K. Photo-induced phenomena in sputtered GeO_2 films. J Non-Cryst Solids 2005;351:54–60.
[3] Su Y, Liang X, Li S, Chen Y, Zhou Q, Yin S, Meng X, Kong M. Self-catalytic VLS growth and optical properties of single-crystalline GeO2 nanowire arrays. Mater Lett 2008;62:1010–3.
[4] Atuchin VV, Kaichev V, Korolkov I, Saraev A, Troitskaia I, Perevalov T, Gritsenko V. Electronic structure of noncentrosymmetric α-GeO2 with oxygen vacancy: ab initio calculations and comparison with experiment. J Phys Chem C 2014;118:3644–50.
[5] Ramana C, Troitskaia I, Gromilov S, Atuchin V. Electrical properties of germanium oxide with α-quartz structure prepared by chemical precipitation. Ceram Int 2012;38:5251–5.
[6] Walker B, Dharmawardhana CC, Dari N, Rulis P, Ching W-Y. Electronic structure and optical properties of amorphous GeO2 in comparison to amorphous SiO_2. J Non-Cryst Solids 2015;428:176–83.
[7] Kresse G, Furthmüller J. Efficiency of ab-initio total energy calculations for metals and semiconductors using a plane-wave basis set. Comput Mater Sci 1996;6:15–50.
[8] Hafner J. Ab-initio simulations of materials using VASP: density-functional theory and beyond. J Comput Chem 2008;29:2044–78.

[9] Perdew JP, Burke K, Ernzerhof M. Generalized gradient approximation made simple. Phys Rev Lett 1996;77:3865.
[10] Fuchs M Exchange-correlation energy: from LDA to GGA and beyond. Faradayweg).
[11] Blöchl PE. Projector augmented-wave method. Phys Rev B 1994;50:17953.
[12] Rangel T, Caliste D, Genovese L, Torrent M. A wavelet-based projector augmented-wave (PAW) method: reaching frozen-core all-electron precision with a systematic, adaptive and localized wavelet basis set. Comput Phys Commun 2016;208:1–8.
[13] Albareda G, Suñé J, Oriols X. Many-particle Hamiltonian for open systems with full coulomb interaction: application to classical and quantum time-dependent simulations of nanoscale electron devices. Phys Rev B 2009;79:075315.
[14] Li J, Sun JL. Application of X-ray diffraction and electron crystallography for solving complex structure problems. Acc Chem Res 2017;50:2737–45.
[15] Fiermans L, Clauws P, Lambrecht W, Vandenbroucke L, Vennik J. Single crystal V_2O_5 and lower oxides. A survey of their electronic, optical, structural, and surface properties. Physica Status Solidi (A) 1980;59:485–504.
[16] Shevchuk V, Usatenko YN, Demchenko PY, Antonyak O, Serkiz RY. Nano-and micro-size V2O5 structures. Chem Metals Alloys 2011:67–71.
[17] Lv B, Qian T, Ding H. Angle-resolved photoemission spectroscopy and its application to topological materials. Nat Rev Phy 2019;1:609–26.
[18] Bussolotti F, Chi D, Goh KJ, Huang YL, Wee AT. 2D semiconductor materials and devices. Elsevier; 2020. p. 199–220.
[19] Fox M. Optical properties of solids. Oxford University Press; 2001.
[20] Fox M. Optical properties of solids. American Association of Physics Teachers; 2002.
[21] Speakman SA. Introduction to x-ray powder diffraction data analysis. Center for Materials Science and Engineering at MIT; 2013.
[22] Reznik D, Olk C, Neumann D, Copley J. X-ray powder diffraction from carbon nanotubes and nanoparticles. Phys Rev B 1995;52:116.
[23] VanHove MA, Weinberg WH, Chan C-M. Low-energy electron diffraction: experiment, theory and surface structure determination, vol. 6. Springer Science & Business Media; 2012.
[24] Tepper B, Richter B, Dupuis A-C, Kuhlenbeck H, Hucho C, Schilbe P, Bin Yarmo M, Freund H-J. Adsorption of molecular and atomic hydrogen on vacuum-cleaved V2O5 (001). Surf Sci 2002;496:64–72.
[25] Shenhui J, Ding D, Quanxing J. Measurement of electromagnetic properties of materials using transmission/reflection method in coaxial line. In: Asia-pacific conference on environmental electromagnetics, 2003. CEEM 2003. Proceedings. IEEE; 2003. p. 590–5.
[26] Liao K-m, Wu Y-q, Qian C, Du G-p. Electrical power systems and computers. Springer; 2011. p. 779–84.
[27] Ellison WJ, Moreau J-M. Open-ended coaxial probe: model limitations. IEEE Trans Instrum Meas 2008;57:1984–91.
[28] Wee FH, Soh PJ, Suhaizal A, Nornikman H, Ezanuddin A. Free space measurement technique on dielectric properties of agricultural residues at microwave frequencies. In: 2009 SBMO/IEEE MTT-S international Microwave and optoelectronics conference (IMOC): IEEE); 2009. p. 183–7.
[29] Baker-Jarvis J, Janezic MD, DeGroot DC. High-frequency dielectric measurements. IEEE Instrum Meas Mag 2010;13:24–31.
[30] Grundmann M. The physics of semiconductors. Springer; 2010. p. 775–6.

CHAPTER

10

Spin-diversified quasiparticle behaviors in rare-rare-earth La- and Eu-adsorbed germanene materials

Vo Khuong Dien[1], Shih-Yang Lin[1], Chi-Hsuan Lee[1], Hsin-Yi Liu[1], Thi My Duyen Huynh[1], Nguyen Thi Han[1], Ngoc Thanh Thuy Tran[2], Thi Dieu Hien Nguyen[1], Wei-Bang Li[1] and Ming-Fa Lin[1]

[1]Department of Physics, National Cheng Kung University, Tainan, Taiwan; [2]Hierarchical Green-Energy Materials (Hi-GEM) Research Center, National Cheng Kung University, Tainan, Taiwan

10.1 Introduction

The first discovery of monolayer graphene through mechanical exfoliation of a graphite surface in 2004 [1] is considered as a milestone in material physics and has triggered many scientists to look for alternative 2D materials [2–11]. Indeed, beyond the graphene, other group-IV systems have been synthesized successfully on specific substrates [12–16], especially for monolayer germanene on Pt (111) [17,18], Au (111) [19,20] and Al (111) [21–25] surfaces. Similar as Carbon atom, the electron configurations of germanium also have four electrons in their outermost (s, p_x, p_y, p_z) orbitals, clearly indicating the rather strong covalent bonding. Depend on the orbital hybridizations, the germanium element may exist in two main difference phases, e.g., the sp^3 bonding in the diamond-like crystals [26,27] and the non-negligible mixing of π & σ bonding due to the buckled germanene crystals. The 2D emergent system displays a lot of unusual quasiparticle properties, as the relativistic massless behaviors of electron near the K and K′, an anomalous quantum Hall effect [28]. Very interestingly, difference graphene, the sp^2-sp^3 mixing makes the honeycomb

Fundamental Physicochemical Properties of Germanene-related Materials
https://doi.org/10.1016/B978-0-443-15801-8.00008-6

lattices of germanene are predicted to be buckled [29–31]. Furthermore, due to the larger atomic number of germanium and silicon as compared to carbon, these materials have a much stronger spin–orbit coupling [32–34]. A small buckling will increase the spin–orbit coupling by orders of magnitude [35]. As results, a small band gap at the Dirac points has been opening.

Adsorption of atoms and molecules provides an effective way strategy in creating dramatic changes in electronic and optical properties of 2D materials [36–46]. Adsorption of hydrogen, alkali and transition metal atoms on 2D material has been investigated extensively presently. Rare-earth (RE) atoms possess an intrinsic large spin-orbit interaction and their magnetic states are well decoupled from the environment [47,48]. These properties have been exploited to design single ion molecular magnets with magnetic bistability on a long timescale. Magnetic remanence and long magnetic lifetimes have also been achieved in single rare-earth (RE) adatoms. To date, there are much less studies of interaction between RE atoms and 2D materials [49,50]. The main challenge is the complicated of the sp-f chemical bonding between group IV elements and RE adatoms [51]. The previous theoretical and experimental have been investigated on the RE-related graphene, however, a comprehensive and comparative work to fully understanding the orbital bonding-enrich phenomena is rather limit up to now.

In this chapter, the geometric, electronic and optical properties of the RE atoms (La and Eu) adsorbed on monolayer Germanene are investigated by using the first-principle calculations. The concentration and arrangement dependents are explored in detail. The orbital-hybridizations in the chemical bonding are achieved by the delicate analysis on the atom-dominated energy band structure, the spatial charge densities & their variations, the atom-/projected van-Hove singularities. The magnetic properties pf the host and guest adatoms can be comprehensive through the spin-split/spin-degeneracy energy band structure, the partial-/net-magnetic moments, the spatial spin density distribution and the polarized-density of states. Especially, the orbital-character of the band-edge states was used to identify the optical excitations, e.g. the threshold frequency, the special prominence peaks in the optical adsorption spectrum, furthermore, the collective optical excitations could also be understood from the dielectric functions, electron energy loss functions, absorptions coefficients and reflection spectrums. Most of theoretical predictions in this work are required the high-solution experimentally techniques.

10.2 Computational details

We used the density functional theory method [52–55] via the Vienna Ab-initio Simmulation Package (VASP) [56] to perform the optimization of crystal structures and the calculation of the electronic properties. The Perdew-Burke-Ernzerhof (PBE) [57] generalized gradient approximation was used for the exchange-correlation functional. The interaction between the valence electrons and ions core was evaluated by the projector augmented wave (PAW) method [58]. The cutoff energy for the expansion of the plane wave basis set is 600 eV for all calculation. The Brillouin zone was integrated with a special k-point mesh of $20 \times 20 \times 1$ and $40 \times 40 \times 1$ in the Monkhorst-Pack sampling technique for the geometric relaxation and electronic/optical calculation, respectively. The convergence condition of the ground-state is set to be 10^{-6} eV between two consecutive simulation steps, and all atoms were allowed to fully relax during the geometry optimization until the Hellmann-Feynman force [59] acting on each atom was smaller than 0.01 eV/Å. Spin-polarized calculations were performed for the geometry optimization and the calculation of the band structure. In the k-point sample, the cutoff energy has been checked for convergence of the calculations.

10.3 Results and discussions

Lanthanum (La) and Europium (Eu) adatoms adsorption monolayer germanene can create unusual geometric structures. The RE-related germanene system is modeled with one RE adatom in six difference hexagonal germanene supercell (Fig. 10.1), namely 2×2, $2\sqrt{3} \times 2\sqrt{3}$, 3×3, 4×4, 5×5 and $3\sqrt{3} \times 3\sqrt{3}$. In which, the corresponding concentration is 12.5%, 5.6%, 4.2%, 3.2%, 2.0% and 1.9%, respectively. Similar as the RE-adsorbed on graphene, the most favorite position is the hollow side, regardless of the RE-type and concentration. This could be verified from the STM measurement.

10.3.1 La adsorption

La-adsorbed on Germanene systems are stable under single side adsorption, with specific distributions and concentrations are shown in Fig. 10.1 and Table 10.1. It is very difficult to form the high concentration adsorption systems since the La radius is very big and Coulomb repulsive interactions between them are significant within a short distance. For pristine germanene, it has a hexagonal primitive unit cell with two Ge atoms per cell. The lattice

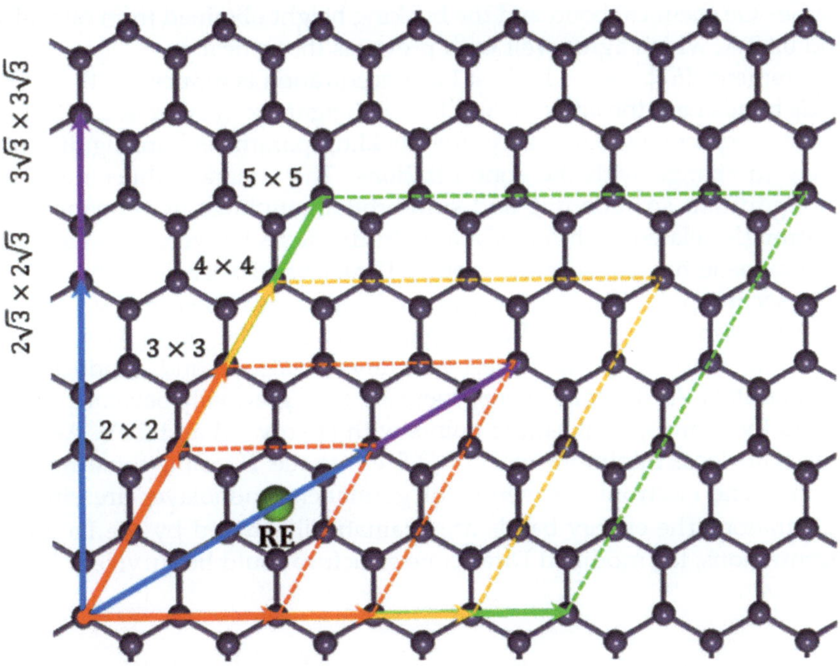

FIG. 10.1 Geometric structures for various RE-adsorbeb germanene.

TABLE 10.1 Calculated Ge−Ge and Ge-RE bond lengths, adatom height, binding energies, energy gap and total magnetic moments per unit cell of Rare-earth-absorbed Germanene.

Adatom	Unit cell	Concentration	Nearest Ge−Ge (Å)	RE height (Å)	Bulkling (Å)	M_{tot} (μ_B)	CT(e)
La	2×2	1:8	2.571	1.395	0.826	0	1.279
	3×3	1:18	2.547	1.482	0.895	0	1.306
	$2\sqrt{3} \times 2\sqrt{3}$	1:24	2.531	1.617	0.869	0	1.319
	4×4	1:32	2.533	1.553	0.894	0	1.325
	5×5	1:50	2.536	1.539	0.897	0	1.327
	$3\sqrt{3} \times 3\sqrt{3}$	1:54	2.531	1.587	0.876	0	1.327
Eu	2×2	1:8	2.512	1.712	0.657	6.926	1.182
	3×3	1:18	2.501	1.801	0.780	7.420	1.219
	$2\sqrt{3} \times 2\sqrt{3}$	1:24	2.498	1.840	0.782	7.214	1.243
	4×4	1:32	2.496	1.811	0.775	7.287	1.245
	5×5	1:50	2.496	1.840	0.778	7.216	1.249
	$3\sqrt{3} \times 3\sqrt{3}$	1:54	2.491	1.879	0.773	7.225	1.251

constant, the Ge−Ge chemical bond and the bucking height obtained from our calculation are 4.03, 2.42 and 0.68 Å, which agree well with previous theoretical predictions [60] and experimental measurements [61]. As the highest La concentration is reduced to the lowest one, the nearest Ge−Ge bonds near the adatom, and the RE-height are vary from 2.571 to 2.531 Å and 1.395 to 1.587 Å, respectively. Especially, the buckling parameters are significantly increase and sensitively to change with the concentrations. The enhanced bucking is expected to induce more significant sp^3 bonding and spin-orbital couplings in germanene. These phenomena are strongly related with the orbital-hybridizations between Ge−La bonds.

Pristine Germanene has an isotropic modify Dirac-cone structure within $|E^{c,v}| \leq 1\,eV$, of which the intersecting point is just located at the K point, as shown in Fig. 10.2A. The negligible energy gap estimated as 0.5 meV. This gap could be enlarged to 3.2 meV under the effects of spin-orbit couplings. The π-electronic linear valence and conduction bands are formed by $4p_z$ orbitals. As the state energy increase, they gradually become parabolic energy dispersions and there are special saddle points with energy −3 and 2 eV. Also, σ-electronic energy bands with local minimum of $E^v = 3.5$ eV at the Γ point, come mainly from $4p_x$ and $4p_y$ orbitals. The electronic properties of germanene monolayer are enrich by tuning the RE concentration. The energy bands are dramatically shifted by the La adsorptions, at the low concentrations, the modified Dirac cone structure could be survived. The Fermi level

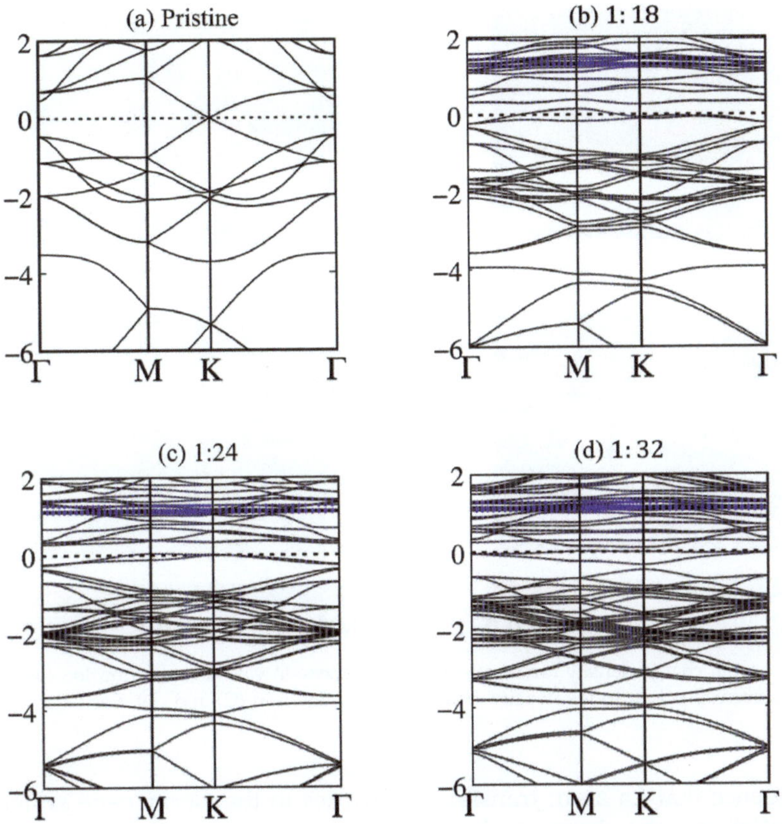

FIG. 10.2 Band structures of La adsorbed germanene for various concentrations: (A) Pristine, (B) 1:18, (C) 1:24 and (D) 1:32. The gray and blue circles represent the dominated of the Ge and Eu atoms, respectively.

is located at the valence cone, all of them are metals/semimetals, for which, free hole appeared between E_D and E_F. La make much concentrations to the certain energy bands with a very weak energy dispersions in the range of 1.2 eV $\leq E^{c,v} \leq$ 1.6 eV, where the amount of concentration is indicated by the radius of blue circles. Roughly speaking, there exist the dramatically modified low-lying energy dispersions, mainly owing to the extremely nonuniform environment in an enlarged unit cell and more important or complicated interlayer orbital hybridizations.

Bader charge analysis, as well as the spatial charge density distributions/charge density variations (Fig. 10.3), can further provide information about the characters, the multi-orbital hybridizations of RE adsorption germanene. In which, the latest one can be achieved by subtracting the charge densities of germanene and Re atoms from that of germanene-RE adsorption systems. The Bader charge analysis accompanied with the charge density

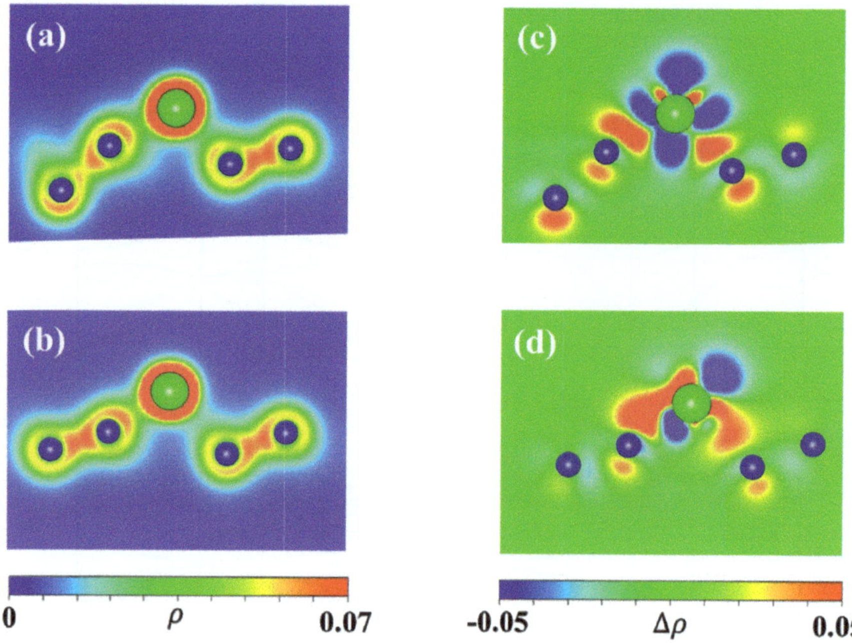

FIG. 10.3 The spatial charge density for La adsorbed germanene in various concentration cases (A) 1:18 and (B) 1:24. Similar plots corresponding to the density differrence are shown in (C) and (D). For La adsorbed germanene.

difference evidence that La atom transfers its charges to the Germanene system. Increasing the La concentrations, the charge exchange is slightly reduced, the main reason is due to the strong repulsive Coulomb interactions between the nearest adatom atoms. There are three kind of orbital bondings in the honeycomb lattices could survive simultaneously. The σ electrionic orbital hybridizations, with a high charge density between two Ge atoms, are slightly affected by La adsorptions, corresponding to the slightly variations of charge density/a minor charge density differences. This responsible for a rigid red-shift of the σ energy bands and the absence of co-dominate of La, Ge in the valence band. In significantly contract, the modifications of the π bondings are absorbed through the variations the charges between Ge and La atoms. These features suggest the existence of a Ge-(4_{pz}) − La-(6s, $5d_{yz}$, $5d_{xy}$, $5d_{z^2}$, $5d_{xz}$, $5d_{z^2-y^2}$), accounting for the (La, Ge) co-domination of conduction bands. In addition, The weak, but important (6s, $5d_{yz}$, $5d_{xy}$, $5d_{z^2}$, $5d_{xz}$, $5d_{z^2-y^2}$) - (6s, $5d_{yz}$, $5d_{xy}$, $5d_{z^2}$, $5d_{xz}$, $5d_{z^2-y^2}$) couplings between two nearest La atoms in sufficient high concentrations.

The main characteristics of DOS are determined by the energy dispersions of band-edge states; furthermore, the orbital-projected DOS directly reflected the features of chemical bondings. The La-adsorbed germanene exhibits many 2D Van Hove singularities, as indicated in Fig. 10.4A–D. A dip, strong symmetric peaks and the shoulder structures, respectively, come from the linear bands, the saddle point and the local extreme points. Germanene processes the low-energy 4_{pz} dominated DOS (the red curve in Fig. 10.4A), including the dip at $E = 0$, the linear E-dependence and the π, peaks in the algorithmically

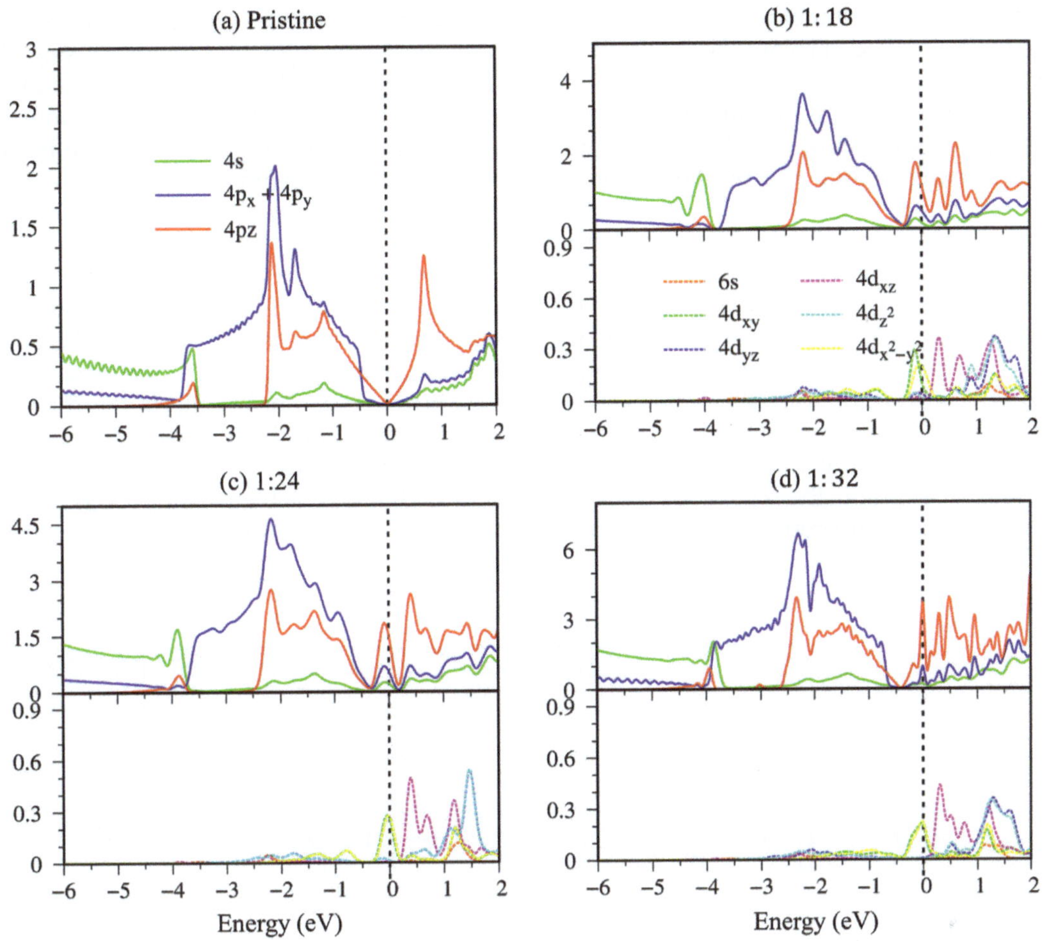

FIG. 10.4 The total and the orbital-projected DOS for La-adsorbed germanene with distinct adsorption cases: (A) pristine, (B) 1:18, (C) 1:24 and (D) 1:32.

form. The special structure in 4_{pz}-related DOS are dramatically changed by the orbital hybridization of d and 4_{pz}. The DOS due to the conduction bands π^*-states is finite at the Fermi level, being a result of the shift of π band. The center of the linear E-dependence, which slightly depended on adatom concentration, is shifted downward. The fit well of the DOS of La and Ge are evident the significant hybridizations of Ge-4_{pz} and La-d orbitals.

When the 2D pristine/adsorbed Germanene are perturbed by an electromagnetic wave, the occupied electronic states are vertically excited to the unoccupied ones. All the charge carriers are capable of dynamically screening this external field and thus generate the induced current density. Within the linear response, the screening ability of charges is well characterized by the transverse dielectric function at the long-wavelength limit $[\varepsilon(q \rightarrow 0, \omega)]$. According to basic electrodynamics, this frequency-dependent function is very useful in understanding the main features of reflectance spectrum, absorption coefficient and energy loss function. The relevant physical quantities are consistent with one another using the same theoretical framework. Furthermore, the single-particle electronic excitations (electron-hole pairs) and collective excitations (plasmon modes) are expected to dominate the couplings of charges and electromagnetic waves. The delicate analyses are done by the direct linking among the orbital-dominated band-edge states (band structures), the joint Hove singularities (the large density of states), and the specific absorption frequency (optical spectra). Each prominent absorption structure will be identified from the corresponding orbital hybridizations.

A pristine germanene clearly exhibits the unusual optical excitation phenomena, as indicated in Fig. 10.5. Its frequency-dependent bare response function, which is due to the vertically interband electron-hole excitations, show four prominent absorption structures, respectively at 0.25, 2.21, 4.23, and 3.80 eV. The featured results only reflects the different orbital hybridizations of 2D germanium honeycomb crystal, respectively, arising from the inter-π-band transitions inside a Dirac-cone structure, the higher-energy ones associated with the saddle M-point, the composite excitations of the σ band & π band from the K and M points. The optical properties of germanene are enriched under La-adsorped, e.g., a strong

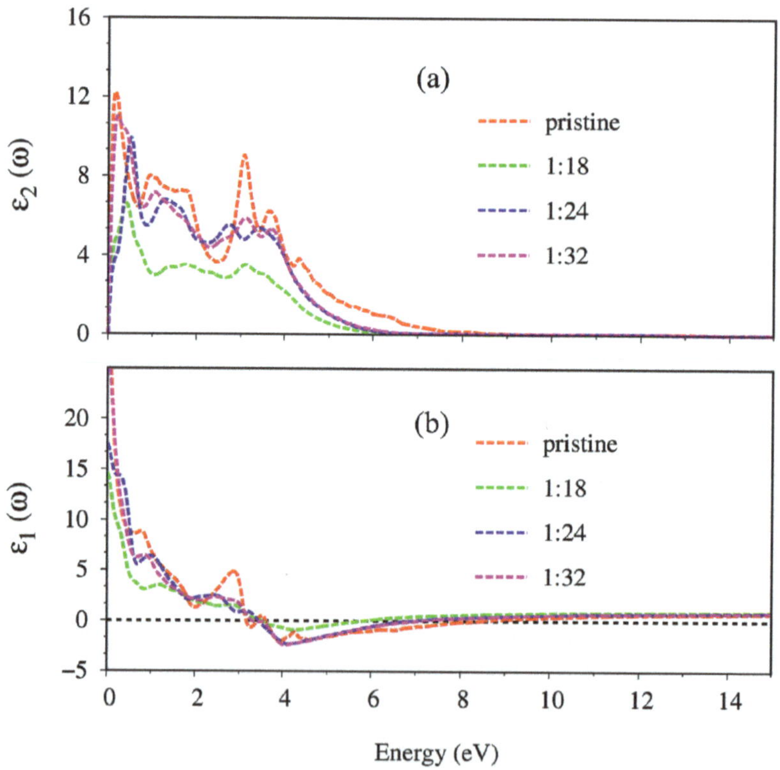

FIG. 10.5 The imaginary part and real part of dielectric functions of pristine/La-doped germanene.

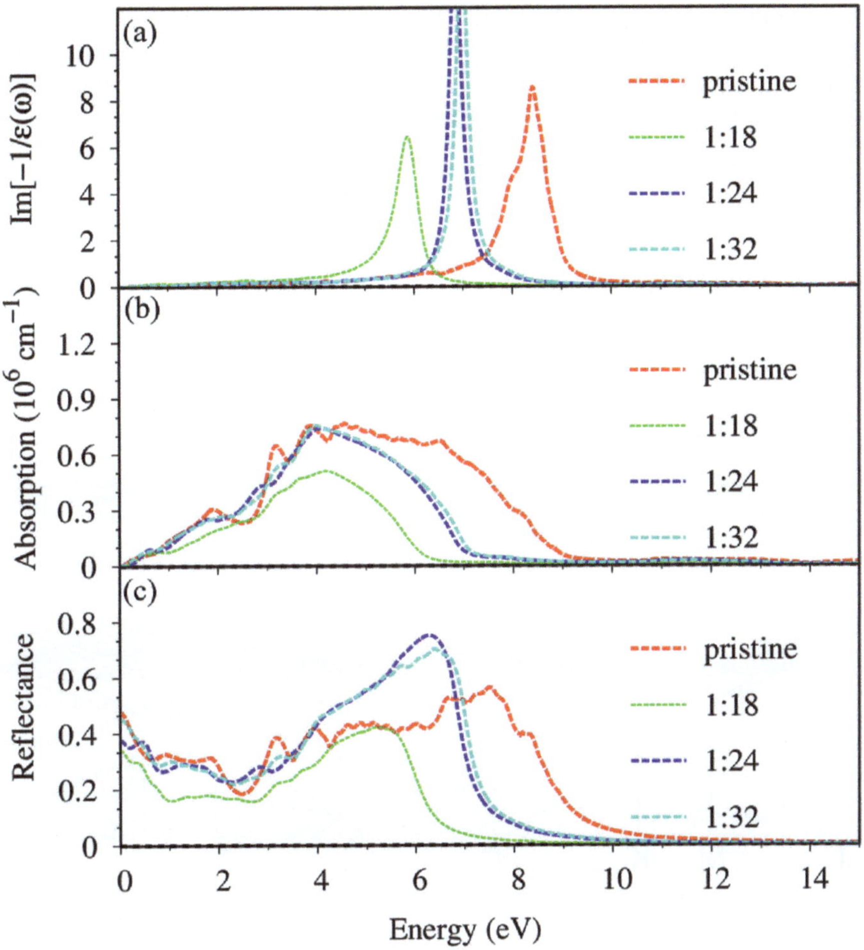

FIG. 10.6 The electron energy loss function, adsorption and reflection coefficient of pristine/La-doped germanene.

spectral response between 0 and 1 eV are significantly decreases for the doped systems, especially for the high concentration cases.

After the charge screenings of all π and σ valence electrons, the energy loss functions (Fig. 10.6A), absorption coefficients (Fig. 10.6B), and reflectance spectra (Fig. 10.6C), clearly reveal the featured plasmon modes, respectively, corresponding to the prominent peaks, the similar ones, and the drastic plasmon-edges. From the calculated results, the π plasmon, the second π-electronic collective excitation, respectively, survive 2.6 and 8.6 eV. Apparently, such energy-dependent plasmons belong to the quantization modes of the coherent charge oscillations under the long wavelength limit, corresponding to the different orbital-dominated valence electrons. There also exist the drastic decline near the plasmon frequencies for the frequency-related reflectance spectra (Fig. 10.6C). Very interestingly, $R(\omega)$ totally vanishes within the specific range 8 eV $< \omega <$ 15 eV. Most of the incident light are transmitted through a single-layer germanene where only a partial perturbation is absorbed by material.

The sensitive dependences of reflectance and transmittance on the layer number and stacking configuration is worthy of the systematic investigations. The plasmon modes are significantly red-shifted for the doped systems. The other properties are also slightly change. All structures possess oscillation mainly in the range 0—10 eV.

10.3.2 Eu adsorption

The optimal Eu adatom adsorptions on a germanene surface are clearly show in Table 10.1. The similar to those of La one, the Eu atom is favorite adsorption at the hollow side, the highest concentration can be achieved through the 1:8 case of single side adsorption. The adatom height are vary from 1.71 to 1.80 Å. They become longer when Eu concentration is decreased. Specifically, the bulking - the non-orthogonal degree of the π and σ bondings are slightly

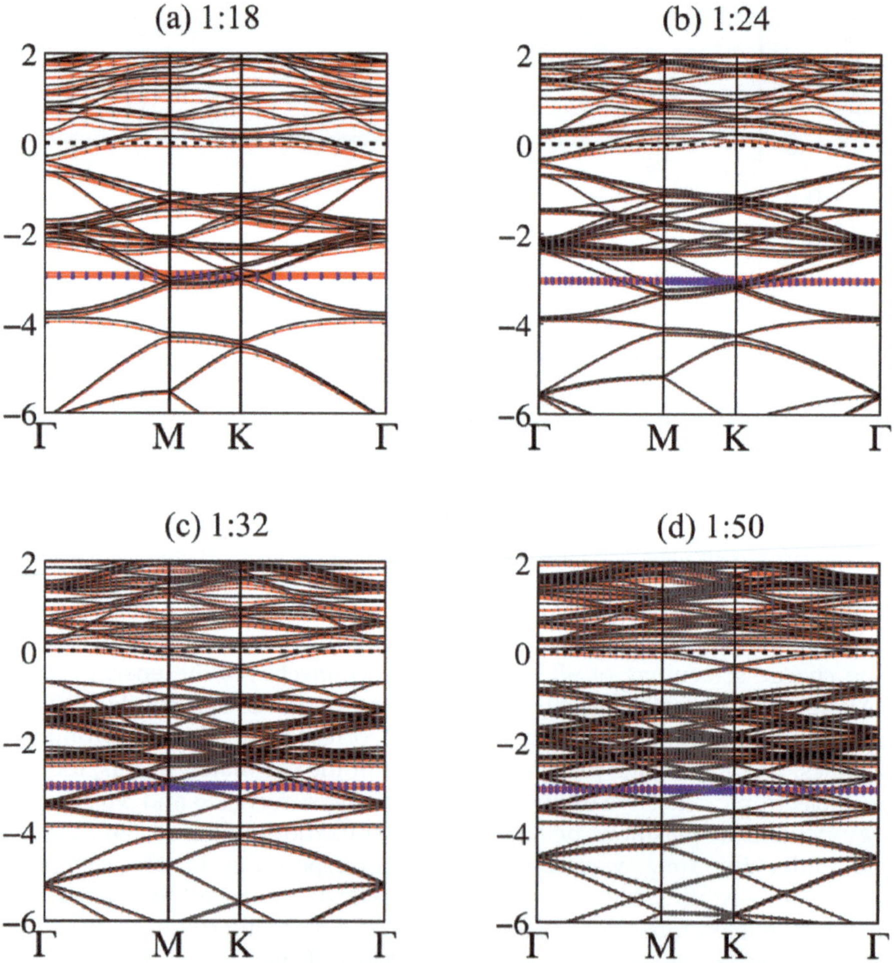

FIG. 10.7 Band structures of Eu adsorbed germanene for various concentrations: (A) 1:18, (B) 1:24, (C) 1:32 and (D) 1:50. The gray and blue circles express the dominated of the Ge and Eu atoms, respectively. While the black and the red solid lines indicate the spin up and spin down energy states.

smaller than those of La adatom adsorption, but it still larger than that of the pristine case. These features suggest that the presence of the sufficiently strong coupling of Eu adatoms and Ge atom, associated with 6s and 4f valence orbitals of Eu. The π bonding, which is extend on the germanene surface, is strongly modify or even completely destroyed due to the critical interlayer atomic interactions (Fig. 10.7).

The Eu adsorptions are able to create a rich and usual band structures, as clearly indicated in Fig. 10.8A–D. The main features of electronic band structures include the highly asymmetric energy spectra near the Fermi energy, E_F, the modified Dirac cone structure with/without an energy spacing, a blue shift of E_F relative to the Dirac point in accordance with n-type doping behavior. Their energy gradually grown with decreasing of Eu concentrations. It is importance to noted that the similar roughly rigid is also revealed in the σ energy bands. Roughly speaking, there exist the dramatically modified low-lying energy dispersions, mainly owing to the extremely non-uniform environment in an enlarged unit cell and more important or complicated interlayer orbital hybridizations. Specially, Eu make much concentrations to the certain energy bands with a very weak energy dispersions in the range of $-2.6\text{ eV} \leq E^{c,v} \leq -2.8\text{ eV}$, where the amount of concentration is indicated by the radius of blue circles. The ferromagnetic configurations, which are clearly characterized by the spin-split valence and conduction energy sub-band and the magnetic moments, are further investigated from the spin-density distributions. All the Eu adsorption cases, as clearly depicts in Fig. 10.8, exhibit an unusual ferromagnetic phenomenon. The Eu adatoms are an able to create the dominant spin density associated with seven 4f orbitals electrons of Eu adatoms.

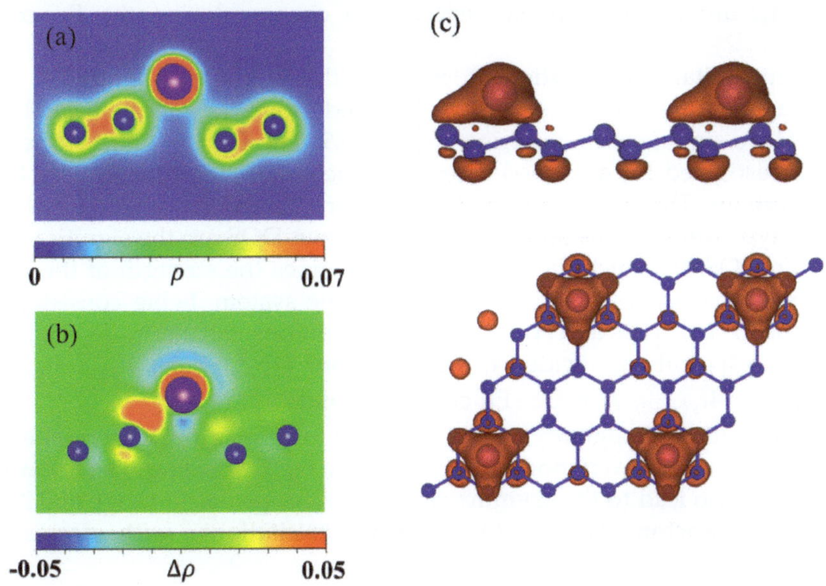

FIG. 10.8 (A) The spatial charge density and (B) charge density difference for Eu adsorbed germanene in 1:18 case. The similar plot for spin density distribution for Eu adsorbed germanene.

The spatial charge densities/charge density differences and effective charge transfer are, respectively can provide the first step orbital-hybridizations in the chemical bonds. Similar to those of La adsorbed, Eu atom transfer its charges to the Germanene system. The average effective charges are lower than that of La adsorbed. The π bondings are strongly modify by Eu adsorption, while the opposite is true for the σ electrionic bands. There are exist significant and unusual charge density differences between the guest and the host atoms. The strongly hybridizations of ($4p_z$) – (6s, $4f_{xyz}$, $4f_{yz^2}$, $4f_{z^3}$, $4f_{xz^2}$, $4f_{z(x^2-3y^2)}$, $4f_{x(x^2-3y^2)}$, $4f_{y(y^2-3x^2)}$) in Ge–Eu chemical bonds have led to modify Dirac cone structures. Specifictly, the observed charge density variations, with the positive and negative $\Delta\rho$, are shown between two Eu atoms for the high Eu concentration cases. The strong (6s, $4f_{xyz}$, $4f_{yz^2}$, $4f_{z^3}$, $4f_{xz^2}$, $4f_{z(x^2-3y^2)}$, $4f_{x(x^2-3y^2)}$, $4f_{y(y^2-3x^2)}$ – (6s, $4f_{xyz}$, $4f_{yz^2}$, $4f_{z^3}$, $4f_{xz^2}$, $4f_{z(x^2-3y^2)}$, $4f_{x(x^2-3y^2)}$, $4f_{y(y^2-3x^2)}$) orbital hybridizations in Eu–Eu bonds are responsible for the wide Eu-dominated valence energy spectra, being consistence with energy band structures and orbital-/atom-decompound density of states.

The magnetic properties of the Eu-adsorption germanene could be comprehended through the spin density distribution and the magnetic moment, in which the net magnetic moment in a unit cell is determined by the competition of spin-up and spin-down components. The germanene-RE systems are ferro-magnetically ordered, and the total magnetic moment is vary from 6.926 μ_B to 7.287 μ_B, as expected, since the Eu atom introduces seven spin-up electrons. As a matter of fact, the spin-up density is the most dominant part and relies on the Eu atom, which is directly reflected by the fully occupied states in the dispersionless energy bands below the Fermi level. Furthermore, because of the strong hybridization with Germanene systems, Ge atoms also manifest partial minor magnetic contributions with values one orders lower than the Eu atom, respectively. As a result, the net magnetic moments of the adsorbed systems are expected to be sensitive to change during the Eu concentration variation.

The significant orbital hybridizations, together with the π and σ bondings, which could survive in Eu-adsorbed germanene, are investigated from the atom-, orbital-, and spin-projected van Hove singularities, referring to Fig. 10.9. The DOS is finite at the Fermi level for all of the investigated Eu adsorbed cases. In which, its magnitude is very large for the flat energy dispertions. The linear E-dependence present the red shift related to the E_F evident a very strong n-type doping. The spin-up and spin-down DOS are the significant difference, especially for Eu-DOS. In which the difference determines the strength of the net magnetic moment, this indicates a presence of magnetism in the system, being consistence with the spin-polarized band structures and spin-density distributions. Furthermore, the lightly merge of the Eu and Ge-$4p_z$ orbitals evidence the important ($4p_z$) – (6s, $4f_{xyz}$, $4f_{yz^2}$, $4f_{z^3}$, $4f_{xz^2}$, $4f_{z(x^2-3y^2)}$, $4f_{x(x^2-3y^2)}$, $4f_{y(y^2-3x^2)}$) in Ge–Eu chemical bonds.

Fig. 10.10A presents the imaginary part of dielectric function $\varepsilon_2(\omega)$ of germanene adsorbed Eu atom. The single particle excitations are enriched by the Eu-doping. The linear energy dispertion near the E_F also lead to the asymmetric square-root peaks. In range of 0.2 eV $< \omega <$ 1 eV, the special structure is due to intraband excitations, with decrease the Eu-

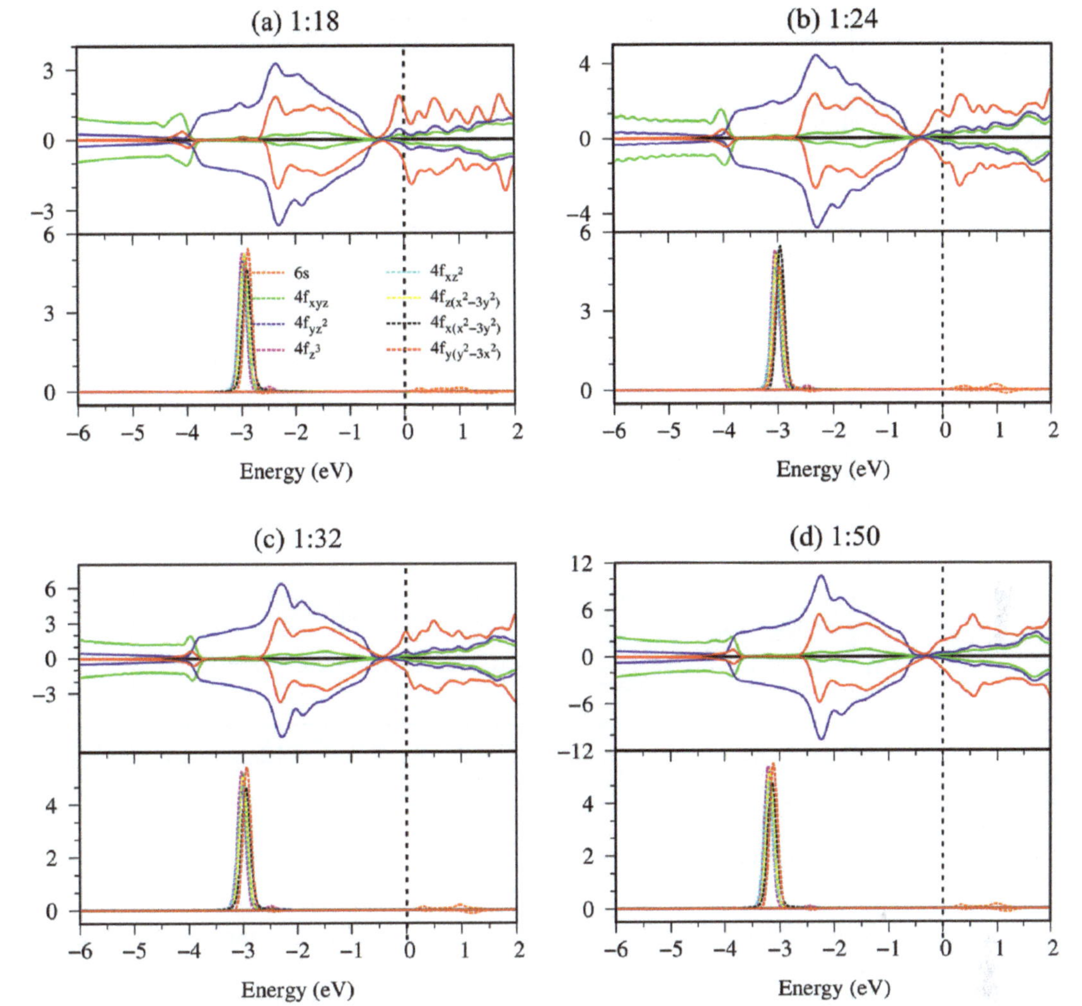

FIG. 10.9 The total and the orbital-projected DOS for Eu-adsorbed germanene with distinct adsorption cases: (A) 1:18, (B) 1:24, (C) 1:32 and (D) 1:50.

concentrations, the identical structure at $\omega \sim 0.25$ eV is increased. Apparently, the variations of the Fermi level will modify the single particle excitations and even results in the creation or destruction of different Plasmon modes. After Eu-adsorbed, the screened response functions, revealing the distinct collective excitations and the various Landau dampings, present a rich and unique phenomenon. The electron energy loss functions, adsorption and reflection spectra are shown in Fig. 10.11A–C. The main features are like those of the pristine one; However, the intensity of the Plasmon mode, the reflection coefficient and the decay

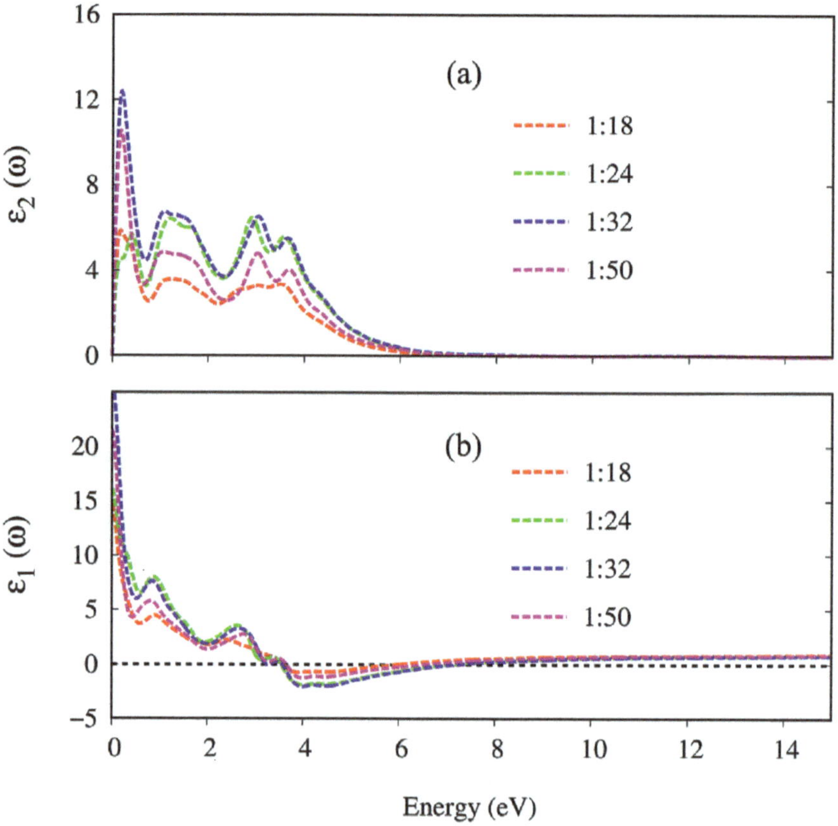

FIG. 10.10 The imaginary part and real part of dielectric functions of Eu-doped germanene.

length-the inverse value of the adsorption coefficients are sensitive to charge with the change of Eu-concentrations.

The rare-earth adsorbed-induced diverse phenomena are worthy of a systematic investigation from the experimental measurements, being useful in understanding the critical orbital hybridizations, spin configurations and the optical excitations. The STM/TEM measurements on GNRs have confirmed the achiral/chiral open boundaries structures, the nano-scaled widths and the layered structures [62,63]. They are available in identifying the planar honeycomb lattice after halogen adsorptions, the top-site position, and the X—C bond lengths. Such measurements, with the spin polarization, could be utilized to examine the metallic band structures, the carbon- and adatom-dominated energy bands near the Fermi level, and the spin-split electronic states. The 1D asymmetric peaks of DOSs in semiconducting GNRs are identified from the STS measurements GNRs [64]. The distinct van Hove singularities, being characterized by FM, AFM and NM metals, require further spin-polarized STS examinations. The transport experiments could test the relation between electron density and rare-earth concentration, and the spin-polarized currents. Moreover, the magneto-optical experiments are suitable for verifying the main features of the spin-split energy bands. The results of

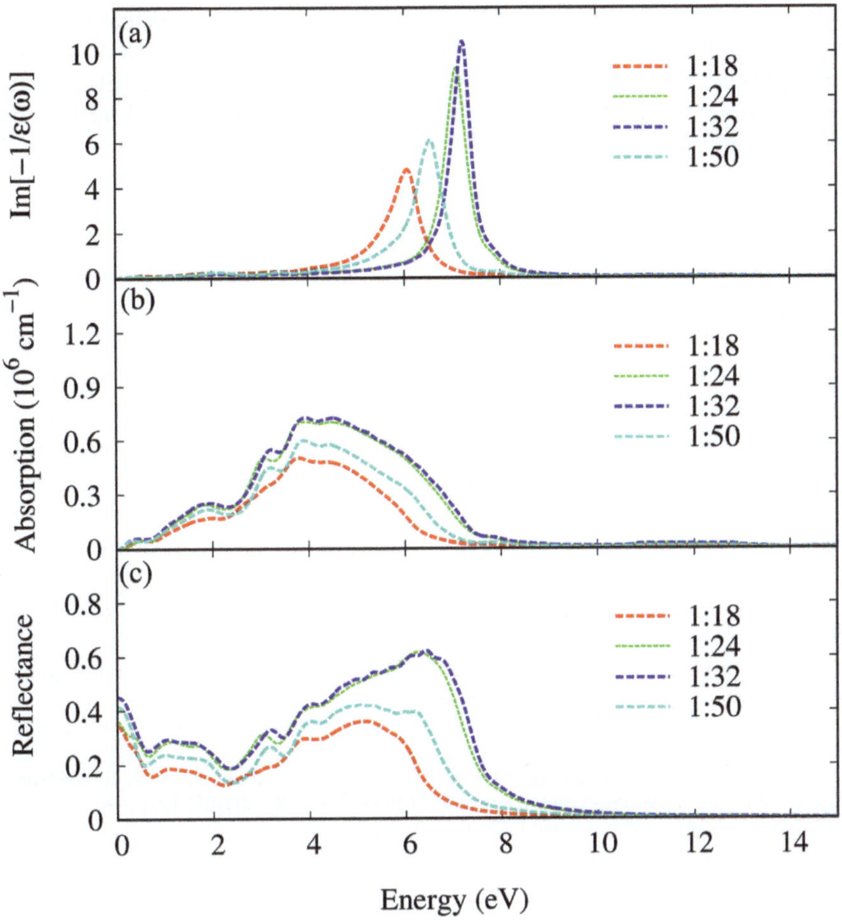

FIG. 10.11 The electron energy loss function, adsorption and reflection coefficient of Eu-doped germanene.

the photoluminescence, adsorption, reflection, or transmission measurements, being supported by the Kramers–Kronig relations [65], can be compared with our dielectric function [66], absorption [67], reflection coefficients [68] and energy loss functions [69]. Other information, such as the diversified optical excitation peaks, the sensitive change of the optical properties [70–72] owing to the rare earth adsorbed on the of the optical properties of germanene can also be carried out.

10.4 Concluding remarks and summary

The geometric, electronic and optical properties of rare-earth adatom adsorbed germanene were investigated by using first-principles calculations. In contract with the pristine germanene as a narrow-gap semiconductor, all of the rare-earth-adsorbed cases that we have investigated fall into the n-type doping, and thus, implying an enhanced the electronic conductivity. On the other hand, The La- and Eu-adsorbed germanene are differ from each other in their fundamental properties. The main differences are the optimal lattice constants, the Re-Ge chemical bond lengths, enhanced bulking degrees, the modified the Dirac cone structures, the spin-degenerate or spin-split energy spectra around the Fermi level, the adatom-enrich charge density distributions, the ferromagnetic or nonmagnetic under different concentrations, the special van Hove singularities, the single- and the collective-optical excitations. The most importantly, the significant hybridizations of (6s, $5d_{yz}$, $5d_{xy}$, $5d_{z^2}$, $5d_{xz}$, $5d_{z^2-y^2}$)/(6s, $4f_{xyz}$, $4f_{yz^2}$, $4f_{z^3}$, $4f_{xz^2}$, $4f_{z(x^2-3y^2)}$, $4f_{x(x^2-3y^2)}$, $4f_{y(y^2-3x^2)}$) & 4pz orbitals and (6s, $5d_{yz}$, $5d_{xy}$, $5d_{z^2}$, $5d_{xz}$, $5d_{z^2-y^2}$) - (6s, $5d_{yz}$, $5d_{xy}$, $5d_{z^2}$, $5d_{xz}$, $5d_{z^2-y^2}$)/(6s, $4f_{xyz}$, $4f_{yz^2}$, $4f_{z^3}$, $4f_{xz^2}$, $4f_{z(x^2-3y^2)}$, $4f_{x(x^2-3y^2)}$, $4f_{y(y^2-3x^2)}$)-(6s, $4f_{xyz}$, $4f_{yz^2}$, $4f_{z^3}$, $4f_{xz^2}$, $4f_{z(x^2-3y^2)}$, $4f_{x(x^2-3y^2)}$, $4f_{y(y^2-3x^2)}$) have been identified in the La–Ge/Eu–Ge and La–La/Eu–Eu chemical bonds. The theoretical predictions on the optimal geometries, the occupied energy dispersions, van Hove singularities, and optical properties could be verified from the high-resolution measurements of scanning tunneling spectrocopy, angle-resolved photoemission spectroscopy, and photoluminescene/reflection/adsorption measurements, respectively.

References

[1] Chen G, Weng W, Wu D, Wu C, Lu J, Wang P, et al. Preparation and characterization of graphite nanosheets from ultrasonic powdering technique. Carbon 2004;42:753–9.

[2] Hashimoto A, Suenaga K, Gloter A, Urita K, Iijima S. Direct evidence for atomic defects in graphene layers. Nature 2004;430:870–3.

[3] Berger C, Song Z, Li T, Li X, Ogbazghi AY, Feng R, et al. Ultrathin epitaxial graphite: 2D electron gas properties and a route toward graphene-based nanoelectronics. J Phys Chem B 2004;108:19912–6.

[4] Khantha M, Cordero N, Molina L, Alonso J, Girifalco L. Interaction of lithium with graphene: an ab initio study. Phys Rev B 2004;70:125422.

[5] Peres N, Araújo M, Bozi D. Phase diagram and magnetic collective excitations of the Hubbard model for graphene sheets and layers. Phys Rev B 2004;70:195122.

[6] Lin J-H, Zhang H, Cheng X-L, Miyamoto Y. Single-layer group IV-V and group V-IV-III-VI semiconductors: structural stability, electronic structures, optical properties, and photocatalysis. Phys Rev B 2017;96:035438.

[7] Demirci S, Avazlı N, Durgun E, Cahangirov S. Structural and electronic properties of monolayer group III monochalcogenides. Phys Rev B 2017;95:115409.

[8] Çınar MN, Sargın GÖ, Sevim K, Özdamar B, Kurt G, Sevinçli H. Ballistic thermoelectric transport properties of two-dimensional group III-VI monolayers. Phys Rev B 2021;103:165422.

[9] Xue D, Betzler K, Hesse H. Dielectric properties of I-III-VI 2-type chalcopyrite semiconductors. Phys Rev B 2000;62:13546.

[10] Yang Z, Hao J. Recent progress in 2D layered III–VI semiconductors and their heterostructures for optoelectronic device applications. Advanc Mater Technol 2019;4:1900108.

[11] Yarema O, Yarema M, Wood V. Tuning the composition of multicomponent semiconductor nanocrystals: the case of I–III–VI materials. Chem Mater 2018;30:1446–61.

[12] Lin C-L, Arafune R, Kawahara K, Tsukahara N, Minamitani E, Kim Y, et al. Structure of silicene grown on Ag (111). Appl Phy Express 2012;5:045802.

[13] Cahangirov S, Audiffred M, Tang P, Iacomino A, Duan W, Merino G, et al. Electronic structure of silicene on Ag (111): strong hybridization effects. Phys Rev B 2013;88:035432.

[14] Voloshina E, Generalov A, Weser M, Böttcher S, Horn K, Dedkov YS. Structural and electronic properties of the graphene/Al/Ni (111) intercalation system. New J Phys 2011;13:113028.

[15] Nam H, Jang S, Lee W. Strategies for the modification of adhesion at graphene/Al interfaces: a first principles study. J Nanosci Nanotechnol 2014;14:7674–8.
[16] Yuhara J, Le Lay G. Beyond silicene: synthesis of germanene, stanene and plumbene. Jpn J Appl Phys 2020;59:SN0801.
[17] Li L, Lu Sz, Pan J, Qin Z, Wang Yq, Wang Y, et al. Buckled germanene formation on Pt (111). Adv Mater 2014;26:4820–4.
[18] Li L. Germanene on Pt (111). In: Fabrication and physical properties of novel two-dimensional crystal materials beyond graphene: germanene, hafnene and PtSe2. Springer; 2020. p. 27–36.
[19] Zhuang J, Gao N, Li Z, Xu X, Wang J, Zhao J, et al. Cooperative electron–phonon coupling and buckled structure in germanene on Au (111). ACS Nano 2017;11:3553–9.
[20] Cantero ED, Solis LM, Tong Y, Fuhr JD, Martiarena ML, Grizzi O, et al. Growth of germanium on Au (111): formation of germanene or intermixing of Au and Ge atoms? Phys Chem Chem Phys 2017;19:18580–6.
[21] Stephan R, Hanf M-C, Derivaz M, Dentel D, Asensio M, Avila J, et al. Germanene on Al (111): interface electronic states and charge transfer. J Phys Chem C 2016;120:1580–5.
[22] Derivaz M, Dentel D, Stephan R, Hanf M-C, Mehdaoui A, Sonnet P, et al. Continuous germanene layer on Al (111). Nano Lett 2015;15:2510–6.
[23] Endo S, Kubo O, Nakashima N, Iwaguma S, Yamamoto R, Kamakura Y, et al. Germanene on Al (111) grown at nearly room temperature. Appl Phys Exp 2017;11:015502.
[24] Liu G, Liu S, Xu B, Ouyang C, Song H, Guan S, et al. Multiple Dirac points and hydrogenation-induced magnetism of germanene layer on Al (111) surface. J Phys Chem Lett 2015;6:4936–42.
[25] Wang W, Uhrberg RI. Coexistence of strongly buckled germanene phases on Al (111). Beilstein J Nanotechnol 2017;8:1946–51.
[26] Keating P. Theory of the third-order elastic constants of diamond-like crystals. Phys Rev 1966;149:674.
[27] Belenkov E, Greshnyakov V. Structure, properties, and possible mechanisms of formation of diamond-like phases. Phys Solid State 2016;58:2145–54.
[28] Mittal S, Orre VV, Leykam D, Chong YD, Hafezi M. Photonic anomalous quantum Hall effect. Phys Rev Lett 2019;123:043201.
[29] Coello-Fiallos D, Tene T, Guayllas J, Haro D, Haro A, Gomez CV. DFT comparison of structural and electronic properties of graphene and germanene: monolayer and bilayer systems. Mater Today Proc 2017;4:6835–41.
[30] Ahamed MI, Anwar N. Germanene. Monoelements: Prop Appl 2020:235–60.
[31] Matthes L, Gori P, Pulci O, Bechstedt F. Universal infrared absorbance of two-dimensional honeycomb group-IV crystals. Phys Rev B 2013;87:035438.
[32] Galitski V, Spielman IB. Spin–orbit coupling in quantum gases. Nature 2013;494:49–54.
[33] Ciccarino CJ, Flick J, Harris IB, Trusheim ME, Englund DR, Narang P. Strong spin–orbit quenching via the product Jahn–Teller effect in neutral group IV qubits in diamond. Npj Quant Mater 2020;5:1–6.
[34] Wepfer G, Collins T, Euwema R. Calculated spin-orbit splittings of some group iv, iii-v, and ii-vi semiconductors. Phys Rev B 1971;4:1296.
[35] Konschuh S, Gmitra M, Fabian J. Tight-binding theory of the spin-orbit coupling in graphene. Phys Rev B 2010;82:245412.
[36] Lin S-Y, Chang S-L, Chen H-H, Su S-H, Huang J-C, Lin M-F. Substrate-induced structures of bismuth adsorption on graphene: a first principles study. Phys Chem Chem Phys 2016;18:18978–84.
[37] Nguyen DK, Lin S-Y, Tran NTT, Liu H-Y, Lin M-F. Rich essential properties of Si-doped graphene. In: Green energy materials handbook. CRC Press; 2019. p. 83–107.
[38] Tran NTT, Gumbs G, Nguyen DK, Lin M-F. Fundamental properties of metal-adsorbed silicene: a DFT study. ACS Omega 2020;5:13760–9.
[39] Averyanov DV, Sokolov IS, Platunov MS, Wilhelm F, Rogalev A, Gargiani P, et al. Competing magnetic states in silicene and germanene 2D ferromagnets. Nano Res 2020;13:3396–402.
[40] Feng W, Liu C-C, Liu G-B, Zhou J-J, Yao Y. First-principles investigations on the berry phase effect in spin–orbit coupling materials. Comput Mater Sci 2016;112:428–47.
[41] Champagne A, Charlier J-C. Physical properties of 2D MXenes: from a theoretical perspective. J Phys: Mater 2020;3:032006.
[42] Mi TY, Triet DM, Tien NT. Adsorption of gas molecules on penta-graphene nanoribbon and its implication for nanoscale gas sensor. Phys Open 2020;2:100014.

[43] Tien NT, Thao PTB, Phuc VT, Ahuja R. Electronic and transport features of sawtooth penta-graphene nanoribbons via substitutional doping. Phys E Low-dimens Syst Nanostruct 2019;114:113572.
[44] Tien NT, Thao PTB, Phuc VT, Ahuja R. Influence of edge termination on the electronic and transport properties of sawtooth penta-graphene nanoribbons. J Phys Chem Solid 2020;146:109528.
[45] Nguyen TT, Mi TY. An ab initio study of small gas molecule adsorption on the edge of N-doped sawtooth penta-graphene nanoribbons. Pap Phys 2021;13. pp. 130003–130003.
[46] Dien VK, Le OK, Chihaia V, Pham-Ho M-P. Monolayer transition-metal dichalcogenides with polyethyleneimine adsorption. J Comput Electron 2021;20:135–50.
[47] Xue D, Hu J, Liang X, Wang H. Perspective on rare earth crystals. Sci China Technol Sci 2020;63:1085–6.
[48] Skomski R, Sellmyer D. Anisotropy of rare-earth magnets. J Rare Earth 2009;27:675–9.
[49] Huang Z-W, Li Z-J, Zheng L-R, Wu W-S, Chai Z-F, Shi W-Q. Adsorption of Eu (III) and Th (IV) on three-dimensional graphene-based macrostructure studied by spectroscopic investigation. Environ Pollut 2019;248:82–9.
[50] Baltic R, Donati F, Singha A, Wäckerlin C, Dreiser J, Delley B, et al. Magnetic properties of single rare-earth atoms on graphene/Ir (111). Phys Rev B 2018;98:024412.
[51] Stekolnikov A, Furthmüller J, Bechstedt F. Adatoms, dimers, and interstitials on group-IV (113) surfaces: first-principles studies of energetical, structural, and electronic properties. Phys Rev B 2003;67:195332.
[52] Parr RG. Density functional theory. Annu Rev Phys Chem 1983;34:631–56.
[53] Han NT, Dien VK, Tran NTT, Nguyen DK, Su W-P, Lin M-F. First-principles studies of electronic properties in lithium metasilicate (Li_2SiO_3). RSC Adv 2020;10:24721–9.
[54] Thi HN, Tran NT, Dien VK, Nguyen DK, Lin MF. Diversified properties in 3D ternary oxide compound: Li2SiO3. In: Lithium-Ion batteries and solar cells. CRC Press; 2021. p. 79–101.
[55] Khuong Dien V, Thi Han N, Nguyen TDH, Huynh TMD, Pham HD, Lin M-F. Geometric and electronic properties of Li_2GeO_3. Front Mater 2020;7:288.
[56] Sun G, Kürti J, Rajczy P, Kertesz M, Hafner J, Kresse G. Performance of the Vienna ab initio simulation package (VASP) in chemical applications. J Mol Struct: Theochem 2003;624:37–45.
[57] Perdew J, Burke K, Ernzerhof M. Perdew, burke, and ernzerhof reply. Phys Rev Lett 1998;80:891.
[58] Blöchl PE. Projector augmented-wave method. Phys Rev B 1994;50:17953.
[59] Politzer P, Murray JS. The Hellmann-Feynman theorem: a perspective. J Mol Model 2018;24:1–7.
[60] Wei W, Dai Y, Huang B, Jacob T. Many-body effects in silicene, silicane, germanene and germanane. Phys Chem Chem Phys 2013;15:8789–94.
[61] Yamamoto K, Horiba K, Taguchi M, Matsunami M, Kamakura N, Chainani A, et al. Temperature-dependent Eu 3 d− 4 f x-ray absorption and resonant photoemission study of the valence transition in $EuNi_2$ (SiO_2GeO_8) 2. Phys Rev B 2005;72:161101.
[62] Tapasztó L, Dobrik G, Lambin P, Biro LP. Tailoring the atomic structure of graphene nanoribbons by scanning tunnelling microscope lithography. Nat Nanotechnol 2008;3:397–401.
[63] Khandelwal M, Kumar A. One-step chemically controlled wet synthesis of graphene nanoribbons from graphene oxide for high performance supercapacitor applications. J Mater Chem 2015;3:22975–88.
[64] Wang W-X, Zhou M, Li X, Li S-Y, Wu X, Duan W, et al. Energy gaps of atomically precise armchair graphene sidewall nanoribbons. Phys Rev B 2016;93:241403.
[65] O'Donnell M, Jaynes E, Miller J. Kramers–Kronig relationship between ultrasonic attenuation and phase velocity. J Acoust Soc Am 1981;69:696–701.
[66] Penn DR. Wave-number-dependent dielectric function of semiconductors. Phys Rev 1962;128:2093.
[67] Roche SM, McConkey KM. Absorption: nature, assessment, and correlates. J Pers Soc Psychol 1990;59:91.
[68] Moses H. Calculation of the scattering potential from reflection coefficients. Phys Rev 1956;102:559.
[69] Nagatomi T, Shimizu R, Ritchie R. Energy loss functions for electron energy loss spectroscopy. Surf Sci 1999;419:158–73.
[70] Singh J. Optical properties of condensed matter and applications. John Wiley & Sons; 2006.
[71] Han NT, Dien VK, Lin M-F. Excitonic effects in the optical spectra of Li_2SiO_3 compound. Sci Rep 2021;11:1–10.
[72] Dien VK, Pham HD, Tran NTT, Han NT, Huynh TMD, Nguyen TDH, et al. Orbital-hybridization-created optical excitations in Li_2GeO_3. Sci Rep 2021;11:1–10.

CHAPTER

11

Germanene on Ag (111)

Shih-Yang Lin[1], Thi Dieu Hien Nguyen[1], Vo Khuong Dien[1], Chi-Hsuan Lee[1], Hsin-Yi Liu[1], Hai Duong Pham[1], Thi My Duyen Huynh[1], Nguyen Thi Han[1], Ngoc Thanh Thuy Tran[2], Wei-Bang Li[1] and Ming-Fa Lin[1]

[1]Department of Physics, National Cheng Kung University, Tainan, Taiwan; [2]Hierarchical Green-Energy Materials (Hi-GEM) Research Center, National Cheng Kung University, Tainan, Taiwan

11.1 Monolayer germanene on Ag (111)

A monolayer germanene on Ag (111) presents the feature-rich geometric and electronic properties, mainly owing the non-uniform atomic bonding environments in the specific Moire super lattices, the greatly enhanced bucklings, and the significant Ge–Ag orbital hybridizations. The substrate effects are taken into consideration by using the sufficiently depth silver layers. The four-layer Ag atoms are utilized in the first-principles calculations, as clearly illustrated in Fig. 11.1. The optimal geometry obviously shows that there exist 72 atoms enclosed under a greatly enlarged unit cell due to the 3×3 and 4×4 mismatched structure in the Ge- and Ag-layers, respectively. They exhibit good agreement with the experimental results [1–5]. The lattice constant on the (x, y) plane is ~11.56 Å. The increasing bond lengths, height differences between two interlayer sub lattices, and the fluctuation in interlayer distances for two neighboring layers directly reflect the above-mentioned mechanisms. The intra layer Ge–Ge/Ge–Ag/Ag–Ag bond lengths are varied within an enlarged unit cell, where they are in the ranges of 2.324–2.517/2.682–3.035/2.918–3.337 Å's. The height differences of the first-layer germanene and silver are, respectively, ~0.530 and 0.486 Å's, in which three Ge-atoms in the former even reach a relatively longer length of ~1.18–1.33 Å's. The strengthened non-orthogonality of π and σ bondings in germanene and the increased sp^3 bondings account for the changes. It should be noticed that the height of buckling drastically decrease in the second-layer silver and the others. The interlayer distances, which correspond to the Ge–Ag layers, the first Ag layers, the second Ag layer, and the third Ag layers, are

Fundamental Physicochemical Properties of Germanene-related Materials
https://doi.org/10.1016/B978-0-443-15801-8.00003-7

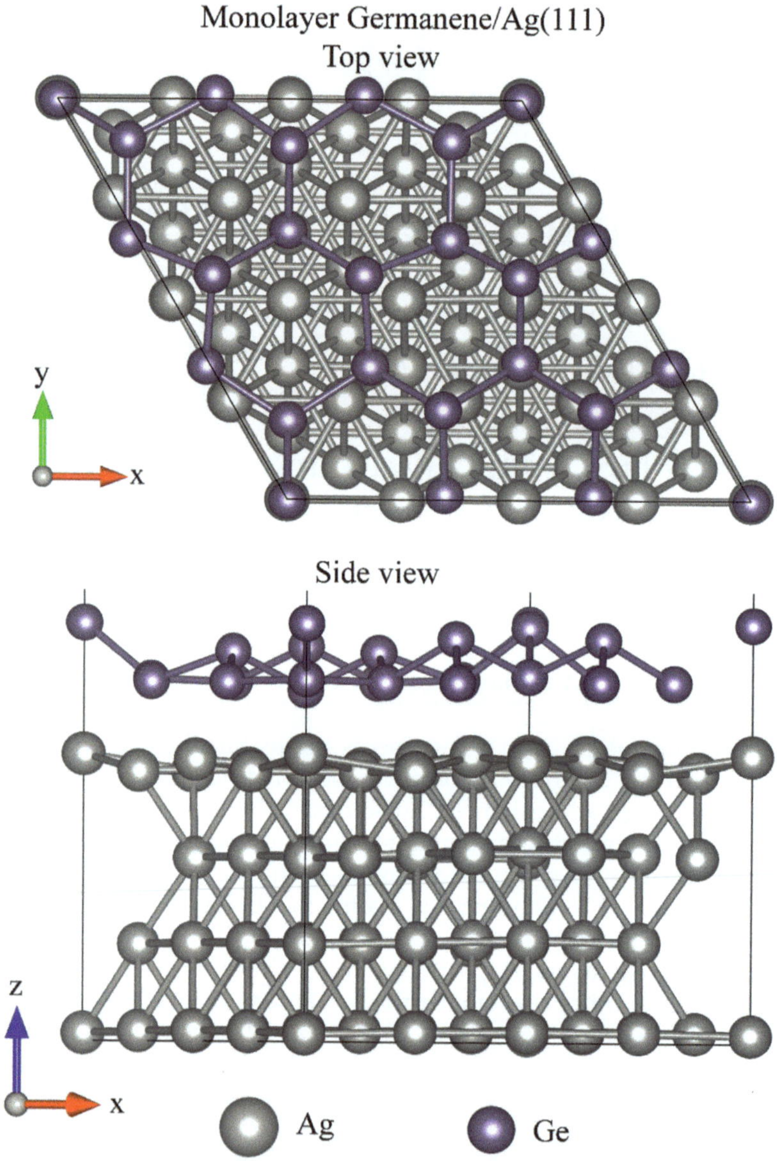

FIG. 11.1 The optimal geometric structure of germanene/Ag (111) with a four-layer substrate in the presence of top and side views, and the enclosed black lines corresponding to a Moire superlattice (3 × 3, 4 × 4). The blue and gray balls, respectively, correspond to the germanium and silver atoms.

about 2.096–3.449, 2.341–2.698, 2.243–2.562, 2.370–2.581 Å's, respectively. Gernerally speaking, they are larger than/comparable to those in bilayer germanene systems. A signifiican fluctuation is shown in the heterojunction so that the Ge–Ag bonds are expected to provide the strong multi-orbital hybridizations. The diversified chemical environments come to exist in the intra layer and interlayer atomic interactions, relating to the specific Moire super lattices; hence, these results become the high barriers in simulating this system through the tight-binding model. The above-mentioned theoretical predictions on the main geometric features could be further examined by the high-resolution STM [6–11], TEM [12–17] and LEED [8,18–23] measurements (details in Chapter 3).

The Ge-adlayer/Ag (111) compound presents an unique band structures, as clearly shown in Fig. 11.1. A pristine germanene possesses more energy subbands for a greatly reduced first Brillouin zone due to a zone-folding effect of the specific 3 × 3 Moire super lattice. The $3p_z$ orbitals across the Fermi level at the Γ point can be identified by the slightly separated Dirac-cone structures. Moreover, they are doubly degenerated. The $3p_z$-π and ($3p_x$, $3p_y$)-σ valence bands might be mixed together through the anti-crossing behaviors with the decreasing state energy (details in Chapter 4). Specifically, 1/3 and 2/3 of electronic valene states are dominated by $3p_z$ and ($3p_x$, $3p_y$), respectively. In detail, the deeper valence band near the Γ point comes from the π-band bottom state.

The pristine zone-folding germanene states are changed drastically under the strong substrate effects, as clearly displayed in Fig. 11.1A–D. Extreme points, saddle points, and plenty of band-edge states are created in the energy-wave-vector space of the presence of hetero junction. The interlayer atomic interactions between the neighboring Ge-/Ag–Ag layers induce asymmetric spectra of electron and hole energy subbands about the Fermi level. This system shows more valence and conduction energy subbands because of many silver atoms in the specific Moire super lattice compared with a pristine monolayer germanene. In other words, there exists a quite complicated electronic structure.

The low-lying energy spectrum reveals four energy subbands with the non-monotonous energy dispersions along the high-symetry directions. They cross the Fermi level; hence, the occupied and unoccupied states come to exist within those subbands. This unusual result obviously suggested the semi-metallic behavior in the low 2D free carrier density, while a monolayer germanene belongs to the narrow-gap semiconductor. Within $|E^{c,v}| < 1.0$ eV, the low-energy valene and conduction bands are almost determined by silver atoms; hence, the substrate is expected to make main contributions to the other essential properties. The importance of the Ag–Ag chemical bondings in the complete band structure of a Ge-adlayer/Ag (111) compound could be illustrated by this phenomenon. Under the Ge–Ag interlayer interactions at the heterojunctions, the Dirac-cone structures at the Fermi level are completely absent. The π chemical bondings are continuously extended on a germanene honeycomb lattice, which are destroyed by the Ge–Ag bond dominated multi-orbital hybridizations. As a result, it is hard to define the band widths of π- and σ-band. Also, the simulation of the tight-binding model by using the first-principles band structure can hardly performed.

The entire band structure could be classified into several energy ranges, according to the germanium and silver dominances (the red/blue and brown/gray circles in Fig. 11.2A and B and Fig. 11.2C and D). The electronic energy spectrum in the ranges of $-10.0 \text{ eV} \leq E^v \leq -6.00$ eV, $-6.00 \text{ eV} \leq E^v \leq -2.50$ eV, and $-2.50 \text{ eV} \leq E^v \leq 5.00$ eV, are dominated by the Ge, Ag, and (Ge, Ag) atoms, respectively.

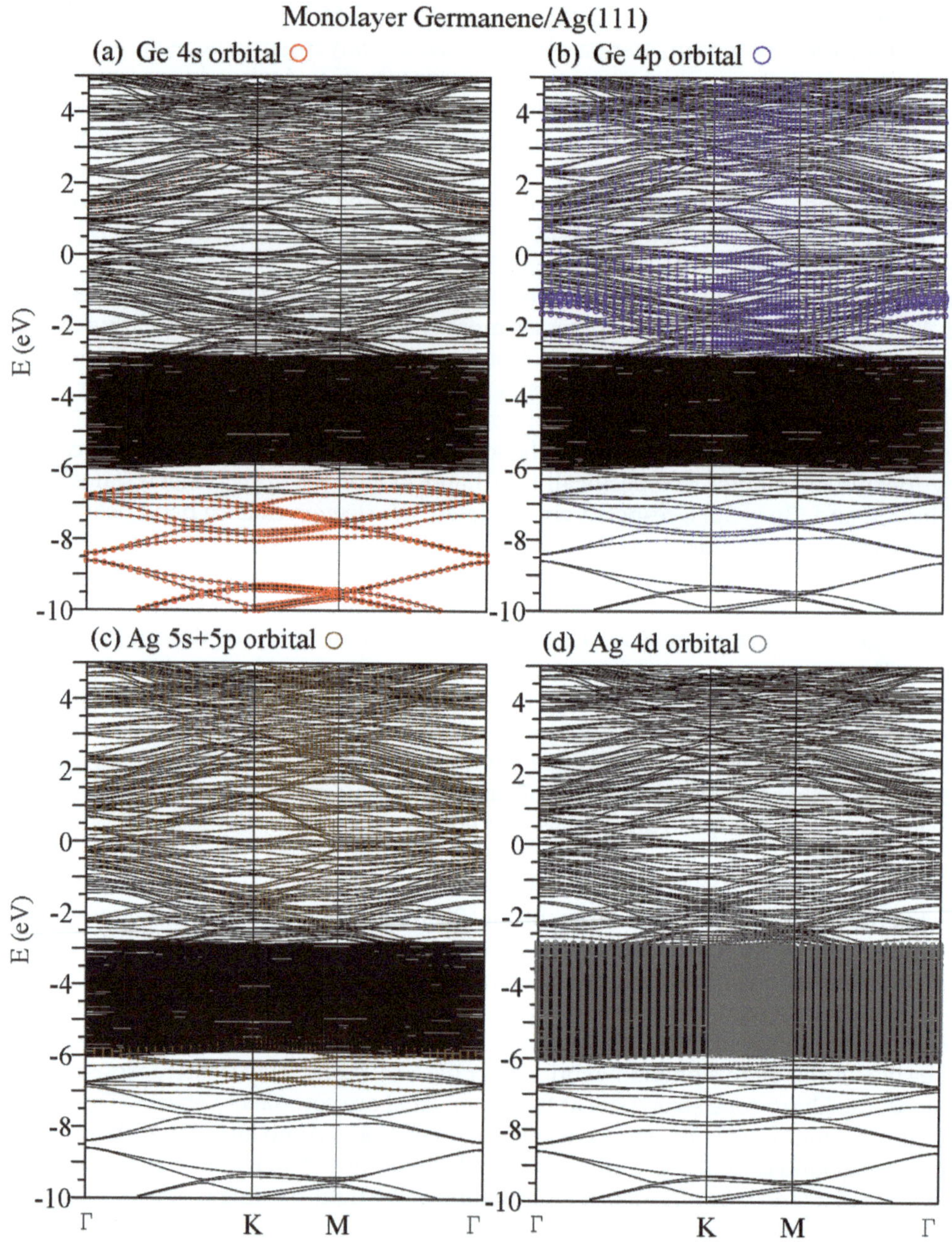

FIG. 11.2 Band structures for the Ge-adlayer/Ag (111) compound, where the (A) Ge 4s orbital, (B) Ge 4p orbital, (C) Ag 5s + 5p orbital, and (d) Ag 4d orbital dominance is shown in different colors.

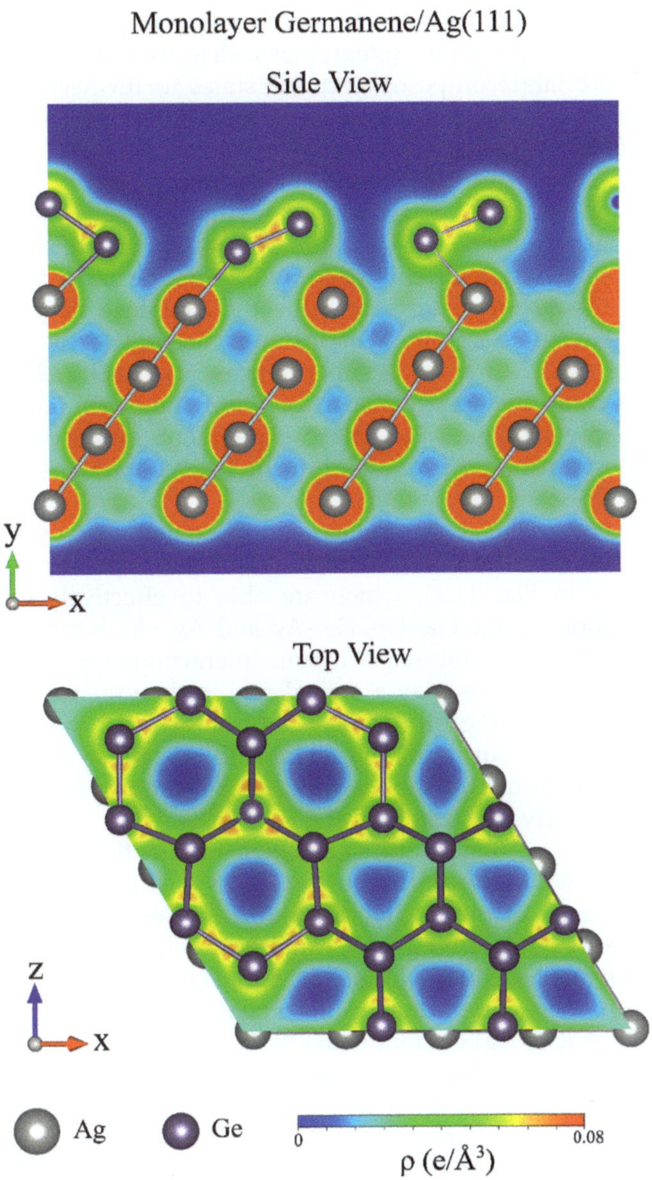

FIG. 11.3 The (A) top view and (B) side view of spatial charge density in germanene/Ag (111).

The similar atom dominances are also exhibited in the AA and AB bilayer cases except for the low-lying valence and conduction bands (discussed later). The d-orbital Ag-layer dominance is very strong in the above-mentioned ranges, as clearly shown in Fig. 11.2D. In addition, among all four silver layers, each silver layer makes a comparable contribution to electronic states, and the first Ag-layer is the only one not significant in contributing to the essential electronic properties. The atom-dominated band structure in Ge monolayer on Ag (111) is closely related to intralayer Ge–Ge & Ag–Ag hopping integrals as well as the interlayer Ge–Ag and Ag–Ag bonds. The atom- and orbital-decomposed electronic states are further examined by the important van Hove singularities in density of states, and will be discussed later in Fig. 11.4. Moreover, the similarity of atom dominances could be found in the bilayer case (discussed later in Figs. 11.6 and 11.10) except for the low-energy bands across the Fermi level.

The complicated band structures and Moire super lattices exist a close relation. Several stable geometries of Ge monolayer on Ag (111) compounds are identified by the experimental examinations [1–5]. They possess the enlarged unit cells on the germanium and silver layers. The zone-folding and substrate effects are directly combined together, leading to the geometry-diversified electronic spectra. For instance, according to the VASP simulation [24] in Fig. 11.2 under the 3×3 & 4×4 super lattices, it is hardly observing the modified structure of the Dirac-cone, the energy dispersions near the Fermi level, the specific Fermi momenta, and the narrow band gap. The ARPES experiments on the complete valence bands (details in Chapter 3) might become a quite difficult way as the non-monotonous **k**-dependences across E_F.

The drastic changes of spatial charge density are presented after the formation of heterojunction in Fig. 11.3. This directly reflect the atom- and orbital-decomposed van Hove singularities (discussed later in Fig. 11.4), which are able to effectively examine the cruicial multi-orbital hybridizations in the Ge–Ge, Ge–Ag and Ag–Ag bonds.

In addition, the interlayer and intralayer atomic interactions simultaneously occur in the Ag–Ag bonds. The Ge–Ag bonds play a critical role in the unusual carreri density, originating from the enhance buckling of Ge-honeycomb lattice and the great deformation on the first Ag-layer (Fig. 11.1). Four germanium atoms (blue balls in a unit cell) of the A and B sublattices have the orbital hybridizations with three silver atoms (white balls) at the specific heterojunction, as clearly illustrated by the (x, z)-plane projection in Fig. 11.3. Interestingly, the first pair in A and B sub lattices exhibits the identical chemical bondings with the first layer Ag-atom, while the third/fourth Ge-atoms correspond to the second/third Ag-ones. Hence, two kinds of orbital hybridizations in Ge–Ag bonds are obtained. The high charge densities are exhibited between Ge and Ag atoms. They are deduced to be transferred from the original (4s, $4p_x$, $4p_y$, $4p_z$), and (5s, 5p, 4d) orbitals (the four and nine orbitals). This mechanism thus become complicated multi-orbital hydizations of four-nine bonding, is further supported by the (Ge, Ag)-co-dominated band structure under certain energy ranges (e.g., $-2.0\ \text{eV} \leq E^v \leq 5.0\ \text{eV}$ in Fig. 11.2B and C). In general, π and σ bondings in Ge-honeycomb lattice are largely affected by the effect of substrate, hence, their well-defined behaviors are completely absent in the essential properties. This viewpoint become not suitable to applicate in Ag-substrate-related germanene systems. The charge density is relatively high in term of the wide and deep red region near Ag-atoms. This mainly comes from 11 valence electrons in each silver. The significant carrier densities are exhibited in the nearest- and second-nearest Ag–Ag atomic interactions (the green/light blue regions). They turn to be the weakest one among the Ag–Ag, Ag–Ge and Ge–Ge bonds.

The complicated band structure in Fig. 11.2 has its main features directly reflected in the 2D van Hove singularities, as clearly illustrated in Fig. 11.4. In general, the band-edge states correspond to the local maximum and local minimun points, saddle-points, and partially flat bands, which are capable of generating the shoulder structues, symmetric logarithmically peaks, asymmetric square-root peaks, and delta-function-like peaks. There exist two

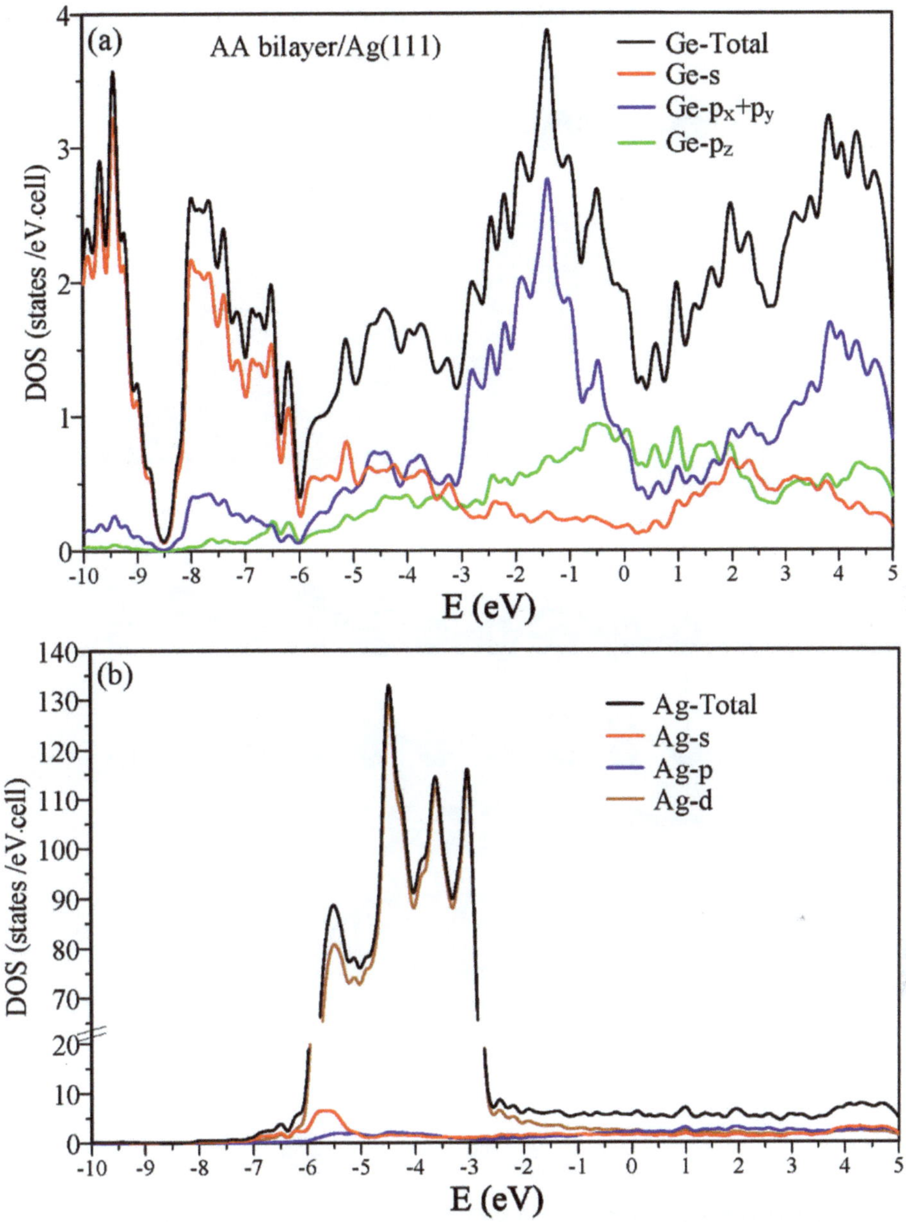

FIG. 11.4 The atom- and orbital-projected density of states in the Ge-adlayer/Ag (111) system: the contributions arising from (A) the total, 3s-, $3p_x$ + $3p_y$- & $3p_z$-Ge orbitals (the black, red, blue & green curves, respectively), and (B) the total, 5s-, 5p- & 4d-Ag orbitals, in which the $5p_x$, $5p_y$, and $5p_z$ make the comparable contributions.

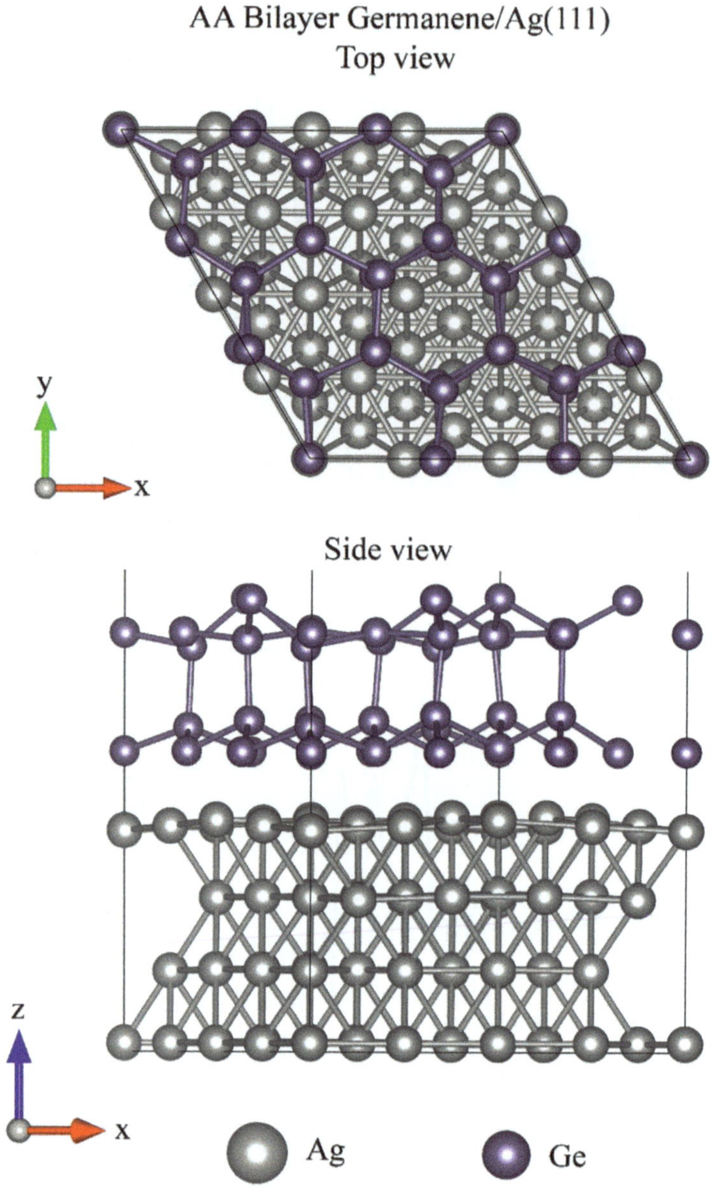

FIG. 11.5 The optimal geometric structure of AA stacking bilayer germanene/Ag (111) with a four-layer substrate in the presence of top and side views, and the enclosed black lines corresponding to a Moire superlattice (3 × 3, 4 × 4). The blue and gray balls, respectively, correspond to the germanium and silver atoms.

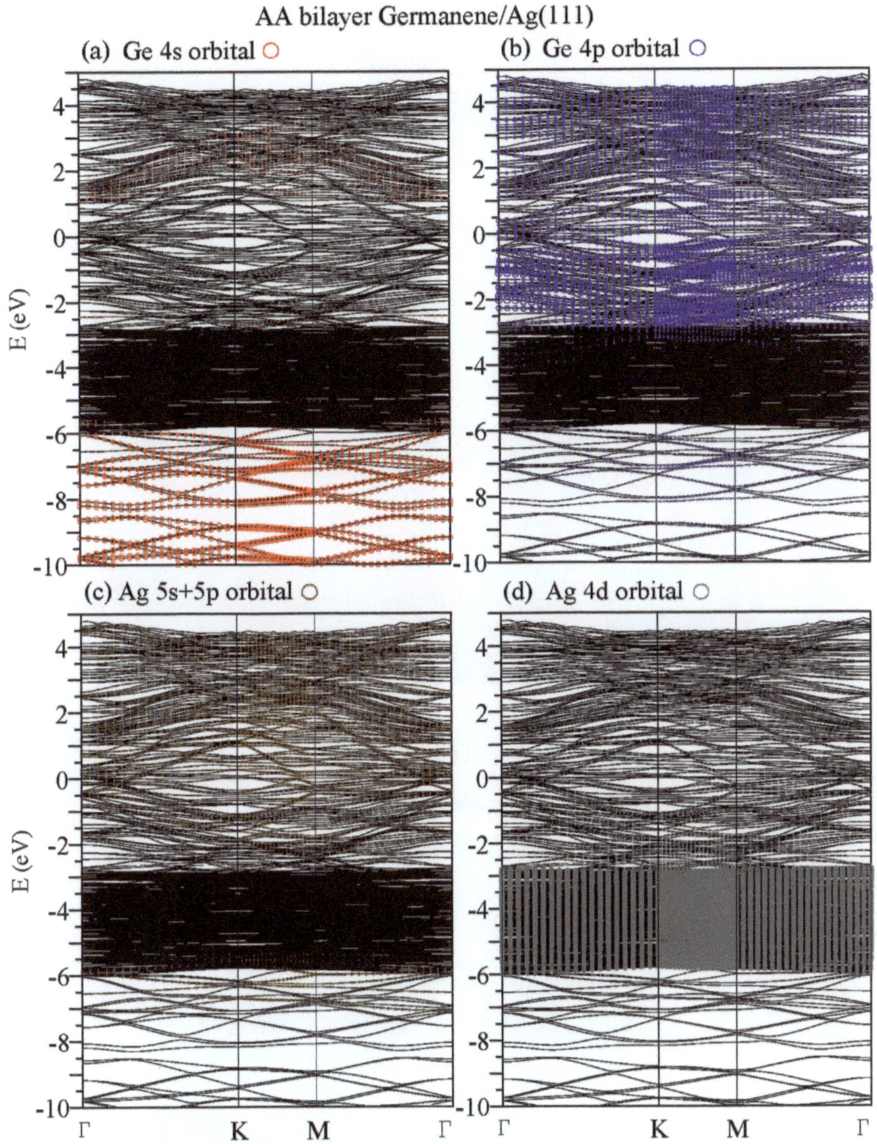

FIG. 11.6 Electronic energy spectra for the AA stacking bilayer Ge/Ag (111) compound, where the (A) Ge 4s orbital, (B) Ge 4p orbital, (C) Ag 5s + 5p orbital, and (D) Ag 4d orbital dominance is shown in different colors.

important van Hove singularities to well distinguish the neighboring peak structures; that is, only the broadening and merged peaks can be observed in the calculated results, such as the Ge- and Ag-atom contributions by the black curves in Fig. 11.4A and B, respectively.

The above-mentioned four types of van Hove singularities become unique due to the strong effects of zone-folding, enhancing the critical points and thus the peaks structures.

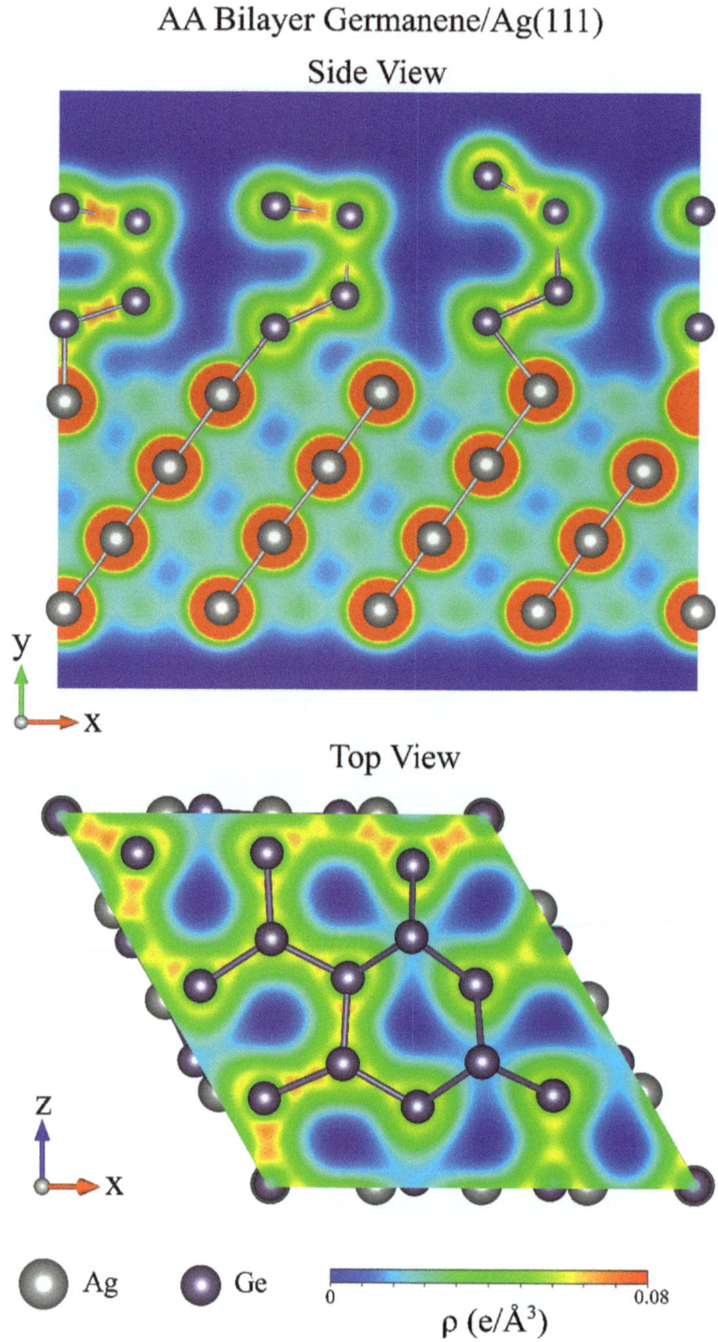

FIG. 11.7 The (A) top view and (B) side view of spatial charge density in AA stacking bilayer germanene on Ag (111).

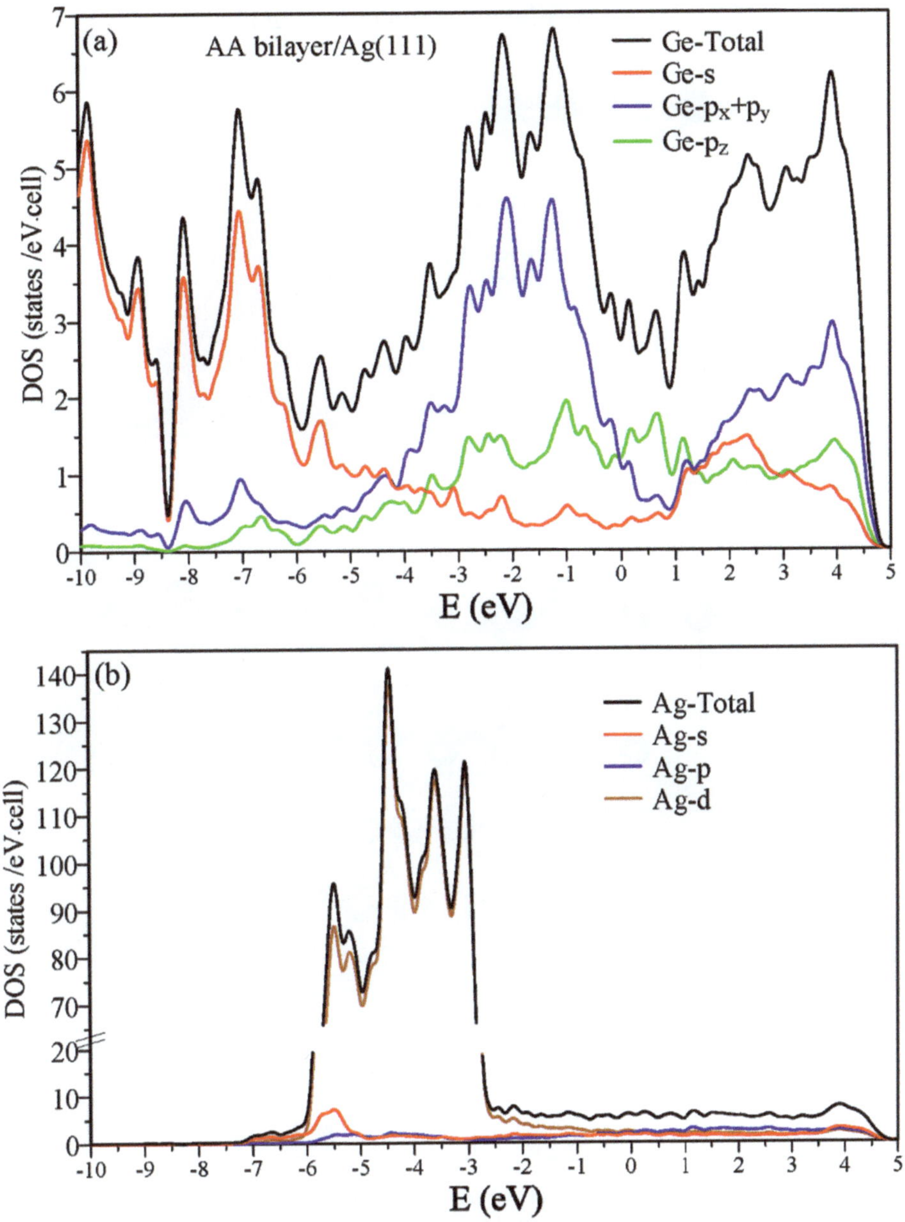

FIG. 11.8 The atom- and orbital-projected density of states in the AA stacking bilayer Ge on Ag (111) system: the contributions arising from (A) the total, 3s-, $3_{px}+3_{py}$- & 3_{pz}-Ge orbitals (the black, red, blue & green curves, respectively), and (B) the total, 5s-, 5p- & 4d-Ag orbitals, in which the 5_{px}, 5_{py}, and 5_{pz} make the comparable contributions.

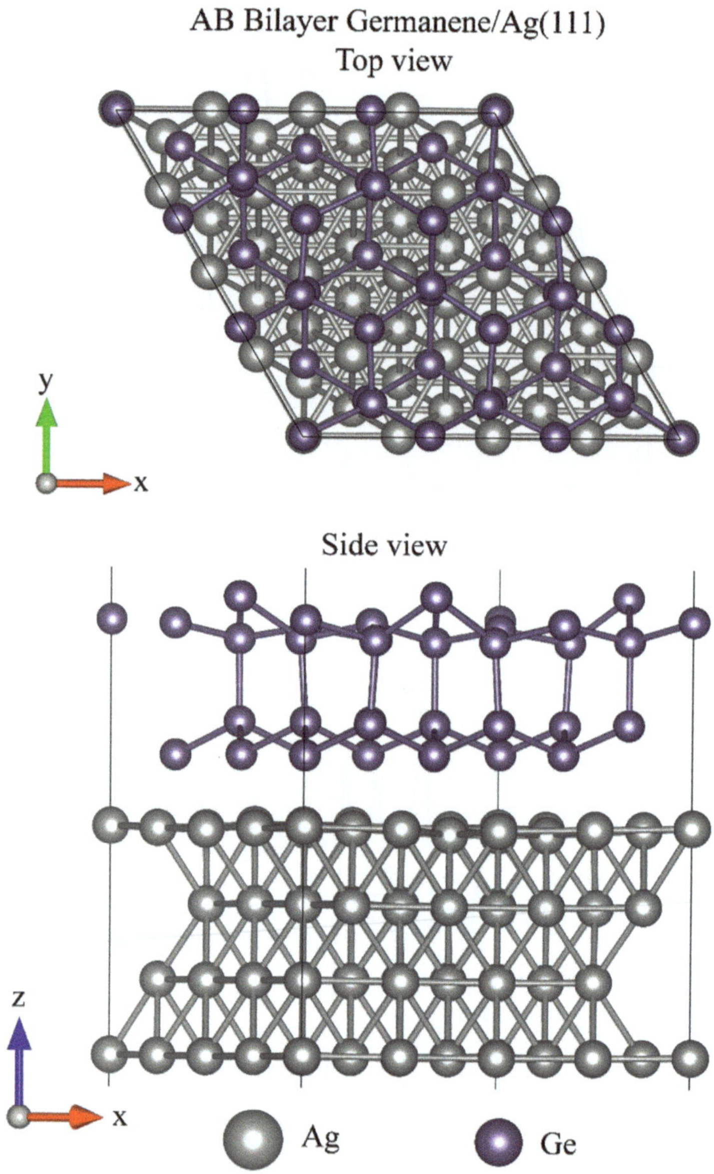

FIG. 11.9 The optimal geometric structure of AB stacking bilayer germanene on Ag (111) with a four-layer substrate in the presence of top and side views, and the enclosed black lines corresponding to a Moire super lattice (3 × 3, 4 × 4). The blue and gray balls, respectively, correspond to the germanium and silver atoms.

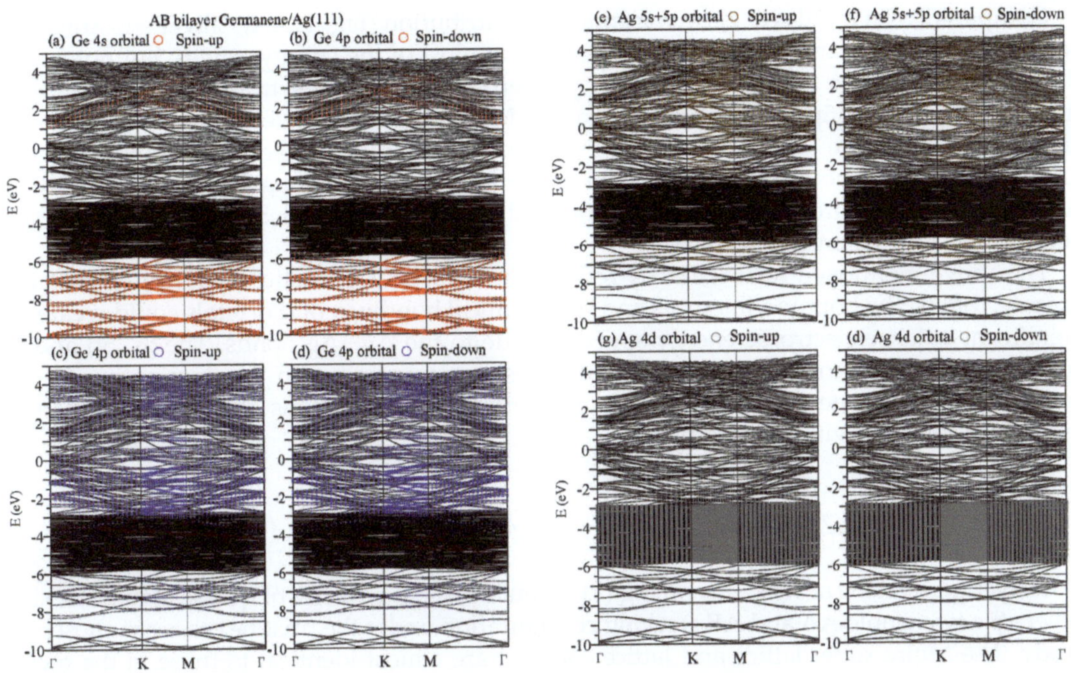

FIG. 11.10 Electronic energy spectra for the AB stacking bilayer Ge/Ag (111) compound with the spin-up (A) Ge 4s orbital, (C) Ge 4p orbital, (E) Ag 5s + 5p orbital, and (G) Ag 4d orbital dominance, and (B), (D), (F), (H) for the spin-down ones.

Density of states is small and possess a finite value at the Fermi level, indicating it's a semi-metal. This property is similar to those of few layer graphenes [25–29] and bernal graphite [30–33]. The V-shape structure near the Fermi level thoroughly disappear, originating from the modified Dirac cone. Moreover, the distribution range of Ge-3p_z is wider than 10 eV (the green curve), being much larger than that (~6 eV) in monolayer germanene (discussed in Chapter 4). The absence of the well-defined π-electronic energy bands & chemical bonding are consistent with the band structure and charge distribution (Figs. 11.2 and 11.3). The σ-electronic orbitals exhibited similar properties, in which their 3s and 3p_x + 3p_y orbitals are presented by the red and blue curves. The (3s, 3p_x, 3p_y, 3p_z)-orbitals of germanium atoms provide the significant contributions at the same energies to the special structures except for $E < -7.0$ eV. In addition, the 3s orbitals dominate the energy spectrum at deeper energies of the Ge monolayer on Ag (111); hence, the third σ-electronic should survive under the strong germanium-substrate interactions.

Compared with monlayer germanium-atom contribution (Fig. 11.4A), siliver substrate present the dominating and co-dominating features in most of the energy ranges, but not for the deeper energy range $E < -7.0$ eV. This property mainly comes from the nine kinds of different orbitals (5s, 5p, 4d) in each silver atom. Many Ag atoms under a specific Moire supelattice also contribute to such property. Furthermore, the nine 4d-electrons make the major and important contributions. They occupied energy spectrum among 5s, 5p and 4d orbitals (the red, blue and brown curves). It is noticed that $5p_x$, $5p_y$ and $5p_z$ are almost equally contributed to the density of states.

The above-mentioned result directly reflects the main characteristics of transition metal elements, that is, the hopping integrals and the chemical bondings of d orbitals could create the significant valence-electronic energy spectrum due to the Ag–Ag bonds. The 5s, 5p and 4d orbitals possess the comparable density of states accompanied with the dominating Ge-$3p_z$ orbitals. This illustrates the complicated multi-orbital hybridizations of the hetero-junctions systems associated with Ge–Ag bonds.

11.2 AA and AB bilayer germanene on Ag (111)

Bilayer germanene subsystems on Ag (111) could present a plenty of quaGe-stable stacking, where the two stable AA and AB stacking configuration under the pristine cases are chosen for study. The Moire super lattice and lattice constant are almost identical to those in the monolayer germanene on Ag (111) substrate system, as clearly indicated in Fig. 11.5. There exist the non-uniform Ge–Ge/Ge–Ag/Ag–Ag bond lengths within the ranges of 2.335–2.398/2.513–2.892/2.863–2.997 Å's. The height differences are revealed in the first and second Ge-layers and the nearest Ag-layer. They are ~0.873/~0.434/~0.335 Å's. The buckling is significantly decreases in the first Ge layer compared with a monolayer sytstem (Fig. 11.1). The interlayer distances lie in the ranges of 2.432–2.456/2.138–3.315/2.236–2.689/2.334–2.533/2.350–2.506 Å's. The sufficiently short distances between atoms and a significant fluctuation in the heterojunction suggests that the interlayer interactions should be the multi-orbital hybridizations in a strong manner. They cannot be the weak van der Waals interactions. Besides the optimal geometric structure, the number of layers, various stacking sequences and different substrate play important roles in the other essential properties.

11.2.1 Bilayer AA germanene on Ag (111)

A bilayer AA germanene presents a unique band structure, as discussed in Chapter 4. The main features include the dominance of spin-orbital couplings on the low-energy valence and conduction bands, an indirect band gap, the spin-up/spin-down degenerate states, the vanished modified Dirac-cone, the oscillatory dispersions, and the undefined π and σ valence bands due to the non-van der Waals forces. However, when this pristine bilayer germanene possess strong interlayer or intra layer interactions with the Ag (111) substrate, the above-mentioned electronic properties are dramatically changed, as indicated in Fig. 11.6. The bilayer AA germanene on Ag (111) compound presents the spin-degenerate valence and conduction bands, as clearly shown in Figs. 11.6A–D. This is similar to the spin-degenerate properties of the Ge monolayer on Ag (111) substrate system (Fig. 11.2). The low-energy valence bands and conuduction bands within the range of $-2.5\ eV \leq E^{c,v} \leq 4.5\ eV$, which cross the Fermi levels (Fig. 11.6C), are dominated by occupied and unoccupied by Ge-4p electronic states with the direction-dependent Fermi momenta. As for the Ag-dominated bands, they present the d-orbit-dominant bands within the range of $-6.0\ eV \leq E^{c,v} \leq -2.5\ eV$, which similar to the Ge-monolayer/Ag (111) case. This indicates that the relatively stable Ag substrate and the Ag–Ag bonds are hardly affected by the most top layer of Ge.

To fully understand the effects of substarte on bilayer AA germanene, the orbital hybridizations of Ge–Ge, Ge–Ag, and Ag–Ag chemical bonds are directly examined from the spatial charge distributions, as clearly displayed in Fig. 11.7. The critical Ge–Ag bonds at the hetero junction come to exist in the unusual form under the (x, z) projection, where six germanium atoms on A and B sub lattices exhibit the very strong multi-orbital hybridizations with silver ones. There exist the higher carrier densities in the Ge–Ag bonds of the bilayer AA germanene/Ag (111) (yellow region). This has greatly emhanced the Ge-atom dominance of the low-lying energy bands with $-2.5\ eV \leq E^{c,v} \leq 4.5\ eV$ (Fig. 11.6B). Part of the π- and σ-electronic orbitals are transferred from the first germanene layer to the hetero junction, leading to the destruction of the well-behaved chemical bondings. Interlayer Ge–Ge atomic interactions also induce the distorted π and σ bondings, as indicated by the red regions between two germanene layers. The significant interlayer Ge–Ag and Ge–Ge orbital hybridizations are responsible for the undefined π and σ states, e.g., the absence of a modified Dirac cone in band structure and a V-shape van Hove singularity (Figs. 11.6 and 11.9A). The multi-orbital hybridizations are identified from the atom- and orbital-decomposed density of states and the atom-dominated band. As a result, Ge–Ag and Ge–Ge bonds, respectively, belong to the ($3s$, $3p_x + 3p_y$, $3p_z$)-($5s$, $5p$, $4d$) and sp^3 bondings. The Ag–Ag bonds behave similarly for both bilayer- and monolayer-related systems, as shown in Figs. 11.7 and 11.3. Their high and wide carrier density exhibits near each Ag atom (red region). In addition, such complicated bondings mainly come from the multi-orbital hybridizations of d orbitls.

The atom- and orbital-projected density of states (Fig. 11.8A and B) is available in determining the orbital hybridizations in Ge–Ge, Ge–Ag and Ag–Ag bonds. The ($3s$, $3p_x + 3p_y$, $3p_z$) and ($5s$, $5p$, $4d$) orbitals arise from germanium and silver atoms, respectively, resulting in the significant number of van Hobe singularities. In general, the important difference among the orbital-decomposed special structures lies in the intensities. The π-band width and the modified *V*-shape across the Fermi level are absent, as shown in Fig. 11.8A. The results which similar to monolayer/Ag (111) also include the 3s-σ-electronic states about $E < -7.0\ eV$ and the 4d-dominated valence states within $-6.0\ eV < E < -2.5\ eV$.

11.2.2 AB stacking germanene on Ag (111)

The Moire superlattice and lattice constant of AB bilayer germanene on Ag (111) are almost identical to those in the AA substrate system, as shown in Fig. 11.9. A bilayer AB germanene on Ag (111) compound presents the spin-split valence and conduction bands, as clearly shown in Figs. 11.10A–D. This is the main difference from the spin-degenerate properties of the Ge monolayer and AA bilayer on substrate system. The spin splittings are entered in the whole energy range. According to the VASP simulations, the magnetic moment within a Moire unit cell, which is competed by the occupied spin-up and spin-down states, is ~1.17 μB. Such calculated value should be sufficiently large for the experimental SQUID verifications [34,35].

The ferromagnetic configuration in bilayer AB germanene on Ag (111) compound is further analyzed from the spatial charge distributions and spin density distribution, as illustrated by Figs. 11.11 and 11.12, respectively. The Ge–Ag and Ge–Ge orbital hybridization of AA systems are found to be similar in AB stacking system (Fig. 11.1). They belong to the (3s, $3p_x + 3p_y$, $3p_z$)-(5s, 5p, 4d) and sp^3 bondings. The Ag–Ag bonds behave similarly for both AA and AB bilayer- and monolayer-related systems. On the other hand, there exists an extremely non-uniform spin arrangement for AB system, being a distinct feature among these three systems. The spin-up electronic states mainly originate from the six germanium atoms away from the hetero junction, while the other Ge and Ag ones only make little distributions. In addition, the nine 4d orbitals of Ag atoms do not have the prominent contributions. Hence, they are unable to create the unusual magnetic phenomena. It is noted that the total summation of magnetic moments from the first-layer germanene and four-layer silver might affect the net value, that is, their spin-dependent phenomena are non-negligible.

The high-resolution ARPES and/or SP-ARPES experiments are suitable in testing the theoretical predictions of the significant substrate effects on electronic energy spectra (details in Chapter 3). The existence of π valence band close to the Fermi level clearly indicates the significant Ge–Ag atomic interactions and the Ge-atom dominance in the enlarged unit cell. The various optimal geometries are able to create the diversified band structures. The current 3 × 3 & 4 × 4 cells of the Ge-adlayer and bilayer AA and AB germanene on substrate Ag (111) present the rich features, including the vanishing Dirac-cone, the non-monotonous energy band dispersions, the semi-metallic behavior, and the silver-dominated bands. The experimental examination could provide part of useful informations about the intralayer and/or interlayer Ge–Ge, Ge–Ag, and Ag–Ag chemical bondings.

The atom- and orbital-projected density of states combined with the atom-dominated and spin-split band structure, as shown in Fig. 11.13A and B. This is suitable in deciding the critical hybridizations of Ge–Ge, Ge–Ag and Ag–Ag orbitals, and thus the ferromagnetic spin configuration. The Ge-/Ag- and spin-dominated density of states near the Fermi level can determine the magnetic behavior. After the organize the atom-, layer- & orbital-decomposed magnetic moment, the spatial spin distribution, and the atom- & orbital-projected density of states, the $3p_z$ orbitals of the specific second-layer six germanium atoms are predicted to play critical roles in the diversified magnetic phenomenon. In summary, the bilayer AB germanene/Ag (111) and AA ones system exhibit the ferromagnetic and non-magnetic behaviors, respectively, where the spin-dependent many- and single-particle interactions need to be included in the reliable simulations of the phenomenological models.

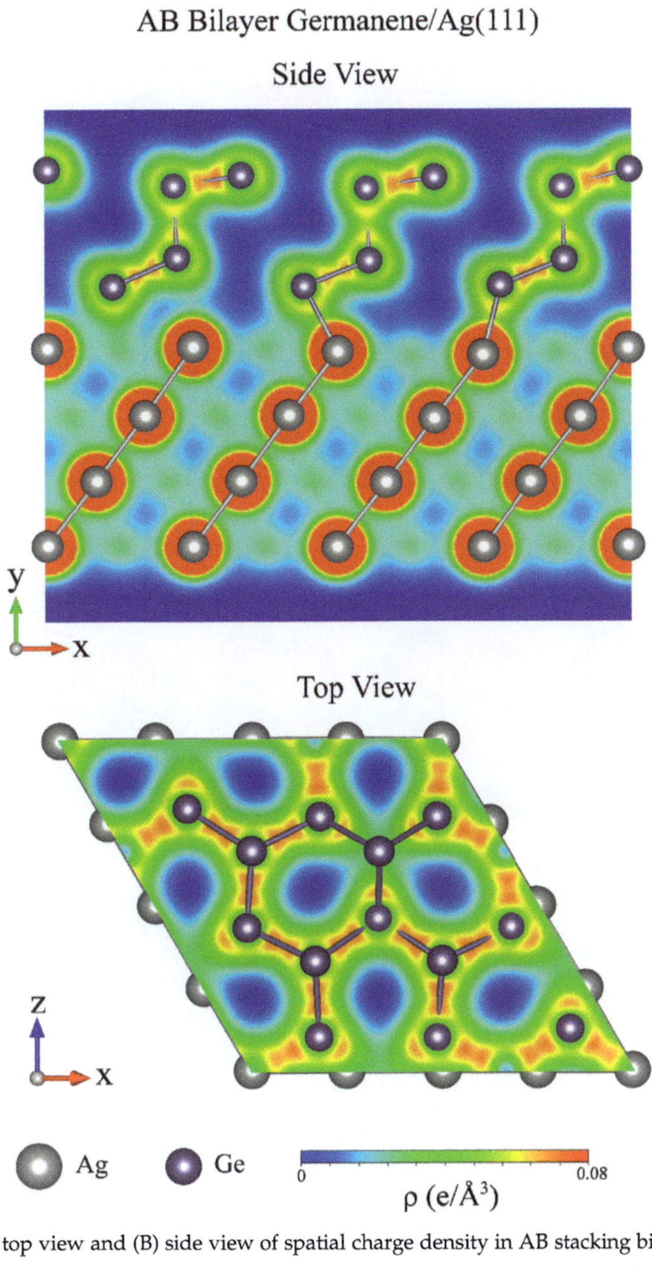

FIG. 11.11 The (A) top view and (B) side view of spatial charge density in AB stacking bilayer germanene on Ag (111).

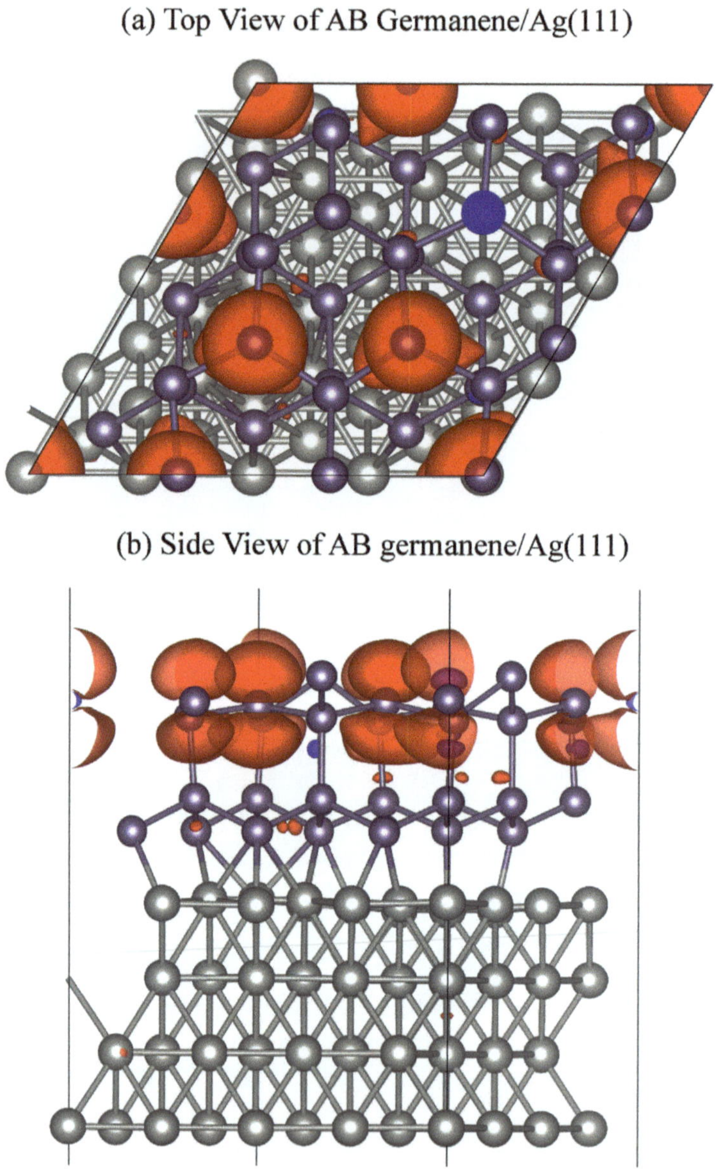

FIG. 11.12 The (A) top view and (B) side view of spin density in AB stacking bilayer germanene on Ag (111).

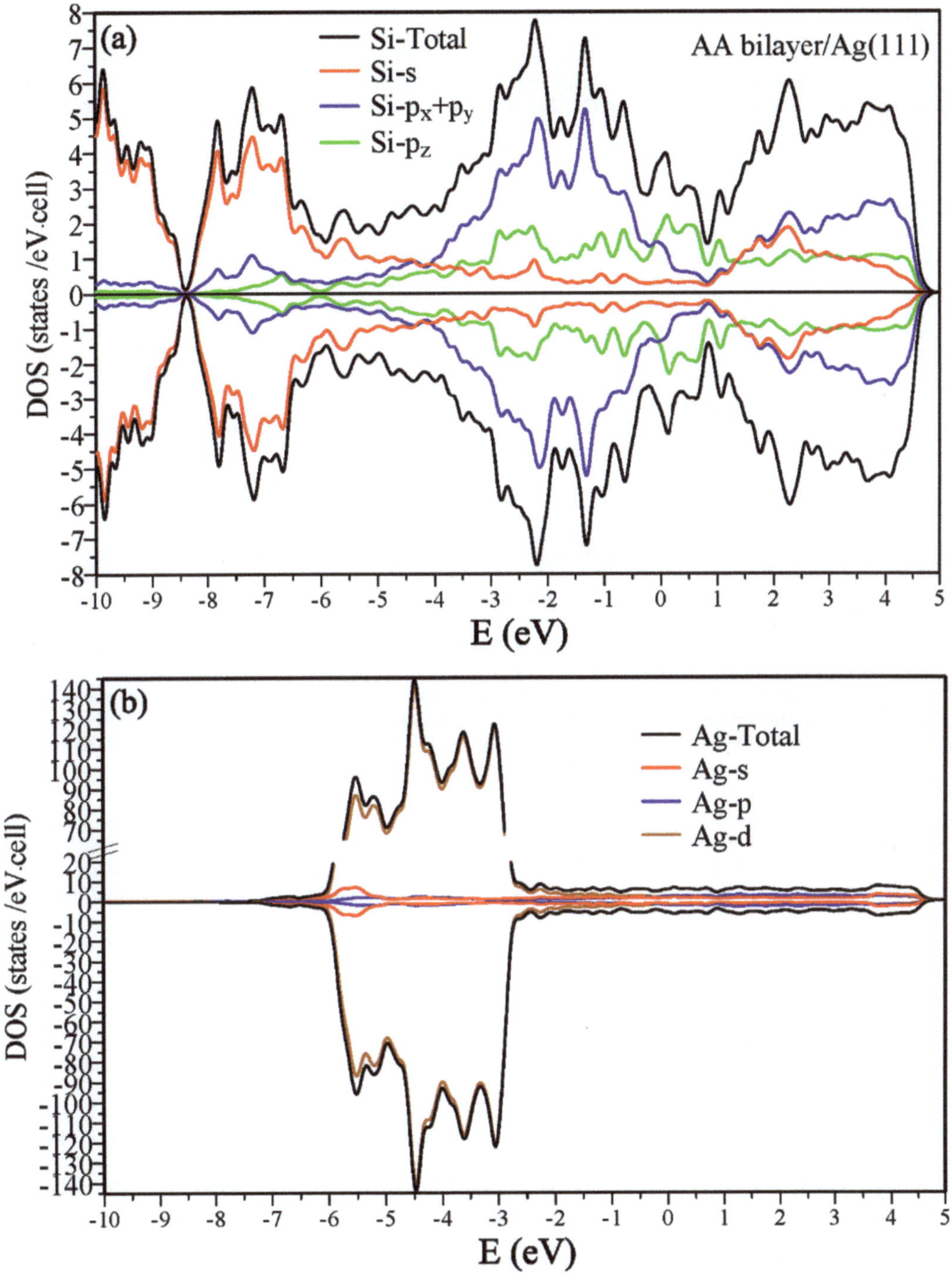

FIG. 11.13 The atom- and orbital-projected density of states in the AB stacking bilayer Ge on Ag (111) system: the contributions arising from (A) the total, 3s-, 3_{px} + 3_{py}- & 3_{pz}-Ge orbitals (the black, red, blue & green curves, respectively), and (B) the total, 5s-, 5p- & 4d-Ag orbitals, in which the 5_{px}, 5_{py}, and 5_{pz} make the comparable contributions.

References

[1] Suzuki S, Iwasaki T, De Silva KKH, Suehara S, Watanabe K, Taniguchi T, et al. Direct growth of germanene at interfaces between van der Waals materials and Ag (111). Adv Funct Mater 2021;31(5):2007038.

[2] Lin CH, Huang A, Pai WW, Chen WC, Chen TY, Chang TR, Yukawa R, Cheng CM, Mou CY, Matsuda I, Chiang TC, Jeng HT, Tang SJ. Single-layer dual germanene phases on Ag (111). Phys Rev Mater 2018;2:24003.

[3] Yuhara J, Shimazu H, Ito K, Ohta A, Araidai M, Kurosawa M, et al. Germanene epitaxial growth by segregation through Ag (111) thin films on Ge (111). ACS Nano 2018;12(11):11632–7.

[4] Ogikubo T, Shimazu H, Fujii Y, Ito K, Ohta A, Araidai M, et al. Continuous growth of germanene and stanene lateral heterostructures. Adv Mater Interfac 2020;7(10):1902132.

[5] Zhuang J, Liu C, Zhou Z, Casillas G, Feng H, Xu X, et al. Dirac signature in germanene on semiconducting substrate. Adv Sci 2018;5(7):1800207.

[6] Zhuang J, Gao N, Li Z, Xu X, Wang J, Zhao J, Dou SX, Du Y. Cooperative electron-phonon coupling and buckled structure in germanene on Au(111). ACS Nano 2017;11:3553–9.

[7] Pierucci D, Sediri H, Hajlaoui M, Girard JC, Brumme T, Calandra M, Velez-Fort E, Patriarche G, Silly MG, Ferro G, Soulière V, Marangolo M, Sirotti F, Mauri F, Ouerghi A. Evidence for flat bands near the Fermi level in epitaxial rhombohedral multilayer graphene. ACS Nano 2015;9:5432–9.

[8] Meng L, Wang Y, Zhang L, Du S, Wu R, Li L, Zhang Y, Li G, Zhou H, Hofer WA, Gao HJ. Buckled silicene formation on Ir (111). Nano Lett 2013;13:685–90.

[9] Papagno M, Rusponi S, Sheverdyaeva PM, Vlaic S, Etzkorn M, Pacilé D, Moras P, Carbone C, Brune H. Large band gap opening between graphene Dirac cones induced by Na adsorption onto an Ir superlattice. ACS Nano 2012;6:199–204.

[10] Dávila ME, Xian L, Cahangirov S, Rubio A, Le Lay G. Germanene: a novel two-dimensional germanium allotrope akin to graphene and silicene. New J Phys 2014;16:95002.

[11] Venema L, Meunier V, Lambin P. Atomic structure of carbon nanotubes from scanning tunneling microscopy. Phys Rev B Condens Matter 2000;61:2991–6.

[12] Egerton RF. Prospects for vibrational-mode EELS with high spatial resolution. Microsc Microanal 2014;20:658–63.

[13] He Z, He K, Robertson AW, Kirkland AI, Kim D, Ihm J, Yoon E, Do LG, Warner JH. Atomic structure and dynamics of metal dopant pairs in graphene. Nano Lett 2014;14:3766–72.

[14] Sato Y, Terauchi M. High-energy resolution electron energy-loss spectroscopy study of interband transitions characteristic to single-walled carbon nanotubes. Microsc Microanal 2014;20:807–14.

[15] Kisielowski C, Freitag B, Bischoff M, Van Lin H, Lazar S, Knippels G, Tiemeijer P, Van Der Stam M, Von Harrach S, Stekelenburg M, Haider M, Uhlemann S, Müller H, Hartel P, Kabius B, Miller D, Petrov I, Olson EA, Donchev T, Kenik EA, Lupini AR, Bentley J, Pennycook SJ, Anderson IM, Minor AM, Schmid AK, Duden T, Radmilovic V, Ramasse QM, Watanabe M, Erni R, Stach EA, Denes P, Dahmen U. Detection of single atoms and buried defects in three dimensions by aberration-corrected electron microscope with 0.5-Å information limit. Microsc Microanal 2008;14:469–77.

[16] Hillerich K, Dick KA, Wen CY, Reuter MC, Kodambaka S, Ross FM. Strategies to control morphology in hybrid group III-V/group IV heterostructure nanowires. Nano Lett 2013;13(3):903–8.

[17] Stehlik S, Orava J, Kohoutek T, Wagner T, Frumar M, Zima V, Hara T, Matsui Y, Ueda K, Pumera M. Carbon nanotube-chalcogenide glass composite. J Solid State Chem 2010;183:144–9.

[18] Hofmann P. The surfaces of bismuth: structural and electronic properties. Prog Surf Sci 2006;81:191–245.

[19] Pfnür H, Tegenkamp C, Vattuone L. Plasmons in one and two dimensions. Springer Handbooks; 2020. p. 557–84.

[20] Gensterblum G, Li-Ming Y, Pireaux JJ, Thiry PA, Caudano R, Lambin P, Lucas AA, Krätschmer W, Fischer JE. High resolution electron energy loss spectroscopy of epitaxial films of C60 grown on GaSe. J Phys Chem Solid 1992;53:1427–32.

[21] Riedl C, Starke U, Bernhardt J, Franke M, Heinz K. Structural properties of the graphene-SiC (0001) interface as a key for the preparation of homogeneous large-terrace graphene surfaces. Phys Rev B Condens Matter 2007;76:245406.

[22] Algdal J, Balasubramanian T, Breitholtz M, Kihlgren T, Walldén L. Thin graphite overlayers: graphene and alkali metal intercalation. Surf Sci 2007;601:1167–75.

[23] Rettner CT, Delouise LA, Auerbach DJ. Effect of incidence kinetic energy and surface coverage on the dissociative chemisorption of oxygen on W(110). J Chem Phys 1986;85:1131–49.
[24] Hafner J. Ab-initio simulations of materials using VASP: Density-functional theory and beyond. Journal of computational chemistry 2008;29(13):2044–78.
[25] Lin C-Y, Ho C-H, Wu J-Y, Do T-N, Shih P-H, Lin S-Y, Lin M-F. In: Lin C-Y, Ho C-H, Wu J-Y, Do T-N, H Shih P-, Lin S-Y, Lin M-F, editors. Diverse quantization phenomena in layered materials. CRC Press; 2019.
[26] Li C, Lu X, Han Y, Tang S, Ding Y, Liu R, Bao H, Li Y, Luo J, Lu T. Direct imaging and determination of the crystal structure of six-layered graphdiyne. Nano Res 2018;11:1714–21.
[27] Averyanov DV, Sokolov IS, Platunov MS, Wilhelm F, Rogalev A, Gargiani P, et al. Competing magnetic states in silicene and germanene 2D ferromagnets. Nano Res 2020;13(12):3396–402.
[28] Qi X, Qu J, Zhang H-B, Yang D, Yu Y, Chi C, Yu Z-Z. $FeCl_3$ intercalated few-layer graphene for high lithium-ion storage performance. J Mater Chem 2015;3:15498–504.
[29] Hattendorf S, Georgi A, Liebmann M, Morgenstern M. Networks of ABA and ABC stacked graphene on mica observed by scanning tunneling microscopy. Surf Sci 2013;610:53–8.
[30] Coletti C, Forti S, Principi A, Emtsev KV, Zakharov AA, Daniels KM, Daas BK, Chandrashekhar MVS, Macdonald AH, Polini M, Starke U. Revealing the electronic band structure of quasi-free trilayer graphene on SiC (0001). Mater Res Soc Symp Proc 2014;1693:155439.
[31] Mak KF, Sfeir MY, Misewich JA, Heinza TF. The evolution of electronic structure in few-layer graphene revealed by optical spectroscopy. Proc Natl Acad Sci U S A 2010;107:14999–5004.
[32] Warner JH, Rümmeli MH, Gemming T, Büchner B, Briggs GAD. Direct imaging of rotational stacking faults in few layer graphene. Nano Lett 2009;9:102–6.
[33] Ho YH, Chiu YH, Su WP, Lin MF. Magneto-absorption spectra of Bernal graphite. Appl Phys Lett 2011;99.
[34] Allman MS, Altomare F, Whittaker JD, Cicak K, Li D, Sirois A, Strong J, Teufel JD, Simmonds RW. Rf-SQUID-mediated coherent tunable coupling between a superconducting phase qubit and a lumped-element resonator. Phys Rev Lett 2010;104.
[35] Finkler A, Segev Y, Myasoedov Y, Rappaport ML, Ne'Eman L, Vasyukov D, Zeldov E, Huber ME, Martin J, Yacoby A. Self-aligned nanoscale SQUID on a tip. Nano Lett 2010;10:1046–9.

CHAPTER

12

Bilayer germanene/germanene composites

Shih-Yang Lin[1], *Thi Dieu Hien Nguyen*[1], *Vo Khuong Dien*[1], *Chi-Hsuan Lee*[1], *Hsin-Yi Liu*[1], *Hai Duong Pham*[1], *Thi My Duyen Huynh*[1], *Nguyen Thi Han*[1], *Ngoc Thanh Thuy Tran*[2], *Wei-Bang Li*[1] *and Ming-Fa Lin*[1]

[1]**Department of Physics, National Cheng Kung University, Tainan, Taiwan;** [2]**Hierarchical Green-Energy Materials (Hi-GEM) Research Center, National Cheng Kung University, Tainan, Taiwan**

The composite bilayer materials are suitable for investigating the essential properties and the feature-rich phenomena without the up-to-date experimental verifications, in which they consist of two different monolayer subsystems. For example, two monolayers of boron-nitride and silicene [1–7], the germanene-silicene van-der-Waals heterostructure [3,8–16]. Based on the group-IV element, the hybridized bilayer systems will become one of the main-stream materials for study. Up to now, the unusual van der Waals heterostructure has been successfully fabricated by the chemical deposition, such as the graphene/germanene/Ru (0001) [17–21]. The experimental results are combined with the first-principles calculations to clearly identify the unique Moire superlattice. The interlayer atomic interactions of germanene and germanene play a critical role in determining the most stable configuration.

This chapter is focused on the germanene/germanene bilayer composites under various stacking using the other group-IV materials as a specific adlayer, especially for the different Moire superlattices. The spin-induced atomic interactions and the van der Waals forces are covered in the delicate first-principles calculations. The ground state energies are delicately analyzed to obtain the optimal geometric structures. Importantly, the π and σ bondings are expected to reveal the modifications and hybridizations.

Fundamental Physicochemical Properties of Germanene-related Materials
https://doi.org/10.1016/B978-0-443-15801-8.00018-9

12.1 Germanene-germanene bilayer systems

A bilayer germanene-germanene composite system presents an highly coincidence of Moire superlattices, being quite different from the normal stacking configurations of silicene/germanene composite [22,23], as shown in Fig. 12.1. In general, there exist a lot of meta-stable geometric structures. Based on the first-principles calculations, two kinds of composite configurations are chosen for a model study, which is AA and AB stackings.

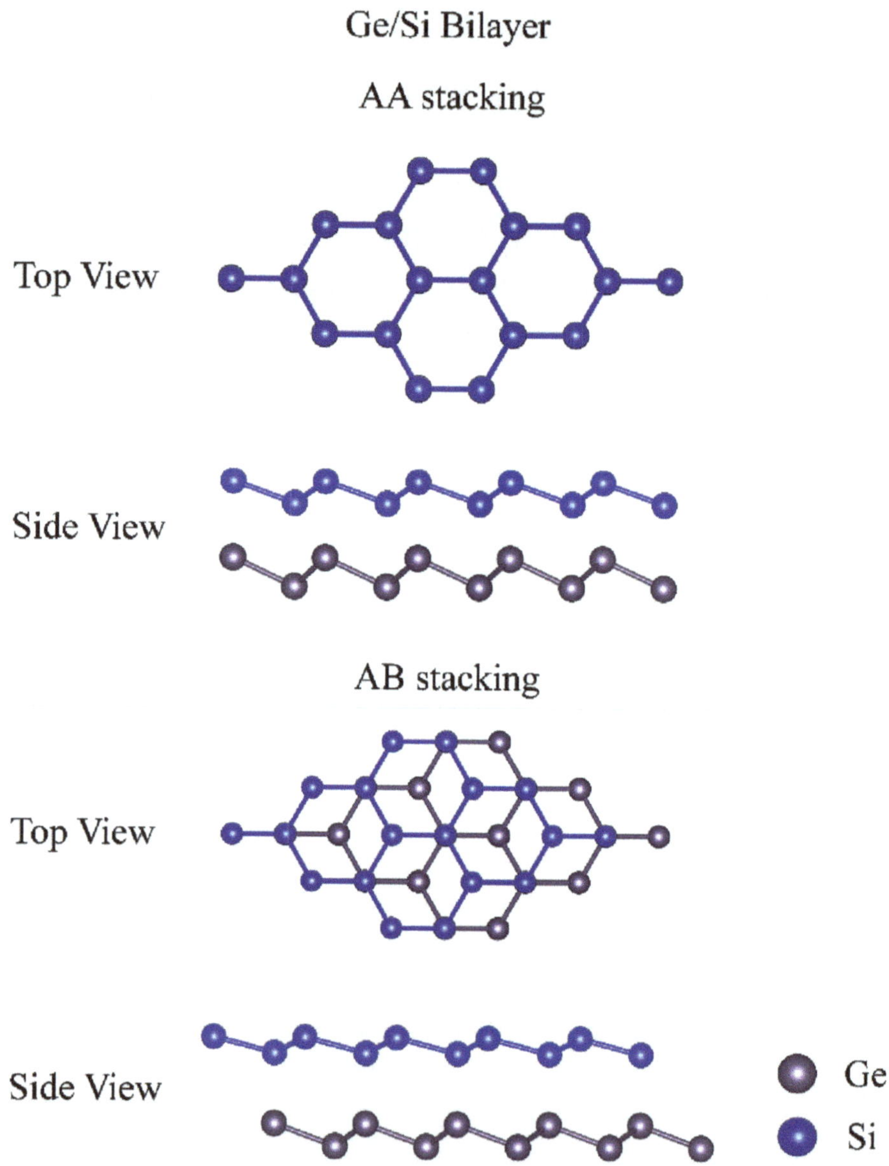

FIG. 12.1 Side view and top view of optimal bilayer germanene-germanene composite system.

The buckling structure and the atom-atom bond length remain almost the same to the pristine systems. This represents the negligible effects on the σ-electronic bondings. In addition, compared with the interlayer orbital hybridizations, the spin-dependent interactions are negligible. The spin-induced splitting phenomenon and magnetic configuration are absent in such composite systems. The experimental measurement by STM can provide the further examination of the geometric structures [24–33].

The bilayer germanene-silicene composite systems present the rich feature of electronic energy spectra, as clearly illustrated in Fig. 12.3. There are comparible germanium-/silicon-dominated electronic states (green/blue balls) due to the same number of atoms in unit cell. An increased number of valence and conduction energy subbands are highly asymmetric about the Fermi level. The low-energy valence and conduction bands are very sensitive to interlayer interation. As for the stacking configuration AA/AB (Fig. 12.2C–F), the low-energy band property is mainly determined by two distinct valleys. They are, respectively, initiated from the K/K′ and M points. Most important, their Fermi levels are situated above and below the Dirac point, since the $3p_z$ orbital energy is deeper than the $4p_z$ one. As a consequence, there exist few free conduction electrons/valence holes with the same free carrier density. This germanene-silicene composite should belong to a semi-metallic system, being generated by the small overlap of valence and conduction subbands. Such phenomenon will be further discussed in the density of states at E_F (Fig. 12.4).

The spatial charge distributions after the germanene-silicene coupling directly show the significant interlayer atomic interactions, as shown in Fig. 12.3. As for AA and AB stackings, the σ-electronic charge densities between two germanium or silicon atoms present the redistribution behaviors, as compared with the monolayer cases (Fig. 12.3A and B). There are many charges exist around atoms, but the opposite is true for the centered region. This clearly indicates that the reduced strengths of σ bondings on germanene and silicene layers (slight red region). These features are consistent with the charge density variations in Ge-Ge and Si–Si bonds (Fig. 12.4). Most important, the interlayer $3p_z$-$4p_z$ single-orbital hybridizations, which indicated by the yellow region in Fig. 12.3C and D, are responsible for two modified Dirac-cone structures across the Fermi level.

The atom- and orbital-decomposed density of states are very useful in determining the critical chemical bondings, as shown in Fig. 12.5A–F. The partially flat energy dispersions, local extrema points, saddle points, respectively, correspond to the shoulders, symmetric peaks in the logarithmically divergent form, square-root divergent peaks, and delta-function-like ones of the 2D band-edge states. The existence of certain number of free conduction electrons and valence holes reflects the Coulomb site-energy difference of Ge-$4p_z$ and Si-$3p_z$ orbitals. Furthermore, a finite density of states at E_F clearly indicates the semi-metallic behavior.

12.2 Sliding germanene/silicene composites

Sliding bilayer germanene-silicene composite system with the shift of stacking configuration is clearly illustrated in Fig. 12.5. Two subsystems are, respectively, represented by the blue and gray balls. The sliding bilayer composite materials are created by the relative shift between two subsystems, where δ is in unit of lattice constant ($\sqrt{3}\,b$ & $b = 2.32$ & 2.48 Å). The typical sliding way is chosen as follows (Fig. 5.1): $\delta_a = 0$ (AA), $\delta_a = 1/8$, $\delta_a = 4/8$, $\delta_a = 8/8$ (AB), $\delta_a = 11/8$, $\delta_a = 12/8$, and along the armchair direction $\delta_z = 1/8$, $\delta_z = 3/8$ along the

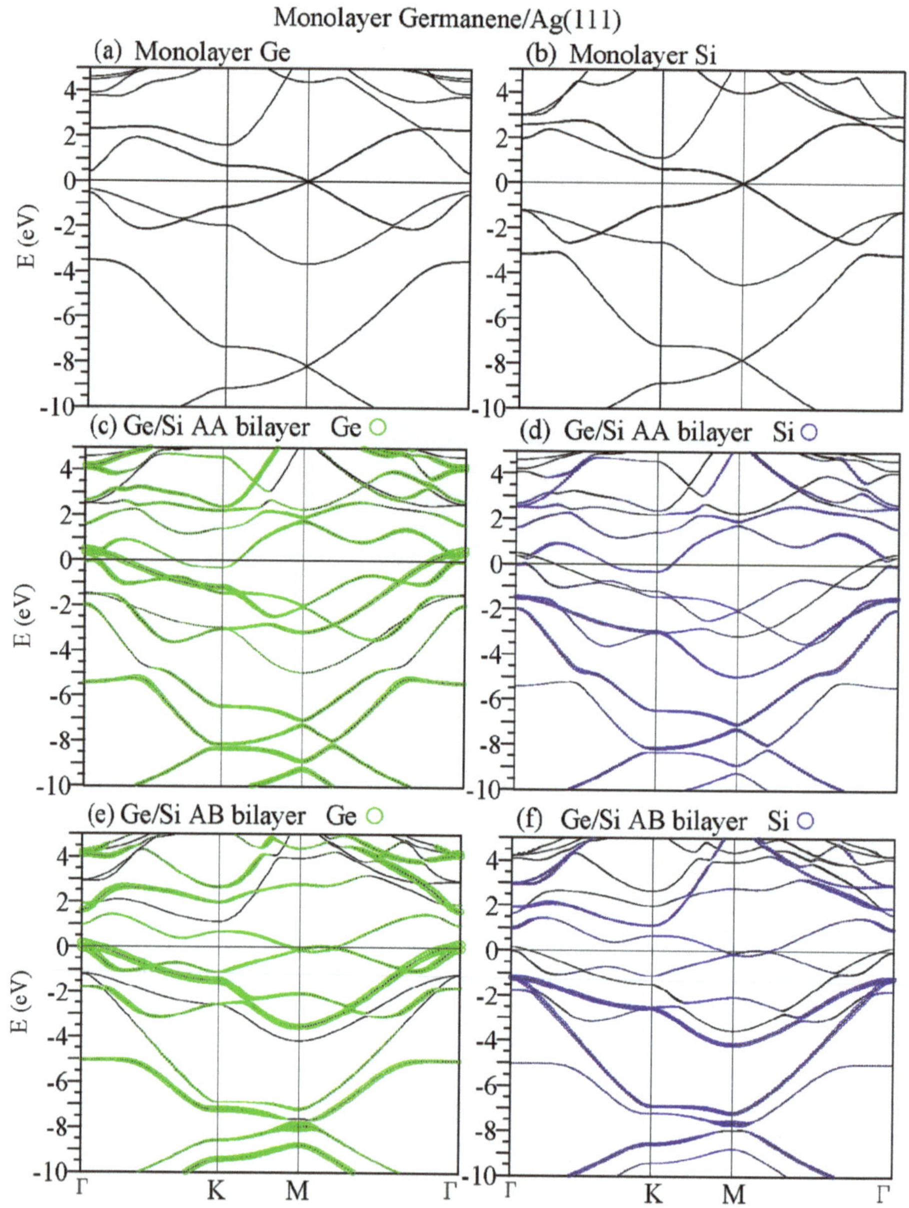

FIG. 12.2 Energy bands of (A) pristine germanene, (B) pristine germanene, (C)/(D) AA stacking Ge/Si bilayer, and (E)/(F) AB stacking Ge/Si bilayer.

zigzag direction. Based on the first-principles calculations and the analyses on ground state energies, this material possesses the smallest Moire super lattice among all the metastable bilayer composites. The Ge-Ge and Si–Si bond lengths are in the ranges of ~2.433–2.492 Å. The lattice constant on the [x, y] plane is slightly varied during the shift process.

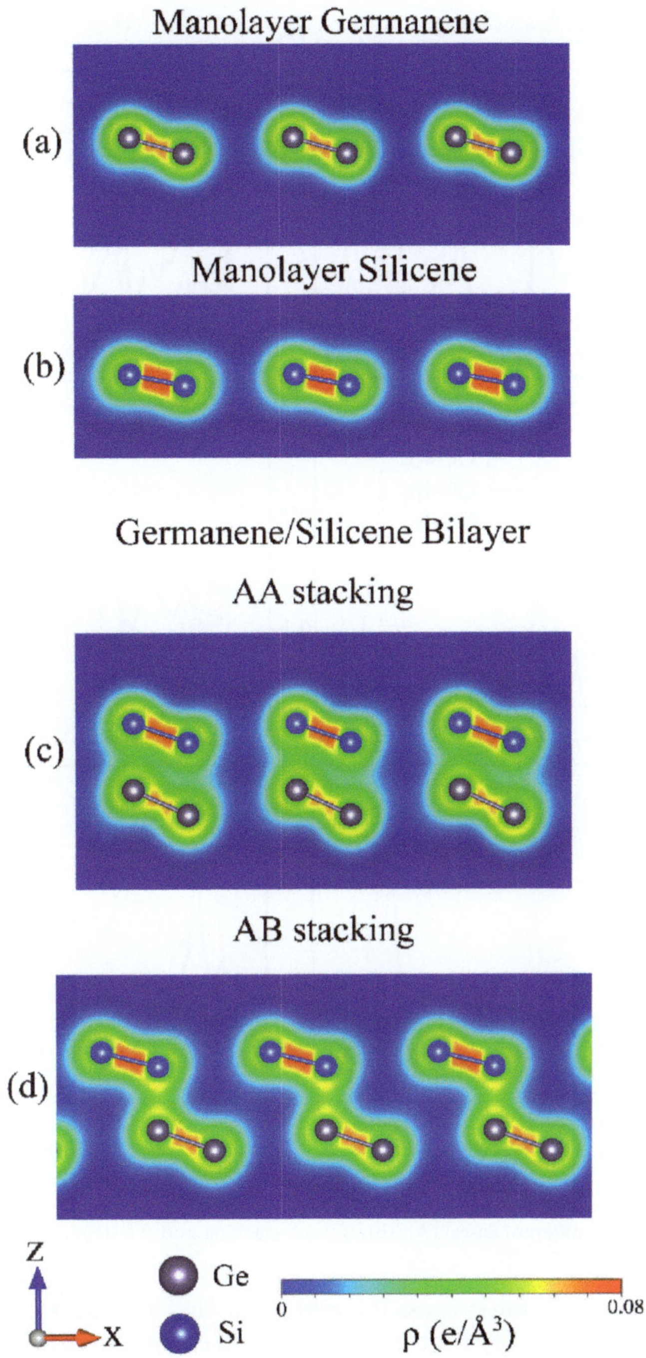

FIG. 12.3 The spatial charge density distributions of (A) pristine germanene (B) pristine germanene, (C) AA stacking Ge/Si bilayer, and (D) AB stacking Ge/Si bilayer.

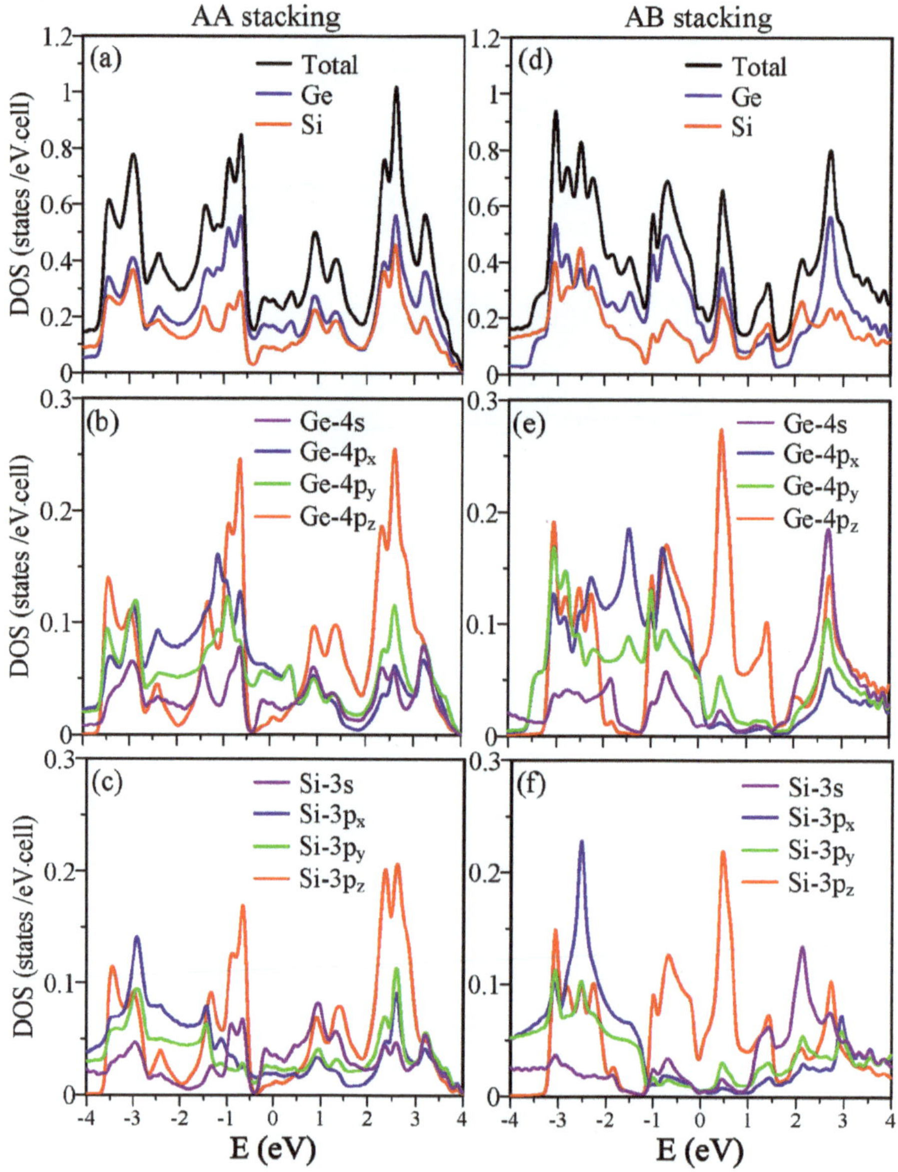

FIG. 12.4 The germanium-, silicon- and (s, p_x, p_y, p_z)-orbital-projected density of states for the bilayer germanene-silicene composites under the different cases: (A)/(B)/(C) AA stacking and (D)/(E)/(F) AB stacking.

A germanene-silicene composite possess the critical $3p_z$-$4p_z$ interlayer orbital hybridizations, which can be further confirmed by the atom-decomposed energy spectra, the spatial charge distributions, and the atom-/orbital-projected density of states (discussed later in Figs. 12.6–12.9).

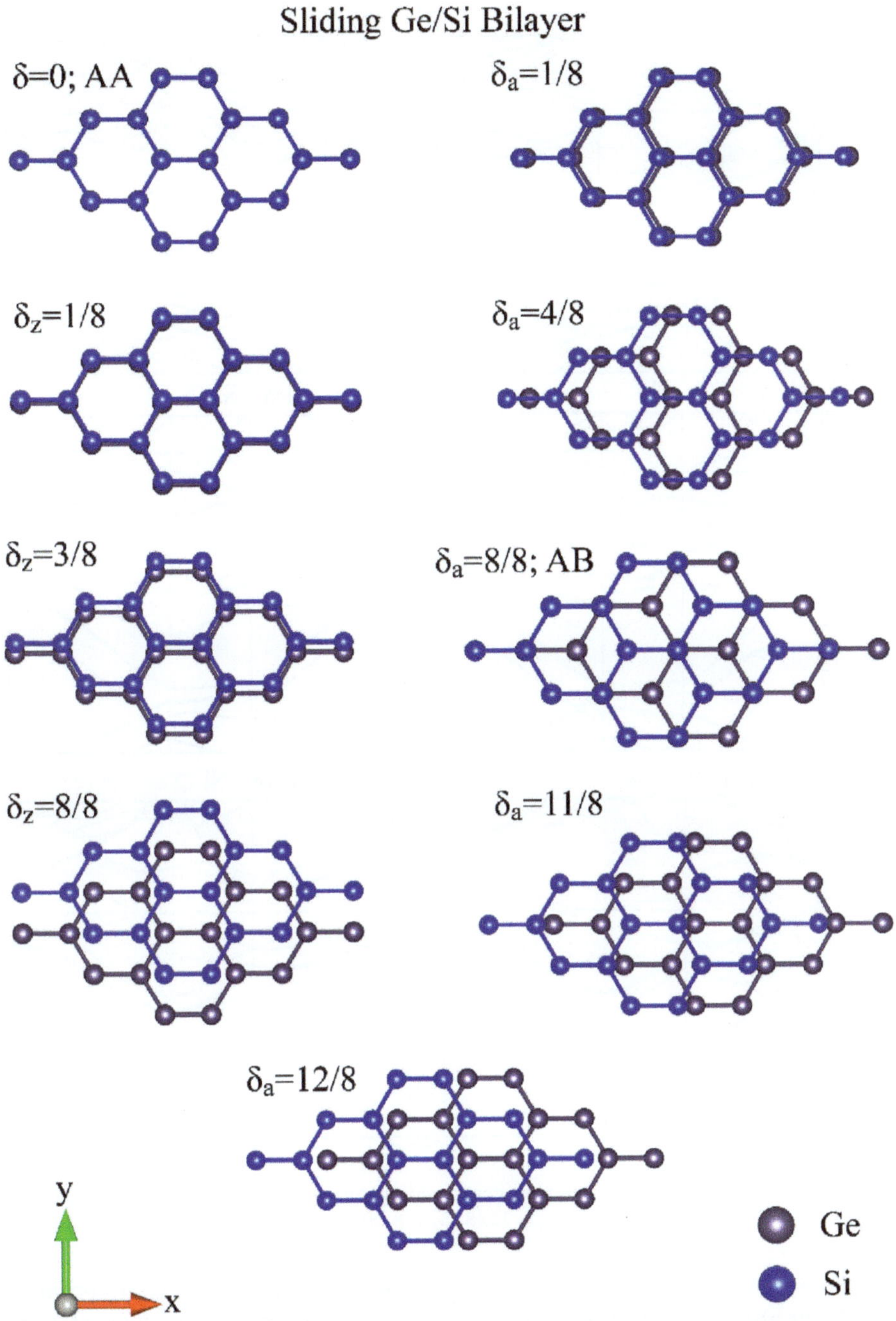

FIG. 12.5 Top view of optimal sliding bilayer germanene-germanene composite system.

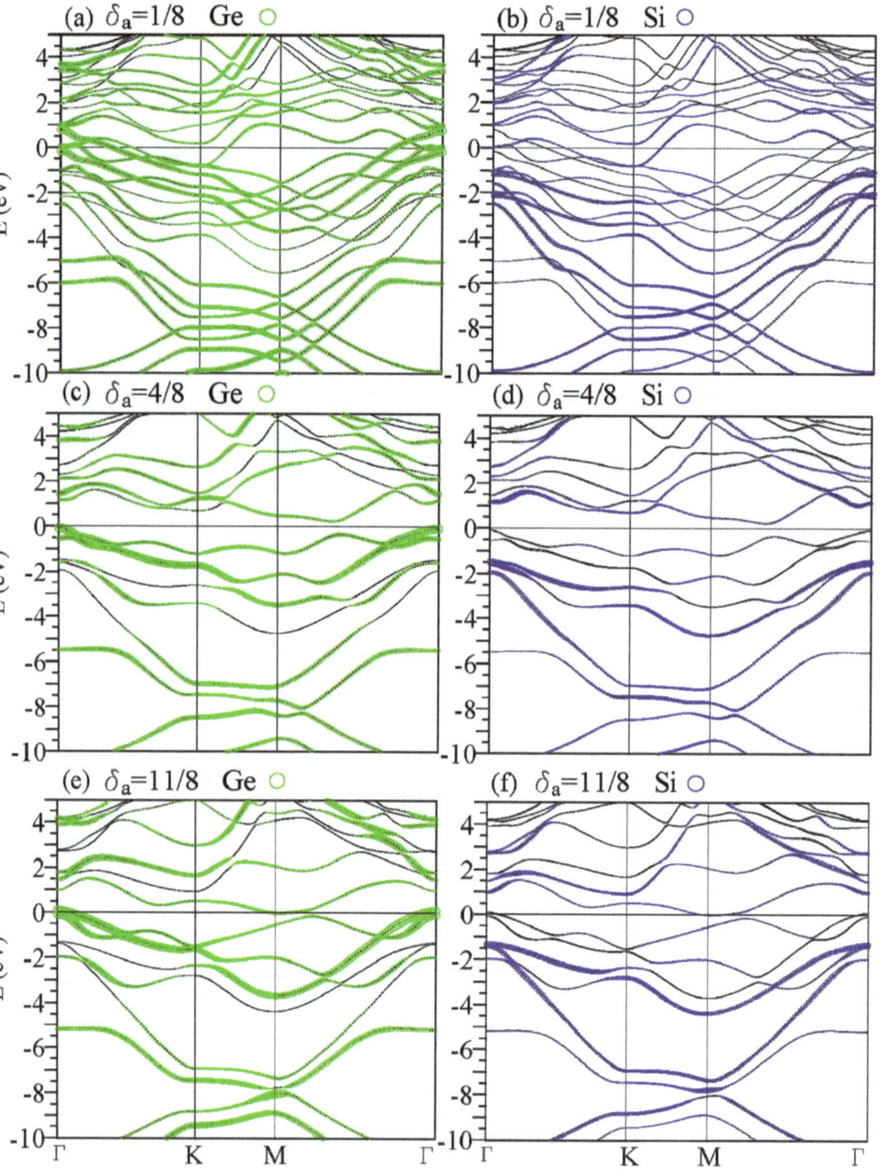

FIG. 12.6 Energy bands of bilayer germanene germanene composite under various shift: (A)/(B) $\delta_a = 1/8$, (C)/(D) $\delta_a = 4/8$, (E)/(F) $\delta_a = 11/8$, (G)/(H) $\delta_a = 12/8$ ($\delta_z = 8/8$), (I)/(J) $\delta_z = 1/8$, and (k)/(L) $\delta_z = 3/8$.

Band structures of germanene-silicene composites are dramatically changed after the significant interlayer atomic interactions, as shown in Fig. 12.6. The composite system presents a very complicated electronic energy spectrum in Fig. 12.6A and B for $\delta_a = 1/8$ with germanium- and silicon-atom dominances, respectively. It should be noted that the Ge and Si atoms make almost equal weight in dominating the energy states. Interestingly, an energy bands

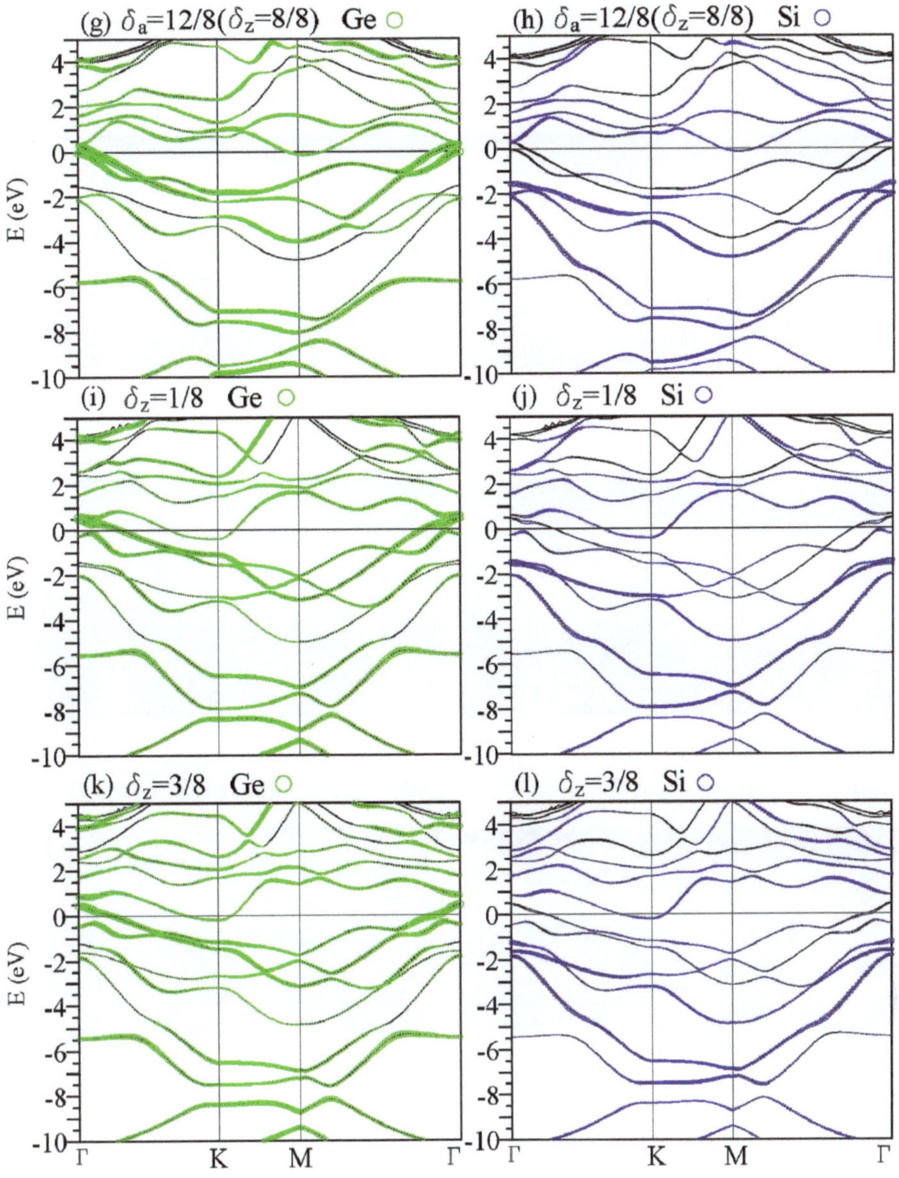

FIG. 12.6 cont'd

gap come to exist when the shift become $\delta_a = 4/8$. Moreover, the lowest unoccupied electronic state and highest occupied state along the MΓ direction are mainly contributed by the germanium honeycomb lattices. As for the other sliding cases, the valence/high conduction energy subbands, the frequent crossings and some anti-crossings come to exist, but not the energy gaps. Apparently, it is hard to clearly identify the π- and σ-band from the complex band structure in these composite systems. These special structures in energy bands could be further verified by the ARPES measurements [27,34–42].

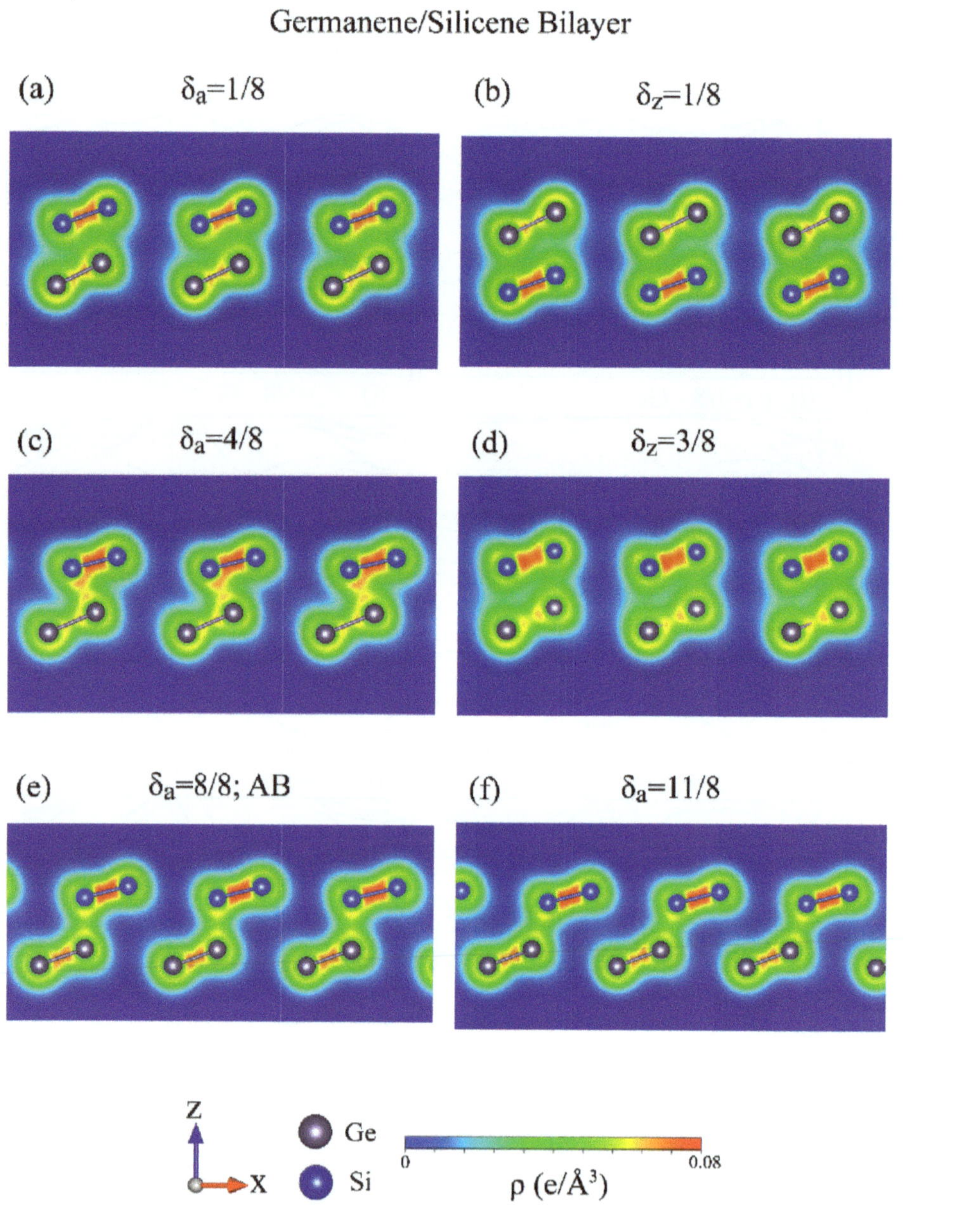

FIG. 12.7 The spatial charge density distributions of germanene germanene composite under various shift: (A) $\delta_a = 1/8$, (B) $\delta_z = 1/8$, (C) $\delta_a = 4/8$, (D) $\delta_z = 3/8$, (E) $\delta_z = 8/8$, and (F) $\delta_a = 11/8$.

The total charge densities for a germanene-silicene composite is shown in Fig. 12.7. They are useful in understanding the critical orbital hybridizations in the intralayer Ge-Si bonds. Apparently, the π and σ bondings could survive in germanium honeycomb lattice as well

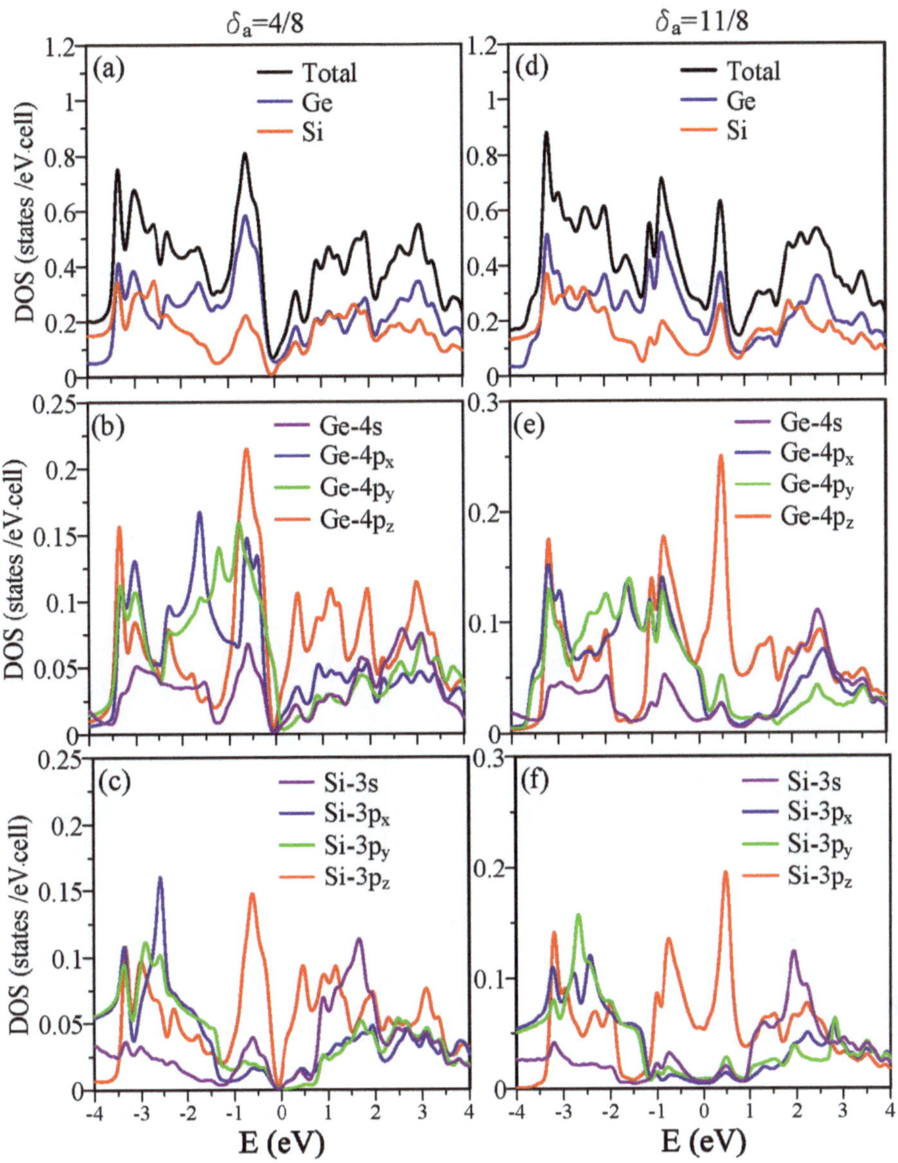

FIG. 12.8 The germanium-, silicon- and (s, p_x, p_y, p_z)-orbital-projected density of states of germanene germanene composite under various shift: (A)/(B)/(C) $\delta_a = 4/8$, (D)/(E)/(F) $\delta_a = 11/8$.

as silicon sublattice, indicating the hardly affected σ bondings feature. As for the interlayer interactions, they are somewhat enhanced through the closing interlayer atomic position, such as $\delta_a = 4/8$ (red region between two layers). This also accounts for the Ge-Si co-dominated energy bands. On the other side, some of the sliding systems shows the relatively weak interlayer interaction, as shown by the free resion between two layers (Fig. 12.7A

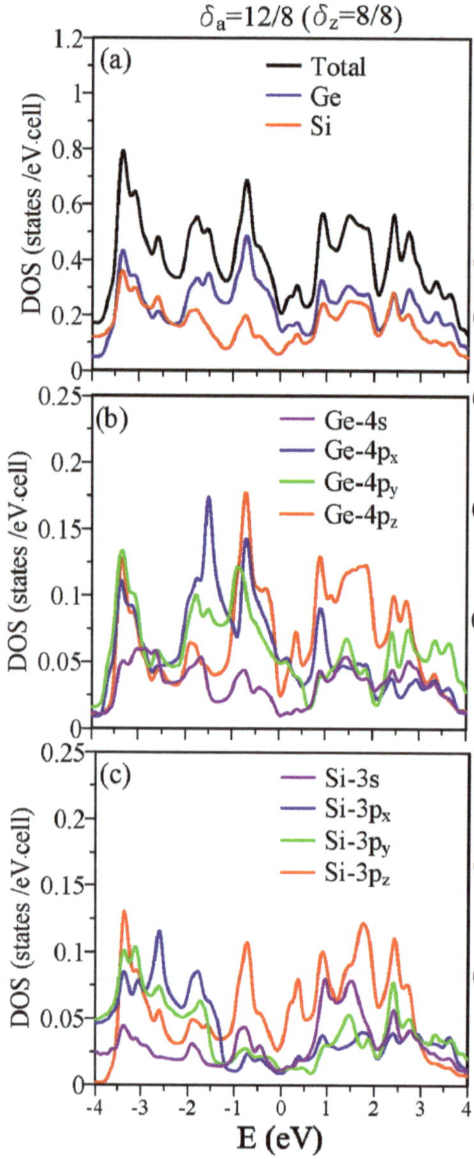

FIG. 12.9 The germanium-, silicon- and (s, p_x, p_y, p_z)-orbital-projected density of states of germanene germanene composite under various shift: (A)/(B)/(C) $\delta_a = 12/8$, (D)/(E)/(F) $\delta_z = 3/8$.

$\delta_a = 1/8$ and 12.7(b) for $\delta_z = 1/8$). As a result, the important van der Waals interactions are a bit weaker in those system and thus the co-dominated electronic states.

The atom- and orbital-decomposed van Hove singularities present the significant differences among the Ge-projected and Si-projected ones, as shown in Figs. 12.8 and 12.9. The former presents the greatly increased $4p_x$ and $4p_y$ states near the Fermi level. The densities of states at E_F approach to zero for some systems, including $\delta_a = 4/8$, which clearly illustrating the semiconducting behaviors. Moreover, these special structures of valence and conduction states are highly asymmetric about E_F for all sliding systems, mainly owing the hybridized σ-electronic states (the (4s, $4p_x$, $4p_y$)/(3s, $3p_x$, $3p_y$) orbitals). There are more van Hove singularities come to exist in the valence energy spectra. The above-mentioned features directly reflect the variation of the coupling effects of the van der Waals interactions and the changed sp^3/sp^2 bondings. The low-energy density of states could be further verified by the high-resolution STS measurements [25,43–47].

References

[1] Kociak M, Henrard L, Stéphan O, Suenaga K, Colliex C. Plasmons in layered nanospheres and nanotubes investigated by spatially resolved electron energy-loss spectroscopy. Phys Rev B Condens Matter 2000;61:13936–44.

[2] Dutta S, Manna AK, Pati SK. Intrinsic half-metallicity in modified graphene nanoribbons. Phys Rev Lett 2009;102:96601.

[3] Bhimanapati GR, Lin Z, Meunier V, Jung Y, Cha J, Das S, Xiao D, Son Y, Strano MS, Cooper VR, Liang L, Louie SG, Ringe E, Zhou W, Kim SS, Naik RR, Sumpter BG, Terrones H, Xia F, Wang Y, Zhu J, Akinwande D, Alem N, Schuller JA, Schaak RE, Terrones M, Robinson JA. Recent advances in two-dimensional materials beyond graphene. ACS Nano 2015;9:11509–39.

[4] Samipour A, Dideban D, Heidari H. First principles study of the ambipolarity in a germanene nanoribbon tunneling field effect transistor. ECS J Solid State Sci Technol 2019;8:M111–7.

[5] Despoja V, Djordjević T, Karbunar L, Radović I, Mišković ZL. Ab initio study of the electron energy loss function in a graphene-sapphire-graphene composite system. Phys Rev B 2017;96.

[6] Crompton KR, Hladky MP, Park HH, Prokes SM, Love CT, Landi BJ. Lithium-ion cycling performance of multi-walled carbon nanotube electrodes and current collectors coated with nanometer scale Al_2O_3 by atomic layer deposition. Electrochim Acta 2018;292:628–38.

[7] Kosynkin DV, Higginbotham AL, Sinitskii A, Lomeda JR, Dimiev A, Price BK, Tour JM. Longitudinal unzipping of carbon nanotubes to form graphene nanoribbons. Nature 2009;458:872–6.

[8] Quang T Le, Cherkez V, Nogajewski K, Potemski M, Dau MT, Jamet M, Mallet P, Veuillen JY. Scanning tunneling spectroscopy of van der Waals graphene/semiconductor interfaces: absence of Fermi level pinning. 2D Mater 2017;4:035019.

[9] Li H, Li S, Naik MH, Xie J, Li X, Wang J, Regan E, Wang D, Zhao W, Zhao S, Kahn S, Yumigeta K, Blei M, Taniguchi T, Watanabe K, Tongay S, Zettl A, Louie SG, Wang F, Crommie MF. Imaging moiré flat bands in 3D reconstructed WSe2/WS2 superlattices. 2020. p. 06113. *arXiv* 2007.

[10] Liu Y, Huang Y, Duan X. Van der Waals integration before and beyond two-dimensional materials. Nature 2019;567:323–33.

[11] Fan Y, Liu X, Wang J, Ai H, Zhao M. Silicene and germanene on InSe substrates: structures and tunable electronic properties. Phys Chem Chem Phys 2018;20:11369–77.

[12] Sharma A, Harnish P, Sylvester A, Kotov VN, Neto AHC. Van der Waals forces and electron-electron interactions in two strained graphene layers. Phys Rev B Condens Matter 2014;89(23):235425.

[13] Ogikubo T, Shimazu H, Fujii Y, Ito K, Ohta A, Araidai M, Kurosawa M, Le Lay G, Yuhara J. Continuous growth of germanene and stanene lateral heterostructures. Adv Mater Interfac 2020;7.

[14] Lin C-Y, Ho C-H, Wu J-Y, Do T-N, Shih P-H, Lin S-Y, Lin M-F. In: Lin C-Y, Ho C-H, Wu J-Y, Do T-N, Shih P-H, Lin S-Y, Lin M-F, editors. Diverse quantization phenomena in layered materials. CRC Press; 2019.

[15] Zhu FF, Chen WJ, Xu Y, Gao CL, Guan DD, Liu CH, Qian D, Zhang SC, Jia JF. Epitaxial growth of two-dimensional stanene. Nat Mater 2015;14:1020–5.

[16] Molle A, Grazianetti C, Tao L, Taneja D, Alam MH, Akinwande D. Silicene, silicene derivatives, and their device applications. Chem Soc Rev 2018;47:6370–87.

[17] Vázquez De Parga AL, Calleja F, Borca B, Passeggi MCG, Hinarejos JJ, Guinea F, Miranda R. Periodically rippled graphene: growth and spatially resolved electronic structure. Phys Rev Lett 2008;100.

[18] Cai T, Song Z, Rodriguez JA, Hrbek J. Preparation and structural characterization of RuS2 nanoislands on Au (111). J Am Chem Soc 2004;126:8886–7.

[19] Bjelkevig C, Mi Z, Xiao J, Dowben PA, Wang L, Mei WN, Kelber JA. Electronic structure of a graphene/hexagonal-BN heterostructure grown on Ru (0001) by chemical vapor deposition and atomic layer deposition: extrinsically doped graphene. J Phys Condens Matter 2010;22:302002.
[20] Sutter P, Hybertsen MS, Sadowski JT, Sutter E. Electronic structure of few-layer epitaxial graphene on Ru (0001). Nano Lett 2009;9:2654–60.
[21] Marchini S, Günther S, Wintterlin J. Scanning tunneling microscopy of graphene on Ru (0001). Phys Rev B Condens Matter 2007;76.
[22] Balendhran S, Walia S, Nili H, Sriram S, Bhaskaran M, et al. Elemental analogues of graphene: silicene, germanene, stanene, and phosphorene. Small 2015;11(6):640–52.
[23] Xu R, Gao N, Li H, Qiu D, Wang Q, Cheng S. Two dimensional monolayer rhombic silicene on the diamond (111) surface. Phys Chem Chem Phys 2018;20(33):21699–704.
[24] Zhuang J, Gao N, Li Z, Xu X, Wang J, Zhao J, Dou SX, Du Y. Cooperative electron-phonon coupling and buckled structure in germanene on Au (111). ACS Nano 2017;11:3553–9.
[25] Pierucci D, Sediri H, Hajlaoui M, Girard JC, Brumme T, Calandra M, Velez-Fort E, Patriarche G, Silly MG, Ferro G, Soulière V, Marangolo M, Sirotti F, Mauri F, Ouerghi A. Evidence for flat bands near the Fermi level in epitaxial rhombohedral multilayer graphene. ACS Nano 2015;9:5432–9.
[26] Meng L, Wang Y, Zhang L, Du S, Wu R, Li L, Zhang Y, Li G, Zhou H, Hofer WA, Gao HJ. Buckled silicene formation on Ir (111). Nano Lett 2013;13:685–90.
[27] Dávila ME, Xian L, Cahangirov S, Rubio A, Le Lay G. Germanene: a novel two-dimensional germanium allotrope akin to graphene and silicene. New J Phys 2014;16:95002.
[28] Bihlmayer G, Sassmannshausen J, Kubetzka A, Blügel S, von Bergmann K, Wiesendanger R. Plumbene on a magnetic substrate: a combined scanning tunneling microscopy and density functional theory study. Phys Rev Lett 2020;124:126401.
[29] Yarotski DA, Kilina SV, Talin AA, Tretiak S, Prezhdo OV, Balatsky AV, Taylor AJ. Scanning tunneling microscopy of DNA-Wrapped carbon nanotubes. Nano Lett 2009;9:12–7.
[30] Gill TG, Fleurence A, Warner B, Prüser H, Friedlein R, Sadowski JT, Hirjibehedin CF, Yamada-Takamura Y. Metallic atomically-thin layered silicon epitaxially grown on silicene/ZrB2. 2D Mater 2017;4:21015.
[31] Yang X, Dou X, Rouhanipour A, Zhi L, Räder HJ, Müllen K. Two-dimensional graphene nanoribbons. J Am Chem Soc 2008;130:4216–7.
[32] Wiesendanger R. Spin mapping at the nanoscale and atomic scale. Rev Mod Phys 2009;81:1495–550.
[33] Wendt S, Schaub R, Matthiesen J, Vestergaard EK, Wahlström E, Rasmussen MD, Thostrup P, Molina LM, Lægsgaard E, Stensgaard I, Hammer B, Besenbacher F. Oxygen vacancies on TiO_2(1 1 0) and their interaction with H_2O and O_2: a combined high-resolution STM and DFT study. Surf Sci 2005;598:226–45.
[34] Usachov D, Fedorov A, Otrokov MM, Chikina A, Vilkov O, Petukhov A, Rybkin AG, Koroteev YM, Chulkov EV, Adamchuk VK, Grüneis A, Laubschat C, Vyalikh DV. Observation of single-spin dirac fermions at the graphene/ferromagnet interface. Nano Lett 2015;15:2396–401.
[35] Senkovskiy BV, Fedorov AV, Haberer D, Farjam M, Simonov KA, Preobrajenski AB, Mårtensson N, Atodiresei N, Caciuc V, Blügel S, Rosch A, Verbitskiy NI, Hell M, Evtushinsky DV, German R, Marangoni T, van Loosdrecht PHM, Fischer FR, Grüneis A. Semiconductor-to-metal transition and quasiparticle renormalization in doped graphene nanoribbons. Adv. Electron. Mater. 2017;3:1600490.
[36] Pogrebna A, Vujičić N, Mertelj T, Borzda T, Cao G, Xu ZA, Chu JH, Fisher IR, Mihailovic D. Spectrally resolved femtosecond reflectivity relaxation dynamics in undoped spin-density wave 122-structure iron-based pnictides. Phys Rev B Condens Matter 2014;89:165131.
[37] De Jong S, Van Heumen E, Thirupathaiah S, Huisman R, Massee F, Goedkoop JB, Ovsyannikov R, Fink J, Dürr HA, Gloskovskii A, Jeevan HS, Gegenwart P, Erb A, Patthey L, Shi M, Follath R, Varykhalov A, Golden MS. Droplet-like Fermi surfaces in the anti-ferromagnetic phase of EuFe2As2, an Fe-pnictide superconductor parent compound. EPL 2010;89:27007.
[38] Coletti C, Forti S, Principi A, Emtsev KV, Zakharov AA, Daniels KM, Daas BK, Chandrashekhar MVS, Macdonald AH, Polini M, Starke U. Revealing the electronic band structure of quasi-free trilayer graphene on SiC (0001). Mater Res Soc Symp Proc 2014;1693:155439.
[39] Levy G, Nettke W, Ludbrook BM, Veenstra CN, Damascelli A. Deconstruction of resolution effects in angle-resolved photoemission. Phys Rev B Condens Matter 2014;90:45150.
[40] Sugawara K, Yamamura N, Matsuda K, Norimatsu W, Kusunoki M, Sato T, Takahashi T. Selective fabrication of free-standing ABA and ABC trilayer graphene with/without Dirac-cone energy bands. NPG Asia Mater 2018;10:e466.

[41] Cuk T, Lu DH, Zhou XJ, Shen ZX, Devereaux TP, Nagaosa N. A review of electron-phonon coupling seen in the high-Tc. Superconductors by angle-resolved photoemission studies (ARPES). Phys Status Solidi Basic Res 2005;242:11–29.
[42] Crepaldi A, Pons S, Frantzeskakis E, Calleja F, Etzkorn M, Seitsonen AP, Kern K, Brune H, Grioni M. Combined ARPES and STM study of Pb/Au (111) Moiré structure: one overlayer, two symmetries. Phys Rev B Condens Matter 2013;87:115138.
[43] Zheng X, Zhang JF, Tong B, Du RR. Epitaxial growth and electronic properties of few-layer stanene on InSb (1 1 1). 2D Mater 2020;7:011001.
[44] Huang H, Wei D, Sun J, Wong SL, Feng YP, Neto AHC, Wee ATS. Spatially resolved electronic structures of atomically precise armchair graphene nanoribbons. Sci Rep 2012;2:983.
[45] Meunier V, Senet P, Lambin P. Scanning tunneling spectroscopy signature of finite-size and connected nanotubes: a tight-binding study. Phys Rev B Condens Matter 1999;60:7792–5.
[46] Matsui T, Kambara H, Niimi Y, Tagami K, Tsukada M, Fukuyama H. STS observations of landau levels at graphite surfaces. Phys Rev Lett 2005;94:226403.
[47] Kano S, Tada T, Majima Y. Nanoparticle characterization based on STM and STS. Chem Soc Rev 2015;44:970–87.

CHAPTER

13

Germanene nanotubes

Hsin-Yi Liu[1], *Shih-Yang Lin*[1], *Vo Khuong Dien*[1], *Chi-Hsuan Lee*[1], *Hai Duong Pham*[1], *Thi My Duyen Huynh*[1], *Nguyen Thi Han*[1], *Ngoc Thanh Thuy Tran*[2], *Thi Dieu Hien Nguyen*[1], *Wei-Bang Li*[1] *and Ming-Fa Lin*[1]

[1]**Department of Physics, National Cheng Kung University, Tainan, Taiwan;** [2]**Hierarchical Green-Energy Materials (Hi-GEM) Research Center, National Cheng Kung University, Tainan, Taiwan**

Tubular nanostructures have been always attracted highly research attention. Carbon nanotubes (CNTs), which are created experimentally [1–7], belong to one dimensional semiconducting nanostructures. Basically, their low dimensionality and the quantum confinement effect are with potentials for various technology Products such as nanocomposites, field emission sources, chemical sensor, photonics, Lithium ion batteries, super capacitors and actuators because of its distinctive mechanical, electronic and optical characteristics [8–13]. The anode of Lithium ion batteries which have been commercialized exhibits limited capacity of graphite is 372 mAh/g theoretically. When a graphitic carbon anode is replaced by tubular nanostructures of C, Si and Ge may increase the anode capacity about two times to 10 times. Carbon nanotubes exhibit reversible capacities of 573 mAh/g at a small current of 0.2°C and 520 mAh/g at a large current of 2°C [14]. Park et al. have investigated that high reversible charge capacity of silicon nanotubes about 3247 mAh/g by increasing the surface area accessible to the electrolyte and deposited a carbon coating on the surface of the nanotubes [15]. Lithium-ion diffusivity of Germanium is 400 times faster than Si and electrical conductivity of Germanium is 10^4 times higher than Si. Besides, theoretical specific capacity of Germanium is very large about 1600 mAh/g, Germanium is also as a promising anode material for lithium-ion batteries. The metals of Si, Ge and Sn exhibit multiple times of capacities when a graphitic anode is replaced. However, the large volume change of IV-grope metals has caused rapid capacity fading during cycling. That has seriously impeded their applicability

Fundamental Physicochemical Properties of Germanene-related Materials
https://doi.org/10.1016/B978-0-443-15801-8.00021-9

of practical lithium-ion cells between 3 and 4.3 V operating voltages. Park et al. prepared Germanium Nanotubes (GeNTs) by method of Kirkendall Effect as anodes of high-rate Lithium-ion batteries. The GeNTs showed exceptionally high rate capability of up to 40°C (40/Ag) with a reversible capacity of 1000 mAh/g maintained over 400 cycles with minimal capacity fading [16]. The growth of germanium nanostructures, which are nanowire (NW) and nanotube (NT) can be controlled by methods of chemical vapor deposition (CVD). Both diameters are with variable diameters from 10 to 500 nm by use of a porous anodic aluminum oxide (AAO) template nanochannels and a wall thickness is controlled between 6 and 18 nm [17]. There are a lot of theoretical studies on the electronic properties of carbon and silicon nanotubes. They are analyzed by the tight-binding model and first-principle method [18–24]. However, simulated calculation of germanium nanotubes published quite a few. Bayani et al. performed carbon and germanium nanotubes applied to field effect transistors (FETs) by simulation of first-principle. The results showed better transistor performance of germanium nanotubes, which output characteristic voltage reached to 0.6 V, compared to carbon nanotubes (0.4 V) [25]. Therefore, it is very significant to realize the electronic properties.

In this chapter we focus on the diversified electronic properties of single wall germanium nanotubes, investigated by first-principles calculations. Achiral germanium nanotubes are taken into consideration to reveal that there are two basic kinds of geometric structures (armchair and zigzag), which can be further modulated by various diameters and chiral vectors. This work presents bond lengths b_1 and b_2, ground state energy as well as total energy E_0, buckling distance Δh, energy band structure, energy gap (E_g), charge density (ρ) and projected density of states (PDOSs) and the both structures dependent on chiral vectors and diameters are fully included in the calculations. A relativistic interaction of a carbon atom's spin with its motion inside a potential is normally weak such as graphene. However, it is significant in a carbon nanotube due to it curvature surface [26]. The curve surface causes asymmetric structure and leads to an intrinsic spin-orbit interaction in carbon nanotubes. The estimating splitting by theoretical calculations is around 100 μeV and the spin-orbit splitting measured by experimental methods are reasonably consistent with theoretical predictions [27–30].

13.1 Achiral germanene nanotubes

The primary symmetry of a nanotube is classified as achiral and chiral. Achiral nanotubes compared to the chiral ones are highly symmetric. They have a mirror image whose structure is identical to the original one. On the contrary, the mirror image of chiral nanotubes shows a spiral symmetry, which are not superposed on to the original one. There are two types of achiral germanium nanotubes; armchair and zigzag nanotubes, their geometric structures of cross-section and side view are shown in Figs. 13.1–13.5, respectively. The width of tube size are dependent on the chirality, which armchair germanium nanotubes (aGeNTs) are (m,m) and zigzag germanium nanotubes (zGeNTs) are (m,0), where m is an integer. A periodical bond length of aGeNTs in a unit cell along the z-direction is 3b (b is a Si–Si bond length) and zGeNTs is $\sqrt{3}$b. The smallest size of aGeNTs starts from Ge (2,2) and then the structures are increased to Ge (10,10). The geometric structure of Ge (2,2) are constructed by two hexagons. The honeycomb structure of Ge (2,2) are due

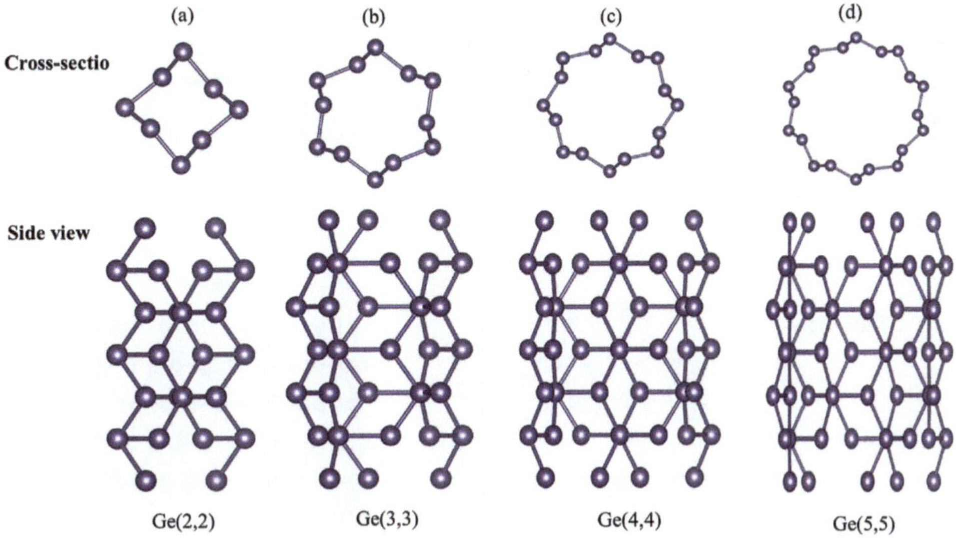

FIG. 13.1 (A)–(D) Geometric structure of armchair germanium nanotubes at side view and cross-section Ge (2,2), Ge (3,3), Ge (4,4) and Ge (5,5).

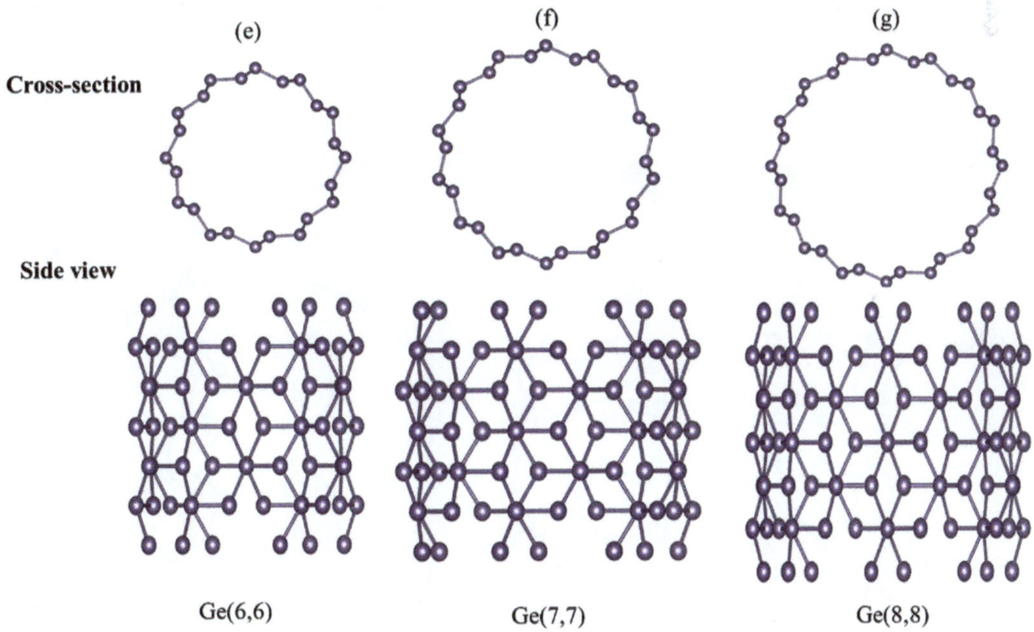

FIG. 13.2 (E)–(G) Geometric structure of armchair germanium nanotubes at side view and cross-section Ge (6,6), Ge (7,7) and Ge (8,8).

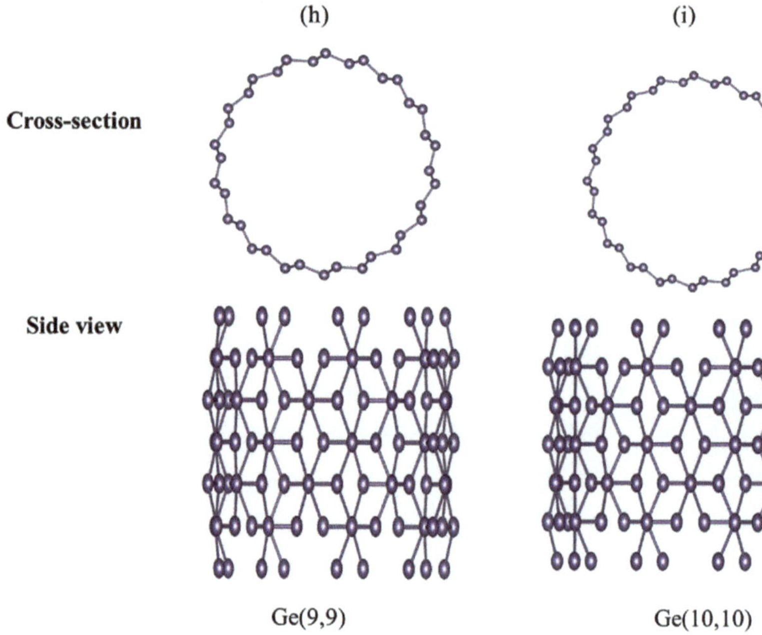

FIG. 13.3 (H) and (I) Geometric structure of armchair germanium nanotubes at side view and cross-section Ge (9,9) and Ge (10,10).

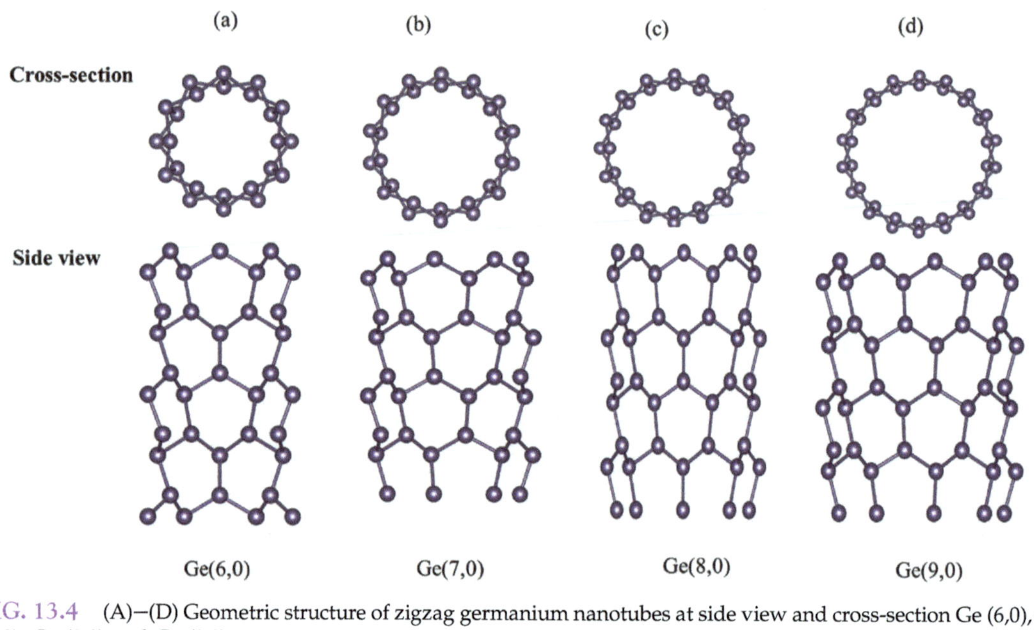

FIG. 13.4 (A)–(D) Geometric structure of zigzag germanium nanotubes at side view and cross-section Ge (6,0), Ge (7,0), Ge (8,0) and Ge (9,0).

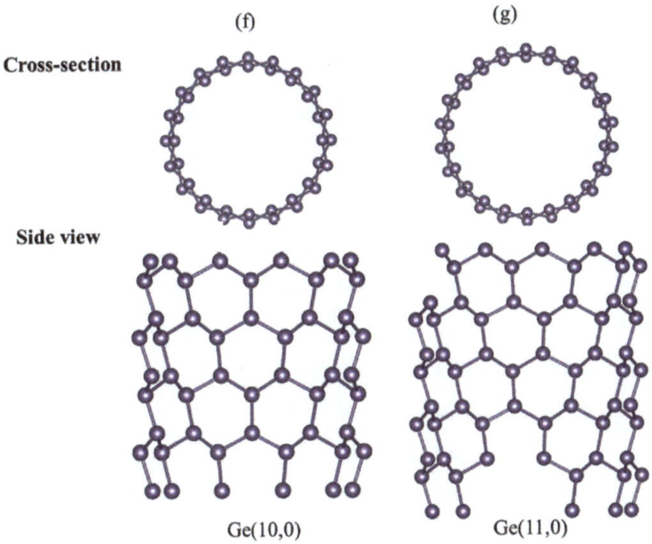

FIG. 13.5 (F) and (G) Geometric structure of zigzag germanium nanotubes at side view and cross-section Ge (10,0) and Ge (11,0).

to the curvature effect highly deformed and generate the buckling distance between each Ge atom. The buckling distance is enhanced by sp^3 hybridization and leads to a gear-likes structure. Also the same with zGeNTs, the structure of zGeNTs start from Ge (6,0) to Ge (11,0), the nanotubes are firstly constructed by six hexagons then are increased to 11 ones. Two kinds of GeNTs, which are aGeNTs and zGeNTs, possess armchair and zigzag edge hexagons, respectively, along the longitudinal direction z surrounded to form a tubular structure. The buckling distance decreases when the chirality and curvature of aGeNTs and zGeNTs decrease. The lattice constant of each GeNT are optimized by moderating adjustments of the structure and the relative lowest ground state energy E_0 is obtained, which are summarized in Tables 13.1 and 13.2. Due to the gearlikes structures, two of the adjacent atoms cause distance difference to the center axis and they result to the two different diameters which define as D_1 and D_2 shown in Tables 13.1 and 13.2; meanwhile, the types of Ge—Ge bonds are labeled as b_1 and b_2 shown in Fig. 13.6 b_1 is a horizontal Ge—Ge bond and b_2 is a vertical one. Both of the bond lengths are altered dependent of tube size especially for the small size of nanotubes. The smallest single wall carbon nanotubes have found by method of HR-TEM is around 4 Å in diameter and the possible structures can be armchair (3,3), zigzag (5,0) and chiral (4,2) [31]. Besides, in a multiwalled carbon nanotube a stable 3 Å of carbon nanotubes can be grown [32]. The smallest size of GeNTs is around 10 nm in diameter and are synthesized by methods of chemical vapor deposition (CVD) [17]. The pipe diameter and thickness of GeNTs can be controlled by use of a porous anodic aluminum oxide (AAO) template nanochannels [17].

TABLE 13.1 Parameters of armchair germanium nanotubes ground state energy E_0, Ge—Ge bond length b_1 and b_2, diameter D_1 and D_2, buckling height Δh and band gap E_g; $d_0 = 2.44$Å.

aGeNTs	Lattice con (Å)	E_0/unit cell (eV)	E_0/atom (eV)	b_1 (Å)	b_2 (Å)	b_1/d0	B_2/d0	D_1 (Å)	D_2 (Å)	Δh (Å)	Eg (eV)
Ge (2,2)	3.81	−31.8343	−3.97928	2.446	2.56	1.002459	1.04918	3.808	5.807	0.9995	0.664
Ge (3,3)	3.97	−47.5743	−3.96453	2.46	2.485	1.008197	1.018443	6.394	7.959	0.7825	0.078
Ge (4,4)	3.99	−63.6873	−3.98046	2.453	2.467	1.005328	1.011066	8.309	9.832	0.7615	0.451
Ge (5,5)	3.99	−79.7941	−3.98971	2.448	2.457	1.003279	1.006967	10.028	11.219	0.5955	0.496
Ge (6,6)	4.00	−95.888	−3.99533	2.448	2.455	1.003279	1.006148	13.309	14.232	0.4615	0.386
Ge (7,7)	4.00	−111.967	−3.99883	2.446	2.452	1.002459	1.004918	14.635	15.676	0.5205	0.324
Ge (8,8)	4.00	−128.037	−4.00117	2.445	2.449	1.002049	1.003689	17.795	18.657	0.431	0.289
Ge (9,9)	4.01	−144.096	−4.00266	2.445	2.45	1.002049	1.004098	19.173	20.137	0.482	0.245
Ge (10,10)	4.00	−160.156	−4.00391	2.446	2.448	1.002459	1.003279	22.323	23.148	0.4125	0.219

TABLE 13.2 Parameters of zigzag germanium nanotubes ground state energy E_0, Ge—Ge bond length b_1 and b_2, diameter D_1 and D_2, buckling distance Δh and band gap E_g; $d_0 = 2.44$Å.

Sample	Lattice con	E_0/unit cell	E_0/atom (eV)	b_1 (Å)	b_2(Å)	b_1/d0	b_2/d0	D_1 (Å)	D_2 (Å)	Buckling distance	E_g (eV)
Ge (6,0)	4.04	−95.2815	−3.970063	2.477	2.466	1.015164	1.010656	6.793	8.503	0.855	0
Ge (7,0)	4	−111.382	−3.977918	2.461	2.466	1.008607	1.010656	8.181	9.659	0.739	0.14
Ge (8,0)	3.99	−127.524	−3.985127	2.457	2.457	1.006967	1.006967	9.7	11.183	0.7415	0.223
Ge (9,0)	4	−143.651	−3.990318	2.454	2.457	1.005738	1.006967	10.804	12.249	0.7225	0.279
Ge (10,0)	4	−159.764	−3.99409	2.451	2.453	1.004508	1.005328	12.256	13.705	0.7245	0.442
Ge (11,0)	4	−175.865	−3.996939	2.45	2.451	1.004098	1.004508	13.416	14.842	0.713	0.442

13.2 Chirality effects

The periodical boundary condition plays an important role in determining metallic or semiconducting behavior of carbon nanotubes. The condition of periodical boundary can

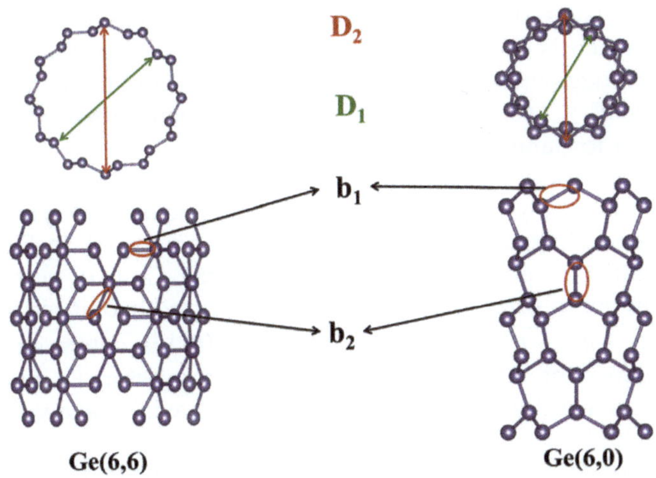

FIG. 13.6 Illustration of germanium nanotubes parameters.

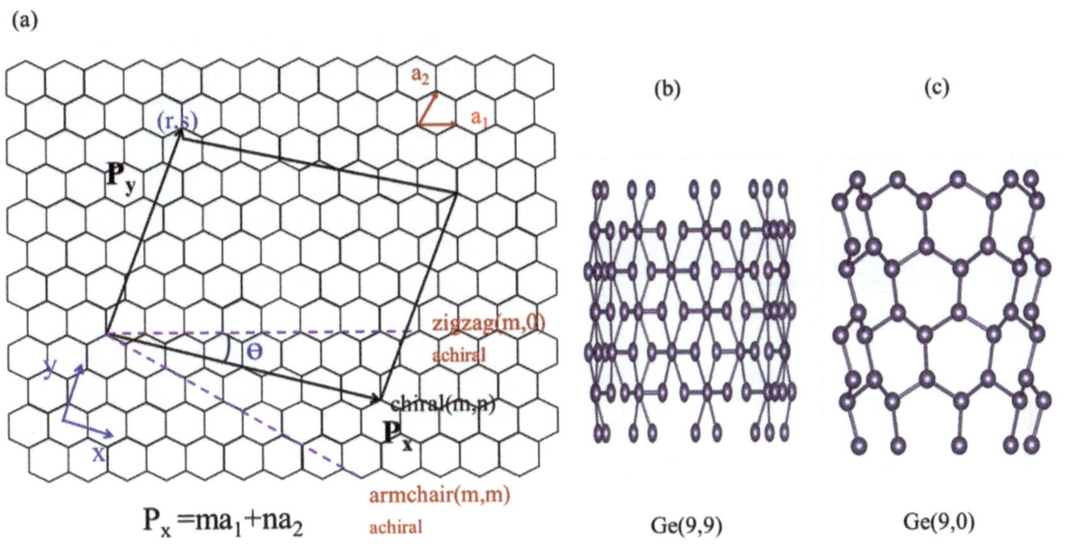

FIG. 13.7 Unrolled honeycomb lattice of a nanotube (A) A single wall armchair Ge (9,9) and zigzag Ge (9,0) carbon nanotube (B) and (C).

also be applied to determining metallic or semiconducting behavior of germanium nanotubes.

A cylindrical shape of nanotube is the one-dimensional structure with axial symmetry and it shows a spiral conformation named chirality. The classification of germanium nanotubes (GeNTs) shown in Fig. 13.7A is characterized by two integers (m,n) and is formed by rolling a germanene sheet from the lattice vector $P_x = ma_1 + na_2$, where a_1 and a_2 are primitive lattice vectors of a 2D sheet. The radius r of a nanotube (m,n) behaves as $b\sqrt{3(m^2+mn+n^2)}/2\pi$. The chiral angle θ which is the angle between the vectors P_x and a_1 can characterize the non-chiral/chiral arrangements of germanium hexagons on a cylindrical surface. It is confined in the range $0° \leq |\theta| \leq 30°$ for all germanium nanotubes. Hence, the chiral angle θ yields an expression for tanθ, which is $\tan\theta = \sqrt{3}/(2m+n)$. The other type of germanium nanotubes are armchair and zigzag ones. The both are classified as achiral nanotubes. (m,m) and (m,0) shown in Fig. 13.7A as a purple dash line are the symbol respectively for the achiral armchair and zigzag systems with $\theta = 30°$ and $\theta = 0°$. In Fig. 13.7B and C are rolled honeycomb lattice of armchair Ge (9,9) and zigzag Ge (9,0) germanium nanotubes, respectively. The number of germanium atoms N_u in a primitive unit cell is given by an equation of $N_u = 4\sqrt{(m^2+mn+n^2)(r^2+rs+s^2)/3}$, where (r,s) is determined by the primitive lattice vector P_y along the nanotube axis. The lattice vector P_y can be described as $P_y = ra_1 + sa_2$ shown as a rectangle in Fig. 13.7A and The axial wave vectors in the 1D first Brillouin zone are $|k_y| \leq \pi/|P_y|$.

The energy band structures of aGeNTs and zGeNTs with the chiral vectors of (2,2) -(10,10) and (6,0) - (11,0) are shown in Figs. 13.8–13.12, respectively. The both have the k_x-dependent energy bands with strong dispersion relations for various armchair and zigzag systems. The 1D dispersion relations of aGeNTs are highly symmetric.

There are two symmetry points in the one dimensional first Brillouin zone (1D BZ) at the Γ (k = 0) point and X point (k = $\pm\pi$/a), where a is the lattice constant. Besides, there is an another important point K between Γ and X with k = $\pm 2\pi$/3a. Germanium nanotubes (GeNTs) possess special 1-D dispersion relations because of buckling angle, honeycomb symmetry, edge structure and quantum confinement. All aGeNTs exhibit semiconducting

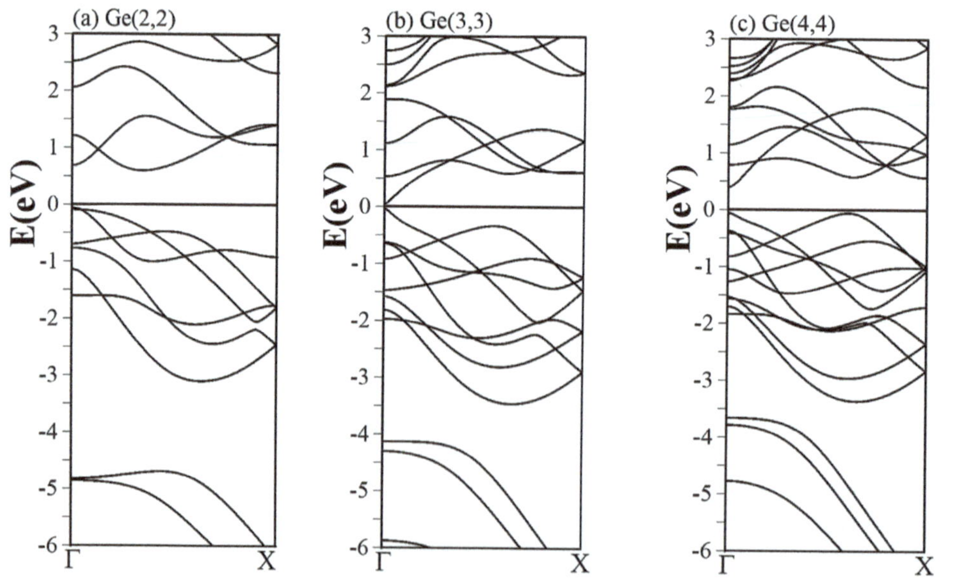

FIG. 13.8 (A)–(C) Band structure of armchair germanium nanotubes Ge (2,2), Ge (3,3) and Ge (4,4).

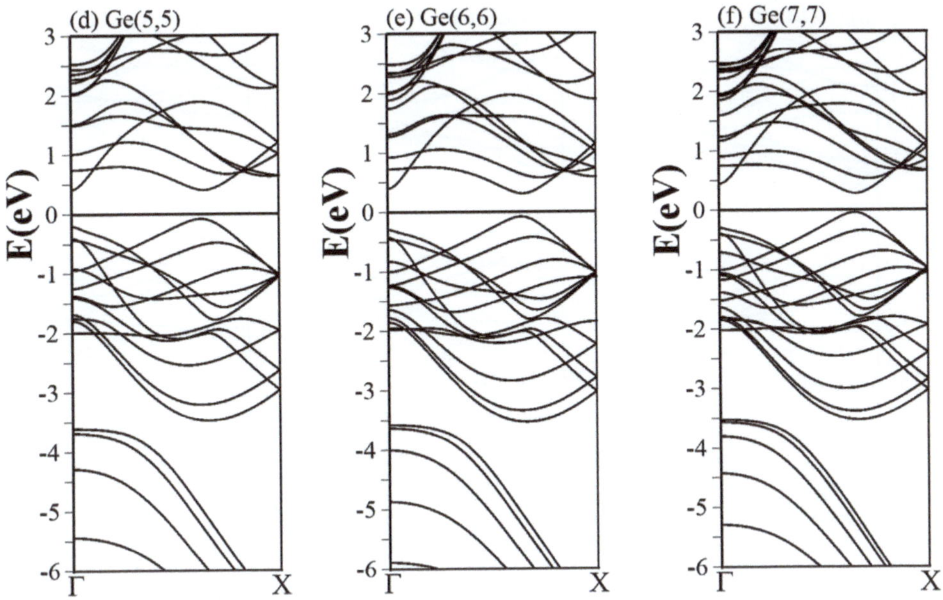

FIG. 13.9 (D)–(F) Band structure of armchair germanium nanotubes Ge (5,5), Ge (6,6) and Ge (7,7).

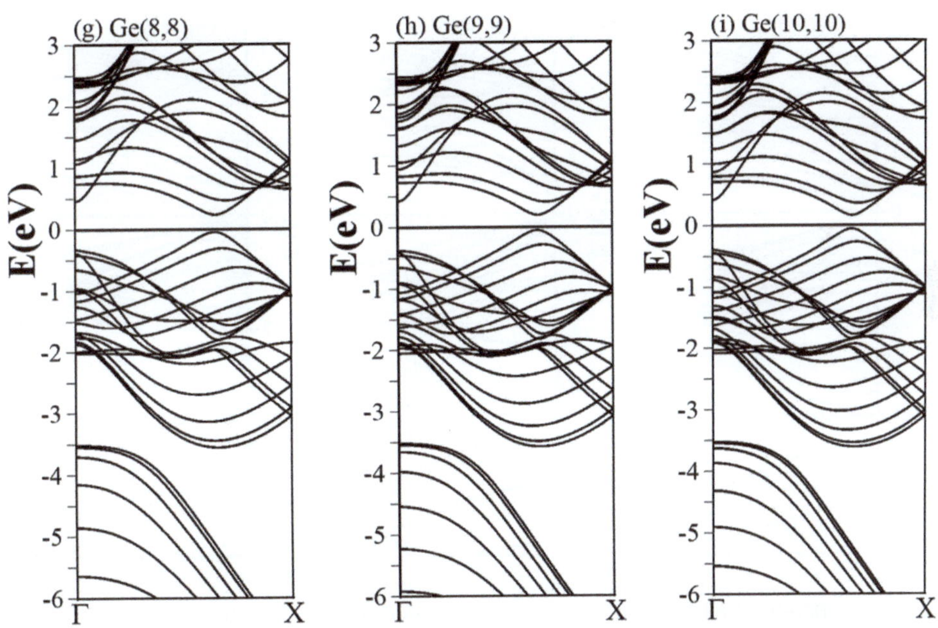

FIG. 13.10 (G)–(I) Band structure of armchair germanium nanotubes Ge (8,8), Ge (9,9) and Ge (10,10).

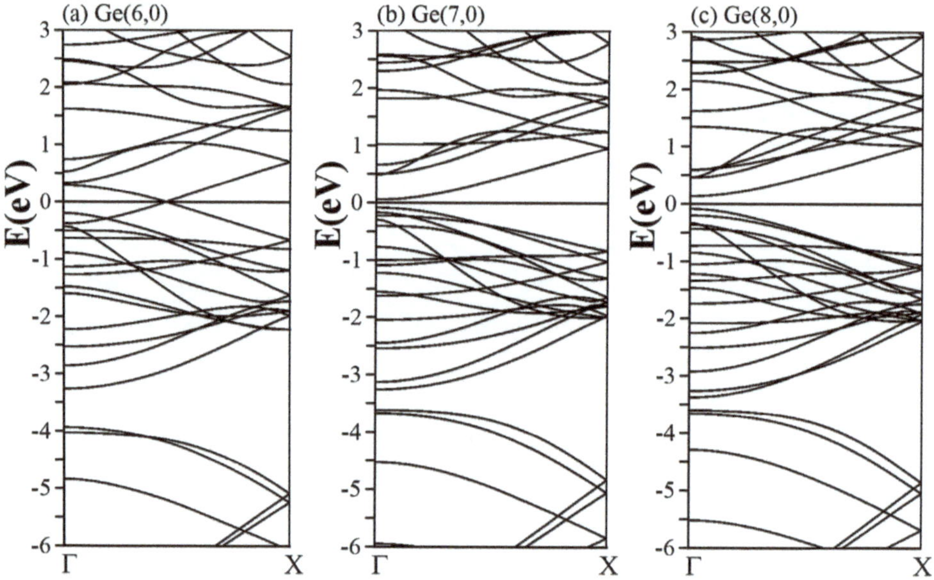

FIG. 13.11 (A)–(C) Band structure of zigzag germanium nanotubes Ge (6,0), Ge (7,0) and Ge (8,0).

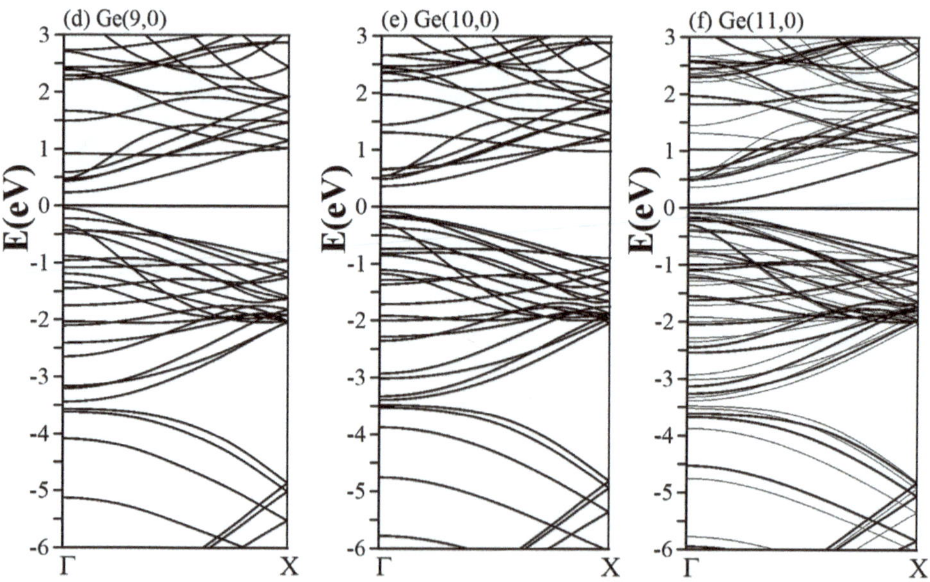

FIG. 13.12 (D)–(F) Band structure of zigzag germanium nanotubes Ge (9,0), Ge (10,0) and Ge (11,0).

characters due to the asymmetric structure comparing with armchair carbon nanotubes. There are two diameters on aGeNTs because of buckling distances. Two diameters will lead to generate two angular momentums and result in energy band gap (E_g) at fermi level (E_F). The number of Energy bands increases with increasing chiral vectors of aGeNTs and most of them are doubly degenerate because of the cylindrical boundary condition. The occupied valence bands are asymmetric to the unoccupied conduction bands around $E_F = 0$, especially for the small size of aGeNTs Ge (2,2) and Ge (3,3). The reasons could be the sp^3 hybridization and periodic boundary conditions (PBCs) [33,34]. Due to the curvature effects, the misorientation of $2p_z$ orbital and the sp^3 orbital hybridization on a cylindrical surface, are not taken into account, the periodical boundary plays a critical factor in determining the metallic or semiconducting behavior. Most of energy bands for armchair germanium nanotubes (aGeNTs) are parabolic dispersion and some sub-bands of them are partially flat. The band-edge states located between $k = 0$ to $k = \pm\pi/a$ (from Γ to X point) present van Hove singularities in density of states (DOSs). The sub-band crossings exhibit dominantly related to the sub-band anti-crossings. Each of sub-bands corresponds to different angular momentum J.

As for armchair germanium nanotubes (aGeNTs) shown in Figs. 13.8A–13.10I corresponding to the chiral vectors of Ge (4,4) to Ge (10,10), present semiconducting behaviors with energy gaps and Dirac cones located at K points. According to the Zone-folding approximation, the Dirac cones of 1D BZ are folded at K point ($k = \pm2\pi/3a$) in agreement with previous results [33–35]. However, the small sized tube of Ge (2,2) exhibits indirected band gap E_g around fermi level, the reason could be due to the curvature effect and leads to the two hexagons and in a circle deformed. The nanotube of Ge (3,3) shows the direct band gap E_g at Γ points and it shows the semi-metal behaviors. The other indirect band gaps of aGeNTs are Ge (4,4) and Ge (5,5).

All of the electronic states are measured from the Fermi level ($E_F = 0$) and the energy bands of occupied valence orbital are asymmetry to the energy bands of unoccupied conduction ones. The energy gap of E_g of Ge (2,2) – Ge (10,10) are summarized in Table 13.1 and illustrated in Fig. 13.8A and B. The nanotubes of Ge (2,2), Ge (4,4) and Ge (5,5) shown in Figs. 13.8A and 13.9D, possess an indirect energy gap of $E_g = 0.664$, 0.451 and 0.496 eV, respectively. In other words, the difference between the highest valence band ($HV\ B$) and the lowest conduction band (LCB) at the located 1D BZ point is not the same. The energy (eV) of the HBV point is at Γ point and the LCB point is around $k = \pm\pi/3a$. For the other indirect E_g energy bands of Si (4,4) and Si (5,5), both of them show the HBV point is around the position of $k = \pm2\pi/3a$ and the LCB point is at Γ point ($k = \pm\pi/a$). However, the values of E_g show irregularity by small values and radii of germanium nanotubes. This features could be due to high buckling distances related to their neighboring atom and result to the high curvature energy of geometric structure [36]. High buckling distance appears a gear-like structure and results to the energy band gap E_g Ref. [37]. There is a direct band gap with increasing chiral vector of Ge (6,6) to Ge (10,10). The anticrossing dispersions take place between valence band and conduction band located around K point in the zone boundary of $k = \pm\pi/3a$. The E_g of Ge (6,6) to Ge (10,10) decreases as $1/r$ scaling law with increasing of chiral vectors and radii r, i.e., the large silicon nanotubes exhibit less buckling and the geometry turns to similar like structure of carbon nanotubes.

The band structures of zigzag germanium nanotubes (zGeNTs) with chiral vectors of (6,0), (7,0), (8,0), (9,0), (10,0) and (11,0) are shown in Figs. 13.11A–13.12F. There are certain important differences between armchair and zigzag systems in energy band structure. The Ge (6,0) exhibits the metallic features and the others (n = 7–11) are semicondutors. The metallic characteristics alter the electronic structure of germanium nanotubes (GeNTs) and they can be due to the mixing of σ^* and π^* bands [24,38–40]. For the band structure of Ge (6,0), there are two flat sub-bands cross the fermi level (E^F) at the position of $k = \pm\pi/2a$.

Other zGeNTS like Ge (7,0) - (11,0), there is no crossing sub-bands at Fermi level E_F. They form direct band gap at Γ point. The value of E_g increase with increasing radius of germanium nanotubes and are illustrated in Fig. In this case of metallic nanotubes, the fundamental differences of electronic properties are observable imposed by curvature effect [36]. In curved system, the σ^* sub-band shifts to upward and π^* sub-band moves to downward. Both sub-bands result the mixing states shown in the energy dispersion. The rehybridization of the σ and π states reflects the consequence of the curvature. In contrast to without curved system, the σ and π orbitals appear orthogonal to each other and therefore cannot be a mixed state [41]. The alternation of electronic structure is dramatically different, especially for those with small radii and chiral vectors of silicon nanotubes. In the case of zigzag (n,0) carbon nanotubes, presents the metallic character when n = 3q (q is an integer). For zSiNTs (n,0) remains semiconducting characters when n = 3q, which m is in the range of 10–24 [24]. There

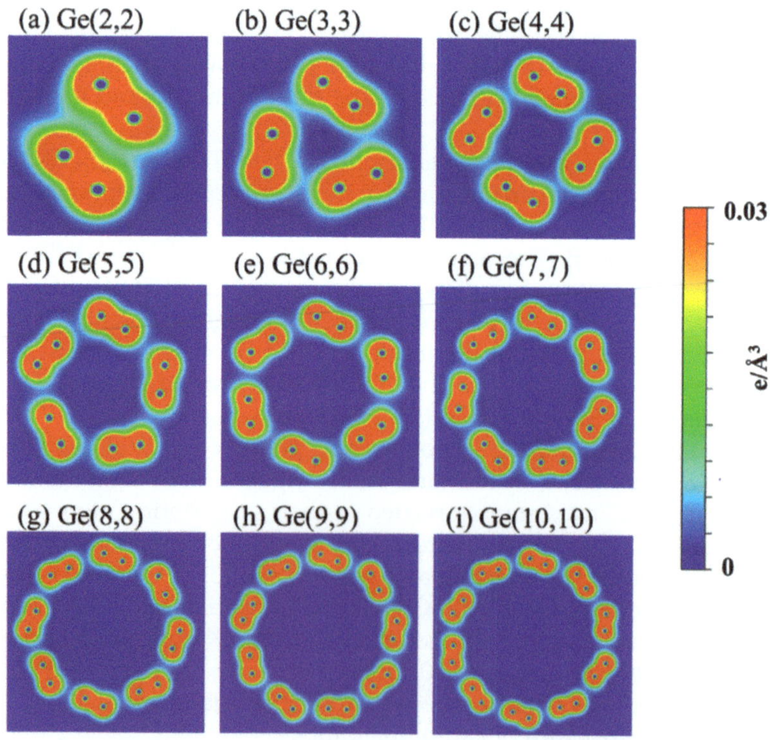

FIG. 13.13 (A)–(I) Charge density of armchair germanium nanotubes from Ge (2,2) to Ge (10,10).

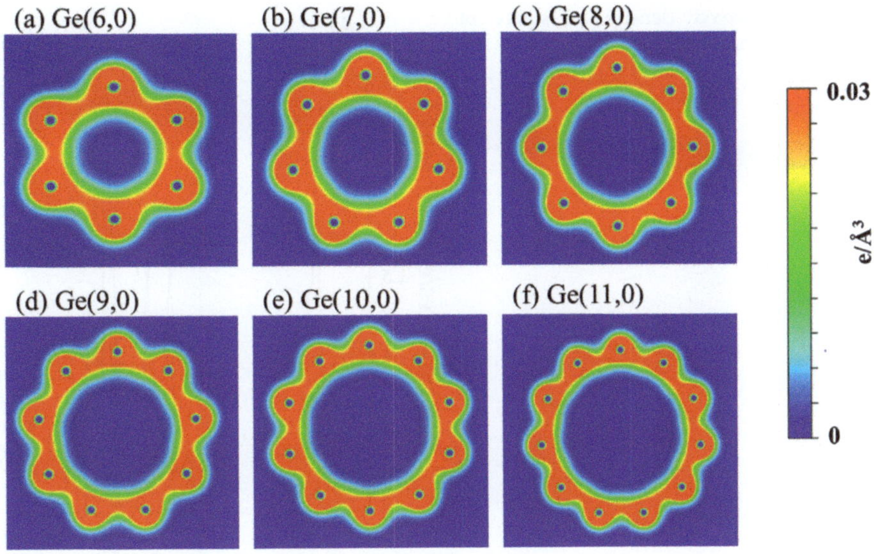

FIG. 13.14 (A)–(F) Charge density of zigzag germanium nanotubes from Ge (6,0) to Ge (11,0).

are quite a few sub-band anti-crossings observable in the results of band structure of zSiNTs; Only Si (4,0) presents an anti-crossing sub-band around −1.1 eV. Most of them are crossing energy dispersions distribute among valence and conduction sub-bands.

Spatial charge density ρ of aGeNTs and zGeNTs from (2,2) to (10,10) and (6,0) to (11,0), respectively, are available in investigating the multi-/single-orbital hybridizations because of various chemical bonds [42]. They are illustrated in Figs. 13.13A–13.14F with the (x,y)-plane projection calculated from first principle calculations. The results of ρ represent clearly the chemical bonding as well as the charge transfer. In both pictures of Figs. 13.13 and 13.14, red color shapes distribute around each germanium atom and appear σ bonds. The (4s, $4p_x$, $4p_y$) or (4s, $4p_x$, $4p_y$, $4p_z$) hybridization of germanium nanotubes is hard to distinguish by the results of spatial charge density ρ. In contrast to monolayer germanene, the results of charge density ρ from GeNTs have the weaker intensity of chemical bonding. The tubular geometric structure seems to effect the distribution of charge density. The results of bond length b_1 from Ge (6,0) to Ge (11,0) are consistent with this phenomenon. The bond length b_1 of Ge (6,0) is 2.477 Å, which is only 1.52% longer than standard Ge–Ge bond length (2.44 Å) then decreases to 2.450 Å when the bond length b_1 is Ge (11,0). For the result of Ge (6,0), a σ bonding is extended to connect tubular silicon atoms and form a circle. There is a yellow region clearly shown in the middle of each atom lead to the weak σ bonding. Comparing to aGeNTs, the special charge density of zGeNTs, which is distributed between two neighboring silicon atoms on the same layer (the red regions), is obviously reduced and transferred.

The orbital-projected density of states (PDOSs) as well as 1D van Hove singularities of energy dependence of GeNTs are clearly shown in Figs. 13.15–13.29, directly reflect the primary features of electronic energy band structure. The partial orbital-projected PDOSs can

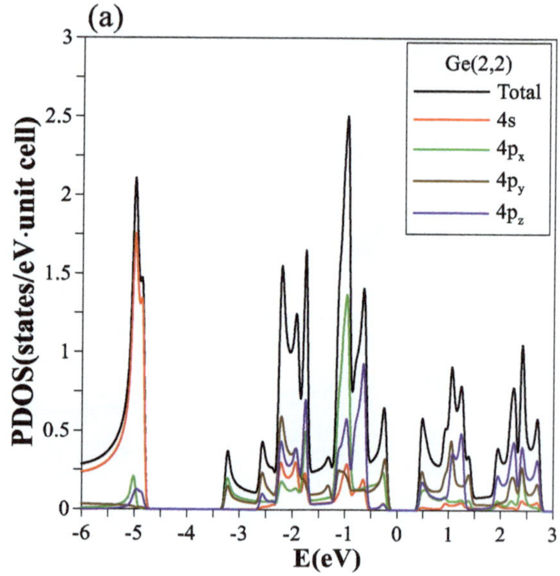

FIG. 13.15 (A) Projected density of states of armchair germanium nanotubes Ge (2,2).

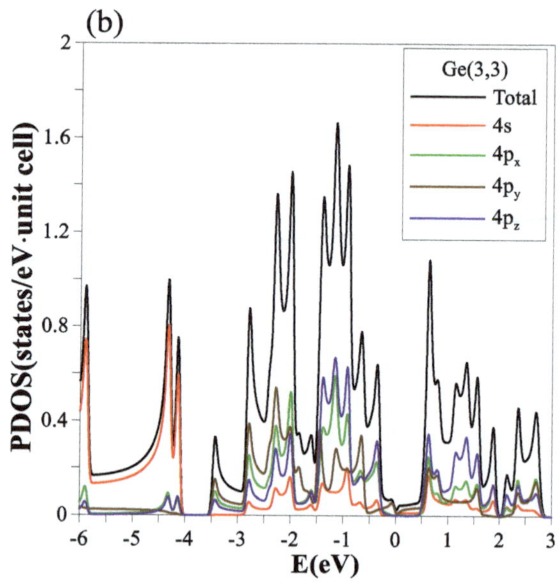

FIG. 13.16 (B) Projected density of states of armchair germanium nanotubes Ge (3,3).

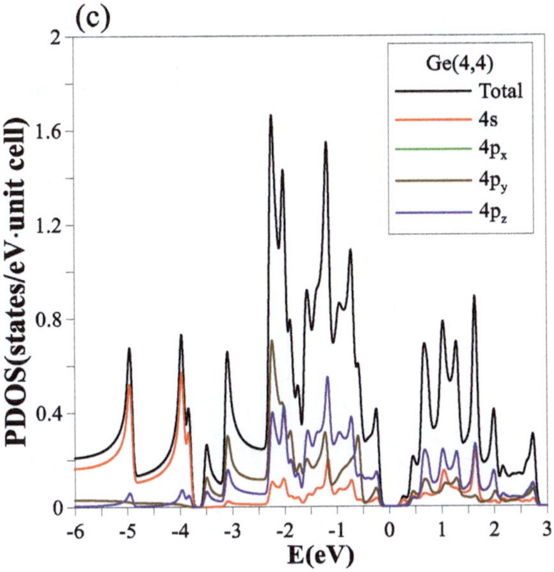

FIG. 13.17 (C) Projected density of states of armchair germanium nanotubes Ge (4,4).

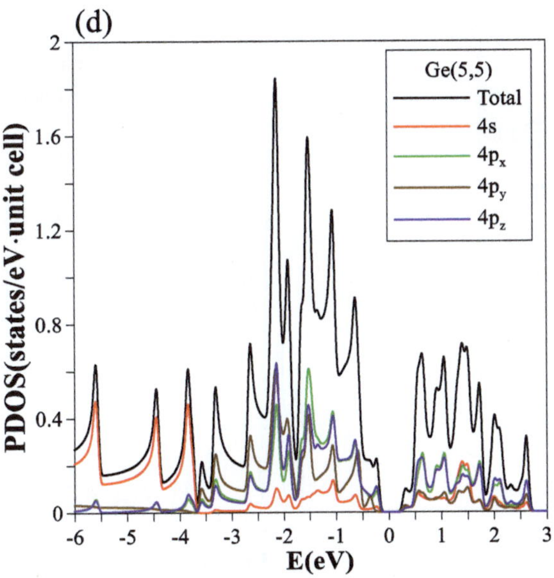

FIG. 13.18 (D) Projected density of states of armchair germanium nanotubes Ge (5,5).

FIG. 13.19 (E) Projected density of states of armchair germanium nanotubes Ge (6,6).

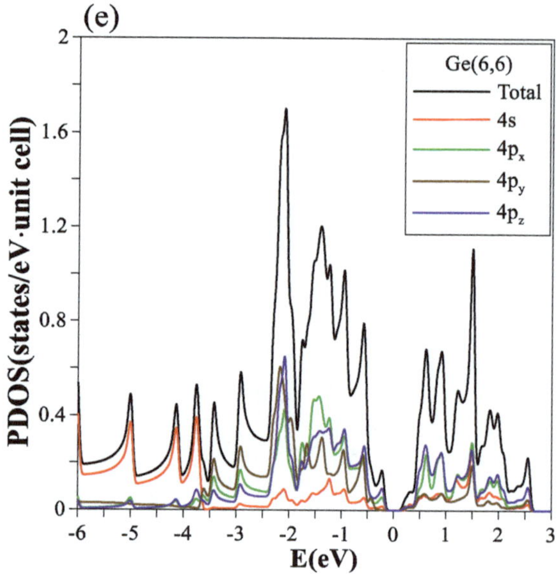

FIG. 13.20 (F) Projected density of states of armchair germanium nanotubes Ge (7,7).

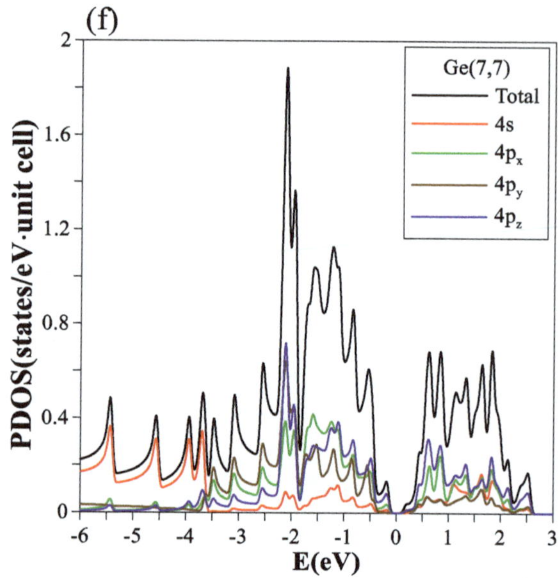

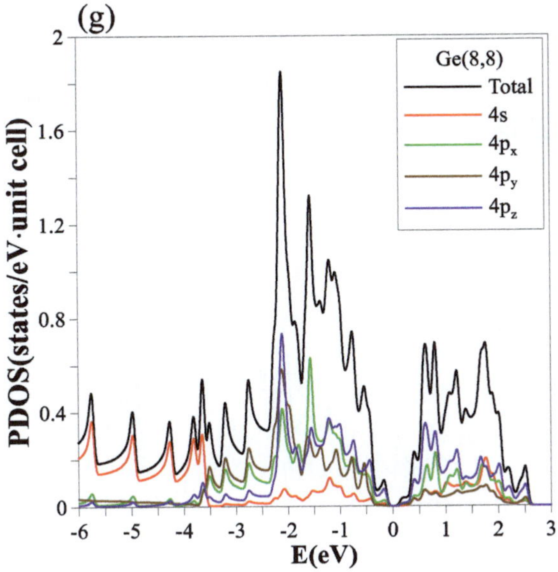

FIG. 13.21 (G) Projected density of states of armchair germanium nanotubes Ge (8,8).

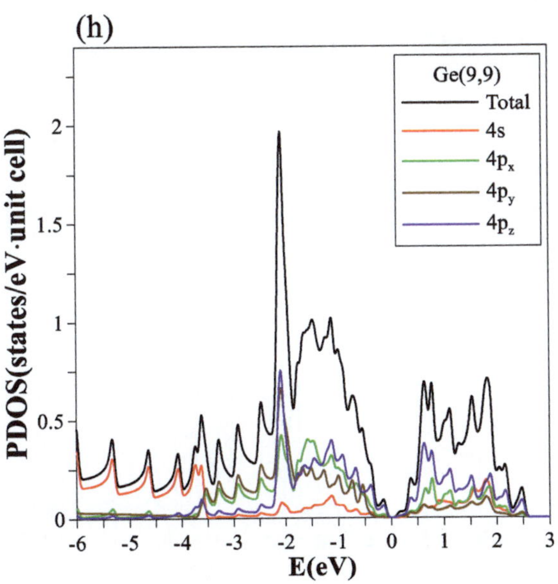

FIG. 13.22 (H) Projected density of states of armchair germanium nanotubes Ge (9,9).

FIG. 13.23 (I) Projected density of states of armchair germanium nanotubes Ge (10,10).

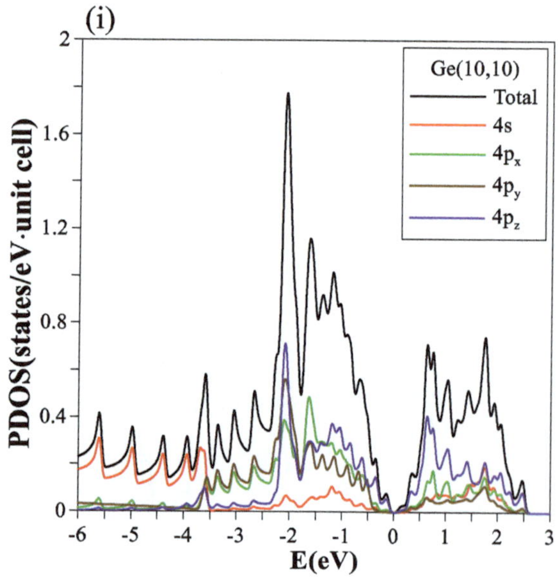

FIG. 13.24 (A) Projected density of states of zigzag germanium nanotubes Ge (6,0).

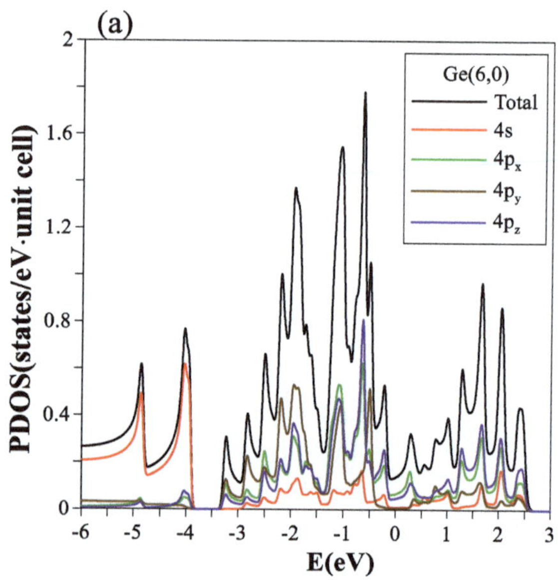

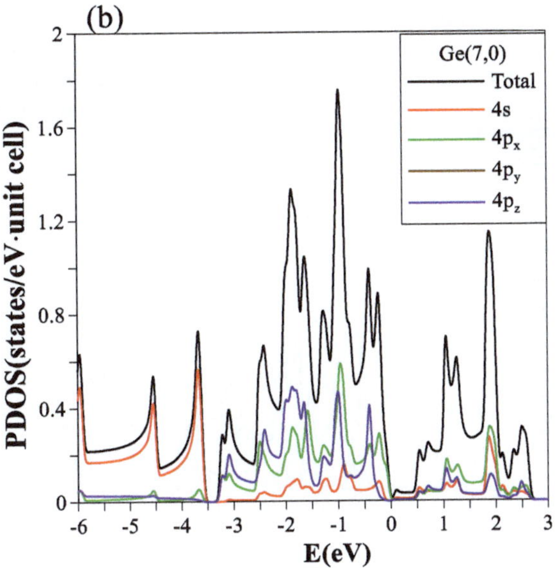

FIG. 13.25 (B) Projected density of states of zigzag germanium nanotubes Ge (7,0).

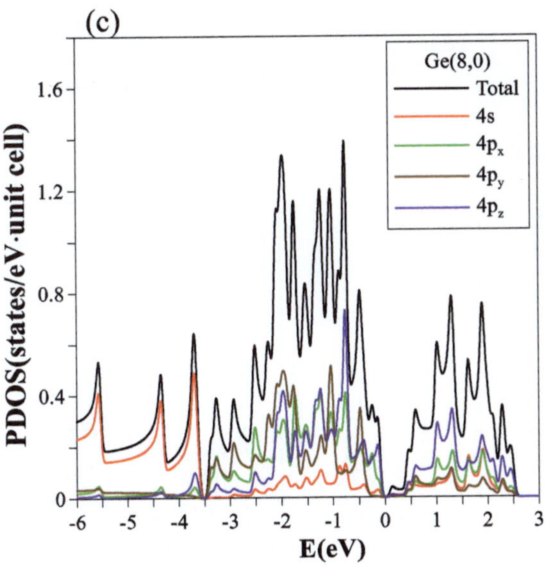

FIG. 13.26 (C) Projected density of states of zigzag germanium nanotubes Ge (8,0).

FIG. 13.27 (D) Projected density of states of zigzag germanium nanotubes Ge (9,0).

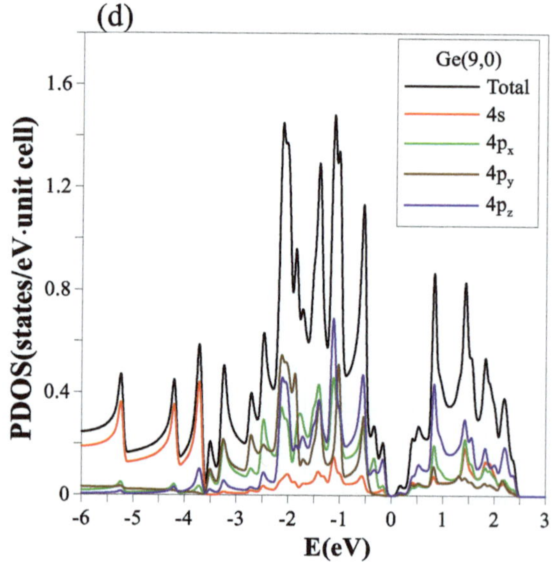

FIG. 13.28 (E) Projected density of states of zigzag germanium nanotubes Ge (10,0).

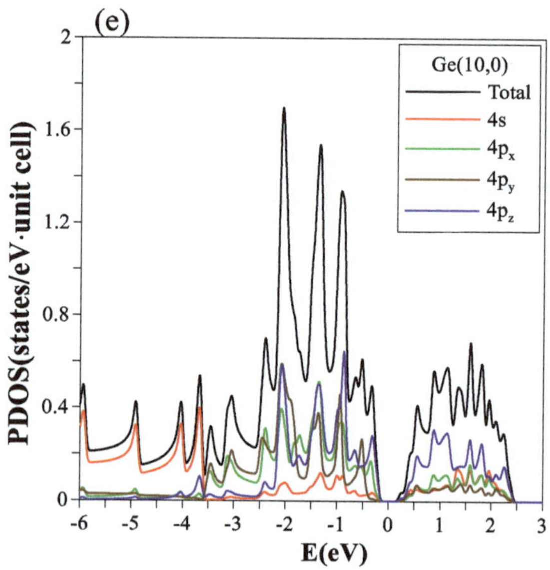

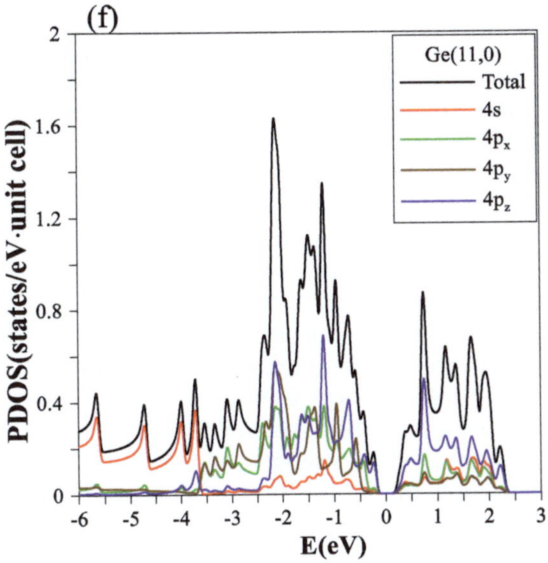

FIG. 13.29 (F) Projected density of states of zigzag germanium nanotubes Ge (11,0).

be investigated fully to realize the orbital-decomposed contributions. For example, the results of 2D materials of monolayer graphene appear zero at $E_F = 0$. The contributions of $2p_z$-orbital present a linear energy dependence at low energy, originating from almost a disappeared energy gap 0 eV and two symmetric logarithmic peaks at 2 and −2.5 eV, respectively, isotropic linear bands and the saddle points of π bands. For the case of monolayer germanene, the contribution at low energy between −1 and 1 eV are not fully dominated by $4p_z$-orbital. The contributions of $4p_x$, $4p_y$ and 4s-orbital are also involved. Other features like a shoulder structure and a prominent peak corresponding to the saddle point of σ subbands at M point. They appear in the middle energy of PDOSs and are mainly contributed by $4p_x$, $4p_y$ and $4p_z$ orbitals. Besides, $4p_z$-orbitals contribute mainly at low and middle energy level from −2 to 2 eV and it yields a distinct peak contributed by hybridization of (4s, $4p_x$, $4p_y$, $4p_z$)-orbitals.

The features of aGeNTs Ge (2,2) – Ge (10,10) in orbital-projected density of states (PDOSs) are illustrated in Figs. 13.15A–13.23I. There are plenty of symmetric and asymmetric peaks which come from the parabolic and partially flat sub-bands. Interestingly, the PDOSs of aGeNTs near the Fermi level E_F located at E = 0. They have a zero value for semiconducting nanotubes. PDOSs of armchair carbon nanotubes calculated by tight binding model show the peaks of mirror image between valence and conduction regions [43]. However, the peaks of aGeNTs occur between valence and conduction regions are not symmetric and it could be due to the unsymmetrical structure. Comparison with 1D PDOSs of a single wall germanium nanotube and 2D DOSs of a germanene monolayer, there are no independent orbitals contributed by $4p_z$-orbitals of aGeNTs [41,44]. Size of nanotubes reflect the strength of curvature effect and they further induce the hybridization of $4p_z$-orbtals. The case of Ge (2,2) shown in Fig. 13.15A appear a $4p_z$ signal initially at ca. −0.3 eV near Fermi level in the valence region.

It is not very significant compared with $4p_x$ and $4p_y$ signals. The small sized of nanotubes due to the increase of curvature lead to inhibition of sp^2 hybridization. When chiral vectors increase from Ge (3,3) to Ge (10,10), the first $4p_z$ peak in the valence region locates at around −0.2 eV to −0.3 eV is highly significant. The reduced curvature effect exhibits the major signals of $4p_z$-orbital and induces the contribution ratio of sp^2 and sp^3 hybridization. Curved geometric structures influence $4p_z$-orbital with interaction of (4s, $4p_x$, $4p_y$)-orbitals and causes the (4s, $4p_x$, $4p_y$) and (4s, $4p_x$, $4p_y$, $4p_z$) hybridization as well as sp^2 and sp^3 bonding.

Figs. 13.24A−13.29F show the multi-orbital projection PDOSs of zGeNT Ge (6,0) - Ge (11,0), respectively. The PDOSs of Ge (6,0) have a non-zero value at Fermi level (E_F) and it is a metallic germanium nanotube due to the mixing of σ^* and π^*. However, the PDOSs of Ge (7,0) − Ge (11,0) possess a zero value at (E_F) and exhibit semiconducting features. Each of the peaks is contributed by 4s-, $4p_x$-, $4p_y$- and $4p_z$-orbitals, it is hard to observe the dominated $4p_z$-orbital at the low energy level. because of curved structures most of them generate either sp^2 or sp^3 σ bonding. Independent $4p_z$ π bonding are not observable in the results of zigzag silicon nanotubes by methods of first principle. The strength of curvature effect induced by the hybridization of $4p_z$-orbitals is noticeable in the results of PDOSs of zGeNTs. The signals of $4p_z$ for Ge (6,0) accumulate at fermi level, then the peak shift to −0.4 eV for Ge (7,0), further the peak shifts to −0.1 eV location for Ge (8,0) and Ge (9,0). Finally, the peak shifts slowly but surely between −0.1 eV and −0.2 eV, when the size of tubes keep increasing Ge (11,0) shown in Fig. 13.29F. It locates at ca. −0.2 eV shown by the nanotubes of Ge (11,0). It is overall very difficult to distinguish the initial, middle (saddle-point structure) and final π/σ bands; thus, both armchair and zigzag germanium nanotubes are significantly different from each other. There are a lot of symmetric peaks, antisymmetric ones and shoulders in the whole energy range, mainly owing to the multi orbital hybridizations, the oscillatory energy dispersions, and the frequent anti crossings.

References

[1] Iijima S. Helical microtubules of graphitic carbon. Nature 1991;354:56.

[2] Journet C, Maser WK, Bernier P, Loiseau A, delaChapelle ML, Lefrant S, Deniard P, Lee R, Fischer JE. Large-scale production of single-walled carbon nanotubes by the electric-arc technique. Nature 1997;388:756.

[3] Thess A, Lee R, Nikolaev P, Dai H, Petit P, Robert J, Xu C, Lee YH, Kim SG, Rinzler AG, Colbert DT, Scuseria GE, Tomanek D, Fischer JE, Smalley RE. Crystalline ropes of metallic carbon nanotubes. Science 1996;273:483.

[4] Yudasaka M, Komatsu T, Ichihashi T, Iijima S. Single-wall carbon nanotube formation by laser ablation using double-targets of carbon and metal. Chem Phys Lett 1997;278:102.

[5] Ebbesen TW, Ajayan PM. Large-scale synthesis of carbon nanotubes. Nature 1992;358:220.

[6] Ebbesen TW, Hiura H, Fujita J, Ochiai Y, Matsui S, Tanigaki K. Patterns in the bulk growth of carbon nanotubes. Chem Phys Lett 1993;209:83.

[7] Seraphin S, Zhou D, Jiao J, Withers JC, Loutfy R. Effect of processing conditions on the morphology and yield of carbon nanotubes. Carbon 1993;31:685.

[8] Terrones M. Science and technology of the twenty-first century: synthesis, properties, and applications of carbon nanotubes. Annu Rev Mater Res 2003;33:419.

[9] Yue GZ, Qiu Q, Gao B, Cheng Y, Zhang J, Shimoda H, Chang S, Lu JP, Zhou O. Generation of continuous and pulsed diagnostic imaging x-ray radiation using a carbon-nanotube-based field-emission cathode. Appl Phys Lett 2002;81:355.

[10] Baughman R, Zakhidov A, de Heer WA. Carbon nanotubes-the route toward applications. Science 2002;297:787.

[11] Crompton KR, Hladky MP, Park HH, Prokes SM, Love CT, Landi BJ. Cycling performance of multi-walled carbon nanotube electrodes and current collectors coated with nanometer scale Al2O3 by atomic layer deposition. Electrochim Acta 2018;292:628.
[12] Frackowiak E, Beguin F. Electrochemical storage of energy in carbon nanotubes and nanostructured carbons. Carbon 2002;40:1775.
[13] Memarian F, Fereidoon A, Khodaei S, Mashhadzadeh AH, Ganji MD. Molecular dynamic study of mechanical properties of single/double wall SiCNTs: consideration temperature, diameter and interlayer distance. Vacuum 2017;139:93.
[14] Chen S, Chen P, Wang Y. Carbon nanotubes grown *in situ* on graphene nanosheets as superior anodes for Li-ion batteries. Nanoscale 2011;3:4323.
[15] Park MH, Kim MG, Joo J, Kim K, Kim J, Ahn S, Cui Y, Cho J. Silicon nanotube battery anodes. Nano Lett 2009;9:3844.
[16] Park MH, Cho YH, Kim K, Kim J, Liu M, Cho J. Germanium nanotubes prepared by using the Kirkendall effect as anodes for high-rate lithium batteries. Angew Chem Int Ed 2011;50:9647.
[17] Li X, Meng G, Xu Q, Kong M, Zhu X, Chu Z, Li AP. Controlled synthesis of germanium nanowires and nanotubes with variable morphologies and sizes. Nano Lett 2011;11:1704.
[18] Chegel R, Behzad S. Bandstructure modulation for Si-h and Si-g nanotubes in a transverse electric field:Tight binding approach. Superlattice Microst 2013;63:79.
[19] Ahmadi N, Shokri AA, Elahi SM. Optical transition of zigzag silicon nanotubes under intrinsic curvature effect. Silicon 2016;8:217.
[20] Behzad S, Chegel R. Magnetic field-induced splitting of optical spectra in silicon nanotubes: tight binding calculations. Silicon 2016;8(43):2016.
[21] Chegel R, Behzad S. Electronic properties of SiNTs under external electric and magnetic fields using the tight-binding method. J Electron Mater 2014;43:329.
[22] Zhang RQ, Lee HL, Li WK, Teo BK. Investigation of possible struc- tures of silicon nanotubes via density-functional tight-binding molecular dynamics simulations and ab initio calculations. J Phys Chem B 2005;109:8605.
[23] Seifert G, Koehler T, Urbassek HM, Hernandez E, Frauenheim T. Tubular structures of silicon. Phys Rev B 2001;63:193409.
[24] Yang X, Ni J. Electronic properties of single-walled silicon nanotubes compared to carbon nanotubes. Phys Rev B 2005;72:195426.
[25] Bayani AH, Dideban D, Akbarzadeh M, Moezi N. Benchmarking performance of a gate-all-around germanium nanotube field effect transistor (GAA-GeNTFET) against GAA-CNTFET. ECS J Solid State Sci Technol 2017;6:M24.
[26] Steele GA, Pei F, Laird EA, Jol JM, Meerwaldt HB, Kouwenhoven LP. Large spin-orbit coupling in carbon nanotubes. Nat Commun 2013;4.
[27] Kuemmeth F, Ilani S, Ralph D, McEuen P. Coupling of spin and orbital motion of electrons in carbon nanotubes. Nature 2008;452:448.
[28] Churchill HOH, Kuemmeth F, Harlow JW, Bestwick AJ, Rashba EI, Flensberg K, Stwertka CH, Taychatanapat T, Watson SK, Marcus CM. Relaxation and dephasing in a two-electron C-13 nanotube double quantum dot. Phys Rev Lett 2009;102:166802.
[29] Jhang SH, Marganska M, Skourski Y, Preusche D, Witkamp B, Grifoni M, van der Zant H, Wosnitza J, Strunk. Spin-orbit interaction in chiral carbon nanotubes probed in pulsed magnetic fields. Phys Rev B 2010;82:041404.
[30] Jespersen TS, Grove-Rasmussen K, Paaske J, Muraki K, Fujisawa T, Nygard J, Flensberg K. Gate-dependent spin-orbit coupling in multielectron carbon nanotubes. Nat Phys 2011;7:348.
[31] Guan L, Suenaga K, Iijima S. Smallest carbon nanotube assigned with atomic resolution accuracy. Nano Lett 2008;8:459.
[32] Zhao X, Liu Y, Inoue S, Suzuki T, Jones RO, Ando Y. Smallest carbon nanotube is 3 Å in diameter. Phys Rev Lett 2004;92:125502.
[33] Hamada N, Sawada S, Oshiyama A. New one-dimensional conductors: graphitic microtubules. Phys Rev Lett 1992;68:1579.
[34] Mintmire J, Dunlap B, White C. Are fullerene tubules metallic? Phys Rev Lett 1992;68:631.

[35] Saito R, Fujita M, Dresselhaus G, Dresselhaus MS. Electronic structure of chiral graphene tubules. Appl Phys Lett 1992;60:2204.
[36] Guelseren O, Yildirim T, Ciraci S. Systematic ab initio study of curvature effects in carbon nanotubes. Phys Rev B 2002;65:153405.
[37] Wang C, Fu X, Guo Y, Guo Z, Xia C, Jia Y. Band gap scaling laws in group IV nanotubes. Nanotechnology 2017;28:115202.
[38] Ouyang M, Huang JL, Cheung CL, Lieber CM. Energy gaps in "metallic" single-walled carbon nanotubes. Science 2001;292:702.
[39] Kane CL, Mele E. Size, shape, and low energy electronic structure of carbon nanotubes. Phys Rev Lett 1997;78:1932.
[40] Blase X, Benedict LX, Shirley EL, Louie SG. Hybridization effects and metallicity in small radius carbon nanotubes. Phys Rev Lett 1994;72:1878.
[41] Lin SY, Chang SL, Thuy Tran NT, Yang PH, Lin MF. H-Si bonding-induced unusual electronic properties of silicene: a method to identify hydrogen concentration. Phys Chem Chem Phys 2015;17:26443.
[42] Jacobsen H. Chemical bonding in view of electron charge density and kinetic energy density descriptors. J Comput Chem 2009;30:1903.
[43] Rao AM, Richter E, Bandow S, Chase B, Eklund PC, Dresselhaus MS, et al. Diameter-selective Raman scattering from vibrational modes in carbon nanotubes. Science 1997;275:187.
[44] Liu HY, Lin SY, Wu JY. Stacking-configuration-enriched essential properties of bilayer graphenes and silicenes. J Chem Phys 2020;153:154707.

CHAPTER

14

Dimension-enriched essential properties of Ge-related materials

Thi My Duyen Huynh[1], Shih-Yang Lin[1], Vo Khuong Dien[1], Chi-Hsuan Lee[1], Hsin-Yi Liu[1], Hai Duong Pham[1], Nguyen Thi Han[1], Ngoc Thanh Thuy Tran[2], Thi Dieu Hien Nguyen[1], Wei-Bang Li[1] and Ming-Fa Lin[1]

[1]Department of Physics, National Cheng Kung University, Tainan, Taiwan; [2]Hierachical Green-Energy Materials (Hi-GEM) Research Center, National Cheng Kung University, Tainan, Taiwan

14.1 Overview

Very interesting, graphene-/silicene-/germanene-/tinene-/plumbene-related materials have stirred a lot of basic [1–8] and applied [9–16] science researches, mainly owing to the rich and unique physical/chemical/material phenomena [1,7,17–31]. The systematic investigations, which are developed under the quasiparticle frameworks in the previous books [32], are successfully conducted on graphite- [33,34], graphene- [5], graphene-nanoribbon- [2,35], carbon nanotube- [36–38], and silicene-based systems [8]. Apparently, there exist certain important similarities [8,21,23,30] and differences [39,40] among them, being attributed to the concise physical and chemical environments. For example, according to the boundary conditions (BCs), a monolayer graphene [41,42], a carbon nanotube [4] and a graphene nanoribbon [35,43], respectively, possess a periodic BC on a 2D plane, a similar one on a hollow cylinder, and open BCs on two open edges. Qualitatively, this will generate the geometry-dominated low-lying valence and conduction energy sub-bands, leading to the featured band gaps. Band gap Eg, which mainly come from the carbon-2pz orbitals in the sufficiently large/wide nanotubes/nanoribbons, are identified to exhibit the unusual (radius/chirality)/(width/chirality) dependences from both theoretical predictions [44,45] and experimental examinations [43,46–51]. (m, n) carbon nanotubes, being characterized by a planar

Fundamental Physicochemical Properties of Germanene-related Materials
https://doi.org/10.1016/B978-0-443-15801-8.00020-7

honeycomb crystal [37,38], are classified into three types: (a) 1D metals for $m = n$ armchair systems [52], (b) $2m + n = 3I$ and m, n narrow-gap semiconductors with Eg inversely proportional to the square of radius due to significant curvature effects [53,54], and others for middle-gap semiconductors in an inverse relation between band gap and radius. On the graphene-nanoribbon side, their band gaps quickly decline as width grows. Only achiral armchair nanoribbons display three type of gap approaches based on the number of dimer lines, in which $Na = 3I + 2$ belong to narrow-gap semiconductors [55], but 3I and $3I + 1$ are middle-gap ones [55–57]. Such rules arise from open boundaries [41,42] and bond length non-uniform fluctuations [58]. The close relations of geometric symmetries and electronic properties are thoroughly test by the simultaneously measurements of high-resolution STM and STS (detail in Chapter 3). Specifically, the dangling zigzag carbon structures, e.g., the local vacancies in graphene-related surfaces [59,60], the two open boundaries of finite-length carbon nanotubes [61,62], the chiral boundaries of graphene nanoribbons [44], are predicted to show the ferromagnetic spin configuration along the same boundary. The above-mentioned quasiparticle phenomena will become diversified for the other outstanding candidates, such as, 1D germanene nanoribbons in the current book under the first-principles calculations [63,64].

The rather active physical/chemical/material environment [10,24,39,65], which consist of the finite-width confinement [24,48], achiral/chiral edge structures [22,25,46], and the planar/deformed honeycomb crystals, are responsible for the diverse quasiparticle behaviors in group-IV-related 1D nanoribbon systems. Very interesting, the close relations of crystal structures among nanoribbon, nanotube and monolayer have been utilized to successfully synthesize the specific low-dimensional materials [66,67]. For example, each graphene nanoribbon could be regarded as a direct cut of graphene along the longitudinal direction [55–57,68]; furthermore, it corresponds to an unrolled carbon nanotube. The former has been grown under the various physical [69–72] and chemical methods [73,74]. The up-to-date experimental growths on graphene nanoribbons clearly show the well-behaved honeycomb planes [75–78], the scrolled/curved structures [4,79–84], the folded ones [85,86], as well as, the multi-walled lattices with different stacking configurations [87–89] and layer numbers [90]. Apparently, the dangling bonds, which survive at surfaces [91,92] and open boundaries [41,42,62], will provide the highly non-uniform/anisotropic environments and thus create the diversified phenomena through the modulations of chemical decorations [93–95]/absorptions [96–98]/intercalations [99,100]/substitutions [101–103], mechanical strains [54,93,104], transverse and longitudinal gate voltages [105], and a perpendicular uniform magnetic field [72]. The diverse quasiparticle behaviors are expected to be clearly revealed in germanene- and silicene-related nanoribbons, mainly due to more complicated couplings of significant mechanisms [56,65,77,92] (discuss later in detail).

The thorough investigations have been done for monolayer graphene [31,41]/silicene [106]/germanene/tinene/plumbene [8] materials both theoretically [6,8,106] and experimentally [10,21,27,28,30,59,60]. From the theoretical viewpoints, the significant orbital hybridizations, which mainly determine the low-energy essential properties [107], are clearly identified to be the perpendicular π and σ bondings in a planar graphene [108], as well as, the dominating sp^2, but observable sp^3 bondings in the other buckled honeycomb crystal [23,109,110]. Both π and π^* conduction sub-bands, which are created by C-$2p_z$/Si-$3p_z$/Ge-$4p_z$ orbitals, can be well distinguished from the deeper σ ones. Of course, the π electronic model is frequently utilized in graphene-related materials with very strong σ bonding as clearly illustrated in the previously published books [32].

The first-principles simulations [64], but not the phenomenological models [63], are very suitable for fully exploring the essential properties of 1D germanene nanoribbons in the absence of a uniform perpendicular magnetic field [3,17,111,112]. The significant coupling effects, which principally come from the finite-width quantum confinement [61,62,78], open edge structures, and a buckled honeycomb crystal [8,31,109,110], will be thoroughly examined to clearly illustrate the intrinsic orbital hybridizations the spin-created single-particle and bi-particle interactions (the spin orbit couplings and the on-site Coulomb interactions) [1,109,113]. Obviously, a pristine germanene nanoribbon is shown to present a highly active chemical environment because of a lot of dangling bonds at a planar surface [2,17] and two open edges [17,111]. The specific passivation of hydrogen atoms on boundaries if able to test the great suppression of the large bond-length fluctuations. All the VASP simulations of physical and chemical quantities are accurately tested to be consistent with one another under a unified quasiparticle framework, as done in the previous and current books. The close relations between theoretical predictions and experimental verifications are fully explored for the essential properties, such as, how to observe the sensitive dependence of Eg on width and edge structure [1,17,55,111,112]. The important differences among germanene-, silicene- and graphene-based nanoribbons are discussed in detail. Weather the tight-binding model/the generalized tight-binding models [18], with the concise physical and chemical pictures, is worthy of the systematic investigations, will be evaluated in detail.

14.2 One-dimensional germanene nanoribbons

14.2.1 Zigzag system

Zigzag germanene nanoribbons (ZGeNRs) with the two edges are defined by the number of zigzag lines labeled N_Z. In this chapter, the pristine (without hydrogen atoms at the ribbon edge) and hydrogen passivated germanene nanoribbons of $N_Z = 8$ are examined as shown in Fig. 14.1. They exhibit the non-uniform geometric configuration illustrated in chemical bonds near the boundaries as listed in Table 14.1. The first Ge—Ge bond are respectively 2.6 and 2.63 Å while the second bonds are 2.69 and 2.693 Å in absence and presence of hydrogen systems, leading to the zigzag edge reconstruction. The Ge—H bond length is approximately 1.5 Å illustrating the identical orbital hybridization due to the edge structures. The ground state energies shown in Table 14.1 in two cases of edge are obviously different presenting spin distributions. In addition, nanoribbon systems remain the buckled structure of germanene monolayer [8,66], in which a mixed sp^2—sp^3 bonding exists due to chemical bonding between the larger atoms.

Electronic properties of ZGeNRs are mainly defined by the boundaries, dandling bonds and quantum confinement. ZGeNRs belong to semimetal in both of absence and presence hydrogen atoms due to the small overlap between conduction and valence bands as shown in Fig. 14.2. In this figure, their energy bands have strong dispersions. The low-ling valence and conduction bands reveal a special valley centered at $k_z = 2/3$. This feature further demonstrates the low-energy electronic states are sampled from those near the Dirac points of monolayer germanene (see Fig. 14.3A) corresponding to the K and K′ valleys. Furthermore, the electronic states are doubly degenerate for the spin polarized configurations in both

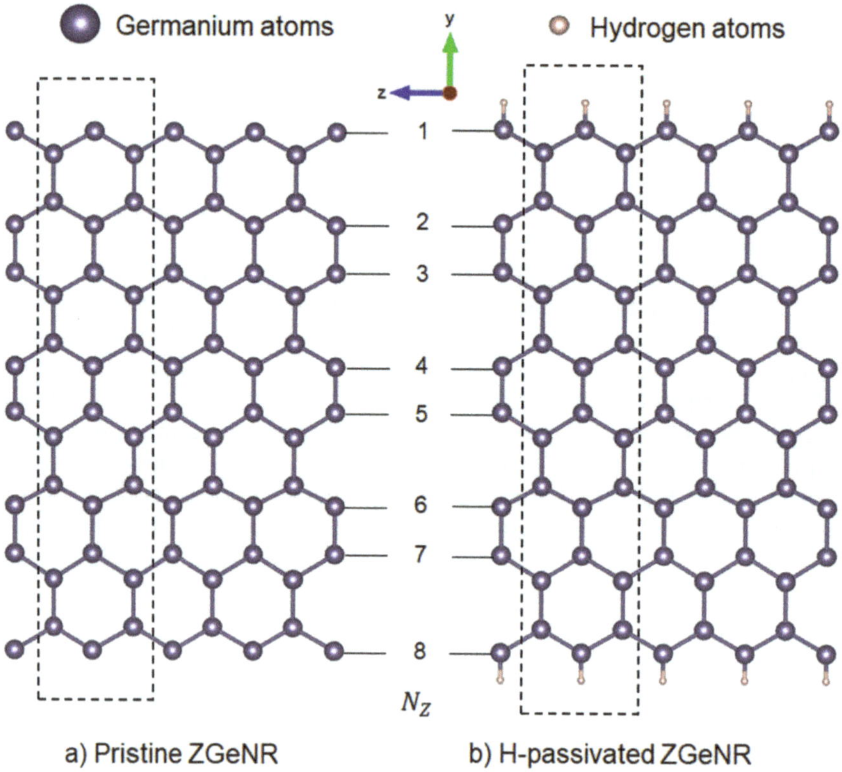

FIG. 14.1 Geometric structure of (A) pristine and (B) hydrogen passivated zigzag nanoribbons.

TABLE 14.1 Bond length (Å), buckle (Δ, Å), ground state energy (eV), binding energy (eV), magnetism (μB), band gap (eV) of zigzag nanoribbons.

ZGeNR	Ground state energy (eV)	Binding energy (eV)	Bond length (Å) Ge–Ge 1st	Bond length (Å) Ge–Ge 2nd	Bond length (Å) Ge–H	Buckle (Å)	Magnetism (μB)	Band gap (eV) SOC	Band gap (eV) Spin polarized
Pristine	−53.5	–	2.6	2.69	–	0.7	0.01	SM	SM
H-passivated	−47.18	−9.135	2.603	2.693	1.5	0.7	0.52	SM	SM

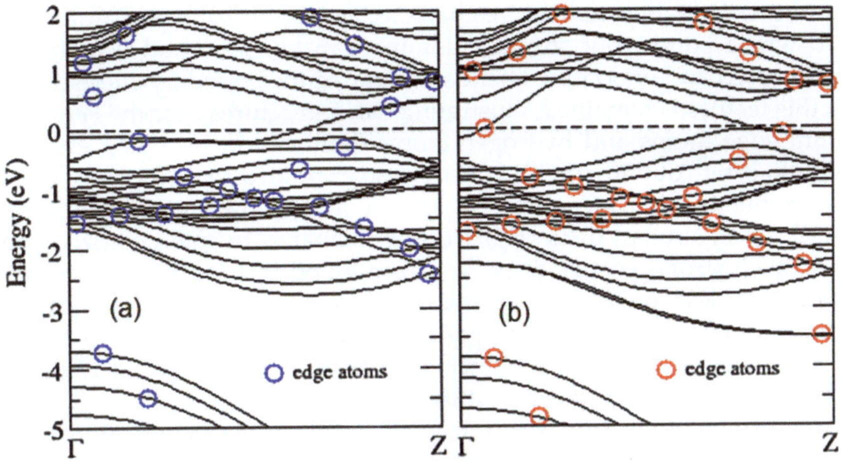

FIG. 14.2 Band energy spectra of (A) pristine and (B) hydrogen passivated zigzag nanoribbons.

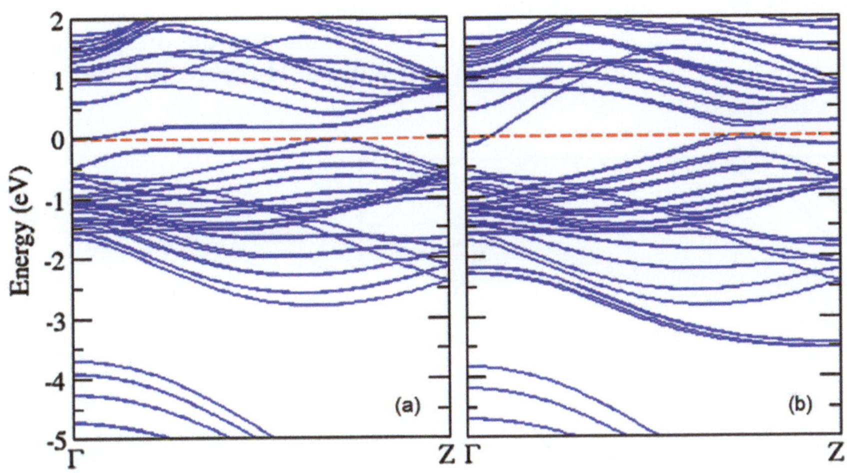

FIG. 14.3 Band energy spectra of (A) pristine and (B) hydrogen passivated zigzag nanoribbons considering spin-orbit coupling effect.

absence and presence of hydrogen passivation, corresponding to the ferromagnetic one. For pristine system, partial flat bands are exhibited within energy range of (−1,1) eV at $k_z > 2/3$ coming from edge atoms as described by blue circles in this zone (Fig. 14.2A). These energy bands vanish in hydrogen passivated case (Fig. 14.2B). The edge atoms contribution in H-passivated is stronger than that of pristine system at higher and deeper conduction and valence bands. Importantly, the hydrogen termination could suppress the edge dangling Ge–Ge bonds leading to induce conduction and valence bands. There exist energy spacing located at $E < 3.3$ eV in absence H case which become smaller in presence H one due to the band-edge valence state with atom dominances (shown by blue and red circles) anticipating the critical boundaries in characterization of band structures. Edge atoms contribute to whole bands, but strongly dominate in valence band near the Fermi level, particularly occupied and unoccupied bands, indicating the boundary effects. After hydrogen passivation, this feature is remained, illustrating band structures near the Fermi level depend on germanium edge atoms and hydrogen atoms dominate in low-lying energies. Simultaneously, the spin-orbit coupling (SOC) effect is also examined to further provide detail information in nanoribbon characteristics. As shown in Fig. 14.4, the band energy spectra including SOC reveal similar/distinct properties compared to spin-polarized configuration. In similarities, such strong energy dispersions are found in conduction and valence bands. The band structures also present the essential features such as partial flat bands located at $k_z > 2/3$, strong dispersion come from parabolic bands, spacing state in range of energy less than $-3/-3.5$ eV in absence/presence H system, e.g., In distinct characteristics, the SOC energy bands show significant features of energy states near the Fermi level. The overlap between conduction and valence bands induces in both cases. An open gap occurs at $k_z = 2/3$ in H-passivated system corresponding to K valleys in germanene monolayer. A pair of symmetric unoccupied conduction and occupied valence bands appear at $k_z < 2/3$. It should be noted that the SOC configuration is mainly affect band characteristic near the Fermi level in nanoribbons due to boundaries.

Features in energy band structures are further examined in the orbital-projected DOS as shown in Fig. 14.5. For $N_z = 8$ system with and without H-passivated, they reveal a plenty

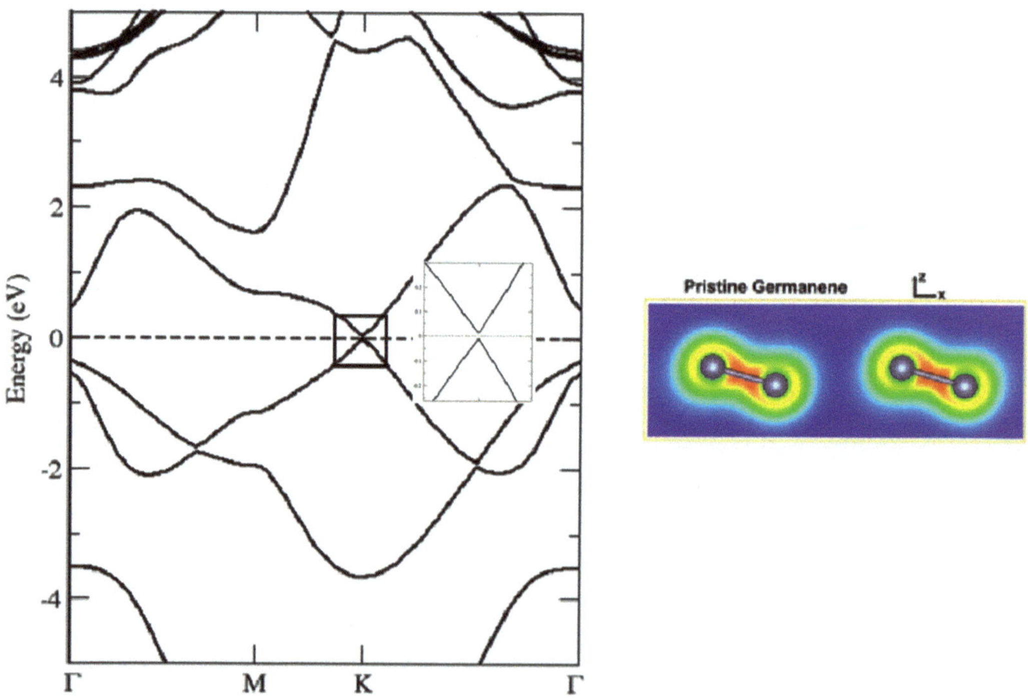

FIG. 14.4 Band energy spectra and (x, z) plane charge density [114] of monolayer germanene.

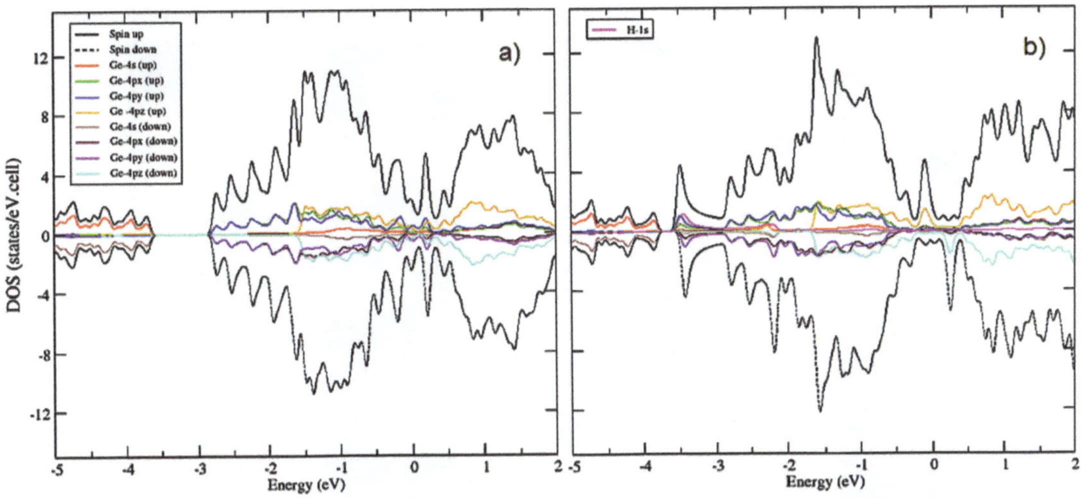

FIG. 14.5 Orbital projected DOS of (A) pristine and (B) hydrogen passivated zigzag nanoribbons.

of asymmetric peaks due to parabolic energy bands in the two open boundaries. They arise from the valence and conduction valleys nearest the Fermi level. Moreover, a delta-function form exists near the Fermi level based on the partial flat energy bands. These strong peaks arising from the pair of edge-localized energy bands, mainly due to the almost dispersion conduction and valence bands in the range of $k_z > 2/3$. At deep energy lower than −2.5 eV, the Ge and H contributions could appear indicating the strong Ge—H bonds. The peaks in DOS of absence/presence H-passivated are associated with the sp^3/π bondings of $(4s,4p_x,4p_y,4p_z)/4p_z$ orbitals, illustrating the orbital hybridizations in these bonds. The π and σ bonds significantly contribute to the low-lying valence energy DOS. Especially, there exists the sp^2-s bonding of the Ge—H bonds come from the presence of $(4s,4p_x,4p_y)$-1s orbitals.

Furthermore, these characteristics related to chemical bonds and energy bands could be verified from the charge density described in Fig. 14.6. The strong bonding strength of Ge—Ge bonds are obviously indicated in (x, y) plane (red regions) as shown in Fig. 14.6A. All the Ge—Ge bonds possess the strong covalent σ bonds compared to germanene monolayer system (see Fig. 14.3B). Specially, Ge—Ge bonds are slightly weakened (see red regions of (x, y) plane in Fig. 14.6B) corresponding to the strong Ge—H bondings in H-passivated system. According to charge transfer, ρ in H-passivated system reveal essential charge transfer to hydrogen compared to that of absence H system (see red regions near H atoms and edge of (y, z) plane).

In general, ZGeNRs of absence and presence hydrogen termination present features of semimetal with ferromagnetic configuration. Their geometry shows non-uniform structures through two ribbon-nearest Ge—Ge bonds leading to buckle honeycomb. The sp^3/sp^2-s bonding near ribbon boundaries is found in pristine/H-passivated system based on boundary terminations. Two spin configurations of spin polarization and SOC reveal similarities and differences as described in band energy spectra. They exhibit strong and special energy

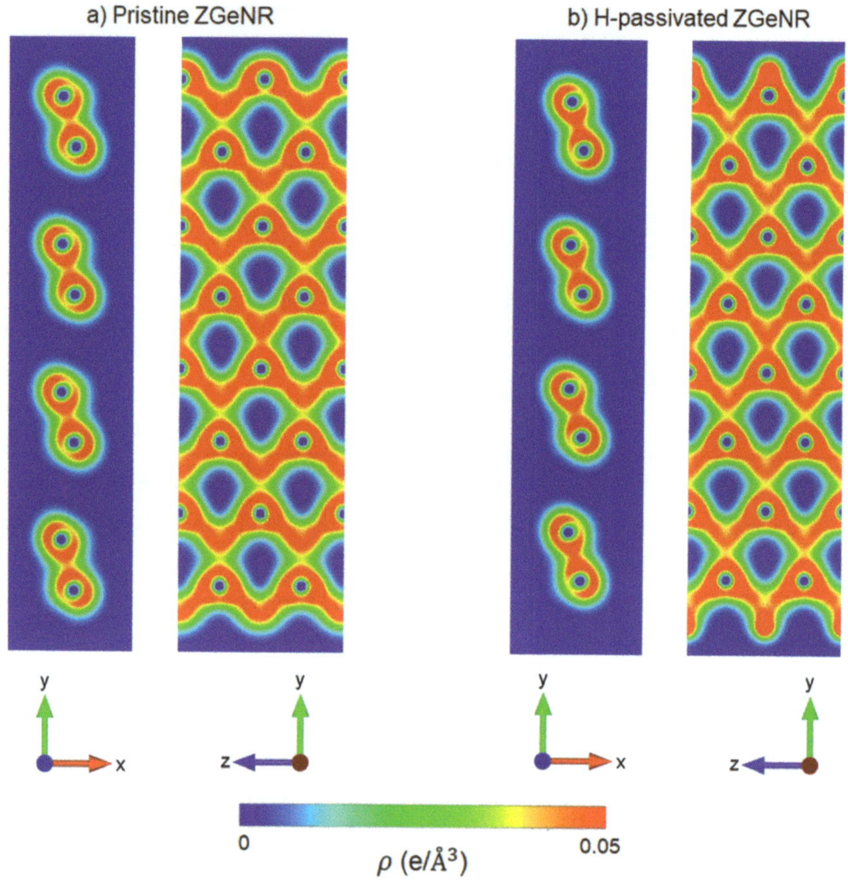

FIG. 14.6 Charge density of $N_z = 8$ (A) pristine and (B) hydrogen passivated zigzag nanoribbons.

dispersion due to parabolic bands and partial flat bands, occupied-unoccupied bands, spacing states. Thus, the boundaries, chemical bonds and quantum confinement could determine characteristic of ZGeNRs.

14.2.2 Armchair system

As mentioned above, various widths are considered to verify dependences between band gap and boundaries. Being similar to graphene nanoribbons [35,55], the widths of armchair germanene nanoribbons (AGeNRs) are determined by the number of dimer lines N_A along y direction as shown in Fig. 14.7. There are three types of system approaches in armchair nanoribbons, which are respectively 3I, 3I + 1 and 3I + 2, with I is integer, in which I = 2 and I = 3 are chosen in the system. The pristine and hydrogen-passivated structures are constructed to illustrate the essential rules of boundaries and bond lengths. Besides, due to spin distribution and boundary terminations, the ground state energies in presence of hydrogen are smaller

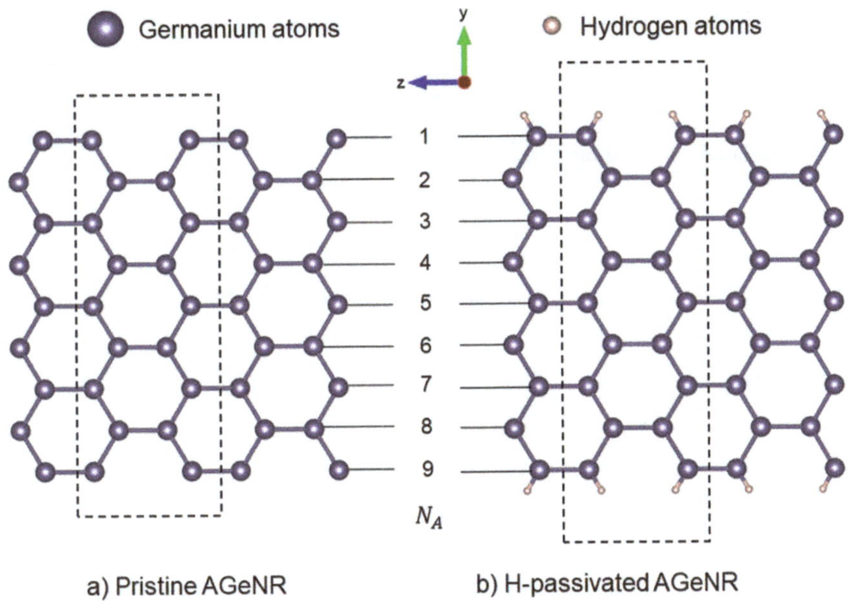

FIG. 14.7 Geometric structure of (A) pristine and (B) hydrogen passivated armchair nanoribbons.

than those of absence systems listed in Table 14.2. For the pristine system, the nearest bonds (or the first bond length) are shorter than that of the second ones, leading to the geometric reconstruction. As a case, the nearest Ge—Ge length of the $N_A = 6$ system is 2.45 Å (see in Table 14.2) while the second Ge—Ge bond length is 2.48 Å, demonstrating the non-uniform of Ge—Ge bond lengths. The similar features can be found in other system as listed in Table 14.2. Compared to zigzag system, the armchair nanoribbon reconstruction is more serious. The C—H bondings are therefore formed to create the stable dangling bonds. In the hydrogen-passivated system, the strong bonding sp^2 of C—H bonds is slightly affected by different widths as shown in Table 14.2, for example, the $N_A = 6$ system has the C—H bond of 1.3 Å while this bond is 1.26 Å in the $N_A = 11$ system. Compared to the pristine structures, the first and second Ge—Ge bonds in all systems are longer than those of pristine one. Take an example of the $N_A = 6$ system, these bonds are respectively 2.59 and 2.62 Å while the pristine case has the bonds of 2.45 and 2.48 Å. Obviously, the terminated boundary affects the chemical bonds illustrated in changing of bond lengths in the two above cases.

Electronic properties of pristine and hydrogen passivated AGeNRs are respectively defined by the edge structure and quantum confinement effects as shown in Figs. 14.8 and 14.9. However, they exist the distinct properties depending on the edge structure and ribbon width. As above mentioned, there are three group of ribbon width approaches; that is 3I, 3I + 1 and 3I + 2 considered in both cases of boundaries. With regarding to pristine ones, energy bands clearly present strong dispersion relations. The small overlap between conduction and valence bands is observed, anticipating semimetal characteristic of all systems. All structures exhibit the partial flat band near the Fermi level that could lead to the delta-function form of the peaks in DOSs. In addition, they possess a pair of linear bands about

TABLE 14.2 Bond length (Å), buckle (Δ, Å), ground state energy (eV), binding energy (eV), magnetism (μB), band gap (eV) of armchair nanoribbons.

System		Ground state energy (eV)	Binding energy (eV)	Bond length (Å) Ge–Ge		Ge–H	Buckle (Å)	Magnetism (μB)	Band gap (eV)	
AGeNR	N_A			1st	2nd				SOC	Spin polarized
Pristine	6	−43.53	–	2.453	2.484	–	0.661	0.89	0.15	0.15
	7	−52.01	–	2.456	2.487	–	0.662	0.73	SM	SM
	8	−58.49	–	2.462	2.494	–	0.664	1	SM	SM
	9	−66.3	–	2.457	2.489	–	0.663	1.15	0.1	SM
	10	−73.75	–	2.459	2.49	–	0.663	0.37	0.07	SM
	11	−81.44	–	2.463	2.493	–	0.664	3.8	SM	SM
H-passivated	6	−53.19	2.31	2.59	2.621	1.3	0.7	0	0.42	0.42
	7	−60.45	0.5	2.573	2.605	1.286	0.69	0	0.36	0.36
	8	−67.81	−1.34	2.557	2.59	1.278	0.69	0	SM	SM
	9	−75.55	−3.28	2.544	2.576	1.272	0.69	0	0.33	0.33
	10	−83.09	−5.17	2.534	2.566	1.267	0.68	0	0.26	0.26
	11	−90.69	−7.07	2.527	2.56	1.265	0.68	0	SM	SM

the Fermi level leading to be gapless. The more energy sub-bands at deeper/higher energy ranged in $|E| > 1$ eV are observed since enhancing the ribbon width. Especially, the NA = 6 system (Figs. 14.8A and 14.9A) reveal a narrow indirect band gap while this gap is absent in orbital-project DOS. This might come from the different number of spin up and spin down. Based on the spin distribution differences of two spins, the spin transition is found in PDOS might cause this feature. With atoms dominance, edge atoms (blue circle in Fig. 14.8) exhibit the strong contribution in whole bands, specially, at range of energy near the Fermi level, illustrating the boundary termination effect.

As an alternative for stable structures, H-passivated armchair systems create the edge terminations that affect the electronic properties of AGeNRs as shown in Fig. 14.9. Among them, 3I, 3I + 1 systems belong to middle-gap semiconductors for $N_A = 6$, 7 and 9, 10 while 3I + 2 systems are semimetal for $N_A = 8$, 11 listed in Table 14.2. As shown in this table, the energy gaps depend on the ribbon width classified in these three groups. The largest energy gap occurs for $N_A = 3I$ and the gapless one for $N_A = 3I + 2$ at the same I value. Furthermore, band gaps reduce quickly when increasing the ribbons width, indicating inverse relations from the well fittings. Therefore, the energy gaps clearly and strongly based on N_A corresponding to three N_A categories. It should be noted that the hydrogen terminations could enhance energy

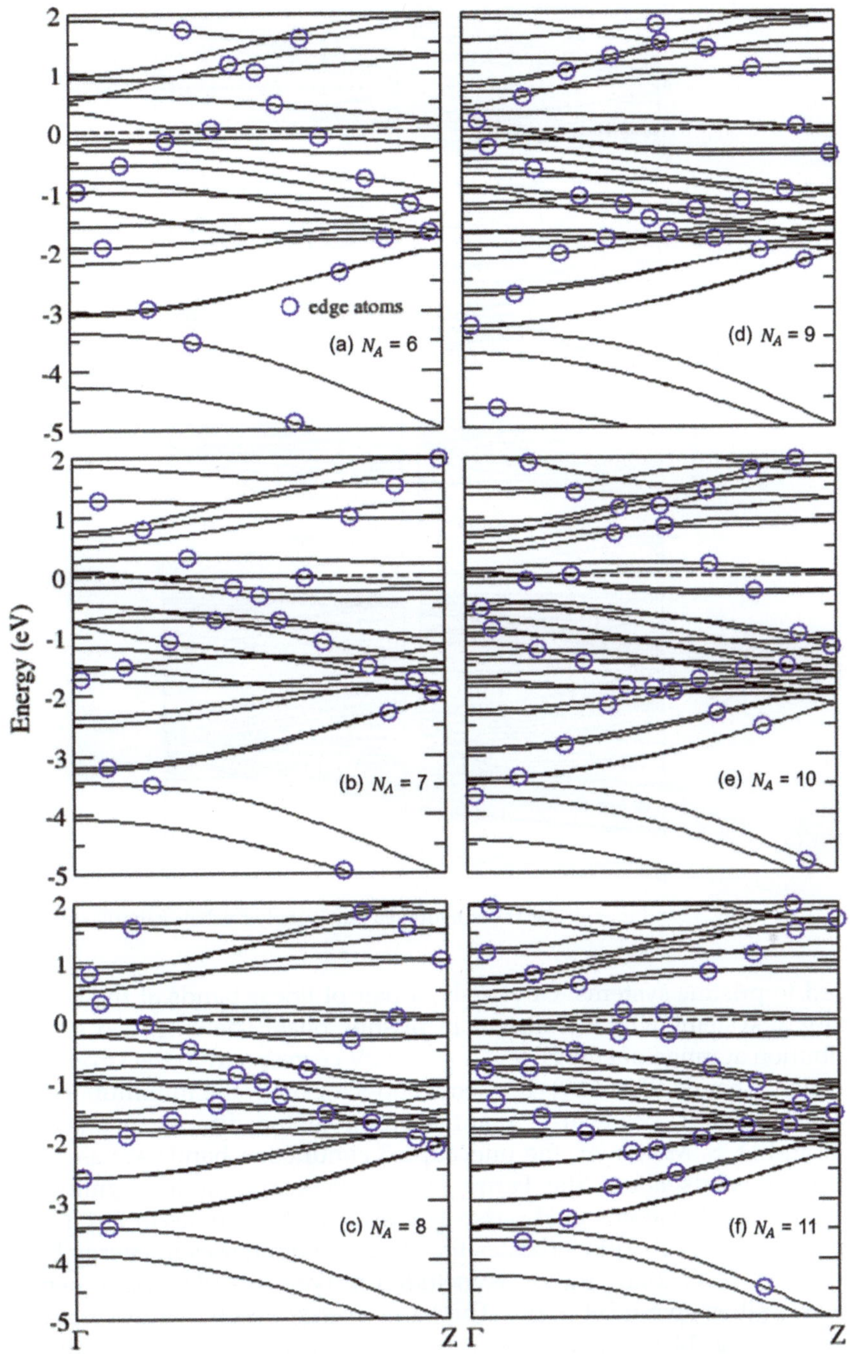

FIG. 14.8 Band energy spectra of pristine armchair nanoribbons.

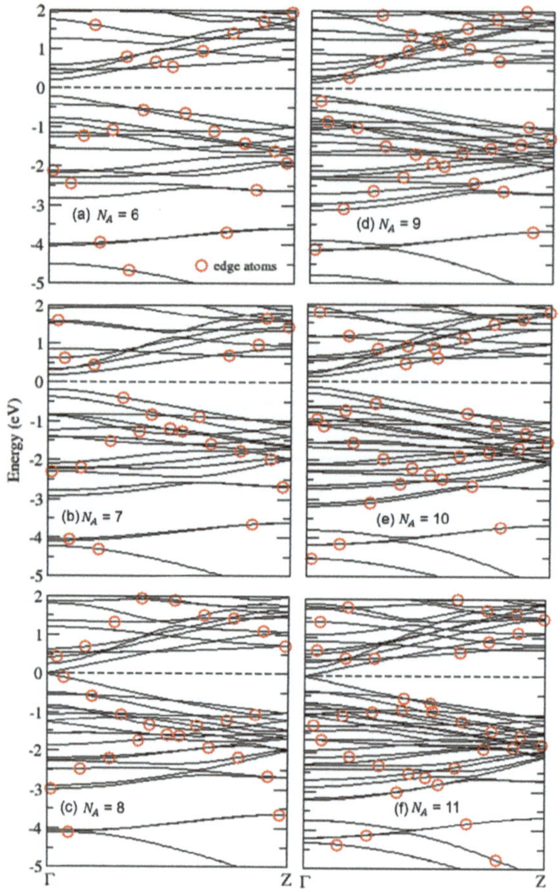

FIG. 14.9 Band energy spectra of hydrogen passivated armchair nanoribbons.

gap compared to pristine systems. Obviously, a pair of linear bands at the Fermi level illustrate opened-gap systems which is absent in pristine ones. The extended states with wide spatial distribution at range of energy less than −3 eV correspond to the strong parabolic dispersions in valence bands (Fig. 14.9). For semiconductor ones, the minimum conduction and maximum valence band locate at Γ point, leading to the direct-gap ones as shown in Fig. 14.9A, B and D, E. Moreover, the unoccupied conduction bands are asymmetric to the occupied valence bands about the Fermi level, demonstrating in asymmetric peaks in PDOS. Furthermore, edge atoms in H-passivated systems (red circles in Fig. 14.9) also dominate in near the Fermi level. This indicates the effect of boundaries in energy bands. As previously mentioned, SOC configuration is included in band calculations to verify boundary dependence as well as spin-orbit effect. The band energy spectra of hydrogen termination ones are shown in Fig. 14.10. Remaining features in zigzag system, SOC configuration only affect energy states near the Fermi level. Although bandgap values are unchanged in SOC systems, the direct band gaps centered by the Fermi level respect to deduce unoccupied

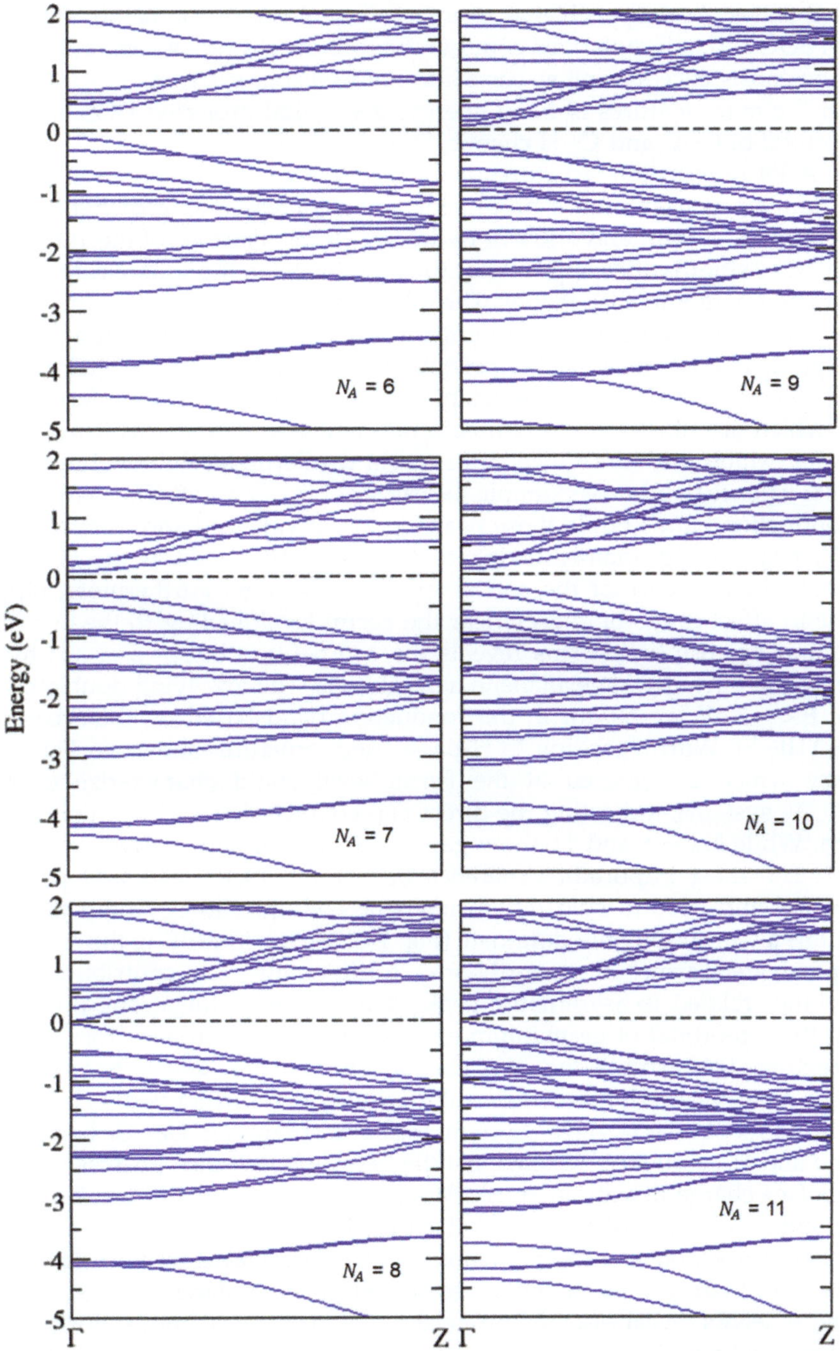

FIG. 14.10 Band energy spectra of hydrogen passivated armchair nanoribbons including spin-orbit coupling (SOC) effect.

and occupied states. At deeper/higher energy valence/conduction bands, the SOC effect is absent. Both SOC and spin-polarized configurations show strong dispersion in energy bands, indicating the consistence as well as correlation of two spin systems.

To reflect the main features of band energy, the orbital projected DOS is considered, in which the orbital of C–C and C–H chemical bonds are described indicating the orbital hybridization. A lot of essential structures in DOSs are revealed as shown in Figs. 14.11 and 14.12. The number of asymmetric prominent peaks increases as growing the nanoribbon width in both cases of with and without hydrogen atoms. Because of the 1D characteristics, most of van Hove singularities appear near the Fermi level in the square root divergent form which come from parabolic dispersions. The orbital projected DOS reveal the main contributions of the outer four orbital of germanium and s orbital of hydrogen atoms corresponding to 4s, $4p_x$, $4p_y$, $4p_z$ and 1s orbitals. For pristine AGeNR systems, the orbital projected DOSs considering spin up and spin down states plotted in Fig. 14.11, showing ferromagnetic configurations which are illustrated in non-zero net magnetic moment as listed in Table 14.2. Few structures exhibit the delta-function form near the Fermi level ranged from 0 to −0.5 eV due to the flat valence bands, for example the cases of $N_A = 7$ and 8. The π and σ peaks occur at whole conduction bands while those peaks locate at valence band in range of $E < 0$ and $E < -2.5$ eV, respectively, mainly owing to the dangling bonds at the boundaries. This feature reveals the existence of the partial sp^3 near the edge. Furthermore, ($4p_x$, $4p_y$) and $4p_z$ compete to affect range of energy near the Fermi level leading to the overlap between conduction and valence band. This reveals the competition of sp^2 and sp^3 hybridization characterizing the buckling in germanene nanoribbons. The low-lying double peaks located in range of $E < -3$ eV are absent in the conduction band based on the broadening effect (see Fig. 14.11B–F). With regarding to H-passivated structures shown in Fig. 14.12, a pair of symmetry which is centered at the Fermi level could characterize energy gap of $N_A = 6$, 7, 9, 10 systems, indicating the direct gap characteristics as mentioned in band energy spectra. While $N_A = 8$ and 11 show gapless structures with a small overlap about the Fermi level. The most important difference between H-passivated and pristine ones is the possess the spin up and spin down magnetic configurations. About the Fermi level, the $4p_z$ orbitals (cyan curve) only dominate (Fig. 14.12A–F), indicating the crucial properties of the H-passivated armchair nanoribbons. Additionally, the contribution of 1s- and ($4p_x$,$4p_y$) orbitals related to asymmetric peaks demonstrating the strong orbital hybridization among three p-orbital of germanium and 1s orbital of hydrogen atoms, as well as the buckled structure. This exhibits the sp^2-s chemical bond in the Ge–H bonds.

The charge density can provide the useful information to further verify chemical bonds as shown in Fig. 14.13. Similar to zigzag system, the bond strength of Ge–Ge bonds is presented with pristine and H-passivated systems. All the Ge–Ge bonds also possess the strong covalent σ bonds (see charge density in (x, z) plane). As shown in this figure, such bondings of Ge–Ge are enhanced in pristine case that correlate to the weaken Ge–Ge bonds in H-passivated case corresponding to red region. ρ in H-passivated systems reveal significant charge transfer to hydrogen as shown in red region of (y, z) plane.

Generally, AGeNRs reveal essential properties due to chemical bonds, buckling, boundary termination and ribbon width. Varying ribbon width by determining three categories 3I, 3I + 1 and 3I + 2 in AGeNRs seriously illustrate the dependence of N_A and band gap as well as energy band characteristics. 3I and 3I + 1 belong to middle-gap semiconductors while 3I + 2

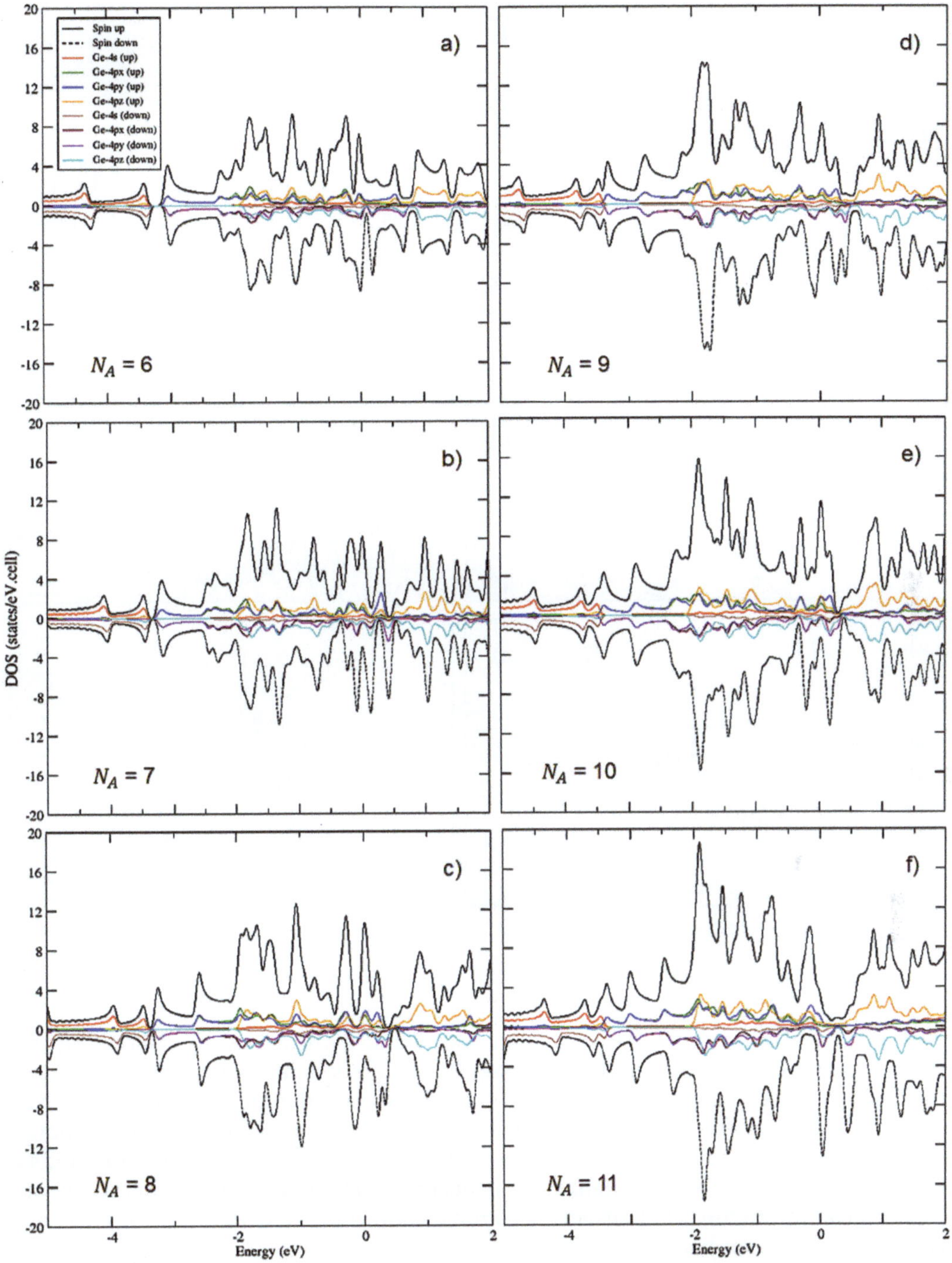

FIG. 14.11 Orbital projected DOS of pristine armchair nanoribbons.

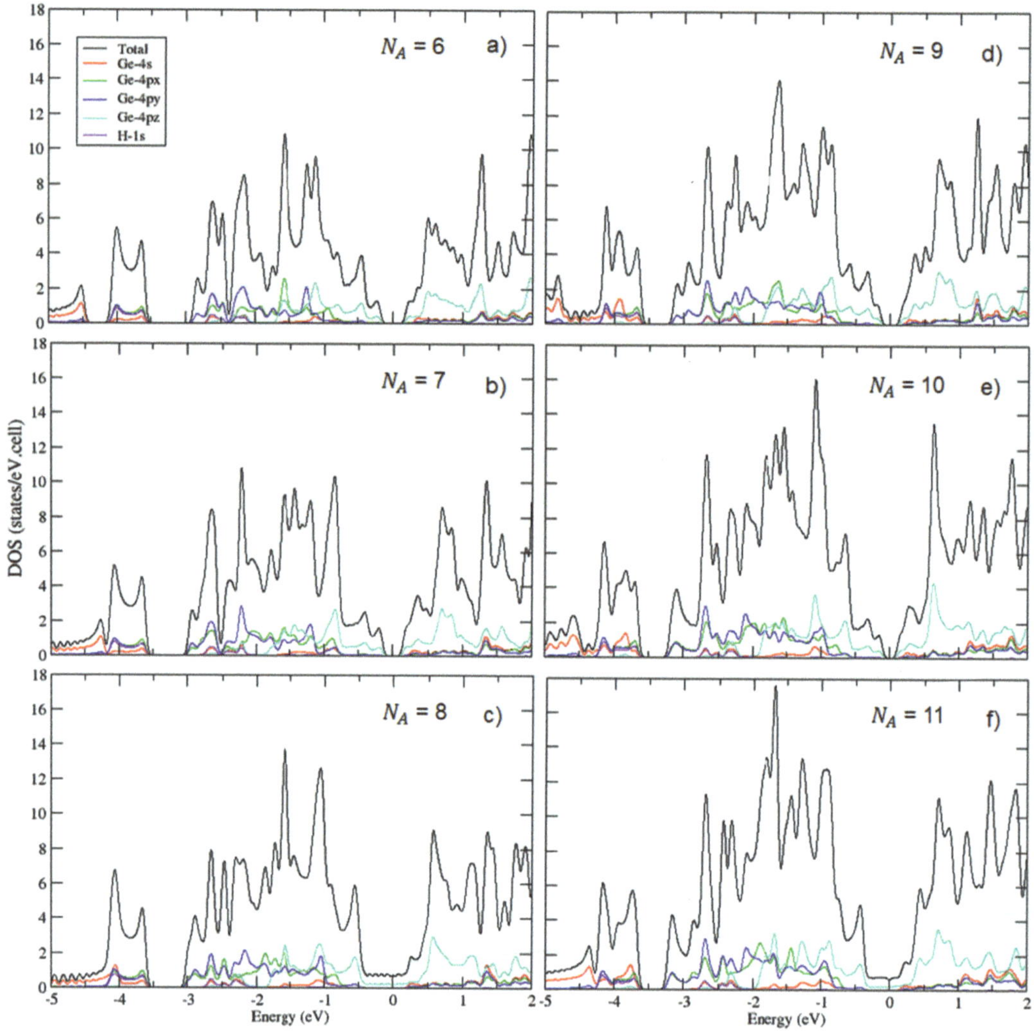

FIG. 14.12 Orbital projected DOS of hydrogen passivated armchair nanoribbons.

present semimetal behavior for H-passivated ones. Band gaps decrease as increasing ribbon width, demonstrating this strong dependence. The differences in features of pristine and H-passivated system also indicate the effect of boundaries and dangling bonds. There exists a transition from semimetal system to direct semiconductors after hydrogen termination. Energy band dispersions follow change, leading the critical characteristics in both bands and PDOSs. In addition, the net magnetization is absent in H-passivated systems. Edge atoms dominate near the Fermi level respect to strong contribute at deeper and higher energy valence and conduction bands.

In summary, both ZGeNRs and AGeNRs exhibit special characteristic of 1D nanoribbons. In ribbon structures, sp^2 and sp^3 bondings compete to create strong hybridizations.

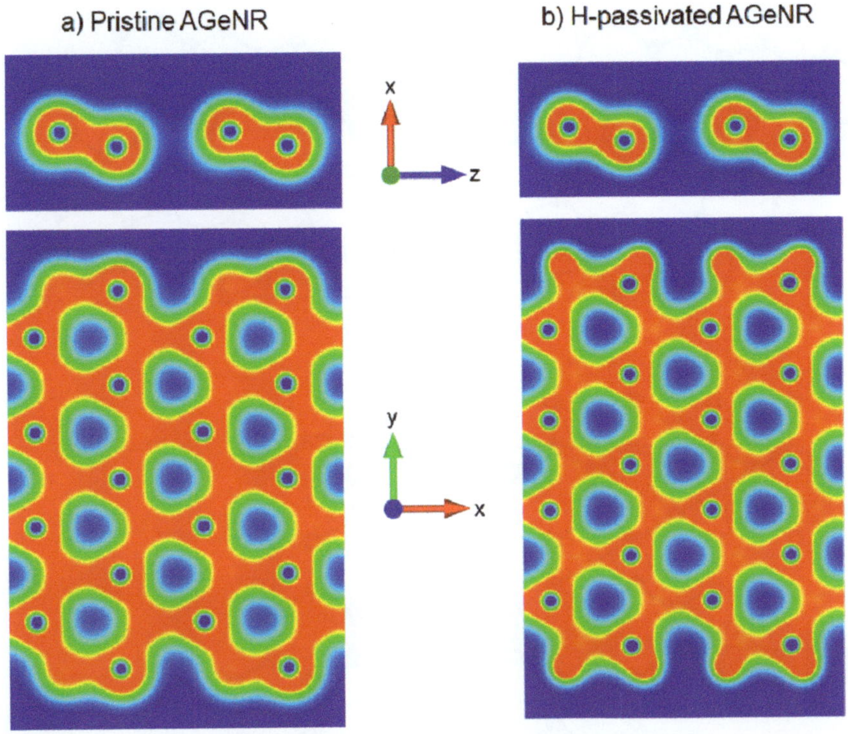

FIG. 14.13 Charge density of (A) pristine and (B) hydrogen passivated armchair nanoribbons with $N_A = 9$.

Zigzag and armchair systems reveal common trend in chemical bonds due to determined boundaries, indicating in non-uniform honeycomb structures as shown in first and second Ge–Ge bonds. However, electronic properties are significantly distinct based on edge termination. The main different point of zigzag and armchair system is semi-metallic/semiconducting behavior due to occupied, unoccupied states and band energy dispersions. The special valleys at $k_z = 2/3$ appear in zigzag systems corresponding to K and K′ valleys in monolayer, while these valleys are absent in armchair ones, anticipating different band dispersions of two cases. Commonly, they all demonstrate diverse quasiparticle phenomena as mentioned in our previous and recent works, based on special width and edge structures.

14.3 Chirality effects

A fundamental shape relevant to nanoribbon behavior of germanene-related research is varying open edges depending on the angle at which it is cut, such as zigzag and armchair as above mentioned. Apart from these two nanoribbons, chirality is one of special edges characterized by chiral vector (m,n) (m, n: integral) [49,51,115] as shown in Fig. 14.14A, create a chiral angle θ. Zigzag and armchair nanoribbons have corresponding chiral of $\theta = 0$ and $\theta = 30°$ (see Fig. 14.4B), respectively. Chirality therefore presents many species in construction edge

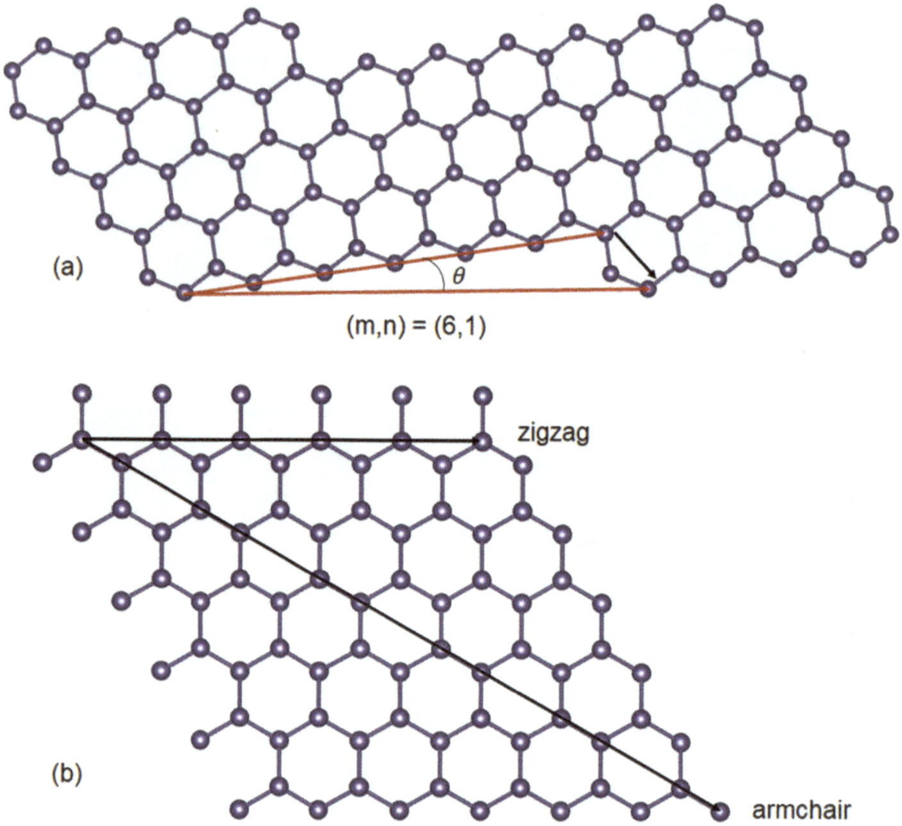

FIG. 14.14 Determination of (A) chiral and (B) zigzag, armchair nanoribbons in monolayer.

terminations, corresponding to value of angle θ or the length of m, n integral. ZGeNRs and AGeNRs are two special cases of chiral germanene nanoribbons. Due to their electronic properties, the most important effect could be observed is band gap dependence. A semiconducting gap system emerge in armchair cases while zigzag edges show semi-metallic behavior, illustrating this effect. The corresponding electronic characteristics have been analyzed, revealing effects of boundaries of these two nanoribbons. However, varying chiral types is absent in this work that should be included in further research.

Based on chirality effect in graphene nanoribbons [25,49,51,104], germanene-related materials are expected to reveal similar, but unique characteristics due to chirality with varying (m,n) vector or angle of θ that possess in this work. Few features of chirality will be reviewed in this section to further provide perspective of chirality and orient for the next research. As a natural phenomenon, chirality extensively exists in nearly all scaled structures, which is applied in many fields, such as chemistry, biochemistry, materials [25,43,47,49–51]. Especially, graphene which have distinctive become representative of 2D and 1D materials. Attachment of chiral components on graphene nanoribbons open a wide range of superior properties and applications. 1D spin-polarized edge states coupled across the width of a chiral graphene nanoribbon [51], which is different from armchair nanoribbons. Controlling chirality at the edges along ZGNRs and AGNRs exhibit a large variation in band gap [35,68], indicating the dependence of edges, particularly chiral effect, and band gaps. A system of different chiral types with changing θ angle [44] have been considered in carbon nanostructures indicate ultimate strength dependence for dependent mechanical properties. Furthermore, attached chiral phenylethyl and p-bromophenylethyl groups [45] onto the edges of graphene nanoribbons create chiral nanographene that might induce chiroptical properties. Generally, the changes of chiral angle show significant follow changes in geometric and electronic properties [22,43,44,46–48,50,51,68,104,105,115–117], which need to be examined in further research of germanene nanoribbons.

14.4 Similarities/differences with carbon/silicon nanoribbons

Because of identical 1D honeycomb nanoribbons, graphene, silicene and germanene nanoribbons present the identical properties in geometry and electronic one due to chemical bonds and quantum confinement effect. They all show essential and unique characteristics of semiconductors/semimetals with considering boundary terminations of hydrogen absence and presence systems. However, silicene and germanene nanoribbons exist the distinct properties compared to graphene ones due to buckled structures that leads to different features of energy bands. In addition, the height fluctuation (buckling) in germanene is larger than that of silicene that might cause the enhanced critical properties in germanene nanoribbons. These characteristics will be discussed below to generally make comparison and further provide detail information in geometric and electronic ones.

14.4.1 Similarities

With regarding to geometry, carbon [2,55–57], silicene [1–3,113] and germanene nanoribbons reveal honeycomb structures from top view with and without hydrogen passivation, for example, systems in Figs. 14.1 and 14.7. To construct 1D nanoribbons, one direction is longitudinal x-axis and a direction make with x direction an angle of 30° corresponding to zigzag and armchair nanoribbons, are considered as shown in Fig. 14.14. Ribbon width is determined by numbers of zigzag line/dimer line (N_Z/N_A) for zigzag/armchair nanoribbons. The difference of ground state energies between absence and presence hydrogen passivation exhibits the distribution of spin polarization. Furthermore, they all reveal non-uniform structures based on the two different nearest (the first and second C–C/Si–Si/Ge–Ge) bondings of nanoribbons. Importantly, strong sp^3 hybridization in monolayer which leads to the common bonds is absent in nanoribbons with respect to the sp^2 or the mixed of sp^2–sp^3 ones. The C–/Si–/Ge–H bonds are identical in each system creating the stable bonds at ribbons, illustrating the existence of strong sp^2–s bonds at edge. The trend of increasing bond length after hydrogen termination is remained in three systems, indicating the effect of boundaries in nanoribbons. Besides, the dangling bonds in pristine system is absent in passivation structures, demonstrating this effect.

In subject of electronic properties, the boundaries, chemical bonds and quantum confinement characterize electronic features of 1D graphene/silicene/germanene nanoribbons.

Two cases of comparison; that is, germanene compared to carbon nanoribbons and germanene with silicene nanoribbons, need to be classified to provide detailed information. For former case, both graphene and germanene nanoribbons reveal strong dispersion in most energy bands come from parabolic ones. The low-lying energy valence and conduction bands of zigzag systems present a special valley centered at $k = 2/3$, corresponding to K and K′ valleys near the Dirac points of monolayer systems. In zigzag systems, energy spacing could also found at deeper energy valence bands, indicating band-edge conduction state is close to the edge-atom-dominated bands, leading to the strong peak in DOS. Simultaneously, all H-passivated armchair systems belong to direct-gap semiconductors instead of indirect-gap semiconductors/semimetal in pristine cases, anticipating the dependence of nanoribbon boundaries. Moreover, varying ribbon width characterizes energy gaps and features of energy bands. For the case of germanene and silicene nanoribbons [1,107], confinement effect also characterizes their energy bands. They all exhibit a semi-metallic behavior as well as the net magnetization come from the different number of spin up and spin down occupied states in zigzag system. There exists a crossing band at the Fermi level located near Γ point for H-passivated structures. The existence of a strong peak near the Fermi level shown in orbital-projected DOS might cause this magnetic configuration. Additionally, when increasing the number of ribbon width, band gaps remarkably change, indicating the width dependence between N_A and Eg which is one of the most important features of armchair systems. It should be noted that they all belong to buckled structures might lead to similar properties in their nanoribbons.

14.4.2 Differences

Obviously, the unique and significant feature of buckling is a critical point of the differences between germanene and carbon nanoribbons. Due to buckled honeycomb structure, germanene nanoribbons reveal the different characteristics in structure parameters and electronic properties. For zigzag systems, graphene ones exhibit the semiconducting anti-ferromagnetic (AFM) [32] when considering spin-polarized configurations while the semi-metallic ferromagnetic (FM) is found in presence and absence hydrogen passivation (Section 14.1.1). Furthermore, the highest occupied state (HOS) and lowest unoccupied state (LUS) exist about $k_z = 2/3$ define the direct energy gaps that is absent in germanene ones. The atom dominances in band energy spectra indicate strong contributions of edge atoms at conduction and valence bands near the Fermi level/low-lying energy valence bands for pristine/H-passivated case. Due to this contribution, edge-localized energy band These features of edge atoms dominance are unclearly presented in germanene nanoribbons corresponding to the contributions in whole bands of edge atoms for two cases (see Fig. 14.4). With armchair systems, AGNRs belong to indirect-gap semiconductors in pristine ones, while all pristine AGeNRs are semimetal due to the small overlap between conduction and valence bands. This finding leads to different characteristics such as non-magnetic/magnetic configurations of pristine AGNRs/AGeNRs. Although H-passivated AGNRs and AGeNRs present semiconducting properties with direct band gaps, the former reveals the largest gaps occurred for $N_A = 3I + 1$ while that of gaps occurs for $N_A = 3I$ in the later.

Even though both silicene and germanene nanoribbons exist buckled structures instead of planar carbon nanoribbons, these two nanoribbons with hydrogen termination exhibit the distinct features due to height fluctuation. Zigzag systems are first discussed based on their magnetization. Although zigzag type of silicene and germanene nanoribbons reveal the magnetic behavior, silicene case complicatedly presents the sensitive magnetism including non-magnetic (NM), ferromagnetic (FM), anti-ferromagnetic (AFM) configurations [1,118] due to changing ribbon width. Some of them exhibit magnetic semiconducting behavior with a narrow direct gap characterized by VBM and CBM at Γ point. Furthermore, ASiNRs [107] obviously exhibit the switch of indirect gap to direct gap when increasing ribbon width, while the direct gap is remained in all AGeNRs with varying N_A. The occupied and unoccupied linear energy states in AGeNRs are absent in ASiNR, leading to the difference of this band gap feature.

14.5 Conclusion

In conclusion, 1D germanene nanoribbons [2,17,111,112], with the specific widths and open edge structures, exhibit the diversified quasiparticle phenomena, being totally different from 2D [8,66] and 3D allotropes [119]. The coupling effects, which mainly arise from the transverse quantum confinement [1], the dangling bonds at boundaries/the hydrogen decorations there [2,120,121], the weak, but significant sp^3 bondings/the dominating sp^2 ones (the non-perpendicular π and σ bondings), and the spin-orbital interactions/the spin-/edge-dependent Coulomb interactions. The first and second mechanisms have generated the observable fluctuations of Ge—Ge bond lengths, especially for those near edge structures. Such highly non-uniform chemical environments, being fully assisted by the intrinsic interactions, are responsible for the geometry-enriched fundamental properties [2,17,121,122]. All of them belong to the middle- or narrow-gap semiconductors, as observed in 1D graphene/silicene nanoribbons [1,2,55]. The band-gap magnitudes gradually decline in the increment of nanoribbon width. Furthermore, armchair systems could be divided into three categories, where the dimer line number of $N_A = 3I + 2$ exhibit the smallest direct gaps. This unique behavior should be associated with the open boundary conditions and the bond-length-dominated hopping integrals. Such information would be very useful for establishing the reliable tight-binding model [18,123]/the generalized tight-binding model under a uniform magnetic field [124,125]. The featured band structures cover the highly asymmetric hole and electron spectra, many low-lying energy dispersions without band overlap across the Fermi level, the frequent crossing and anti-crossing behaviors, and the absence of the well separated π and σ energy sub-bands. Very interesting, the spatial charge density distributions clearly show the position-dependent π, σ and sp^3 bondings simultaneously, with the high position dependences. A lot of van Hove singularities, which are induced by the band-edge states of 1D energy sub-bands, exhibit the significant characteristics of $4p_z$-, ($4s$, $4p_x$, $4p_y$)- and ($4s$, $4p_x$, $4p_y$, $4p_z$) within the different energy ranges.

On the experimental side, STM and TEM (details in Chapter 3.1) are, respectively, available in verifying the position-dependent Ge—Ge bond lengths/the achiral and chiral edge structures and the buckled structures (the top and side views). Such observations can examine the sensitive passivation effects of the frequent hydrogen decorations [126,127]. The geometry-dominated electronic properties are directly measured by the high-precision STS (Chapter 3.2), revealing the featured band gaps and low energy van Hove singularities of occupied valence and unoccupied conduction sub-bands. Such measurements, being fully supported by the VASP simulations in this work, are sufficient in identifying the critical orbital hybridizations at the distinct energy ranges. As for the predicted 1D band structures, the ARPES observations could be utilized to verify the geometry-enriched band gaps, as well as the parabolic/oscillatory/linear/partially flat valence dispersion relations. Whether a buckled structure strongly affects the well definition of momentum transfer along the longitudinal direction should be very carefully done during the detailed analyses. However, the straightforward tests on the charge/spin density distributions seem to be lack of the high-resolution tools up to now. A close combination of the theoretical and experimental research might overcome this open issue. Most important, the strong relations among the VASP simulations, the phenomenological models and the delicate experiments need to be greatly strengthened for fully comprehending the diverse quasiparticle behaviors in germanene-related materials. That is, a unified quasiparticle framework is consistent with one another for three approaches, as finished in the previously published books [32]. In addition, the similar studies could be generalized to 1D germanene nanotube-related systems [2,3,17,111,112,122,126,127], with the periodical boundary conditions, buckled honeycomb structures and curvature effects. Such low-dimensional systems are expected to present the diverse quasiparticle phenomena in terms of geometric [22,35,41,61,62], electronic [1,2,17,19,31,36,38,56,96], optical [1,5,17], magnetic [1,59,101,110,122,125,127] and transport properties [23,56,57,65,98].

References

[1] Ferri M, Fratesi G, Onida G, Debernardi A. Ab initio study of the structural, electronic, magnetic, and optical properties of silicene nanoribbons. Phys Rev B 2019;99.

[2] Pandya A, Sangani K, Jha PK. Band gap determination of graphene, h-boron nitride, phosphorene, silicene, stanene, and germanene nanoribbons. J Phys D-Appl Phys 2020;53.

[3] Hattori A, Yada K, Araidai M, Sato M, Shiraishi K, Tanaka Y. Influence of edge magnetization and electric fields on zigzag silicene, germanene and stanene nanoribbons. J Phys Condens Matter 2019;31.

[4] de Lima JDM, Gomes E, da Silva FF, Moraes F, Teixeira R. Geometric effects on the electronic structure of curved nanotubes and curved graphene: the case of the helix, catenary, helicoid, and catenoid. Eur Phys J Plus 2021;136.

[5] Radsar T, Khalesi H, Ghods V. Graphene properties and applications in nanoelectronic. Opt Quant Electron 2021;53.

[6] Cai B, Zhang SL, Hu ZY, Hu YH, Zou YS, Zeng HB. Tinene: a two-dimensional Dirac material with a 72 meV band gap. Phys Chem Chem Phys 2015;17:12634–8.

[7] Mahdavifar S, shayesteh SF, Tagani MB. Electronic and mechanical properties of Plumbene monolayer: a first-principle study. Phys E Low-dimens Syst Nanostruct 2021;134.

[8] Yuhara J, Le Lay G. Beyond silicene: synthesis of germanene, stanene and plumbene. Jpn J Appl Phys 2020;59.

[9] Martin C, Kostarelos K, Prato M, Bianco A. Biocompatibility and biodegradability of 2D materials: graphene and beyond. Chem Commun 2019;55:5540–6.

[10] Fusco L, Gazzi A, Peng GT, Shin Y, Vranic S, Bedognetti D, Vitale F, Yilmazer A, Feng X, Fadeel B, Casiraghi C, Delogu LG. Graphene and other 2D materials: a multidisciplinary analysis to uncover the hidden potential as cancer theranostics. Theranostics 2020;10:5435–88.

[11] Fu YF, Hansson J, Liu Y, Chen SJ, Zehri A, Samani MK, Wang N, Ni YX, Zhang Y, Zhang ZB, Wang QL, Li MX, Lu HB, Sledzinska M, Torres CMS, Volz S, Balandin AA, Xu XF, Liu J. Graphene related materials for thermal management. 2D Mater 2020;7.

[12] Lavin-Lopez MP, Valverde JL, de Lucas-Consuegra A, Calcerrada AB, Paton-Carrero A, Romero A. Influence of the synthesis method on electrical storage capacity of graphene-related materials. Mater Sci Technol 2019;35:361–7.

[13] Netkueakul W, Korejwo D, Hammer T, Chortarea S, Rupper P, Braun O, Calame M, Rothen-Rutishauser B, Buerki-Thurnherr T, Wick P, Wang J. Release of graphene-related materials from epoxy-based composites: characterization, quantification and hazard assessment in vitro. Nanoscale 2020;12:10703–22.
[14] Catania F, Marras E, Giorcelli M, Jagdale P, Lavagna L, Tagliaferro A, Bartoli M. A review on recent Advancements of graphene and graphene-related materials in biological applications. Appl Sci 2021;11.
[15] Goldie SJ, Bush S, Cumming JA, Coleman KS. A statistical approach to Raman analysis of graphene-related materials: implications for quality control. ACS Appl Nano Mater 2020;3:11229–39.
[16] Woo YS. Transparent conductive electrodes based on graphene-related materials. Micromachines 2019;10.
[17] Shiraz AK, Goharrizi AY, Hamidi SM. The electronic and optical properties of armchair germanene nanoribbons. Phys E Low-dimens Syst Nanostruct 2019;107:150–3.
[18] Chegel R, Hasani M. Electronic and thermal properties of silicene nanoribbons: third nearest neighbor tight binding approximation. Chem Phys Lett 2020;761.
[19] Li R. Electronic properties of hybrid graphene/nitrogene/graphene hetero-trilayers. Phys E Low-dimens Syst Nanostruct 2020;123.
[20] Olaniyan O, Moskaleva LV. First-principles study of the electronic properties and thermal expansivity of a hybrid 2D carbon and boron nitride material. C-J Carbon Res 2021;7.
[21] Qiao H, Liu HT, Huang ZY, Hu R, Ma Q, Zhong JX, Qi X. Tunable electronic and optical properties of 2D monoelemental materials beyond graphene for promising applications. Energy Environ Mater 2021.
[22] Yang N, Li C, Tang Y. Effects of chirality and stacking on the thermal expansion effects of graphene. Mater Res Exp 2020;7.
[23] Zhang QQ, Jia PZ, Chen XK, Zhou WX, Chen KQ. Thermal transport properties in monolayer group-IV binary compounds. J Phys Condens Matter 2020;32.
[24] Silvestri A, Criado A, Prato M. Concluding remarks: chemistry of 2-dimensional materials: beyond graphene. Faraday Discus 2021;227:383–95.
[25] Zhao B, Yang SH, Deng JP, Pan K. Chiral graphene hybrid materials: structures, properties, and chiral applications. Adv Sci 2021;8.
[26] Immanuel S, Dar MA, Sivasubramanian R, Karim MR, Kim DW, Gul R. Progress and prospects on the fabrication of graphene-based nanostructures for energy storage, energy conversion and biomedical applications. Chem–Asian J 2021;16:1365–81.
[27] Li ZL, Chen YP, Ma TY, Jiang YZ, Chen J, Pan HG, Sun WP. 2D metal-free nanomaterials beyond graphene and its analogues toward electrocatalysis applications. Adv Energy Mater 2021.
[28] Zhang EQ, Xing ZK, Wan DA, Gao HR, Han YD, Gao YS, Hu HF, Cheng ZZ, Liu TG. Surface-enhanced Raman spectroscopy chips based on two-dimensional materials beyond graphene. J Semiconduct 2021;42.
[29] Lu LX, Yang Q, Xu Q, Sun YJ, Tang SS, Tang XB, Liang H, Yu YD. Two-dimensional materials beyond graphene for the detection and removal of antibiotics: a critical review. Crit Rev Environ Sci Technol 2021.
[30] Lan CY, Shi Z, Cao R, Li C, Zhang H. 2D materials beyond graphene toward Si integrated infrared optoelectronic devices. Nanoscale 2020;12:11784–807.
[31] Barraza-Lopez S, Xia FN, Zhu WJ, Wang H. Beyond graphene: low-symmetry and anisotropic 2D materials. J Appl Phys 2020;128.
[32] Ngoc Thanh Thuy Tran S-YL, Lin C-Y, Lin M-F. Geometric and electronic properties of graphene-related systems. In: Chemical bonding schemes; 2018.
[33] Tuinstra F, Koenig JL. Raman spectrum of graphite. J Chem Phys 1970;53:1126.
[34] Cousins CSG, Heggie MI. Elasticity of carbon allotropes. III. Hexagonal graphite: review of data, previous calculations, and a fit to a modified anharmonic Keating model. Phys Rev B 2003;67.
[35] Gospodarev IA, Grishaev VI, Manzhelii EV, Sirenko VA, Syrkin ES, Feodosyev SB. The effect of size quantization on the electron spectra of graphene nanoribbons. Low Temp Phys 2020;46.
[36] Soncini C, Bondino F, Magnano E, Bhardwaj S, Kumar M, Cepek C, Pedio M. Electronic properties of carbon nanotubes as detected by photoemission and inverse photoemission. Nanotechnology 2021;32.
[37] Haddon RC. Carbon nanotubes. Acc Chem Res 2002;35:997.
[38] Popov VN. Carbon nanotubes: properties and application. Mater Sci Eng R-Rep 2004;43:61–102.
[39] Wang JN, Zhang YL, Liu Y, Zheng WH, Lee LP, Sun HB. Recent developments in superhydrophobic graphene and graphene-related materials: from preparation to potential applications. Nanoscale 2015;7:7101–14.

[40] Ambrosi A, Chua CK, Bonanni A, Pumera M. Electrochemistry of graphene and related materials. Chem Rev 2014;114:7150–88.
[41] Wang HD, Kurata K, Fukunaga T, Zhang X, Takamatsu H. Width dependent intrinsic thermal conductivity of suspended monolayer graphene. Int J Heat Mass Tran 2017;105:76–80.
[42] Nakanishi T, Koshino M, Ando T. Transmission through a boundary between monolayer and bilayer graphene. Phys Rev B 2010;82.
[43] Liu XF, He XQ, Sun LG, Wang JB, Yang D, Shi X. A chirality-dependent peridynamic model for the fracture analysis of graphene sheets. Mech Mater 2020;149.
[44] Thamaraikannan S, Sunny MR, Pradhan SC. Chirality dependent mechanical properties of carbon nanostructures. Mater Res Exp 2019;6.
[45] Nishitani S, Sekiya R, Haino T. Chirality-embedded nanographenes. Angew Chem-Int Ed 2020;59:669–73.
[46] Wang YK, Wang ZY, Wang QJ, Zhou SQ, Han QY, Gao W, Ren KL, Qi JX, Dong J. Active control and biosensing application of induced chirality between symmetric metal and graphene nanostructures. J Phys Chem C 2019;123:24754–62.
[47] Aniban X, Hartwig B, Wuttke A, Mata RA. Dispersion forces in chirality recognition—a density functional and wave function theory study of diols. Phys Chem Chem Phys 2021;23:12093–104.
[48] Yang L, Guo YJ. Dynamics of water confined in a graphene nanochannel: dependence of friction on graphene chirality. Nanotechnology 2020;31.
[49] Wang XM, Zheng ZB, Xu NS, Wang WL, Chen HJ, Deng SZ. A nano-imaging study of graphene edge plasmons with chirality-dependent dispersions. Adv Opt Mater 2021;9.
[50] Nissler R, Kurth L, Li H, Spreinat A, Kuhlemann I, Flavel BS, Kruss S. Sensing with chirality-pure near-infrared fluorescent carbon nanotubes. Anal Chem 2021;93:6446–55.
[51] Tao CG, Jiao LY, Yazyev OV, Chen YC, Feng JJ, Zhang XW, Capaz RB, Tour JM, Zettl A, Louie SG, Dai HJ, Crommie MF. Spatially resolving edge states of chiral graphene nanoribbons. Nat Phys 2011;7:616–20.
[52] Charlier A, McRae E, Heyd R, Charlier MF, Moretti D. Classification for double-walled carbon nanotubes. Carbon 1999;37:1779–83.
[53] Gulseren O, Yildirim T, Ciraci S. Systematic ab initio study of curvature effects in carbon nanotubes. Phys Rev B 2002;65.
[54] Ding JW, Yan XH, Cao JX, Wang DL, Tang Y, Yang QB. Curvature and strain effects on electronic properties of single-wall carbon nanotubes. J Phys Condens Matter 2003;15:L439–45.
[55] Son YW, Cohen ML, Louie SG. Energy gaps in graphene nanoribbons. Phys Rev Lett 2006;97.
[56] Treske U, Ortmann F, Oetzel B, Hannewald K, Bechstedt F. Electronic and transport properties of graphene nanoribbons. Phys Status Solidi A-Appl Mater Sci 2010;207:304–8.
[57] Wakabayashi K, Takane Y, Yamamoto M, Sigrist M. Electronic transport properties of graphene nanoribbons. New J Phys 2009;11.
[58] Hossain MS, Huynh DH, Jiang LM, Rahman S, Nguyen PD, Al-Dirini F, Hossain F, Bahk JH, Skafidas E. Investigating enhanced thermoelectric performance of graphene-based nano-structures. Nanoscale 2018;10:4786–92.
[59] Yazyev OV. Emergence of magnetism in graphene materials and nanostructures. Rep Prog Phys 2010;73.
[60] Tang Q, Zhou Z, Chen ZF. Graphene-related nanomaterials: tuning properties by functionalization. Nanoscale 2013;5:4541–83.
[61] Rochefort A, Salahub DR, Avouris P. Effects of finite length on the electronic structure of carbon nanotubes. J Phys Chem B 1999;103:641–6.
[62] Zhu HY, Klein DJ, Schmalz TG, Rubio A, March NH. Geometric boundary effects on the electronic properties of finite carbon nanotubes. J Phys Chem Solid 1998;59:417–23.
[63] Hellman A, Baerends EJ, Biczysko M, Bligaard T, Christensen CH, Clary DC, Dahl S, van Harrevelt R, Honkala K, Jonsson H, Kroes GJ, Luppi M, Manthe U, Norskov JK, Olsen RA, Rossmeisl J, Skulason E, Tautermann CS, Varandas AJC, Vincent JK. Predicting catalysis: understanding ammonia synthesis from first-principles calculations. J Phys Chem B 2006;110:17719–35.
[64] Loschen C, Bromley ST, Neyman KM, Illas F. Understanding ceria nanoparticles from first-principles calculations. J Phys Chem C 2007;111:10142–5.
[65] Azad ZA, Shokri A, Khezrabad MSA. Principle study on the quantum transport in an armchair silicene nanoribbon-based p-n diode. Mater Chem Phys 2021 First-;257.

[66] Mas-Balleste R, Gomez-Navarro C, Gomez-Herrero J, Zamora F. 2D materials: to graphene and beyond. Nanoscale 2011;3:20–30.
[67] Dubertret B, Heine T, Terrones M. The rise of two-dimensional materials. Acc Chem Res 2015;48:1–2.
[68] Wang HS, Chen LX, Elibol K, He L, Wang HM, Chen C, Jiang CX, Li C, Wu TR, Cong CX, Pennycook TJ, Argentero G, Zhang DL, Watanabe K, Taniguchi T, Wei WY, Yuan QH, Meyer JC, Xie XM. Towards chirality control of graphene nanoribbons embedded in hexagonal boron nitride. Nat Mater 2021;20:202.
[69] Fedotov PV, Rybkovskiy DV, Chernov AI, Obraztsova EA, Obraztsova ED. Excitonic photoluminescence of ultra-narrow 7-armchair graphene nanoribbons grown by a new "bottom-up" approach on a Ni substrate under low vacuum. J Phys Chem C 2020;124:25984–91.
[70] Chen ZP, Narita A, Mullen K. Graphene nanoribbons: on-surface synthesis and Integration into electronic devices. Adv Mater 2020;32.
[71] Zhou XH, Yu G. Modified engineering of graphene nanoribbons prepared via on-surface synthesis. Adv Mater 2020;32.
[72] Aristov VY, Chaika AN, Molodtsova OV, Aristova IM, Potorochin DV. Nanostructured graphene on beta-SiC/Si(001): atomic and electronic structures, magnetic and transport properties (brief review). Jetp Lett 2021;113:176–93.
[73] Goltl F, Murray EA, Chen BWJ, Jacobberger RM, Arnold MS, Mavrikakis M. Exploring driving forces for length growth in graphene nanoribbons during chemical vapor deposition of hydrocarbons on Ge(001) via kinetic Monte Carlo simulations. Appl Surf Sci 2020;527.
[74] Shekhirev M, Lipatov A, Torres A, Vorobeva NS, Harkleroad A, Lashkov A, Sysoev V, Sinitskii A. Highly selective gas sensors based on graphene nanoribbons grown by chemical vapor deposition. Acs Appl Mater Interfaces 2020;12:7392–402.
[75] Senkovskiy BV, Usachov DY, Fedorov AV, Haberer D, Ehlen N, Fischer FR, Gruneis A. Finding the hidden valence band of N = 7 armchair graphene nanoribbons with angle-resolved photoemission spectroscopy. 2D Mater 2018;5.
[76] Gigli L, Manini N, Benassi A, Tosatti E, Vanossi A, Guerra R. Graphene nanoribbons on gold: understanding superlubricity and edge effects. 2D Mater 2017;4.
[77] Hamalainen SK, Sun ZX, Boneschanscher MP, Uppstu A, Ijas M, Harju A, Vanmaekelbergh D, Liljeroth P. Quantum-confined electronic states in atomically well-defined graphene nanostructures. Phys Rev Lett 2011;107.
[78] Karakachian H, Nguyen TTN, Aprojanz J, Zakharov AA, Yakimova R, Rosenzweig P, Polley CM, Balasubramanian T, Tegenkamp C, Power SR, Starke U. One-dimensional confinement and width-dependent bandgap formation in epitaxial graphene nanoribbons. Nat Commun 2020;11.
[79] Savin AV, Dmitriev SV, Korznikova EA, Kistanov AA. Multilayered scrolls of carbon nanoribbons. Mater Phys Mech 2018;35:155–66.
[80] Savin AV, Mazo MA. Simulation of scrolled packings of graphone nanoribbons. Phys Solid State 2017;59:1260–6.
[81] Savin AV, Korznikova EA, Lobzenko IP, Baimova YA, Dmitriev SV. Symmetric scrolled packings of multilayered carbon nanoribbons. Phys Solid State 2016;58:1278–84.
[82] Savin AV, Mazo MA. Two-dimensional model of scrolled packings of molecular nanoribbons. Phys Solid State 2018;60:826–35.
[83] Zhang J, Fahrenthold EP. Conductance of curved 3M-1 armchair graphene nanoribbons. J Phys Chem C 2019;123:21805–12.
[84] Niu WH, Ma J, Soltani P, Zheng WH, Liu FP, Popov AA, Weigand JJ, Komber H, Poliani E, Casiraghi C, Droste J, Hansen MR, Osella S, Beljonne D, Bonn M, Wang HI, Feng XL, Liu JZ, Mai YY. A curved graphene nanoribbon with multi-edge structure and high intrinsic charge carrier mobility. J Am Chem Soc 2020;142:18293–8.
[85] Song J, Xia XM, Chen JT, Xia D, Xue QZ, Li Q, Dong MD. Folding 2D graphene nanoribbons into 3D nanocages induced by platinum nanoclusters. J Phys Chem C 2020;124:10495–501.
[86] Fonseca AF, Galvao DS. Self-tearing and self-peeling of folded graphene nanoribbons. Carbon 2019;143:230–9.
[87] Ko DJ, Choi J, Yan BY, Hwang TJ, Jin XZ, Kim JM, Piao YZ. A facile and scalable approach to develop electrochemical unzipping of multi-walled carbon nanotubes to graphene nanoribbons. J Mater Chem 2020;8:22045–53.

[88] Zheng QF, Guo Y, Liang Y, Shen Q. Graphene nanoribbons from electrostatic-force-controlled electric unzipping of single- and multi-walled carbon nanotubes. ACS Appl Nano Mater 2020;3:4708–16.
[89] Shukla A, Bhat SD, Pillai VK. Simultaneous unzipping and sulfonation of multi-walled carbon nanotubes to sulfonated graphene nanoribbons for nanocomposite membranes in polymer electrolyte fuel cells. J Membr Sci 2016;520:657–70.
[90] Ujjain SK, Ahuja P, Bhatia R, Attri P. Printable multi-walled carbon nanotubes thin film for high performance all solid state flexible supercapacitors. Mater Res Bull 2016;83:167–71.
[91] Gao Y, Xu D, Cui T, Li D. Stability of hydrogen-terminated graphene edges. Phys Chem Chem Phys 2021.
[92] Kleshchonok A, Gutierrez R, Cuniberti G. Contact effects and quantum interference in engineered dangling bond loops on silicon surfaces. Nanoscale 2015;7:13967–73.
[93] Zhang SD, Cao C, Zeng BW, Long MQ. The effects of strain and electric field on the half-metallicity of pristine and O-H/C-N-decorated zigzag graphene nanoribbons. J Phys Condens Matter 2020;32.
[94] Aierken Y, Leenaerts O, Peeters FM. Principles study of the stability and edge stress of nitrogen-decorated graphene nanoribbons. Phys Rev B 2018 First-;97.
[95] Guo G, Xie ZX, Tang J, Zhang Y, Xu GB. Fluorine edge decoration on zigzag silicene nanoribbons. Superlattices Microstruct 2020:139.
[96] Abdelsalam H, Saroka VA, Teleb NH, Ali M, Osman W, Zhang Q. Electronic and adsorption properties of extended chevron and cove-edged graphene nanoribbons. Phys E Low-dimens Syst Nanostruct 2021;126.
[97] Li Y, He HJ, Liu ZC, Lai Z, Wang Y. A facile method for preparing three-dimensional graphene nanoribbons aerogel for uranium(VI) and thorium(IV) adsorption. J Radioanal Nucl Chem 2021;328:289–98.
[98] Zitoune H, Adessi C, Benchallal L, Samah M. Quantum transport properties of gas molecules adsorbed on Fe doped armchair graphene nanoribbons: a first principle study. J Phys Chem Solid 2021;153.
[99] Sharma H, Sandha KS. Impact of intercalation doping on the conductivity of multi-layer graphene nanoribbon (MLGNR) in on-chip interconnects. J Circ Syst Comput 2020;29.
[100] Kaur M, Gupta N, Kumar S, Raj B, Singh AK. RF analysis of intercalated graphene nanoribbon-based global-level interconnects. J Comput Electron 2020;19:1002–13.
[101] Yang SY, Li J, Li SS. Antiferromagnetic-ferromagnetic transition in zigzag graphene nanoribbons induced by substitutional doping. Chin Phys B 2018;27.
[102] Friedrich N, Brandimarte P, Li JC, Saito S, Yamaguchi S, Pozo I, Pena D, Frederiksen T, Garcia-Lekue A, Sanchez-Portal D, Pascual JI. Magnetism of topological boundary states induced by boron substitution in graphene nanoribbons. Phys Rev Lett 2020;125.
[103] Kawai S, Nakatsuka S, Hatakeyama T, Pawlak R, Meier T, Tracey J, Meyer E, Foster AS. Multiple heteroatom substitution to graphene nanoribbon. Sci Adv 2018;4.
[104] Motamedi M, Aghili AH. Stress-strain behavior of two-layer graphene with different chirality. J Nanoanal 2021;8.
[105] Venkataiah C, Prakash V, Mallikarjuna K, Prasad TJ. Investigating the effect of chirality, oxide thickness, temperature and channel length variation on a threshold voltage of MOSFET, GNRFET, and CNTFET. J Mech Cont Math Sci 2019:232–44.
[106] Chabi S, Kadel K. Two-dimensional silicon carbide: emerging direct band gap semiconductor. Nanomaterials 2020;10.
[107] Tan XY, Wu DD, Liu QB, Fu HH, Wu RQ. Spin caloritronics in armchair silicene nanoribbons with sp(3) and sp(2)-type alternating hybridizations. J Phys Condens Matter 2018;30.
[108] Joshi RK, Yoshimura M, Kumar A. Graphene. J Nanomater 2010;2010.
[109] Tabert CJ, Nicol EJ. Dynamical polarization function, plasmons, and screening in silicene and other buckled honeycomb lattices. Phys Rev B 2014;89.
[110] Tabert CJ, Carbotte JP, Nicol EJ. Magnetic properties of Dirac fermions in a buckled honeycomb lattice. Phys Rev B 2015;91.
[111] Ayatollahi A, Roknabadi MR, Behdani M, Shahtahmassebi N, Sanyal B. Density functional investigations on the adsorption characteristics of nucleobases on germanene nanoribbons. Phys E Low-dimens Syst Nanostruct 2020;117.
[112] Lebedeva OS, Lebedev NG, Lyapkosova IA. Elastic conductivity of silicene and germanene nanoribbons. St Petersburg Polytech Univ J-Phys Math 2019;12:38–49.

[113] da Silva JG, Enders BG, Silva GME, Fonseca ALD. Electron-phonon coupling in armchair silicene nanoribbons. Phys Lett 2019;383.

[114] Pham HD, Nguyen TDH, Vo KD, Huynh TMD, Lin MF. Rich essential properties of boron, carbon, and nitrogen substituted germanenes. APEX 2020;13.

[115] Kapoor P, Sharma M, Ahluwalia PK. Chirality dependent structural, electronic and mechanical properties of pristine Ag, Au and Pt nanotubes: a DFT study. Phys E Low-dimens Syst Nanostruct 2021;131.

[116] Guo YH, Liu MM, Zhu XJ, Zhu LR, He CA. Catalytic asymmetric synthesis of silicon-stereogenic dihydrodibenzosilines: silicon central-to-axial chirality relay. Angew Chem-Int Ed 2021;60:13887–91.

[117] Ben-Moshe A, da Silva A, Muller A, Abu-Odeh A, Harrison P, Waelder J, Niroui F, Ophus C, Minor AM, Asta M, Theis W, Ercius P, Alivisatos AP. The chain of chirality transfer in tellurium nanocrystals. Science 2021;372:729.

[118] Song LL, Yang ZH, Liu L, Yuan LW, Zhao H, Chen X, Zhang Y, Zheng XH. Realizing stable half-metallicity in zigzag silicene nanoribbons with edge dihydrogenation and chemical doping. J Phys Condens Matter 2021;33.

[119] Okano M, Mochida K. Electrochemical approach to germenium and silicenium ions. Chem Lett 1991:819–22.

[120] Rojas KIM, Villagracia AC, Moreno JL, David M, Arboleda NB. Ca and K decorated germanene as hydrogen storage: an ab initio study. Int J Hydrogen Energy 2018;43:4393–400.

[121] Xiao P, Fan XL, Liu LM. Tuning the electronic properties of half- and full-hydrogenated germanene by chlorination and hydroxylation: a first-principles study. Comput Mater Sci 2014;92:244–52.

[122] Ildarabadi F, Farghadan R. Edge magnetization and spin-valley-caloritronics in germanene and stanene nanoribbons. J Magn Magn Mater 2021;529.

[123] Bhandari AS, Verma A. A tight-binding analysis of models, sheet width, and strain for 2D monoatomic germanium sheets. Mater Exp 2019;9:273–80.

[124] Dhakal U, Rai D. Magnetic field control of current through model graphene nanosheets. Phys Lett A 2019;383:2193–200.

[125] Chegel R, Behzad S. Chirality dependence of dipole matrix element of carbon nanotubes in axial magnetic field: a third neighbor tight binding approach. Opt Commun 2014;313:406–15.

[126] Arjmand T, Tagani MB, Soleimani HR. The effect of buckling on I-V characteristics of symmetric and asymmetric zigzag germanene nanoribbons: a first-principle calculation. J Phys D-Appl Phys 2017;50.

[127] Monshi MM, Aghaei SM, Calizo I. Edge functionalized germanene nanoribbons: impact on electronic and magnetic properties. RSC Adv 2017;7:18900–8.

CHAPTER

15

Alkali-adsorbed germanene nanoribbons

Thi My Duyen Huynh[1], Shih-Yang Lin[1], Vo Khuong Dien[1], Chi-Hsuan Lee[1], Hsin-Yi Liu[1], Hai Duong Pham[1], Nguyen Thi Han[1], Ngoc Thanh Thuy Tran[2], Thi Dieu Hien Nguyen[1], Wei-Bang Li[1] and Ming-Fa Lin[1]

[1]Deparment of Physics, National Cheng Kung University, Tainan, Taiwan; [2]Hierachical Green-Energy Materials (Hi-GEM) Research Center, National Cheng Kung University, Tainan, Taiwan

15.1 Overview

Alkali and germanium atoms, which, respectively, belong to group-I and group-IV elements in a periodic table, play the critical roles in generating a lot of significant condensed-matter system, such as, the well-known lithium- [1–3]/sodium-ion-base batteries [4–6] and semiconductor industries [7–15]. They are able to create very active chemical environments, mainly owing to the atomic configurations of the outmost single orbital and the half-occupied ($4s,4p_x,4p_y,4p_z$) orbitals, respectively. The up-to-date basic research show that the highly non-uniform and anisotropic chemical bonds have been observed in the efficient cathode/electrolyte/anode materials of lithium-ion-related materials, e.g., ternary LiTiO [16]/LiGeO [17,18]/LiFeO [19] compounds, with many atoms in different Moire superlattice [20]. Very interesting, two kinds of chemical bonds, Li–Ti/Li–Ge/Li–Fe and Li–O ones, exhibit the observable bond-length fluctuations. The optimal crystal symmetries, being fully assisted by the featured energy spectra and wave functions, the obviously deformed charge density distributions, as well as the atom-/orbital-/spin-projected van Hove singularities, have successfully achieved the concise physical and chemical pictures for all diversified quasiparticle phenomena [21–24]. By the delicate VASP simulations and phenomenological analyses, the systematic investigated are thoroughly conducted on three kinds of core materials in lithium-/sodium-/

Fundamental Physicochemical Properties of Germanene-related Materials
https://doi.org/10.1016/B978-0-443-15801-8.00014-1

aluminum-/iron-based batteries, as clearly indicated in the published book. One of the important results are critical multi-orbital hybridizations of 2s-($3d_{z^2}$,$3d_{x^2-y^2}$,$3d_{xy}$,$3d_{yz}$,$3d_{xz}$) and 2s-(4s,$4p_x$,$4p_y$,$4p_z$) and 2s-(2s,$2p_x$,$2p_y$,$2p_z$)/2s-($3d_{z^2}$, $3d_{x^2-y^2}$,$3d_{xy}$,$3d_{yz}$,$3d_{xz}$) and 2s-(2s,$2p_x$,$2p_y$,$2p_z$) for the LiTiO, LiGeO, and LiFeO materials, respectively. As for the germanium- and silicon-related materials, their semiconducting behaviors have created the recent industry revolutions and will dominate the next-generation electronic developments in applied chips (detail in Chapter 22). Their basic science researches are fully explored in the current and previous books [25], respectively, where certain important differences between these two kinds of main-stream systems are thoroughly examined from all the relevant quantities.

Chemical adsorptions [26–32], substitutions [33–35], intercalations/de-intercalations [20,36–39] and edge passivation [40,41], as systematic studies done for bulk graphite [42], layered graphene systems [43–47], carbon nanotubes [48–54] and graphene nanoribbons [26,40,55–61], are very powerful in diversifying the fundamental quasiparticle behaviors both theoretically and experimentally. Very interesting, graphite alkali-atom intercalations compounds, which were successfully synthesized, at least, 40 years ago, exhibit the excellent electrical conductivities, comparable with that of a copper metal [5,39,62]. The intercalation-induced high free carrier densities have generated a lot of dramatic transformations about the low-energy physical properties, e.g., the blue shift of the threshold adsorption frequency [39,62], and the coherent charge oscillations of conduction electrons (the plasmon modes of free carriers). The interlayer alkali-carbon atomic interactions are thoroughly identified to be the critical mechanism. As a result of the weak, but significant van der Waals interaction between graphitic layers, the guest-atom intercalation and de-intercalations are highly efficiently for alkali atoms during the charging and discharging processes of ion-based batteries. Such compounds frequently serve as an anode material of a commercialized battery [1–6,36,38,63]. On the graphene side, alkalization [30,31,55,58], hydrogenation [41,64–66], oxidation [37] and halogenation [67], are very popular in driving the unusual phenomena, the unique functionalities, and the highly potential applications [15,43,68,69]. The various single-/multi-orbital hybridizations and spin-dependent ferromagnetic configurations are clearly examined through the VASP simulations [70] and the phenomenological models. For example, the potassium-adsorbed graphene systems, which are detected by the ARPES measurements (detail in Chapter 3) show a roughly rigid shift about the Dirac-cone energy spectrum below the Fermi level. The spectral distribution width is further utilized to determine the Coulomb decay rates of the excited quasi-particle states, e.g., the anisotropic and non-monotonous wave vector-/energy-dependences. The inelastic scatterings are totally different between 2D graphene and electron gas. The collective excitations from free conduction electrons are successfully verified from the high-precision measurements of electron energy loss spectroscopy (detail in Chapter 3). Apparently, a lot of unusual quasiparticle behaviors are closely related to the n-/p-type doping ability; that is, the close relations between free carrier densities and guest-atom adsorption cases (their concentrations and distributions are responsible for the other fundamental properties). Each alkali-adsorbed graphene nanoribbon is an outstanding model system in determining the critical charge transfer, since its free carriers per unit length is proportional to the Fermi momenta under the 1D quantum confinement. A plenty of alkali-chemisorption configurations have identified the 100% charge transfer between guest-alkali and host-carbon atoms through the delicate VASP simulations. Whether the similar relations come to exist in alkali-adsorbed germanene nanoribbons should be investigated in detail.

1D germanene nanoribbons, with/without chemical modifications, are expected to exhibit the rich quasiparticle phenomena, as clearly revealed in partner materials of graphene nanoribbons [26,32,33,40,55,56,58–61,67,71,72]. Each single-layer honeycomb crystal could be regarded as a cut 2D germanene along the longitudinal direction, or a broken germanene nanotube along the tubular axis. Two close approaches are frequently utilized in the theoretical predictions [28,56,73,74] and experimental syntheses/examinations [67,75–77]; that is, there exist strong relations among the electronic states in three kinds of low-dimensional system. The very active chemical/physical environments [13] can cover a buckled surface full of the dominating σ bondings/the weak, but significant sp^3 bondings [78], a finite width with two open boundaries, achiral/chiral edge structures in the absence/presence of hydrogen decorations [79–83], and the interlayer/intralayer guest-host hybridized multi-orbital interactions after chemical adsorption [26,27,31]/passivation [40,41]/substitutions [13,33–35]. The alkali-adsorbed germanene nanoribbons are chosen for a model study, since it is relatively easy to observe the diversified quasiparticle phenomena through the concise chemical and physical mechanism, such as, the sensitive width and edge-structure dependences of band gaps under the alkali surface adsorptions and/or the hydrogen edge decorations. The highly non-uniform and anisotropic chemical-bond distributions have created the extra multi-orbital mixings between host and guest adatoms. The significant couplings, which might consist of the nearest-neighboring intralayer hopping integrals in host- and guest-atom layers without/with open boundary limits, the prominent interlayer atomic interactions, the spin-orbital interactions (the single-particle form in Ref. [84]), the spin-dependent on-site electron-electron under a ferromagnetic configuration (the Hubbard-like Coulomb interactions [24,77,85]). Their strong effects on all relevant quantities are thoroughly examined and analyzed by the VASP simulations. The reliable establishments of phenomenological models will be discussed in detail.

15.2 Alkali-adsorbed structural optimization

From a viewpoint of geometric structure, each honeycomb system has four possible positions, in which another atom can adsorb to create an adsorbed structure. In case of alkali adsorbed germanene, there are four types of adsorptions namely hollow, bridge, top and valley site as shown in Fig. 15.1. In Fig. 15.1A, top view of adsorbed positions is determined in a hexagonal cell. The hollow site atoms locate at center of hexagonal cell, in which the bond length of adsorbed atoms and germanium is the same. The bridge site atoms locate at middle of Ge–Ge bonds. Because of the common side view of these two types in (x, z) plane, Fig. 15.1D and E show other side views to further illustrate the correlation positions. Due to buckled structure of 2D germanene, two nearest germanium atoms cannot locate in the same plane, leading to the top and valley positions for adsorptions/substitutions/doping. As clearly shown in Fig. 15.1B and C, top site adsorbed atoms are on the top of top germanium while valley site ones locate on the top of bottom germanium atoms. Furthermore, in top view with considering hexagons (see Fig. 15.1A), blue and red rectangle are additionally defined to figure out the top and valley sites in adsorbed positions, respectively. Based on these possible positions, adsorbed atoms could be determined varying adatoms, concentrations or distributions [26,27,55,83] in single or double side. Two typical types of germanene nanoribbons correspond to zigzag and armchair structures, determining by the hexagons arranged along the edges from 2D germanene (mentioned in Chapter 14). Due to the structure, the four adsorbed types could be applied in nanoribbon systems.

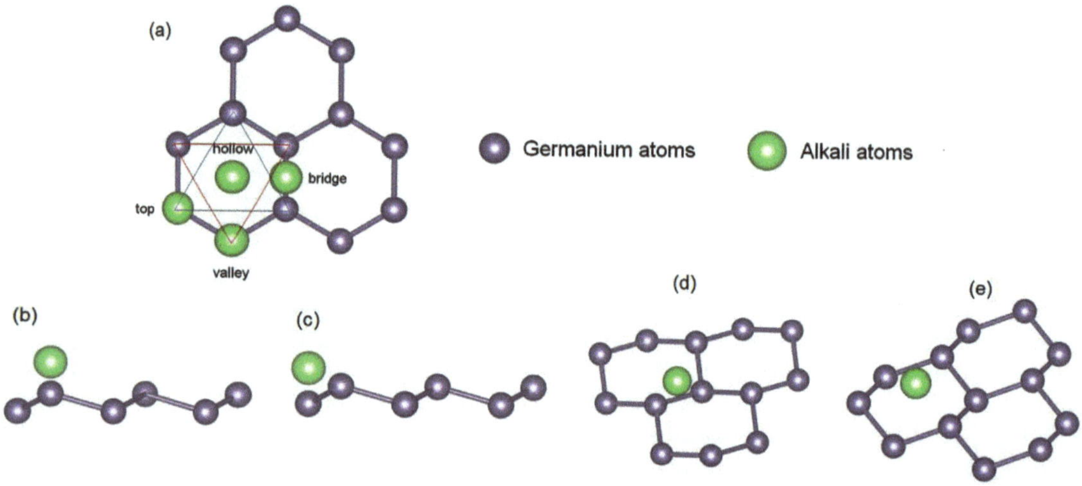

FIG. 15.1 (A) Top view and side view of adsorbed position in (B) top site, (C) valley site, (D) bridge site and (E) hollow site of alkali-adsorbed germanene nanoribbons.

To determine the favorable position in adsorption of germanene nanoribbons, the optimal calculations are considered to find the minimum ground state energies. Fig. 15.2 shows ground state energies of top, valley, bridge and hollow sites in $N_A = 9$ system of lithium adsorbed armchair germanene nanoribbons. The concentration of these systems is 1:20 (5%), in which a lithium atom adsorbs four positions (hollow, bridge, top and valley) in a unit cell. As shown in this figure, valley sites have the minimum ground state energies compared to others. The valley site is therefore considered in this work for zigzag and armchair germanene nanoribbons that will be discussed later.

The essential properties of alkali-adsorbed germanene nanoribbons are investigated for the distinct adatoms including Li, Na, K in two typical nanoribbons (zigzag and armchair).

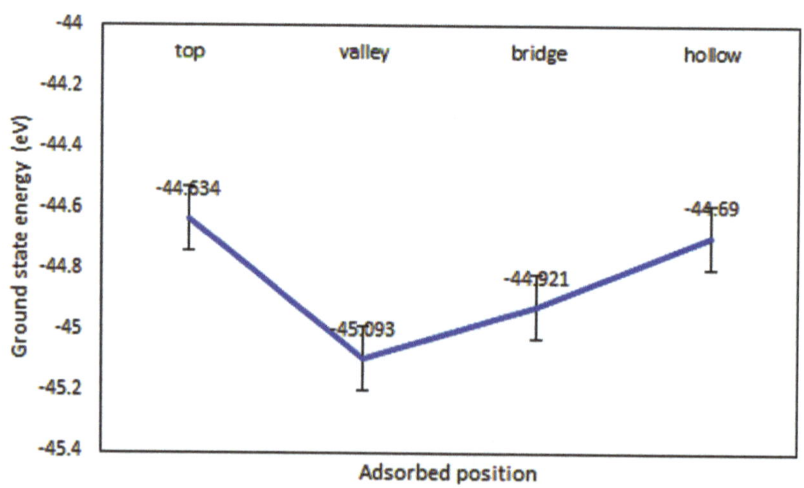

FIG. 15.2 Optimized ground state energies (eV) of four adsorbed positions in armchair germanene nanoribbons.

The top and side views of alkali adsorbed zigzag and armchair germanene nanoribbons are described in Fig. 15.3, in which their widths are characterized by the number of zigzag and dimer lines, respectively. The valley alkali adsorbed sites are defined due to minimum ground state energies of the structural optimization. In these systems, alkali adsorbed at Ge4 and Ge3 of $N_Z = 8$ and $N_A = 12$, respectively, with 1:18 (5.6%) and 1:29 (3.45%) concentrations in single valley site. The nearest Ge–Ge and A-Ge bonds strongly depend on the distinct adatoms and edge as shown in Table 15.1. The smallest bond of Ge–Ge is found

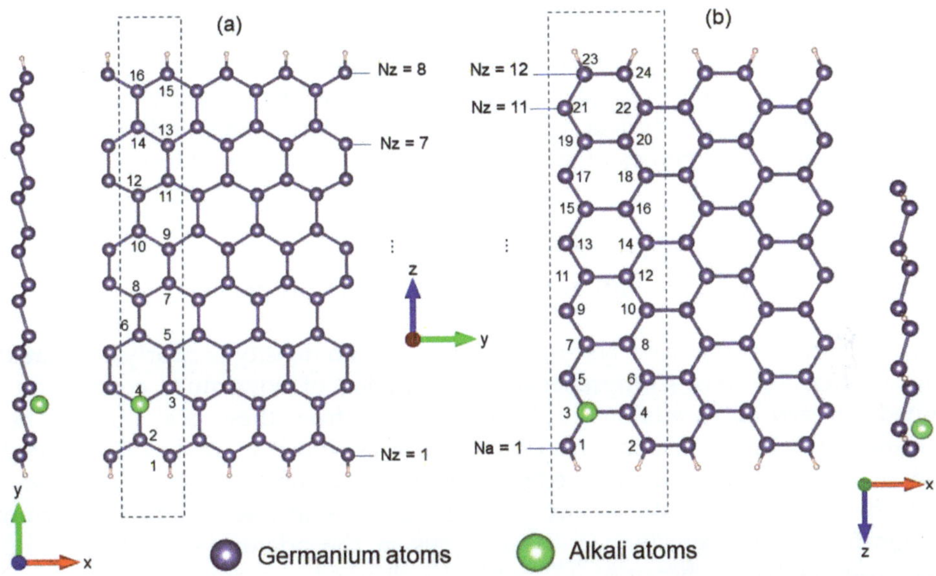

FIG. 15.3 Geometric structures of alkali adsorbed (A) $N_Z = 8$ zigzag and (B) $N_A = 12$ armchair germanene nanoribbons in single valley site.

TABLE 15.1 Calculated nearest Ge–Ge and A–Ge bond length, height of passivated germanium, binding energies, magnetic moment and band gaps of $N_Z = 8$ and $N_A = 12$ alkali-adsorbed germanene nanoribbons.

		Bond length (Å)				
System	**Configuration**	**Nearest**	**A-Ge**	**Height (Å)**	**Magnetic moment**	**Eg (eV)**
Zigzag	$(4)_{single}$; Li	2.631	2.643	1.433	0.26	SM
	$(4)_{single}$; Na	2.572	2.583	1.401	0.24	SM
	$(4)_{single}$; K	2.6	2.734	1.782	0.26	SM
Armchair	$(3)_{single}$; Li	2.636	2.66	1.421	0	SM
	$(3)_{single}$; Na	2.523	2.553	1.359	0	SM
	$(3)_{single}$; K	2.584	2.667	1.73	0.01	SM

in K-adsorbed structure while it presents a largest bond of K–Ge corresponding to 2.6 and 2.782 Å, 2.58 and 2.73 Å in zigzag and armchair case, respectively. This might cause from the adatom radius that K has the biggest radius in group of Li, Na and K, indicating in growing this bond lengths when increasing adatom radius from Li to K. Moreover, adsorbed structures reveal the larger bonds of A-Ge compared to Ge–Ge bonds. Furthermore, the buckling (height fluctuation) exhibits the same features when going from Li to K. The strongest height is found in K-adsorbed configuration due to the size of adsorbed atoms. The height of these structures is dramatically larger than those of pristine and H-passivated cases (see Chapter 14), demonstrating the critical effects of alkali adsorption on structural characteristics. Besides, magnetic moment shown in Table 15.1 anticipates the ferromagnetic/nonmagnetic configuration of zigzag/armchair nanoribbons. These values significant change after alkali adsorption compared to pristine and H-passivated cases; especially, armchair nanoribbons transfer from magnetic to nonmagnetic configurations, indicating the adsorption affect to magnetic properties of germanene nanoribbons. This feature will be discussed in DOS properties in detail.

15.3 Alkali-adsorbed electronic properties

In alkali-adsorbed germanene systems, the free electron density is clearly calculated using first-principles calculations. Comprehensive information of how much conduction charge contributed by each alkali adatom could be provided from these evaluations. The Bader charge transfer is used to describe this characteristic as shown in Table 15.2, depending on the adatom kind, relative position, edge structure and ribbon width. The alkali atoms contribute the outmost s-orbital electrons as free carriers in the system. Furthermore, the charge transfer is very sensitive to the changes in the adatom kind, for example, Li-adsorbed case has Bader charge of 0.74 while K-adsorbed one presents the value of 0.16e in zigzag nanoribbons. This feature also depends on edge boundaries, indicating in 0.16 and 0.12e for K-adsorbed zigzag and armchair nanoribbons, respectively. This indicates

TABLE 15.2 Free electron densities for various adatom kind in $N_A = 12$ and $N_Z = 8$ alkali-adsorbed germanene nanoribbons.

System	Configuration	n (10^7 e/cm)	n/Adatom concentration (e)	Bader charge transfer (e)
Zigzag	$(4)_{single}$; Li	3.69	1.85	0.74
	$(4)_{single}$; Na	3.58	1.79	0.53
	$(4)_{single}$; K	3.58	1.79	0.16
Armchair	$(3)_{single}$; Li	3.77	1.89	0.73
	$(3)_{single}$; Na	3.58	1.79	0.54
	$(3)_{single}$; K	3.59	1.79	0.12

the charge transfer cannot be utilized to evaluate the free carrier density in adatom adsorption. In addition, the free electron densities listed in Table 15.2 calculated using $n = 2k_F/\pi$, where k_F is the fermi momentum, reflect the same orbital hybridizations in A-Ge bonds under various adatom adsorption. The free electron densities slightly change going from Li to Na and K corresponding to 3.69 and 3.58e, indicating this feature is independent on adatom kind (the distribution configurations). The same property is also found in edge boundary dependent, both zigzag and armchair cases show similar free electron densities; illustrating the boundary termination does not affect free electron densities. Generally, the free electron densities might depend on the concentration configurations by varying concentrated values that is absent in this work.

Apart from the effects of geometric structures, chemical bonds, finite width confinement, edge boundaries on critical properties of 1D systems; the spin configuration is an important factor which affect the carrier density. Take Li-absorbed system as a representative, Fig. 15.4 describes the spin configurations of $N_Z = 8$ and $N_A = 12$ zigzag and armchair nanoribbons. These spin structures strongly depend on edge boundaries, corresponding to zigzag and armchair nanoribbons. Free carriers in the spin-down of zigzag system (Fig. 15.4A), due to the ferromagnetic ordering at one edge without adatom. This demonstrates that the A-Ge interactions seriously frustrate the spin configurations. Furthermore, the absence of spin distribution near adatom exhibits the magnetic suppression. In contrast, the spin degenerate electronic states become split with lithium adatom near armchair edge (Fig. 15.4B), associated

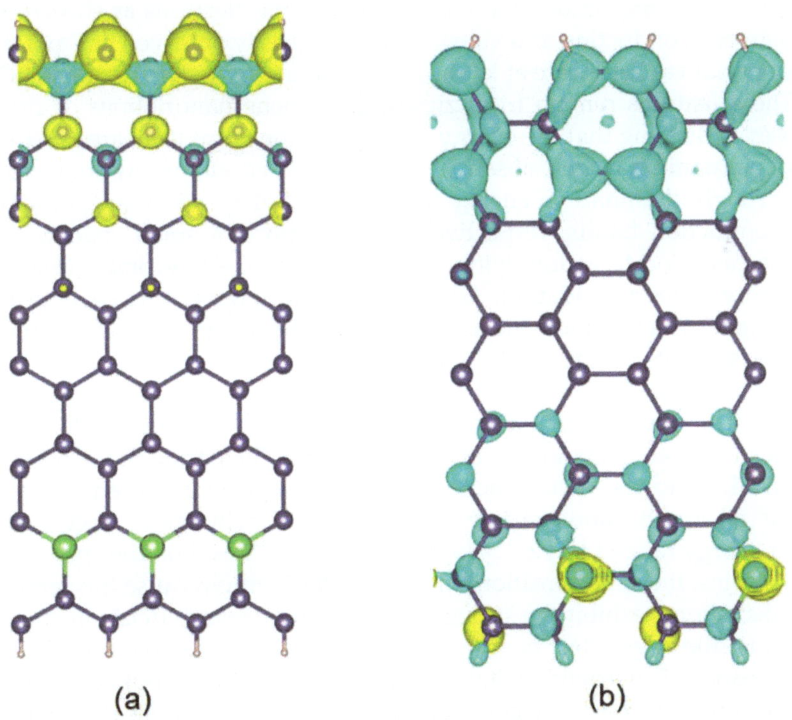

FIG. 15.4 Spin configurations of Li-adsorbed (A) zigzag and (B) armchair nanoribbons.

with edge atoms away and near adatom. There are more free carriers in spin up/spin down near the edge without/with adatom, indicating the spin configurations are enhanced by the A-Ge interactions. In short, the spin configurations with ferromagnetic/non-magnetic obviously depend on adatom and edge boundaries.

The main features of energy band structures strongly depend on relative position, adatom and edge structure. Fig. 15.5 clearly shows energy spectra with alkali atom dominance of zigzag and armchair nanoribbons that directly induced by the finite-size confinement effect, leading to a lot of 1D energy bands. As mentioned in Chapter 14, the edge boundaries can diversify band structures and spin configurations. The alkali-adsorbed systems exhibit metallic band, in which conduction bands are easily modulated by adatom (green and maroon circles). Besides, valence bands reveal energy dispersions affected by the adatom adsorption with unchanged germanium atom dominance. These characteristics might come from the close relations between the quantum confinement indicated in nanoribbon systems and the orbital hybridizations in A-Ge bonds. The bandgap features are determined by the strong competition between spin arrangement and quantum confinement. All structures reveal semi-metallic features after alkali-adsorption with distinct band dispersions for zigzag and armchair nanoribbons. The translation from direct-gap semiconductors of armchair germanene nanoribbons to semi-metallic ones indicates the strong A-Ge bonding, as well as essential effects of alkali adsorption in germanene nanoribbons. Furthermore, energy bands present parabolic dispersions in both conduction and valence bands, as well as few of partially flat bands.

For the zigzag system, the partial flat bands exist near the fermi level around $k_z > 2/3$ in valence band, in which the conduction band only has free electrons as shown in Fig. 15.5A. The overlap between conduction and valence bands is clearly observed at Γ point and around $k_z > 2/3$. The special valleys locate at $k_z = 2/3$ corresponding to Dirac cone in 2D germanene monolayer. These features remain from zigzag germanene nanoribbons of pristine and H-passivated cases, indicating that the energy bands near the fermi level are mostly contributed from the edge germanium atoms, absent of alkali-adsorbed atoms. As shown in this figure, alkali atoms strongly dominate in energy ranged from -1 to -3 eV and more than 1 eV in valence and conduction bands, respectively. Furthermore, a small spacing (smaller than that of pristine case) locate at low-lying energy less than -3 eV, indicating the effects of band-edge valence states and adatom. Adatom dominance (green circles) is clearly observed in deeper and higher valence and conduction band, respectively, when changing alkali atoms follow the increase of atoms radius (Li to Na, to K).

With regarding to armchair system (Fig. 15.5B), most of characteristics exhibited in zigzag cases are absent. The small overlap between conduction and valence bands is only observed at Γ point instead of Γ and Z points as in zigzag system. The partial flat band near the Fermi level that could lead to the delta-function form of the peaks is absent. In addition, they possess a pair of linear bands about the Fermi level. The more energy sub-bands at deeper/higher energy ranged in $|E| > 1$ eV are observed. Based on the spin distribution differences of two spins, the spin transition is found in PDOS might cause this feature. Similar to zigzag case, alkali atoms exhibit the strong contribution (maroon circles) in energy of $|E| > 1$ eV, demonstrate the edge effect near the fermi level, as well as adatom effect at low-lying and high-lying bands. The extended states with wide spatial distribution at range of energy less than -3 eV and more than 1 eV correspond to the strong parabolic dispersions in valence and conduction bands. Especially, the small spacing locating at energy less than -3 eV is

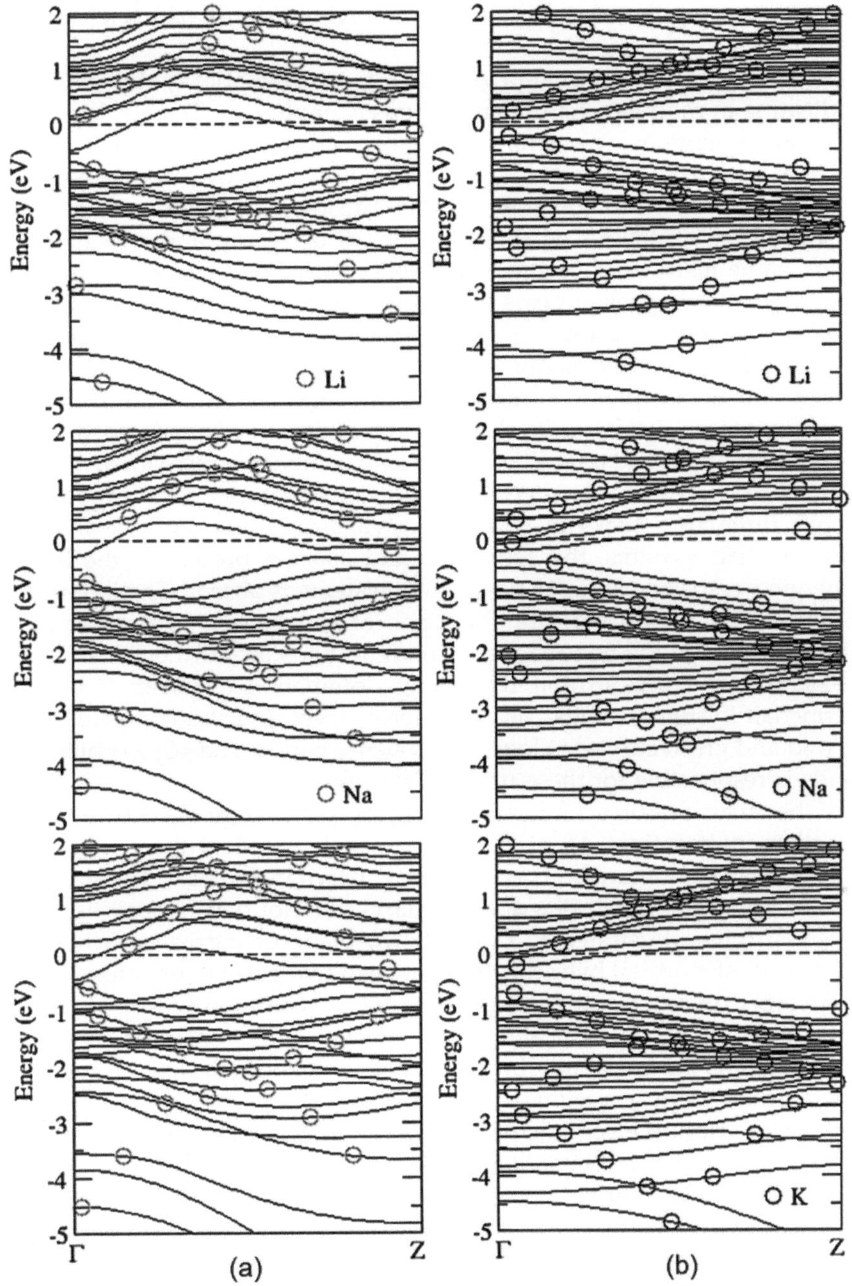

FIG. 15.5 Band energy spectra of alkali-adsorbed (A) $N_Z = 8$ zigzag and (B) $N_A = 12$ armchair germanene nanoribbons.

totally absent, illustrating the adatom effects in low-lying valence bands (clearly indicated in PDOS).

The critical characteristics of orbital-projected DOS are dramatically changed, due to dimensionality, indicating in monolayer and nanoribbons (Chapter 14). The alkali-adsorbed germanene nanoribbons exhibits many asymmetric peaks, as well as few symmetric ones as shown in Fig. 15.6. They arise from the parabolic and partially flat energy dispersions, corresponding to square-root and delta-function divergent forms. The low-lying peaks are additionally diversified by distinct edge structures as shown in PDOS of zigzag and armchair systems. For zigzag system, some features of pristine nanoribbons are remained that the spacing due to the pair of anti-symmetric peak appears in valence band of Li- and Na-adsorbed zigzag systems (Fig. 15.6A). However, the spacing in conduction bands is absent after alkali adsorptions. In addition, adatoms induce the asymmetric peaks near the fermi level, due to alkali dominated occupied and unoccupied conduction bands. Zigzag systems can clearly present the delta-function form near the fermi level, especially in Li- and Na-adsorbed ones, based on the edge-localized energy bands, sensitive to the spin configurations. Moreover, the ferromagnetic configurations are obviously revealed in zigzag structures, for example, Na-adsorbed system exhibits three/two peaks in conduction/valence bands near the fermi level. With regarding to armchair systems, the zero DOS and spacing present at the fermi level and low-lying energies, respectively, in H-passivated systems is absent. The intensity is reduced in the spin-split energy bands, indicating in a pair of symmetric peaks across the fermi level (Fig. 15.6B). Additionally, the non-magnetic configurations are further illustrated due to the merged peaks. All structures exhibit the strong domination of Ge-4p_z (cyan/blue curves in zigzag/armchair systems) orbital near the fermi level and the competed dominance of 4s-(4p_x,4p_y) (red and green curves) at low-lying valence bands. The s-2pz orbital hybridization in A-Ge bonds is responsible for these unusual features.

To reflect the main features of band and DOS, the charge density is very useful in providing further information. As a representative, charge density of Li-adsorbed zigzag and armchair germanene nanoribbons is examined shown in Fig. 15.7. The significant hybridization in A-Ge bond can create the spatial charge distribution. Apparently, the σ and π simultaneously survive, illustrated in (x, y)/(x, z) plane of zigzag/armchair system. The high charge density of σ orbital hybridization between two germanium atoms is independent with alkali-adsorbed effect. Furthermore, the strong distribution of s-sp_z bond (red region around Li—Ge bond) is clearly indicated charge densities, demonstrating the strong Li—Ge interactions. Generally, the essential orbital hybridization and free electron contribution of alkali atoms are independent of various adatoms.

In general, geometric and electronic properties dramatically change when alkali atoms are absorbed on germanene nanoribbons due to chemical bonds, height fluctuation, orbital hybridization, energy bands, charge transfer and free electron carrier. The A-Ge bonds reveal the strong orbital hybridizations and critical effects on band energy and orbital projected DOS due to s-sp_z bonds. Furthermore, the charge transfer indicating contribution of each alkali atoms strongly depends on adatom kind that reduces as increasing radius of atom, while free electron densities are independent on adsorbed atom kind. These featured behaviors characterize alkali adsorbed germanene nanoribbons are sensitive dependence on adatom concentration and configuration, adsorbed atom types (valley sites) and height due to buckling honeycomb structure.

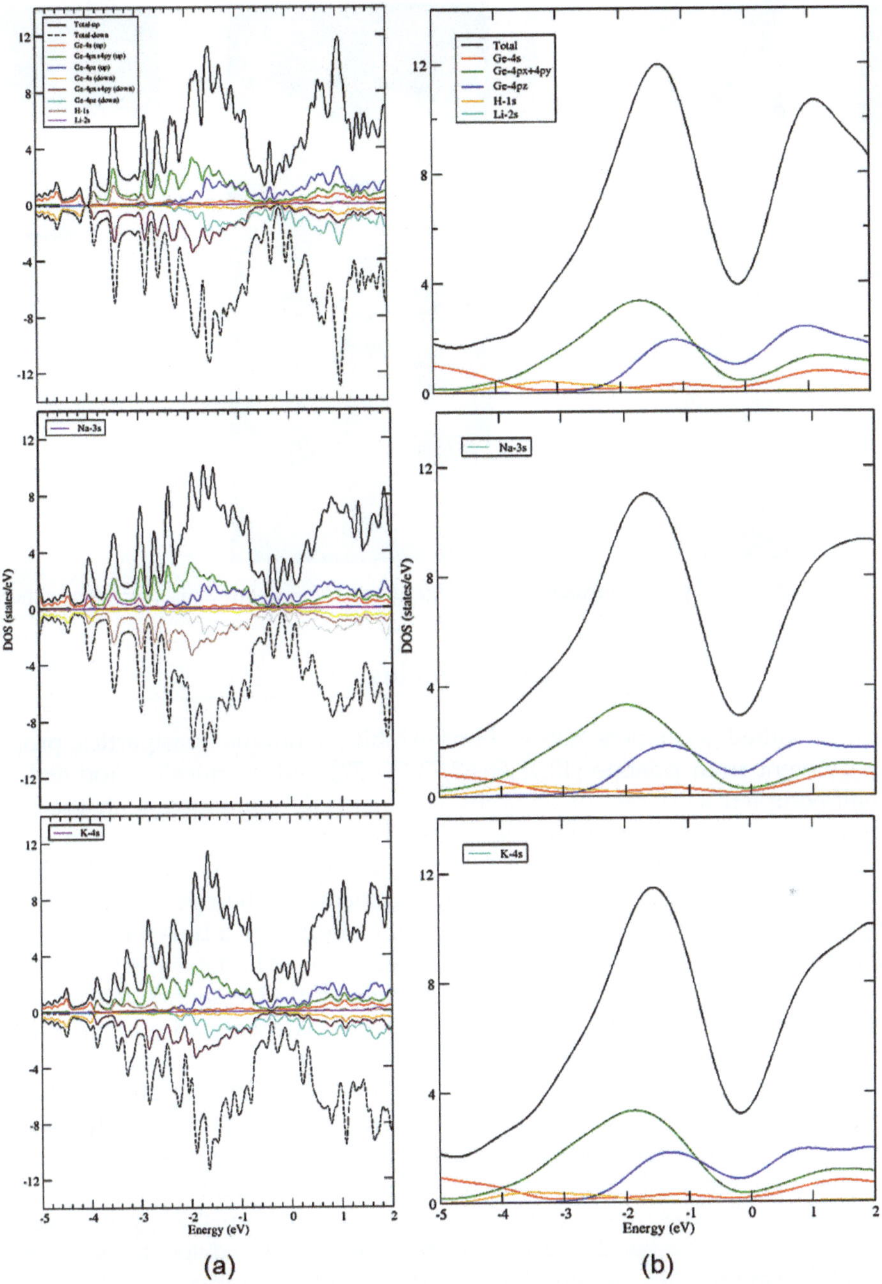

FIG. 15.6 Orbital-projected DOS of alkali-adsorbed (A) $N_Z = 8$ zigzag and (B) $N_A = 12$ armchair germanene nanoribbons.

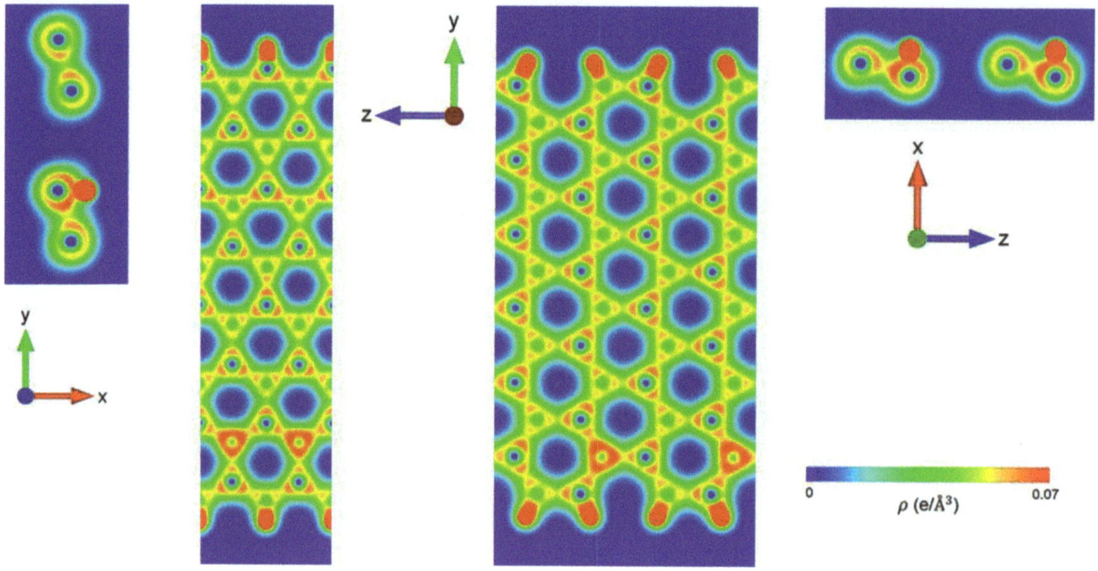

FIG. 15.7 Charge density of Li-adsorbed $N_Z = 8$ zigzag and $N_A = 12$ armchair germanene nanoribbons.

15.4 Conclusion

The alkali-adsorbed germanene nanoribbons exhibit the unique quasiparticle properties, being totally different from pristine [13,28,56,67,73,75–77] and chemically modified cases. The outmost half-occupied s-orbitals of adatoms have directly combined with the highly non-uniform and active chemical environment of Ge-host 1D systems. The creation of the prominent/minor interlayer chemical bonding with the Ge-4p_z/Ge-(4s, 4p_x, 4p_y) orbitals, which are fully assisted by the finite-width quantum confinement (the open boundary conditions; [58,67,75]), the chirality of open edge structures, the coexistent intralayer $\pi/\sigma/sp^3$ bonding, the single-orbital hybridizations in A—A bonds, and the spin-dependent single- or many-particle interactions. Their strong couplings can diversify all essential properties and thus greatly promote the highly potential applications [86–92]. The featured quasiparticle behaviors cover the very sensitive dependences on the alkali-adatom concentrations and configuration, the position-dependent valley sites and heights, the blue shift of the Fermi level (the generation of free conduction electrons in the semiconductor-metal transitions)/the highly asymmetric energy spectra about valence holes and conduction electrons/a lot of energy sub-bands in the first Brillouin zone with the distinct dispersion relations/the frequent sub-band non-crossings, crossings and anti-crossings/many critical points under the vanishing group velocities/the non-well-behaved π- and σ-band widths, the extreme-/saddle-/dispersionless-/linear-point-related van Hove singularities, the simultaneous display of the intralayer and interlayer orbital hybridizations in the spatial charge density distributions, and the adatom- and edge-enriched spin arrangements. Specifically, the linear free carrier density due to the alkali-adatom charge transfer is directly examined from the calculated Fermi momenta ($n = 2k_F/\pi$). This feature shows the wear change when varying adatom, indicating the independent on adatom kind (distribution configuration). That might depend on concentration when changing concentration that is not shown in this chapter. The various concentration configurations will be further examined in the next research. Generally, the 1D conduction electrons will further lead to an outstanding transport property and an optical threshold frequency. They are thus expected to play the critical roles in electronic devices and optical modulators.

The alkali-adsorbed germanene nanoribbons are very excellent candidates in establishing the phenomenological models (details in Chapter 2.2) because of the diversified quasiparticle behaviors mainly arising from the concise mechanisms. For example, the tight-binding model/the generalized tight-binding model, which is in the absence/presence of electric and magnetic fields, should cover the s-s single-orbital hybridizations of A—A bonds, the (4s, $4p_x$, $4p_y$, $4p_z$)- (4s, $4p_x$, $4p_y$, $4p_z$) bonding in the buckled honeycomb Ge-lattices [13,78], the interlayer dominant s-$4p_z$/minor s-sp^3 atomic interactions, as well as, the spin-orbital coupling or the on-site spin-dependent Coulomb potentials, The Hubbard-like Hamiltonian will be delicately established by comparing the low-lying energy sub-bands in the current VASP simulations. strong cooperation between the numerical and model approaches is able greatly promote the progress of physical/chemical/material sciences, as done for the previously published books. When these two models are further linked with the static and dynamic Kubo formulas, and the random-phase approximation (RPA)/self-energy method, the Hall effects/optical-magneto-optical transitions, electron-hole and plasmon excitations/the Coulomb decay rates could be thoroughly through the delicate calculations. The above-mentioned viewpoints are worthy of the systematic investigations under the near-further research.

On the experimental side, the high-resolution detections are required to present the concise physical/chemical/material pictures for the alkalization phenomena of 1D germanene nanoribbons. Their examinations are also suitable for hydrogenation (details in Chapter 7), oxidization, and halogenation. Such strategy is very useful in fully understanding the diversified effects of chemical modifications. However, the successful observations on the sample syntheses (the various techniques in Refs. [93,94]), low-dimensional optimal symmetries (STM/TEM, details in Chapter 3.1), occupied energy spectra below the Fermi level (ARPES, Chapter 3.3), and energy-dependent van Hove singularities (STS, Chapter 3.2) are absent up to now. The close cooperation among the experimental measurements, the first-principles simulations, and the phenomenological models will be rather helpful in solving the open measurement issues. How to present the spatial charge density distributions about the alkali layer and buckled Ge-honeycomb lattices, as well as the adatom-dominated spin configurations in zigzag germanene nanoribbons, should be very interesting research topics. The 1D ferromagnetic metals might have the unusual transport properties and thus display potential applications for electronic nanodevices. The similar VASP studies on hydrogenations, oxidizations and halogenations can reveal the totally different orbital hybridizations and thus generate the diverse quasiparticle properties.

References

[1] Yang Y, Yan C, Huang JQ. Research progress of solid electrolyte interphase in lithium batteries. Acta Phys Chim Sin 2021;37.

[2] Ye YK, Hu ZX, Liu JH, Lin WC, Chen TW, Zheng JX, Pan F. Research progress of theoretical studies on polarons in cathode materials of lithium-ion batteries. Acta Phys Chim Sin 2021;37.

[3] Wang H, An HW, Shan HM, Zhao L, Wang JJ. Research progress on interfaces of all-solid-state batteries. Acta Phys Chim Sin 2021;37.

[4] Tao F, Liu Y, Ren XY, Jiang AJ, Wei HJ, Zhai XL, Wang F, Stock HR, Wen SF, Ren FZ. Carbon nanotube-based nanomaterials for high-performance sodium-ion batteries: recent advances and perspectives. J Alloys Compd 2021;873.

[5] Majumder T, Das D, Jena S, Mitra A, Das S, Majumder SB. Electrophoretic deposition of metal-organic framework derived porous copper oxide anode for lithium and sodium ion rechargeable cells. J Alloys Compd 2021;879.

[6] Hakim C, Sabi N, Saadoune I. Mixed structures as a new strategy to develop outstanding oxides-based cathode materials for sodium ion batteries: a review. J Energy Chem 2021;61:47–60.

[7] Dugarte-Dugarte A, Pineda NR, Nieves L, Henao JA, de Delgado GD, Delgado JM. The crystal structure of Cu(2) GeSe(3)( )and the structure-types of the I-2-IV-VI3 family of semiconducting compounds. Acta Crystallogr Sect B-Struct Sci Cryst Eng Mater 2021;77:158–67.

[8] Pontes RB, Mancano RR, da Silva R, Cotica LF, Miwa RH, Padilha JPE. Electronic and optical properties of hydrogenated group-IV multilayer materials. Phys Chem Chem Phys 2018;20:8112–8.

[9] Dimple JN, Rawat A, Ahammed R, Mohanta MK, De Sarkar A. Emergence of high piezoelectricity along with robust electron mobility in Janus structures in semiconducting Group IVB dichalcogenide monolayers. J Mater Chem 2018;6:24885–98.

[10] Imai M. Semiconducting ternary Si clathrates. Jpn J Appl Phys 2020;59.

[11] Pena-Pedraza H, Lopez-Rivera SA, Martin JM, Delgado JM, Power C. Crystal and phonon structure of ZnSiP2, a II-IV-V-2 semiconducting compound. Mater Sci Eng B-Adv Funct Solid-State Mater 2012;177:1465–9.

[12] Jungwirth T, Novak V, Marti X, Cukr M, Maca F, Shick AB, Masek J, Horodyska P, Nemec P, Holy V, Zemek J, Kuzel P, Nemec I, Gallagher BL, Campion RP, Foxon CT, Wunderlich J. Demonstration of molecular beam epitaxy and a semiconducting band structure for I-Mn-V compounds. Phys Rev B 2011;83.

[13] Arjmand T, Tagani MB, Soleimani HR. The effect of buckling on I-V characteristics of symmetric and asymmetric zigzag germanene nanoribbons: a first-principle calculation. J Phys D-Appl Phys 2017;50.

[14] Liu J, Li XB, Wang D, Liu H, Peng P, Liu LM. Single-layer Group-IVB nitride halides as promising photocatalysts. J Mater Chem 2014;2:6755–61.

[15] Qiao H, Liu H T, Huang Z Y, Hu R, Ma Q, Zhong J X and Qi X 2021 Tunable electronic and optical properties of 2D monoelemental materials beyond graphene for promising applications Energy Environ Mater.

[16] Jiang JH. Synthesis of $Li_2Ti_{1.5}O_4$ and its selectivity to Li^+ exchange. Asian J Chem 2013;25:841–3.

[17] Dien VK, Han NT, Nguyen TDH, Huynh TMD, Pham HD, Lin MF. Geometric and electronic properties of Li_2GeO_3. Front Mater 2020;7.

[18] Dien VK, Pham HD, Tran NTT, Han NT, Huynh TMD, Nguyen TDH, Ming FL. Orbital-hybridization-created optical excitations in Li_2GeO_3. Sci Rep 2021;11.

[19] Catti M, Montero-Campillo M. Principles modelling of lithium iron oxides as battery cathode materials. J Power Sources 2011 First-;196:3955–61.

[20] Walter J, Shioyama H. Boron trichloride graphite intercalation compound studied by selected area electron diffraction and scanning tunneling microscopy. J Phys Chem Solid 1999;60:737–41.

[21] Yoshioka N, Mizukami W, Nori F. Solving quasiparticle band spectra of real solids using neural-network quantum states. Commun Phys 2021;4.

[22] Verresen R, Moessner R, Pollmann F. Avoided quasiparticle decay from strong quantum interactions. Nat Phys 2019;15:750.

[23] Liu FL, Lundgren R, Titum P, Pagano G, Zhang JH, Monroe C, Gorshkov AV. Confined quasiparticle dynamics in long-range interacting quantum spin chains. Phys Rev Lett 2019;122.

[24] Fandan R, Pedros J, Guinea F, Bosca A, Calle F. Effect of quasiparticle excitations and exchange-correlation in coulomb drag in graphene. Commun Phys 2019;2.

[25] Ngoc Thanh Thuy Tran S-YL, Lin C-Y, Lin M-F. Geometric and electronic properties of graphene-related systems. In: Chemical bonding schemes; 2018.

[26] Brito WH, Miwa RH. Adsorption and diffusion of gold adatoms on graphene nanoribbons: an ab initio study. Phys Rev B 2010;82.

[27] Pang Q, Zhang CL, Li L, Fu ZQ, Wei XM, Song YL. Adsorption of alkali metal atoms on germanene: a first-principles study. Appl Surf Sci 2014;314:15–20.

[28] Ayatollahi A, Roknabadi MR, Behdani M, Shahtahmassebi N, Sanyal B. Density functional investigations on the adsorption characteristics of nucleobases on germanene nanoribbons. Phys E Low-dimens Syst Nanostruct 2020;117.

[29] Abdelsalam H, Saroka VA, Teleb NH, Ali M, Osman W, Zhang Q. Electronic and adsorption properties of extended chevron and cove-edged graphene nanoribbons. Phys E Low-dimens Syst Nanostruct 2021;126.

[30] Wang HW, Luo WW, Tian ZF, Ouyang CY. First principles study of alkali and alkaline earth metal ions adsorption and diffusion on penta-graphene. Solid State Ionics 2019;342.

[31] Sun ML, Tang WC, Ren QQ, Wang SK, JinYu DYH, Zhang YJ. First-principles study of the alkali earth metal atoms adsorption on graphene. Appl Surf Sci 2015;356:668–73.

[32] Saffarzadeh A. Modeling of gas adsorption on graphene nanoribbons. J Appl Phys 2010;107.

[33] Yang SY, Li J, Li SS. Antiferromagnetic-ferromagnetic transition in zigzag graphene nanoribbons induced by substitutional doping. Chin Phys B 2018;27.

[34] Friedrich N, Brandimarte P, Li JC, Saito S, Yamaguchi S, Pozo I, Pena D, Frederiksen T, Garcia-Lekue A, Sanchez-Portal D, Pascual JI. Magnetism of topological boundary states induced by boron substitution in graphene nanoribbons. Phys Rev Lett 2020;125.

[35] Kawai S, Nakatsuka S, Hatakeyama T, Pawlak R, Meier T, Tracey J, Meyer E, Foster AS. Multiple heteroatom substitution to graphene nanoribbon. Sci Adv 2018;4.

[36] Li Y, Lu YX, Adelhelm P, Titirici MM, Hu YS. Intercalation chemistry of graphite: alkali metal ions and beyond. Chem Soc Rev 2019;48:4655–87.

[37] Du Y, Zhuang JC, Wang JO, Li Z, Liu H, Zhao J, Xu X, Feng H, Chen L, Wu K, Wang X, Dou SX. Quasi-freestanding epitaxial silicene on Ag(111) by oxygen intercalation. Sci Adv 2016;2.

[38] Gopalakrishnan V, Sundararajan A, Omprakash P, Panemangalore DB. Review-energy storage through graphite intercalation compounds. J Electrochem Soc 2021;168.

[39] Kaur M, Gupta N, Kumar S, Raj B, Singh AK. RF analysis of intercalated graphene nanoribbon-based global-level interconnects. J Comput Electron 2020;19:1002–13.

[40] Son YW, Cohen ML, Louie SG. Energy gaps in graphene nanoribbons. Phys Rev Lett 2006;97.

[41] Gao Y, Xu D, Cui T, Li D. Stability of hydrogen-terminated graphene edges. Phys Chem Chem Phys 2021.

[42] Sonia FJ, Jangid MK, Ananthoju B, Aslam M, Johari P, Mukhopadhyay A. Understanding the Li-storage in few layers graphene with respect to bulk graphite: experimental, analytical and computational study. J Mater Chem 2017;5:8662–79.

[43] Lan CY, Shi Z, Cao R, Li C, Zhang H. 2D materials beyond graphene toward Si integrated infrared optoelectronic devices. Nanoscale 2020;12:11784–807.

[44] Mas-Balleste R, Gomez-Navarro C, Gomez-Herrero J, Zamora F. 2D materials: to graphene and beyond. Nanoscale 2011;3:20–30.

[45] Joshi RK, Yoshimura M, Kumar A. Graphene. J Nanomater 2010;2010.

[46] Fusco L, Gazzi A, Peng GT, Shin Y, Vranic S, Bedognetti D, Vitale F, Yilmazer A, Feng X, Fadeel B, Casiraghi C, Delogu LG. Graphene and other 2D materials: a multidisciplinary analysis to uncover the hidden potential as cancer theranostics. Theranostics 2020;10:5435–88.

[47] Gigli L, Manini N, Benassi A, Tosatti E, Vanossi A, Guerra R. Graphene nanoribbons on gold: understanding superlubricity and edge effects. 2D Mater 2017;4.

[48] Haddon RC. Carbon nanotubes. Acc Chem Res 2002;35:997.

[49] Popov VN. Carbon nanotubes: properties and application. Mater Sci Eng R Rep 2004;43:61–102.

[50] Chegel R, Behzad S. Chirality dependence of dipole matrix element of carbon nanotubes in axial magnetic field: a third neighbor tight binding approach. Opt Commun 2014;313:406–15.

[51] Charlier A, McRae E, Heyd R, Charlier MF, Moretti D. Classification for double-walled carbon nanotubes. Carbon 1999;37:1779–83.

[52] Ding JW, Yan XH, Cao JX, Wang DL, Tang Y, Yang QB. Curvature and strain effects on electronic properties of single-wall carbon nanotubes. J Phys Condens Matter 2003;15:L439–45.

[53] Rochefort A, Salahub DR, Avouris P. Effects of finite length on the electronic structure of carbon nanotubes. J Phys Chem B 1999;103:641–6.

[54] Litinskii AO, Kamnev VV. Electron-energy characteristics of double-walled carbon nanotubes doped with alkali metal atoms. Russ J Phys Chem B 2013;7:62–70.

[55] Lin YT, Lin SY, Chiu YH, Lin MF. Alkali-created rich properties in graphene nanoribbons: chemical bondings. Sci Rep 2017;7.

[56] Pandya A, Sangani K, Jha PK. Band gap determination of graphene, h-boron nitride, phosphorene, silicene, stanene, and germanene nanoribbons. J Phys D-Appl Phys 2020;53.
[57] Zhang J, Fahrenthold EP. Conductance of curved 3M-1 armchair graphene nanoribbons. J Phys Chem C 2019;123:21805–12.
[58] Mao YL, Hao WP, Wei XL, Yuan JM, Zhong JX. Edge-adsorption of potassium adatoms on graphene nanoribbon: a first principle study. Appl Surf Sci 2013;280:698–704.
[59] Gospodarev IA, Grishaev VI, Manzhelii EV, Sirenko VA, Syrkin ES, Feodosyev SB. The effect of size quantization on the electron spectra of graphene nanoribbons. Low Temp Phys 2020;46.
[60] Treske U, Ortmann F, Oetzel B, Hannewald K, Bechstedt F. Electronic and transport properties of graphene nanoribbons. Phys Status Solidi A 2010;207:304–8.
[61] Wakabayashi K, Takane Y, Yamamoto M, Sigrist M. Electronic transport properties of graphene nanoribbons. New J Phys 2009;11.
[62] Sharma H, Sandha KS. Impact of intercalation doping on the conductivity of multi-layer graphene nanoribbon (MLGNR) in on-chip interconnects. J Circ Syst Comput 2020;29.
[63] Immanuel S, Dar MA, Sivasubramanian R, Karim MR, Kim DW, Gul R. Progress and prospects on the fabrication of graphene-based nanostructures for energy storage, energy conversion and biomedical applications. Chem–Asian J 2021;16:1365–81.
[64] Liu XM, Li G, Lipatov A, Sun T, Pour MM, Aluru NR, Lyding JW, Sinitskii A. Chevron-type graphene nanoribbons with a reduced energy band gap: solution synthesis, scanning tunneling microscopy and electrical characterization. Nano Res 2020;13:1713–22.
[65] Yuan LH, Gong JJ, Wang DB, Zhang CR, Zhang ML, Su JY, Kang L. Hydrogen storage capacity of alkali metal atoms decorated porous graphene. Acta Phys Sin 2020;69.
[66] Singh R. Spin-orbit coupling in graphene, silicene and germanene: dependence on the configuration of full hydrogenation and fluorination. Bull Mater Sci 2018;41.
[67] Monshi MM, Aghaei SM, Calizo I. Edge functionalized germanene nanoribbons: impact on electronic and magnetic properties. RSC Adv 2017;7:18900–8.
[68] Li ZL, Chen YP, Ma TY, Jiang YZ, Chen J, Pan HG, Sun WP. 2D metal-free nanomaterials beyond graphene and its analogues toward electrocatalysis applications. Adv Energy Mater 2021.
[69] Radsar T, Khalesi H, Ghods V. Graphene properties and applications in nanoelectronic. Opt Quant Electron 2021;53.
[70] Hafner J. Ab-initio simulations of materials using VASP: density-functional theory and beyond. J Comput Chem 2008;29:2044–78.
[71] Savin AV, Mazo MA. Simulation of scrolled packings of graphone nanoribbons. Phys Solid State 2017;59:1260–6.
[72] Tao CG, Jiao LY, Yazyev OV, Chen YC, Feng JJ, Zhang XW, Capaz RB, Tour JM, Zettl A, Louie SG, Dai HJ, Crommie MF. Spatially resolving edge states of chiral graphene nanoribbons. Nat Phys 2011;7:616–20.
[73] Shiraz AK, Goharrizi AY, Hamidi SM. The electronic and optical properties of armchair germanene nanoribbons. Phys E Low-dimens Syst Nanostruct 2019;107:150–3.
[74] Bhandari AS, Verma A. A tight-binding analysis of models, sheet width, and strain for 2D monoatomic germanium sheets. Mater Exp 2019;9:273–80.
[75] Ildarabadi F, Farghadan R. Edge magnetization and spin-valley-caloritronics in germanene and stanene nanoribbons. J Magn Magn Mater 2021;529.
[76] Lebedeva OS, Lebedev NG, Lyapkosova IA. Elastic conductivity of silicene and germanene nanoribbons. St Petersburg Polytech Univ J-Phys Math 2019;12:38–49.
[77] Hattori A, Yada K, Araidai M, Sato M, Shiraishi K, Tanaka Y. Influence of edge magnetization and electric fields on zigzag silicene, germanene and stanene nanoribbons. J Phys Condens Matter 2019;31.
[78] Pham HD, Nguyen TDH, Vo KD, Huynh TMD, Lin MF. Rich essential properties of boron, carbon, and nitrogen substituted germanenes. APEX 2020;13.
[79] Rojas KIM, Villagracia AC, Moreno JL, David M, Arboleda NB. Ca and K decorated germanene as hydrogen storage: an ab initio study. Int J Hydrogen Energy 2018;43:4393–400.
[80] Zhang SD, Cao C, Zeng BW, Long MQ. The effects of strain and electric field on the half-metallicity of pristine and O-H/C-N-decorated zigzag graphene nanoribbons. J Phys Condens Matter 2020;32.
[81] Aierken Y, Leenaerts O, Peeters FM. First-principles study of the stability and edge stress of nitrogen-decorated graphene nanoribbons. Phys Rev B 2018;97.

[82] Guo G, Xie ZX, Tang J, Zhang Y, Xu GB. Fluorine edge decoration on zigzag silicene nanoribbons. Superlattice Microst 2020;139.
[83] Xiao P, Fan XL, Liu LM. Tuning the electronic properties of half- and full-hydrogenated germanene by chlorination and hydroxylation: a first-principles study. Comput Mater Sci 2014;92:244–52.
[84] Dresselhaus G. Spin-orbit coupling effects in zinc blende structures. Phys Rev 1955;100:580–6.
[85] Ho JH, Lu CL, Hwang CC, Chang CP, Lin MF. Coulomb excitations in AA- and AB-stacked bilayer graphites. Phys Rev B 2006;74.
[86] Singh S, Garg K, Sareen A, Mehla R, Kaur I. Doped armchair germanene nanoribbon exhibiting negative differential resistance and analysing its nano-FET performance. Org Electron 2018;54:261–9.
[87] Sharma V, Srivastava P, Jaiswal NK. Edge-oxidized gemanene nanoribbons for nanoscale metal interconnect applications. IEEE Trans Electron Dev 2018;65:3893–900.
[88] Bayani AH, Dideban D, Moezi N. Impact of uniaxial compressive strain on physical and electronic parameters of a 10 nm germanene nanoribbon field effect transistor. Superlattice Microst 2016;100:198–208.
[89] Sharma V, Srivastava P. Probing gold-doped germanene nanoribbons for nanoscale interconnects under DFT-NEGF framework. J Electron Mater 2020;49:3938–46.
[90] Sharma V, Srivastava P, Jaiswal NK. Prospects of asymmetrically H-terminated zigzag germanene nanoribbons for spintronic application. Appl Surf Sci 2017;396:1352–9.
[91] Hsu CH, Fang YM, Wu SQ, Huang ZQ, Crisostomo CP, Gu YM, Zhu ZZ, Lin H, Bansil A, Chuang FC, Huang L. Quantum anomalous hall insulator phase in asymmetrically functionalized germanene. Phys Rev B 2017;96.
[92] Nejad HF, Dideban D, Ketabi A, Vali M, Bayani AH, Heidari H. Tuning the analog and digital performance of germanene nanoribbon field effect transistors with engineering the width and geometry of source, channel and drain region in the ballistic regime. Mater Sci Semicond Process 2018;80:18–23.
[93] Wang JN, Zhang YL, Liu Y, Zheng WH, Lee LP, Sun HB. Recent developments in superhydrophobic graphene and graphene-related materials: from preparation to potential applications. Nanoscale 2015;7:7101–14.
[94] Qin ZH. Recent progress of graphene-like germanene. Acta Phys Sin 2017;66.

CHAPTER

16

RE-T-silicon/-germanium intermetallic materials

Shih-Yang Lin[1], Thi Dieu Hien Nguyen[1], Vo Khuong Dien[1], Chi-Hsuan Lee[1], Hsin-Yi Liu[1], Hai Duong Pham[1], Thi My Duyen Huynh[1], Nguyen Thi Han[1], Ngoc Thanh Thuy Tran[2], Wei-Bang Li[1] and Ming-Fa Lin[1]

[1]Department of Physics, National Cheng Kung University, Tainan, Taiwan; [2]Hierarchical Green-Energy Materials (Hi-GEM) Research Center, National Cheng Kung University, Tainan, Taiwan

16.1 $Sm_2T_3Ge_5$ 3D compounds

Novel materials with density wave behavior have attracted a bunch of researches for its widely applications. As a result of this investigation, the family of compounds $RE_2Ru_3Ge_5$ (RE = Ce, Sm) that exhibit rich charge density wave feature are the main focus in this chapter [1–8]. They can be study through the X-ray diffraction [9–15], the charge transport measurements [16–24], the magnetic measurements [12,25–30], and the heat capacity [2,5,31–33], the CDW state in these compounds.

According to the first-principles simulations [34,35], certain meta-stable crystal structures of three-component $Sm_2T_3Ge_5$ compounds are chosen as model systems to clearly study the diversified orbital hybridizations [36–40]. The high-resolution numerical evaluations are used for the important steps, such as the sufficient cut-off energies, the large number of wave-vector points in the greatly reduced first Brillouin zone, and the almost vanishing net forces on all atoms. $Sm_2T_3Ge_5$ can present the rich and unique crystal structures, being achieved from the VASP simulations. This 3D materials possess Moire superlattice with many atoms in a unit cell; hence, the numerical calculations need to have the highly precise in order to optimize the structures. To fully explore the spin-dependent magnetic configurations, f orbital is taken into consideration with the single- and many-particle intrinsic

Fundamental Physicochemical Properties of Germanene-related Materials
https://doi.org/10.1016/B978-0-443-15801-8.00006-2

interactions. Such simulation method principally covers the frequent electron-ion scatterings due to the atom-related crystal potentials [41], and the exchange & correlation energies originating from the spatial charge distribution. The three-component $Sm_2T_3Ge_5$ compounds present the highly non-uniform physical & chemical environments, according to the VASP results in Fig. 16.1. The unit cell contains 8 Sm atoms, 12 Ru atoms, and 20 Ge atoms. This complicated system belong to the orthorhombic lattices with the different lattice constants along three perpendicular. Furthermore, $Sm_2T_3Ge_5$ possess the position-dependent and anisotropic atomic bondings. However, the fluctuations of their bond lengths are significant and observable. The bond length measure from Sm-/Ru-/Ge range in 2.198–2.473/2.385–2.824 Å. Each bond length directly reflects the binding strength of distinct orbitals.

The electronic properties of $Sm_2T_3Ge_5$ display the unusual electronic energy spectra and wave functions, as shown in Fig. 16.2. The main features include the strong wave vector dependences along the high-symmetry points of the reduced first Brillouin zone [42–44], a significant number of valence & conduction bands due to Ge/Ru/Sm atoms in a specific unit cell & their active orbitals (Fig. 16.2B–D), the asymmetry of valence and conduction states about the Fermi level under the complicated intrinsic interactions, the distinct atomic dominances at the various energy ranges, and the metallic/semiconducting behavior in the absence/presence of band gap. Specially, the Ge-, Ru-, and Sm-dominated band are found to be in the range of $E < -7.5$ eV, -5.0 eV $< E < 2.0$ eV, and 2.0 eV $< E < 5.0$ eV. The absence of Ge-dominated bands in -1 eV $< E < 1$ eV means that its low-energy π electronic states are no-longer exist. Instead, the low-lying states are mainly come from the Ru atoms within -5.0 eV $< E < 2.0$ eV. Importantly, the Ge/Ru/Sm co-dominating bands reveal at the range of 2.0 eV $< E < 4.0$ eV. Such feature will be further discus for the complicated multi-orbital hibridizations. However, the direct experimental examinations by the high-resolution ARPES are absent up to now.

A real-space distribution of charge density is useful in comprehending the critical orbital hybridizations in various bond, as shown by Fig. 16.3 The VASP-based simulations provide the total charge densities of $Sm_2T_3Ge_5$. The dramatic transformations of electron clouds come to exist everywhere, especially for the neighboring regions from the atom center. There exist a lot of multi-orbital mixings, namely, (4s, $4p_x$, $4p_y$, $4p_z$), ($3d_{z^2}$, $3d_{x^2-y^2}$, $3d_{xz}$, $3d_{zy}$, $3d_{yx}$), and (f_{z^2}, f_{xz^2}, $f_{yz^2}f_{xyz}$, $f_{z(x^2-y^2)}$, $f_{x(x^2-3y^2)}$, $f_{y(y^2-3x^2)}$) in Ge-, Ru- and Sm bonds, respectively. It should be noticed that they might have the same contributions. As a consequence, the π and σ bonding due to the (4s, $4p_x$, $4p_y$, $4p_z$) orbitals totally disappear in $Sm_2T_3Ge_5$. When the active multi-orbital hybridizations are combined with the orbital-decomposed van Hove singularities, their detailed energy dependences could be clarified in the further analyses (Fig. 16.4).

16.2 $Ce_2T_3Ge_5$ bulk systems

Another the three-component $Ce_2T_3Ge_5$ compounds also present the highly non-uniform physical & chemical environments, as indicated in Fig. 16.5. This complicated system, similar to the $Sm_2T_3Ge_5$, also belong to the orthorhombic lattices. $Ce_2T_3Ge_5$ possess the position-dependent and anisotropic atomic bondings, very close to the $Sm_2T_3Ge_5$. However, the fluctuations of their bond lengths are significant and observable. The bond length measure from Ce-/Ru-/Ge range in 2.374–2.736/2.925–2.824 Å. Those bond lengths directly reflect the slightly changed binding strength of distinct orbitals.

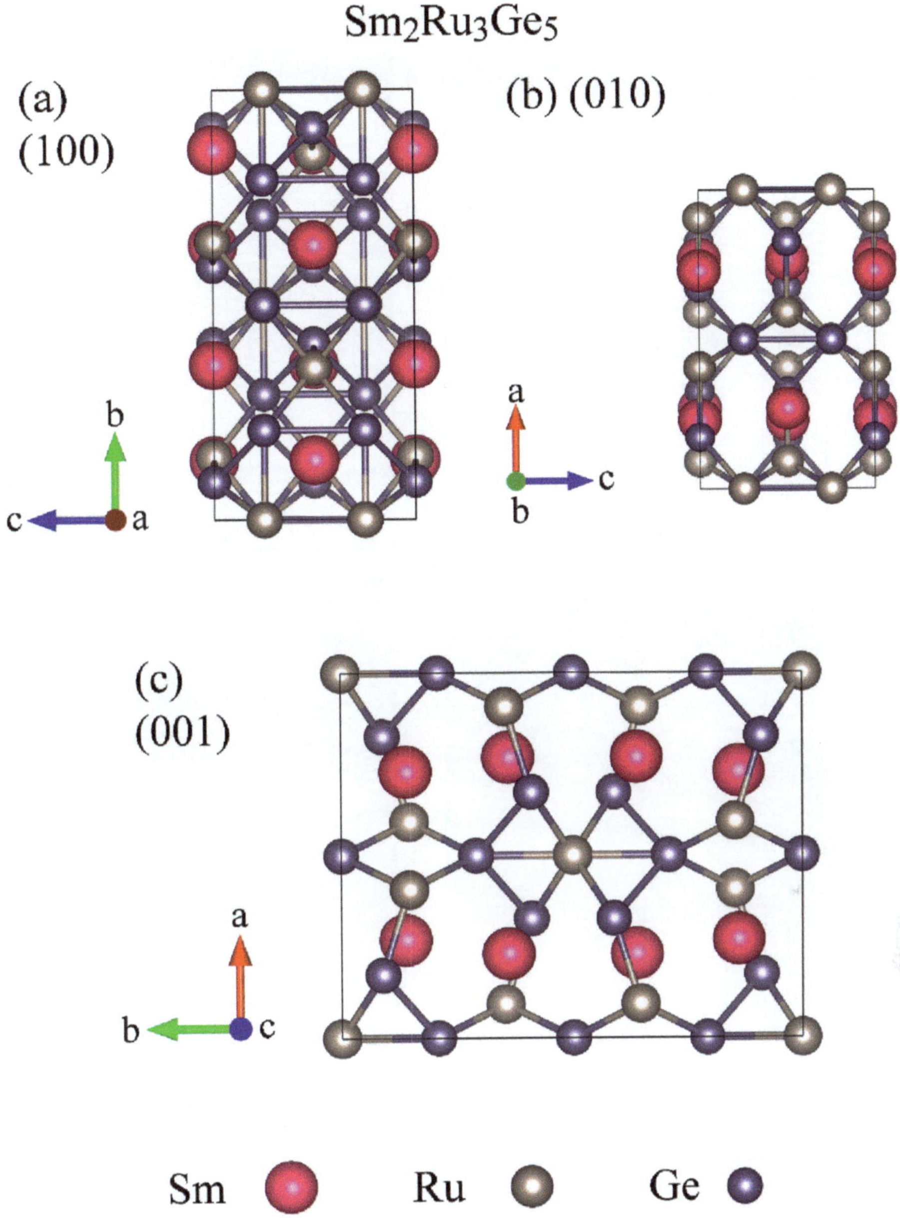

FIG. 16.1 The 3D orthorhombic structure of $Sm_2Ru_3Ge_5$ in the (A) 100, (B) 010, and (C) 001 surface.

The electronic properties of $Ce_2T_3Ge_5$ also display the unusual electronic energy spectra and wave functions, as shown in Fig. 16.6. The main features, which is similar to $Sm_2T_3Ge_5$, include the strong wave vector dependences along the high-symmetry points of the reduced first Brillouin zone, a significant number of valence & conduction bands due to Ce/Ru/Ge

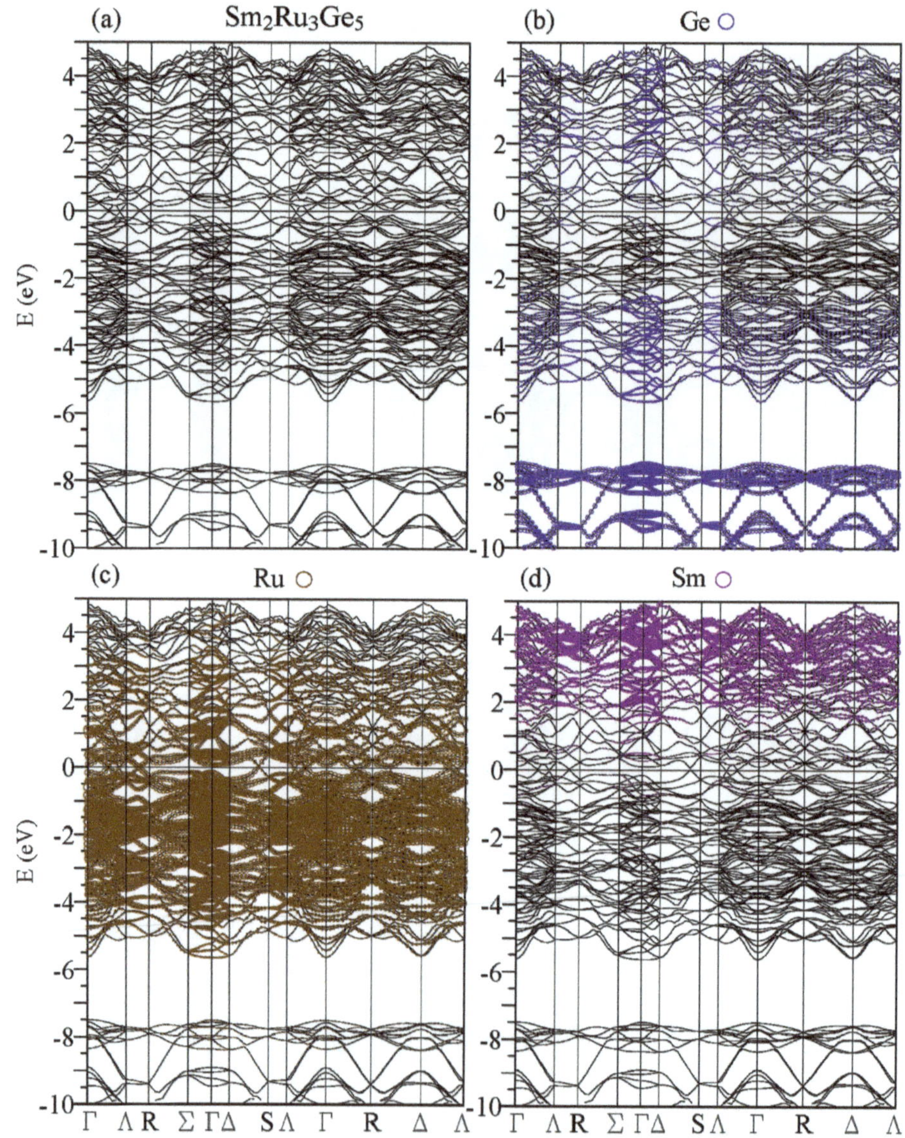

FIG. 16.2 Electronic energy spectra for quaternary $Sm_2Ru_3Ge_5$ compounds: (A) Energy bands, (B) Ge-, (C) Ru-, and (D) Sm-dominated bands. Three kinds of atom dominances are revealed by the colored circles in each energy subbands.

atoms in a specific unit cell & their active orbitals, the asymmetry of valence and conduction states about the Fermi level under the complicated intrinsic interactions, However, the Ce contributed bands are almost vanish in the range of $E > 0.5$ eV, being the main difference from $Sm_2T_3Ge_5$. The Ge-dominated bands mainly exist at deeper energy ($E < -7.5$ eV),

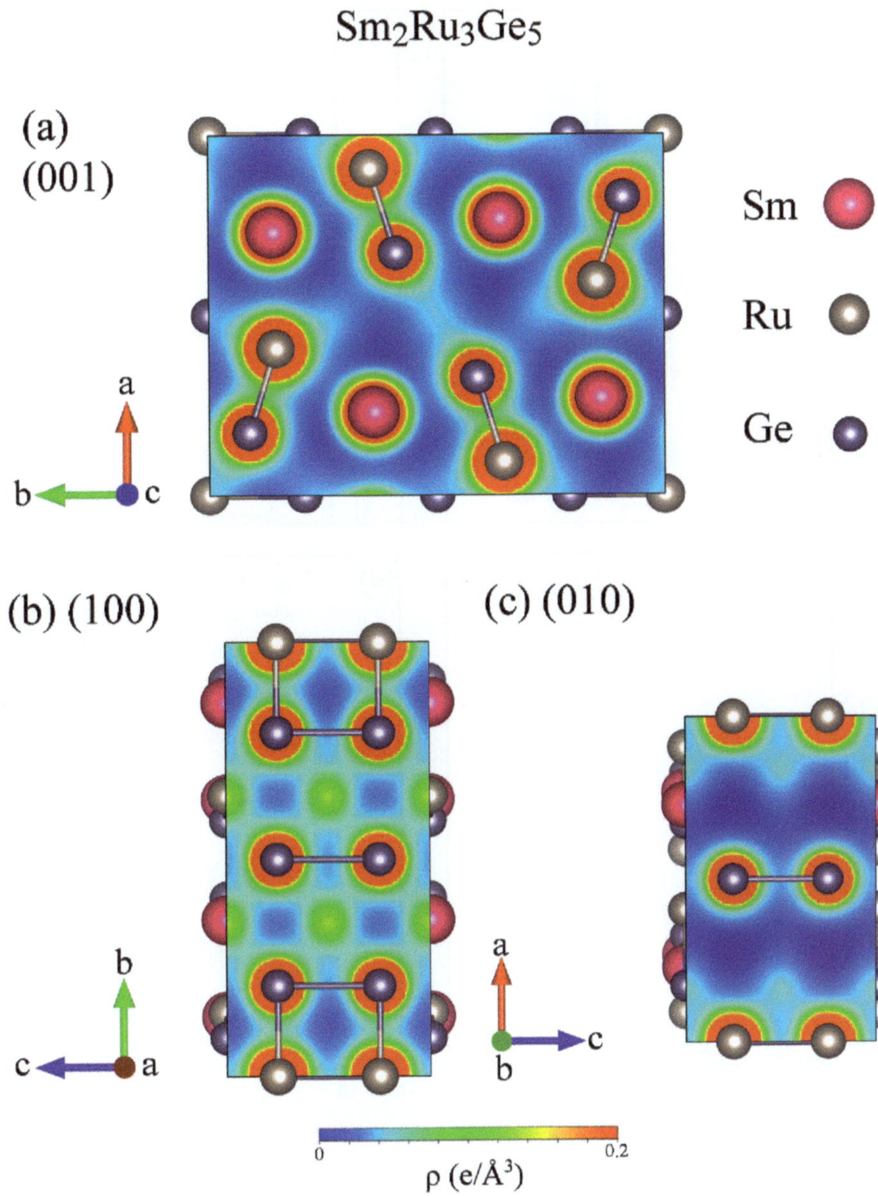

FIG. 16.3 The spatial charge density distributions of $Sm_2Ru_3Ge_5$ in (A) 100, (B) 010, and (C) 001 surface.

such properties are also obtained by $Sm_2T_3Ge_5$. Specially, the Ce-dominated bands, within the range of -0.5 eV < 0.7 eV, are localized with a very small bandwidth, such phenomenon is absent in $Sm_2T_3Ge_5$. However, the bandwidth, energy dispersion, and energy range of Ru-dominated bands of $Ce_2T_3Ge_5$ are keep similar to the $Sm_2T_3Ge_5$.

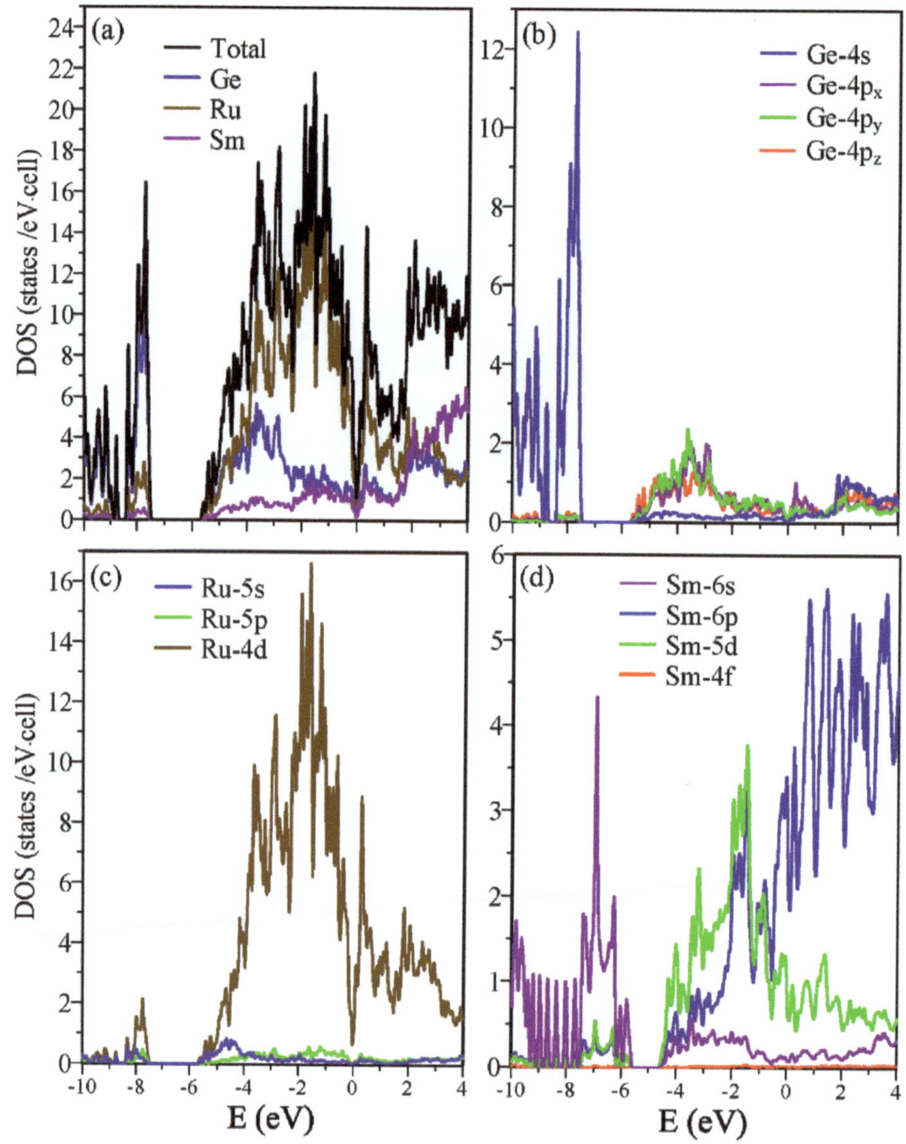

FIG. 16.4 The atom-/orbital-/spin-decomposed density of states of $Sm_2Ru_3Ge_5$ under the various cases: (A) Atom-, (B) Ge-orbital-, (C) Ru-orbital-, and (D) Sm-orbital-decomposed DOSs.

The high-precision DOS calculations, which are conducted on the atom-, orbital-contributions, are able to provide the detailed information about the critical mechanisms of quasiparticle charges (Fig. 16.7) and electronic states, as shown in Fig. 16.8. The feature-rich van Hove singularities, being fully supported by the position-dependent chemical bonds (Fig. 16.5), the band-edge states & atom dominances (Fig. 16.6), and the spatial charge density

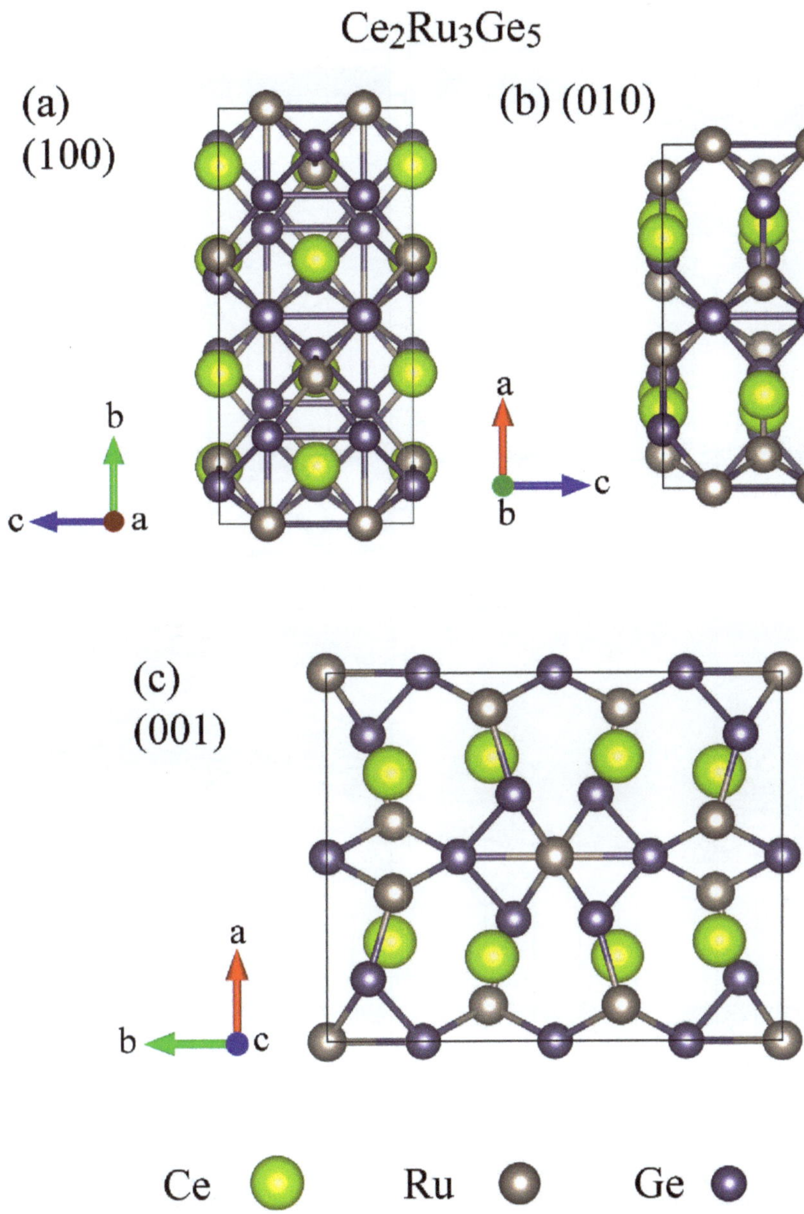

FIG. 16.5 The 3D orthorhombic structure of $Ce_2Ru_3Ge_5$ in the (A) 100, (B) 010, and (C) 001 surface.

distributions (Fig. 16.7), are the most significant evidences. Their prominent intensities are thoroughly examined from the dominating contributions of multi-orbitals in Ce-/Ru-/Ge bonds.

Furthermore, the emerged peaks due to few orbitals might survive within certain energy ranges. As a result, one can get the active orbitals in chemical bondings and their strengths.

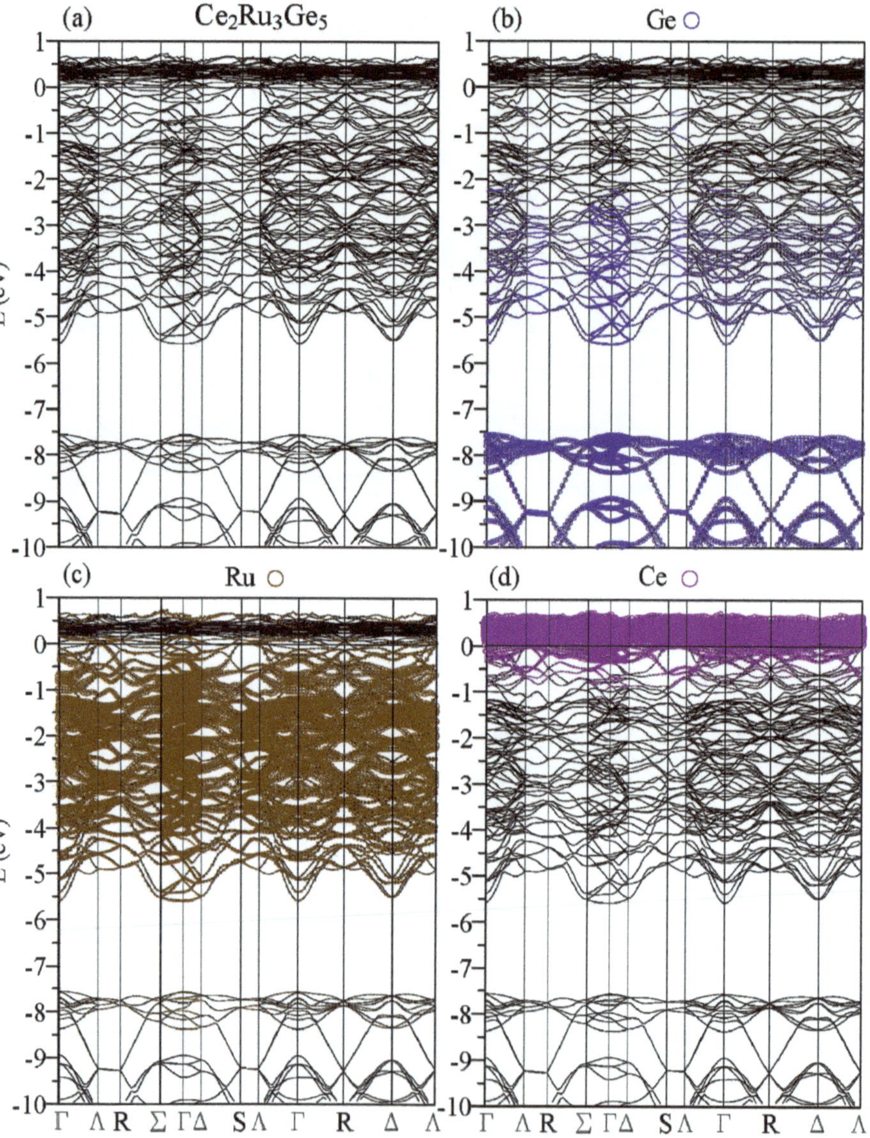

FIG. 16.6 Electronic energy spectra for quaternary $Ce_2Ru_3Ge_5$ compounds: (A) Energy bands, (B) Ge-, (C) Ru-, and (D) Ce-dominated bands. Three kinds of atom dominances are revealed by the colored circles in each energy subbands.

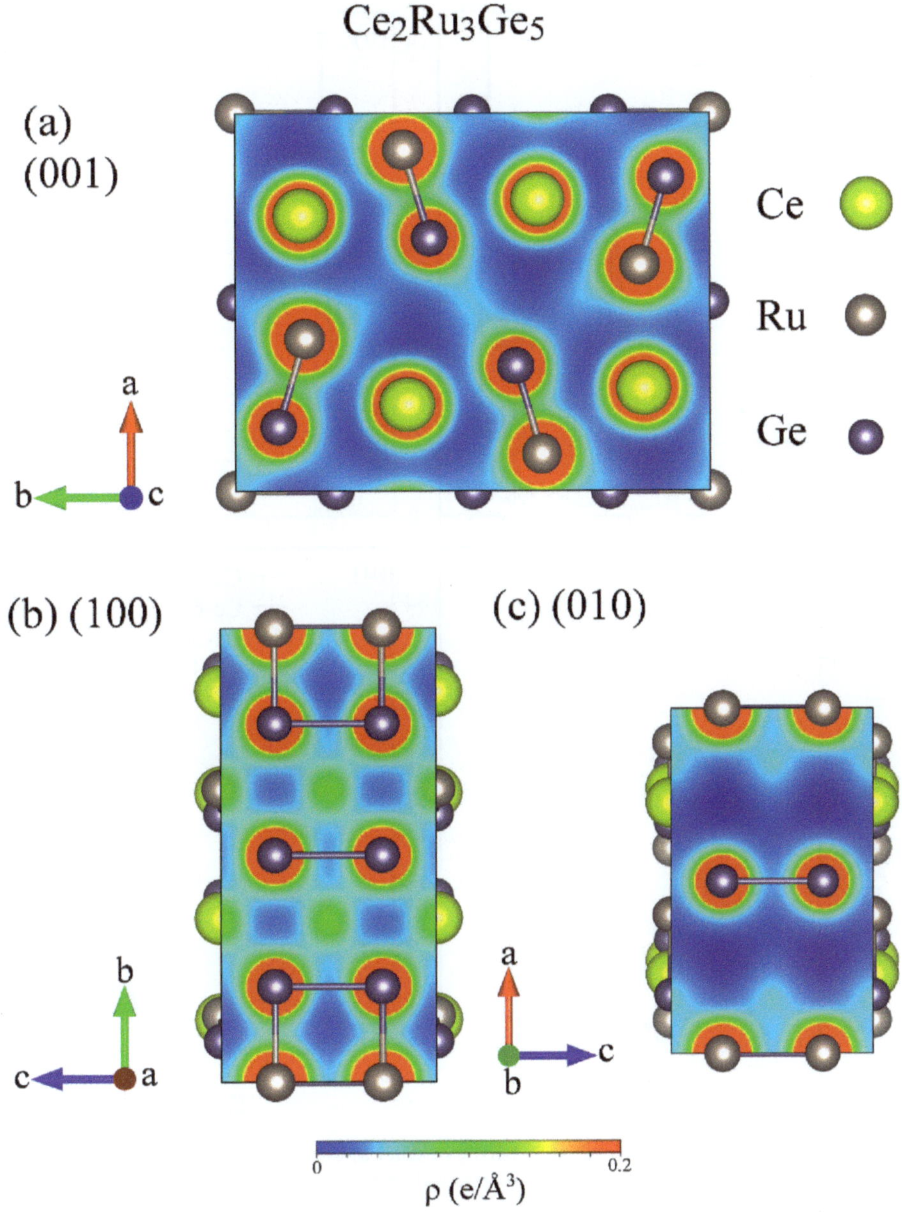

FIG. 16.7 The spatial charge density distributions of $Ce_2Ru_3Ge_5$ in (A) 100, (B) 010, and (C) 001 surface.

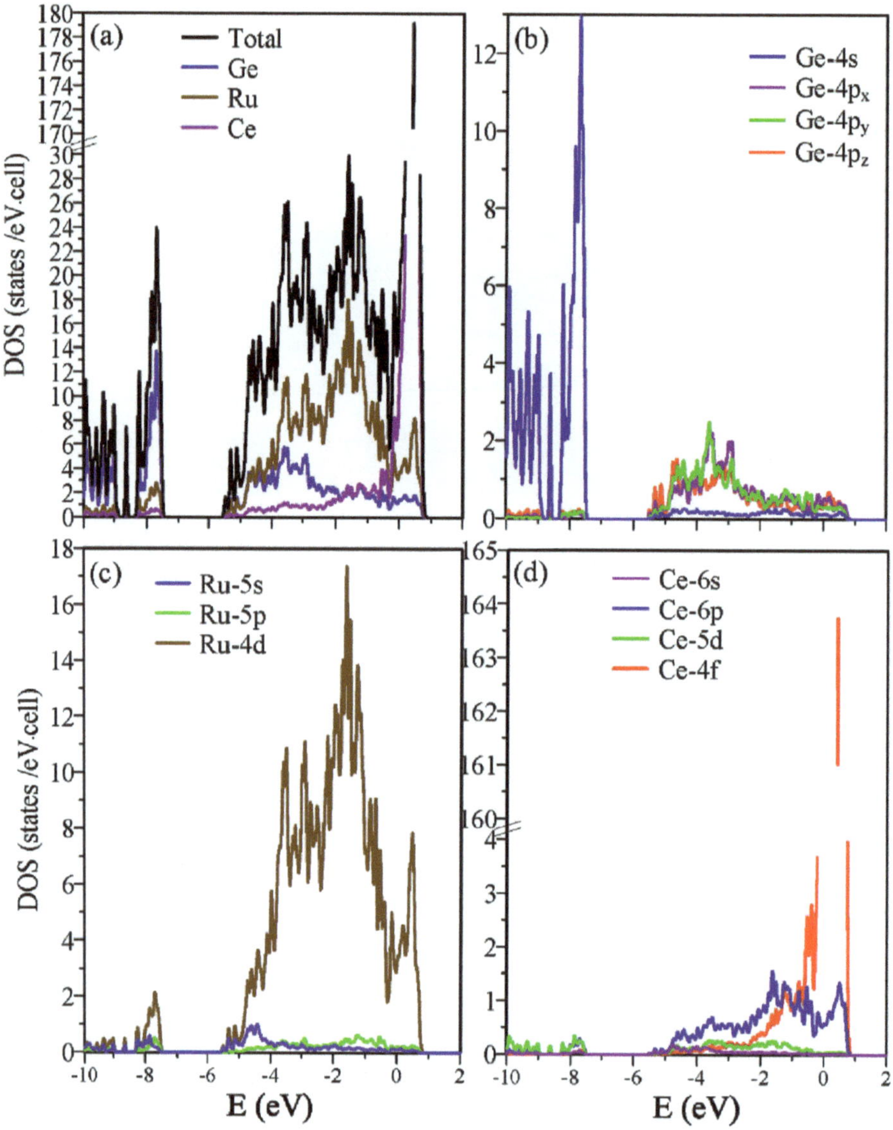

FIG. 16.8 The atom-/orbital-/spin-decomposed density of states of $Ce_2Ru_3Ge_5$ under the various cases: (A) Atom-, (B) Ge-orbital-, (C) Ru-orbital-, and (D) Sm-orbital-decomposed DOSs.

References

[1] Chabot B, Parthé E. $Ce_2Co_3Si_5$ and $R_2Ni_3Si_5$ (R ≡ Ce, Dy, Y) with the orthorhombic $U_2Co_3Si_5$-type structure and the structural relationship with the tetragonaL $Sc_2Fe_3Si_5$-type structure. J Less Common Met 1984;97:285–90.

[2] Thomas EL, Kim MS, Sokolov DA, Bennett MC, Aronson MC, Chan JY. Structure, and magnetic and transport behavior of twinned $Ce_2Rh_3(Pb,Bi)_5$. J Solid State Chem 2007;180:2356–62.

[3] Galadzhun YV, Hoffmann R-D, Pöttgen R, Adam M. Complex three-dimensional $[Au_3In_5]$ polyanions in $Ln_2Au_3In_5$ (Ln=Ce, Pr, Nd, Sm). J Solid State Chem 1999;**148** 425–32.

[4] Seropegin YD, Shapiev BI, Gribanov AV, Bodak OI. Isothermal cross-section of the Ce-Ru-Ge phase diagram at 600°C. J Alloys Compd 1999;288:155–8.

[5] Bugaris DE, Malliakas CD, Han F, Calta NP, Sturza M, Krogstad MJ, Osborn R, Rosenkranz S, Ruff JPC, Trimarchi G, Bud'ko SL, Balasubramanian M, Chung DY, Kanatzidis MG. Charge density wave in the new polymorphs of RE 2 Ru 3 Ge 5 (RE = Pr, Sm, Dy). J Am Chem Soc 2017;139:4130–43.

[6] Morozkin AV, Seropegin YD. Sm-Ru-Ge system at 1070 K. J Alloys Compd 2004;365:168–72.

[7] Kuo CN, Hsu CJ, Tseng CW, Chen WT, Lin SY, Liu WZ, et al. Charge density wave like behavior with magnetic ordering in orthorhombic $Sm_2Ru_3Ge_5$. Phys Rev B 2020;101:155140.

[8] Cao K, Skowron ST, Biskupek J, Stoppiello CT, Leist C, Besley E, Khlobystov AN, Kaiser U. Imaging an unsupported metal–metal bond in dirhenium molecules at the atomic scale. Sci Adv 2020;**6**. eaay5849.

[9] Kanno R, Shirane T, Kawamoto Y, Takeda Y, Takano M, Ohashi M, Yamaguchi Y. Synthesis, structure, and electrochemical properties of a new lithium iron oxide, $LiFeO_2$, with a corrugated layer structure. J Electrochem Soc 1996;143:2435–42.

[10] Vartanyants IA, Singer A. Coherence properties of third-generation synchrotron sources and free-electron lasers. Synchrotron Light Source Free Las Accel Phys Instrum Sci Appl 2020;12:987–1029.

[11] Thess A, Lee R, Nikolaev P, Dai H, Petit P, Robert J, Xu C, Lee YH, Kim SG, Rinzler AG, Colbert DT, Scuseria GE, Tománek D, Fischer JE, Smalley RE. Crystalline ropes of metallic carbon nanotubes. Science 1996;273:483–7.

[12] Freccero R, Choi SH, Solokha P, De Negri S, Takeuchi T, Hirai S, Mele P, Saccone A. Synthesis, crystal structure and physical properties of $Yb_2Pd_3Ge_5$. J Alloys Compd 2019;783:601–7.

[13] Siqueira KPF, Carvalho GB, Dias A. Influence of the processing conditions and chemical environment on the crystal structures and phonon modes of lanthanide orthotantalates. Dalton Trans 2011;40:9454–60.

[14] Jeong HK, Yun PL, Lahaye RJWE, Park MH, Kay HA, Ick JK, Yang CW, Chong YP, Ruoff RS, Young HL. Evidence of graphitic AB stacking order of graphite oxides. J Am Chem Soc 2008;130:1362–6.

[15] Sakurai Y, Arai H, Okada S, Yamaki JI. Low temperature synthesis and electrochemical characteristics of $LiFeO_2$ cathodes. J Power Sources 1997;68:711–5.

[16] Berciu M, Bhatt RN. Spin waves in disordered III-V diluted magnetic semiconductors by a modified RPA approach. Phys Rev B Condens Matter 2002;66:852071–8520713.

[17] Kumar A, Poumirol JM, Escoffier W, Goiran M, Raquet B, Broto JM. Electronic properties of graphene, few-layer graphene, and bulk graphite under very high magnetic field. Int J Nanosci 2011;10:43–7.

[18] Han MY, Özyilmaz B, Zhang Y, Kim P. Energy band-gap engineering of graphene nanoribbons. Phys Rev Lett 2007;98:206805.

[19] Huang B, Li Z, Liu Z, Zhou G, Hao S, Wu J, Gu BL, Duan W. Adsorption of gas molecules on graphene nanoribbons and its implication for nanoscale molecule sensor. J Phys Chem C 2008;112:13442–6.

[20] Cummings AW, Valenzuela SO, Ortmann F, Roche S. Graphene spintronics 2D mater. Prop Dev 2017;9:197–218.

[21] Dreher M, Gunder D, Zorb S, Witte G. Van der Waals bound organic semiconductor/2D-material hybrid heterosystems: intrinsic epitaxial alignment of perfluoropentacene films on transition metal dichalcogenides. Chem Mater 2020;32:9034–43.

[22] Senkovskiy BV, Fedorov AV, Haberer D, Farjam M, Simonov KA, Preobrajenski AB, Mårtensson N, Atodiresei N, Caciuc V, Blügel S, Rosch A, Verbitskiy NI, Hell M, Evtushinsky DV, German R, Marangoni T, van Loosdrecht PHM, Fischer FR, Grüneis A. Semiconductor-to-Metal transition and quasiparticle renormalization in doped graphene nanoribbons. Adv Electron Mater 2017;3:1600490.

[23] Yusoff ARBM, Nazeeruddin MK. Organohalide lead perovskites for photovoltaic applications. J Phys Chem Lett 2016;7:851–66.

[24] Chu SW, Baek SJ, Kim DC, Seo S, Kim JS, Park YW. Charge transport in graphene doped with diatomic halogen molecules (I 2, Br 2) near Dirac point. Synth Met 2012;162:1689–93.

[25] Matsui T, Kambara H, Niimi Y, Tagami K, Tsukada M, Fukuyama H. STS observations of landau levels at graphite surfaces. Phys Rev Lett 2005;94:226403.
[26] Červenka J, Katsnelson MI, Flipse CFJ. Room-temperature ferromagnetism in graphite driven by two-dimensional networks of pointdefects. Nat Phys 2009;5:840–4.
[27] Pöttgen R, Lang A, Hoffmann RD, Künnen B, Kotzyba G, Müllmann R, Mosel BD, Rosenhahn C. The stannides YbPtSn and Yb 2 Pt 3 Sn 5. In: Pöttgen R, Lang A, Hoffmann R-D, Künnen B, Kotzyba G, Müllmann R, Mosel BD, Rosenhahn C, editors. Zeitschrift für Krist. - cryst. Mater. vol. 214; 1999. p. 143–50.
[28] Lai YH, Ho JH, Chang CP, Lin MF. Magnetoelectronic properties of bilayer Bernal graphene. Phys Rev B Condens Matter 2008;77:85426.
[29] Lin YT, Lin SY, Chiu YH, Lin MF. Alkali-created rich properties in graphene nanoribbons: chemical bondings. Sci Rep 2017;7.
[30] Tao C, Jiao L, Yazyev OV, Chen YC, Feng J, Zhang X, Capaz RB, Tour JM, Zettl A, Louie SG, Dai H, Crommie MF. Spatially resolving edge states of chiral graphene nanoribbons. Nat Phys 2011;7:616–20.
[31] Alekseeva OK, Pushkareva IV, Pushkarev AS, Fateev VN. Graphene and graphene-like materials for hydrogen energy. Nanotechnologies Russ 2020;15:273–300.
[32] Cho J, Kim YJ, Park B. Novel $LiCoO_2$ cathode material with Al_2O_3 coating for a Li ion cell. Chem Mater 2000;12:3788–91.
[33] Li C, Zhang HP, Fu LJ, Liu H, Wu YP, Rahm E, Holze R, Wu HQ. Cathode materials modified by surface coating for lithium ion batteries. Electrochim Acta 2006;51:3872–83.
[34] Ong SP, Wang L, Kang B, Ceder G. Li - Fe - P - O_2 phase diagram from first principles calculations. Chem Mater 2008;20:1798–807.
[35] Nakayama M, Yamada S, Jalem R, Kasuga T. Density functional studies of olivine-type $LiFePO_4$ and $NaFePO_4$ as positive electrode materials for rechargeable lithium and sodium ion batteries. Solid State Ionics 2016;286:40–4.
[36] Han NT, Dien VK, Thuy Tran NT, Nguyen DK, Su WP, Lin MF. First-principles studies of electronic properties in lithium metasilicate (Li_2SiO_3). RSC Adv 2020;10:24721–9.
[37] Lin S-Y, Thanh Thuy Tran N, Chang S-L, Su W-P, Lin M-F. Structure-and adatom-enriched essential properties of graphene nanoribbons. CRC press; 2018.
[38] Pham HD, Nguyen TDH, Dien Vo K, Huynh TMD, Lin MF. Rich essential properties of boron, carbon, and nitrogen substituted germanenes. Appl Phys Express 2020;13.
[39] Khuong Dien V, Thi Han N, Nguyen TDH, Huynh TMD, Pham HD, Lin MF. Geometric and electronic properties of Li_2GeO_3. Front Mater 2020;7.
[40] Chen S-C, Wu J-Y, Lin M-F. Novel magnetic quantization of bismuthene *arXiv*. 2017.
[41] Perdew JP, Burke K, Ernzerhof M. Generalized gradient approximation made simple. Phys Rev Lett 1996;77:3865–8.
[42] Chadi DJ, Cohen ML. Special points in the brillouin zone. Phys Rev B 1973;8:5747–53.
[43] Baldereschi A. Mean-value point in the Brillouin zone. Phys Rev B 1973;7:5212–5.
[44] Monkhorst HJ, Pack JD. Special points for Brillouin-zone integrations. Phys Rev B 1976;13:5188–92.

CHAPTER

17

Li-Ge-O electrolyte of Li^+-based batteries

Vo Khuong Dien[1], Shih-Yang Lin[1], Chi-Hsuan Lee[1], Hsin-Yi Liu[1], Hai Duong Pham[1], Thi My Duyen Huynh[1], Nguyen Thi Han[1], Ngoc Thanh Thuy Tran[2], Thi Dieu Hien Nguyen[1], Wei-Bang Li[1] and Ming-Fa Lin[1,2]

[1]Department of Physics, National Cheng Kung University, Tainan, Taiwan; [2]Hierarchical Green-Energy Materials (Hi-GEM) Research Center, National Cheng Kung University, Tainan, Taiwan

17.1 Introduction current applications

Lithium-ion batteries (LIBs) dominate in commercial purposes due to their high performance energy resources, e.g., high power, energy densities, and high reliability, as well as other certain fundamental merits including affordable price, long cycle life, and friendly environment [1]. The recent experimental progress shows many applications of LIBs such as electric vehicles (EV) and hybrid EV and mobile devices, which require the optimum in producing efficient energy [1,2,3]. A Li-ion battery principally consists of a cathode (positive electrode), an anode (negative electrode), and an ionically conductive Li^+-containing electrolyte, in which the third component is closely related to the unusual transport of the positive lithium ions (Li^+) between two electrodes. Depending on the combined alternative materials of three components, Li-ion batteries could provide various performance, e.g., the specific energy density in the values of 100−250 Wh/Kg [4], the volumetric energy density from 250 to 680 Wh/L [5,6], the specific power density in the range of 300−1500 W/kg [6], and the faster charging time (80% of charge of states in 15 min) [7]. The critical mechanisms of LIBs are characterized by the specific charging and discharging process which is based on the exchange of Li+-ions. During the charging process, Li+-ions commonly move from the cathode material and transport externally to intercalate into the anode

Fundamental Physicochemical Properties of Germanene-related Materials
https://doi.org/10.1016/B978-0-443-15801-8.00009-8

through the electrolyte. When discharging, the electrons flow from the anode to the cathode through an outer circuit creating an electric current, mainly converted chemical energy into electrical energy. Most importantly, the electrolyte plays an important role during the charging discharging process, however, recently electrolyte used in Li-ion batteries is not compatible with latterly developed high-voltage positive electrodes, which are one of the most effective ways of increasing the energy density. Nowadays, solid electrolytes are now rapidly emerging as promising alternatives given their wider electrochemical window of stability [10]. As potential electrolyte candidates, Li-Ge-O compounds exhibit the large ionic conductivity ($1.5 \times 10^{-5}\,\Omega\,cm^{-1}$ for Li_2GeO_3), in which electronic conductivity is negligible with a high ionic transference number [10]. Furthermore, Li_2GeO_3 shows wide cycling stability with a reserved charge capacity of $725\,mAhg^{-1}$ after 300 cycles at $50\,mAg^{-1}$ [8]. As for the battery safety, Li_2GeO_3 is suitable for the selection of solid electrolytes according to the reasonable decreasing in interface resistance. Recently, a lot of high-resolution experimental measurements on essential physical properties of anodes/cathodes/electrolytes in the Li$^+$-based batteries, e.g., optimized geometry structures, band structures, optical properties and transport. The X-ray diffraction and the low-energy electron diffraction (LEED; [9]) are available in investigating the 3D lattice symmetries, especially Li-X-O related systems. The angle-resolved photoemission spectroscopy (ARPES; [10]) is a powerful experimental technique directly measuring the single particle spectral function, depending on wave vectors and frequency. For example, ARPES could measure the Dirac cone structure of AA/AB bilayer graphene, the band energy dispersion along the synthetic dimension, band gap, and so on. There are many methods developed for measuring optical properties, for example, the complex permittivity, which is based on the constraints in specific frequencies, materials, applications could be measured by transmission/reflection line method [11], open-ended coaxial probe method [12], free space method [13], resonant method [14]. Other optical properties, such as energy loss functions, reflectance, absorptance, refractivity, excitation could be found through spectroscopy measurements, for example, analysis of the specular reflection from the polished surface of a solid can give the complex index of refraction using the Kramers-Kronig method.

Although the high potential of Lithium Germanate compound in optoelectronic and battery applications is expected, the fundamental properties, such as electronic structure and optical properties of Li_2GeO_3 were not been completely comprehended. (i) Density functional theory (DFT) is a good starting point to understand the electronic properties of solid-state substances. However, it cannot correctly estimate the fundamental band gap of materials [15], especially for large gap semiconductors or insulators. For example, the previous standard DFT calculation indicated the band gap of 3.8 eV for Li_2GeO_3 [16], it is not consistent with the optical measurement of Trukhin et al. [17]. (ii) Up to now, the optical characters of Li_2GeO_3 have been detected by the photoluminescence measurements [17]. Other experiments or calculations on reflection, absorption coefficients, and energy loss functions to thoroughly understand the optical properties of Li_2GeO_3 are absent. Furthermore, the excitonic effects - the mutual interactions between excited electrons and holes, are usually observed in materials within suitable conditions, e.g., large band gap or low temperature. This electron-hole pair strongly influences on optical excitations and thus worthy to include in calculations. However, the theoretical studies to point out the effect of exciton on optical properties of Li_2GeO_3 have not achieved. (iii) The significant orbital hybridizations that survive in chemical bonds are rather complicated and responsible for the critical fundamental properties [18–20]. However, the investigations to identify such critical bonding mechanism in lithium-based materials are still rather limited. Furthermore, the systematic connection of the orbital hybridizations with the electronic properties, optical excitations and excitonic effects is a question that has not been answered up to now. The main reasons are the presence of large unit cell, non-uniform chemical environments, and thus, the extremely complicated orbital hybridizations in these ternary compounds [21].

In this work, the theoretical framework, being based on the significant orbital hybridizations in chemical bonds [16,22,23], is developed by examining the essential properties in the ternary lithium-based materials. This strategy is based on the accurate first-principles calculations on an optimal lattice symmetry with position-dependent chemical bondings, the atom-dominated band structure at different energy ranges, the spatial charge densities due to various orbitals, and the atom- and orbital-projected van Hove singularities related to orbital overlaps. The energy-decomposed single-/multi-orbital hybridizations will be utilized to account for the optical threshold frequency, a lot of prominent absorption structures, a very strong plasmon response in terms of the dielectric functions, energy loss functions, reflectance spectra, and absorption coefficients under the distinct electric polarizations, the excitonic states with rather strong energy binding. The current study is of paramount importance not only for fundamental physics but also for technical applications. Most predicted results in this work require high-resolution experimental examinations.

17.2 Methods and computational details

We used the Vienna Ab-initio Simulation Package (VASP) to perform the optimization of the structure and calculation the electronic and optical properties. For the ground states calculations, the Perdew-Burke-Ernzerhof (PBE) generalized gradient approximation [24] was used for the exchange-correlation functional. Projector augmented wave (PAW) pseudopotentials are used to describe the electronic wave functions in the core region [25]. The cutoff energy for the expansion of the plane wave basis was set to 500 eV. The Brillouin zone was integrated with a special k-point mesh of $15 \times 15 \times 15$ in the Monkhorst-Pack sampling technique [26] for geometric optimization. The convergence condition is set to be 10^{-8} eV between two consecutive simulation steps, and all atoms were allowed to fully relax during geometric optimization until the Hellmann-Feynman force acting on each atom was smaller than 0.01 eV/Å.

Subsequent to the DFT results, the single-particle Green's function and the screened Coulomb interactions (G0W0) approach [27] using 250 eV energy cutoff for the response functions and the Brillouin zone was integrated with a special k-points mesh of $10 \times 10 \times 10$ in the Γ-center sampling technique to obtain the corrected quasi-particle density of states and electronic band structure. The single-particle optical excitation of Li_2GeO_3 can be described by the Kubo formula [28]:

$$\varepsilon_2(\omega) = \frac{8\pi^2 e^2}{\omega^2} \sum_{vck} |e.\langle vk|v|ck\rangle|^2 \delta(\omega - E_{ck} - E_{vk}),$$

Where the intensity of each excitation peak and the available transition channels is directly related to the velocity matrix element, $|e.\langle vk|v|ck\rangle|^2$, and joined of density of states $\delta(\omega - E_{ck} - E_{vk})$, respectively. Regarding the optical response beyond the independent particle approach, the electron-hole interaction was taken into account. The connection of the exciton energies Ω_S and corresponding electron-hole amplitude $|e.\langle 0|v|S\rangle|^2$ of the correlated electron-hole excitations S is obtained by solving the Bethe-Salpeter equation (BSE) [28].

$$\varepsilon_2(\omega) = \frac{8\pi^2 e^2}{\omega^2}\sum_{vck}|e.\langle 0|v|S\rangle|^2\delta(\omega - \Omega_s).$$

In which, the k-point sampling, energy cutoff, and number of bands setting the same as in the G0W0 calculation. In this calculation, the 20 highest occupied valence bands (VBs) and eight lowest unoccupied conduction bands (CBs) are included as a basis for the excitonic states with a photon energy region from 0 to 25 eV. In addition, the broadening parameter γ, which arises from various de-excitation mechanisms, was set at 0.1 eV. All parameters in this study have been checked for convergence of the calculations.

17.3 Results and discussions

17.3.1 Li-Ge-O ternary electrolyte compound of Li-ion-based batteries

The ternary compound, Li_2GeO_3 with 24 atoms within a conventional unit cell, is chosen for a model study in illustrating the geometric, electronic and optical properties. The optimal lattice, as clearly shown in Fig. 17.1, belongs to an orthorhombic structure. Each Li/Ge atom is surrounded by four O atoms in a tetrahedral form. There exist 32 Li-/16 Ge—O bonds, in which the former and the latter display large fluctuations about their lengths (~1.952–2.167 Å and ~1.757–1.870 Å, respectively). The calculated lattice constants along the x-, y- and z-directions, (9.612, 5.462, 4.874 Å), are very close to the X-ray diffraction measurements, (9.602, 5.502, 4.849 Å) in Ref. [8] and (9.632, 5.479, 4.842 Å) in Ref. [29]. Obviously, the greatly non-uniform chemical/physical environment indicates the importance of orbital

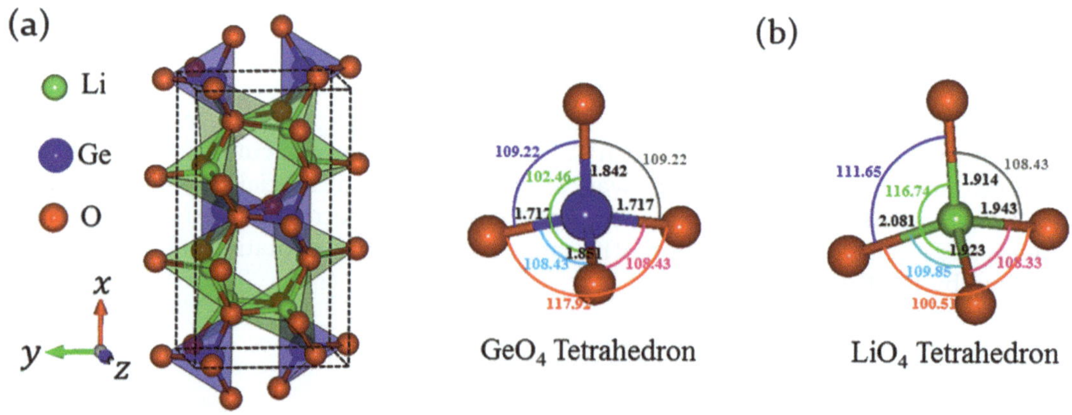

FIG. 17.1 (A) The optimal geometric structure of the Li_2GeO_3 ternary compound, (B) oxygen atoms around each Li/Ge one.

hybridizations in chemical bondings. Furthermore, it might be very useful for supporting the ion transport in electrolyte/electrode materials [21]. This behavior will be responsible for the unusual electronic properties and the highly anisotropic optical transitions.

Li_2GeO_3 exhibits a rich and unique band structure. The main features in Fig. 17.2A show a lot of valence and conduction sub-bands in a wide energy range (-26 eV $< E^{c,v} <$ 12 eV), the high asymmetry of hole and electron energy spectra about the Fermi level ($E_F = 0$), the

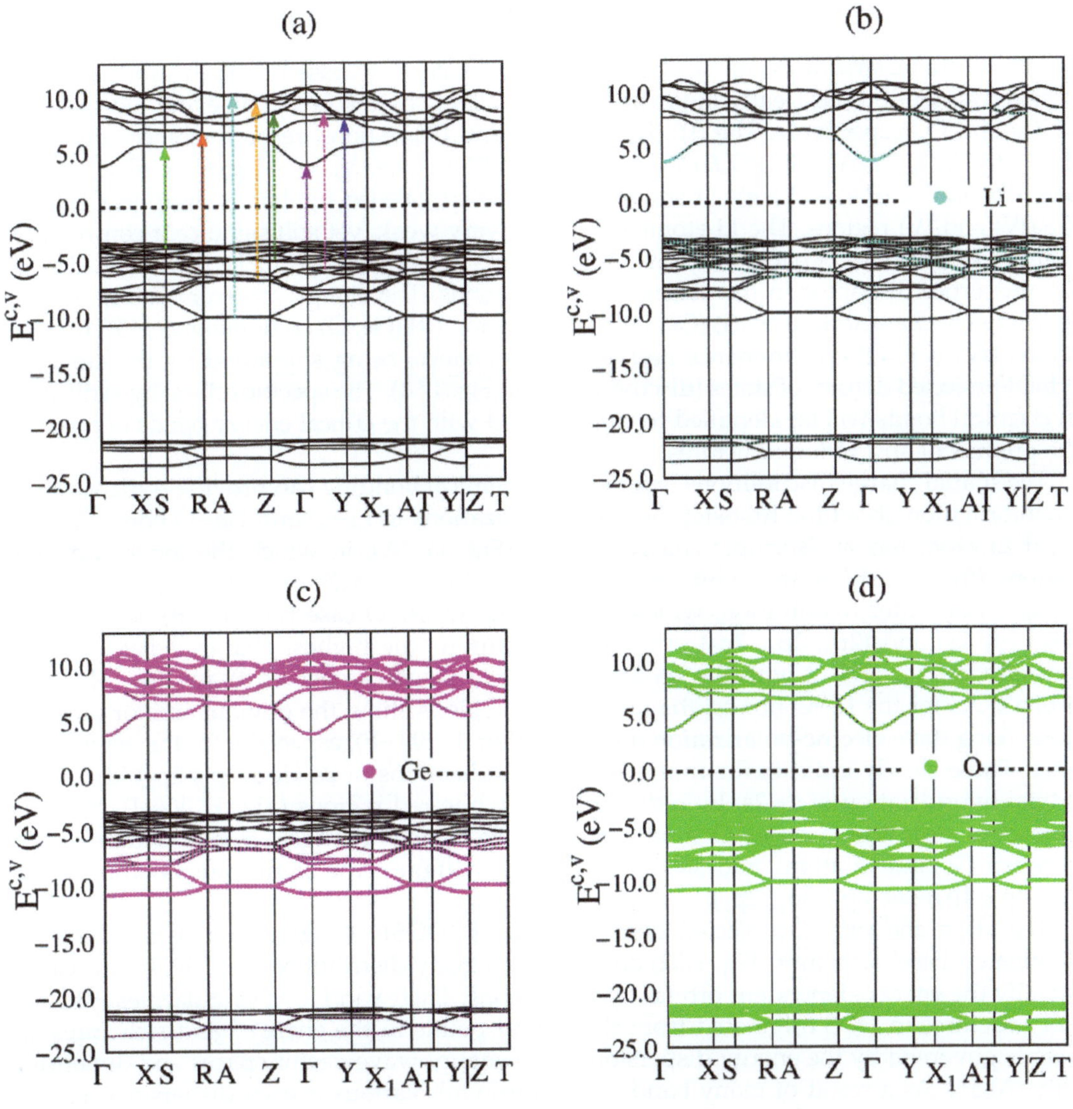

FIG. 17.2 (A) The quasi-particle band structure along the high-symmetry points in the wave-vector space, the vertical-colored arrows indicate the optical excitations. The domination of (B) Li, (C) Ge and (D) O atoms (the cyan, pink and red circles) in the electronic band structure. The Fermi level is denoted by the dash-black line.

parabolic/oscillatory/linear/partially flat energy dispersions with various band-edge states, and the atom-dominated electronic states are frequently present in the different energy regions. An indirect band gap of $E_g^i = 6.9\,eV$ related to the highest occupied state and the lowest unoccupied one, respectively, at the Z and Γ points. The quasi-particle band gap in this work is systematically larger than the corresponding standard DFT band gap (the gray curves with the energy gap of 3.8 eV) [16], which suggest an enhancement of electron-electron interactions [27] (The detail explanations as shown in supporting information). In addition, the low screening leads to excitonic effects well below the quasiparticle band gap. This point will be discussed in the next section.

The electronic structures can be classified into five categories based on the various atom and orbital contributions, as clearly indicated in Fig. 17.2B–D by the green, blue and red circles for Li, Ge and O, respectively. The (I), (II), (III), (IV) and (V) regions, respectively, correspond to $E^c > 7.6$ eV, 3.2 eV $< E^c < 7.6$ eV, -6.0 eV $< E^v < -3.8$ eV, -10.6 eV $< E^v < -6.0$ eV, and -24.5 eV $< E^v < -21.8$ eV. It should be noted that valence sub-bands disappear between the (IV) and (V) regions. The Li atoms only make very weak, yet still significant contributions within a whole range of band structures (small green circles in Fig. 17.2B), i.e., the unusual essential properties disappear in the absence of Li–O bonds. The energy-spectrum sub-groups are qualitatively characterized by (I) (Ge, O) co-dominance, (II) (Ge, O) co-dominance, (III) O dominance, (IV) (Ge, O) co-dominance and (V) O dominance, being supported by the atom and orbital-projected density of states (discussed later in Fig. 17.4). The specific orbital hybridizations in chemical bonds will be identified to be associated with the critical energy bands in revealing the prominent absorption structures.

The spatial charge distributions before/after chemical bonding can provide some very useful information about the first-step orbital hybridizations in Li–O and Ge–O bonds. An isolated Li atom has an isotropic charge density (Fig. 17.3A), in which the inner and outer regions (the dark-blue and white-red parts) arise from 1 to 2s orbitals, respectively. The similar, but wider distribution, which corresponds to the O case (Fig. 17.3B), is associated with (1s, 2s) and ($2p_x$, $2p_y$, $2p_z$) orbitals. Furthermore, the highest charge density appears around Ge (Fig. 17.3C), with the separate ranges of (4s, those below it) and ($4p_x$, $4p_y$, $4p_z$) orbitals. As for the Li–O bonds, the outer/inner regions show the obvious/minor deformations along three electric-polarization directions (Fig. 17.3D–F), especially for the neighboring ones. These clearly indicate the multi-orbital hybridizations of 2s-($2p_x$, $2p_y$, $2p_z$)/the single-orbital hybridization of 2–2s. In addition to the white-red regions (Fig. 17.3G–I), the dark-blue ones near the Ge and O atoms present observable changes, dense charge density between Ge and O atoms and short Ge–O bonds suggest the significant bonding of (4s, $4p_x$, $4p_y$, $4p_z$)-(2s, $2p_x$, $2p_y$, $2p_z$).

The atom and orbital-projected density of states (DOS), being supported by the atom-dominated band structure (Fig. 17.2) and charge density distribution (Fig. 17.3), can clearly identify the energy-dependent orbital hybridizations. Their van Hove singularities - the special structures in DOS originated from the critical points in the energy band structures, are mainly governed by the energy dispersion in the energy-wave-vector spaces and dimensionality [30,31]. As a result of many band-edge states with various energy dispersions, such as the extreme, saddle and dispersionless ones, Fig. 17.4A displays various prominent asymmetric peaks and shoulders (the black curves). The van Hove singularities of different atoms

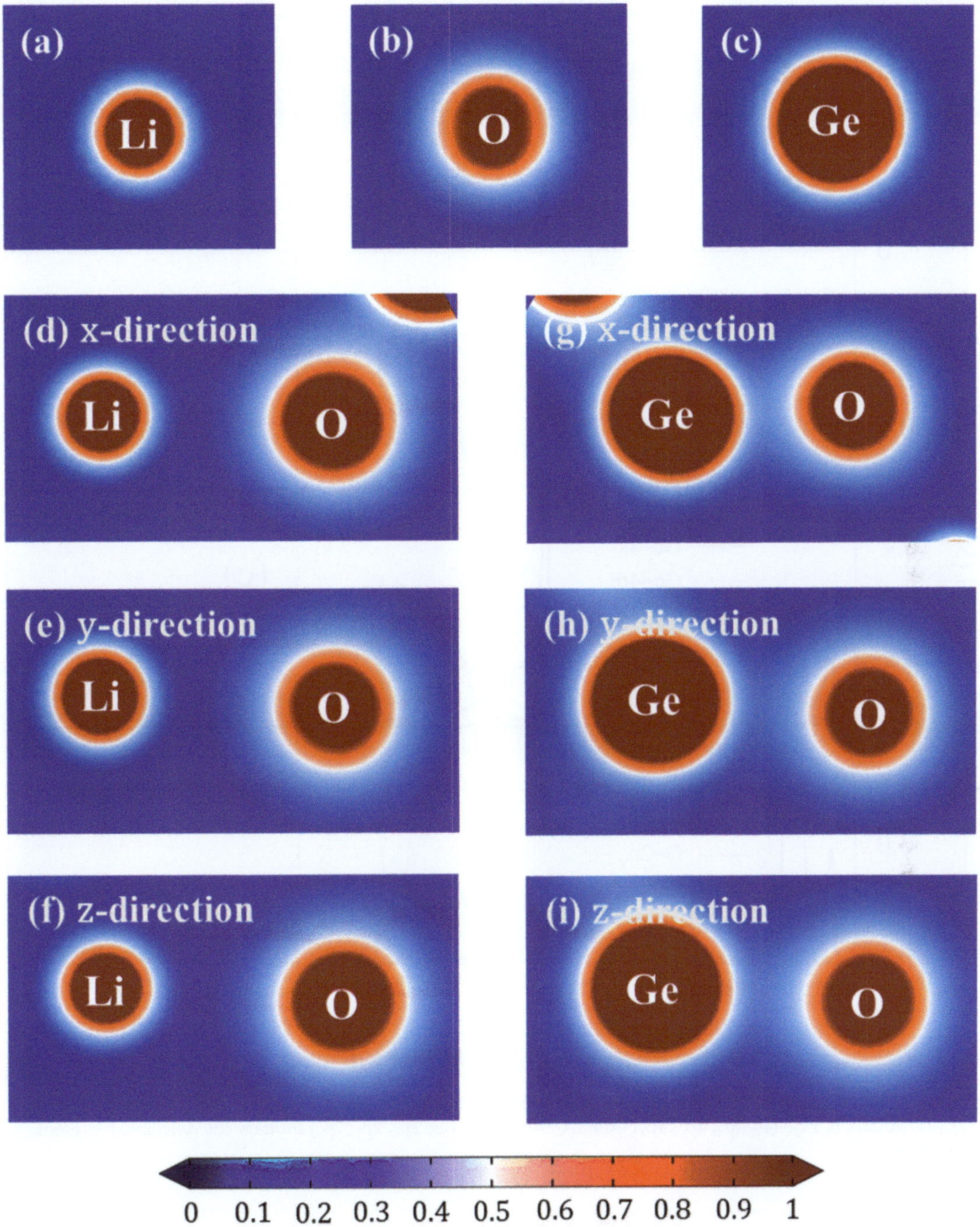

FIG. 17.3 The spatial charge density distributions for (A)/(B)/(C) an isolated Li/Ge/O atom and (D)/(E)/(F) the shortest Li—O & (G)/(H)/(I) Ge—O bond along x-/y-/z-direction, respectively.

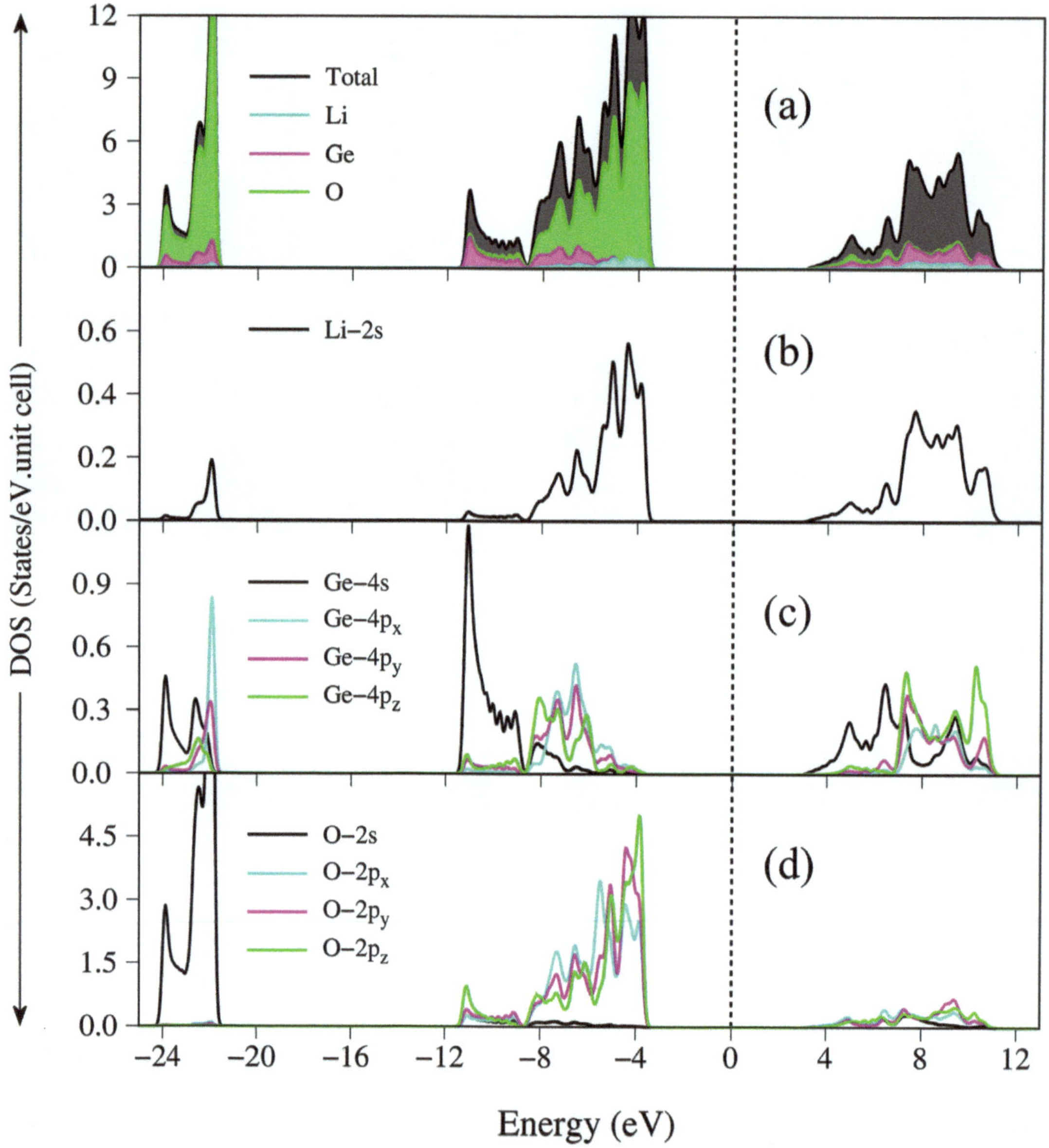

FIG. 17.4 The density of states for different components: (A) total magnitude with Li-, Ge-, O-atom contributions, (B) Li-2s orbitals, (C) Ge-(4s, $4p_x$, $4p_y$, $4p_z$) orbitals, and (D) O- (2s, $2p_x$, $2p_y$, $2p_z$) orbitals, the dash line represents the Fermi level.

and orbitals can merge (Fig. 17.4A–D), clearly indicating the specific orbital hybridizations. According to their strong co-relations, the five subgroups of band structure are further examined to be dominant through (I) 2s-(2s, $2p_x$, $2p_y$, $2p_z$) & (4s, $4p_x$, $4p_y$, $4p_z$)-(2s, $2p_x$, $2p_y$, $2p_z$), (II) 2s-($2p_x$, $2p_y$, $2p_z$) & 4s-($2p_x$, $2p_y$, $2p_z$), (III) 2s-($2p_x$, $2p_y$, $2p_z$) & (4s, $4p_x$, $4p_y$, $4p_z$)-($2p_x$, $2p_y$, $2p_z$), (IV) 2s-($2p_x$, $2p_y$, $2p_z$) & 4s-($2p_x$, $2p_y$, $2p_z$) and (V) the single orbital 2–2s & 4–2s. Such identifications of significant orbital bonding would become more delicate under strong optical responses.

17.3.2 Optical properties of Li-Ge-O compounds

After the perturbation of an electromagnetic (EM) wave, electrons are vertically excited from the occupied to unoccupied states. The excited valence holes and conduction electrons simultaneously come into existence during the optical excitations; furthermore, they tend to combine through the attractive Coulomb potentials under the suitable condition, e.g., a large band gap for the suppression of temperature broadening. The coupled quasiparticles, the stable excitons, might strongly affect the main features of the optical absorption spectra since they make important contributions to many-body effects. This mean that they take part in the various-order excitation processes using the Coulomb scatterings, leading to the dramatic transformations of photons. The excitonic effects, which are closely related to the critical orbital hybridizations, are the focus of this study.

As a consequence of the electromagnetic wave absorption, the dielectric function is complex and reflects the main features of electronic properties and thus orbital-hybridizations. The imaginary part, $\varepsilon_2(\omega)$, as clearly showed in Fig. 17.5A, presents the excitation characteristics of available channels in the absence of a screening effect of all valence charges. Because of the non-uniform environment in the orthorhombic structures, the optical absorption of Li_2GeO_3 is three-fold degenerate and manifests a strong polarization effect. An optical gap (E_g^o), which corresponds to the threshold absorption frequency, is about 5.95/5.92/5.87 eV under the many-body effects for the x-/y-/z-directions of electric polarizations (the black/red/blue curves). However, it is ~ 7.00 eV purely through the single-particle optical excitations for $\boldsymbol{E} \parallel x$ (the blue curve in Fig. 17.5C). Compared with band gap of $E_g^i = 6.9$ eV, the great red shift clearly indicates the very strong Coulomb couplings between the excited holes and electrons. The composite quasiparticles should be quite stable, so that they are expected to survive at room temperature. The theoretical prediction is relatively in good accord with the previous experiment result (the cyan curve in Fig. 17.5A) [17].

In addition to the optical gap, the main profile of the single particle optical excitations also exists various pronounced peaks or shoulder structures (the distinct colored arrows in Fig. 17.5C). In which, the appearance of these certain optical spectral structures is associated with the transitions from band-edge states/Van Hove singularities of electronic band structure/orbital-projected density of states. Furthermore, the formation of the robust excitonic states around the respective extrema of the band-edge state also creates extra prominence peaks below the band gap (the black arrows in Fig. 17.5C) and largely modify optical transition intensities. The frequency-dependent vertical transitions and orbital hybridizations are identified (detail in Table 17.1), the corresponding excitation of each channel is marked with arrow heads in Fig. 17.2A and the colored triangles in Fig. 17.4. For example, the

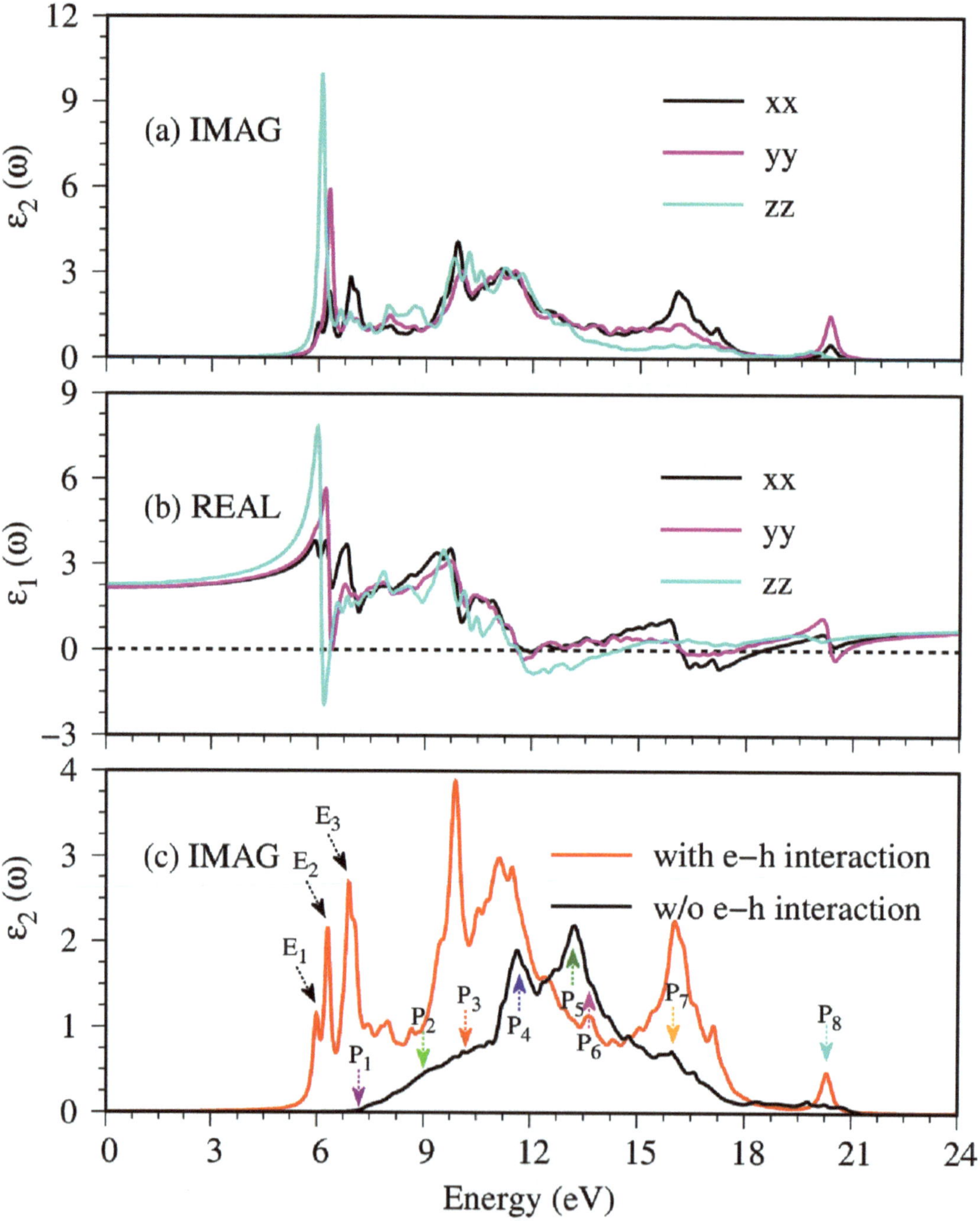

FIG. 17.5 (A) The imaginary and (B) real parts of dielectric functions with the excitonic effects under three electronic-polarization directions (The black, pink and cyan lines). (C) Comparison of the imaginary parts of the dielectric function of x-direction with and without excitonic effects. The independent-particle optical absorption and the three first exciton peaks are, respectively, made with colored and black arrows.

TABLE 17.1 Calculated the prominent absorption structures and the leading transition of each peaks. E_i denotes *i*th excition, while P_i denotes *i*th interband transition peaks.

Peaks	Colors	Energy (eV)	Specific orbital hybridizations in Ge–O bonds
E_1	Black	6.0	-
E_2	Black	6.3	-
E_3	Black	6.9	-
P_1	Purple	7.0	Ge (4s) – O ($2p_x$,$2p_y$,$2p_z$)
P_2	Ligh-green	9.0	Ge (4s) – O ($2p_x$,$2p_y$ $2p_z$)
P_3	Red	10.2	Ge (4s) – O ($2p_x$,$2p_y$ $2p_z$)
P_4	Blue	11.4	Ge (4s) – O ($2p_x$,$2p_y$ $2p_z$)
P_5	Green	13.0	Ge ($4p_x$, $4p_y$, $4p_z$) – O ($2p_x$,$2p_y$ $2p_z$)
P_6	Purple	14.0	Ge ($4p_x$, $4p_y$, $4p_z$) – O ($2p_x$,$2p_y$ $2p_z$)
P_7	Yellow	15.8	Ge ($4p_x$, $4p_y$, $4p_z$) – O ($2p_x$,$2p_y$ $2p_z$)
P_8	Cyan	20.3	Ge ($4p_x$, $4p_y$, $4p_z$) – O ($2p_x$,$2p_y$ $2p_z$)

excitation from the valence flat band to the conduction minimum extremes at the Γ point, leading to the threshold structure of the independent excitation spectrum (purple arrow in Fig. 17.5C). The shoulder structures, which arise from the several extreme-related transitions with equivalent frequency (light-green, red and cyan arrows in Fig. 17.5C as examples), while the sharp peaks (blue and green arrows) correspond to the optical excitations between flat bands, for which, their inequivalent intensities are associated with the relevant hybridizations. Additionally, due to increasing electron-hole overlaps, the absorption coefficient above the band gap is significantly enhanced. Using this strategy, the specified mechanisms for the various absorption structures could be achieved. This viewpoint has been successfully generalized to other emergence materials.

For the real part $\varepsilon_1(\omega)$ (Fig. 17.5B), their prominent absorption structures can be understood from those of $\varepsilon_2(\omega)$ by the Kramers-Kronig relation $\left(\varepsilon_1(\omega) = 1 + \frac{2}{\pi} P \int_0^\infty \frac{\omega'^2 \varepsilon_2(\omega)}{\omega'^2 - \omega^2} d\omega'\right)$ under the event of optical excitations [32]. Apparently, the special structures in $\varepsilon_1(\omega)$ and $\varepsilon_2(\omega)$ might be similar or different, being strongly co-related by the principal-value integration on the complex plane. Below the threshold frequency, $\varepsilon_1(\omega)$ weakly depends on ω, in which its value roughly lies within the range of 2.1–2.2 for x-, y-polarizations and the z one. This will determine the low-frequency reflectance spectrum (Fig. 17.6B) and the vanishing absorption coefficient (Fig. 17.6C). Very interesting, $\varepsilon_1(\omega)$ is very sensitive to the changes of frequency during the creation of the excited holes and electrons. It can vanish at weak Landau damping, in which its zero point and the small $\varepsilon_2(\omega)$ might appear simultaneously,

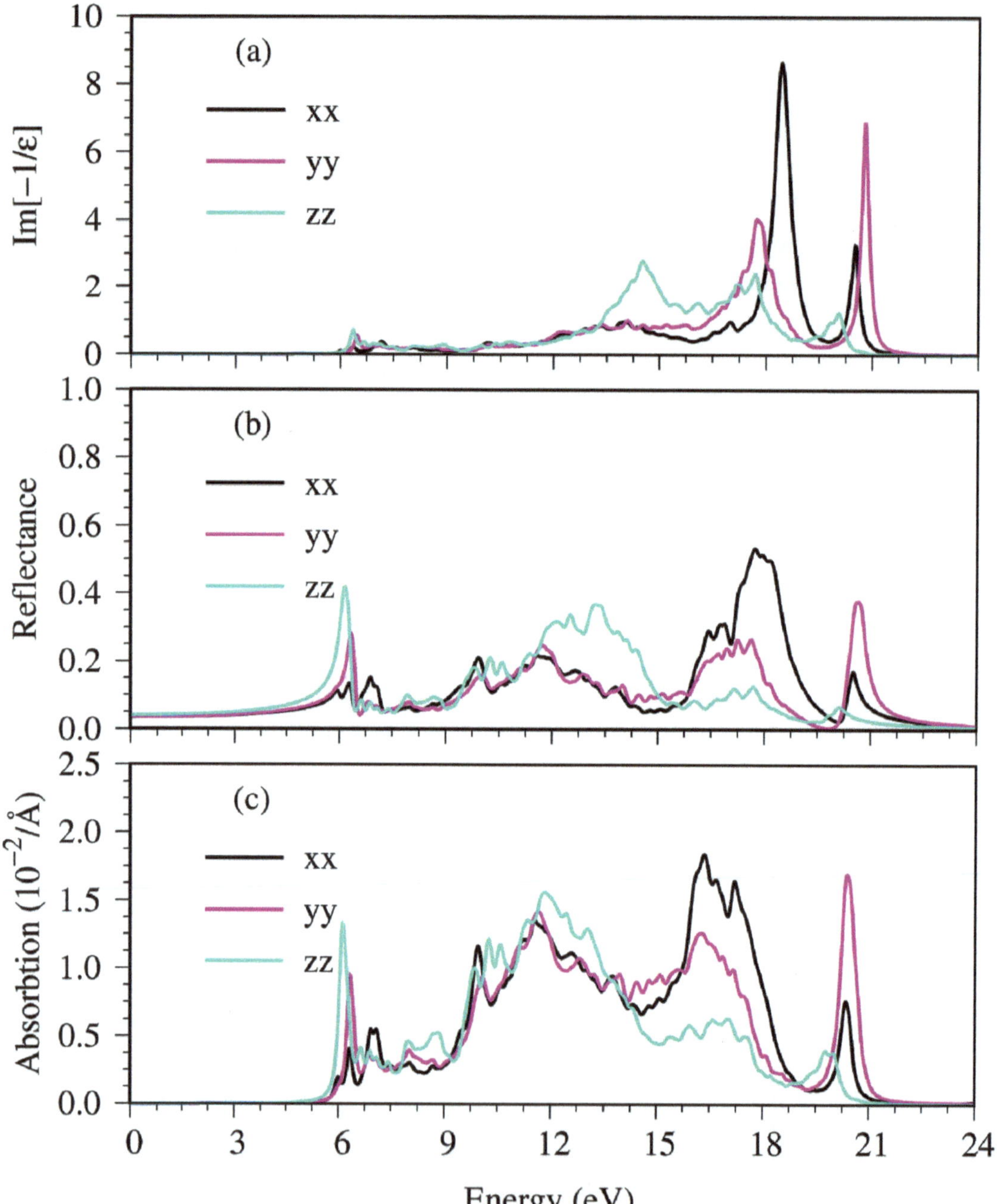

FIG. 17.6 The various optical properties: (A) energy loss functions, (B) reflectance spectra, and (C) absorption coefficients.

e.g., the vanishing $\varepsilon_1(\omega)$ at 18.5/20/13.6 eV under $E \parallel x/E \parallel y/E \parallel z$. In addition, the zero points for the z-polarization (the blue curve) at 6 eV become meaningless because of the combination with very strong electron-hole excitations.

The energy loss function (ELF), being defined as $Im[1/\varepsilon(\omega)]$ is the screened response function due to the significant valence charges of the Li, Ge and O atoms. The charge screenings can determine the coherent carrier oscillations at the long wave-length limit during the optical transitions. In general, the collective excitations (a plasmon mode) are revealed as a sufficiently strong peak with ELF higher than 1, as clearly indicated in Fig. 17.6A. The strongest peak comes into existence at $\omega_P = 18.5/20/13.6$ eV for the x-/y-/z-direction electric polarizations, being attributed to the significant Li-2s, O-($2p_x$, $2p_y$, $2p_z$) & Ge-(4s, $4p_x$, $4p_y$, $4p_z$) orbitals. The O-2s contribution for the most prominent plasma wave is ignored, since its dominating valence states only appear below −22 eV (the black curve in Fig. 17.4D). In addition, a few minor plasmon peaks are accompanied by serious Landau dampings. It should be noted that two manners could be utilized to identify the collective excitations, but the current peak in ELF is much better than the zero points in $\varepsilon_1(\omega)$. The absence of the latter and/or the combination with a very large $\varepsilon_2(\omega)$ is the main reason.

When an electromagnetic wave is normally incident on Li_2GeO_3. The total electric field will be reflected by the surface, absorbed by the valence electrons and transmitted through a finite-width sample. The reflectance $R(\omega)$ $\left(\text{which is characterized as } \left|\sqrt{\frac{\varepsilon(\omega)-1}{\varepsilon(\omega)+1}}\right|^2\right)$ and the absorption coefficient $\alpha(\omega)$ $\left(\text{defined as } \sqrt{2}\,\omega\left[\sqrt{\varepsilon_1^2(\omega)+\varepsilon_2^2(\omega)} - \varepsilon_1(\omega)\right]^{1/2}\right)$ are two characteristic optical phenomena of solid-state materials, directly reflecting the main features of the single-particle (Fig. 17.5A and B) and the collective excitations (Fig. 17.6A). As shown in Fig. 17.6B and C, for $\omega < E_g^o$, a small reflectance weakly depends on the frequency and is roughly given by $\left|\sqrt{\frac{\varepsilon_1(0)-1}{\varepsilon_1(0)+1}}\right|^2$. The absorption coefficient $\alpha(\omega) = 0$ because of the vanishing electron-hole excitations, leading to the non-decay EM-wave propagation and thus a very efficient transmission. However, within the active region of valence-electron excitations $\left(E_g^o < \omega < 21\ eV\right)$, the sensitive and significant frequency dependences come into existence. Reflectance is enhanced and displays a large fluctuation under the various inter-band excitation channels, in which a drastic change of the plasmon edge appears at ω_P, e.g., ~ 40% variation under the $E \parallel y$ case (the red curve). Moreover, since absorption coefficient $\alpha(\omega)$ is very large, especially that related to the plasmon mode, and its inverse is about 50–300 Å, the EM waves are easily absorbed by Li_2GeO_3 through the rich electronic excitations.

In addition to X-ray diffractions, only very few experimental examinations on electronic and optical properties exist. In general, the wave-vector dependences of occupied valence states can be directly tested by angle-resolved photoemission spectroscopy (ARPES) [10]. A lot of unusual energy sub-bands (Fig. 17.2) due to a large and complex unit cell would create high barriers in the ARPES measurements. Scanning tunneling spectroscopy (STS) [33] is available for the clear identification of van Hove singularities near the Fermi level. However, STS might be suitable only under a thin-film sample because of very weak quantum currents. Very interestingly, the optical spectroscopy methods of reflectance [34], absorption [35] and transmission [36] are reliable in verifying the frequency-dependent optical properties. They can provide significant information about the initial excitonic peaks, the greatly reduced threshold excitation frequency, many prominent absorption structures, and the strongest plasmon mode at $\omega_P = 6.0$ eV. Up to now, one photo-luminescence measurement has verified the optical gap situated at $\sim$ 6.0 eV [17], being close to the current prediction of $\sim$ 5.95 eV. Apparently, the diverse optical properties in lithium-based compounds are worthy of systematic investigations, both experimentally and theoretically.

17.4 Conclusion

The significant orbital hybridizations in chemical bonds, being based on first-principles calculations, are thoroughly identified for the Li_2GeO_3 compound in terms of the geometric, electronic and optical properties. They will play important roles in fully understanding the diversified essential properties of optical/electrode materials in ion-related materials. This material presents unusual features, i.e., a large unit cell with a highly non-uniform chemical environment, the Li-, Ge- or O-dominated energy bands, orbital-induced spatial charge densities, atom- and orbital-decomposed van Hove singularities. As a result, the band-edge states, which might create the prominent optical responses, are well characterized by the specific orbital interactions through the developed theoretical framework. The featured optical transitions cover a red-shift optical gap ($E_g^o = 5.95$ eV) much lower than an indirect one ($E_g^i = 6.9$ eV), a very long transmission length/low reflectance for $\omega < E_g^o$, various pronounced single-particle absorption structures (short decay lengths), and the strongest plasmon peak/a quickly decreasing edge in the energy loss function/reflectance spectrum at $\omega_P = 17-18$ eV, and sensitive changes due to the electric-field directions. The many-body excitonic effects have strongly modified the single-particle inter-band excitations. Under current investigations, the developed theoretical framework can be further generalized to include other emergence materials.

Acknowledgments

This work is supported by the Hi-GEM Research Center and the Taiwan Ministry of Science and Technology under grant number MOST 108-2212-M-006-022-MY3, MOST 109- 162811-M-006-505 and MOST 108-3017-F-006-003.

References

[1] Marom R, Amalraj SF, Leifer N, Jacob D, Aurbach D. J Mater Chem 2011;21:9938–54.
[2] Barghamadi M, Kapoor A, Wen C. J Electrochem Soc 2013;160:A1256.
[3] Hussein AA-H, Batarseh I. IEEE Trans Veh Technol 2011;60:830–8.
[4] Jia C, Pan F, Zhu YG, Huang Q, Lu L, Wang Q. Sci Adv 2015;1:e1500886.
[5] Pang Q, Liang X, Kwok CY, Kulisch J, Nazar LF. Adv Energy Mater 2017;7:1601630.
[6] Yokota S. Mechanical engineering reviews **1**. DSM0014-DSM0014. 2014.
[7] Goto K, Nakagawa T, Nakamura O, Kawata S. IEEE (Inst Electr Electron Eng) Trans Biomed Eng 2001;48:830–3.
[8] Rahman MM, Sultana I, Yang T, Chen Z, Sharma N, Glushenkov AM, et al. Angew Chem 2016;128:16293–7.
[9] VanHove MA, Weinberg WH, Chan C-M. Low-energy electron diffraction: experiment, theory and surface structure determination. Springer Science & Business Media; 2012.

[10] Lv B, Qian T, Ding H. Nat Rev Phys 2019;1:609–26.
[11] Caloz C, Itoh T. Application of the transmission line theory of left-handed (LH) materials to the realization of a microstrip LH line. In: IEEE antennas and propagation society international symposium. IEEE Cat. No. 02CH37313); 2002. p. 412–5.
[12] Marsland T, Evans S. Dielectric measurements with an open-ended coaxial probe. In: IEE proceedings H (microwaves, antennas and propagation); 1987. p. 341–9.
[13] Ghodgaonkar DK, Varadan VV, Varadan VK. IEEE Trans Instrum Meas 1989;38:789–93.
[14] Yu KB, Ogourtsov S, Belenky V, Maslenikov A, Omar AS. IEEE Trans Microw Theor Tech 2000;48:2159–64.
[15] International JPP. J Quan Chem 1985;28:497–523.
[16] Khuong Dien V, Thi Han N, Nguyen TDH, Huynh TMD, Pham HD, Lin M-F. Front Mater 2020;7.
[17] Trukhin AN, Rogulis U, Spingis M. J Lumin 1997;72–74:890–2.
[18] Nguyen DK, Tran NTT, Chiu Y-H, Gumbs G, Lin M-F. Sci Rep 2020;10:12051.
[19] Nguyen DK, Tran NTT, Nguyen TT, Lin M-F. Sci Rep 2018;8:17859.
[20] Lin S-Y, Lin Y-T, Tran NTT, Su W-P, Lin M-F. Carbon 2017;120:209–18.
[21] Zhao Q, Stalin S, Zhao C-Z, Archer LA. Nat Rev Mater 2020;1–24.
[22] Han NT, Dien VK, Thuy Tran NT, Nguyen DK, Su W-P, Lin M-F. RSC Adv 2020;10:24721–9.
[23] Nguyen TDH, Pham HD, Lin S-Y, Lin M-F. RSC Adv 2020;10:14071–9.
[24] Perdew JP, Burke K, Ernzerhof M. Phys Rev Lett 1996;77:3865–8.
[25] Kresse G, Joubert D. Phys Rev B 1999;59:1758–75.
[26] Wisesa P, McGill KA, Mueller T. Phys Rev B 2016;93:155109.
[27] Hybertsen MS, Louie SG. Phys Rev B 1986;34:5390.
[28] Rohlfing M, Louie SG. Phys Rev B 2000;62:4927–44.
[29] Yin C, Xiang H, Li C, Porwal H, Fang L. J Am Ceram Soc 2018;101:4608–14.
[30] Cohen ML, Louie SG. Fundamentals of condensed matter physics. Cambridge University Press; 2016.
[31] Kittel C. Solid state physics, 3. Shell Development Company; 1955.
[32] Fox AM, Fox DPAM. Optical properties of solids. Oxford University Press; 2001.
[33] Feenstra RM. Surf Sci 1994;299:965–79.
[34] Kortüm G. Reflectance spectroscopy: principles, methods, applications. Springer Science & Business Media; 2012.
[35] BV DC. Hydraulics. 1962.
[36] Suzuki S, Takamura M, Yamamoto H. Jpn J Appl Phys 2016;55:06GF08.

CHAPTER

18

Insight into electronic and optical properties of inorganic Ge-based halide perovskites ($CsGeI_3$)

Nguyen Thi Han[1], *Shih-Yang Lin*[1], *Vo Khuong Dien*[1], *Chi-Hsuan Lee*[1], *Hsin-Yi Liu*[1], *Thi My Duyen Huynh*[1], *Ngoc Thanh Thuy Tran*[2], *Wei-Bang Li*[1] *and Ming-Fa Lin*[1,2]

[1]Department of Physics, National Cheng Kung University, Tainan, Taiwan; [2]Hierarchical Green-Energy Materials (Hi-GEM) Research Center, National Cheng Kung University, Tainan, Taiwan

18.1 Introduction

Nowadays, perovskite solar cell devices have become one of the mainstream systems in the basic materials science and engineering applications which they can provide an eco-friendly energy, their efficiency and a special route to convert photon energy into electricity [1–9]. The principal architecture of a perovskite solar cell includes the absorber layer sandwich between the electron-transporting and the hole-transporting layers [10–12] (Fig. 18.1). In which, the first component is very important since it absorbs the electromagnetic wave (EM) and generates the electron-hole pairs [13]. Apparently, the geometric, electronic properties and optical excitation mechanisms of these components are regarded as very important common characteristics. These factors should be critical conditions in determining the efficiency of light-generated current and searching for optimal match of three kinds of core components. The crucial mechanism of solar cells based on photovoltaic effects involves two key processes [14,15] (Fig. 18.2). In the first process, the electron-hole pairs (excitons) are generated in the absorber layer upon the absorption of a suitable photon. In the second process, charge carriers (electrons and holes) are produced upon the action of the electric field existing at the interface between the absorber and the charge-transporting layers. When the electron separates from the hole and injected into the electron-transporting layer and then migrates to the anode.

Fundamental Physicochemical Properties of Germanene-related Materials
https://doi.org/10.1016/B978-0-443-15801-8.00015-3

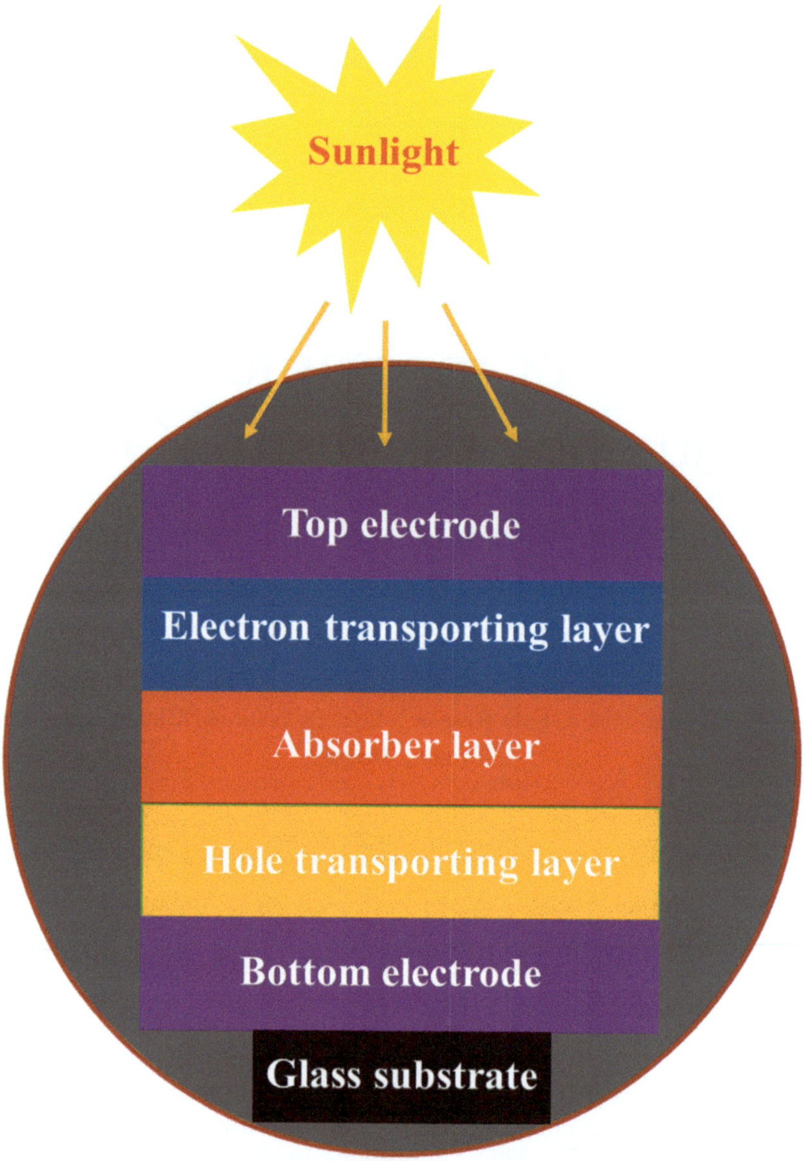

FIG. 18.1 Principle architecture of a perovskite solar cell.

Simultaneously, the hole injects into the hole-transporting layer and subsequently migrates to the cathode. The electrons and holes are collected by working and counter electrodes respectively and transported to the external circuit to produce current.

Generally, the electron-transporting and the hole-transporting layers of solar cell panel are fabricated by the different semiconductors (or the same semiconductor with different types of conduction), while semiconductor materials used for the absorber layer must be dissimilar to

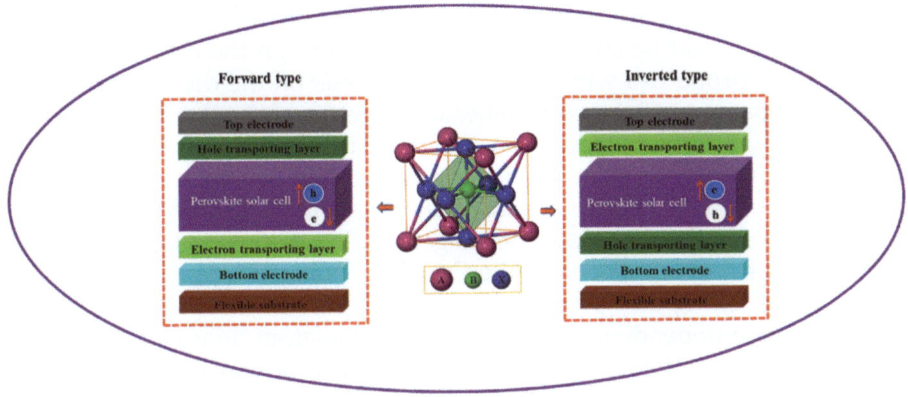

FIG. 18.2 Two general device structures of perovskite solar cell.

the charge transporting layers in order to produce the built-in electric field and to carry the electric current [3]. Recently, the metal halide perovskites with the general formula ABX_3 (where, A = cation, B = metal ion, and X = halogen anion) have been indicated to be a novel class of promising semiconductor materials in photovoltaic applications [16–20].

Some presentative systems for this kind of materials, such as $CaTiO_3$, $BiVO_3$, $CsSnCl_3$, $CsPbBr_3$ and so on are investigated by many methods [21–27]. Among them, $CsGeI_3$ has attracted much more and more attentions. For example, many studies have investigated $CsGeI_3$ which makes it possible to enhance enrich properties of solar cell applications [28–33]. In addition, inorganic halide perovskite $CsGeI_3$ is stable up to 350°C, with devices recording efficiency to only 0.2% [34]. The effects of pressure on the structure and band gap of $CsGeI_3$ was also indicated [29,35].

On the theoretical side, investigations indicated that the first-principles calculation method is very efficient and reliable in determining the rich essential properties of the 3D compounds, such as the geometric, electronic and optical properties of the perovskite materials [23,29]. However, the critical but very complicated orbital hybridizations that survive in the chemical bonds and the optical excitation mechanism with related to the high-potential functionalities of the solar cell devices have not been achieved so far. Furthermore, they also have not dealt with the unique excitonic effects, the mutual interactions of electron-hole upon the optical excitations. Fortunately, the previous results are calculated by DFT based on VASP package are fully understanding of the diversified physical and chemical phenomena. Indeed, systematic studies have been successfully conducted on layered 2D graphene/silicene [36,37] with or without chemical modifications (absorptions and substitutions) [38–41] and the 3D ternary cathode/anode/electrolyte compounds [42–47]. Through delicate analyses, the critical mechanisms and a concise picture of the geometric, electronic, magnetic and optical properties could be firmly obtained. As result, the concise calculations might be suitable for exploring the fundamental properties of a lot of complicated perovskites compound for solar cell applications. Obviously, the significant single/multi-orbital hybridizations in the various chemical bonds are obtained from the optimal lattice symmetry, the atom-dominated band structures, the spatial charge densities and their changes after chemical modifications, and the atom and orbital-projected density of states. The optical excitation mechanism, which is generated upon the absorption of EM could be successfully identified through the close combination of the geometric, electronic properties with the delicate analysis on the dielectric function, energy loss function, absorption coefficient and reflection spectrum [43,44].

The developed theoretical framework, which based on the significant orbital hybridizations in chemical bonding, is utilized to identify the crucial features in solar cell materials. In this chapter, we have performed the theoretical investigation by using the first-principles calculations and delicate analysis to get a fully comprehending the essential properties of germanium-based halide perovskite $CsGeI_3$ for solar cell applications. Also, we have connected electronic and optical properties of $CsGeI_3$ perovskite. The optimized geometric structure with position-dependent chemical bonding, the atom dominated energy spectrum at various energy ranges, the spatial charge density distribution due to different orbitals, and the atom- and orbital-projected density of states are associated with the overlap of orbitals. Furthermore, our results took a comprehensive insight into various orbital hybridizations to interpret the onset frequency, many prominent absorption structures, the prominence plasmon modes in terms of real part and imaginary one of the dielectric functions, reflectance spectra, absorption coefficients under three identical directions and the very important electron-hole interactions. The current study is of paramount importance not only for fundamental physics but also for technical applications.

Most of theoretical predictions in this chapter can be tested by the highly resolved experimental methods. For example, X-ray diffractions (XRD) is usually used to measure the lattice symmetries of 3D materials [28,48,49], while the morphology and the atom concentration can be detected by the scanning electron microscopy (SEM) [50] and Atomic force microscope (AFM) [51]. The high resolution of the nano-scale top and side views can be achieved by the Scanning tunneling microscopy (STM) [52] and tunneling electron microscopy (TEM) [53] respectively. Besides, the occupied electronic states can be accurate examination by Angle-resolved photoemission spectroscopy (ARPES) [54]. The high-resolution scanning tunneling spectroscopy (STS) are rather efficient method in thoroughly identify the whole valence and conduction van Hove singularities [55]. Furthermore, the optical measurements, such as absorption, transmission and reflectance spectroscopies combine with Kramer–Kronig relations can give information of the optical excitations [56]. The presented theoretical framework will be fully comprehending the diverse phenomena and potential applications for solar cell and other optoelectronic devices.

18.2 Computation methods

In this chapter, we have investigated the first-principal calculations via the Vienna Ab-initio Simulation Package (VASP) [43–46,57–59] to exhibit the optimization of the geometric and the electronic and optical properties for $CsGeI_3$ solar cell. The Perdew-Burke-Ernzerhof (PBE) generalized gradient approximation was used for the exchange-correlation functional [60–63]. The projector augmented wave (PAW) method was utilized for characterization of the ion and valence electron interactions [64]. The cutoff energy for the expansion of the plane wave basis was set to 500 eV. The Brillouin zone was integrated with a special k-point mesh of $25 \times 25 \times 25$ in the Γ-centered sampling technique for the structural optimization. The convergence condition of the ground-state is set to be 10^{-8} eV between two consecutive simulation steps, and all atoms could fully relax during the geometric optimization until the Hellmann-Feynman force acting on each atom was smaller than 0.01 eV.

Optical properties of $CsGeI_3$ is investigated by many-body perturbation theory (called GW method where G is the Green' function and W the screened coulomb potential) [65–67]. In which the quasiparticle energies are obtained within approximation for the self-energy [68–70], the cutoff energy for the response function was set to 200 eV, 10 × 10 × 10 Γ-centered k points sampling was used to represent reciprocal space. The wannier interpolation procedure performed in the WANNIER90 code [71,72] was used to plot the quasi-band structure. These can be described by the Kubo formula [73,74]:

$$\varepsilon_2(\omega) = \frac{8\pi^2 e^2}{\omega^2} \sum_{vc\mathbf{k}} |ev\mathbf{k}|\mathbf{v}|c\mathbf{k}|^2 \delta(\omega - (E_{c\mathbf{k}} - E_{v\mathbf{k}})),$$

where the first part, $|ev\mathbf{k}|\mathbf{v}|c\mathbf{k}|^2$ is the square of the electric dipole moment, which is responsible for the strength of the excitation peaks, and the second part, $\delta(\omega - (E_{c\mathbf{k}} - E_{v\mathbf{k}}))$, is the joined of the density of states, which corresponds to the available excitation transition channels. In addition to the independent particle excitations, the present of stable exciton may strongly affect in the optical properties. To evaluate this bound state of the electron-hole pair, we solve the standard Bethe-Salpeter equation (BSE) [75].

$$\varepsilon_2(\omega) = \frac{8\pi^2 e^2}{\omega^2} \sum_{vc\mathbf{k}} |e0|\mathbf{v}|S|^2 \delta(\omega - \Omega_S),$$

Where, the k-point sampling, energy cutoff, number of bands, were set to the same values as in the GW calculation. In addition, the 6 lowest unoccupied conduction bands (CBs) and the 11 highest occupied valence bands (VBs) are included as a basis for the excitonic states with a photon energy region from 0 to 12 eV. The fully chemical and physical pictures, the physics of the selective optical absorption and the effect of the robust exciton are the main study focus of the current chapter.

18.3 Rich electronic properties

The $CsGeI_3$ compound displays a unique geometric structure, this material crystallizes in the rhombohedral structure, as clearly illustrated in Fig. 18.3A. This reveal the lattice parameters $x = y = z = 6.11$ Å, $\alpha = \beta = \delta = 88.6°$ for R-3m symmetry (Table 18.1). The primitive cell of the crystal consists of five atoms in the unit cell (Table 18.2).These results are consistent not only with the X-ray powder diffraction data [40,41] but also with the previous theoretical reports. The shortest and longest bond lengths are listed in Table 18.3. The $CsGeI_3$ halide perovskites cover corner-sharing (GeI_6) octahedra and create a 3D framework with a Cs component surrounded by 12 I anions (Fig. 18.3B). Thus, they can conduct a non-uniform chemical environment. This will display directly the complicated multi-orbital hybridizations in chemical bonding and are responsible for the unique optical excitations.

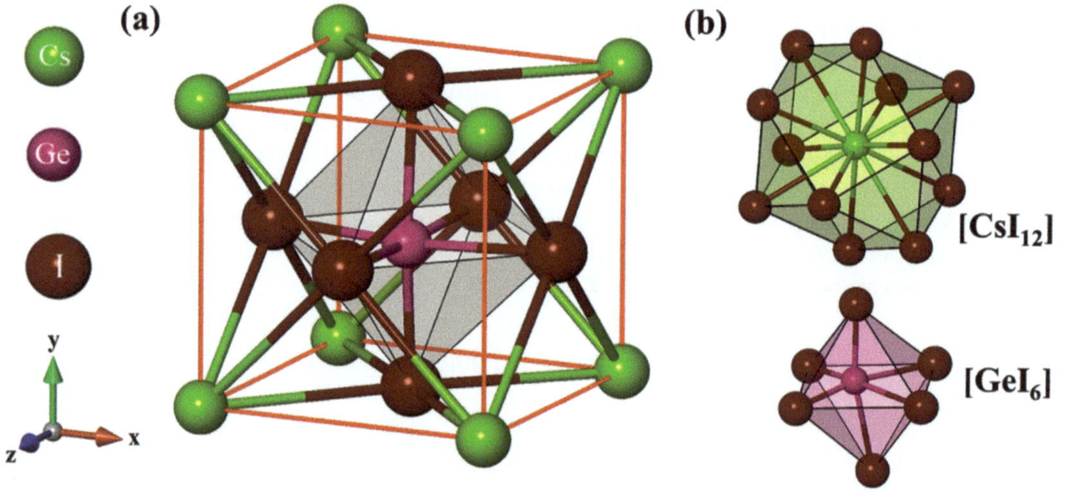

FIG. 18.3 (A) The optimal lattice structure of $CsGeI_3$ trigonal symmetry with lattice constants $x = y = z = 6.11$ Å. Unit cell is a solid red (B) two kinds of chemical bonding: Cs—I and Ge—I bonds.

TABLE 18.1 Compare lattice constant of perovskite $CsGeI_3$ between this work and previous reports.

		Lattice constant	
Space group	Crystal system	This work	Previous reports (DFT and XRD)
R-3m	Trigonal	$a = b = c = 6.10$ Å	$a = b = c = 5.98$ Å
		$\alpha = \beta = \gamma = 88.61$Å	$\alpha = \beta = \gamma = 88.61$Å

TABLE 18.2 The bond lengths in $CsGeI_3$ perovskite for the total number of 12 Cs—I and 6 Ge—I bonds.

Atom-atom	No. of bonds	Bond length (Å)
Cs—I	12	2.79—3.30
Ge—I	6	4.30—4.40

Interestingly, Fig. 18.5A—D exhibits the diversified electronic properties in unique band structure for $CsGeI_3$ perovskite. Indeed, the valence bands are highly asymmetric to the conduction ones about the Fermi level $E_F = 0$. The presence of many valence and conduction bands along high-symmetry points as named M, R, Γ, X, M, Γ points (Fig. 18.4). Their energy dispersions exist the parabolic, oscillatory and camel's back shape relations.

TABLE 18.3 Prominent absorption structures: frequencies, label indicators and identified orbital hybridizations without electron-hole interaction in Ge—I bonds.

Energy (eV)			
With e-h interaction	**W/o e-h interaction**	**Label**	**Specific orbital hybridizations in Ge—I bonds**
2.0	2.0 (threshold)	E_1	Ge ($4p_x$, $4p_y$)- I ($5p_x$, $5p_y$)
3.5	3.5	A	Ge (4s, $4p_x$, $4p_y$, $4p_z$)-I ($5p_x$, $5p_y$)
4.2	4.2	B	Ge (4s, $4p_x$, $4p_y$, $4p_z$)-I ($5p_x$, $5p_y$)
5.2	5.2	C	Ge (4s, $4p_x$, $4p_y$, $4p_z$)-I ($5p_x$, $5p_y$, $5p_z$)
6.0	6.0	D	Ge (4s, $4p_x$, $4p_y$, $4p_z$)-I ($5p_x$, $5p_y$, $5p_z$)
6.5	6.5	E	Ge (4s, $4p_x$, $4p_y$, $4p_z$)-I ($5p_x$, $5p_y$, $5p_z$)
7.0	7.0	F	Ge (4s, $4p_x$, $4p_y$, $4p_z$)-I ($5p_x$, $5p_y$, $5p_z$)

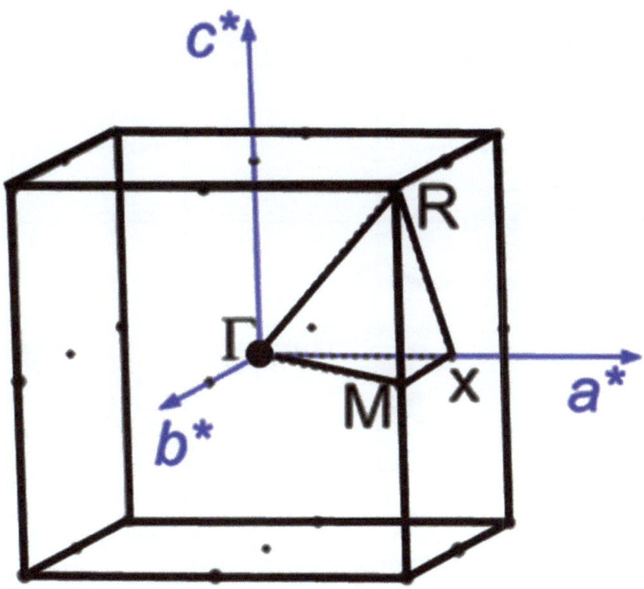

FIG. 18.4 The first brillouin zone with high-symmetry points of $CsGeI_3$ perovskite.

Furthermore, the energy sub-bands with non-crossing, crossing and anti-crossing phenomena come into existence frequently. The critical points in the energy-wave-vector space will create the unique van Hove singularities in density of states (DOS). These behaviors related to the complex orbital hybridizations in Cs-I and Ge-I chemical bonding and can be responsible for strong absorption structures in optical responses. According to the atom-dominated energy spectrum, two groups of energy sub-bands can be observed: (i) the valence bands related

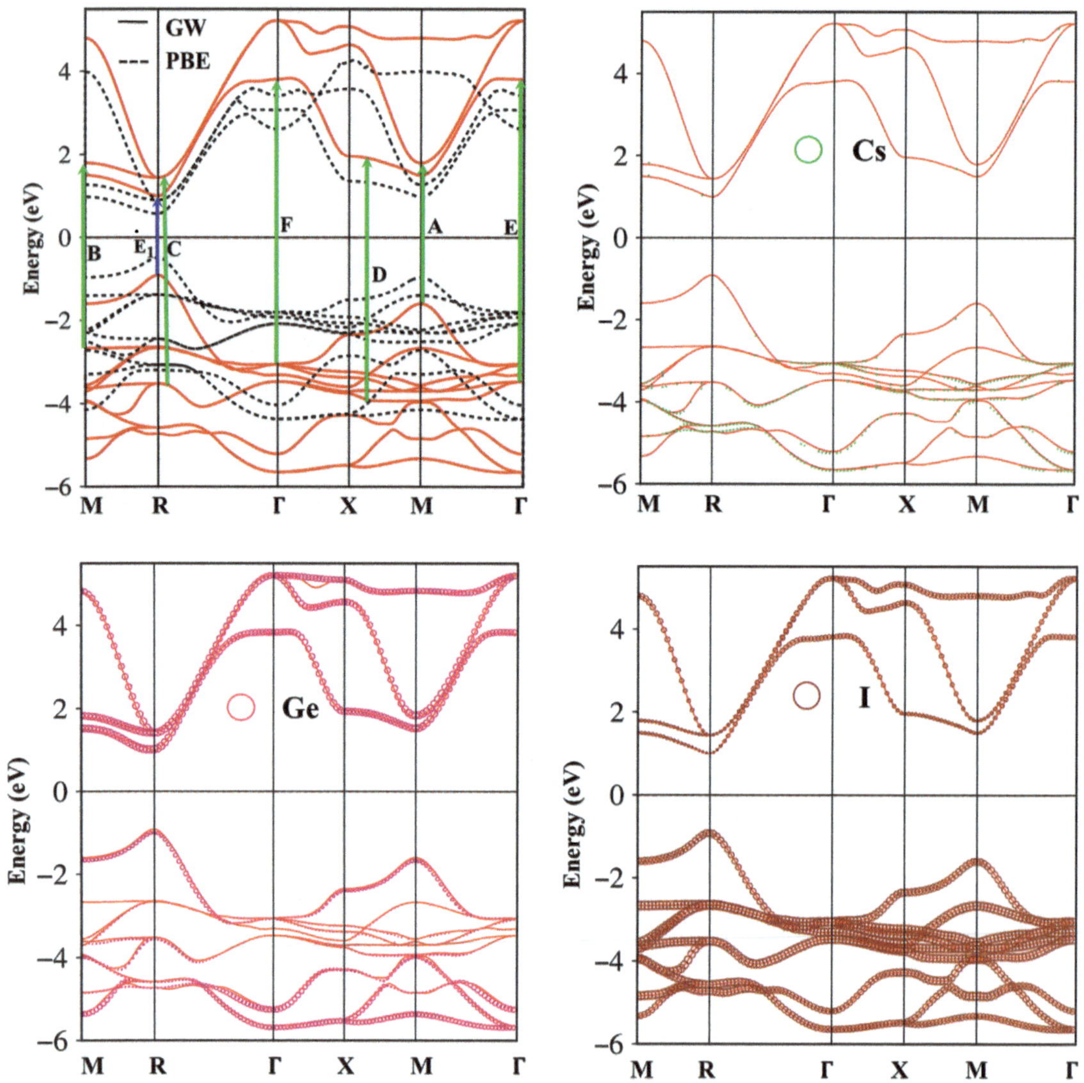

FIG. 18.5 (A) Quasiparticle band structure along the high-symmetry points in the wave-vector space, with (B) Cs- (C) Ge- and (D) I-atom dominances (green, light purple and brown circles respectively).

to energy ranges of -6.00 eV< $E^v < -1.00$ eV due to the strongly dominated of I atom, and (ii) the conduction band related to energy ranges of 1.00 eV $<E^c < 5$ eV mostly due to Ge contributions. Apparently, the Cs only has 6s valence orbital, its contributions to all energy sub-bands are very weak, but very important (Fig. 18.5B), e.g., several features will be disappeared within the absence of Cs atoms.

Most importantly, the quasiparticle band structure of this system belongs to a direct-gap semiconductor of $E_g^d = 2.0$ eV for the highest occupied and lowest unoccupied states in between the MΓ direction (solid red in Fig. 18.5A). Indeed, GW method is closer to reported experimental values of 1.63 eV. While using density function theory, the band gap value is1.04 eV for DFT method (dash black in Fig. 18.5A). It is well consistent with previous report of 1.06 eV for PBE sol + SOC [28] (smaller than GW method). Obviously, the slightly quasi-particle corrections to the DFT energy gap evidence e-e Coulomb interactions are insignificant. The low screening leading the hole-electron interactions are weak and excitonic states are right below the threshold frequency.

Interestingly, the charge density distributions of $CsGeI_3$ perovskite which clearly showed in Fig. 18.6A–C, provides complex chemical environments and useful first-step information of multi-orbital hybridizations. It should be noticed that the chemical bonds are achieved by combining them with the atom- and orbital-projected density of states (later discussions in Fig. 18.7A–D), and the atom-dominated band structures (Fig. 18.5B and D). The isolated of Cs, Ge, I atoms are contributed by (6s) orbital (4s, $4p_x$, $4p_y$, $4p_z$) orbitals and (5s, $5p_x$, $5p_y$, $5p_z$) orbitals, respectively, as evident Fig. 18.6A. After combine to form the chemical bonding, the strong orbital overlaps are easy to identify in the Ge-I bonds (Fig. 18.6C). This obviously deformed of this charge due to (4s, $4p_x$, $4p_y$, $4p_z$)-($5p_x$, $5p_y$, $5p_z$) orbital hybridizations. As for the Cs-I bonds (Fig. 18.6B), the weak but important orbital hybridizations between Cs and I atoms as expose on the slight deformation of the charge. Specifically, the I-5s orbitals are relatively far away that of the Cs-atom; therefore, they are almost negligible for the chemical bonds. According to the above-mentioned features, the Cs-I and Ge-I bonds are predicted to exhibit 6s-($5p_x$, $5p_y$, $5p_z$) and (4s, $4p_x$, $4p_y$, $4p_z$)-($5p_x$, $5p_y$, $5p_z$) orbital hybridizations, respectively.

Density of states (DOS) and Project-density of states (PDOS) are depicted in Fig. 18.7A–D, being fully comprehended the occupied and unoccupied states simultaneously. Apparently, its special structures, van Hove singularities, mainly originate from the band-edge states with the zero group velocities (Fig. 18.4A). Such critical points in the energy-wave-vector space will include the local extreme points, the saddle points, and the dispersionless energy sub-bands. Apparently, the van Hove singularities are greatly diversified under the different dimensions, e.g., the delta-function-like prominent peaks for the zero-dimension (0D) discrete energy levels, the square-roots asymmetric peaks/plateau structures of the one-dimension (1D) parabolic/linear energy sub-bands, and the shoulder structures/V-shapes/logarithmic peaks/square-roots asymmetric peaks in the two- dimensions parabolic/Dirac cone/ saddle-point/constant-energy-loop band structures. Moreover, the higher dimension in three-dimensions (3D) materials makes them only indicate symmetric/asymmetric peaks and shoulder structures, mainly due to the absence of quantum confinement; the shoulder structures arise from the 3D parabolic energy dispersions. For instance, van Hope singularity of 3D $CsGeI_3$ perovskite are exhibited by a lot of symmetric/asymmetric/shoulder peaks which are created by the critical point in band structure.

Very interestingly, the atom- and orbital-projected van Hove singularities of $CsGeI_3$ perovskite could be directly utilized to settle the important multi-orbital hybridizations in all Cs-I and Ge-I bonds. Three kinds of atoms (the green, purple and brown curves) have the significant contributions and the similar van Hove singularities in the whole energy range. The band gap value is obtained for 2 eV with a vanishing density of states around the Fermi level and split the very high symmetric occupied and unoccupied states. Apparently, the Cs contributions are lowest as evident the green curves in Fig. 18.7B. Besides, the Ge and I atoms

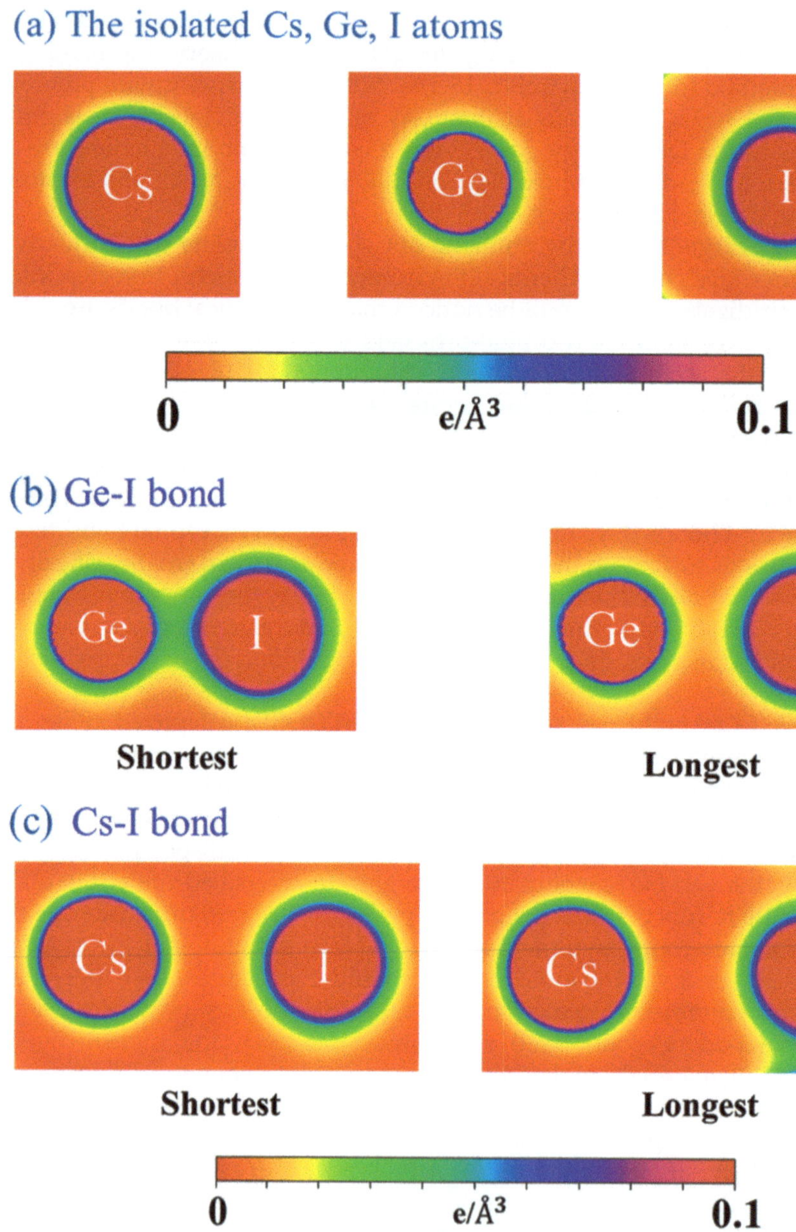

FIG. 18.6 (A), The isolated Cs/Ge/I atom (B)/(C) the charge density distributions related to the significant orbital hybridizations in Cs–I and Ge–I bonds, respectively along the x-/y-/z-identical directions for the shortest and longest bonds.

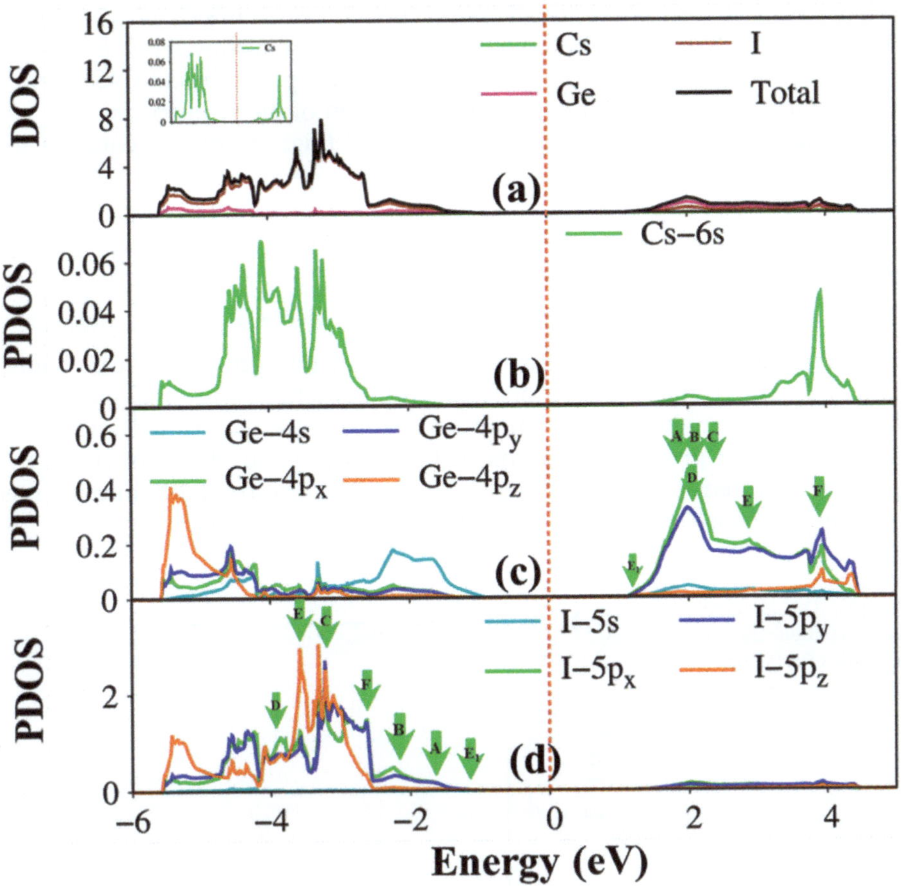

FIG. 18.7 (A), Density of states under Cs-/Ge-/I- atom contributions, (B)/(C)/(D) with projected density of states calculations, respectively Cs- (6s), Ge- (4s, $4p_x$, $4p_y$, $4p_z$) and I- (5s, $5p_x$, $5p_y$, $5p_z$).

(Fig. 18.7C–D) depicted to present entire both occupied and unoccupied states. Apparently, I-($5p_x$, $5p_y$, $5p_z$) (the cyan, green, blue and orange curves in Fig. 18.7D) orbitals and Cs-(6s) orbital strongly hybridize with the four Ge-(4s, $4p_x$, $4p_y$, $4p_z$) one (the cyan, green, blue, and orange curves in Fig. 18.7C), since their van Hove singularities have merged each other. Among their available orbitals, the I-5s orbitals only make a minor contribution at any energies and then vanishing at some energy region. The high-resolution STS are rather efficient method in thoroughly examination the whole valence and conduction van Hove singularities.

18.4 Enrich optical properties

It is known that the electrons in system is vertical promoted from the valence to conduction states under the perturbation of an electromagnetic wave (EM) or photon. As a direct consequence of photon absorption, the dielectric function is complex and supplies important information to fully understand the optical transition mechanisms. Fig. 18.8A showed the imaginary part $\varepsilon_2(\omega)$ of the dielectric function, the optical properties of $CsGeI_3$ are identical under three identical directions. Due to the conservation of the momentum, the optical gap, also called the threshold frequency, are around 2.0 eV (Fig. 18.9B) and equal to the fundamental direct gap E_g^d. However, when the presence of excitonic effects is taken into account (Fig. 18.9A), the optical gap is slightly reduced by 0.5 eV, leading to 1.5 eV. The slightly redshift shows that the electron-hole interactions are weak, and thus, the exciton states are easy to break to create free electrons and holes.

In addition to threshold frequency $\left(E_g^0\right)$, $\varepsilon_2(\omega)$ also survives extra six prominent peaks. The orbital character of these peaks can be understood through the delicate connection of van Hop singularities/critical points in the band structure with the intensity and the frequency of each channels. The excitation mechanism was confirmed and indicated by the alphabet Labels on green arrowheads in the energy band structure (Fig. 18.5A), and the orbital-projected density of states (Fig. 18.7C and D), the orbital hybridizations related to these optical responses are successfully achieved Fig. 18.9B and evidenced in Table 18.3.

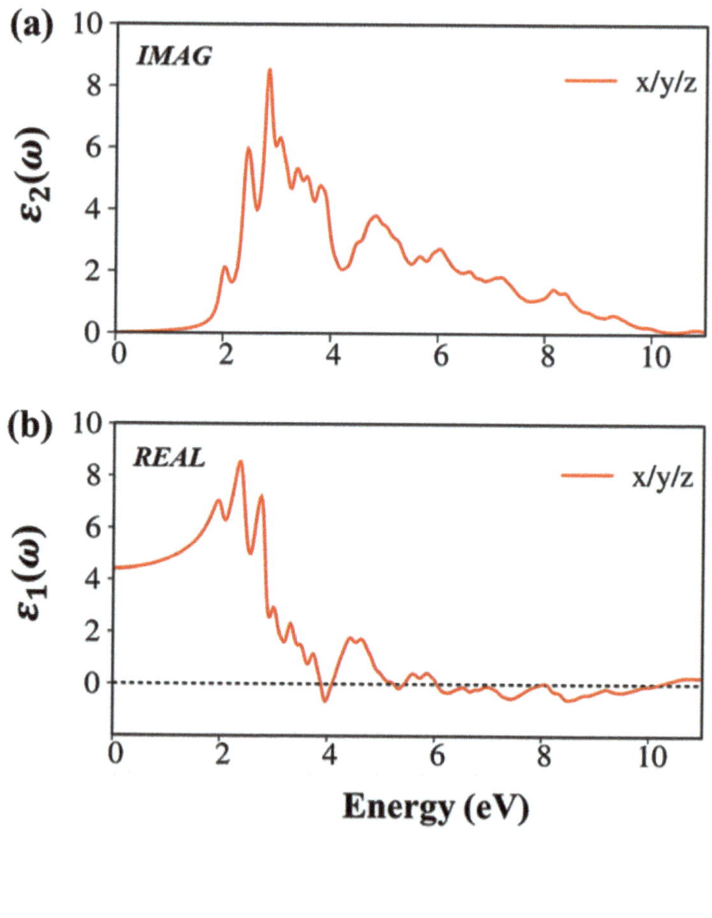

FIG. 18.8 (A), The imaginary-part and (B) real-part dielectric functions with the excitonic effects under x-/y-/z-identical directions (red curves).

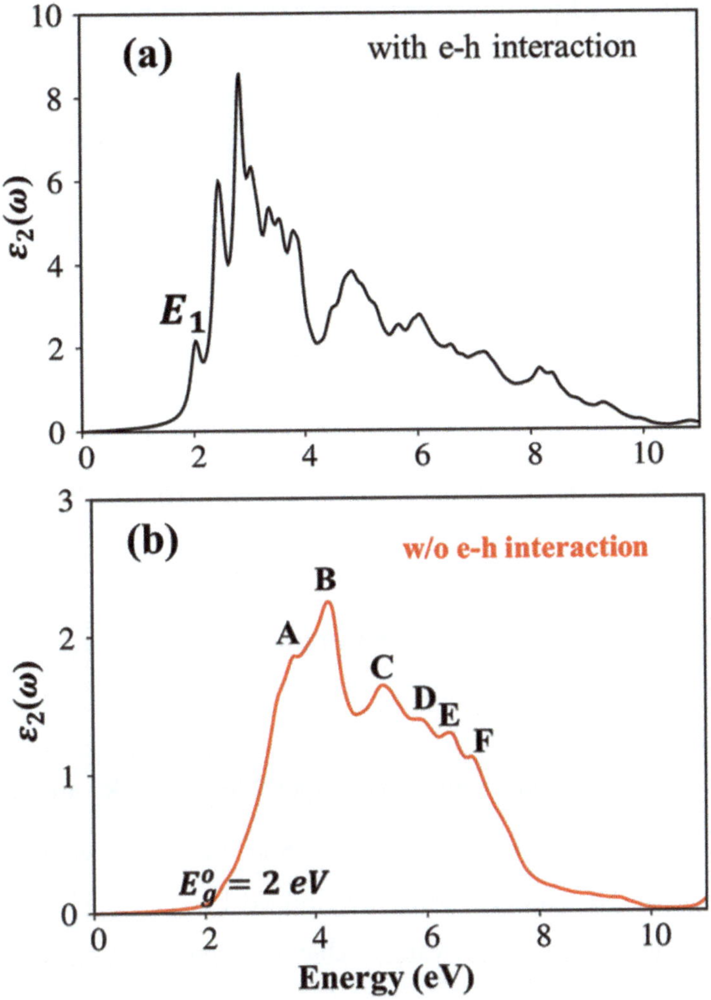

FIG. 18.9 Comparison of the imaginary part of dielectric function with and w/o electron-hole interactions (or called excitonic effect) under three identical directions.

The real part $\varepsilon_1(\omega)$ of the dielectric function (Fig. 18.8B) relates to the imaginary part $\varepsilon_2(\omega)$ through the Kramers- Kronig relationship [76–78]. This close relation will appear the pair pronounced peaks of the real part exist simultaneously with the prominent structures of the imaginary one. Obviously, $\varepsilon_1(\omega)$ is weakly dependent on the energy in the inactive region. For example, dielectric constant at zero point ε_1 (0) are about 4.5 and almost unchanged for frequencies less or equal to the optical gap, through which, the vanishing range of absorption coefficients and low energy reflectance spectrum could be examined. Contrarily, during the excitation channels, the real part is very sensitive to a change of EM. Besides, the real part almost is zero at some frequencies. The simultaneous existence of the zero points of the real part with very weak single-particle transitions showed and corresponding to the weak damping of the plasmon mode (detail in Fig. 18.10 later). For example, the insignificant Landau

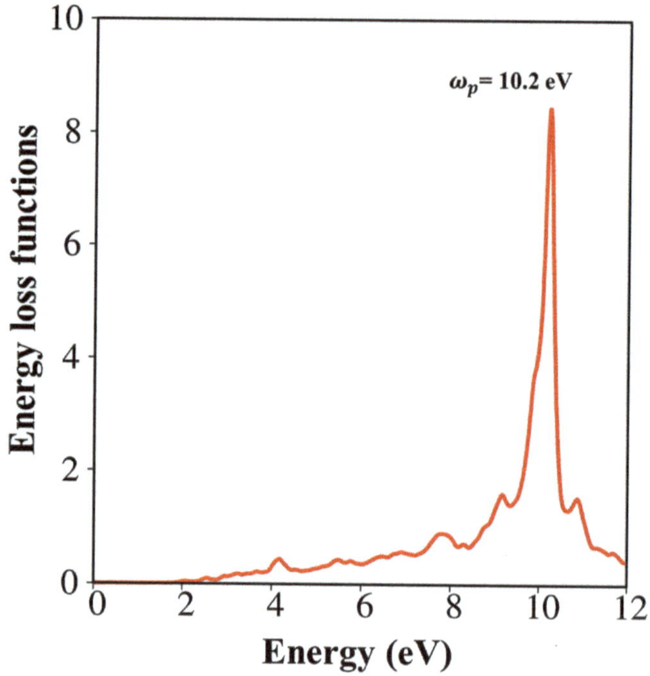

FIG. 18.10 Energy loss functions with exitonic effect in optical properties of $CsGeI_3$ compound.

damping by the single partial excitation at the real part is 10.2 eV. The zero points at 3.8 eV combine with an inter-band transition peak, and therefore, correspond to the weak Landau damping.

The energy loss function (ELF) is the screened responses function due to the various valence charge for Cs, Ge and I atoms in unit cell, being determined by $\mathrm{Im} = \left|\frac{-1}{\varepsilon(\omega)}\right|$ [79]. The collective excitation (Plasmon mode) of a certain valence charge is confirmed by each spectral peak in ELF. As obviously indicated in Fig. 18.10, the peak at almost un-damping of the plasmon resonances are about $\omega_p = 10.2$ eV. This value is rather suitable with the zero points in the real part $\varepsilon_1(\omega)$ and imaginary part $\varepsilon_2(\omega)$ of dielectric functions. This strongest plasmon mode is due to contributions of Cs-6s, Ge- (4s, $4p_x$, $4p_y$, $4p_z$) and I-($5p_x$, $5p_y$, $5p_z$) orbitals. In addition to the most pronounced peak, a few plasmon modes with relatively weak intensity are also exist.

In order to comprehend about the absorption coefficient [80] $\alpha(\omega)$ and the reflectance spectra [81] ε_1 of $CsGeI_3$ compound which are clearly performed by both Figs. 18.11 and 18.12, expose optical phenomena and reflect the principal characteristics of single and coherent excitations. When photon is smaller than the onset frequency, the reflectance spectra R(ω) is weakly dependent on the frequency while the absorption coefficient $\alpha(\omega)$ is zero due to the lacking the electronic excitation contributions. Beyond the thresh-hold frequency, both ε_1 and R(ω) are changing significantly and sensitively. As for the absorption coefficients, the fast increase is contributed by the various inter-band transitions. The inverse values of the

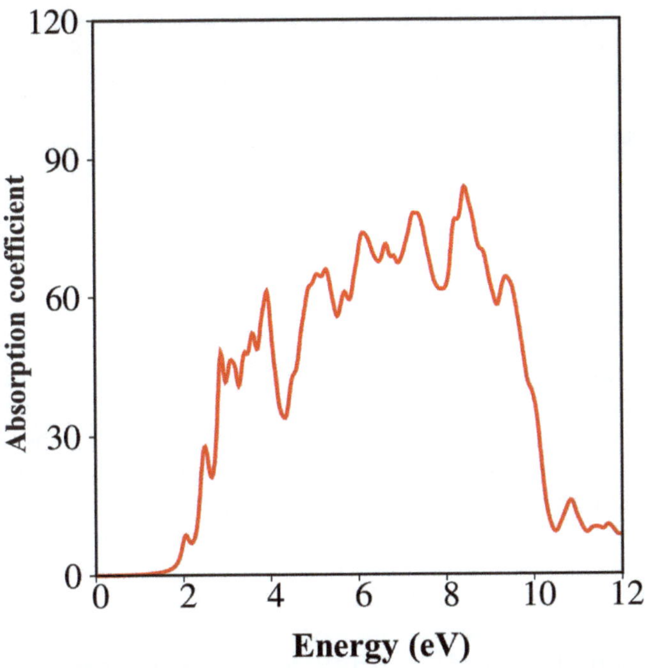

FIG. 18.11 Absorption coefficients with exitonic effect in optical properties of $CsGeI_3$ compound.

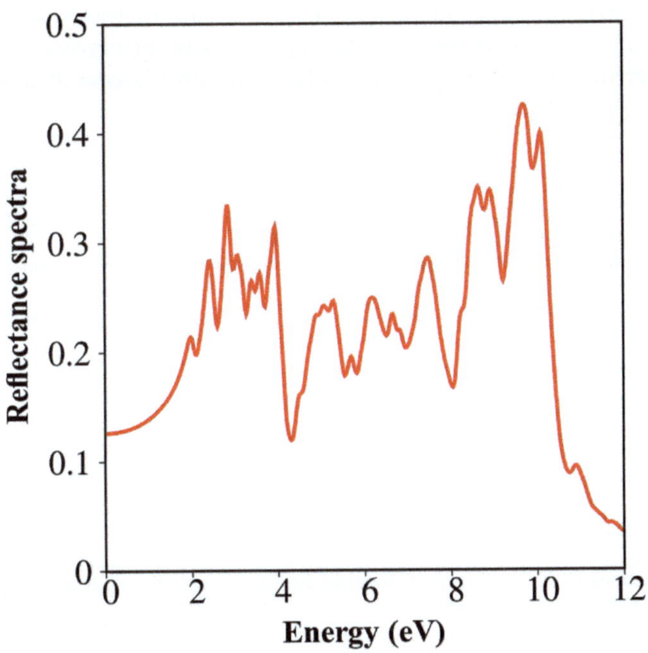

FIG. 18.12 Reflectance spectra with exitonic effect in optical properties of $CsGeI_3$ compound.

absorption coefficient are 100 Å, indicating that the photon energy propagate in the medium are easily absorbed by the enrich electronic excitations. Besides, reflectance spectra $R(\omega)$ is significantly enhanced and exhibited huge fluctuation. In the above consideration, the reflective index is $\sim\left|\frac{\sqrt{\varepsilon_1(0)}-1}{\sqrt{\varepsilon_1(0)}+1}\right|^2 = 0.16$ eV. Due to the collective excitations, the reflectance spectra with 40% variation at the resonance frequency ω_p are indicated. The optical properties of $CsGeI_3$ perovskite could be investigated and identified by experimental methods.

18.5 Conclusion remarks

In summary, we have performed the unique geometric, electronic and optical properties of $CsGeI_3$ by using first-principles calculations (based on VASP package). In this chapter, many clearly physical and chemical pictures of $CsGeI_3$ perovskite solar cell are achieved. Furthermore, this material exposed a lot of featured properties, such as extremely non-uniform physical and chemical environments, unique band structure, direct energy gap, charge density distribution, van Hop singularity. These results are calculated and related to the optical response under the distinct orbital hybridization. The diversified properties and huge application of this material and other perovskite materials could be fully comprehended.

Especially, the enrich optical properties involve under three identical directions, the presence of excitonic states with a reduced optical gap for $E_g^O = 1.5$ eV, modified single particle a high absorption coefficient, six prominent single-particle excitation peaks, and a lot of pronounced peaks in ELF at the plasmon resonance frequency $\omega_p = 10.2\,eV$.

The main optical features of $CsGeI_3$ could be examined by the experimental methods. The presented theoretical framework could be explored for potential applications of perovskite solar cell materials and other optoelectronic materials. The $CsGeI_3$ compound is one of the worthy outstanding candidates for large applications.

References

[1] Lewis NS. Research opportunities to advance solar energy utilization. Science 2016;351.
[2] Rong Y, Hu Y, Mei A, Tan H, Saidaminov MI, Seok SI, et al. Challenges for commercializing perovskite solar cells. Science 2018;361.
[3] Wang R, Mujahid M, Duan Y, Wang ZK, Xue J, Yang Y. A review of perovskites solar cell stability. Adv Funct Mater 2019;29:1808843.
[4] Lewis NS. Toward cost-effective solar energy use. Science 2007;315:798–801.
[5] Atwater HA, Polman A. Plasmonics for improved photovoltaic devices. Mater sustain Energy: A Collect Peer-Rev Res Rev Art Nat Publish Group 2011:1–11.
[6] Chopra K, Paulson P, Dutta V. Thin-film solar cells: an overview. Prog Photovoltaics Res Appl 2004;12:69–92.
[7] Polman A, Knight M, Garnett EC, Ehrler B, Sinke WC. Photovoltaic materials: present efficiencies and future challenges. Science 2016;352.
[8] Yang G, Tao H, Qin P, Ke W, Fang G. Recent progress in electron transport layers for efficient perovskite solar cells. J Mater Chem 2016;4:3970–90.
[9] Pham HD, Yang TCJ, Jain SM, Wilson GJ, Sonar P. Development of dopant-free organic hole transporting materials for perovskite solar cells. Adv Energy Mater 2020;10:1903326.
[10] Grånäs O, Vinichenko D, Kaxiras E. Establishing the limits of efficiency of perovskite solar cells from first principles modeling. Sci Rep 2016;6:1–6.

[11] Fujiwara H, Kato M, Tamakoshi M, Miyadera T, Chikamatsu M. Optical characteristics and operational principles of hybrid perovskite solar cells. Phys Status Solidi (A) 2018;215:1700730.
[12] Hussain I, Tran HP, Jaksik J, Moore J, Islam N, Uddin MJ. Functional materials, device architecture, and flexibility of perovskite solar cell. Emergent Mater 2018;1:133–54.
[13] Gregg BA. Excitonic solar cells. ACS Publications; 2003.
[14] Jung HS, Park NG. Perovskite solar cells: from materials to devices. Small 2015;11:10–25.
[15] Zhang J, Gao X, Deng Y, Zha Y, Yuan C. Comparison of life cycle environmental impacts of different perovskite solar cell systems. Sol Energy Mater Sol Cell 2017;166:9–17.
[16] Song X, Hodes G, Zhao K, Liu S. Metal-free organic halide perovskite: a new class for next optoelectronic generation devices. Adv Energy Mater 2021;11:2003331.
[17] Lyu M, Yun JH, Chen P, Hao M, Wang L. Addressing toxicity of lead: progress and applications of low-toxic metal halide perovskites and their derivatives. Adv Energy Mater 2017;7:1602512.
[18] Hoefler SF, Trimmel G, Rath T. Progress on lead-free metal halide perovskites for photovoltaic applications: a review. Monatshefte für Chemie-Chem Month 2017;148:795–826.
[19] Zhang Z, Li Z, Meng L, Lien SY, Gao P. Perovskite-based tandem solar cells: get the most out of the sun. Adv Funct Mater 2020;30:2001904.
[20] Mesquita I, Andrade L, Mendes A. Perovskite solar cells: materials, configurations and stability. Renew Sustain Energy Rev 2018;82:2471–89.
[21] Lemanov V, Sotnikov A, Smirnova E, Weihnacht M, Kunze R. Perovskite CaTiO3 as an incipient ferroelectric. Solid State Commun 1999;110:611–4.
[22] Diallo P, Jeanlouis K, Boutinaud P, Mahiou R, Cousseins J. Improvement of the optical performances of Pr3+ in $CaTiO_3$. J Alloys Compd 2001;323:218–22.
[23] Dutta DP, Jayakumar O, Tyagi A, Girija K, Pillai C, Sharma G. Effect of doping on the morphology and multiferroic properties of $BiFeO_3$ nanorods. Nanoscale 2010;2:1149–54.
[24] Yang J, He Q, Suresha S, Kuo C, Peng C, Haislmaier R, et al. Orthorhombic $BiFeO_3$. Phys Rev Lett 2012;109:247606.
[25] Huang L-y, Lambrecht WR. Electronic band structure, phonons, and exciton binding energies of halide perovskites $CsSnCl_3$, $CsSnBr_3$, and $CsSnI_3$. Phys Rev B 2013;88:165203.
[26] Huang L-y, Lambrecht WR. First-principles calculations of phonons and Raman spectra in monoclinic $CsSnCl_3$. Phys Rev B 2015;91:075206.
[27] Shen Z, Loo W, Kuok M, Tang S. High pressure phase transition studies of $CsSnCl_3$. J Mol Struct 1994;326:73–80.
[28] Tang L-C, Chang C-S, Huang JY. Electronic structure and optical properties of rhombohedral $CsGeI_3$ crystal. J Phys Condens Matter 2000;12:9129.
[29] Liu D, Li Q, Jing H, Wu K. Pressure-induced effects in the inorganic halide perovskite $CsGeI_3$. RSC Adv 2019;9:3279–84.
[30] Huang L-y, Lambrecht W. Phase stability, electronic structure and phonons in CsGeI 3. In: APS March Meeting Abstracts; 2016. S8. 007.
[31] Lu X, Zhao Z, Li K, Han Z, Wei S, Guo C, et al. First-principles insight into the photoelectronic properties of Ge-based perovskites. RSC Adv 2016;6:86976–81.
[32] Zhang Q, Mushahali H, Duan H, Lee M-H, Jing Q. The linear and nonlinear optical response of $CsGeX_3$ (X= Cl, Br, and I): the finite field and first-principles investigation. Optik 2019;179:89–98.
[33] Liu H, Zhang H. Structural, electronic and optical properties of $CsMI_3$ (M = Ge, Sn, Pb) perovskite from first principles. Am Sci Res J Eng, Technol Sci 2020;71:87–98.
[34] Krishnamoorthy T, Ding H, Yan C, Leong WL, Baikie T, Zhang Z, et al. Lead-free germanium iodide perovskite materials for photovoltaic applications. J Mater Chem 2015;3:23829–32.
[35] Seo D-K, Gupta N, Whangbo M-H, Hillebrecht H, Thiele G. Pressure-induced changes in the structure and band gap of $CsGeX_3$ (X = Cl, Br) studied by electronic band structure calculations. Inorg Chem 1998;37:407–10.
[36] Matthes L, Pulci O, Bechstedt F. Optical properties of two-dimensional honeycomb crystals graphene, silicene, germanene, and tinene from first principles. New J Phys 2014;16:105007.
[37] Tran NTT, Lin S-Y, Glukhova OE, Lin M-F. Configuration-induced rich electronic properties of bilayer graphene. J Phys Chem C 2015;119:10623–30.
[38] Tran NTT, Nguyen DK, Lin S-Y, Gumbs G, Fa-Lin M. Fundamental properties of transition-metals-adsorbed graphene. 2019. arXiv preprint arXiv:1905.01010.

[39] Nguyen DK, Lin Y-T, Lin S-Y, Chiu Y-H, Tran NTT, Fa-Lin M. Fluorination-enriched electronic and magnetic properties in graphene nanoribbons. Phys Chem Chem Phys 2017;19:20667–76.
[40] Pablo-Pedro R, Magana-Fuentes MA, Videa M, Kong J, Li M, Mendoza-Cortes JL, et al. Understanding disorder in 2D materials: the case of carbon doping of Silicene. Nano Lett 2020;20:6336–43.
[41] Lopez-Bezanilla A. Substitutional doping widens silicene gap. J Phys Chem C 2014;118:18788–92.
[42] Han NT, Tran NTT, Dien VK, Nguyen DK, Lin M-F. Diversified properties in 3D ternary oxide compound. Lithium-Ion Batteries and Solar Cells: Phys, Chem, Mater Propert 2021.
[43] Han NT, Dien VK, Lin M-F. Excitonic effects in the optical spectra of Li_2SiO_3 compound. Sci Rep 2021;11:1–10.
[44] Dien VK, Pham HD, Tran NTT, Han NT, Huynh TMD, Nguyen TDH, et al. Orbital-hybridization-created optical excitations in Li_2GeO_3. Sci Rep 2021;11:1–10.
[45] Han NT, Dien VK, Tran NTT, Nguyen DK, Su W-P, Lin M-F. First-principles studies of electronic properties in lithium metasilicate (Li_2SiO_3). RSC Adv 2020;10:24721–9.
[46] Dien VK, Han NT, Nguyen TDH, Huynh TMD, Pham HD, Lin MF. Geometric and electronic properties of Li_2GeO_3. Front Mater 2020;7:288.
[47] Li W-B, Lin S-Y, Tran NTT, Lin M-F, Lin K-I. Essential geometric and electronic properties in stage-n graphite alkali-metal-intercalation compounds. RSC Adv 2020;10:23573–81.
[48] Epp J. X-ray diffraction (XRD) techniques for materials characterization. In: Materials characterization using nondestructive evaluation (NDE) methods. Elsevier; 2016. p. 81–124.
[49] Wu X, Song W, Li Q, Zhao X, He D, Quan Z. Synthesis of lead-free $CsGeI_3$ perovskite colloidal nanocrystals and electron beam-induced transformations. Chem–Asian J 2018;13:1654–9.
[50] Montiel KA, Yang C, Andreasen CH, Gottlieb MS, Pfefferkorn MR, Wilson LG, et al. Lead-free perovskite thin film solar cells from binary sources. In: 2019 IEEE 46th photovoltaic Specialists Conference. PVSC); 2019. p. 1183–6.
[51] Houari M, Bouadjemi B, Matougui M, Haid S, Lantri T, Aziz Z, et al. Optoelectronic properties of germanium iodide perovskites $AGeI_3$ (A = K, Rb and Cs): first principles investigations. Opt Quant Electron 2019;51:1–14.
[52] Qiu L, He S, Ono LK, Qi Y. Progress of surface science studies on ABX3-based metal halide perovskite solar cells. Adv Energy Mater 2020;10:1902726.
[53] Chen L-J. Synthesis and optical properties of lead-free cesium germanium halide perovskite quantum rods. RSC Adv 2018;8:18396–9.
[54] Hermanson J. Final-state symmetry and polarization effects in angle-resolved photoemission spectroscopy. Solid State Commun 1977;22:9–11.
[55] Hipps K. Scanning tunneling spectroscopy (STS). In: Handbook of applied solid state spectroscopy. Springer; 2006. p. 305–50.
[56] Lucarini V, Saarinen JJ, Peiponen K-E, Vartiainen EM. Kramers-Kronig relations in optical materials research. Springer Science & Business Media; 2005.
[57] Hafner J. Ab-initio simulations of materials using VASP: density-functional theory and beyond. J Comput Chem 2008;29:2044–78.
[58] Sun G, Kürti J, Rajczy P, Kertesz M, Hafner J, Kresse G. Performance of the Vienna ab initio simulation package (VASP) in chemical applications. J Mol Struct: Theochem 2003;624:37–45.
[59] Thi HN, Tran NTT, Dien VK, Nguyen DK, Lin M-F. Diversified properties in 3D ternary oxide compound: Li2SiO3. In: Lithium-ion Batteries and solar cells. CRC Press; 2021. p. 79–101.
[60] Perdew JP, Burke K, Wang Y. Generalized gradient approximation for the exchange-correlation hole of a many-electron system. Phys Rev B 1996;54:16533.
[61] Peng H, Perdew JP. Rehabilitation of the Perdew-Burke-Ernzerhof generalized gradient approximation for layered materials. Phys Rev B 2017;95:081105.
[62] Madsen GK. Functional form of the generalized gradient approximation for exchange: the PBE α functional. Phys Rev B 2007;75:195108.
[63] Maximoff SN, Ernzerhof M, Scuseria GE. Current-dependent extension of the Perdew–Burke–Ernzerhof exchange-correlation functional. J Chem Phys 2004;120:2105–9.
[64] Blöchl PE. Projector augmented-wave method. Phys Rev B 1994;50:17953.
[65] Bruneval F, Sottile F, Olevano V, Del Sole R, Reining L. Many-body perturbation theory using the density-functional concept: beyond the G W approximation. Phys Rev Lett 2005;94:186402.
[66] Friedrich C, Schindlmayr A. Many-body perturbation theory: the GW approximation. NIC Series; 2006. p. 335. vol. 31.

[67] Wing D, Haber JB, Noff R, Barker B, Egger DA, Ramasubramaniam A, et al. Comparing time-dependent density functional theory with many-body perturbation theory for semiconductors: screened range-separated hybrids and the G W plus Bethe-Salpeter approach. Phys Rev Mater 2019;3:064603.
[68] Rohlfing M, Louie SG. Electron-hole excitations in semiconductors and insulators. Phys Rev Lett 1998;81:2312.
[69] Grossman JC, Rohlfing M, Mitas L, Louie SG, Cohen ML. High accuracy many-body calculational approaches for excitations in molecules. Phys Rev Lett 2001;86:472.
[70] Rohlfing M, Louie SG. Electron-hole excitations and optical spectra from first principles. Phys Rev B 2000;62:4927.
[71] Mostofi AA, Yates JR, Lee Y-S, Souza I, Vanderbilt D, Marzari N. wannier90: a tool for obtaining maximally-localised Wannier functions. Comput Phys Commun 2008;178:685–99.
[72] Pizzi G, Vitale V, Arita R, Blügel S, Freimuth F, Géranton G, et al. Wannier90 as a community code: new features and applications. J Phys Condens Matter 2020;32:165902.
[73] Crépieux A, Bruno P. Theory of the anomalous Hall effect from the Kubo formula and the Dirac equation. Phys Rev B 2001;64:014416.
[74] Kawabata A, Kubo R. Electronic properties of fine metallic particles. II. Plasma resonance absorption. J Phys Soc Jpn 1966;21:1765–72.
[75] Vinson J, Rehr J, Kas J, Shirley E. Bethe-Salpeter equation calculations of core excitation spectra. Phys Rev B 2011;83:115106.
[76] Hutchings DC, Sheik-Bahae M, Hagan DJ, Van Stryland EW. Kramers-Krönig relations in nonlinear optics. Opt Quant Electron 1992;24:1–30.
[77] Carcione JM, Cavallini F, Ba J, Cheng W, Qadrouh AN. On the Kramers-Kronig relations. Rheol Acta 2019;58:21–8.
[78] Grosse P, Offermann V. Analysis of reflectance data using the Kramers-Kronig relations. Appl Phys A 1991;52:138–44.
[79] Nagatomi T, Shimizu R, Ritchie R. Energy loss functions for electron energy loss spectroscopy. Surf Sci 1999;419:158–73.
[80] Thomasson S-I. On the absorption coefficient. Acta Acustica united Acustica 1980;44:265–73.
[81] Vrhel MJ, Gershon R, Iwan LS. Measurement and analysis of object reflectance spectra. Color Res Appl 1994;19:4–9.

CHAPTER

19

Chemical modification of Sb thin film

Chi-Hsuan Lee[1], *Shih-Yang Lin*[1], *Vo Khuong Dien*[1], *Hsin-Yi Liu*[1], *Hai Duong Pham*[1], *Thi My Duyen Huynh*[1], *Nguyen Thi Han*[1], *Ngoc Thanh Thuy Tran*[2], *Thi Dieu Hien Nguyen*[1], *Wei-Bang Li*[1] *and Ming-Fa Lin*[1]

[1]Department of Physics, National Cheng Kung University, Tainan, Taiwan; [2]Hierarchical Green-Energy Materials (Hi-GEM) Research Center, National Cheng Kung University, Tainan, Taiwan

Surface states of a topological insulator [1–9] are typically represented by Dirac cones in the energy gap of the bulk electronic band structure. The electron states of the linear crossing bands possess time-reversal symmetry and hence guarantee non dissipative transport against the presence of impurity atoms and geometric perturbations. Potential applications of the topological surface states include spin-polarized transport, quantum computing, and the more subtle manifestation of Majorana fermions [10]. However, the topological insulator has to have large enough dimensions so that the conducting surface states can be unambiguously separated from the bulk states and have all the characters of topological conduction. This requirement in size sometimes means tens of nanometers in thickness that a very thin film does not have.

Thin films, nonetheless, have the advantage of being easier to synthesize and convenient for making miniaturized devices, but the reduced thickness also means increased coupling between the two surfaces, which may reach the point of destroying the Dirac cones and opening a band gap in the surface states. Furthermore, a pristine film synthesized without the contamination of impurity atoms is sometimes difficult to achieve. Even if that is possible, the sample still has to face the possibility of exposure to foreign atoms in some applications [11]. Knowing better how the thickness and adatoms affect the surface electronic structure of the thin film is thus important for studies in and potential applications of topological conduction.

Fundamental Physicochemical Properties of Germanene-related Materials
https://doi.org/10.1016/B978-0-443-15801-8.00024-4

In this regard, experiments performed by Bian et al. have shown that pristine Sb films as thin as four bilayers can be successfully synthesized over a Bi-terminated Si (111) surface [12]. Their angle-resolved photoemission spectroscopy indicates that a Dirac cone is formed close to the Fermi level. However, their calculation based on density functional theory (DFT) also shows that a free-standing Sb film consisting of four bilayers opens a band gap in the surface states due to tunneling-induced coupling between the top and bottom surfaces. Theoretically, one can try to close the gap by increasing the thickness of the film. However, if the thickness of four bilayers is to be maintained, the gap will also disappear by applying hydrogen termination of the Sb film. The latter result clearly points to the crucial role played by the substrate and, in a more general sense, adsorbed atoms in forming and shaping topological surface states of thin films that would otherwise be topologically trivial and therefore warrants a wider investigation in this direction.

19.1 Bilayer Sb systems

In our study on the consequences of the adsorption of foreign atoms, we first wanted to know how different configurations of H adsorption affect the surface states of the four-bilayer Sb film. We used spin-polarized DFT calculations as implemented in the Vienna Ab initio Simulation Package codes [13,14] as our tools. The important spin-orbit interaction was included in the calculation and local-density approximation selected for the exchange-correlation potential. To accommodate various configurations involving H adsorption, an expanded unit cell of 32 Sb atoms was used. Cutoff energy was set at 300 eV and a $9 \times 9 \times 1$ sampling of **k** points in the first Brillouin zone was also adopted. Relaxation of the atomic positions in the Sb films with and without impurity atoms was always executed before relevant energy bands were calculated.

The left part of Fig. 19.1 shows the energy bands of the pristine four-bilayer Sb film along the $\overline{\Gamma} - \overline{M}$ direction, with projected Sb bulk states superimposed in shaded regions. Around the $\overline{\Gamma}$ point, two bands are found between the shaded areas and thus represent surface states

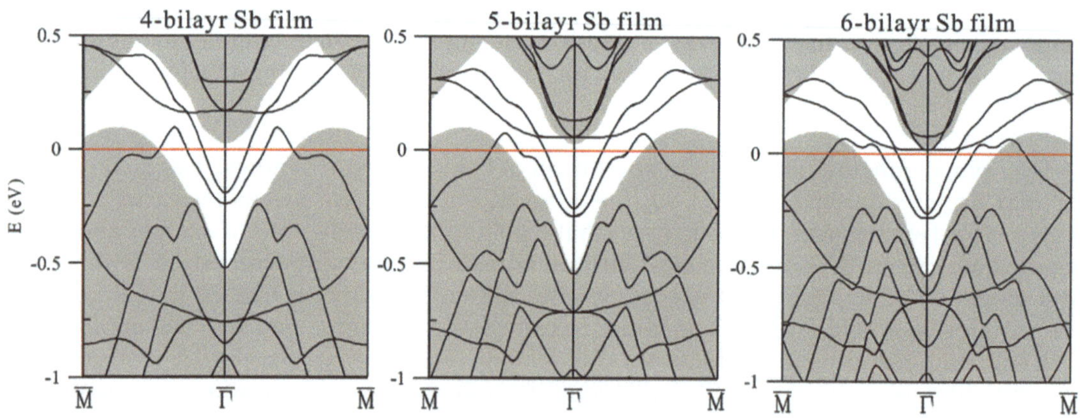

FIG. 19.1 Energy bands of the four-, five-, and six-bilayer Sb films (from left to right) along the $\overline{\Gamma} - \overline{M}$ direction, with projected bulk Sb states superimposed in the shaded area. The band gap shrinks as the thickness of the film increases.

of the Sb film. However, an energy gap of 47.5 meV due to quantum tunneling between the two faces of the film separates the two doubly degenerate bands. If we increased the thickness of the film to five and six bilayers (as represented by the corresponding band structures in the middle and right parts of Fig. 19.1) and thus reduced the tunneling, the gap shrank to 31.1 and 15.8 meV, respectively, in accordance with the reported trend in Ref. [12]. As with the four-bilayer film, each of the two surface bands was doubly degenerate and occupied by electrons with both spin directions.

19.2 The effect of hydrogen concentration on the adsorption

With the adsorption of H atoms, the conduction of the Sb film was drastically different. Three configurations of H adsorption were considered in our calculation. In the first case, a single H atom was attached to a Sb atom in the bottom surface layer of the film in the expanded unit cell of 32 Sb atoms. Shown in Fig. 19.2A, the bond length between the Sb and H is found to be 1.736 Å, and the binding energy is 1.687 eV per H atom per unit cell. Such a firm adsorption indicated strong interaction between the adatom and the host. Most importantly, a Dirac cone was clearly formed as a result, as shown in Fig. 19.2B, where two bands mostly within the gap between the shaded areas cross each other at 0.440 eV below the Fermi level. Analysis of the composition of the wave functions revealed that the two bands were almost entirely composed of electrons from the top surface without the H adatoms. In fact, only in **k** points approaching the Brillouin zone boundary did electrons from the bottom Sb surface with H adatoms make an appreciable contribution to the two bands. This strong suppression of quantum tunneling [12] between the two surfaces of the film destroyed symmetry along the normal to the film and lifted degeneracy of the bands of the pristine film, forming Dirac cones as a results. Moreover, the conduction over the uncontaminated surface layer was then protected by time-reversal symmetry. An analysis of the

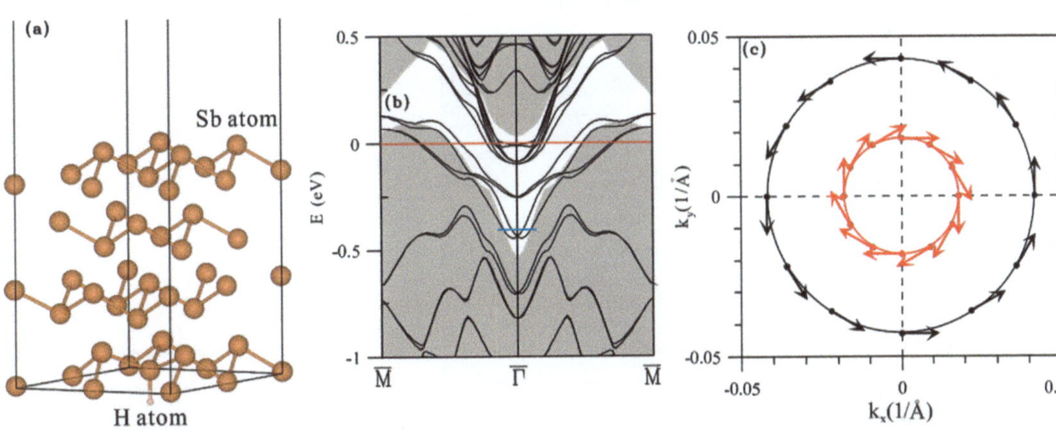

FIG. 19.2 (A) A hydrogen atom is adsorbed on the bottom face of the four-bilayer Sb film in a unit cell of 32 Sb atoms. (B) Energy bands of the configuration shown in (A) have the gap closed and a Dirac cone formed. (C) Spin texture of the configuration is revealed in the k_x - k_y plane by two concentric equi-energy circles corresponding to the energy specified by the horizontal bar in (B).

spin texture [15] of the topological surface states was performed by extracting **k** points of energy equal to −0.406 eV. In Fig. 19.2C the **k** points corresponding to the energy form two concentric equi-energy circles centered at $\overline{\Gamma}$ in the k_x - k_y plane with the spin at each point locked perpendicularly to the direction of the crystal momentum. The fact that each spin angular momentum had no component along the k_z axis and the spin direction at point **k** was totally opposite to that of −**k** assured the complete disappearance of 180° elastic backscattering of the surface electrons and facilitated non dissipative conduction.

If there were more than one H atom adsorbed on the surface per unit cell, the binding energy for each adsorbed H atom increased to 1.907 eV for two H and dropped to 1.872 eV for four H per unit cell in our calculation, and the optimal configurations after the relaxation process had the Sb−H bonds deviated from the normal to the surface. Shown in Figs. 19.3 and 19.4 are the configurations for two H and four H adsorptions in the expanded unit cell as well as the energy bands and spin texture, with each having a Dirac cone between the shaded regions. The figures also show that time-reversal symmetry is still preserved for the surface states, but the projections of constant energy surfaces over the k_x - k_y plane are no longer circles but more like distorted ellipses, reflecting the asymmetry introduced by the additional H adatoms. The spin is no longer necessarily orthogonal to **k** and appreciable out-of-plane components also appear from the calculation.

19.3 3*d* transition metal adsorption

Naturally, a question arose about whether adsorption of atoms other than H would also produce Dirac cones and non dissipative conduction on the Sb film. To address the issue, we further included all 3*d* transition metals (TM) as individual adsorbates in our investigation. In each case, a single TM atom was adsorbed on the top surface of the four-bilayer Sb film in the same expanded unit cell. Each configuration was relaxed, and the corresponding energy bands and spin texture analyzed.

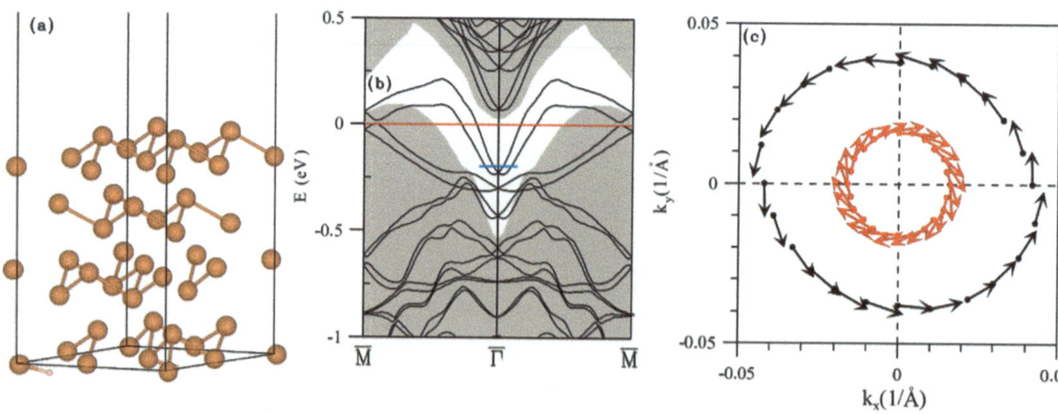

FIG. 19.3 (A) Two H atoms are, respectively, adsorbed on the center and edge Sb atoms of the bottom face of the four-bilayer Sb film. (B) Energy bands of the configuration shown in (A). (C) Spin texture corresponding to the energy specified by the horizontal bar in (B).

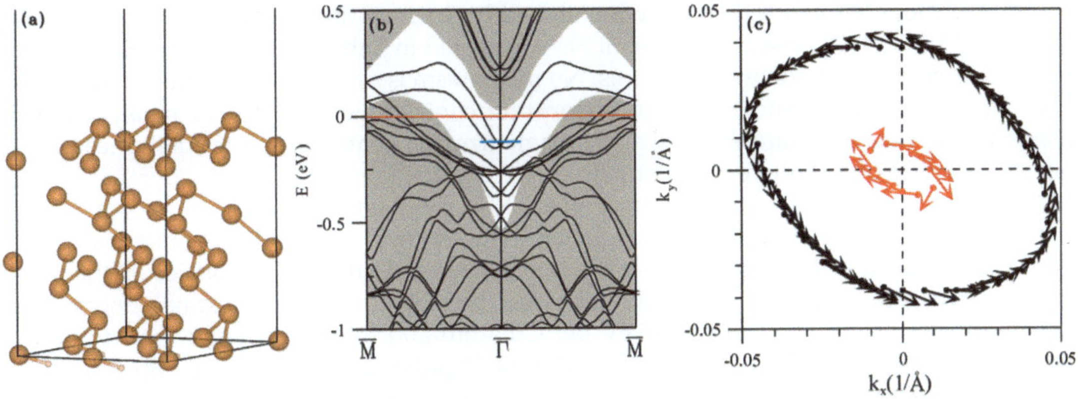

FIG. 19.4 Similar plots as Fig. 19.2, but every Sb atom of the bottom face of the four-bilayer Sb film is connected with a hydrogen atom.

Table 19.1 provides information about the binding energy, charge transfer, and magnetic moment for each of the 3*d* TM adsorbates from the calculation. As in the case of H adsorption, binding energy was derived by subtracting the total energy of the doped Sb film from the sum of total energies of the pristine Sb film and the free TM atom. It is obvious from the listed high binding energies that almost all TM-doped Sb films are robust structures, with the only exception of Zn. Bader charge analysis [16] also points to a trend, with the only exception of Cr, that higher binding energy is generally associated with larger charge transfer between the TM and Sb film, regardless of the sign of charge. Thus, the Sc-adsorbed film has the highest binding energy of 7.890 eV due to the transfer of more than one electron (1.0781 e) from the Sb film to the Sc atom, while Zn is least bonded to the film as a result of the smallest transfer of charge. We also consider the charge transfer of other atoms to compete the results of 3*d* TM atoms (not shown in Table 19.1). The fluorine atom can gain a larger charge (0.681 e) from the

TABLE 19.1 Binding energies E_b, charge transfer ΔQ, and magnetic moments M_{tot} of Sb-film adsorbed by transition-metal atoms.

	Sc	**Ti**	**V**	**Cr**	**Mn**
E_b (eV/atom)	7.890	4.981	3.910	2.747	3.090
ΔQ (e)	1.0780	−0.8140	−0.5005	−0.3101	−0.1140
M_{tot} (μ_B)	0.0003	0.6004	1.2350	4.0493	3.1366
	Fe	Co	Ni	Cu	Zn
E_b (eV/atom)	4.430	6.062	5.113	3.145	0.807
ΔQ (e)	0.2032	0.3211	0.2917	0.0991	−0.0488
M_{tot} (μ_B)	2.0496	0.0021	0.0000	0.0000	0.0000

Sb film. On the other hand, the above-mentioned hydrogen adsorption has a middle charge transfer ($\Delta Q = 0.2143$) close to the case of Fe. Also listed in Table 19.1 are magnetic moments per unit cell for the TM-adsorbed Sb films. No magnetism is associated with the adsorption of a Ni, Cu, or Zn atom, and Sc or Co adsorption contributes only a very small magnetic moment. The other TM atoms, however, magnetize the unit cell substantially, leading to serious violation of time-reversal symmetry.

We first examined the electronic structures of Sb films with nonmagnetic adsorption. A typical example was provided by a single Cu atom, shown in Fig. 19.5A, adsorbed on the top surface of the four-bilayer Sb film in the same expanded unit cell. Relaxation of the whole structure places the Cu atom in a position having three Sb atoms as nearest neighbors. Binding energy was calculated to be 3.145 eV per Cu atom and thus indicated a robust binding. Two Dirac cones (Fig. 19.5B), both centered at the $\overline{\Gamma}$ point and below the Fermi level, were spotted as a result of the Cu adsorption. The one lower in energy was found overwhelmingly to comprise wave functions from electrons in the surface layer without the Cu adatom, and only in **k** points close to the Brillouin zone boundary and of energy above the Fermi level did electrons from the surface with the Cu adatom make an appreciable contribution to the energy bands associated with the cone, proving, once again, suppression of quantum tunneling occurred. The Dirac cone with higher energy had more wave functions from Cu and thus represented more of the surface with Cu adsorption.

It is, thus, feasible, by adjusting the Fermi level, to promote non dissipative conduction along, for example, the $\overline{\Gamma} - \overline{\mathrm{M}}$ direction in the surface layer without the adsorbed Cu. The relevant energy bands are spin polarized and invariant under time reversal. Analysis of spin texture at constant energy confirms the time-reversal symmetry by two projected concentric equi-energy circles centered at the $\overline{\Gamma}$ point on the k_x - k_y plane and spin-locked chiral structure. A typical example is shown in Fig. 19.5C, where the two concentric circles correspond to the energy of 0.360 eV below the Fermi level. Each spin is orthogonal to its associated **k** vector and has no out-of-plane component. For the Dirac cone of higher energy,

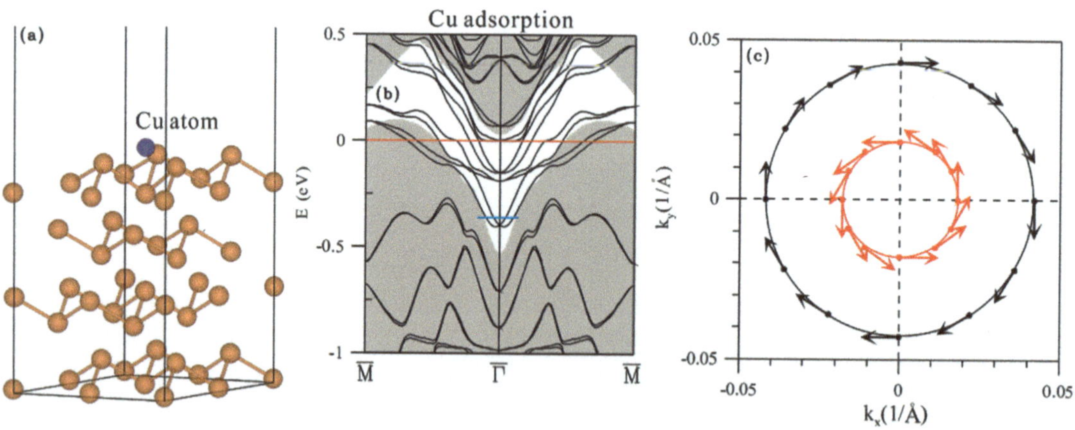

FIG. 19.5 (A) A Cu atom is adsorbed on the top surface of the four -bilayer Sb film in a unit cell of 32 Sb atoms. (B) Energy bands along the $\overline{\Gamma} - \overline{\mathrm{M}}$ direction for the adsorption. Two Dirac cones are formed as a result. (C) Spin texture corresponding to the energy indicated by the horizontal bar in (B).

the spin texture is warped by appreciable spin components out of the plane, but time-reversal symmetry still remains, giving a picture of topological conduction on both surfaces of the film. Fig. 19.6 gives a panorama of energy bands along some symmetry directions of the Brillouin zone. It shows that the contributions to energy bands from the adsorbed Cu atoms, as indicated by the size of circles, are mostly low in energy and of d orbital characters. An exception is found for a band crossing the Fermi level in either the $\overline{\Gamma} - \overline{K}$ or $\overline{\Gamma} - \overline{K}'$ direction comprising a Cu s orbital.

This kind of adsorption-assisted formation of Dirac cones also occurs in the cases of Zn and Ni. Binding energy is a weaker one of only −0.807 eV for Zn, and two Dirac cones are also found (Fig. 19.7A) with the one lower in energy associated mostly with electrons from the Sb atoms of the pristine surface layer. Spin texture analyzed at the energy of −0.225 eV is shown in Fig. 19.7B, revealing the same spin-locked chiral structure at each **k** point. Fig. 19.8A and B are energy bands and spin texture for the Ni-adsorbed Sb film. The Dirac cone corresponding to the pristine surface is higher in energy, and the spin texture is also

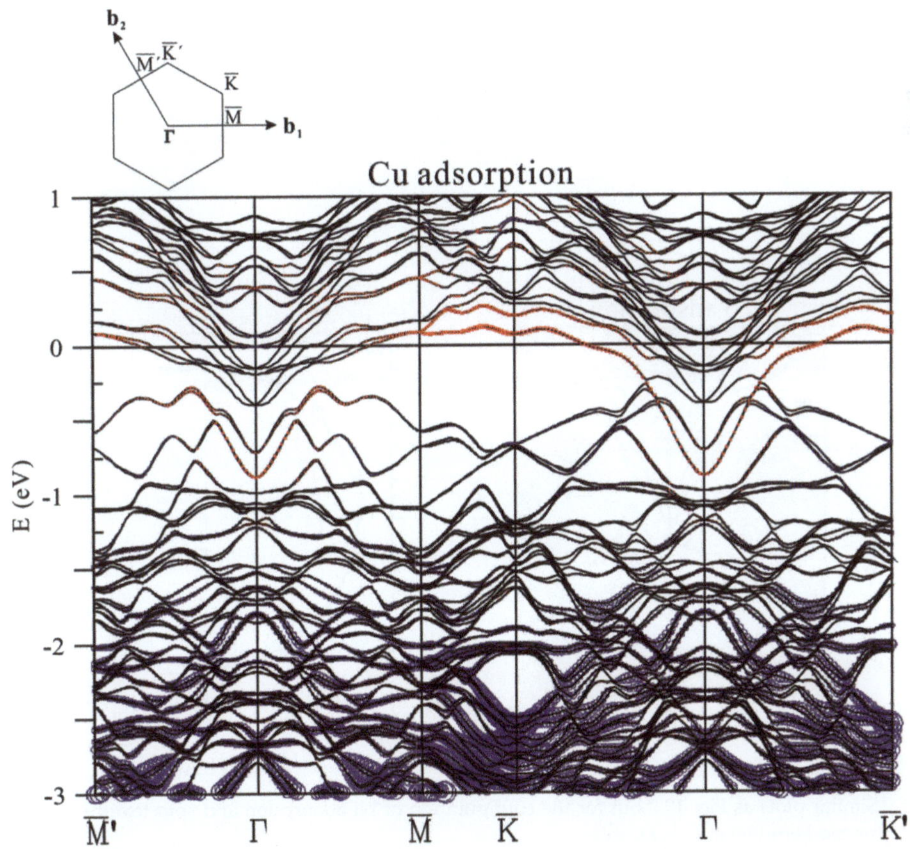

FIG. 19.6 Band structure of the geometrical configuration shown in Fig. 19.5A along symmetry directions of the brillouin zone.

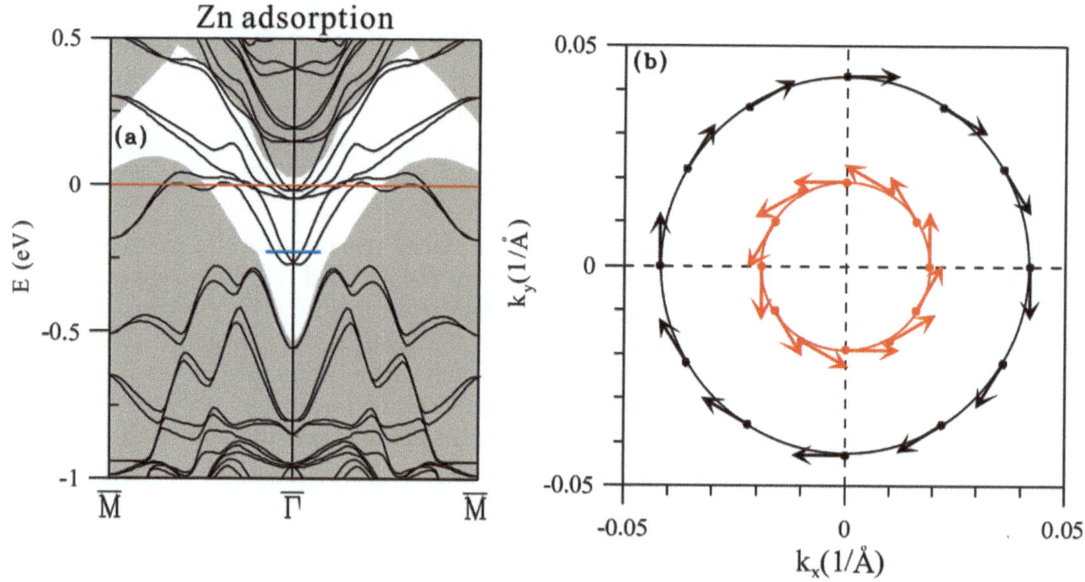

FIG. 19.7 (A) energy bands along the $\overline{\Gamma} - \overline{M}$ direction for the configuration of Zn adsorption (one Zn atom per unit cell) on the four -bilayer Sb film. (B) Spin texture corresponding to the energy indicated by the horizontal bar in (A).

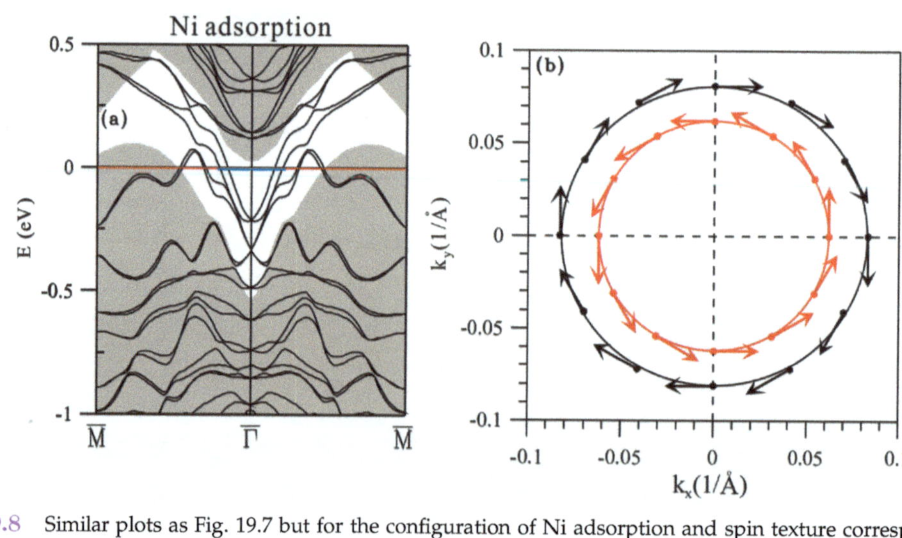

FIG. 19.8 Similar plots as Fig. 19.7 but for the configuration of Ni adsorption and spin texture corresponding to the energy near the Fermi level (−0.005 eV).

analyzed at a higher energy of only 0.005 eV below the Fermi level. Again, spin angular momenta lie completely in the k_x - k_y plane, and time-reversal symmetry is observed for the two bands associated with the Dirac cone on all **k** points. Fig. 19.9A illustrates the distribution of total density of states (DOS) of the Sb film with the Ni adatom. The contribution of the Ni atom can be extracted from Fig. 19.9B, which shows that the local DOS (LDOS) of Ni consists overwhelmingly of 3d orbitals, and its presence at the Fermi level is negligible. Fig. 19.9C is the LDOS taken from a Sb atom on the surface without Ni adsorption. Summed together, the wave functions from surface Sb atoms dominate the bands crossing the Fermi level. It is also a possible scenario that impurity adsorption occurs on both faces of the Sb film.

The four-bilayer Sb film is not a stable structure for such kind of double adsorptions, due to relatively strong quantum tunneling between the two faces. However, in a six-bilayer film deposited on a substrate (simulated by H adsorption as mentioned in Ref. [12]) and with a Cu atom attached to the top surface, we find that Dirac cones are also formed and time-reversal symmetry still observed. The spin texture is warped by large out-of-plane components, but the spin is still orthogonal to **k**, and topological conduction is still in effect on both surfaces.

Finally, we considered the adsorption of magnetic atoms. There have been theoretical and experimental studies on the adsorption or doping of magnetic atoms on bulk topological insulators such as Bi_2Se_3 [17,18] and Bi_2Te_3 [19], where topological characters remained intact despite the disturbance of the magnetic impurities. For the Sb thin film, the situation was different. In the very robust Sc adsorption, the extremely small magnetic moment did not destroy the time-reversal symmetry of the energy bands, as is shown in Fig. 19.10A. But spin texture corresponding to energy at 0.06 eV below the Fermi level (Fig. 19.10B) is greatly distorted from concentric circles. Very large out-of-plane components of spin angular momenta can be found from the figure.

For the other 3*d* TM atoms with large magnetic moments, which are Ti, V, Cr, Mn, and Fe, topological conduction completely disappeared. Three typical examples are selected for discussion.

In the adsorption of a Mn atom per expanded unit cell of the four-bilayer Sb film, a higher magnetic moment of 3.1366 μB and a binding energy of 3.090 eV are obtained from the calculation. Shown in Fig. 19.11A, there are bands converging at the $\overline{\Gamma}$ point, but no Dirac cone is formed. The larger magnetic moment even causes more splitting of energy bands and skewed crossing bands. Analysis of the spin texture also clearly indicates breaking of the time-reversal symmetry on equi-energy curves, with large out-of-plane components at some **k** points. A magnetic moment of 2.0496 μB is associated with Fe adsorption, which has a larger binding energy of 4.430 eV (Fig. 19.11B). There is also serious violation of time-reversal symmetry for the relevant energy bands, providing further evidence that magnetism and topological surface states do not coexist in the situation of magnetic atoms absorbed on the Sb thin film.

Although the magnetic moment of Co adsorption is an order of magnitude larger than that of Sc but still insignificant compared with the other magnetic TM atoms, no Dirac cones are formed with bands extending above the Fermi level, as is evident from the band structure in Fig. 19.11C, making topological conduction impossible.

In conclusion, our calculations have shown that a four bilayer Sb thin film can be turned into a conductor with energy bands possessing time-reversal symmetry by the adsorption of

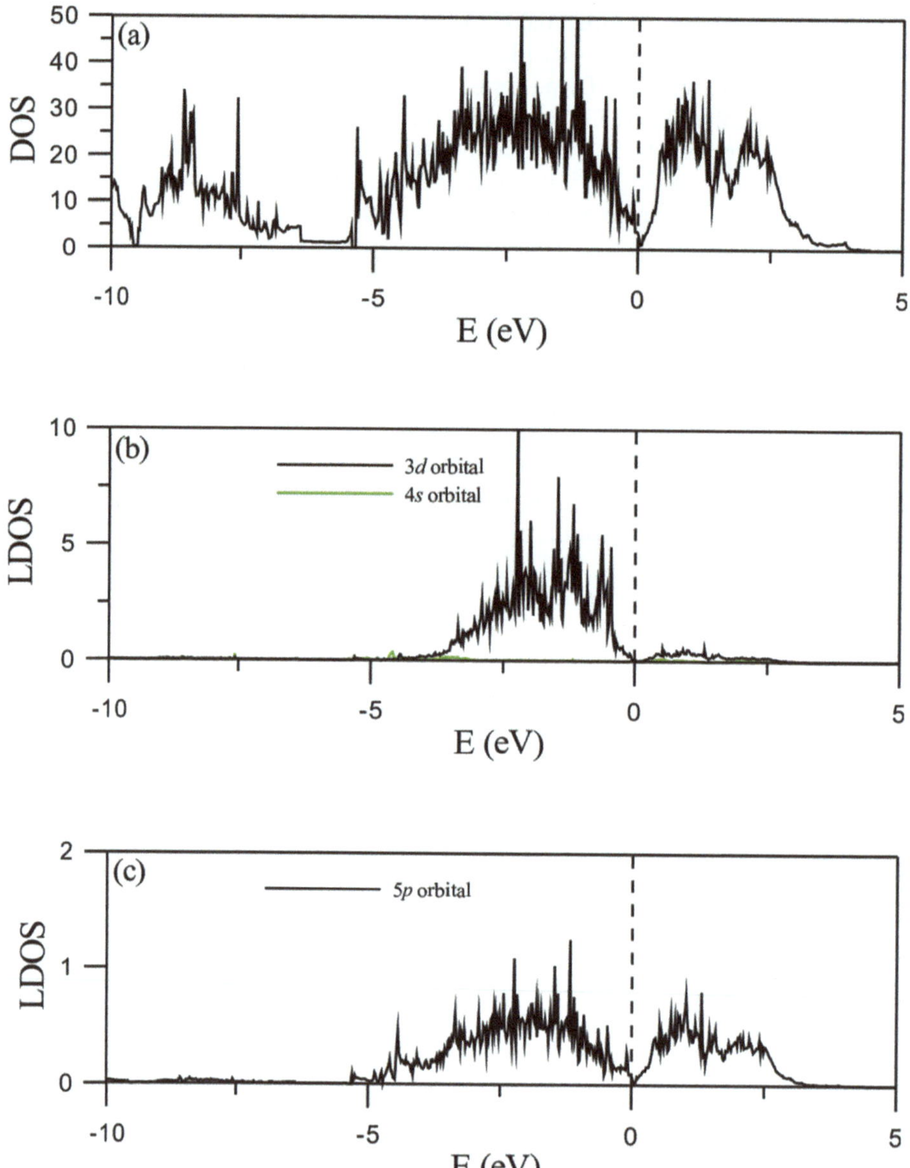

FIG. 19.9 (A) Total DOS of Ni adsorption by the four -bilayer Sb film. LDOS contributed by (B) the Ni adatom and (C) the Sb atom on the pristine surface layer.

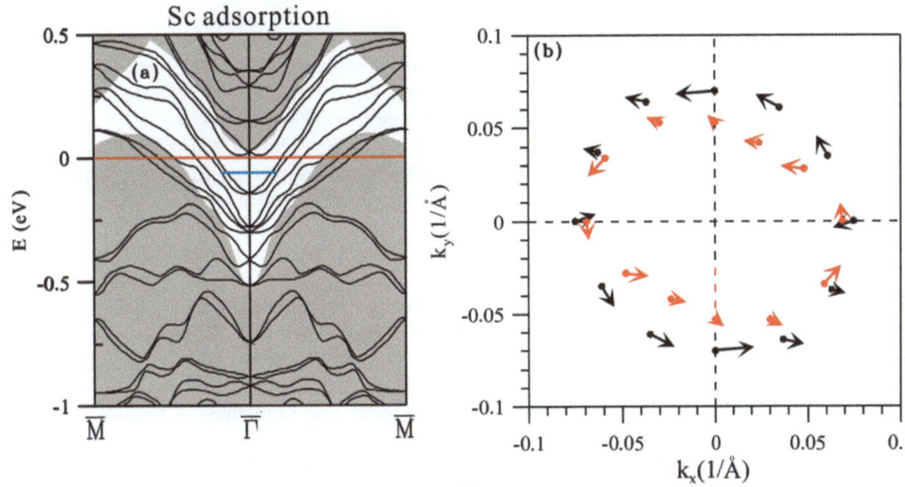

FIG. 19.10 (A) Energy bands along the $\overline{\Gamma}-\overline{M}$ direction for the configuration of Sc adsorption (one Sc atom per unit cell) on the four-bilayer Sb film. (B) Spin texture corresponding to the energy indicated by the horizontal bar in (A).

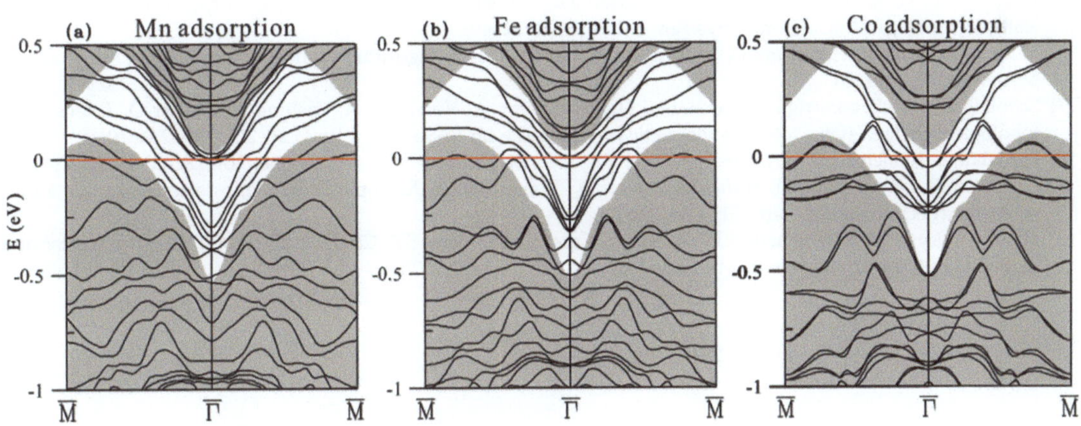

FIG. 19.11 Energy bands along the $\overline{\Gamma}-\overline{M}$ direction for (A) Mn, (B) Fe, and (C) Co adsorptions by the four-bilayer Sb film.

H and nonmagnetic 3*d* TM atoms on its surfaces. Adsorption of magnetic TM atoms other than Sc, however, destroys time-reversal symmetry and topological conduction. We have also shown that a nonmagnetic impurity adsorbed on one surface makes possible the formation of Dirac cones and non dissipative transport by suppressing quantum tunneling between the two faces of the thin film. Instead of growing bulky topological insulators with tens or even hundreds of layers of atoms, one has the option of using the very thin films of only a few layers and obtaining topological conduction at the surfaces of the films. This should promote the application of thin films as vehicles of topological conduction in future research.

References

[1] Kane CL, Mele EJ. Z2 topological order and the quantum spin Hall effect. Phys Rev Lett 2005;95:146802. Kane C L and Mele E J 2005 Quantum Spin Hall Effect in Graphene Phys. Rev. Lett. 95 226801.
[2] Bernevig BA, Zhang SC. Quantum spin Hall effect. Phys Rev Lett 2006;96:106802.
[3] Fu L, Kane CL. Topological insulators with inversion symmetry. Phys Rev B 2007;76:045302.
[4] Fu L, Kane CL, Mele EJ. Topological insulators in three dimensions. Phys Rev Lett 2007;98:106803.
[5] Moore JE, Balents L. Topological invariants of time-reversal-invariant band structures. Phys Rev B 2007;75:121306.
[6] Murakami S. Phase transition between the quantum spin Hall and insulator phases in 3D: emergence of a topological gapless phase. New J Phys 2007;9:356.
[7] Roy R. Topological phases and the quantum spin Hall effect in three dimensions. Phys Rev B 2009;79:195322.
[8] Hasan MZ, Kane CL. *Colloquium*: topological insulators. Rev Mod Phys 2010;82:3045.
[9] Qi XL, Zhang SC. Topological insulators and superconductors. Rev Mod Phys 2011;83:1057.
[10] Wilczek F. Majorana returns. Nat Phys 2009;5:614.
[11] Wang X, Bian G, Miller T, Chiang T-C. Fragility of surface states and robustness of topological order in Bi_2Se_3 against oxidation. Phys Rev Lett 2012;108:096404.
[12] Bian G, Wang X, Liu Y, Miller T, Chiang T-C. Interfacial protection of topological surface states in ultrathin Sb films. Phys Rev Lett 2012;108:176401.
[13] Kresse G, Joubert D. From ultrasoft pseudopotentials to the projector augmented-wave method. Phys Rev B 1999;59:1758.
[14] Kresse G, Furthmüller J. Efficient iterative schemes for ab initio total-energy calculations using a plane-wave basis set. Phys Rev B 1996;54:11169.
[15] Basak S, Lin H, Wray LA, Xu S-Y, Fu L, Hasan MZ, Bansil A. Spin texture on the warped dirac-cone surface states in topological insulators. Phys Rev B 2011;84(R):121401.
[16] Tang W, Sanville E, Henkelman G. A grid-based bader analysis algorithm without lattice bias. J Phys Condens Matter 2009;21:084204.
[17] Scholz MR, Sánchez-Barriga J, Marchenko D, Varykhalov A, Volykhov A, Yashina LV, Rader O. Tolerance of topological surface states towards magnetic moments: Fe on Bi_2Se_3. Phys Rev Lett 2012;108:256810.
[18] Honolka J, Khajetorians AA, Sessi V, Wehling TO, Stepanow S, Mi J-L, Iversen BB, Schlenk T, Wiebe J, Brookes NB, Lichtenstein AI, Hofmann P, Kern K, Wiesendanger R. In-plane magnetic anisotropy of Fe atoms on Bi_2Se_3 (111). Phys Rev Lett 2012;108:256811.
[19] Henk J, Flieger M, Maznichenko IV, Mertig I, Ernst A, Eremeev SV, Chulkov EV. Topological character and magnetism of the Dirac state in Mn-doped Bi_2Te_3. Phys Rev Lett 2012;109:076801.

CHAPTER

20

Substrate effects of two-dimensional materials on few-layer antimony

Chi-Hsuan Lee[1], Shih-Yang Lin[1], Vo Khuong Dien[1], Hsin-Yi Liu[1], Hai Duong Pham[1], Thi My Duyen Huynh[1], Nguyen Thi Han[1], Ngoc Thanh Thuy Tran[2], Thi Dieu Hien Nguyen[1], Wei-Bang Li[1] and Ming-Fa Lin[1]

[1]Department of Physics, National Cheng Kung University, Tainan, Taiwan; [2]Hierarchical Green-Energy Materials (Hi-GEM) Research Center, National Cheng Kung University, Tainan, Taiwan

Topological insulators [1−10] have conducting surface states that are separated from the insulating bulk states. The surface states form Dirac cones with time-reversal symmetry in the energy gap of the bulk electronic bands. The directions of their associated crystal momenta are locked to the electron spins in a way that prevents back scattering of electrons irrespective of the presence of non-magnetic impurities or geometric disturbances. Topological insulators are very valuable in applications requiring non-dissipative or spin-polarized transports. In this regard, an important question concerns the physical sizes of topological insulators. If a topological insulator is to be integrated into a nanoscale circuit, it has to be shrunk in at least one dimension. However, there is a limit on how far the topological insulator can be reduced. Below a certain thickness quantum tunneling between the two surfaces of a free-standing film would open a gap and destroy the Dirac cone [10].

A thin film deposited on a substrate, however, may circumvent the effect of the tunneling between the two surfaces of the film. Although a free-standing four-bilayer Sb (111) film has a large tunneling-induced energy gap, the same film supported by a Bi-terminated silicon substrate forms a Dirac cone at the Brillouin zone center, as was demonstrated by the angle-resolved photoemission experiment [11]. The thin Sb film [11−13] in contact with heterogeneous atoms in the substrate can be simulated theoretically as an adsorption of impurity atoms on one film surface. According to calculations based on density functional theory

Fundamental Physicochemical Properties of Germanene-related Materials
https://doi.org/10.1016/B978-0-443-15801-8.00007-4

(DFT), there is substantial charge transfer between the adsorbed hydrogen or non-magnetic transition-metal atoms and the film, and the energy gap is closed as a result [11,14]. In order to better understand the interaction between thin films of a topological insulator and their substrate, we set out to investigate the electronic structure of the Sb film deposited on a layer of heterogeneous atoms. Bonding between the film and the layer and the interface states formed as a result should offer useful information about the feasibility of producing topological conduction for such configurations, which are more realistic in practice and can be used in more general applications.

20.1 Essential properties of Sb (111) substrates

Our first calculation dealt with Sb (111) films in contact with a single layer of Bi. We used the VASP [15,16] codes for this and subsequent calculations. General features included spin–orbit interaction, local density approximation for exchange–correlation functionals, and a cutoff energy of 400 eV for expanded plane waves. Sampling of the Brillouin zone was usually done with a $9 \times 9 \times 1$ mesh of **k** points. Relaxation of atomic positions to minimize the strain was always executed before the calculation of energy bands. The vacuum layer for the relaxation process has a height of 18 Å perpendicular to the surface.

Graphene and hexagonal boron nitride (BN) monolayers are two-dimensional (2-D) materials with the same honeycomb structures, but the electronic properties of the two are hugely different. Graphene has the p_z orbital perpendicular to the atomic plane and crossing linear energy bands near the Fermi level [17]. It leads to unconventional quantum Hall effects observed in transport properties. In contrast, the h-BN layer is an insulator with a wide energy gap [18]. Furthermore, it is also possible to construct a wide variety of hybrid 2-D structures out of the C and BN components, because of the similar lattice parameters between graphene and h-BN. From the previous Experimental and theoretical studies, the electronic properties of graphene can be turned into either n type or p type by replacing individual C atoms with N or B atoms respectively [19]. The single-layer materials, such as Bi, graphene, boron nitride, and boron-doped graphene of different concentration, are chosen to be the candidates of substrates with few-layer Sb (111) films. The effect of defect in single-layer Bi is also studied.

20.2 Substrate effects of few-layer antimony

Fig. 20.1A shows the geometrical structure of a four-bilayer Sb (111) film in contact with a single layer of Bi, consisting of 32 Sb and 8 Bi atoms in the 2×2 supercell. As the experiment [11] has concluded, Fig. 20.1B displays that a Dirac cone at approximately 0.33 eV below the Fermi level is formed in the gap of the projected bulk energy bands represented by shaded areas. The states of the cone are mostly made of electrons from the surface which is not in contact with Bi (open black circle in Fig. 20.1B). The contribution of single-layer Bi film is mainly observed at the bands near the Fermi level, as shown with green solid circle. Not surprisingly, Dirac cones also form in Sb films having more bilayers. Fig. 20.1C presents the band structure of a four-bilayer Sb (111) film in contact with a Bi layer, and Fig. 20.1D stands for that of a five-bilayer Sb film. It is clear that increasing Sb layers does not change the fact that a Dirac cone exits in the gap region of bulk Sb. Apart from that, even if the Bi layer has vacancy defects, the whole structure is still able to produce Dirac cones and topological

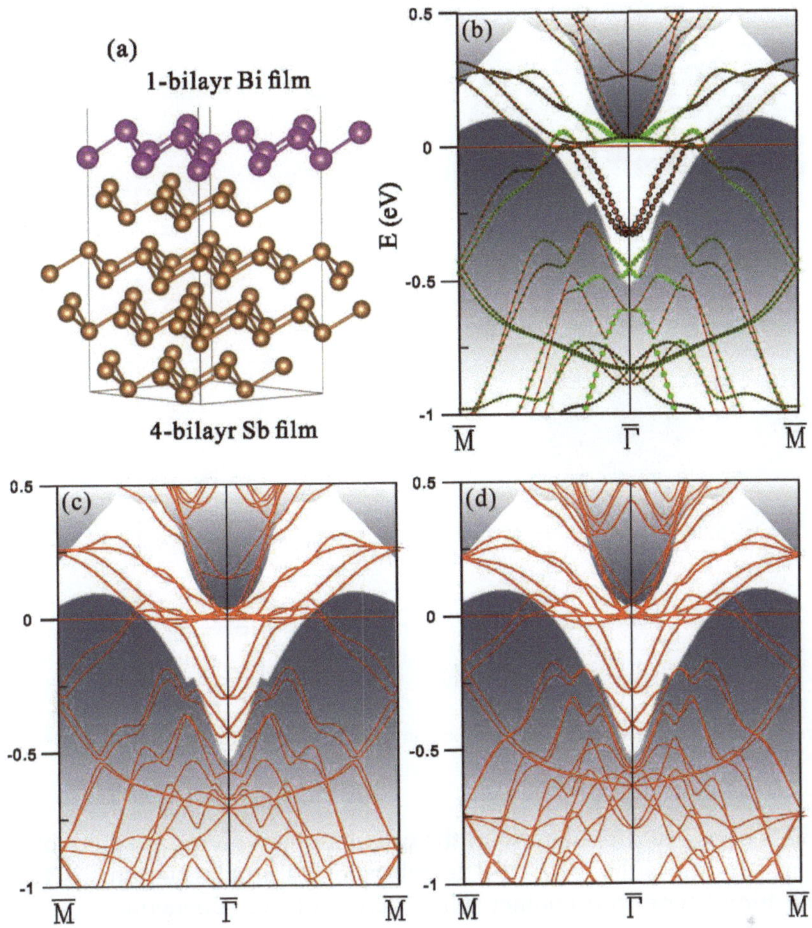

FIG. 20.1 Energy bands along the $\overline{\Gamma} - \overline{\mathrm{M}}$ direction of a (B) three-bilayer, (C) four bilayer, (D) five-bilayer Sb (111) film with a single layer of Bi atoms, and (A) the geometrical configuration of four-bilayer Sb (111) film in contact with a single layer of Bi atoms. Solid green (open black) circles indicate contribution from interface (pristine surface) electrons.

conduction. The feature also stands in the two types of vacancy defects. In order to study the difference between the two type of defects, the above-mentioned larger unit cell, simultaneously containing 24 Sb atoms from the three-bilayer Sb (111) film and 7 Bi atoms from defective Bi monolayer, is considered.

In Fig. 20.2, a layer of Bi that is greatly distorted due to some absent Bi atoms is shown over the Sb (111) film. A Dirac cone is still preserved in the gap of projected bulk bands of Sb. Also shown is the spin texture [20] in the k_x - k_y plane at 0.28 eV below the Fermi level. Two equi-energy circles presenting spin-locked crystal momentum vectors confirm time-reversal symmetry.

Another distribution of defects results in the less distorted Bi layer shown in Fig. 20.3, where a Dirac cone and its associated spin texture at −0.19 eV are also drawn. It is obvious that defects in the Bi layer are not a factor affecting topological conduction of the composite

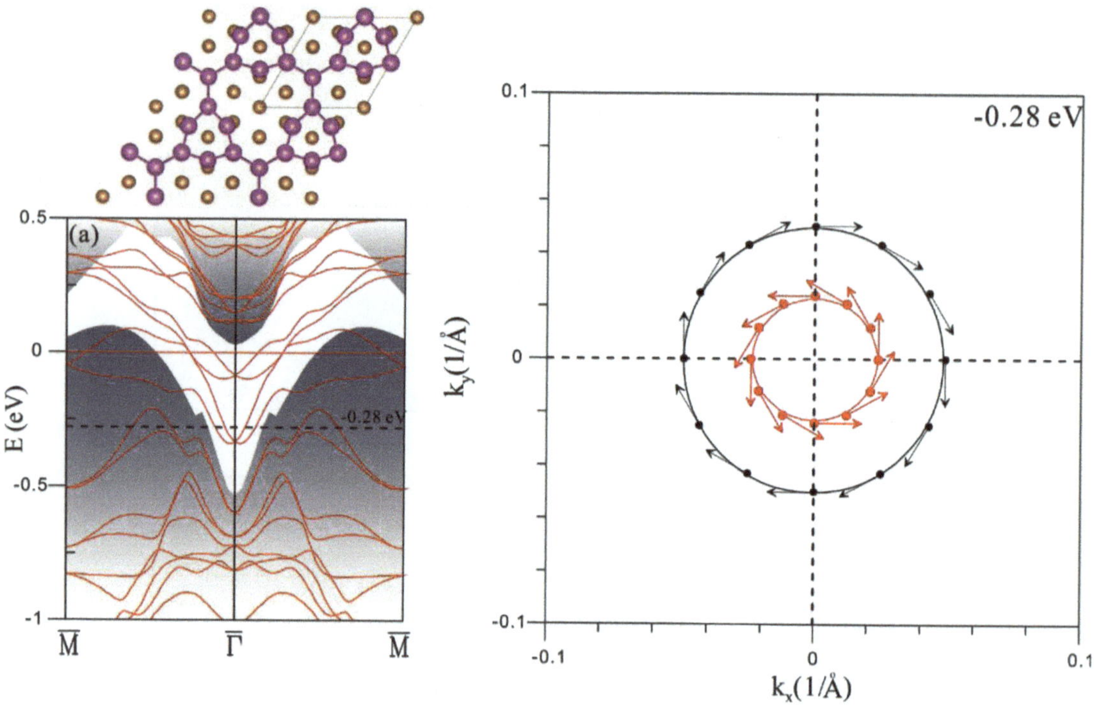

FIG. 20.2 Energy bands and spin texture of a three-bilayer Sb (111) film in contact with a more distorted Bi layer with a vacancy defect.

structure. This is certainly consistent with the calculation involving isolated impurity adsorptions of Sb films. These results reveal the flexible role Sb thin films assume in topological conduction if the films come into contact with a layer of heterogeneous atoms. It should be stressed that we continue to use the term 'topological conduction' here and thereafter as an extension of the description of the bulk topological insulator Sb.

We would like to exploit further the capability of the three-bilayer Sb (111) film, which is probably the lower limit in thickness for most applications in topological transport, by placing it in contact with other materials. Graphene in particular is made up of a single layer of carbon atoms, and the deposition of such light atoms on the Sb film was the target of our next investigation. We employed the same large unit cell as was used in the investigation of the distorted Bi adlayer contact, consisting of 24 Sb atoms from the three-bilayer Sb film and 24 C atoms from graphene. There were, however, several possible configurations for the composite structure. An Sb atom can occur directly over a C atom (top site), perch on the middle point of the bond between 2 C atoms (bridge site), or sit right on the center of the hexagon of C atoms (hollow site). Calculation indicated that the configuration with Sb atoms at hollow sites is energetically most favorable, followed by the bridge-site contact. Binding energies (per unit cell) for the three configurations, given in Table 20.1, differ little from one another. Band structures, however, show more distinct features for each of the three configurations. Overall, the Dirac cone of a free-standing graphene is folded back to the Brillouin zone center as a

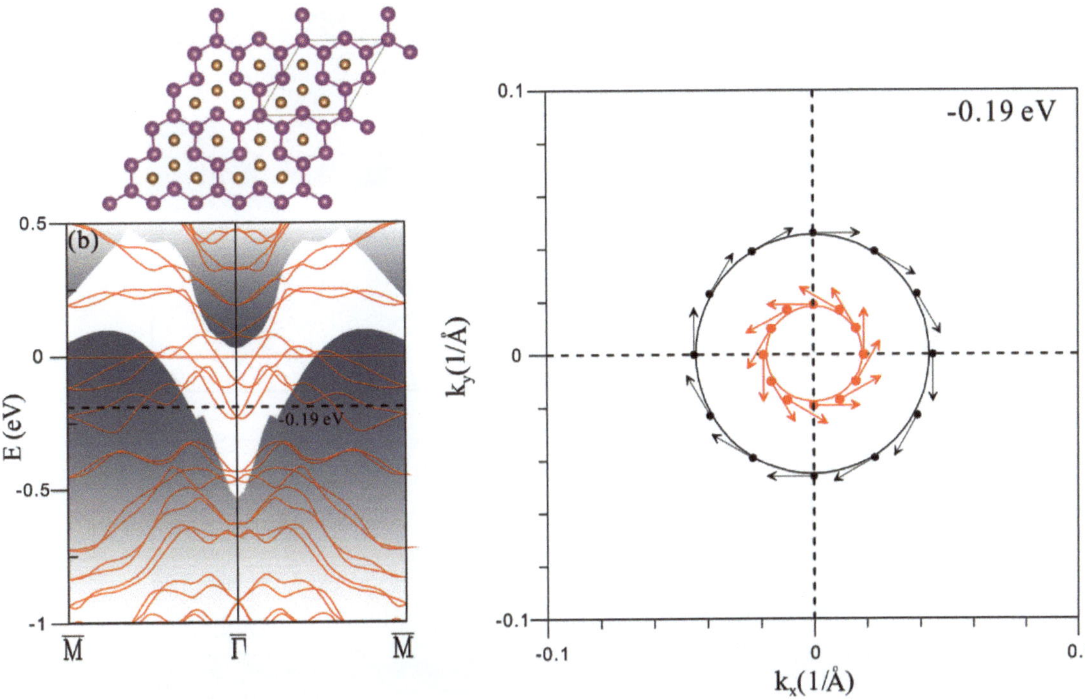

FIG. 20.3 Similar plots as Fig. 20.2, except that the Bi layer with a vacancy defect is less distorted.

TABLE 20.1 Binding energies E_b and charge transfers of a three-bilayer Sb (111) film in contact with a single layer of atoms.

Layer contact	E_b (eV)	Electrons transferred to the layer contact (e)
Graphene (bridge sites)	−0.467	0.124
Graphene (hollow sites)	−0.512	0.138
Graphene (top sites)	−0.444	0.136
BC_3 (bridge sites)	−1.452	0.630
BC_3 (hollow sites)	−0.991	0.462
BC_3 (top sites)	−1.394	0.550
BN (top sites)	−0.492	0.084

result of the deposition, as is shown in Fig. 20.4A–C, corresponding to bridge, hollow, and top site binding, respectively. Although the binding with the Sb film is relatively weak, it is enough to generate an energy gap between the two tips of the cones due to uneven

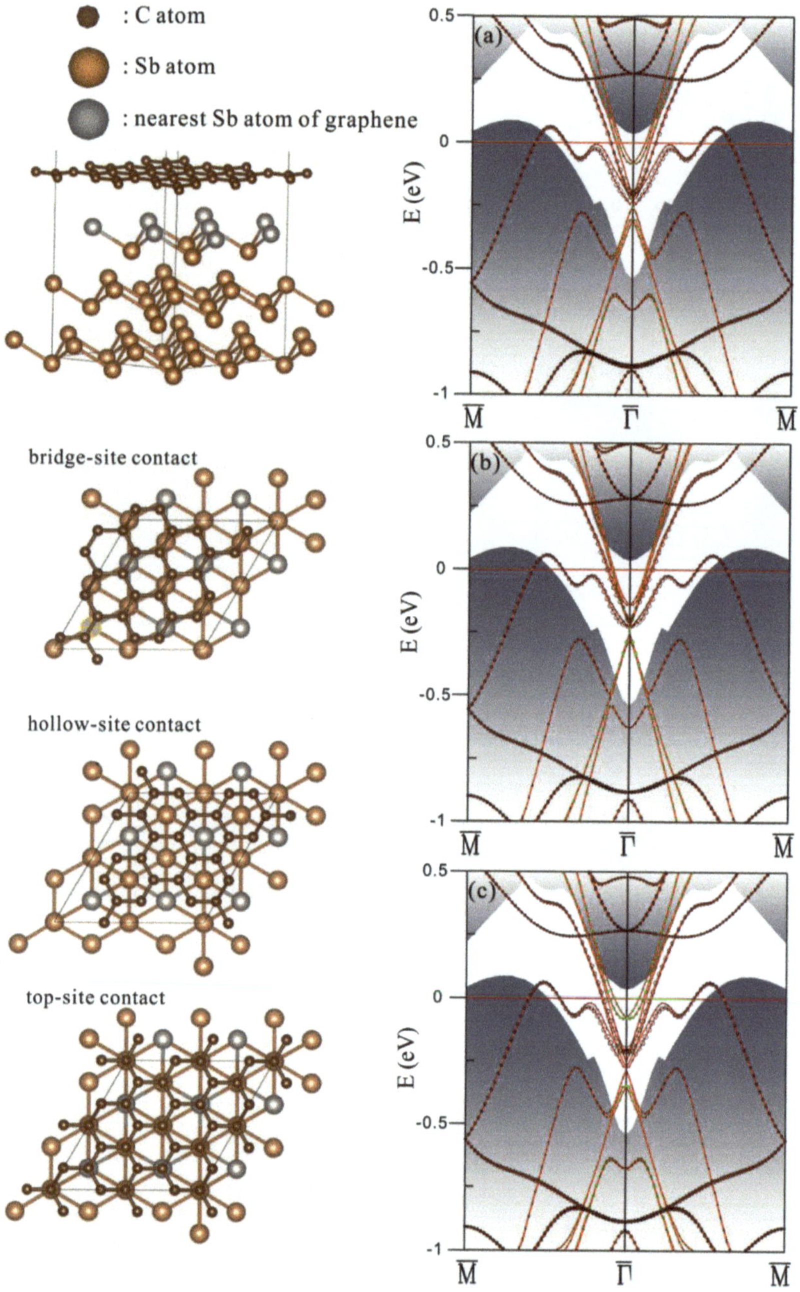

FIG. 20.4 Configurations and energy bands of a three-bilayer Sb (111) film in (A) bridge-site (B) hollow-site (C) top-site contact with graphene. Solid green (open black) circles indicate contribution from interface (pristine surface) electrons.

interaction between the C and Sb atoms, with some C atoms being closer to Sb than others. Interestingly, it is the two weaker bindings that produce Rashba spin splitting in the energy bands, as in Fig. 20.4A and C. The bands with green circles are largely made of electrons from interface Sb and C atoms. Apparently, Sb atoms at bridge sites or on top of C atoms interact with C more strongly and result in more splitting of energy bands. Bands with black circles are mostly from Sb atoms on the other surface of the film.

If one substitutes boron atoms for some of the C atoms in the graphene layer, different electronic structures are expected due to the change in bonding and different number of valence electrons. As an example we display five calculated energy bands in Fig. 20.5, along the $\overline{\Gamma} - \overline{\mathrm{M}}$ direction in the first Brillouin zone, of a three-bilayer Sb (111) film in top-site contact with a B-doped graphene sheet, with the latter evolving from a lower concentration of B (1/24) to BC_3. The increase in B atoms alters the shapes and distributions of bands and realigns the position of the Fermi level accordingly. Dirac cones, nonetheless, are present in all five configurations. In the case of BC_3 contact, for example, two Dirac cones are present and two energy values are chosen for spin texture analysis. For 0.07 eV above the Fermi level two equi-energy circles in the $k_x - k_y$ plane are shown in Fig. 20.6A, with the inner one corresponding to states mostly contributed by Sb atoms (blue circles in left panel) at the interface and the outer one representing states largely from the other surface of the Sb film. Both energy circles are spin-polarized and have spin direction perpendicular to the electron momentum. At 0.14 eV below the Fermi level, two equi-energy circles on the $k_x - k_y$ plane are also present (Fig. 20.6B). However, spin directions are reversed at **k** points in each energy circle of the cone relative to those of Fig. 20.6A, and states are mostly made of orbitals contributed by the Sb atoms on the pristine surface. This clearly shows that transport through the film can either occur via the mixture of the interface states due to the BC_3 contact and pristine surface states of Sb atoms, or from the latter only. By gating the film with a voltage it is possible to selectively choose either mode of transport.

Fig. 20.7 shows the evolution of energy bands for hollow-site contact. The Fermi level tends to be lower in energy for the same sequence of B substitutions, implying that the substrate contains fewer electrons. In fact, analysis of the Bader charge transfer [21] between the BC_3 layer and the Sb film has indicated that the hollow-site contact has the fewest electrons going into the Sb film, followed by the top-site contact. Listed in Table 20.1, the bridge-site contact induces the largest charge transfer and hence has the strongest binding between the film and the BC_3 layer. The main reason why bridge-site and top-site contacts have larger charge transfers is simply due to the closer contact between atoms in the two configurations.

We present the full band structure of the three-bilayer Sb (111) film in bridge-site contact with BC_3 in Fig. 20.8. The size of blue (black) circles indicates the contribution interface (pristine surface) electrons make to the wave function of point **k**. For the Dirac cone near the Fermi level, wave functions corresponding to **k** points close to the $\overline{\Gamma}$ point are largely made of electrons from Sb atoms of the interface layer. At **k** points some distance away from the origin, one band continues to be dominated by the interface electrons, while the other switches to the electrons from the pristine surface. Around 0.25 eV below the Fermi energy, two bands also cross each other at the $\overline{\Gamma}$ point. Both bands are heavily made of electrons from the Sb atoms of the pristine surface. The total density of states of the system is drawn in Fig. 20.9A. Decomposed into individual atoms, local density of states (LDOS) of boron atoms in Fig. 20.9B indicate that they contribute negligibly to the valence bands of the whole structure, independent

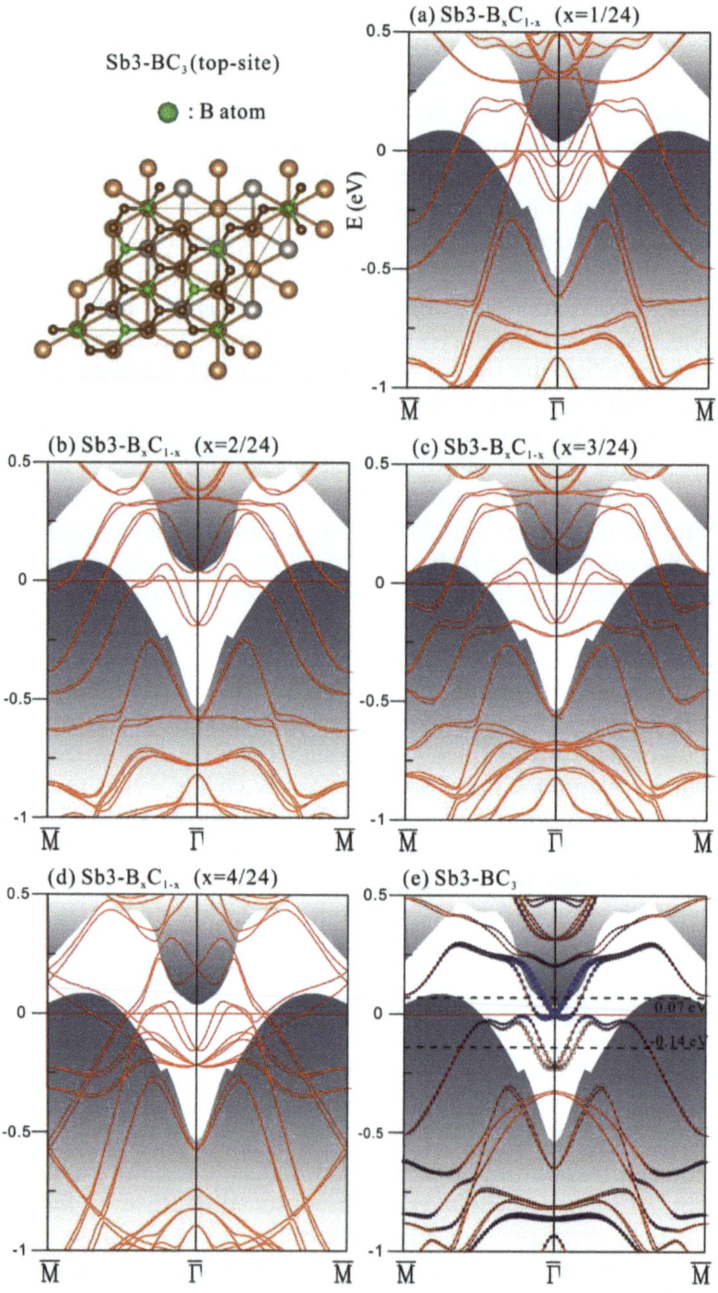

FIG. 20.5 (A)–(E) evolution of energy bands of a three-bilayer Sb (111) film in top-site contact with B-doped graphene from the concentration of 1/24 to BC_3. Solid blue (open black) circles in (E) indicate contribution from interface (pristine surface) electrons.

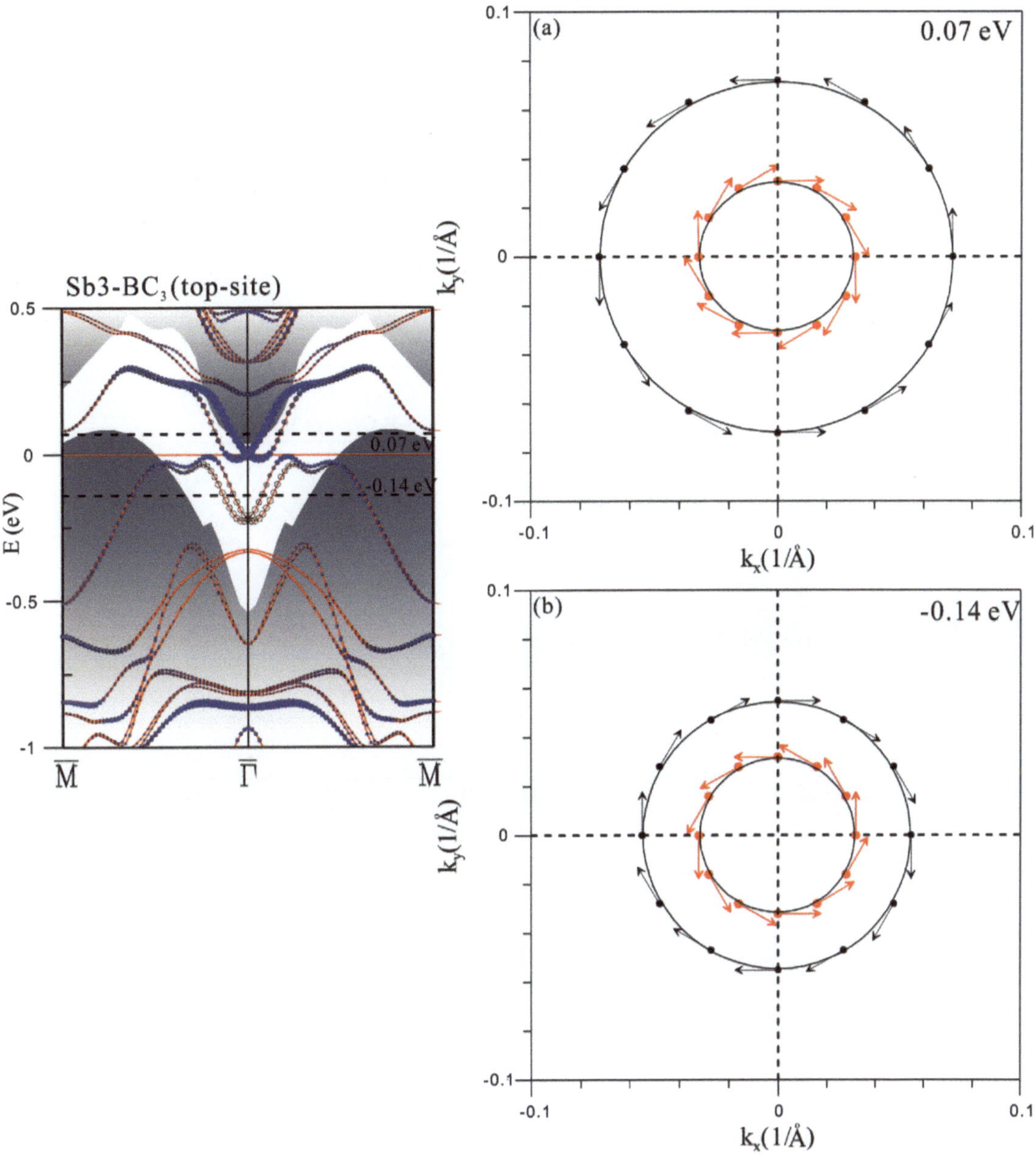

FIG. 20.6 Spin texture of the three-bilayer Sb (111) film in top-site contact with a single layer of BC_3 is shown on the $k_x - k_y$ plane, corresponding to energy of (A) 0.07 eV above the Fermi level and (B) 0.14 eV below the Fermi level.

of whether they are closer to (red curve) or farther away from (black curve) the surface Sb atoms. Fig. 20.9C shows the contributions from an Sb atom in close contact (red curve) with a boron atom and an Sb in the pristine surface layer. They provide interface and surface states to the valence bands just as Fig. 20.7 has explained.

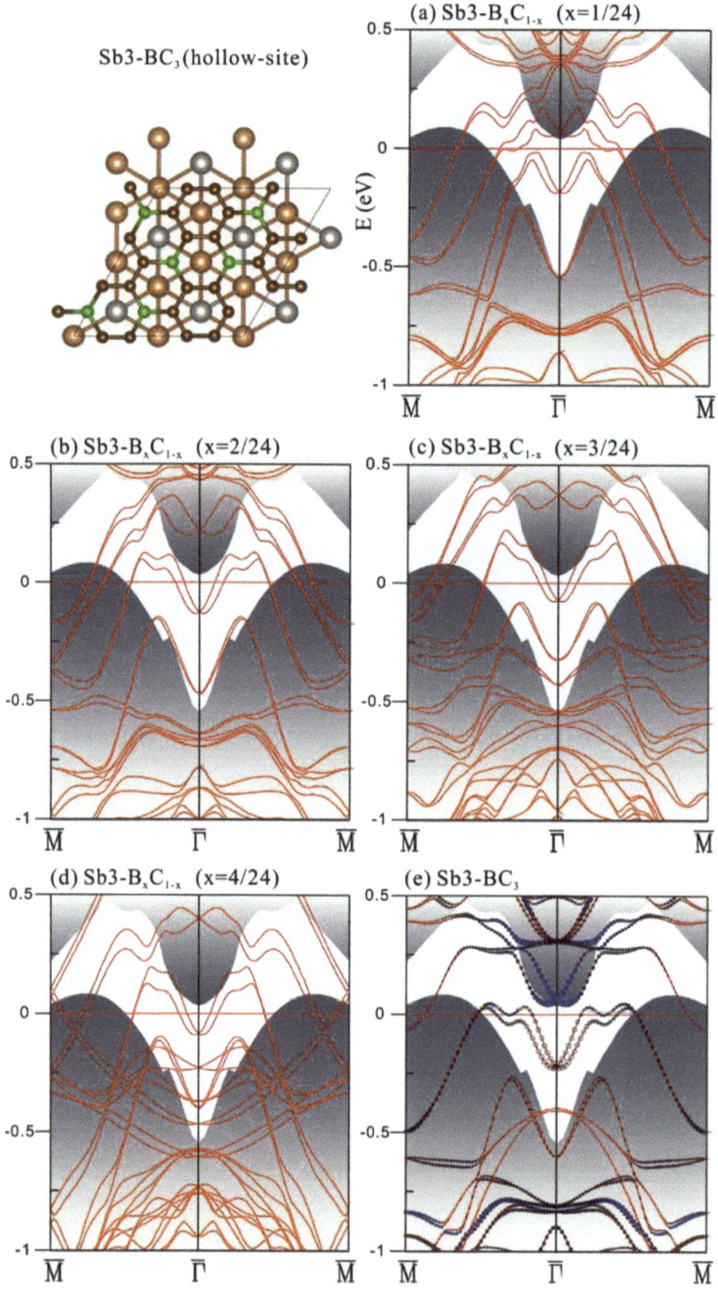

FIG. 20.7 Similar plots as Fig. 20.5 with the three-bilayer Sb (111) film in hollow-site contact with B-doped graphene.

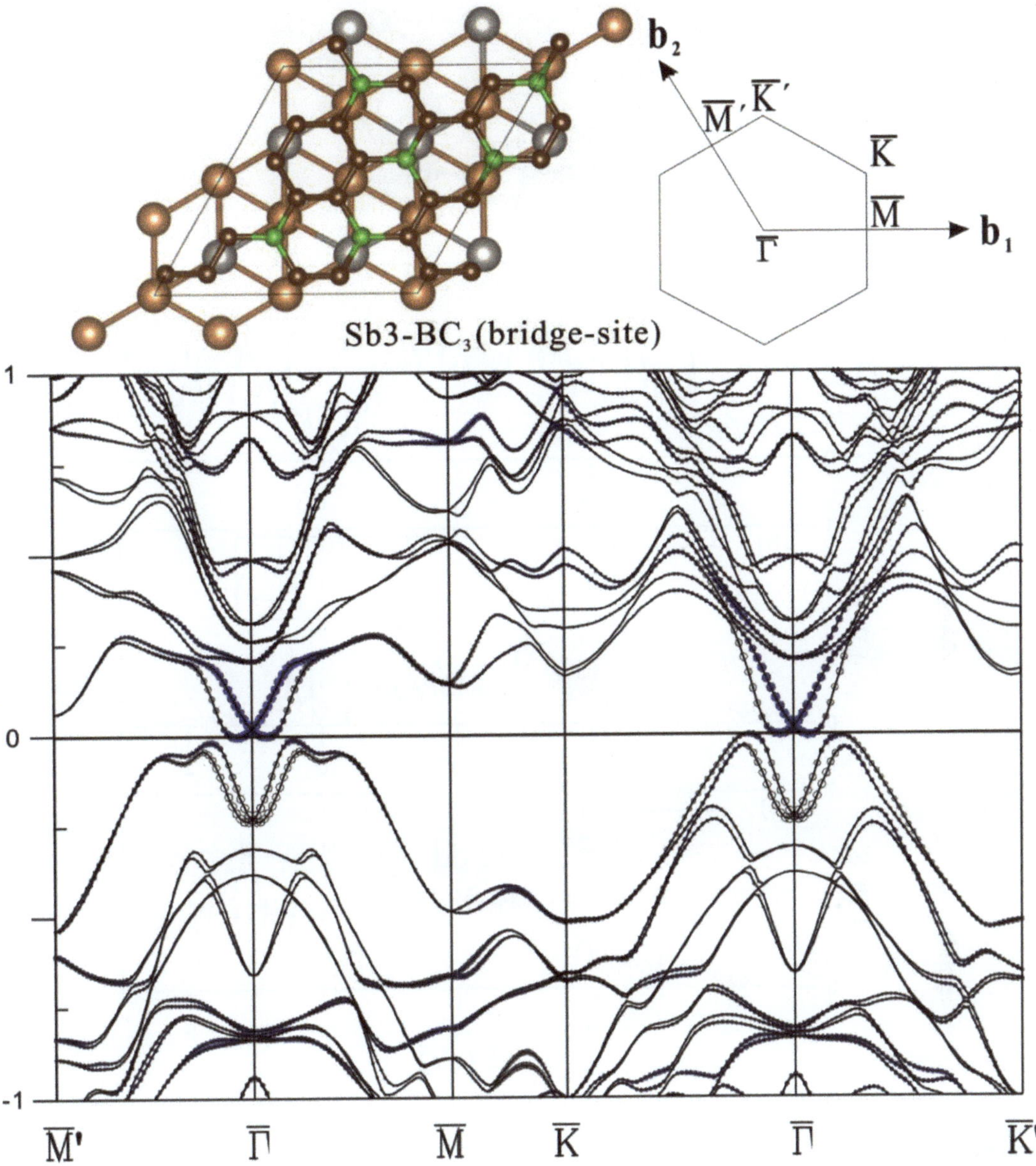

FIG. 20.8 Band structure of a three-bilayer Sb (111) film in bridge-site contact with a single layer of BC_3 along symmetry directions in the brillouin zone. Blue (black) circles indicate contribution from interface (pristine surface) electrons.

Three energy values are chosen for the analysis of spin-dependent transport of the three-bilayer Sb film in bridge-site contact with BC_3. At 0.1 eV above the Fermi level (Fig. 20.10A) two equi-energy circles of opposite helix are obtained on the k_x - k_y plane, each having spin direction locked perpendicularly to the momentum. At 0.02 eV, which is exactly the energy

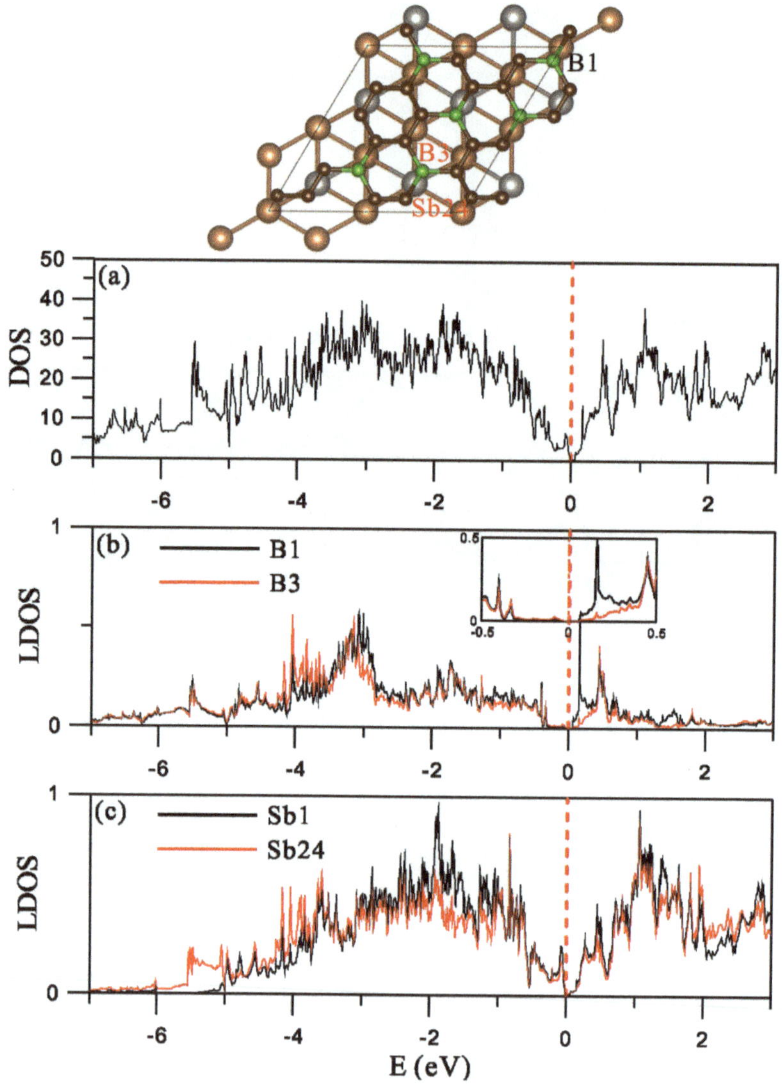

FIG. 20.9 (A) Total density of states derived from Fig. 20.8. (B) LDOS of B atoms B1 and B3, with the latter being a nearest neighbor of an Sb atom. (C) LDOS of Sb atoms Sb1 and Sb24. Sb1 is in the pristine surface layer, while Sb24 is a nearest neighbor of B3.

level of the Dirac point (Fig. 20.10B), only one circle is obtained. With only one mode of transport, the current along the $\overline{\Gamma} - \overline{\mathrm{M}}$ direction is 100% spin-polarized with no possibility of back scattering, and the conductance of one quantum unit (e^2h^{-1}) is only half that of the two-mode cases. If the energy is again shifted downward to, say, 0.008 eV above the Fermi level (Fig. 20.10C), another two-mode transport appears with complete spin polarization. It is thus possible to switch back and forth between a completely spin-polarized and unpolarized transport by gating the film with continuously adjustable voltage.

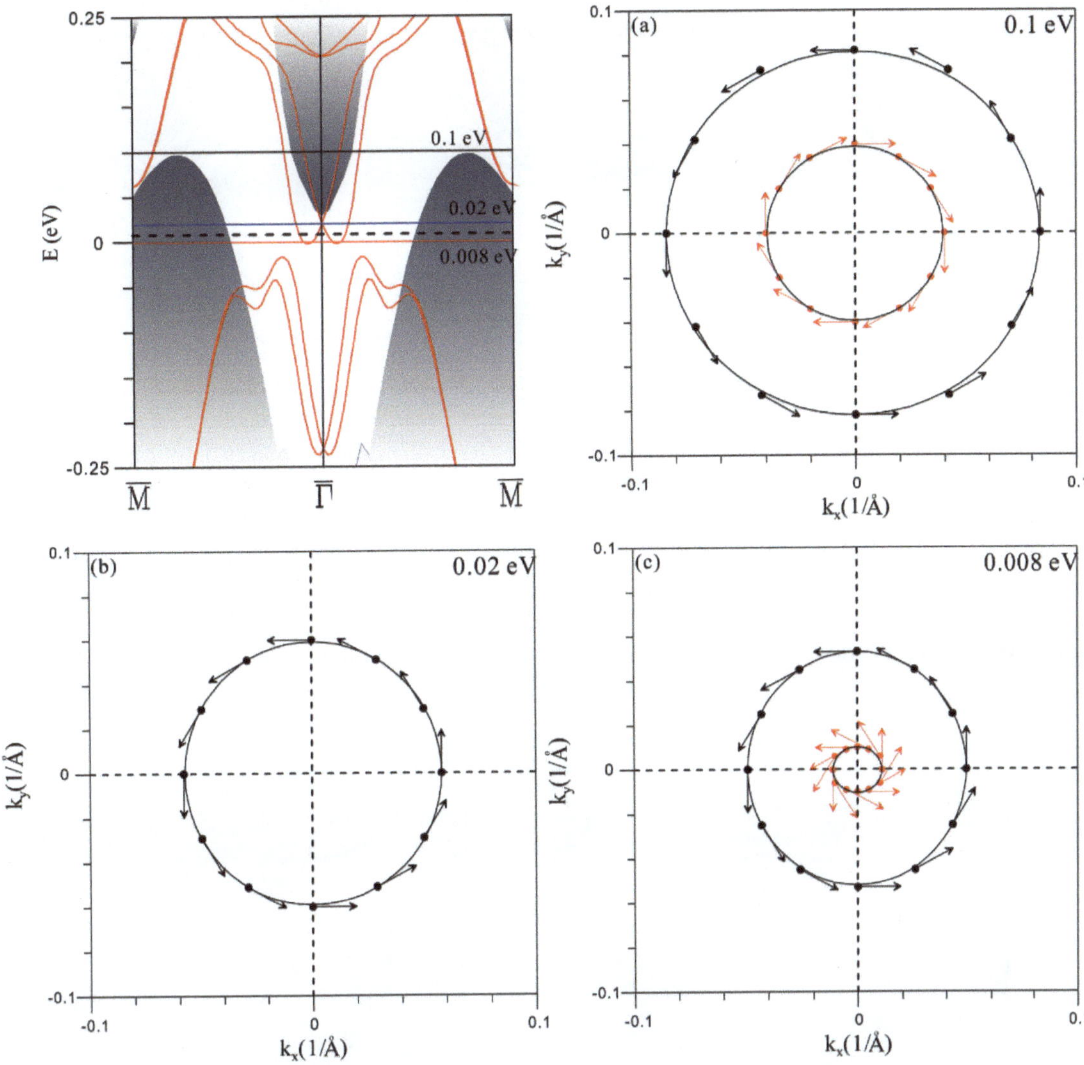

FIG. 20.10 Spin texture of the three-bilayer Sb (111) film in bridge-site contact with a single layer of BC_3 is shown on the $k_x - k_y$ plane, corresponding to energy of (A) 0.1 eV (B) 0.02 eV and (C) 0.008 eV above the Fermi level. There is only one mode of transport in (B).

We also investigated the dependence of topological conduction on the thickness of Sb film by increasing the number of bilayers from four to six. Their band structures along the $\overline{\Gamma} - \overline{M}$ direction are drawn in Fig. 20.11A–C for top-site contact of BC_3 respectively. It is clear that Dirac cones and topological conduction are also present with thicker structures, with the upper cone more related to interface states and the lower one to surface states of the pristine face. There is therefore certain flexibility in synthesizing Sb films for the purpose of topological conduction.

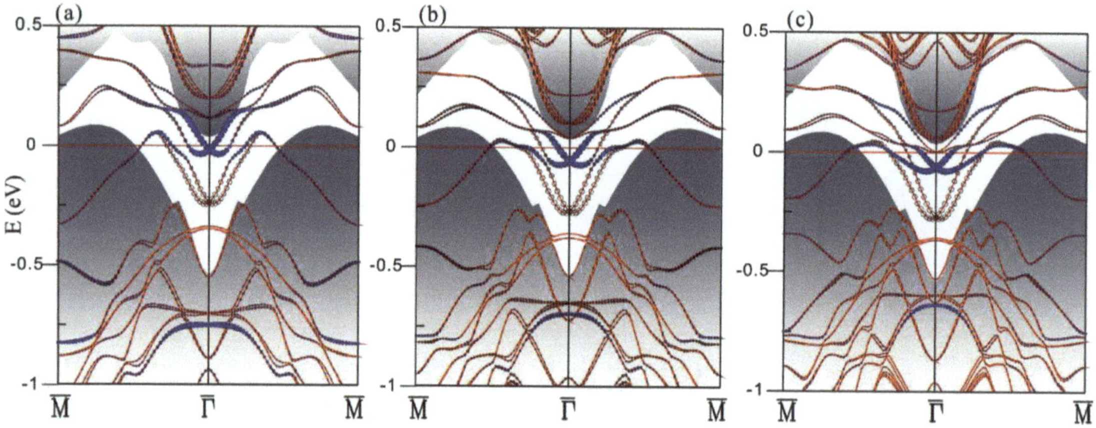

FIG. 20.11 Energy bands of an Sb (111) film with (A) four bilayers, (B) five bilayers, and (C) six bilayers in top-site contact with a single layer of BC_3. Blue (black) circles indicate contribution from interface (pristine surface) electrons.

As a final note, we briefly mention the calculation of three-bilayer Sb (111) film deposited on a single layer of BN (Fig. 20.12). Binding energy is almost as weak as that of graphene, and charge transfer is significantly smaller, as listed in Table 20.1. As a result of this contact, no Dirac cones are formed and degeneracy of the two surface states is not broken.

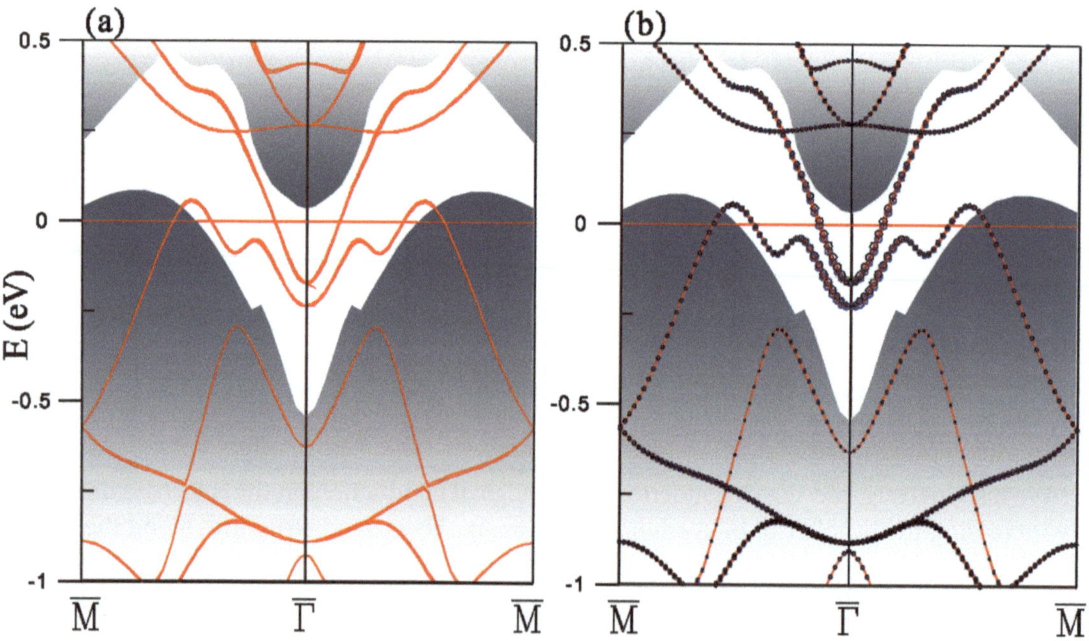

FIG. 20.12 Band structures of a three-bilayer Sb (111) film in top-site contact with boron nitride on the (A) boron and (B) nitrogen atom. Blue (black) circles indicate contribution from interface (pristine surface) electrons of the Sb film.

In summary, our calculations have found that an Sb (111) film as thin as three-bilayer can have topological conduction if the film is in contact with a layer of various heterogeneous atoms, including Bi, pure and B-doped graphene. In particular, if the film is in bridge-site contact with BC_3, topological conduction can be switched between one-mode and two-mode transport by adjusting gate voltage, with the former being 100% spin-polarized and the latter being either unpolarized or completely polarized. These, together with calculations on thicker films, point to a convenient way to exploit topological transport.

References

[1] Kane CL, Mele EJ. Z_2 topological order and the quantum spin Hall effect. Phys Rev Lett 2005;95:146802. Kane C L and Mele E J 2005 Quantum Spin Hall Effect in Graphene Phys. Rev. Lett. 95 226801.

[2] Bernevig BA, Zhang SC. Quantum spin Hall effect. Phys Rev Lett 2006;96:106802.

[3] Fu L, Kane CL. Topological insulators with inversion symmetry. Phys Rev B 2007;76:045302.

[4] Fu L, Kane CL, Mele EJ. Topological insulators in three dimensions. Phys Rev Lett 2007;98:106803.

[5] Moore JE, Balents L. Topological invariants of time-reversal-invariant band structures. Phys Rev B 2007;75:121306.

[6] Murakami S. Phase transition between the quantum spin Hall and insulator phases in 3D: emergence of a topological gapless phase. New J Phys 2007;9:356.

[7] Roy R. Topological phases and the quantum spin Hall effect in three dimensions. Phys Rev B 2009;79:195322.

[8] Hasan MZ, Kane CL. *Colloquium*: topological insulators. Rev Mod Phys 2010;82:3045.

[9] Qi XL, Zhang SC. Topological insulators and superconductors. Rev Mod Phys 2011;83:1057.

[10] Zhang Y, et al. Crossover of the three-dimensional topological insulator Bi_2Se_3 to the two-dimensional limit. Nat Phys 2010;6:584.

[11] Bian G, Wang X, Liu Y, Miller T, Chiang T-C. Interfacial protection of topological surface states in ultrathin Sb films. Phys Rev Lett 2012;108:176401.

[12] Bian G, Miller T, Chiang T-C. Passage from spin-polarized surface states to unpolarized quantum well states in topologically nontrivial Sb films. Phys Rev Lett 2011;107:036802.

[13] Hsieh D, et al. Observation of unconventional quantum spin textures in topological insulators. Science 2009;323:919.

[14] Lee CH, Yang CK. Topological surface states revealed by Sb thin films adsorbed with impurity atoms. Phys Rev B 2013;87:115306.

[15] Kresse G, Joubert D. From ultrasoft pseudopotentials to the projector augmented-wave method. Phys Rev B 1999;59:1758.

[16] Kresse G, Furthmüller J. Efficient iterative schemes for ab initio total-energy calculations using a plane-wave basis set. Phys Rev B 1996;54:11169.

[17] Latil S, Henrard L. Charge carriers in few-layer graphene films. Phys Rev Lett 2006;97:036803.

[18] Watanabe K, Taniguchi T, Kanda H. Direct-bandgap properties and evidence for ultraviolet lasing of hexagonal boron nitride single crystal. Nat Mater 2004;3:404.

[19] Xie J, Zhang ZY, Yang DZ, Xue DS, Si MS. Theoretical prediction of carrier mobility in few-layer BC2N. J Phys Chem Lett 2014;5:4073.

[20] Basak S, Lin H, Wray LA, Xu S-Y, Fu L, Hasan MZ, Bansil A. Spin texture on the warped Dirac-cone surface states in topological insulators. Phys Rev B 2011;84(R):121401.

[21] Tang W, Sanville E, Henkelman G. A grid-based Bader analysis algorithm without lattice bias. J Phys Condens Matter 2009;21:084204.

CHAPTER

21

Concluding remarks

Shih-Yang Lin[1], Vo Khuong Dien[1], Chi-Hsuan Lee[1], Hsin-Yi Liu[1], Hai Duong Pham[1], Thi My Duyen Huynh[1], Nguyen Thi Han[1], Ngoc Thanh Thuy Tran[2], Thi Dieu Hien Nguyen[1], Wei-Bang Li[1] and Ming-Fa Lin[1]

[1]Department of Physics, National Cheng Kung University, Tainan, Taiwan; [2]Hierarchical Green-Energy Materials (Hi-GEM) Research Center, National Cheng Kung University, Tainan, Taiwan

This work, which is devoted to the development of quasiparticle framework, is able to clearly illustrate the diverse phenomena through the outstanding candidates of the germanium-/germanene-related materials [1–10]. The delicate simulations and analyses are thoroughly done under the first-principles method [11], being fully assisted by the phenomenological models [12]. A plenty of critical factors have been successfully covered in the numerical calculations: dimensionalities [13–15], stacking configurations [16–19], layer numbers [20–23], buckled/planar honeycomb lattices [4,24–27], quantum confinements/chiralities of nanotubes and nanoribbons [28–32], chemical adsorptions [33–37] and substitutions [38–41], composites of two monolayer subsystems [42,43], heterojunctions/substrate effects [33,44,45], Ge-based battery electrolytes [46–48], solar cells [49–51], intermetallic compounds [52–54] and ferromagnetics [55,56], and few-layer antimonene without/with substrate [57–59]. Very obviously, the various chemical environments is capable of providing a sufficient platform in predicting the rich and unique physical/chemical/material properties, identifying the diverse phenomena and thus realizing the highly potential applications [60–64]. The strong relations between two theoretical approaches are fully discussed in Chapters 4–20. The current VASP predictions can account for the up-to-date experimental measurements, while most of them need the systematic examinations. The similar studies will be generalized to the binary/ternary/quaternary semiconductor compounds, such as, the unusual II–V/II–VI/III–V/III–VI compounds [65–77]. This is very helpful in greatly promoting the semiconductor-based chips with the specific operation functionalities.

The theoretical quasiparticle framework is further extended through the high-accurate VASP simulations for the germanene-/germanium-related emergent materials. The calculated

Fundamental Physicochemical Properties of Germanene-related Materials
https://doi.org/10.1016/B978-0-443-15801-8.00022-0

quantities, which include the optimal crystal structures, atom-dominated/spin-split band structures, spatial carrier densities, net magnetic moments, spin configurations, atom-, orbital- and spin-decomposed density of states, optical dielectric functions, reflectance spectra, screened response functions, and absorption coefficients, agree with one another within the quasiparticle viewpoints [78–82]. The critical mechanisms, corresponding to the concise physical and chemical pictures, are delicately achieved from the featured results. The diverse mutli-/single-orbital hybridizations, the unusual spin-orbital couplings and/or the bi-particle Coulomb interactions frequently come to exist in the outstanding systems. The clear identifications of the significant chemical bondings cover s–sp^3 (e.g., H–Ge/Alkali-Ge bonds; details in Chapter 7 [83–85]), pure sp^3–sp^3 (3D Ge–Ge bonds; Chapter 13 [83,86,87]), hybridized sp^3–sp^3 (Ge–O bonds [88–90]), {s, p_x, p_y, p_z}-{$3d_z^2$, $3d_{x^2-y^2}$, $3d_{xz}$, $3d_{zy}$, $3d_{yx}$} (Ge-T bonds; Chapter 8 [91–94]), {s, p_x, p_y, p_z}-{f_z^2, f_{xz}^2, $f_{yz}^2 f_{xyz}$, $f_{z(x^2-y^2)}$, $f_{x(x^2-3y^2)}$, $f_{y(y^2-3_x^2)}$} (Ge–Re bonds; Chapter 10 [95,96]), the non-orthogonality/orthogonality of pure π and σ bondings (2D Ge–Ge/C–C bonds; Chapter 7 [89,97]), the impure π and σ bondings with the non-perpendicular/perpendicular relations (in the buckled/non-buckled binary compounds of B-, C-, and N-substituted germanene systems; Chapter 6). Very interesting, each prominent absorption peak/plasmon mode can be assigned to the specific orbitals of the initial and final states (e.g., details in Chapter 7). The important differences among the group-IV hydrogenated materials are illustrated thoroughly (details in Chapter 7). These rich quasiparticle behaviors will play the critical roles in establishing the well-behaved phenomenological models [98,99], e.g., the tight-binding model/the generalized tight-binding model for the hydrogenations effects. On the drawback side, this numerical approach might present the serious problems in certain important research topics, especially for giant Moire superlattices due to the significant bond-length fluctuations (in multi-component compounds/heterojunctions/chemical modifications and magnetic-/electric-field modulations). Of course, the well-known quasiparticles of phonons and photons need to be covered in the extended quasiparticle framework under the near-future studies. The separate and composite quasiparticle modes are worthy of the systematic researches.

As for Ge-related main-stream materials in Chapters 14–20, the full measurement examinations on the essential properties are almost absent up to now, leading to the urgent requirements in greatly promoting the simultaneous progress of experiments and theories. Of course, a lot of crystal structures have been successfully identified from the high-resolution measurements of STM/TEM/LEED and X-ray, respectively, corresponding 2D/1D [27,30,100–104] and 3D systems [105,106]. The former and the latter cover few-layer germanene on substrates/adatom-adsorbed germanene/germanene nanoribbons and Re–T–Ge intermetallic compounds/LiGeO/$CsGeI_3$. STS [20,107,108] seems to present quite large uncertainties for the low-dimensional systems in observing the band property near the Fermi level, since is very difficult to largely reduce the substrate effects. The updated combination of STM and STS might be available in understanding the unusual relations of geometric and electronic properties in germanene nanotubes [7] and nanoribbons [109–112]. The scanning electron microscopy and single crystal X-ray diffraction are available for testing the crystal structure of rare-earth metal/Ge compounds [95]. The ARPES measurements for layered germanene on substrates clearly reveal the strong modified band structures below the Fermi level, being in great sharp contrast with the free-standing cases of the theoretical predictions [94,97,112–114]. The critical pictures arise from significant orbital hybridizations at two subsystems and heterojunction between them. Their high-precision abilities would be almost disappear in crystal structures with three dimensions, an obvious deformations of nanotube axis, and a large Moire superlattice. The conservation of momentum transfer and the number of low-lying valence subbands will determine the existence of well-characterized energy spectra. Optical reflectance, absorption and photoluminescence spectroscopies are mainly conducted on conducted on LiGeO. There exist several ambiguous absorption structures within an insufficiently wide frequency range (details in Chapter 17), in which they are not assigned to the suitable band-edge states. The last one is very useful in understanding a red shift of threshold absorption frequency. Such precise observation could be generalized to the germanium-based semiconductors in realizing the stability of excitonic effects (the stable quasiparticles [115–117]) and thus the optical applications.

Monolayer germanene materials, as indicated in the previous chapter, exhibit diverse low-lying energy bands because of buckled honeycomb lattices, multi- or single-orbital hybridization, and significant or negligible spin-orbital coupling. The generalized tight-binding model [118–125], with reliable parameters in the presence of electric and magnetic fields, could be further developed to explore the varied fundamental properties fully through direct combination with the single- [17] and many-particle theories [126], such as optical absorption spectra and selection rules [127,128], magneto-electronic properties [129,130], Coulomb excitations and decay rates [17,131,132], and quantum Hall effects [133–136]. More, layered structures, stacking configurations and layer numbers play critical roles in greatly diversifying the fundamental properties and enhancing the potential applications. Up to now, there are a lot of experimental [137–140] and theoretical [2,25–30,57,141–144] researches on the graphene-related materials. The monolayer and sliding bilayer germanenes in this work are expected to display more complicated phenomena, as a result of the strong couplings among the buckled structures, stacking configurations, multi-orbital hybridizations, intralayer and interlayer hopping integrals, and spin-orbital interactions [145]. The sublattice-, layer-, orbital- and spin-decomposed contributions to the fundamental properties and the significant cooperative/competitive relations with the displacements will be the studying focuses, e.g., the semimetal-semiconductor transition [146] and the transformation of non-magnetic and magnetic configurations [147]. The first-principles calculation is frequently utilized to explore the electronic properties. It is very suitable in studying the total ground state energy, the multi-/single-orbital hybridizations and the spin-related configurations under the optimal geometric structures; therefore, it can account for the concise physical and chemical pictures in explaining the diversified phenomena. The geometric, electronic and magnetic properties of mono- and bilayer germanene materials and the differences among the theoretical models are worthy of the systematic studies. Furthermore, the various physical, chemical and material properties of IV-group materials need to be conducted on the further investigations.

In past years, the adsorption on germanene has gradually sprung up due to the buckled honeycomb structure and unusual electronic properties [148–150]. The studies cover wide issue, such as, gas adsorption [151] or transition metal [152]. On the basis of the first-principles calculations, we display the optimal structures, spatial charge distributions, charge variations, band structures, and density of states of boron-, carbon-, and nitrogen-adsorbed germanene. The heights of buckling are drastically changed to be three times more than the pristine germanene. For the single-side adsorbed and double-side adsorbed germanene. And the host-guest bond length of Ge–B/Ge–C/Ge–N for single-cases are 2.095 ÅÅ/

1.935 ÅÅ/1.956 ÅÅ, respectively; and for the double-cases of the top adatom (the bottom adatom) are 2.453 ÅÅ/2.082 ÅÅ/1.984 ÅÅ (2.457 ÅÅ/2.081 ÅÅ/1.982 ÅÅ), respectively; and the ground state energy are −13.129 eV/−14.460 eV/−14.796 eV (−18.768 eV/−21.983 eV/−22.506 eV). The spatial charge distributions show that boron atoms get stronger interaction with the horizon-neighboring germanium atoms, but the nitrogen has the stronger interactions with the germanium atoms blow. Whether single-ones or double-side ones, the nitrogen-adsorbed germanene shows the most stability among all. The calculated results all reveal that there exist very strong but complicated [2s, 2px, 2py, 2pz]-[4s, 4px, 4py, 4pz] sp^3-orbital hybridizations. Even the double-side adsorbed cases exhibit like bilayer structures.

Very interesting, boron-, carbon- and nitrogen-substituted germanene systems exhibit the diverse orbital hybridizations and spin configurations, as thoroughly examined from the first-principles simulations [63] for the highly non-uniform chemical bonds, the atom- and spin-dominated band structures within the specific energy ranges, the spatial charge/spin densities and their variations after the chemical substitutions, and the atom-, orbital- and spin-projected density of states. The featured quasiparticle properties cover the position-dependent chemical bonds under the various substitution cases [63], the diversified energy spectra and wave functions (in terms of semiconductors/metals, direct/indirect band gap, asymmetry of hole and electron bands, distinct energy dispersions, more band-edge states, zone-folding effect, existence/absence of π and σ bands, non-crossing/crossing/anti-crossing behaviors [76]), the diversified orbital hybridizations (the sp^2 or sp^3 ones from top and side views [63]), and various van Hove singularities due to different critical points [75]. Through the detailed analyses of all relevant quantities, the specific orbital hybridizations are the dominant π and σ bondings and the minor sp^3 ones in a pristine monolayer germanene, the [4s, $4p_x$, $4p_y$,$4p_z$]-[4s, $4p_x$, $4p_y$,$4p_z$] and [4s, $4p_x$, $4p_y$,$4p_z$]-[2s, $2p_x$, $2p_y$,$2p_z$] mixings, respectively, in Ge−Ge, Ge−B/Ge−C/Ge−N and B−B/C−C/N−N bonds of buckled binary compounds, as well as, the almost perpendicular π and σ bondings for non-buckled systems. Also noticed that the hopping integrals and the one-site Coulomb potentials very differ from one another among these complex orbital mixings under the distinct atomic configurations. The high-precision measurements of STM/TEM, ARPES and STS (details in Chapter 3), are necessary in, respectively, examining the theoretical predictions about the optimal geometries, occupied energy bands, and density of states across the Fermi level.

Hydrogenations of monolayer group-IV systems clearly present the diverse quasiparticle phenomena, greatly promoting the development of theoretical framework [153]. Through the delicate analyses, both charge- and spin-dependent mechanisms are clearly identified to account for the unusual quasiparticle behaviors. The former covers the perpendicular π and σ bondings [154], sp^2, and sp^3 bondings which they are verified from graphene [154], silicene/germanene/tinene/plumbene [26,155−157], and hydrogenated group-IV materials [158,159], respectively. Furthermore, the latter exhibit through the spin-orbital couplings [160] and/or on-site Coulomb interactions [161]. The rich crystal symmetries are directly reflected in the planar or buckled honeycomb lattice, IV−IV bond lengths, their observable fluctuations, IV−H bonding angles, H-adatom hollow sites and heights. Most of hydrogenated group-IV require STM and TEM (details in Chapter 3) examinations except for graphene and graphene [162,163]. These featured geometries, which are largely enhanced by the various chemisorption cases, have successfully created the unique band structures, wave functions, spatial charge density distributions/their variations after hydrogenations, as well as, van Hove singularities. The VASP simulations clearly reveal many important characteristics, e.g., as, a zero-gap semiconductor arising from the isotropic Dirac-cone structure, the small-, middle- and large-gap semiconductors due to the spin-based interactions and/or hydrogenation effects, the perturbed/modified/destroyed π and σ valence subbands. the high asymmetry of hole and electron spectra about the Fermi level, linear/parabolic/oscillatory/partially flat dispersion relations, the non-crossing/crossing/anticrossing subband behaviors, a lot of band-edge states, the diverse charge distributions corresponding to the various chemical bondings, as well as, the merged van Hove singularities associated with orbital hybridizations. Optical properties, transverse dielectric functions, energy loss spectra, absorption coefficients and reflectance spectra, and absorption coefficients [164,165], clearly show the unusual interband electron-hole excitations and coherent carrier oscillations, in which the different plasmon modes come to exist in the specific energy ranges. The pronounced absorption peaks and excitonic effects are successfully combined with the joint van Hove singularities. Each prominent structure is thoroughly examined to be a special set of orbital hybridizations from the initial and final band-edge states. This is completely assisted by the featured band structures. The lowest/lower excitonic peaks are closely related to the stable or quasi-stable electron-hole bound states, depending on the dielectric screening of the total valence charges [166,167]. The red shifts about the threshold absorption frequency indicates that optical gaps are smaller than band gaps. In addition, the excitonic theory of emergent 2D materials could be developed in the further studies, while their numerical calculations might present a giant barrier associated with the Coulomb-field screening effects (the evaluation of the longitudinal dielectric function) actions for the different energy-range valence carriers (the distinct plasmons [153]). The prominent absorption structures and the excitonic effects have been successfully licked with the joint van Hove singularities (details in Table 7.2). Each pronounced peak is clearly identified to a specific set of orbital hybridizations from the initial to final band-edge states, being fully supported by the featured band structures. The lowest/lower excitonic peaks are due to the stable or quasi-stable electron-hole bound states. The red shifts of threshold absorption frequency indicates the fact that optical gaps are less than energy gaps. Specifically, the 2D excitonic theories need to be established in the near-future research, even under a super-high barrier of numerical evaluations related to the Coulomb-field carrier screenings. As to the phenomenological models, they are well developed for pristine graphene/silicene/germmanene/tinene/plumbene [168–172], while the opposite are true for hydrogenated systems [173,174]. For example, the generalized tight-binding model has successfully generated the diverse magnetic quantization phenomena on the former [175,176], but not about the latter. On the experimental side, graphene is fully examined for its essential properties [154], and graphene is conducted on geometric structures [154] and transport behaviors [177]. Most of theoretical predictions must be test in the further measurements, in which the geometric, electronic and optical properties could be verified, respectively, from STM/TEM (details in Chapter 3), ARPES/STS, and Reflectance/absorption/transmission/photoluminescence spectroscopies. Apparently, the chemical reactions, e.g., hydrogenation [178], halogenation [179,180], oxidization [180], and alkalization [181], are very powerful in generating the diverse quasiparticle behaviors and enhancing the highly potential applications. Such approach is fully consistent with the development of quasiparticle framework.

In general, the transition metal (Fe, Co, and Ni)-doped germanene can greatly diverse electronic and magnetic properties. Although all systems exhibit optimal structures at the hollow site, each TM atom-adsorbed germanene system exhibits distinctly different properties. While Fe-doped germanene exhibits a semiconductor nature with a very large magnetic moment at the four possible absorption sites, the Co–Ge compound demonstrates a metallic nature with a smaller magnetic moment. However, the Co absorption also presents semiconducting nature but does not possess magnetism in any of the absorption sites. Based on the results, TM-Ge bondings exhibit a very complex multi-orbital hybridization, where the orbitals come from the TM ($3d_{x2-y2}$, $3d_{xy}$, $3d_{yx}$, $3d_{xz}$, $3d_{z2}$) atoms interact with the Ge (4s, $4p_x$, $4p_y$, $4p_z$) atoms. This hybridization can be easily observed through the results of the charge distribution and projected atom-, spin-, and orbital-decomposed density of states. Moreover, there is clearly a charge transfer from Ge to TM atoms, which weakens the strength of Ge–Ge bonds and shows that TM–Ge bonds are formed mainly performing ionic character. Very interestingly, comparing TM atoms doped-germanene with such systems in silicene and graphene [182–184], the absorbed energy values showed that germanene might be the best candidate to accommodate TM atoms among all three 2D materials. Moreover, the recent results about the successful synthesis of adatom-doped graphene [185–187] as well as the creation of pristine germanene in the free-standing state [155,188,189] have paved the way for germanene and functionalized germanene compounds to be applied to the manufacture of practical devices.

$\alpha - GeO_2$ shows rich and unusual properties compound, which belongs to trigonal structures. The main band structures possess a large band gap of $E_g^i = 3.2193$ eV, while the asymmetry properties present about the Femi level of occupied and unoccupied states. The concise essential quasiparticle properties are fully explored under the first-principles calculations [190,191]. It is noted that the Ge atoms dominated in unoccupied electron states and the O atom are prominent in occupied hole states. The appearance of van Hove density of states that consistent with the band-edge states of band spectrum, and change density with Ge–O bond orbital hybridizations. Furthermore, the excitonic effects, mainly arise from the hole-electron pairs by the electromagnetic fields are studied under the optical properties [192]. Dielectric functions, absorption coefficient, energy loss spectra, and reflectivity index are provided in this book chapters. All these are combined in terms of orbital hybridizations. There are many prominent shoulders/peaks that come into existence, can identify from the band-edge states band structures and orbital-decomposed densities of states. Based on this work, the further theoretical strategy could be expanded into 1D, 2D, and 3D systems [193–197].

The rare-earth adatom adsorption on a 2D germanene exhibit a diverse and unique chemical bondings. There are Ge–Ge, Ge–X and X–X chemical bondings, in which, the latter is presented at high concentrations. The significant hybridization of the RE and Ge atoms leading to the enhanced of the bulked height, and thus, lead to distort of the π-bondings. The rich and diverse electronic properties include n-type doping behaviors with the significantly blue-shift of the Fermi level, the electron transfers are gradually changing with kinds and adatoms concentrations, the dominated and co-dominated of RE and Ge atoms in the conductions and valence bands, the active orbital-hybridizations-related to the orbital hybridizations. There have many weak energy dispersions and thus create strong van-Hove singularities in the density of states. Very interestingly, the Eu-adsorbed germanene exhibits the ferromagnetic ordered, which clearly reflects on the slitting of spin-up/spin-down states on the energy band structures, the polarized density of states and the spin density distributions. In addition, the optical excitations of germanene are also changed under RE-chemisorption, the main feature cover the unique single-particle excitations, the plasmon modes—collective excitations of charges in term of energy loss functions, absorption, reflection spectrums. The close connections of the initial and final states and their orbital hybridizations are successfully achieved.

The 1D germanium nanotubes, with a unique cylindrical symmetry, are very suitable for studying the basic and applied sciences. The geometric structure, electronic properties of armchair and zigzag nanotubes are investigated by method of first-principles. The both nanotubes exhibit the diverse geometries and electronic properties, as a result of the significant chemical bonding of outer four orbitals and spin-orbital couplings. The band structure dependent of chiralities and radius, charge density and orbital-projected density of state are useful in exploring the orbital-dependent essential properties. They possess buckled honeycomb lattices of 3.81−4.04 Å, the ground state energy E_0 and buckling height Δh decrease with increase of diameters. The bond length b_1 and b_2 of germanium nanotubes increase when the diameter of the tube turns small. The results of band structure from germanium and carbon nanotubes are quite different. Comparing to armchair carbon nanotubes (aCNTs), armchair germanium nanotubes (aGeNTs) exhibit direct and indirect band gap due to their unsymmetrical structure. The armchair germanium nanotubes from (2, 2) to (10, 10) possess semiconducting behaviors but the aCNTs behave gapless metallic features. The energy dispersion of aGeNTs and aCNTs are reflected in PDOSs as many anti-symmetric and symmetric peaks. There are five kinds of low-lying peak structures characteristic of the main features of essential properties. The predicted geometric structure, energy bands and PDOSs are reliable by experimental tools of STM, ARPES and STS, respectively. Besides, ARPES and STS is particularly for the spin resolution. The 1D armchair carbon and germanium nanotubes, with a unique cylindrical symmetry, are very suitable for studying the basic and applied sciences. The chirality- and radius dependent semiconductors and metals are available in functionalized nanodevices, such as electronic and optoelectronic devices. The intrinsic 1D band structures, with the decoupled angular momenta, dominate the quantized electrical and optical properties, e.g., the geometry-dependent resistance, capacitance, inductance, and optical conductance [198−200]. At present, the Feld-effect transistors (FETs) based on the semiconducting carbon nanotubes are widely developed, mainly owing to the advantage of high carrier mobility under the low scatterings. The first FETs composed of single- or multi-walled carbon nanotubes were reported in 1998 [201,202], which present superior electrical conductance at room temperature using the gate-voltage modulation. A very high carrier mobility ($>105\ cm^2/Vs$) is identified to be related to the diffusive transport of holes under the electron-phonon scatterings [199,200]. The nano-scaled radii allow the gate's ability to control the potential of the conducting channel in the ultimate thin FETs, while suppressing short-channel effects [203,204]. Moreover, the semiconducting carbon nanotubes have the highly potential applications in optoelectronic devices, including electroluminescent light emitters [101,205], supercapacitors [206] and photodetectors [207−209]. The band-edge states in 1D nanotubes, with the high PDOSs, can create the very prominent optical excitations [198].

The excited electrons and holes on a cylindrical surface are further driven toward each other by using an appropriate bias between source and drain of the FET. The electron-hole recombination will emit the strong electroluminescence. The highly efficient photon-emission process has been extensively utilized in light-emitting diodes (LEDs) [210–212]. The application of photodetector is based on the electric current generated by the resonant excitations. On the other hand, the metallic carbon nanotubes could be selected as candidates in future interconnect materials of integrated electronic devices [213,214].

1D germanene nanoribbons are outstanding candidates in fully comprehending the dimensionality-enriched quasiparticle behaviors, since their fundamental properties are mainly determined by the multi-coupling effects among the finite-width quantum confinement [215], the dangling bonds at two open edges [216], their hydrogen decorations, the weak, but important sp^3 bondings in a buckled honeycomb crystal [217], and the spin-orbital interactions [216] or the edge- and spin-dependent electron-electron ones (the Hubbard-like Coulomb interactions in Ref. [218]). The main features cover the highly non-uniform Ge–Ge bond lengths (the position-related hopping integrals in a unit cell), the geometry-dominated band gaps/a large asymmetry of hole and electron energy spectra about the Fermi level/the various energy dispersion relations/the frequent crossings and/or anti-crossings/a lot of band-edge states with zero group velocities, the coexistence of the π, σ and sp^3 bondings, as well as, the merged or non-merged van Hove singularities of different orbitals at the specific energy ranges. Apparently, all the calculated results are consistent with one another within a unified quasiparticle framework, as clearly illustrated in the previous books [219]. The theoretical predictions on the top- and side-view morphologies could be, respectively, verified by the high-resolution STM [220] and TEM measurements [221]. The width- and chirality-characterized band gaps will be thoroughly examined by the STS observations [222]. The featured density of states, which crosses the occupied and unoccupied states, is directly detected by the STS experiments [223]. This is capable of identifying the concise orbital hybridizations for three kinds of coexistent orbital hybridizations. However, the spatial charge/spin density distributions cannot be observed in these low-dimensional materials. How to utilize the direct combinations of theoretical and experimental researches in overcoming on detection difficulties is one of the next-step studying focuses. Most important, the first-principles simulations, the phenomenological models, and the high-precision observations need to establish rather strong relations among them. For example, the measured ARPES valence sub-bands/band gaps and the VASP calculations can provide a sufficient information about all intrinsic interactions in the tight-binding model/the generalized tight-binding model [224]. The above-mentioned quasiparticle phenomena will be greatly diversified by the various crystal structures, e.g., those in cylindrical/curved/rolled/folded germanene systems.

The alkali chemisorption on a 1D germanene nanoribbon clearly illustrates the diversified quasiparticle properties, being totally different from those in a pristine system [225–228]. By the delicate VASP simulations and analyses, the critical mechanisms are thoroughly identified through all the consistent physical and chemical quantities, covering the highly non-uniform crystal structures in the presence of intralayer and interlayer chemical bonds, the atom-dominated 1D valence and conduction energy sub-bands with the blue-shift Fermi level (the n-type alkali-adatom doping in Refs. [181,229,230]), the active orbital-related charge density distributions, the ferromagnetic, anti-ferromagnetic and non-magnetic spin configurations, and the atom-, orbital- and spin-projected van Hove singularities (the merged and non-merged special structures). Even under the complicated physical and chemical environments after the chemical modifications, the concise pictures are achieved under a unified quasiparticle framework. This is capable of providing a sufficient information about the reliable establishment of phenomenological models. The critical mechanisms need to be included in the Hamiltonian of the tight-binding model [224], where they correspond to the finite-width quantum confinement [226,227,231,232], the chirality of an edge structure [233–235], the hydrogen passivation on open boundaries [226,236], the s-s orbital hybridizations in A-A bonds/the interlayer dominant s-4pz and minor s-(4s, 4px, 4py) ones, and the spin-dependent single- or many-particle interactions. A suitable strategy should directly reflect the alkali-induced conduction electrons; that is, it is reliable in accounting for the sensitive dependences of adatom concentrations and arrangements about 1D free carrier densities. Furthermore, the Hubbard-like tight-binding model would be very useful in presenting the physical pictures of the spatial spin configurations. Obviously, the various experimental observations are absent up to date, such as, the growth methods, the optimal crystal structures under STM/TEM measurements [223], the occupied valence sub-bands by ARPES, the geometry-characterized band gaps/the van Hove singularities using STS detections [223]. The close relations among the first-principles method, the phenomenological models, and the high-precision experimental measurements are able to greatly promote the much progress of the basic and applied sciences, since such viewpoints are initiated from a unified quasiparticle framework, as fully expressed in the previous and current books. How to verify the orbital-deformed charge/spin density distributions is a very interesting research topic. The systematic investigations could be done for hydrogenations [237–240], oxidizations [241] and halogenations [242] of germanene-based materials.

The 3D ternary Li_2GeO_3 compounds, a candidate for electrolyte materials of lithium-ion batteries, are predicted to exhibit the rich, and unique electronic and optical properties. The accurate analyses, which are made from the exact first-principles calculations but not for the phenomenological methods, are available in identifying the significant single-/multi-orbital hybridizations in Li–O and Ge–O bonds. Specifically, the 3D Li_2GeO_3 compound presents the Moire superlattice and anisotropy geometric, the complicated band structure with various atom dominations, the sizable indirect gap of 6.9 eV, the spatial charge distribution, and a lot of van Hove singularities due to the critical points in energy spectrum. As a consequence, the important single-/multi-orbital orbital hybridizations in Ge–O and Li–O bonds could be well characteristics. The optical properties include the strong polarizations of the optical response, the strong red-shift of the optical gap due to the excitonic effects, various single-/collective excitations and their orbital-hybridizations, the high-optical transparency within a bond energy range, a short optical decay length. The calculated results clearly indicate that the 3D ternary Li_2GeO_3 could be served as a candidate anode/electrolyte component in Lithium-based batteries as well as optal-electronic devices. Our prediction provides certain meaningful information about the critical physical/chemical pictures in ternary germanium-based compound. Such states-of-the-art analysis is very useful for fully comprehended the diversified properties in other emerging materials. Furthermore, the theoretical predictions on the optimal geometries, the rich occupied bands, many van Hove singularities, and optical properties could be verified from the high-resolution measurements of the X-ray diffraction, angle-resolved photoemission spectroscopy, scanning tunneling spectroscopy, and photoluminescence/reflection/adsorption measurements, respectively.

In Chapter 18, based on the state-of-the-art first-principles calculations and delicate analysis, many clearly physical/chemical phenomena of CsGeI3 perovskite solar cell have been achieved. This material exhibited a lot of unusual properties as 12 Cs—I bonds and 6 Ge—I bonds, leading to an extremely non-uniform physical and chemical environments, unique quasiparticle band structure with a direct energy gap of 2 eV, the charge density distribution, a lot of symmetric, asymmetric peaks and shoulder structures van Hove singularities. Consequently, the orbital-hybridizations in the chemical bonds have been obtained. About the optical excitations, the presence of excitonic effects includes the slight redshift of the optical gap due to the weak Coulomb interactions, the modified single-particle inter-band transitions, the low reflectance and high transparency of electromagnetic waves for $\omega < E_g^o$ and high absorption coefficient. Especially, the initial and the finial states and the orbital characters of six prominent single-particle excitation peaks, as well as the most pronounced peaks in energy loss function at the Plasmon resonance frequency of 10.2 eV have been well identified.

The main electronic, optical features of $CsGeI_3$ could be examined by the experimental methods [243—246]. The presented theoretical framework could be explored for potential applications of other perovskites solar cells and optoelectronic materials [247,248]. The CsGeI3 compound is one of the worthy outstanding candidates for large applications.

Our calculations have shown that adsorptions of non-magnetic impurity atoms of hydrogen, copper, or zinc on a four-bilayer Sb thin film actually facilitate the formation of Dirac cones that preserve time-reversal symmetry. But magnetic atoms such as iron and manganese do just the opposite. The results suggest the counterintuitive concept of achieving topological conduction by doping nonmagnetic foreign atoms on thin films of topological insulators.

An Sb (111) film as thin as three-bilayer can have topological conduction if the film is in contact with a layer of various heterogeneous atoms by us the density functional calculation. The substrate with defects, such as Bi monolayer, still induces the topological conduction of the composite structure. Similar features can also be observed in a three-bilayer Sb film in contact with B-doped graphene from the concentration of 1/24 to BC_3. When the Sb film is in bridge-site contact with BC_3, topological conduction can be switched between one-mode and two-mode transport by adjusting gate voltage, with the former being 100% spin-polarized and the latter being either unpolarized or completely polarized.

References

[1] Bhimanapati GR, Lin Z, Meunier V, Jung Y, Cha J, Das S, Xiao D, Son Y, Strano MS, Cooper VR, Liang L, Louie SG, Ringe E, Zhou W, Kim SS, Naik RR, Sumpter BG, Terrones H, Xia F, Wang Y, Zhu J, Akinwande D, Alem N, Schuller JA, Schaak RE, Terrones M, Robinson JA. Recent advances in two-dimensional materials beyond graphene. ACS Nano 2015;9:11509—39.

[2] Zhu FF, Chen WJ, Xu Y, Gao CL, Guan DD, Liu CH, Qian D, Zhang SC, Jia JF. Epitaxial growth of two-dimensional stanene. Nat Mater 2015;14:1020—5.

[3] Tabert CJ, Nicol EJ. Valley-spin polarization in the magneto-optical response of silicene and other similar 2D crystals. Phys Rev Lett 2013;110.

[4] Lin CH, Huang A, Pai WW, Chen WC, Chen TY, Chang TR, Yukawa R, Cheng CM, Mou CY, Matsuda I, Chiang TC, Jeng HT, Tang SJ. Single-layer dual germanene phases on Ag(111). Phys Rev Mater 2018;2:24003.

[5] Zhuang J, Gao N, Li Z, Xu X, Wang J, Zhao J, Dou SX, Du Y. Cooperative electron-phonon coupling and buckled structure in germanene on Au(111). ACS Nano 2017;11:3553–9.

[6] Stephan R, Hanf MC, Derivaz M, Dentel D, Asensio MC, Avila J, Mehdaoui A, Sonnet P, Pirri C. Germanene on al(111): interface electronic states and charge transfer. J Phys Chem C 2016;120:1580–5.

[7] Snehha P, Nagarajan V, Chandiramouli R. Germanene nanotube electroresistive molecular device for detection of NO_2 and SO_2 gas molecules: a first-principles investigation. J Comput Electron 2019;18:308–18.

[8] Derivaz M, Dentel D, Stephan R, Hanf MC, Mehdaoui A, Sonnet P, Pirri C. Continuous germanene layer on Al(111). Nano Lett 2015;15:2510–6.

[9] Yuhara J, Shimazu H, Ito K, Ohta A, Araidai M, Kurosawa M, Nakatake M, Le Lay G. Germanene epitaxial growth by segregation through Ag(111) thin films on Ge(111). ACS Nano 2018;12:11632–7.

[10] Suzuki S, Iwasaki T, De Silva KKH, Suehara S, Watanabe K, Taniguchi T, Moriyama S, Yoshimura M, Aizawa T, Nakayama T. Direct growth of germanene at interfaces between van der Waals materials and Ag(111). Adv Funct Mater 2021;31.

[11] Argaman N, Makov G. Density functional theory: an introduction. Am J Phys 2000;68:69–79.

[12] Goringe CM, Bowler DR, Hernández E. Tight-binding modelling of materials. Rep Prog Phys 1997;60:1447–512.

[13] Yera R, Rossi N, Méndez CG, Huespe AE. Topology design of 2D and 3D elastic material microarchitectures with crystal symmetries displaying isotropic properties close to their theoretical limits. Appl Mater Today 2020;18:100456.

[14] Saito R, Fujita M, Dresselhaus G, Dresselhaus MS. Electronic structure of graphene tubules based on C60. Phys Rev B 1992;46:1804–11.

[15] Thess A, Lee R, Nikolaev P, Dai H, Petit P, Robert J, Xu C, Lee YH, Kim SG, Rinzler AG, Colbert DT, Scuseria GE, Tománek D, Fischer JE, Smalley RE. Crystalline ropes of metallic carbon nanotubes. Science (80-.) 1996;273:483–7.

[16] Hattendorf S, Georgi A, Liebmann M, Morgenstern M. Networks of ABA and ABC stacked graphene on mica observed by scanning tunneling microscopy. Surf Sci 2013;610:53–8.

[17] Ho JH, Lu CL, Hwang CC, Chang CP, Lin MF. Coulomb excitations in AA- and AB-stacked bilayer graphites. Phys Rev B—Condens Matter Mater Phys 2006;74:85406.

[18] Guinea F, Castro Neto AH, Peres NMR. Electronic states and Landau levels in graphene stacks. Phys Rev B—Condens Matter Mater Phys 2006;73:245426.

[19] Lin CY, Huang BL, Ho CH, Gumbs G, Lin MF. Geometry-diversified coulomb excitations in trilayer AAB stacking graphene. Phys Rev B 2018;98.

[20] Zheng X, Zhang JF, Tong B, Du RR. Epitaxial growth and electronic properties of few-layer stanene on InSb (1 1 1). 2D Mater 2020;7:011001.

[21] Lu CL, Chang CP, Huang YC, Ho JH, Hwang CC, Lin MF. Electronic properties of AA- and ABC-stacked few-layer graphites. J Phys Soc Japan 2007;76:024701.

[22] Xiao YM, Xu W, Peeters FM. Infrared to terahertz absorption window in mono- and multi-layer graphene systems. Opt Commun 2014;328:135–42.

[23] Lin SY, Liu HY, Nguyen DK, Tran NTT, Pham HD, Chang SL, Lin CY, Lin MF. Silicene-based layered materials: essential properties. IOP Publishing; 2020.

[24] Berdiyorov GR, Neek-Amal M, Peeters FM, Van Duin ACT. Stabilized silicene within bilayer graphene: a proposal based on molecular dynamics and density-functional tight-binding calculations. Phys Rev B—Condens Matter Mater Phys 2014;89:24107.

[25] Lee J, Tian WC, Wang WL, Yao DX. Two-dimensional pnictogen honeycomb lattice: structure, on-site spin-orbit coupling and spin polarization. Sci Rep 2015;5:11512.

[26] Li L, Lu SZ, Pan J, Qin Z, Wang YQ, Wang Y, Cao GY, Du S, Gao HJ. Buckled germanene formation on Pt(111). Adv Mater 2014;26:4820–4.

[27] Meng L, Wang Y, Zhang L, Du S, Wu R, Li L, Zhang Y, Li G, Zhou H, Hofer WA, Gao HJ. Buckled silicene formation on Ir(111). Nano Lett 2013;13:685–90.

[28] Lin J, Raji ARO, Nan K, Peng Z, Yan Z, Samuel ELG, Natelson D, Tour JM. Iron oxide nanoparticle and graphene nanoribbon composite as an anode material for high-performance li-ion batteries. Adv Funct Mater 2014;24:2044–8.

[29] Rurali R, Coluci VR, Galvão DS. Prediction of giant electroactuation for papyruslike carbon nanoscroll structures: first-principles calculations. Phys Rev B—Condens Matter Mater Phys 2006;74:85414.

[30] Tao C, Jiao L, Yazyev OV, Chen YC, Feng J, Zhang X, Capaz RB, Tour JM, Zettl A, Louie SG, Dai H, Crommie MF. Spatially resolving edge states of chiral graphene nanoribbons. Nat Phys 2011;7:616–20.
[31] Cai J, Ruffieux P, Jaafar R, Bieri M, Braun T, Blankenburg S, Muoth M, Seitsonen AP, Saleh M, Feng X, Müllen K, Fasel R. Atomically precise bottom-up fabrication of graphene nanoribbons. Nature 2010;466:470–3.
[32] Lee RS, Kim HJ, Fischer JE, Thess A, Smalley RE. Conductivity enhancement in single-walled carbon nanotube bundles doped with K and Br. Nature 1997;388:255–7.
[33] Vanin M, Mortensen JJ, Kelkkanen AK, Garcia-Lastra JM, Thygesen KS, Jacobsen KW. Graphene on metals: a Van Der Waals density functional study. Phys Rev B—Condens Matter Mater Phys 2010;81:81408.
[34] Yan J, Fuhrer MS. Correlated charged impurity scattering in graphene. Phys Rev Lett 2011;107:377381.
[35] Clair S, Pons S, Seitsonen AP, Brune H, Kern K, Barth JV. STM study of terephthalic acid self-assembly on Au(111): hydrogen-bonded sheets on an inhomogeneous substrate. J Phys Chem B 2004;108:14585–90.
[36] Wendt S, Schaub R, Matthiesen J, Vestergaard EK, Wahlström E, Rasmussen MD, Thostrup P, Molina LM, Lægsgaard E, Stensgaard I, Hammer B, Besenbacher F. Oxygen vacancies on TiO_2(1 1 0) and their interaction with H_2O and O_2: a combined high-resolution STM and DFT study. Surf Sci 2005;598:226–45.
[37] Bhuvaneswari R, Nagarajan V, Chandiramouli R. Adsorption studies of trimethyl amine and n-butyl amine vapors on stanene nanotube molecular device—a first-principles study. Chem Phys 2018;501:78–85.
[38] Fasel R, Osterwalder J. Alkali-metal adsorption geometries on metal surfaces from photoelectron-diffraction experiments. Surf Rev Lett 1995;02:359–86.
[39] Li M, Hong M, Tang X, Sun Q, Lyu WY, Xu SD, Kou LZ, Dargusch M, Zou J, Chen ZG. Crystal symmetry induced structure and bonding manipulation boosting thermoelectric performance of GeTe. Nano Energy 2020;73:104740.
[40] Akturk OÜ, Tomak M. Bismuth doping of graphene. Appl Phys Lett 2010;96:81914.
[41] An H, Li Y, Long P, Gao Y, Qin C, Cao C, Feng Y, Feng W. Hydrothermal preparation of fluorinated graphene hydrogel for high-performance supercapacitors. J Power Sources 2016;312:146–55.
[42] Virojanadara C, Watcharinyanon S, Zakharov AA, Johansson LI. Epitaxial graphene on 6H-SiC and Li intercalation. Phys Rev B—Condens Matter Mater Phys 2010;82:205402.
[43] Fiori S, Murata Y, Veronesi S, Rossi A, Coletti C, Heun S. Li-intercalated graphene on SiC(0001): an STM study. Phys Rev B 2017;96:125429.
[44] Goler S, Coletti C, Piazza V, Pingue P, Colangelo F, Pellegrini V, Emtsev KV, Forti S, Starke U, Beltram F, Heun S. Revealing the atomic structure of the buffer layer between SiC(0 0 0 1) and epitaxial graphene. Carbon NY 2013;51:249–54.
[45] Wang YP, Cheng HP. Absence of a Dirac cone in silicene on Ag(111): first-principles density functional calculations with a modified effective band structure technique. Phys Rev B—Condens Matter Mater Phys 2013;87:245430.
[46] Hu Z, Sheng J, Chen J, Sheng G, Li Y, Fu XZ, Wang L, Sun R, Wong CP. Enhanced Li ion conductivity in Ge-doped $Li_{0.33}La_{0.56}TiO_3$ perovskite solid electrolytes for all-solid-state Li-ion batteries. New J Chem 2018;42:9074–9.
[47] Tsiulyanu D, Stratan I, Ciobanu M. Influence of glassy backbone on the photoformation and properties of solid electrolytes Ag: as-S-Ge. Chalcogenide Lett 2020;17:9–14.
[48] Sarangika HNM, Dissanayake MAKL, Senadeera GKR. Efficiency enhancement in dye-sensitized solar cells irrespective of the electrolyte medium by nanostructured tri-layer TiO_2 photoanode. Ionics 2020;26:4747–55.
[49] Liu B, Hu J, Jia L, Liu J, Ren X, Zhang X, Guo X, Frank, Liu S. Ge quantum-dot enhanced c-Si solar cell for improved light trapping efficiency. Sol Energy 2018;167:102–7.
[50] Barrigón E, Ochoa M, García I, Barrutia L, Algora C, Rey-Stolle I. Degradation of Ge subcells by thermal load during the growth of multijunction solar cells. Prog Photovoltaics Res Appl 2018;26:102–11.
[51] Baran V, Cat Y, Sertel T, Ataser T, Sonmez NA, Cakmak M, Ozcelik S. A comprehensive study on a stand-alone germanium (Ge) solar cell. J Electron Mater 2020;49:1249–56.
[52] Sasaki T, Kato T, Fukushima J, Hayashi Y, Takizawa H. High-pressure synthesis and crystal structure of a novel intermetallic compound Mn(Al,Ge)5. J Alloys Compd 2019;806:58–62.
[53] Lekhal A, Benkhelifa FZ, Mecabih S, Abbar B, Bouhafs B. Structural and electronic properties of non-magnetic intermetallic YAuX (X = Ge and Si) in hexagonal and cubic phases. Bull Mater Sci 2016;39:195–200.
[54] Komatsu T, Mesuda M, Yashima T. Aromatization of butane on Pt-Ge intermetallic compounds supported on HZSM-5. Appl Catal Gen 2000;194:333–9.

[55] Usachov D, Fedorov A, Otrokov MM, Chikina A, Vilkov O, Petukhov A, Rybkin AG, Koroteev YM, Chulkov EV, Adamchuk VK, Grüneis A, Laubschat C, Vyalikh DV. Observation of single-spin Dirac fermions at the graphene/ferromagnet interface. Nano Lett 2015;15:2396–401.

[56] Fayazi M, Liu B, Lei L, Shuai G, Odunmbaku O, Wang S, He Y, Boi FS. Ferromagnetic hysteresis and structural recrystallization in turbostratic graphite. Mater Res Exp 2019;6:105612.

[57] Shao Y, Liu ZL, Cheng C, Wu X, Liu H, Liu C, Wang JO, Zhu SY, Wang YQ, Shi DX, Ibrahim K, Sun JT, Wang YL, Gao HJ. Epitaxial growth of flat antimonene monolayer: a new honeycomb analogue of graphene. Nano Lett 2018;18:2133–9.

[58] Lei T, Liu C, Zhao JL, Li JM, Li YP, Wang JO, Wu R, Qian HJ, Wang HQ, Ibrahim K. Electronic structure of antimonene grown on Sb_2Te_3 (111) and Bi_2Te_3 substrates. J Appl Phys 2016;119.

[59] Jałochowski M, Krawiec M. Antimonene on Pb quantum wells. 2D Mater 2019;6.

[60] Goriparti S, Miele E, De Angelis F, Di Fabrizio E, Proietti Zaccaria R, Capiglia C. Review on recent progress of nanostructured anode materials for Li-ion batteries. J Power Sources 2014;257:421–43.

[61] Dahn JR, von Sacken U, Juzkow MW, Al-Janaby H. Rechargeable $LiNiO_2$/carbon cells. J Electrochem Soc 1991;138:2207–11.

[62] Zhong H, Xu K, Liu Z, Xu G, Shi L, Fan Y, Wang J, Ren G, Yang H. Charge transport mechanisms of graphene/semiconductor Schottky barriers: a theoretical and experimental study. J Appl Phys 2014;115:13701.

[63] Lin MC, Gong M, Lu B, Wu Y, Wang DY, Guan M, Angell M, Chen C, Yang J, Hwang BJ, Dai H. An ultrafast rechargeable aluminium-ion battery. Nature 2015;520:325–8.

[64] Geim AK, Novoselov KS. The rise of graphene. Nanosci Technol A Collect Rev Nat J 2009;6:11–9.

[65] Shen G, Chen D. One-dimensional nanostructures and devices of II-V group semiconductors. Nanoscale Res Lett 2009;4:779–88.

[66] Bhogal S, Mukherjee D, Banerjee S, Islam AM, Daggubati R, Paul TK. Current trends and future perspectives in the treatment of pulmonary hypertension: WHO group II-V. Curr Probl Cardiol 2018;43:217–31.

[67] Indolia RS. Relationship of refractive index with optical energy gap and average energy gap for II-VI and III-V group of semiconductors. Int J Pure Appl Phys 2017;13:185–91.

[68] Zhou J. Correction to: recent progress on 2D group II-VI binary chalcogenides ZnX and CdX (X = S, Se, Te): from a theoretical perspective. Adv Theory Simul 2019;2(8):1900061. https://doi.org/10.1002/adts.201900061.

[69] Alyörük MM. Piezoelectric properties of monolayer II–VI group oxides by first-principles calculations. Phys Status Solidi Basic Res 2016;253:2534–9.

[70] Zhou J. Recent progress on 2D group II-VI binary chalcogenides ZnX and CdX (X = S, Se, Te): from a theoretical perspective. Adv Theory Simul 2019;2:1900061.

[71] Cornet C, Chen L, Skibitzki O, Pedesseau L, Létoublon A, Stervinou J, Bernard R, Levallois C, Piron R, Perrin M, Schubert MA, Moréac A, Durand O, Schroeder T, Bertru N, Even J, Léger Y. Strong electron-phonon interaction in 2d vertical homovalent III-V singularities. ACS Nano 2020;14:13127–36.

[72] Elçioğlu EB, Didari A, Okutucu Özyurt T, Mengüç MP. Tunable near-field radiative transfer by III-V group compound semiconductors. J Phys D Appl Phys 2019;52.

[73] Hillerich K, Dick KA, Wen CY, Reuter MC, Kodambaka S, Ross FM. Strategies to control morphology in hybrid group III-V/group IV heterostructure nanowires. Nano Lett 2013;13:903–8.

[74] Cornet C, Charbonnier S, Lucci I, Chen L, Létoublon A, Alvarez A, Tavernier K, Rohel T, Bernard R, Rodriguez JB, Cerutti L, Tournié E, Léger Y, Bahri M, Patriarche G, Largeau L, Ponchet A, Turban P, Bertru N. Zinc-blende group III-V/group IV epitaxy: importance of the miscut. Phys Rev Mater 2020;4.

[75] Li X, Li L, Wu M. Various polymorphs of group III–VI (GaSe, InSe, GaTe) monolayers with quasi-degenerate energies: facile phase transformations, high-strain plastic deformation, and ferroelastic switching. Mater Today Phys 2020;15.

[76] Uematsu T, Wajima K, Sharma DK, Hirata S, Yamamoto T, Kameyama T, Vacha M, Torimoto T, Kuwabata S. Narrow band-edge photoluminescence from AgInS2 semiconductor nanoparticles by the formation of amorphous III–VI semiconductor shells. NPG Asia Mater 2018;10:713–26.

[77] Wang H, Qin G, Yang J, Qin Z, Yao Y, Wang Q, Hu M. First-principles study of electronic, optical and thermal transport properties of group III-VI monolayer MX (M = Ga, In; X = S, Se). J Appl Phys 2019;125.

[78] Kresse G, Furthmüller J. Efficient iterative schemes for ab initio total-energy calculations using a plane-wave basis set. Phys Rev B—Condens Matter Mater Phys 1996;54:11169–86.

[79] Benrekia AR, Benkhettou N, Nassour A, Driz M, Sahnoun M, Lebègue S. Structural, electronic and optical properties of cubic $SrTiO_3$ and $KTaO_3$: ab initio and GW calculations. Phys B Condens Matter 2012;407:2632–6.

[80] Rani P, Dubey GS, Jindal VK. DFT study of optical properties of pure and doped graphene. Phys E Low-dimen Syst Nanostruct 2014;62:28–35.
[81] Marqués Arias M. Vienna ab-initio simulation package. University of Edinburgh; 2009.
[82] Sun G, Kürti J, Rajczy P, Kertesz M, Hafner J, Kresse G. Performance of the Vienna ab initio simulation package (VASP) in chemical applications. J Mol Struct Theochem 2003;624:37–45.
[83] Lin X, Lu J, Zhu H. The stability and electronic properties of a new allotrope of silicene and silicon nanotube. Superlattices Microstruct 2017;101:480–7.
[84] Medeiros PVC, Mascarenhas AJS, De Brito Mota F, De Castilho CMC. A DFT study of halogen atoms adsorbed on graphene layers. Nanotechnology 2010;21:485701.
[85] Lalmi B, Oughaddou H, Enriquez H, Kara A, Ś V, Ealet B, Aufray B. Epitaxial growth of a silicene sheet. Appl Phys Lett 2010;97:223109.
[86] Vacacela Gomez C, Pisarra M, Gravina M, Riccardi P, Sindona A. Plasmon properties and hybridization effects in silicene. Phys Rev B 2017;95:85419.
[87] Duo I, Fujishima A, Comninellis C. Electron transfer kinetics on composite diamond (sp^3)-graphite (sp^2) electrodes. Electrochem Commun 2003;5:695–700.
[88] Kang MH, Lee D, Sung J, Kim J, Kim BH, Park J. Structure and chemistry of 2D materials. Compreh Nanosci Nanotechnol 2019;1–5:55–90.
[89] Liu G, Liu SB, Xu B, Ouyang CY, Song HY. First-principles study of the stability of free-standing germanene in oxygen atmosphere. J Appl Phys 2015;118.
[90] Donarelli M, Ottaviano L. 2D materials for gas sensing applications: a review on graphene oxide, mos2, ws2 and phosphorene. Sensors (Switzerland) 2018;18.
[91] Sosa AN, Miranda Á, Pérez LA, Trejo A, Cruz-Irisson M. CO and CO_2 adsorption performance of transition metal-functionalized germanene. Mater Lett 2021;300.
[92] Sun M, Ren Q, Wang S, Zhang Y, Du Y, Yu J, Tang W. Magnetism in transition-metal-doped germanene: a first-principles study. Comput Mater Sci 2016;118:112–6.
[93] Sosa AN, de Santiago F, Miranda Á, Trejo A, Salazar F, Pérez LA, Cruz-Irisson M. Alkali and transition metal atom-functionalized germanene for hydrogen storage: a DFT investigation. Int J Hydrogen Energy 2021;46:20245–56.
[94] Zhou Q, Ju W, Yong Y, Liu Y, Li J. Quantum capacitance of supercapacitor electrodes based on germanene influenced by vacancy and co-doping: a first-principles study. Comput Mater Sci 2021;188.
[95] Bugaris DE, Malliakas CD, Han F, Calta NP, Sturza M, Krogstad MJ, Osborn R, Rosenkranz S, Ruff JPC, Trimarchi G, Bud'ko SL, Balasubramanian M, Chung DY, Kanatzidis MG. Charge density wave in the new polymorphs of RE2Ru3Ge5 (RE = Pr, Sm, Dy). J Am Chem Soc 2017;139:4130–43.
[96] Morozkin AV, Seropegin YD. Sm-Ru-Ge system at 1070 K. J Alloys Compd 2004;365:168–72.
[97] Hoat DM, Nguyen DK, Ponce-Pérez R, Guerrero-Sanchez J, On VV, Rivas-Silva JF, Cocoletzi GH. Opening the germanene monolayer band gap using halogen atoms: an efficient approach studied by first-principles calculations. Appl Surf Sci 2021;551.
[98] Chen SC, Wu JY, Lin CY, Lin MF. Theory of magnetoelectric properties of 2D systems. IOP Publishing; 2017.
[99] Lin C-Y, Ho C-H, Wu J-Y, Do T-N, Shih P-H, Lin S-Y, Lin M-F. Diverse quantization phenomena in layered materials. CRC Press; 2019.
[100] Oka T, Aoki H. All optical measurement proposed for the photovoltaic Hall effect. J Phys Conf Ser 2011;334:012060.
[101] Niimi Y, Matsui T, Kambara H, Tagami K, Tsukada M, Fukuyama H. Scanning tunneling microscopy and spectroscopy of the electronic local density of states of graphite surfaces near monoatomic step edges. Phys Rev B—Condens Matter Mater Phys 2006;73:85421.
[102] Morishita S, Senga R, Lin Y-C, Sawada H, Suenaga K. Highly depth-sensitive TEM imaging of graphene by using monochromatic electron source at low accelerating voltage. Microsc Microanal 2018;24:1610–1.
[103] Huang H, Wei D, Sun J, Wong SL, Feng YP, Neto AHC, Wee ATS. Spatially resolved electronic structures of atomically precise armchair graphene nanoribbons. Sci Rep 2012;2:983.
[104] Talirz L, Söde H, Cai J, Ruffieux P, Blankenburg S, Jafaar R, Berger R, Feng X, Müllen K, Passerone D, Fasel R, Pignedoli CA. Termini of bottom-up fabricated graphene nanoribbons. J Am Chem Soc 2013;135:2060–3.

[105] Stolyarova E, Kwang TR, Ryu S, Maultzsch J, Kim P, Brus LE, Heinz TF, Hybertsen MS, Flynn GW. High-resolution scanning tunneling microscopy imaging of mesoscopic graphene sheets on an insulating surface. Proc Natl Acad Sci U S A 2007;104:9209–12.
[106] Fujita D, Kumakura T, Onishi K, Sagisaka K, Ohgi T, Harada M. Sprout-like growth of carbon nanowires on a carbon-doped Ni(1 1 1) surface. Surf Sci 2004;566–568:361–6.
[107] Pierucci D, Sediri H, Hajlaoui M, Girard JC, Brumme T, Calandra M, Velez-Fort E, Patriarche G, Silly MG, Ferro G, Soulière V, Marangolo M, Sirotti F, Mauri F, Ouerghi A. Evidence for flat bands near the Fermi level in epitaxial rhombohedral multilayer graphene. ACS Nano 2015;9:5432–9.
[108] Passoni M, Donati F, Li Bassi A, Casari CS, Bottani CE. Recovery of local density of states using scanning tunneling spectroscopy. Phys Rev B—Condens Matter Mater Phys 2009;79.
[109] Wang Y, Zheng J, Ni Z, Fei R, Liu Q, Quhe R, Xu C, Zhou J, Gao Z, Lu J. Half-metallic silicene and germanene nanoribbons: towards high-performance spintronics device. Nano 2012;7:1250037.
[110] Wang Y, Ding Y. Mechanical and electronic properties of stoichiometric silicene and germanene oxides from first-principles. Phys Status Solidi Rapid Res Lett 2013;7:410–3.
[111] Rahman MH, Chowdhury EH, Redwan DA, Hong S. Computational characterization of thermal and mechanical properties of single and bilayer germanene nanoribbon. Comput Mater Sci 2021;190.
[112] Samipour A, Dideban D, Heidari H. First principles study of the ambipolarity in a germanene nanoribbon tunneling field effect transistor. ECS J Solid State Sci Technol 2019;8:M111–7.
[113] Dávila ME, Xian L, Cahangirov S, Rubio A, Le Lay G. Germanene: a novel two-dimensional germanium allotrope akin to graphene and silicene. New J Phys 2014;16:95002.
[114] Bayani AH, Dideban D, Vali M, Moezi N. Germanene nanoribbon tunneling field effect transistor (GeNR-TFET) with a 10 nm channel length: analog performance, doping and temperature effects. Semicond Sci Technol 2016;31.
[115] Miyake T, Saito S. Quasiparticle band structure of carbon nanotubes. AIP Conf Proc 2005;772:1037–8.
[116] Bostwick A, Ohta T, Seyller T, Horn K, Rotenberg E. Quasiparticle dynamics in graphene. Nat Phys 2007;3:36–40.
[117] Kotov VN, Uchoa B, Pereira VM, Guinea F, Castro Neto AH. Electron-electron interactions in graphene: current status and perspectives. Rev Mod Phys 2012;84:1067–125.
[118] Kaloni TP, Modarresi M, Tahir M, Roknabadi MR, Schreckenbach G, Freund MS. Electrically engineered band gap in two-dimensional GE, SN, and PB: a first-principles and tight-binding approach. J Phys Chem C 2015;119:11896–902.
[119] Bravo S, Correa J, Chico L, Pacheco M. Tight-binding model for opto-electronic properties of penta-graphene nanostructures. Sci Rep 2018;8.
[120] Do TN, Gumbs G, Shih PH, Huang D, Chiu CW, Chen CY, Lin MF. Peculiar optical properties of bilayer silicene under the influence of external electric and magnetic fields. Sci Rep 2019;9.
[121] Kahl K, Kintscher N. Geometric multigrid for the tight-binding Hamiltonian of graphene. SIAM J Numer Anal 2018;56:499–519.
[122] Chiu YH, Lai YH, Ho JH, Chuu DS, Lin MF. Electronic structure of a two-dimensional graphene monolayer in a spatially modulated magnetic field: Peierls tight-binding model. Phys Rev B—Condens Matter Mater Phys 2008;77.
[123] Konschuh S, Gmitra M, Fabian J. Tight-binding theory of the spin-orbit coupling in graphene. Phys Rev B—Condens Matter Mater Phys 2010;82:245412.
[124] Fan Y, Liu X, Wang J, Ai H, Zhao M. Silicene and germanene on InSe substrates: structures and tunable electronic properties. Phys Chem Chem Phys 2018;20:11369–77.
[125] Jiang L, Marconcini P, Hossian MS, Qiu W, Evans R, Macucci M, Skafidas E. A tight binding and →k. →p study of monolayer stanene. Sci Rep 2017;7.
[126] Shyu FL, Lin MF. Loss spectra of graphite-related systems: a multiwall carbon nanotube, a single-wall carbon nanotube bundle, and graphite layers. Phys Rev B—Condens Matter Mater Phys 2000;62:8508–16.
[127] Lin CY, Wu JY, Ou YJ, Chiu YH, Lin MF. Magneto-electronic properties of multilayer graphenes. Phys Chem Chem Phys 2015;17:26008–35.
[128] Do TN, Shih PH, Chang CP, Lin CY, Lin MF. Rich magneto-absorption spectra of AAB-stacked trilayer graphene. Phys Chem Chem Phys 2016;18:17597–605.
[129] Tegenkamp C, Pfnür H, Langer T, Baringhaus J, Schumacher HW. Plasmon electron-hole resonance in epitaxial graphene. J Phys Condens Matter 2011;23.

[130] Ho YH, Chiu YH, Lin DH, Chang CP, Lin MF. Magneto-optical selection rules in bilayer Bernal graphene. ACS Nano 2010;4:1465–72.
[131] Shih PH, Chiu YH, Wu JY, Shyu FL, Lin MF. Coulomb excitations of monolayer germanene. Sci Rep 2017;7.
[132] Lin CY, Huang BL, Gumbs G, Lin MF. Geometry-diversified coulomb excitations in trilayer AAB stacking graphene. Phys Rev 2018;98:195442.
[133] Zhang Y, Tan YW, Stormer HL, Kim P. Experimental observation of the quantum Hall effect and Berry's phase in graphene. Nature 2005;438:201–4.
[134] Zhang L, Zhang Y, Camacho J, Khodas M, Zaliznyak I. The experimental observation of quantum Hall effect of lD3 chiral quasiparticles in trilayer graphene. Nat Phys 2011;7:953–7.
[135] Jiao N, Kang S, Han K, Shen X, Wang W. Shape effects on graphene magnetoplasmons. Phys Rev B 2019;99.
[136] Kiryu T, Koshino M. Quantum Hall effect in three-dimensional graphene. Phys Rev B 2019;99.
[137] Yaokawa R, Ohsuna T, Morishita T, Hayasaka Y, Spencer MJS, Nakano H. Monolayer-to-bilayer transformation of silicenes and their structural analysis. Nat Commun 2016;7:1–6.
[138] Li H, Hui-Xia F, Meng S. Silicene: from monolayer to multilayer—a concise review. Chin Phys B 2015;24.
[139] Lee JK, Lee SC, Ahn JP, Kim SC, Wilson JIB, John P. The growth of AA graphite on (111) diamond. J Chem Phys 2008;129.
[140] Geim AK, MacDonald AH. Graphene: exploring carbon flatland. Phys Today 2007;60:35–41.
[141] Dávila ME, Le Lay G. Few layer epitaxial germanene: a novel two-dimensional Dirac material. Sci Rep 2016;6:20714.
[142] Liu H, Neal AT, Zhu Z, Luo Z, Xu X, Tománek D, Ye PD. Phosphorene: an unexplored 2D semiconductor with a high hole mobility. ACS Nano 2014;8:4033–41.
[143] Yang QQ, Liu RT, Huang C, Huang YF, Gao LF, Sun B, Huang ZP, Zhang L, Hu CX, Zhang ZQ, Sun CL, Wang Q, Tang YL, Zhang HL. 2D bismuthene fabricated: via acid-intercalated exfoliation showing strong nonlinear near-infrared responses for mode-locking lasers. Nanoscale 2018;10:21106–15.
[144] Pumera M, Sofer Z. 2D monoelemental arsenene, antimonene, and bismuthene: beyond black phosphorus. Adv Mater 2017;29.
[145] Lin SY, Chang SL, Thuy Tran NT, Yang PH, Lin MF. H-Si bonding-induced unusual electronic properties of silicene: a method to identify hydrogen concentration. Phys Chem Chem Phys 2015;17:26443–50.
[146] Tran NTT, Lin SY, Glukhova OE, Lin MF. Configuration-induced rich electronic properties of bilayer graphene. J Phys Chem C 2015;119:10623–30.
[147] Do TN, Shih PH, Gumbs G, Huang D, Chiu CW, Lin MF. Diverse magnetic quantization in bilayer silicene. Phys Rev B 2018;97.
[148] Roome NJ, Carey JD. Beyond graphene: stable elemental monolayers of silicene and germanene. ACS Appl Mater Interfaces 2014;6:7743–50.
[149] Matthes L, Pulci O, Bechstedt F. Massive Dirac quasiparticles in the optical absorbance of graphene, silicene, germanene, and tinene. J Phys Condens Matter 2013;25.
[150] Kaneko S, Tsuchiya H, Kamakura Y, Mori N, Ogawa M. Theoretical performance estimation of silicene, germanene, and graphene nanoribbon field-effect transistors under ballistic transport. Appl Phys Exp 2014;7.
[151] Xia W, Hu W, Li Z, Yang J. A first-principles study of gas adsorption on germanene. Phys Chem Chem Phys 2014;16:22495–8.
[152] Kaloni TP. Tuning the structural, electronic, and magnetic properties of germanene by the adsorption of 3d transition metal atoms. J Phys Chem C 2014;118:25200–8.
[153] Lin C-Y, Wu J-Y, Chiu C-W, Lin M-F. Coulomb excitations and decays in graphene-related systems. CRC Press; 2019.
[154] Geim AK, Novoselov KS. The rise of graphene. In: Nanoscience and technology: a collection of reviews from nature journals. World Scientific; 2009. p. 11–9.
[155] Molle A, Grazianetti C, Tao L, Taneja D, Alam MH, Akinwande D. Silicene, silicene derivatives, and their device applications. Chem Soc Rev 2018;47:6370–87.
[156] Cai B, Zhang S, Hu Z, Hu Y, Zou Y, Zeng H. Tinene: a two-dimensional Dirac material with a 72 meV band gap. Phys Chem Chem Phys 2015;17:12634–8.
[157] Yuhara J, He B, Matsunami N, Nakatake M, Le Lay G. Graphene's latest cousin: plumbene epitaxial growth on a "nano WaterCube". Adv Mater 2019;31:1901017.
[158] Pumera M, An Wong CH. Graphane and hydrogenated graphene. Chem Soc Rev 2013;42:5987–95.

[159] Houssa M, Scalise E, Sankaran K, Pourtois G, Afanas'Ev VV, Stesmans A. Electronic properties of hydrogenated silicene and germanene. Appl Phys Lett 2011;98.
[160] Liu CC, Feng W, Yao Y. Quantum spin Hall effect in silicene and two-dimensional germanium. Phys Rev Lett 2011;107.
[161] Pujari BS, Gusarov S, Brett M, Kovalenko A. Single-side-hydrogenated graphene: density functional theory predictions. Phys Rev B—Condens Matter Mater Phys 2011;84.
[162] Wang Y, Xu X, Lu J, Lin M, Bao Q, Özyilmaz B, Loh KP. Toward high throughput interconvertible graphane-to-graphene growth and patterning. ACS Nano 2010;4:6146–52.
[163] Guillaume SO, Zheng B, Charlier JC, Henrard L. Electronic properties and STM images of doped bilayer graphene. Phys Rev B—Condens Matter Mater Phys 2012;85.
[164] Han NT, Dien VK, Lin MF. Excitonic effects in the optical spectra of Li_2SiO_3 compound. Sci Rep 2021;11.
[165] Dien VK, Pham HD, Tran NTT, Han NT, Huynh TMD, Nguyen TDH, Fa-Lin M. Orbital-hybridization-created optical excitations in Li_2GeO_3. Sci Rep 2021;11.
[166] Yang L, Deslippe J, Park CH, Cohen ML, Louie SG. Excitonic effects on the optical response of graphene and bilayer graphene. Phys Rev Lett 2009;103.
[167] Ju L, Wang L, Cao T, Taniguchi T, Watanabe K, Louie SG, Rana F, Park J, Hone J, Wang F, McEuen PL. Tunable excitons in bilayer graphene. Science (80-.) 2017;358:907–10.
[168] Mak KF, Sfeir MY, Wu Y, Lui CH, Misewich JA, Heinz TF. Measurement of the optical conductivity of graphene. Phys Rev Lett 2008;101.
[169] Chowdhury S, Jana D. A theoretical review on electronic, magnetic and optical properties of silicene. Rep Prog Phys 2016;79.
[170] Matthes L, Pulci O, Bechstedt F. Optical properties of two-dimensional honeycomb crystals graphene, silicene, germanene, and tinene from first principles. New J Phys 2014;16.
[171] Chaudhary RP, Saxena S, Shukla S. Optical properties of stanene. Nanotechnology 2016;27.
[172] Yu XL, Huang L, Wu J. From a normal insulator to a topological insulator in plumbene. Phys Rev B 2017;95.
[173] Lu P, Wu L, Yang C, Liang D, Quhe R, Guan P, Wang S. Quasiparticle and optical properties of strained stanene and stanane. Sci Rep 2017;7.
[174] Cudazzo P, Attaccalite C, Tokatly IV, Rubio A. Strong charge-transfer excitonic effects and the Bose-Einstein exciton condensate in graphane. Phys Rev Lett 2010;104.
[175] Lin CY, Chen RB, Ho YH, Lin MF. Electronic and optical properties of graphite-related systems. 2017.
[176] Lin C-Y, Ho C-H, Wu J-Y, Do T-N, Shih P-H, Lin S-Y, Lin M-F. In: Lin C-Y, Ho C-H, Wu J-Y, Do T-N, Shih P-H, Lin S-Y, Lin M-F, editors. Diverse quantization phenomena in layered materials. CRC Press; 2019.
[177] Peres NMR. The transport properties of graphene. J Phys Condens Matter 2009;21.
[178] Huang HC, Lin SY, Wu CL, Lin MF. Configuration- and concentration-dependent electronic properties of hydrogenated graphene. Carbon N Y 2016;103:84–93.
[179] Tran NTT, Nguyen DK, Glukhova OE, Lin MF. Coverage-dependent essential properties of halogenated graphene: a DFT study. Sci Rep 2017;7:1–13.
[180] Tran NTT, Lin SY, Glukhova OE, Lin MF. π-Bonding-dominated energy gaps in graphene oxide. RSC Adv 2016;6:24458–63.
[181] Lin YT, Lin SY, Chiu YH, Lin MF. Alkali-created rich properties in graphene nanoribbons: chemical bondings. Sci Rep 2017;7.
[182] Sahin H, Peeters FM. Adsorption of alkali, alkaline-earth, and 3d transition metal atoms on silicene. Phys Rev B—Condens Matter Mater Phys 2013;87.
[183] Kaloni TP, Singh N, Schwingenschlögl U. Prediction of a quantum anomalous Hall state in Co-decorated silicene. Phys Rev B—Condens Matter Mater Phys 2014;89.
[184] Wanno B, Tabtimsai C. A DFT investigation of CO adsorption on VIIIB transition metal-doped graphene sheets. Superlattices Microstruct 2014;67:110–7.
[185] Toh RJ, Poh HL, Sofer Z, Pumera M. Transition metal (Mn, Fe, Co, Ni)-doped graphene hybrids for electrocatalysis. Chem Asian J 2013;8:1295–300.
[186] Kakaei K, Rahnavardi M. Synthesis of nitrogen-doped reduced graphene oxide and its decoration with high efficiency palladium nanoparticles for direct ethanol fuel cell. Renew Energy 2021;163:1277–86.
[187] Van Nang L, Van Duy N, Hoa ND, Van Hieu N. Nitrogen-doped graphene synthesized from a single liquid precursor for a field effect transistor. J Electron Mater 2016;45:839–45.

[188] Liu N, Bo G, Liu Y, Xu X, Du Y, Dou SX. Recent progress on germanene and functionalized germanene: preparation, characterizations, applications, and challenges. Small 2019;15.
[189] Hanf MC, Marjaoui A, Stephan R, Zanouni M, Diani M, Sonnet P. Undulated silicene and germanene free-standing layers: why not? J Phys Condens Matter 2020;32.
[190] Liu QJ, Liu ZT, Feng LP, Tian H. First-principles study of structural, elastic, electronic and optical properties of rutile GeO_2 and α-quartz GeO_2. Solid State Sci 2010;12:1748–55.
[191] Hafner J. Ab-initio simulations of materials using VASP: density-functional theory and beyond. J Comput Chem 2008;29:2044–78.
[192] Franz M. Solid-state physics: U-turns strictly prohibited. Nature 2010;466:323–4.
[193] Dresselhaus MS, Dresselhaus G, Saito R. Physics of carbon nanotubes. Carbon N Y 1995;33:883–91.
[194] Yang X, Dou X, Rouhanipour A, Zhi L, Räder HJ, Müllen K. Two-dimensional graphene nanoribbons. J Am Chem Soc 2008;130:4216–7.
[195] Meng X, Tongay S, Kang J, Chen Z, Wu F, Li SS, Xia JB, Li J, Wu J. Stable p- and n-type doping of few-layer graphene/graphite. Carbon N Y 2013;57:507–14.
[196] Büyükyazi M, Mathur S. 3D nanoarchitectures of α-$LiFeO_2$ and α-$LiFeO_2$/C nanofibers for high power lithium-ion batteries. Nano Energy 2015;13:28–35.
[197] Das T, Tosoni S, Pacchioni G. Structural and electronic properties of bulk and ultrathin layers of V_2O_5 and MoO_3. Comput Mater Sci 2019;163:230–40.
[198] Lin MF, Shung KWK. Plasmons and optical properties of carbon nanotubes. Phys Rev B 1994;50:17744–7.
[199] Pennington G, Goldsman N. Semiclassical transport and phonon scattering of electrons in semiconducting carbon nanotubes. Phys Rev B—Condens Matter Mater Phys 2003;68:45426.
[200] Dürkop T, Getty SA, Cobas E, Fuhrer MS. Extraordinary mobility in semiconducting carbon nanotubes. Nano Lett 2004;4:35–9.
[201] Tans SJ, Verschueren ARM, Dekker C. Room-temperature transistor based on a single carbon nanotube. Nature 1998;393:49–52.
[202] Martel R, Schmidt T, Shea HR, Hertel T, Avouris P. Single- and multi-wall carbon nanotube field-effect transistors. Appl Phys Lett 1998;73:2447–9.
[203] Shlafman M, Tabachnik T, Shtempluk O, Razin A, Kochetkov V, Yaish YE. Self aligned hysteresis free carbon nanotube field-effect transistors. Appl Phys Lett 2016;108:163104.
[204] Myodo M, Inaba M, Ohara K, Kato R, Kobayashi M, Hirano Y, Suzuki K, Kawarada H. Large-current-controllable carbon nanotube field-effect transistor in electrolyte solution. Appl Phys Lett 2015;106:213503.
[205] Misewich JA, Martel R, Avouris P, Tsang JC, Heinze S, Tersoff J. Electrically induced optical emission from a carbon nanotube FET. Science (80-.) 2003;300:783–6.
[206] Yuksel R, Sarioba Z, Cirpan A, Hiralal P, Unalan HE. Transparent and flexible supercapacitors with single walled carbon nanotube thin film electrodes. ACS Appl Mater Interfaces 2014;6:15434–9.
[207] Liang S, Ma Z, Wu G, Wei N, Huang L, Huang H, Liu H, Wang S, Peng LM. Microcavity-integrated carbon nanotube photodetectors. ACS Nano 2016;10:6963–71.
[208] Zhang TF, Li ZP, Wang JZ, Kong WY, Wu GA, Zheng YZ, Zhao YW, Yao EX, Zhuang NX, Luo LB. Broadband photodetector based on carbon nanotube thin film/single layer graphene Schottky junction. Sci Rep 2016;6:38569.
[209] He X, Léonard F, Kono J. Uncooled carbon nanotube photodetectors. Adv Opt Mater 2015;3:989–1011.
[210] Mueller T, Kinoshita M, Steiner M, Perebeinos V, Bol AA, Farmer DB, Avouris P. Efficient narrow-band light emission from a single carbon nanotube p-n diode. Nat Nanotechnol 2010;5:27–31.
[211] Pyatkov F, Fütterling V, Khasminskaya S, Flavel BS, Hennrich F, Kappes MM, Krupke R, Pernice WHP. Cavity-enhanced light emission from electrically driven carbon nanotubes. Nat Photonics 2016;10:420–7.
[212] Wang S, Zeng Q, Yang L, Zhang Z, Wang Z, Pei T, Ding L, Liang X, Gao M, Li Y, Peng LM. High-performance carbon nanotube light-emitting diodes with asymmetric contacts. Nano Lett 2011;11:23–9.
[213] Mittal J, Lin KL. Carbon nanotube-based interconnections. J Mater Sci 2017;52:643–62.
[214] Singh K, Raj B. Temperature-dependent modeling and performance evaluation of multi-walled CNT and single-walled CNT as global interconnects. J Electron Mater 2015;44:4825–35.
[215] Hämäläinen SK, Sun Z, Boneschanscher MP, Uppstu A, Ijäs M, Harju A, Vanmaekelbergh D, Liljeroth P. Quantum-confined electronic states in atomically well-defined graphene nanostructures. Phys Rev Lett 2011;107.
[216] Dresselhaus G. Spin-orbit coupling effects in zinc blende structures. Phys Rev 1955;100:580–6.

[217] Arjmand T, Bagheri Tagani M, Rahimpour Soleimani H. The effect of buckling on I-V characteristics of symmetric and asymmetric zigzag germanene nanoribbons: a first-principle calculation. J Phys D Appl Phys 2017;50.
[218] Meservey R, Tedrow PM. Spin-polarized electron tunneling. Phys Rep 1994;238:173–243.
[219] Tran NTT, Lin SY, Lin CY, Lin MF. Geometric and electronic properties of graphene-related systems: chemical bonding schemes. CRC Press; 2017.
[220] Uchida H, Huang D, Grey F, Aono M. Site-specific measurement of adatom binding energy differences by atom extraction with the STM. Phys Rev Lett 1993;70:2040–3.
[221] Meyer RR, Kirkland AI, Saxton WO. A new method for the determination of the wave aberration function for high resolution TEM: 1. Measurement of the symmetric aberrations. Ultramicroscopy 2002;92:89–109.
[222] Matsui T, Kambara H, Niimi Y, Tagami K, Tsukada M, Fukuyama H. STS observations of landau levels at graphite surfaces. Phys Rev Lett 2005;94:226403.
[223] Kano S, Tada T, Majima Y. Nanoparticle characterization based on STM and STS. Chem Soc Rev 2015;44:970–87.
[224] Bhandari AS, Verma A. A tight-binding analysis of models, sheet width, and strain for 2D monoatomic germanium sheets. Mater Exp 2019;9:273–80.
[225] Pandya A, Sangani K, Jha PK. Band gap determination of graphene, h-boron nitride, phosphorene, silicene, stanene, and germanene nanoribbons. J Phys D Appl Phys 2020;53.
[226] Monshi MM, Aghaei SM, Calizo I. Edge functionalized germanene nanoribbons: impact on electronic and magnetic properties. RSC Adv 2017;7:18900–8.
[227] Ildarabadi F, Farghadan R. Edge magnetization and spin-valley-caloritronics in germanene and stanene nanoribbons. J Magn Magn Mater 2021;529.
[228] Hattori A, Yada K, Araidai M, Sato M, Shiraishi K, Tanaka Y. Influence of edge magnetization and electric fields on zigzag silicene, germanene and stanene nanoribbons. J Phys Condens Matter 2019;31.
[229] Pang Q, Zhang CL, Li L, Fu ZQ, Wei XM, Song YL. Adsorption of alkali metal atoms on germanene: a first-principles study. Appl Surf Sci 2014;314:15–20.
[230] Sun M, Tang W, Ren Q, Wang S, Jinyu DY, Zhang Y. First-principles study of the alkali earth metal atoms adsorption on graphene. Appl Surf Sci 2015;356:668–73.
[231] Mao Y, Hao W, Wei X, Yuan J, Zhong J. Edge-adsorption of potassium adatoms on graphene nanoribbon: a first principle study. Appl Surf Sci 2013;280:698–704.
[232] Gospodarev IA, Grishaev VI, Manzhelii EV, Sirenko VA, Syrkin ES, Feodosyev SB. The effect of size quantization on the electron spectra of graphene nanoribbons. Low Temp Phys 2020;46.
[233] Thamaraikannan S, Sunny MR, Pradhan SC. Chirality dependent mechanical properties of carbon nanostructures. Mater Res Express 2019;6.
[234] Nishitani S, Sekiya R, Haino T. Chirality-embedded nanographenes. Angew Chem Int Ed 2020;59:669–73.
[235] Wang HS, Chen L, Elibol K, He L, Wang H, Chen C, Jiang C, Li C, Wu T, Cong CX, Pennycook TJ, Argentero G, Zhang D, Watanabe K, Taniguchi T, Wei W, Yuan Q, Meyer JC, Xie X. Towards chirality control of graphene nanoribbons embedded in hexagonal boron nitride. Nat Mater 2021;20:202–7.
[236] Yuan LH, Gong JJ, Wang DB, Zhang CR, Zhang ML, Su JY, Kang L. Hydrogen storage capacity of alkali metal atoms decorated porous graphene. Wuli Xuebao/Acta Phys Sin 2020;69.
[237] Yao Q, Zhang L, Kabanov NS, Rudenko AN, Arjmand T, Rahimpour Soleimani H, Klavsyuk AL, Zandvliet HJW. Bandgap opening in hydrogenated germanene. Appl Phys Lett 2018;112.
[238] Liu L, Ji Y, Liu L. First principles calculation of electronic, phonon and thermal properties of hydrogenated germanene. Bull Mater Sci 2019;42.
[239] Song L, Yang Z, Liu L, Yuan L, Zhao H, Chen X, Zhang Y, Zheng X. Realizing stable half-metallicity in zigzag silicene nanoribbons with edge dehydrogenation and chemical doping. J Phys Condens Matter 2021;33.
[240] Singh R. Spin-orbit coupling in graphene, silicene and germanene: dependence on the configuration of full hydrogenation and fluorination. Bull Mater Sci 2018;41.
[241] Du Y, Zhuang J, Wang J, Li Z, Liu H, Zhao J, Xu X, Feng H, Chen L, Wu K, Wang X, Dou SX. Quasi-freestanding epitaxial silicene on Ag(111) by oxygen intercalation. Sci Adv 2016;2:e1600067.
[242] Sun M, Wang S, Yu J, Tang W. Hydrogenated and halogenated blue phosphorene as Dirac materials: a first principles study. Appl Surf Sci 2017;392:46–50.

[243] Wu X, Song W, Li Q, Zhao X, He D, Quan Z. Synthesis of lead-free CsGeI3 perovskite colloidal nanocrystals and electron beam-induced transformations. Chem Asian J 2018;13:1654–9.
[244] Yue S, McGuire SC, Yan H, Chu YS, Cotlet M, Tong X, Wong SS. Synthesis, characterization, and stability studies of Ge-based perovskites of controllable mixed cation composition, produced with an ambient surfactant-free approach. ACS Omega 2019;4:18219–33.
[245] Walters G, Sargent EH. Electro-optic response in germanium halide perovskites. J Phys Chem Lett 2018;9:1018–27.
[246] Ming W, Shi H, Du MH. Large dielectric constant, high acceptor density, and deep electron traps in perovskite solar cell material CsGeI3. J Mater Chem 2016;4:13852–8.
[247] Docampo P, Bein T. A long-term view on perovskite optoelectronics. Acc Chem Res 2016;49:339–46.
[248] Pazoki M, Edvinsson T. Metal replacement in perovskite solar cell materials: chemical bonding effects and optoelectronic properties. Sustain Energy Fuels 2018;2:1430–45.

CHAPTER

22

Potential applications

Ngoc Thanh Thuy Tran[1], *Shih-Yang Lin*[2], *Vo Khuong Dien*[2], *Chi-Hsuan Lee*[2], *Hsin-Yi Liu*[2], *Hai Duong Pham*[2], *Thi My Duyen Huynh*[2], *Nguyen Thi Han*[2], *Thi Dieu Hien Nguyen*[2], *Wei-Bang Li*[2] *and Ming-Fa Lin*[2]

[1]Hierarchical Green-Energy Materials (Hi-GEM) Research Center, National Cheng Kung University, Tainan, Taiwan; [2]Department of Physics, National Cheng Kung University, Tainan, Taiwan

Ever since graphene, the first two-dimensional (2D) material, was successfully fabricated [1], the scientific activity related to this substance has continued unabated. Significant effort on exploiting other graphene-related materials reportedly continues for the purpose of extending potential device applications [2–4]. Following the early exploration of graphene, materials with a honeycomb lattice of Group-IV elements were paid considerable attention [2,5–9]. Germanene, as a two-dimensional form of germanium, was successfully synthesized in the laboratory in 2014 [7]. It displays several remarkable properties arising from graphene's hexagonal symmetry and Dirac cone structure, e.g., high carrier mobilityjavascript:void (0); [10,11]; and excellent optical absorption [12,13]. In contrast to the flat geometrical structure of graphene with sp^2 hybridization of carbon atoms, germanene possesses buckled structure as a consequence of the mixed sp^3 hybridizations of Ge atoms [14]. This allows germanene to surpass graphene in some aspects, such as greater spin–orbit coupling (SOC) and quantum spin Hall effect (QSHE) [15,16], more accessible tunability of the band gap [17], and other related properties. Theoretical calculation indicates that the band gap produced by the spin-orbit coupling is about 24.5 meV [6]. This implies that germanene is an ideal candidate for the observation of the QSHE and has potential applications in spintronic devices [18]. To further expand the range of applications, germanene's properties can be modulated by changing the geometric structures, doping, and applying external electric/magnetic fields.

In a lot of approaches to change the nature of germanene intrinsic, chemical adsorption show is one of the most effective methods. The dopant-created chemical bonds will play an

Fundamental Physicochemical Properties of Germanene-related Materials
https://doi.org/10.1016/B978-0-443-15801-8.00017-7

important role in determining the most stable configuration along with possessing a non-uniform environment and thus create diverse essential properties. The distinct magnetism, which is induced by the host atoms/molecules and guest ones, are closely related to the emergent spintronic and can provide the full information on potential applications. They cover the paramagnetism, diamagnetism, non-magnetism ferromagnetism, and anti-ferromagnetism. These phenomena mainly originate from the orbital currents of valence and conduction electrons, being created by a uniform magnetic field.

As we know, due to the emission of greenhouse gases, global warming has become increasingly severe in recent years, in which carbon dioxide (CO_2) is the main artificial greenhouse gas. Hence, the development of technology for CO_2 separation, capture and storage has been an important issue. Interestingly, recent research [19,20] suggests that lithium- and nitrogen-substituted germanene has a strong effect on the adsorption of CO_2. Theoretical calculations show that the CO_2 capacity of lithium-substituted germanene can reach 12.57 mol/kg [20]. This means that the chemically modified germanene has potential applications in the removal of CO_2.

Enhancing the free carrier density or opening of the bandgap is considered as a significant important role for broadening the applications of germanene-based materials. The previous research results have shown that the transition metal doped- 2D systems will create plenty of novel features [21–24]. They have been reported these systems exhibit magnetism [25,26] and other interesting properties that are promising for electronics [27], optical devices [28,29], and batteries [30]. Specifically, in Chapter 8, we have shown the transition metal (Fe, Co, Ni)-absorbed germanene systems. As a result, iron-/cobalt-doped germanenes exhibited a semiconductor behavior with a finite bandgap value. The atom-, orbital- and spin-projected density of states clearly demonstrate the bands at near Fermi level mainly contributing by Ge-4p_z and Fe-3 d_{z^2} or Co-3d_{yz}, 3d_{zx} orbitals, respectively. The distance between the highest valence state and the lowest occupied state is largely enhanced due to interactions between the Ge and Fe/Co atoms, which is caused by breaking the sublattice symmetry as compared to pristine germanene. As a result, a large indirect band gap of 0.22 and 0.36 eV for Fe- and Co-doped germanene, respectively, are achieved. The obtained values of band gap are 11–18 times larger than that of pristine germanene (~0.02 eV), which could be interesting for many kinds of germanene-based electronic devices. Besides, in the case of Fe-/Co-doped germanene, the system exhibited the total magnetic moment of 1.77 and 1.0 μ_B, respectively, in which the TM atoms contribute a large amount in the magnetic strength. Apparently, such results of our calculations can motivate experimentalists to further develop germanene and functionalized germanene as well as has great potential for Ge-based spintronic device applications.

According to the delicate theoretical calculations, the diverse electronic, magnetic, and optical properties are predicted to survive in 2D germanene adsorbed rare-earth atoms. For example, the narrow-gap germanene adsorbed La and Eu atoms exhibit the non-magnetic and paramagnetic phenomena, respectively. Being reflects on the splitting of spin-up, spin-down band structures, spin-density distribution/net magnetic moments, and spin-polarized van Hove singularities, and thus, these systems are available for spintronic applications. In addition, the specific excitation frequencies due to the prominent inter-/intra-band transitions in the adsorb systems are suitable for optical detectors, sensors, and attenuators. How to combine the generalized tight-binding model and the first-principles calculations are the key points of view in constructing the exact and reliable theoretical framework in the near future. In which, the strong cooperations or competitions among the host-atom-, guest atom induced spin moments, and magnetic field will be fully explored.

To date, solar cell has been extensively investigated in both experiment and theoretical methods by many researchers worldwide [31–35]. Perovskites are a class of compounds defined by the formula AMX_3 where X represents a halogen, such as I−, Br−, Cl−. and A and M are cations of different sizes [36–38]. Apparently, the Perovskite materials exposed to the non-uniform and complex orbital-hybridizations, are highly suitable for studying the complicated physical, chemical, and material phenomena. For example, the $CsGeI_3$ compounds have been chosen in this book to illustrate the basic behaviors by using first-principles calculations. A lot of physical and chemical phenomena of this compound are obviously achieved in here. Indeed, $CsGeI_3$ compound exhibits 12 Cs–I and 6 Ge–I bonds with higher than 5% fluctuations and thus, leading to a non-uniform physical-/chemical environment. Besides, it possesses a unique quasiparticle band structure, direct gap semiconductor of 2 eV, a complex charge density distribution due to various orbitals contributions, strong Van Hove singularities. These results are corresponding to the optical responses under single- and collective excitations & their close connections with the orbital hybridizations, the low reflectance and high efficiency of optical absorption in the visible region (a very short decay length of ~100 Å). Especially, the presence of a very weak binding of exciton-hole pairs, and therefore, it is very easy to produce a current of electricity upon sunlight adsorption [39–42]. Consequently, inorganic Ge-based halide perovskite could be promising candidate for potential applications as solar cell, electro-optics [43–45]. The presented theory framework in this book will be explored to fully comprehending other solar cell and electro-optics materials.

On the other hand, an Sb thin film will be open an energy gap due to the quantum tunneling from the two surface states. However, a four-bilayer Sb thin film can be turned into a conductor by the adsorption of H and nonmagnetic 3*d* TM atoms on its surfaces in our calculations. Furthermore, energy states from the two surfaces possess time-reversal symmetry. We have the option of using the very thin films of only a few layers and obtaining topological conduction at the surfaces of the films. We also can choose different absorbed atoms to influence the surface state. This should promote the application of thin films as vehicles of topological conduction in future research.

A thin film deposited on a substrate may be a valid method to decrease the effect of the tunneling between the two surfaces of the film. It is worth mentioning that the thin film is easy to grow on the substrate. On the other hand, the surface states from the pristine surface and interface states of surface and substrate can be used to realize the electronic and spintronic properties of topological thin films. Furthermore, we also can consider the doping or defect in the substrates to influence the interface states.

In summary, a wide range of applications for germanene and functionalized germanene in various fields have been reported. Diverse germanene nanomaterials in the form of nanowires, nanofilms, nanotube, and mesoporous structures have been used in electrodes for Li-ion batteries, solar-cells, gas sensors, semiconductor devices, optoelectronic detectors, infrared optics, and aspects other as aforementioned. These practical results once again confirm that germanene is one of the potential materials in the nanomaterials industry.

References

[1] Novoselov KS, Geim AK, Morozov SV, Jiang D-e, Zhang Y, Dubonos SV, et al. Electric field effect in atomically thin carbon films. Science 2004;306(5696):666–9.
[2] Tran NTT, Gumbs G, Nguyen DK, Lin M-F. Fundamental properties of metal-adsorbed silicene: a DFT study. ACS Omega 2020;5(23):13760–9.
[3] Yu X, Cheng H, Zhang M, Zhao Y, Qu L, Shi G. Graphene-based smart materials. Nat Rev Mater 2017;2(9):1–13.
[4] Borenstein A, Hanna O, Attias R, Luski S, Brousse T, Aurbach D. Carbon-based composite materials for supercapacitor electrodes: a review. J Mater Chem 2017;5(25):12653–72.
[5] Oughaddou H, Enriquez H, Tchalala MR, Yildirim H, Mayne AJ, Bendounan A, et al. Silicene, a promising new 2D material. Prog Surf Sci 2015;90(1):46–83.
[6] Acun A, Zhang L, Bampoulis P, Farmanbar M, van Houselt A, Rudenko A, et al. Germanene: the germanium analogue of graphene. J Phys Condens Matter 2015;27(44):443002.
[7] Dávila M, Xian L, Cahangirov S, Rubio A, Le Lay G. Germanene: a novel two-dimensional germanium allotrope akin to graphene and silicene. New J Phys 2014;16(9):095002.
[8] Derivaz M, Dentel D, Stephan R, Hanf M-C, Mehdaoui A, Sonnet P, et al. Continuous germanene layer on Al (111). Nano Lett 2015;15(4):2510–6.
[9] Li L, Lu S, Pan J, Qin Z, Wang Y, Wang Y, et al. Buckled germanene formation on Pt (111). Adv Mater 2014;26(28):4820–4.
[10] Ye X-S, Shao Z-G, Zhao H, Yang L, Wang C-L. Intrinsic carrier mobility of germanene is larger than graphene's: first-principle calculations. RSC Adv 2014;4(41):21216–20.
[11] Bishnoi B, Ghosh B. Spin transport in silicene and germanene. RSC Adv 2013;3(48):26153–9.
[12] Matthes L, Pulci O, Bechstedt F. Optical properties of two-dimensional honeycomb crystals graphene, silicene, germanene, and tinene from first principles. New J Phys 2014;16(10):105007.
[13] Bechstedt F, Matthes L, Gori P, Pulci O. Infrared absorbance of silicene and germanene. Appl Phys Lett 2012;100(26):261906.
[14] Liu N, Bo G, Liu Y, Xu X, Du Y, Dou SX. Recent progress on germanene and functionalized germanene: preparation, characterizations, applications, and challenges. Small 2019;15(32):1805147.
[15] Azizi F, Rezania H. Effects of spin-orbit coupling and magnetic field on electronic properties of Germanene structure. Synth Met 2021;273:116707.
[16] Zhang R-w, Ji W-x, Zhang C-w, Li S-s, Li P, Wang P-j. New family of room temperature quantum spin Hall insulators in two-dimensional germanene films. J Mater Chem C 2016;4(10):2088–94.
[17] Ni Z, Liu Q, Tang K, Zheng J, Zhou J, Qin R, et al. Tunable bandgap in silicene and germanene. Nano Lett 2012;12(1):113–8.
[18] Kane CL, Mele EJ. Quantum spin Hall effect in graphene. Phys Rev Lett 2005;95(22):226801.
[19] Monshi MM, Aghaei SM, Calizo I. Doping and defect-induced germanene: a superior media for sensing H_2S, SO_2, and CO_2 gas molecules. Surf Sci 2017;665:96–102.
[20] Aghaei SM, Monshi M, Torres I, Banakermani M, Calizo I. Lithium-functionalized germanene: a promising media for CO_2 capture. Phys Lett 2018;382(5):334–8.
[21] Luo M, Shen YH, Yin TL. Ab initio study of electronic and magnetic properties in TM-doped germanene. J Supercond Nov Magnetism 2017;30(4):1019–24.
[22] Pang Q, Zhang CL, Li L, Fu ZQ, Wei XM, Song YL. Adsorption of alkali metal atoms on germanene: a first-principles study. Appl Surf Sci 2014;314:15–20.
[23] Chandiramouli R, Nagarajan V. Adsorption studies of NH3 molecules on functionalized germanene nanosheet - a DFT study. Chem Phys Lett 2016;665:22–30.
[24] Pang Q, Li L, Zhang LL, Zhang CL, Song YL. Functionalization of germanene by metal atoms adsorption: a first-principles study. Can J Phys 2015;93(11):1310–8.
[25] Sharma DK, Kumar S, Auluck S. Magnetism by embedding 3d transition metal atoms into germanene. J Phys D Appl Phys 2018;51(22).
[26] Sun ML, Ren QQ, Wang S, Zhang YJ, Du YH, Yu J, et al. Magnetism in transition-metal-doped germanene: a first-principles study. Comput Mater Sci 2016;118:112–6.
[27] Ye M, Quhe R, Zheng JX, Ni ZY, Wang YY, Yuan YK, et al. Tunable band gap in germanene by surface adsorption. Physica E 2014;59:60–5.

[28] Dhar N, Jana D. Effect of beryllium doping and vacancy in band structure, magnetic and optical properties of free standing germanene. Curr Appl Phys 2017;17(12):1589–600.
[29] Dhar N, Jana D. Magnetic and optical properties of carbon and silicon decorated free standing buckled germanene: a DFT approach. J Phys Chem Solid 2018;115:332–41.
[30] Panigrahi P, Pal Y, Hussain T, Ahuja R. Application of germanene monolayers as efficient anchoring material to immobilize lithium polysulfides in Li-S batteries. Appl Surf Sci 2021:558.
[31] Lee TD, Ebong AU. A review of thin film solar cell technologies and challenges. Renew Sustain Energy Rev 2017;70:1286–97.
[32] Wang R, Mujahid M, Duan Y, Wang ZK, Xue J, Yang Y. A review of perovskites solar cell stability. Adv Funct Mater 2019;29(47):1808843.
[33] A review on computational modelling of individual device components and interfaces of perovskite solar cells using DFT. In: Jayan KD, Sebastian V, editors. AIP conference proceedings. AIP Publishing LLC; 2019.
[34] Mohajeri A, Omidvar A. Fullerene-based materials for solar cell applications: design of novel acceptors for efficient polymer solar cells—a DFT study. Phys Chem Chem Phys 2015;17(34):22367–76.
[35] Shi J, Yun S. First-principles DFT calculations for perovskite solar cells. Counter electrodes for dye-sensitized and perovskite solar cells, vol. 2; 2018. p. 487–509.
[36] Zhou T, Zhang Y, Wang M, Zang Z, Tang X. Tunable electronic structures and high efficiency obtained by introducing superalkali and superhalogen into AMX3-type perovskites. J Power Sources 2019;429:120–6.
[37] Gao P, Grätzel M, Nazeeruddin MK. Organohalide lead perovskites for photovoltaic applications. Energy Environ Sci 2014;7(8):2448–63.
[38] Targhi FF, Jalili YS, Kanjouri F. MAPbI3 and FAPbI3 perovskites as solar cells: case study on structural, electrical and optical properties. Results Phys 2018;10:616–27.
[39] Bagher AM, Vahid MMA, Mohsen M. Types of solar cells and application. Am J Opt Photon 2015;3(5):94–113.
[40] Grätzel M. Conversion of sunlight to electric power by nanocrystalline dye-sensitized solar cells. J Photochem Photobiol Chem 2004;164(1–3):3–14.
[41] Dincer F, Akgol O, Karaaslan M, Unal E, Sabah C. Polarization angle independent perfect metamaterial absorbers for solar cell applications in the microwave, infrared, and visible regime. Prog Electromag Res 2014;144:93–101.
[42] Yu Q, Wang Y, Yi Z, Zu N, Zhang J, Zhang M, et al. High-efficiency dye-sensitized solar cells: the influence of lithium ions on exciton dissociation, charge recombination, and surface states. ACS Nano 2010;4(10):6032–8.
[43] Muhammad N, Fu T, Liu Q, Tang X, Deng Z-L, Ouyang Z. Plasmonic metasurface absorber based on electro-optic substrate for energy harvesting. Materials 2018;11(11):2315.
[44] Gaith MS, Al Hayek I. Mechanical Evaluation of the Hetrojunction Layers in Solar Photovoltaic Cells.
[45] Nann T, Skinner WM. Quantum dots for electro-optic devices. ACS Nano 2011;5(7):5291–5.

CHAPTER

23

Open issues and future challenges

Thi Dieu Hien Nguyen[1], *Shih-Yang Lin*[1], *Vo Khuong Dien*[1], *Chi-Hsuan Lee*[1], *Hsin-Yi Liu*[1], *Hai Duong Pham*[1], *Thi My Duyen Huynh*[1], *Nguyen Thi Han*[1], *Ngoc Thanh Thuy Tran*[2], *Wei-Bang Li*[1] *and Ming-Fa Lin*[1]

[1]Department of Physics, National Cheng Kung University, Tainan, Taiwan; [2]Hierarchical Green-Energy Materials (Hi-GEM) Research Center, National Cheng Kung University, Tainan, Taiwan

This book has developed the germanium-based main-stream materials, clearly illustrating a grand framework of composite quasiparticles through the modulation engineering of condensed-matter physics [1–4], physical chemistry and material sciences. The up-to-date theoretical researches show that quantum quasiparticles could be classified into three subgroups (I) The atom- and orbital-based chemical bondings, such as, 2D pure π & π bondings, 2D impure π, π & π ones, and 3D (s, p_x, p_y, p_z)-(s′, p_x', p_y', p_z'), respectively in 2D graphene/silicene [5]/[6–8], 2D silicon carbide/III-VI compounds [9], and 3D silicon carbide/layered III-VI semiconductor (SiC/[10]). (II) The partial orbital hybridizations in optical transitions and Coulomb excitations. The prominent absorption peaks and the excitonic structures are assigned from the energy-dependent mixings of the initial and final states. Furthermore, the π and π plasmon modes in graphene-related systems, respectively, appear at the middle and high frequencies (<10 eV and >20 eV; [11,12]). (III) The stable and intermediate crystal structures during the charging/discharging processes of battery ion transports [13,14], e.g., the time-resolved configurations due to the continuous variation of lithium ion. A lot of III-VI-related semiconductor compounds have clearly revealed the diverse quasiparticle phenomena in which the various crystal phases, energy spectra, spatial orbital deformations, merged van Hove singularities, and featured absorption structures can provide an outstanding platform in finishing the strategy engineering. More rich composite quasiparticles are expected to appear in other semiconductor compounds, e.g., the optical modulation devices due to threshold gap [15]. In addition to VASP simulations, the direct combinations between the first-principles method and molecular dynamics will be useful in clarifying the

Fundamental Physicochemical Properties of Germanene-related Materials
https://doi.org/10.1016/B978-0-443-15801-8.00005-0

time revolution of a stable crystal structure and the various crystal phases. The time-dependent LDA is available to present the diverse (momentum, frequency) Coulomb -excitation phase diagrams with the rich electron-hole pair excitations and plasmon modes [16]. A grand quasiparticle framework will be extended the coupling and/or couplings among electrons, phonons, and photons. Their strong partnerships should be coved in the current and near further researches of the basic and applied sciences. All the above-mentioned phenomena, which mainly arise from the modulation engineering of physical, chemical, and material environments, can be united together only through a grand quasiparticle framework.

A lot of open issues and challenges, being associated with the intrinsic and extrinsic limits, principally originate from the close relations among the high-precision measurements [17–19], the first-principles simulations [6,20–25] and the phenomenological models [16,26,27]. For example, it is almost mission impossible to synthesize and examine all the quasi-stable crystal structures of bilayer group-IV materials except for graphenes [2], mainly owing to very complex intralayer/interlayer multi-orbital hybridizations [28], prominent buckling structures [29], and ferromagnetic/anti-ferromagnetic spin configurations [30]. The simultaneous identifications of geometric and electronic properties by the nanoscaled STM and STS [19,31–33] might be rather difficult to be achieved under the up-to-date techniques. The VASP simulations are available in fully exploring the various ground state energies [5,6] and the featured electronic [5,6], optical [34] and magnetic properties [35,36]. However, the stacking-dependent tight-binding models cannot be obtained from the well-fitting with the VASP calculations, since the low-lying valence and conduction bands are predicted to exhibit the abnormal energy dispersion relations [5,26]. As a result, the multi-orbital hopping integrals, which are fully assisted by the prominent spin-orbital couplings and the buckled environment, are not clearly examined in the up-to-date theoretical predictions [26]. On the other side, both approaches are consistent with each other for the essential properties in graphene-related layered structures [details in books], such as, those of bulk graphites [5,26], few-layer graphenes [5,26], 1D carbon nanotubes [12,37], and graphene nanoribbons [38,39]. This is dominated by the single-$2p_z$-orbital intralayer & interlayer atomic interactions. How to efficiently deal with the diverse quasi-particle behaviors in non-graphene group-IV materials will become the near-future studying focuses, especially for silicon- and germanium-related semiconductors [30,40,41].

All monolayer group-IV materials have the hexagonal honeycombs crystals, in which their buckled structures of A and B sublattices are getting obvious in the increment of atomic number, or the π bonding is greatly reduced by its competitive partner of π. The well-behaved stacking configurations of ABA [42], ABC [7] and AAB [43,44] have been clearly identified for few-layer graphene systems through the high-resolution examinations of STM [19,33,45] and TEM [46,47]. Such crystal structures could survive in thin films and even bulk graphites, i.e., the graphitic/graphene honeycomb lattices keep planar geometries under the very strong π bondings and the weak, but significant van der Waals [48]. As for other group-IV few-layer systems, these characteristics are drastically changed by the obvious hybridizations of π and π orbitals, more complex buckled structures, and the enhanced spin-orbital couplings, e.g., the rich geometric symmetries and band structures of bilayer germanene (details in Chapter 4) and silicene systems (book [6]). The crossover behavior of two totally different orbital hybridizations, the dominance interchange of 2D-π and 3D-π chemical bondings, is expected to appear quickly in the increment of layer number. This will be achieved by the direct combination of the first-principles method and molecular dynamics [20,49].

In addition to bulk and layered structures [50–52], many low-dimension materials can generate the diverse crystal symmetries [8,53,54]. For example, the condensed-matter systems cover 1D coaxial carbon/silicon nanotubes [55,56], graphene/silicene/germanene nanoribbons [57–59]/[60,61]/[41], 0D carbon fullerenes [62], carbon disks [63], carbon onions [64], carbon rings [32], carbon chains [65], carbon nanoclusters [66], as well as the carbon-composite systems [67]. Specifically, each hollow carbon nanotube clearly displays three types of geometric structures, namely, almost straight [68], toroidal (Fig. 23.1A; [69,70]) and coiled forms [71]. There also exist 3D single- and multi-walled carbon nanotube bundles (Fig. 23.1B; [72,73]). Four types of 1D finite-width carbon honeycomb structures, planar (Fig. 23.2A; [29]), curved (Fig. 23.2B; [74,75]), scrolled (Fig. 23.2C; [76]) and folded (Fig. 23.2D; [77]) geometries, have been successfully synthesized in experimental laboratories. Whether the other group-IV-related systems could survive through the experimental syntheses [78,79] requires the theoretical predictions of VASP simulations [20,80] and the high-precision STM/TEM/LEED examinations [19,31,32,46,47,81], e.g., the existence/absence of germanium/tin/lead nanotubes (buckling-, radius- and chirality-defined crystals), tinene/plumbene nanoribbons (buckling-, width-, & edge-characterized geometries), and tin/lead quantum dots. The orbital- and spin-enriched intrinsic interactions should play the critical roles in determining the various crystal symmetries and thus are responsible for the diverse quasiparticle phenomena. Their concise pictures can be clearly identified from the relevant physical quantities (details in Chapter 2; [5,6]).

The time-dependent crystal structures frequently come into existence during the charging and discharging processes in ion-based batteries [82,83]. For example, the lithium-ion transports are able to enhance/reduce π concentrations and dramatically alter their distribution configurations in cathode/electrolyte/anode materials [such as, the ternary or quaternary lithium oxides in Refs. [23,84]/[85–87], obviously leading to the dynamic behaviors. Generally speaking, the theoretical predictions [20,80] and experimental examinations [88–90] are only focused on very few stable or quasi-stable geometric structures, e.g., the crystal symmetries of the initial and final states during the chemical reactions. How to predict the continuous variations of crystal structures and identify/observe the dominating inter-mediate configurations is a main-stream topic in the current studies. This is very useful in thoroughly understanding the main features of quasiparticle transports. How to directly link geometric symmetries/asymmetries [50,91], electronic properties, as well as, stationary ion currents, is worthy of systematic investigations. Evaluate whether their close relations could account for the critical mechanisms of ion transport properties [92].

The lithium-, phosphorus- and sulfur-related condensed-matter systems play the critical roles in a plenty of commercial applications, e.g., the high-performance lithium sulfide batteries [93–95]. These 3D compounds could be classified into binary (π/LiP/ P_4S_x π in Refs. [96]/[97]/[98]), ternary of Li_3PS_4 [99], and quaternary of LPSO [100]. These three elements are able to build the various 3D allotropes with the totally from crystal symmetries. Each compound is expected to be a Moire superlattice with the position-dependent chemical bonds, clearly indicating the highly non-uniform chemical environment. Furthermore, their special functionalities could be greatly modulated by the changes of their relative concentrations and distribution configurations; therefore, there exist the highly potential applications [101,102] (R). Apparently, they are outstanding candidates in exploring the diversified quasiparticle phenomena and testing the development of theoretical framework for the current

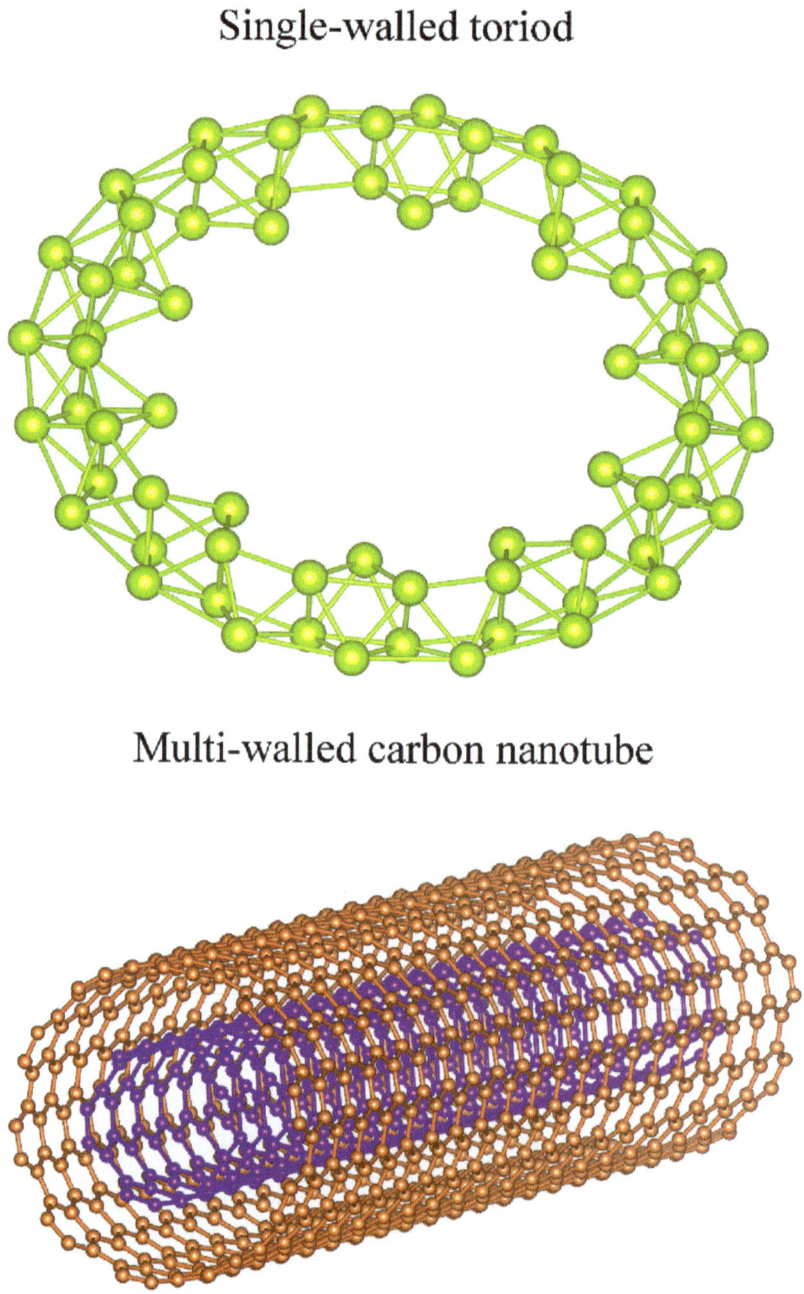

FIG. 23.1 (A) A single-walled toroidal [321] and (B) multi-walled carbon nanotube.

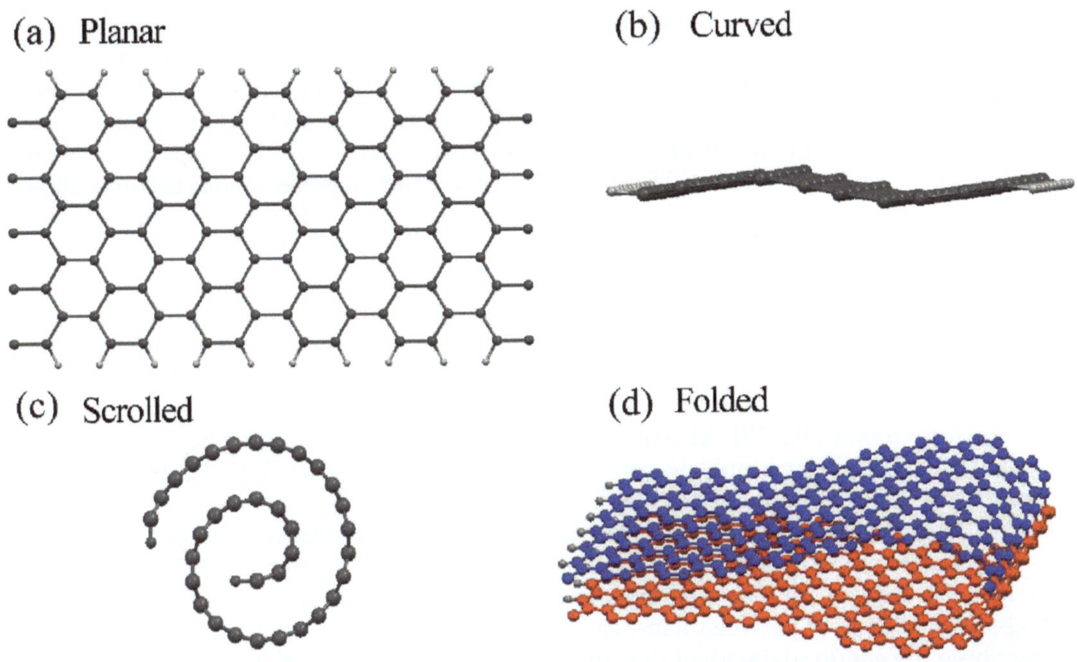

FIG. 23.2 Four types of 1D finite-width graphene nanoribbons with (A) planar, (B) curved, (C) scrolled and (D) folded honeycomb lattices.

book. It would be very interesting to systemically comprehend the close relations among the component-dependent crystal structures, the active chemical bondings and the other fundamental properties (detail in books [5,6,103]). The diverse orbital hybridizations in various chemical bonds of the same system/the different compounds could be further identified from the spatial charge density distributions, compared with those of isolated atoms [23,104]. The rich and unique quasiparticle phenomena, which are expected to be clearly revealed in various Li-, P- and S-based compounds, are under the thoroughly current investigations. The similar studies could be further generalized to aluminum- [99] and iron-related-ion batteries [105]. For example, the intercalations or de-intercalations. Which mainly arise from large-molecule ions of $AlCl_4^{-1}/Al_2Cl_7^{-1}$ between graphitic layers [106], serve as the high-performance anode of the former. The stable crystal structures are greatly diversified by their distributions on the honeycomb lattices and the periodically stacking direction [5], as well as the bonding angles and lengths of large-molecule ions [5]. In short, the crystal symmetries of the different ion batteries could be summarized together, the significant orbital hybridizations & spin configurations are clearly identified from the relevant physical/chemical quantities, and then the critical mechanisms are responsible for the diverse quasiparticle phenomena. This is the theoretical framework in the previous and current books [6].

Rather obviously, electronic and magneto-electronic of main-stream materials display the diverse physics/chemistry/material phenomena and the frequent applications, as done in the previous experiments and predictions. However, the quasiparticle energy spectra of certain specific systems cannot be examined by the high-resolution angle-resolved photoemission spectroscopy and/or scanning tunneling spectroscopy due to the intrinsic limits. For example, the VASP simulations are able to predict band structures of 3D compounds for Moire superlattices within very small first Brillouine zones, such as, those of ternary/quaternary/pentanary lithium oxides in π-ion-based batteries with many weakly dispersive valence subbands. Furthermore, the direct ARPES examinations and analyses would become mission impossible in the absence of momentum transfer conservation along the perpendicular direction and without the featured occupied energy dispersions near the high-symmetry points. The similar 3D compounds frequently come to exist in lithium sulfide batteries [93–95], aluminum- [107–109] and iron-related batteries [105], as well as, their partners with chemical adsorptions or substitutions. Very interestingly, the momentum non-conservation frequently appears in the curved low-dimension crystals, e.g., the disappearance of the well-defined wave vector along the symmetric axis of a very long carbon nanotube, leading to the absence of the reliable ARPES measurements [19,110]. Also noticed that the ARPES measurement have been successfully conducted on planar graphene nanoribbons (details in Chapter 3; [111]), but not folded [77], curved [74,75] and scrolled ones [76].

Both STM and STS measurements are frequently utilized to characterize geometric structures and van Hove singularities simultaneously (details in Chapter 3; [19,32,33,112]). The latter are operated under the well-known quantum tunneling effects, in which the gate-voltage-induced current between nanoscale device and sample is very weak, but significant in describing the energy-dependent density of states across the Fermi level. They provide the total contributions from the active atoms/orbitals in chemical bonds, being compared with the atom-, orbital and spin-decomposed density of states through the delicate VASP calculations. Apparently, such examinations are suitable for 2D few-layer structures [5,113], 1D nanotubes & nanoribbons [74,114,115], and 0D quantum dots [116], such as, very successful identifications on the close relations between geometric symmetries (radii & chiralities) and band properties (metals, narrow-and middle-gap semiconductors) for single-walled carbon nanotubes [117–119], and similar observations about edge structures and electronic properties for graphene nanoribbons [120–122]. That is, band property at the Fermi level (semiconductor, semimetal or metal; [57]) and valence & conduction van Hove singularities [21,103], show the strong evidences in consistence with the theoretical predictions. As a result, the active single- or multi-orbital hybridizations and the magnetic spin configurations, which correspond to the specific energy ranges can be achieved from the delicate analyses. These quasiparticle properties are further reflected in optical transitions [123–125], Coulomb electronic excitations [16,126], and quantum transports [127]. All the physical, chemical, and material phenomena will be closely unified together under the quasiparticle framework [5,6,103] (Fig. 23 3).

The first-principles simulations [128,129], being supported by the method of molecular dynamics [130], are very powerful in fully exploring most of quasiparticle properties. However, they cannot investigate the giant Moire superlattices with a lot of atoms/molecules in a primitive unit cell beyond several hundreds, in which the highly modulated environments frequently appear under the various cases: an achievable perpendicular magnetic field (the magnetic quantization phenomena due to the periodical vector-potential-dependent Peierls phases; [131]), the spatially modulated electric and/or magnetic external fields [132], the low-concentration adatom chemisorption [24], the dilute guest-atom substitution [133], the

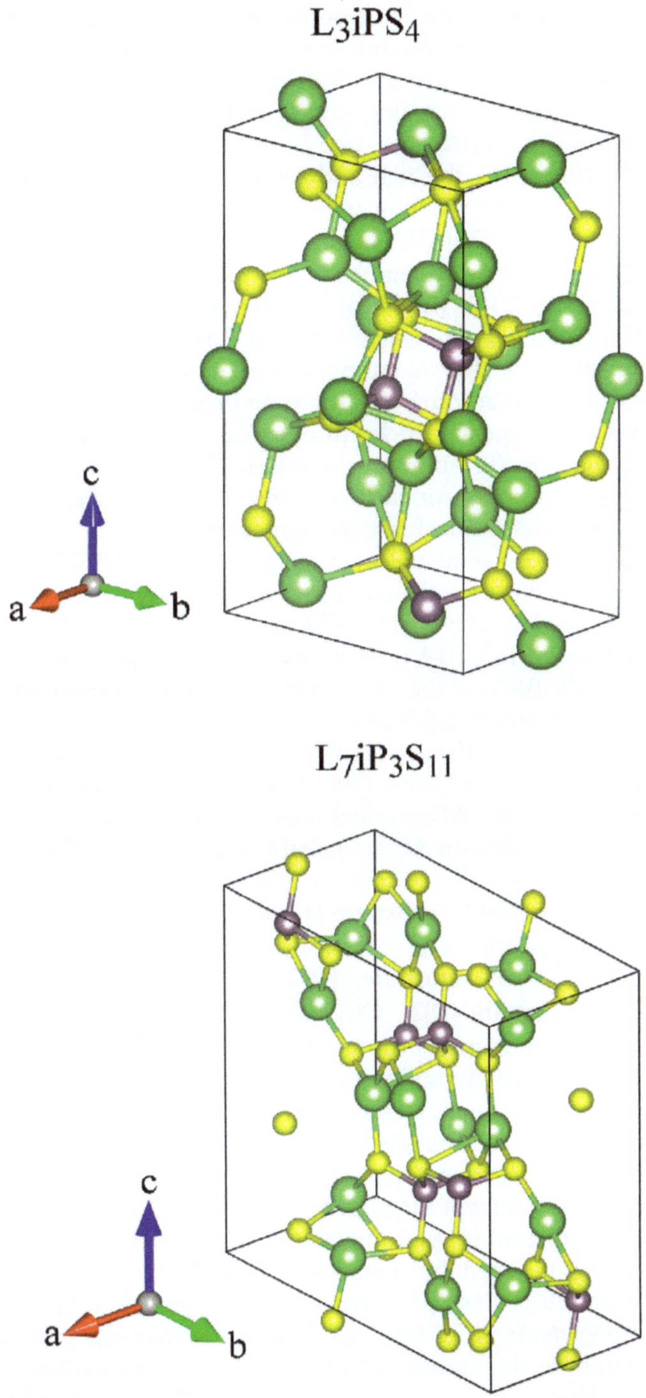

FIG. 23.3 The most stable crystal structure of a 3D ternary LiPS compound.

few-defect modulation structure, the long-period mechanical deformation [134,135], non-commensurate cathode/electrolyte/anode boundaries [136]/[137], the slow modulation of sample and substrate (an incommensurate heterojunction in Ref. [134]), a too thick film, as well as, the highly asymmetric stacking configurations in multi-layer structures [138], multi-walled nanotubes [139,140] & nanotube bundles [73,141,142], and multi-shell onions [143]. As to electronic properties, the rich and unique quasiparticle behaviors are expected to reveal in many emergent binary/ternary/quaternary/pentanary materials and their allotropes, such as, those in 3D/2D [36]/1D/0D $B_xC_yN_z$ [144]/[145]/[146]/[147], 3D Li-, P, & S-related compounds [148], dimension-enriched II-VI/III-V semiconductor compounds with the different substrates (Fig. 23 4; [28,149]), silicone (a silicon-oxygen bone attached with functional groups; [150]), and n- or π-type dopings for graphene-related systems (various intercalations/de-intercalations due to atoms, large molecules or ions; [5]). A plenty of allotropes and varieties are outstanding candidates in providing the systematic investigations about the verifications of the quasiparticle framework. The critical orbital hybridizations in chemical bonds and the concise spin configuration (how to identify sees the details of Chapter 2; books), need to propose for each condensed-matter system; that is, the significant mechanisms are unified under a quasiparticle framework.

In addition to traditional 3D semiconductor compounds [151,152], 2D/1D/0D allotropes and varieties [153,154] have become the emergent materials with the rich and unique quasiparticle phenomena. Through very successfully experimental syntheses, these condensed-matter systems have covered II-V, II-VI, III-V, and III-VI compounds, respectively, such as, BeN_2/BeP_2 [155,156]/[157], BeO on SiC & Ag (111) substrates [158,159]/MgO on Ni (100), Ag (001) & Pd (100) substrates [160]/CaO, BN on Cu (111), Ru (111), and Pt (111) substrates/[161]/AlN on Ag (111) substrate [162]/GaN on graphene, Al_2O_3 (0001) and Si (111) substrates [163–166], and GaS on GaAs (100) [167]/GaSe on graphene [168], GaAs (001) [169], GaN/Sapphire [170], and Mica substrates [171]/GaTe on graphene/InS, InSe on SiO_2/Si substrate [172] & In_2Te_3 on Mylar substrate (polymer; [173]). Furthermore, there also exist ternary and quaternary compounds, e.g., AlGaN on GaN & SiO_2/Si subtrates [174], GaInN [175], AlAsSb on InAs substrate [176], +AlInN on GaN & sapphire substrates [177,178], AlGaP on GaP substrate [179], $Ga_xIn_{1-x}P_yAs_{1-y}$ on GaSe/InP substrate [180], $Al_xIn_{1-x}P_yAs_{1-y}$ on GaAs/InP substrate [181], π on GaAs subtracte, as well as $Al_xGa_yIn_{1-x-y}As$ on InP substrate [182]. Obviously, a plenty of semiconducting materials can greatly diversify the fundamental properties [168,169], largely enhance the specific functionalities [168,169] and extensively widen potential applications [40].

The diversified quasiparticle phenomena, which are closely related to the strong coupling effects due to various composites/crystal symmetries [183], multi-components [184,185], substrates [28,186], excitons [187], carrier static/dynamic screenings [188,189], and coherent atom vibrations [12], are under the current through the solid cooperation of theoretical predictions (NCKU; [190]) and experimental measurements (NYCU; [191]). For example, each semiconductor compound, being induced by a successful synthesis of multi-component system grown on a specific substrate, will display the essential properties totally different from those of two independent subsystems. In addition to the original chemical bondings, its heterojunction is expected to generate the significant multi-orbital hybridizations from the left- and right-hand atoms [6]. Very interestingly, a more complicated chemical environment is

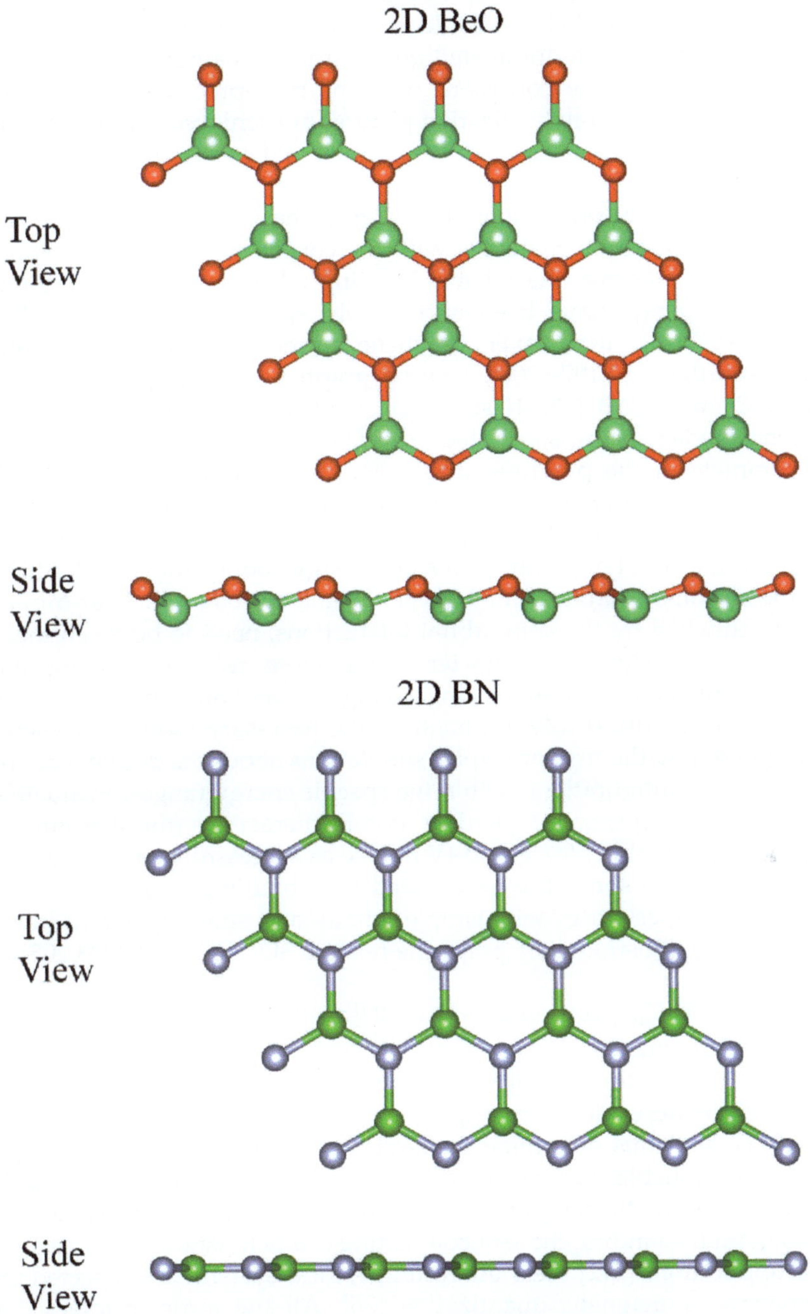

FIG. 23.4 The optimal geometric structures of 2D semiconducting (A) II-VI and III-V compounds with a suitable substrate.

responsible for a large Moire superlattice due to the slow bond-length modulation, thus leading to the total changes of electronic [5], optical [192], Coulomb-excitations [193], magnetic [194], and transport properties [92]. It should be noticed that a suitable substrate width plays a critical role in revealing the co-dominated quasiparticle behaviors of both subsystems, e.g., a three- or four-layer substrate consistent with the high-precision ARPES measurements [195]. While the concise orbital-hybridization pictures are achieved for all the conductor compounds, the very useful information can further promote the great developments of engineering applications [196].

The up-to-date theoretical predictions clearly show that the phenomenological models are developed for the main-stream materials with the concise orbital hybridizations and spin-orbital couplings. For example, the reliable hopping integrals/on site Coulomb potentials, and spin-orbital couplings have been successfully specified for certain 2D condensed-matter systems, such as, monolayer graphene/silicence/germanene/tinene/plumbene/phosphorene/bismuthene [4,197–200], layered graphene systems with AAA [201], ABA [42,53], ABC [202] and AAB [43,44] stackings, bilayer phosphorene materials. However, the tight-binding models, being associated with few-layer inorganic group-IV systems, are not reliable/complete in the previous studies [16,188], mainly owing to the highly non-uniform/anisotropic π bondings, the greatly enhanced interlayer atomic interactions, and the complicated spin-orbital couplings. The single-orbital Hamiltonian should be insufficient in accounting for the low-lying multi-constant-energy loops from the VASP calculations [20,80], e.g., the silicon-$3p_z$ orbitals for bilayer silicene systems [6]. Four-orbital hybridizations, which are enriched by the spin-orbital interactions, need to be taken into account for non-graphene systems. How to characterize the close relations among the interlayer orbital-dependent interactions, the stacking configurations, and the buckled structures of different layers plays a critical role in obtaining the necessary model parameters. This can be partially supported by the first-principles simulations about the dominances of orbital hybridizations and spin configurations within the specific energy ranges. Evaluate whether the empirical formulas of the position-dependent atomic interactions (the strengths of orbital hybridizations) could be obtained for group-IV atoms, as successfully conducted for graphene-related systems [5,203]. In short, the generalized tight-binding models are very convenient only for giant Moire superlattices with simple orbital- and spin-dependent intrinsic interactions [204], as clearly illustrated in graphene-related 3D [132]/2D [203,205]/1D systems [206,207].

Binary and ternary $B_xC_yN_z$ compounds [208,209], with 3D, 2D and 1D structures, might be very suitable in examining the diverse quasiparticle behaviors from the phenomenological models [16,26]. In general, the four orbitals of (2s, $2p_x$, $2p_y$, $2p_z$) generate the active chemical bondings, the position-dependent hopping integrals [26,210] and the different on-site ionization energies [211], and thus determine the electronic/optical/magnetic/transport/thermal properties [26]. The reliable parameters of the generalized tight-binding model could be achieved from the systematic comparisons with the VASP simulations [20,80]. Their values are very useful in understanding the semiconducting characteristics, being frequently utilized in commercial applications [212]. Very interestingly, these systems are expected to present the diverse phenomena of magnetic quantization [26]. All the intrinsic interactions and the external fields are included in Hamiltonian matrix elements simultaneously. Furthermore, the magneto-tight-binding functions are built on the distinct sublattices, e.g., six categories of enlarged carbon-$2p_z$ tight-binding functions for trilayer AAA-/ABA-/ABC-/AAB-stacked graphene systems under a perpendicular magnetic field. This Hermitian matrix has a giant dimension under a periodic Peierls phase associated with vector potential, while it could be solved more efficient through the exact method (details in books [16,26]). The highly degenerate magneto-electronic energy spectra and the unusual oscillations of magneto-wave functions will be clearly revealed in the exact diagonalization of a magnetic Hamiltonian. The zero-point number in a specific sublattice, with a regular localization behavior, could be regarded as the quantum number. 3D/1D and 2D would display Landau subbands and levels, respectively, since the energy dispersions of the former mainly arise from the perpendicular orbital hybridizations interactions/the finite-width quantum confinements [213]/[214]. The main features of magneto-electronic properties are worthy of the further studies, e.g., the number of Landau-level/Landau-subband subgroups & their quantum-number definitions, the sensitive magnetic-field dependences [215], the non-crossing/crossing/anti-crossing behaviors [26], the composite effects due to the superposition magnetic and electric fields [216,217], as well as, the diversified quantization from dimensionalities [218], crystal structures [219] and core components [197,219].

The first-principles simulations are efficient in fully exploring vertically optical transitions for main-stream materials with small primitive unit cells, since the initial and final wave functions are required during the vertical excitations of dynamic charge screenings [188]. Such calculations can deal with the many-body excitonic effects in emergent semiconductors [220]. The systematic investigations, being mainly focused on electron-hole excitations [221] and plasmon modes [12], will be conducted for layered graphene systems with AAA/ABA/ABC/AAB stackings, carbon nanotubes [69,222]/carbon nanotube bundles [73,141], and planar/curved/folded/scrolled graphene nanoribbons [223]/[74,75]/[77]/[76]. In addition, such thorough researches are almost absent up to now. The strong coupling effects from geometric symmetries (hexagons, hollow cylinders and finite-width rectangles), chemical modifications (chemisorptions/substitutions; [149]/[24,133]), transverse electric fields (gate voltages/Coulomb potentials in Ref. [224]), and excitons (the screened electron-hole pairs; [221]) are simultaneously taken into account in the VASP calculations. The close relations among crystal structures, electronic energy spectra & wave functions, and optical transitions need to be clarified through the delicate VASP calculations and analyses. For example, band and optical gaps, which, respectively, play the critical roles in electronic and optical properties, dominate their potential applications [225]. Their sensitive dependences on stacking configurations, layer numbers, electric-field strengths, oxidization/halogenation/alkalizations ([226]/[22,91]/[227,228]) in the presence/absence will be done in the near-future studies. The rich and unique quasiparticle behaviors, which are expected to clearly reveal in 1D carbon nanotubes and graphene nanoribbons, such as, the totally different dependences about radius & chirality [229,230] and width & edge structure [121,231], respectively.

The similar researches could be generalized to silicon- [40,232] and germanium-related low-dimensional materials [30]. The layered crystals can present the dimension crossover phenomena in terms of fundamental physical/chemical/material properties, as expected for graphene- [5], BC_3- [233] and BN-related systems [234,235] during the variation of layer number (from few-layer, thin film and then bulk structures). While the hopping integrals and on-site Coulomb potentials are solved under the layer-dependent tight-binding models [128,236], many unusual features will be clearly revealed in electronic energy spectra, Bloch wave functions, as well as optical absorption structures. For example, monolayer graphene is a zero-gap semiconductor with a vanishing density of states at the Fermi level, bilayer semi-metallic systems have weak band overlaps, and bulk graphites are typical semimetals, where free carrier densities of conduction electrons and valence holes are very sensitive to the stacking configurations AAA [201], ABA [42,53], ABC [237,238], AAB [43,44] and turbostratic ones [239]. Furthermore, the diverse energy dispersions, the linear, parabolic, Sombrero-shape, oscillatory, and partially flat ones, frequently come to exist in 2D and 3D graphene-based materials. A lot of van Hove singularities, being due to the vanishing group velocities of wave packets, are directly reflected in carrier static and dynamic screening responses in the presence of electromagnetic/Coulomb field perturbations. A time-dependent electromagnetic wave, which is normally incident on a finite-width/an infinite graphite-like sample, partially penetrates into this layered structure and the other directly reflects. Of course, the incident, reflected and refracted beams satisfy two boundary conditions (continuities for the normal direction of electric displacement and the tangential one of magnetic field; details in electrodynamics of Jackson; [240]). The former is immediately screened by all the valence and conduction electrons through the well-defined transverse dielectric function. The reliable real- and imaginary parts need to be evaluated under the full band-structure effects (the simultaneous consideration of all intralayer and interlayer atomic interactions; [241]). When the interband electron-hole absorptions are not very strong under specific frequencies, the propagating waves can reach another boundary. The similar scattering appears and thus the refracted beam becomes the initial transmission. The reflected part will experience the repeated behaviors. The analytic formulas, being conducted on the electric fields of the totally reflected and transmitted waves, could be derived according to the reliable frequency- and layer-dependent dielectric function. The reflectance spectra, absorption coefficients and transmission intensities have been under the current investigations [242]. Such rich results might be very helpful in generating the various optical devices [243].

Coulomb excitations clearly indicate static and dynamic charge screenings [16,188,213,216], belong to the main-stream research topics in basic sciences. Both single-particle excitations and plasmon modes, which are confined in the low- and middle-frequency ranges, have been systematically investigated for graphene-related systems using the modified random-phase approximations [244], such as, 3D AA-, AB- and ABC-stacked bulk graphites [213,237], 2D few-layer graphene systems with AAA/ABA/ABC/AAB stackings [201]/[53,245]/[246]/[43,44], 1D coaxial carbon nanotubes [55]/2D aligned carbon nanotubes [247]/3D carbon nanotube bundles [73], and 0D carbon tori [248]. Both atomic and Coulomb interactions, with the intralayer and interlayer ones, are included in the delicate calculations and analyses The coupling effects of band structures [16,126], thermal excitations [249], dopings [203], gate voltages (transverse electric fields; [224,250,251]), perpendicular magnetic fields [250,252,253], dimensionalities [254,255], are clearly identified to create the diverse (momentum/angular momentum, frequency)-excitation phase diagrams, where there exist the different electron-hole excitation regions [220,256] and the unusual dispersion relations of plasmon frequencies [249,257]. The critical temperatures, momenta, gate voltages, and magnetic field strengths can determine the absence of disappearance of single-particle and collective excitation modes. The concise physical pictures are proposed for the various intraband and interband excitation channels, especially for those associated with the band-edge and Fermi-momentum states [220]. The coherent oscillations of band-related carriers are deduced to be responsible for the very interesting many-body Coulomb-field screening behaviors [241]. The modified/extra viewpoints of collective excitations are clearly illustrated in the intrinsic and extrinsic emergent materials (details in books [16,26]).

The static and dynamic screening behaviors, which are revealed under the classical [240] and quantum pictures [256], exhibit the diverse quasiparticle phenomena, such as, the Friedel oscillations under a static Coulomb-field perturbation [258,259], magneto-plasmon modes [202,260] Coulomb decay rates of excited electrons and holes [188], as well as, position- and time-dependent plasma waves [261]. A lot of low-energy physical quantities are dominated by the main features of Fermi surface (Fermi momenta of k_F s; [262]). For example, at T = 0 K, a charged impurity can create an external Coulomb potential which is effectively screened by valence and conduction electrons in any condensed-matter system. Whether the long-range Coulomb interaction would quickly decline around its source is determined by the density of free carriers. This has been fully explored for 1D metallic [231] and semiconducting carbon nanotubes [12], where the former display very short screening lengths of $\sim 1 - 5$ Å and the latter only exhibit the classical distance dependence through the reduced amplitude. Another an interesting evidence of static charge screening in armchair systems (the former) is the long-distance rapid charge oscillations in the 1D form of $\sin(2k_F)/\ r$ ($\sin(2k_F)/\ r^2$ and $\sin(2k_F)/\ r^3$, respectively for 2D and 3D electron gases; [189]). The Fermi surface would become more complicated through the effective modulations about the Fermi-momentum states, such as, the layer numbers, stacking configurations, the n- or π-type doping density and the gate voltages for layered graphene systems [263], leading to the unusual multi-structures of the highest occupied loops [250,264]. The fermi Fermi-momentum-dependent discontinuous electron distributions are able to generate the singular structure in the static polarizations. As a result, the composite Friedel oscillations, directly reflecting the multi-Fermi-loop structures, are expected to exhibit the quasi-beating phenomena. Pristine and n-type-doped trilayer bilayer graphene systems, with AAA [201]/[42,53] ABA/ABC [42,246]/AAB [43,44] stackings AA [265]/AB stackings [213], are outstanding candidates in fully exploring the diversified screening behaviors from the semiconducting, semi-metallic and metallic materials. The similar studies on bilayer silicene/germanene materials are able to provide another unusual charge screenings because of the enhanced π bondings, i.e., the orbital-dependent Fermi surfaces will make the significant contributions the singular dielectric function and thus produce variously complex superpositions of the normal-mode Friedel oscillations. It is very interesting in fully examining the prominent long-distance charge oscillations on 2D surfaces through the STS measurements [31–33,112].

The dynamic charge screenings come to exist at any times and positions when a condensed-matter system is observed by an external ion beam [240]. They are predicted behave as the coherent [11] and/or non-coherent charge oscillations. In general, the time- and position-dependent total Coulomb potentials and induced charge densities can characterize the featured plasma waves. The 2D layered structures [113,266], 1D cylindrical nanotubes [267], finite-width nanoribbons [268], and spherical onions [64] are very suitable for generating the diverse propagations of charge density waves. The momentum-, frequency- and subsystem-dependent response functions has been deduced to dominate the quasiparticle screening phenomena

[269], e.g., the dielectric function tensors of layered graphene systems under the linear response. All the intralayer and interlayer interactions, which mainly arise from orbital hybridizations and Coulomb potentials [188,213], are taken into account simultaneously through the modified random-phase approximation. Very interestingly, electronic properties and their electronic excitations are fully explored under the generalized tight-binding model (discusses later for magneto-plasmon modes; [12,270]); that is, the current studies are reliable in understanding magneto-plasma waves [43,260]. The experimental measurements on the unusual propagation of plasma waves are very sensitive to the initial Coulomb-field perturbation. The theoretical model perturbation is required to be consistent with the external case. For example, a rather rapid electron beam perpendicularly passes a bilayer AA-/AB-stacked graphene [237,271] a trilayer AAA-/ABA-/ABC-/AAB-stacked system; [201]/[42,53]/[246]/[43,44] using a very short time. This is expected to induce the anisotropic and/or isotropic propagations, the separation or superposition of electron-hole and collective excitations, and the existence/coexistence or absence of the Friedel charge oscillations. Specifically, 1D carbon nanotubes provide the perpendicular and parallel cases, in which an ion beams might propagate along the nanotube axis, or perpendicularly penetrate the nanotube center. Both of cases will reveal the totally different responses of dynamic charge screenings. The distinct dimensionalities, crystal structures, orbital hybridizations, spin-orbital couplings, gate voltages, and magnetic fields play the critical roles in observing plasma waves [261].

After the Coulomb- or electromagnetic-field perturbations, the excited quasiparticles can further decay by the electron-electron [240] and electron-photon interactions. At very low temperatures, the latter could be ignored in the experimental measurements. The Coulomb electron-hole and plasmon excitations become the most efficient decay channels [270], being clearly indicated by the various (momentum/angular momentum, frequency)-excitation phase diagrams [16]. The theoretical predictions, being associated with the Coulomb excitations and decay rates, are done by the phenomenological models [16] and the time-dependent first-principles simulations [21,129,272]. However, the thorough and concise physical pictures could be obtained only from the modified random-phase approximation [273] and self-energy method [274,275]. The similar viewpoint, which is based on the layer-dependent sublattices [54], has been successfully utilized in the generalized tight-binding model. That is to say, three kinds of theoretical models are directly combined together in the model calculations and analyses. The quasiparticle framework presents a unification of electronic properties, Coulomb excitations and decay rates. For example, 2D few-layer graphenes and 1D coaxial multi-wall carbon nanotubes, with distinct stacking configurations, represent potential candidates in the systematic investigations about quasiparticle energy spectra [276] and lifetimes [277]. These predictions could be directly examined by the high-resolution ARPES measurements through the wave-vector-dependent spectrum with observable broadening widths [110].

The Coulomb electronic excitations, which appear in the presence of a perpendicular magnetic field, will be greatly diversified by the strong competitions in the Lorentz forces [278]. The magnet-electronic Coulomb excitations could only be explored by the phenomenological models, but not the first-principles simulations. The coherent electron oscillations, being localized around the ion sites, are able to establish the longitudinal Coulomb field. This further leads to the propagating plasma waves, where the group velocity is characterized by the first derivative of plasmon frequency versus momentum transfer. Very interesting, the application of a uniform magnetic force can generate the transverse cyclotron motion and thus induce the abnormal momentum dispersion of plasmon frequency (the unusual group velocity; [11,12]). Moreover, the sensitive 2D wave-vector dependences of electronic states disappear under a full magnetic quantization; that is, all the neighboring states in a constant-energy loop are magnetically quantized into the dispersionless Landau levels (LLs in Refs. [42,238]). These LLs have the delta-function-like density of states so that the inter-LL electron-hole excitations are very prominent and might be accompanied with the sufficiently strong plasmon responses in energy loss functions [279]. As a result, there exist a lot of discrete magneto-plasmon modes with the irregular momentum dispersion relations. Their existences are rather sensitive to the changes of momenta and magnetic field strengths [11,12]. The up-to-date theoretical predictions clearly show that the diverse magnet-electronic Coulomb excitation behaviors in monolayer [16], AA- and AB-stacked bilayer graphene systems. These studies include certain critical factors, the strongly coupling effects due to longitudinal and transverse forces temperatures, dopings and stackings. Very interestingly, while the free carrier density is enough high under the extrinsic case [265,269], the highest-frequency magneto-plasmon mode is predicted to become a 2D plasmon one, as clearly revealed in a 2D electron gas [189]. That is, their momentum-dependent frequencies behave as the square-root form, mainly owing to the 2D bare Coulomb potential of $2\pi e^2/q$ [240] and a whole oscillation of all free conduction electrons/valence holes. The similar studies will be generalized to trilayer graphene systems with AAA [201], ABA [42,53], ABC [237] and AAB stackings [43,44], as well as, monolayer $B_xC_yN_z$ compounds/single-walled $B_xC_yN_z$ nanotubes [144]. One could evaluate whether the emergent few-layer systems are good candidates in creating the diversified magneto-plasmon quasiparticle phenomena, e.g., bilayer silicene [217]/germanene [126]/tinene [4]/plumbene materials [217,280].

The main features of Landau levels and subbands are responsible for the magneto-electrical conductivities, in which their close relations are greatly diversified by the various 1D, 2D and 3D crystal structures. Whether quantum transport properties come to exist strongly depends on the open/cylindrical boundary conditions in nanoribbons/nanotubes [34,122], the significant k_z-dependence I layered bulk systems (e.g., AA-, AB- and ABC-stacked graphites; [213,265,281]), magnetic-field strengths [250,282], gate voltages [283], n- or π-type free carrier dopings [24,284], thermal excitations [249], and impurity disorders [285]. The well-known Landauer-Buttiker and static Kubo formulas [286] are, respectively, suitable for 1D and 2D/3D. The featured transport coefficients are mainly dominated by the occupation or non-occupation of the band-edge states (or the dispersionless Landau states), such as, the plateau and well-like structures in carbon nanotubes [56,287]/few-layer graphene systems [53,113] during the variation of the longitudinal/perpendicular magnetic field. The systematic investigations will be conducted on 2D $B_xC_yN_z$ [288], 3D BC_3 [233], as well as hydrogenated/oxidized/halogenated/alkalized graphene systems [289,290]/[226]/[22]/[291,292]. Specifically, a 2D material is able to present the static scatterings of the initial and final Landau levels, in which both of them possess the well-defined symmetries/the superposition of them about the spatial probability distributions. The magneto-selection rules, being achieved from all the inter-LL scattering events, dominate quantum Hall transport properties. The generalized tight-binding model [16,26] of the quasiparticle framework is available for the near-future studies.

Ferro-magnetic, anti-ferromagnetic and non-magnetic spin configurations in low-dimensional emergent materials [2D/1D/0D condensed-matter systems; [36]/[35,293]/[55], which are strongly coupled with the multi-orbital hybridizations, can present the unusual quasiparticle phenomena. The theoretical predictions and high-precision experimental examinations can be finished almost simultaneously, such as, the direct comparisons from the VASP calculations [20,80] and spin-polarized measurements of STS [31,33,112] and ARPES [110,276]. The previous works clearly show that ferromagnetism and anti-ferromagnetism could be generated by the magnetic atoms [219,294] and the specific morphologies [219]. Furthermore, transition metals and rare-earth elements, respectively, possess 10-d and fourteen-f orbitals, providing the very active chemical environments (the highly non-uniform ones; [295]). Obviously, their orbital- and spin-decomposed physical/chemical/material quantities would be very complicated, but rather significant in creating the diverse magnetisms. Such adatoms and guest atoms, which, respectively, corresponds to chemical absorptions [296,297] and substitutions [298–300], can totally modify the layered systems, e.g., the dramatic transformations of essential properties in chemical modifications for few-layer group-IV [298–300] and group-V materials [301]. The geometric, electronic, and optical properties are under the current studies. Most importantly, both charge- and spin-related intrinsic interactions will be clarified simultaneously from the position-dependent chemical bonds (the crystal symmetry; [302,303]), the spin-split/degenerate & atom-dominated energy spectrum within the specific energy ranges [24,298,304], the atom- & orbital-projected magnetic moment [298], the spatial charge & spin density distributions before and after chemical reactions, the merged van Hove singularities in the presence of atom, orbital & spin decomposition [302,303], and the spin- and orbital-hybridization-enriched absorption structures [24,133,298].

Both chemical adsorptions and substitutions are frequently utilized to clearly observe the diversities of fundamental properties [149,303,305,306], greatly enhance the specific functionalities [306–308], and largely widen the potential applications [289,309,310]. For example, the oxidization of germanium-/silicon-/tin-/lead-based emergent materials [311]/[312]/[313]/[314] frequently come to exist as the 3D, 2D, 1D and 0D crystal structures for the various allotropes. Furthermore, the multi-component compounds [binary/ternary/quaternary/pentanary forms] could be further generated by the active participations of extra elements/molecules/ions, such as, the well-known lithium germanium oxides with the distinct π-ion concentrations and distributions during the charging and discharging processes [82]. Such semiconductor compounds might present the unusual physical/chemical/material properties and thus play the critical roles in the applied sciences [225]. As for the dimensionality-, morphology- and component-dependent semiconductor oxide compounds [315], the close relations among the crystal symmetries, other essential properties and specific functionalities could be well established from the current quasiparticle framework by the delicate first-principles simulations [6,24], being fully supported through the high-resolution experimental examinations [88,90]. The linking mechanisms, the active orbital hybridizations within the distinct energy ranges and their partners of spin-split configurations, need to be clearly identified from the highly non-uniform chemical environments [24], the rich electronic properties [24], as well as the featured optical transitions [241].

The heavy group-IV- (tin and lead in Refs. [314,316]) and halide-based compounds (chloride, bromine and iodine; [317]) have been successfully synthesized and verified to be typical solar cells with the high efficiencies of energy convention, e.g., the up-to-adate performance of ~ 26% . In general, these multi-component condensed-matter systems possess the Perovskite crystal structures, being surrounded by the organic-inorganic or inorganic absorbers. For example, the latter might be methylammonium ($CH_3NH_3^+$; [318]) or cesium ion (Cs^+; [319]). Recently, silicon-based tandem cells show the maximum efficiency about 29%. Very interesting, the two cations possess the prominent chemical bondings with the active halogen anions. That is, the ionic bonds appear between the large-molecule positive and halogen negative. This should be totally different from the weak chemical bondings (the interlayer atomic interactions of the saturated ions and carbon-honeycomb lattice) under the significant intercalations/de-intercalations of the large-molecule negative ions within the graphitic layers, as clearly illustrated in the chemical reactions of aluminum-/iron-ion-based batteries due to $AlCl_4^-$ & $Al_2Cl_7^-/FeCl_3^-$ [320]/[321–323]. In addition, there also exist the intercalation layers of guest molecules. Most important, the complicated orbital hybridizations are induced by three kinds of ions, especially for the large distributions near the large molecule ions before and after the active participations. The concise physics & chemistry pictures could be clearly presented under a unified quasiparticle framework, becoming one of the studying focuses on the current investigations [5,6].

Recently, silicone, polymerized siloxanes (polysiloxanes; [324]), is frequently utilized as a super-high-performance anode in lithium-ion-based batteries. That is, it can replace the critical roles of π-ion intercalations and de-intercalatons through the greatly reduced time of the chemical reactions [82]. This well-known material belongs to the inter-mediate polymer compounds which is made up of a giant Moire superlattice with a straight inorganic Si–O-bone tightly attached to organic benzene-like functional groups. Its chemical formula is π, where R represents the benzene-related organic molecules, and they have the prominent covalent bondings with silicon atoms. Very interestingly, there exist the position-dependent chemical bonds, being fully assisted by the multi-orbital hybridization. These crystal structures, with a lot of active atoms and orbitals, are outstanding candidates in testing, unifying and extending the development of quasiparticle framework, since they are expected to reveal very complicated essential properties under the first-principles simulations, e.g., the unusual energy dispersions [325], van Hove singularities [325] and spatial charge density distributions [40]. The well-characterized orbital mixings in different chemical bonds would be a near-future challenge.

In summary, various condensed-matter systems, which are generated by the chemical [326] and physical modifications, provide the diverse crystal symmetries in the basic and applied sciences. The quasiparticle framework is further developed through the phenomenological models [128,210,211] and the first-principles simulations [21,149,292,305,327], being based on the featured materials. The close relations among the different physical/chemical/material properties and the significant orbital hybridizations/spin configurations will be established for the main-stream materials under the systematic investigations, as clearly illustrated in the previously published books [5,6].

References

[1] Zhang G, Du S, Wu K, Gao H-J. Two-dimensional materials research. Science 2018;360:15–8.
[2] Vogt P. Silicene, germanene and other group IV 2D materials. Beilstein-Institut; 2018.

[3] Lin C-H, Huang A, Pai WW, Chen W-C, Chen T-Y, Chang T-R, Yukawa R, Cheng C-M, Mou C-Y, Matsuda I. Single-layer dual germanene phases on Ag (111. Phys Rev Mater 2018;2:024003.
[4] Matthes L, Pulci O, Bechstedt F. Massive Dirac quasiparticles in the optical absorbance of graphene, silicene, germanene, and tinene. J Phys Condens Matter 2013;25.
[5] Tran NTT, Lin S-Y, Lin C-Y, Lin M-F. Geometric and electronic properties of graphene-related systems: chemical bonding schemes. CRC Press; 2017.
[6] Shih-Yang Lin H-YL, Khanh Nguyen D, Thanh Thuy Tran N, Duong Pham H, Chang S-L, Lin C-Y, Lin M-F. Silicene-based layered materials: essential properties. IOP Publishing; 2020.
[7] Lin SY, Liu HY, Nguyen DK, Tran NTT, Pham HD, Chang SL, Lin CY and Lin MF. Stacking-configuration-enriched fundamental properties in bilayer silicenes.
[8] Liu H-Y, Lin S-Y, Wu J-Y. Stacking-configuration-enriched essential properties of bilayer graphenes and silicenes. J Chem Phys 2020;153:154707.
[9] Terhell J. Polytypism in the III–VI layer compounds. Prog Cryst Growth Char 1983;7:55–110.
[10] Maboudian R, Carraro C, Senesky DG, Roper CS. Advances in silicon carbide science and technology at the micro-and nanoscales. J Vac Sci Technol: Vacuum, Surface Films 2013;31:050805.
[11] Lin M, Huang C, Chuu D. Plasmons in graphite and stage-1 graphite intercalation compounds. Phys Rev B 1997;55:13961.
[12] Lin M-F, Shung KW-K. Plasmons and optical properties of carbon nanotubes. Phys Rev B 1994;50:17744.
[13] Zhu X, Ong CS, Xu X, Hu B, Shang J, Yang H, Katlakunta S, Liu Y, Chen X, Pan L. Direct observation of lithium-ion transport under an electrical field in Li x CoO 2 nanograins. Sci Rep 2013;3:1–8.
[14] Fong KD, Self J, Diederichsen KM, Wood BM, McCloskey BD, Persson KA. Ion transport and the true transference number in nonaqueous polyelectrolyte solutions for lithium ion batteries. ACS Cent Sci 2019;5:1250–60.
[15] Klimmer S, Ghaebi O, Gan Z, George A, Turchanin A, Cerullo G, Soavi G. All-optical polarization and amplitude modulation of second-harmonic generation in atomically thin semiconductors. Nat Photonics 2021:1–6.
[16] Chiun-Yan Lin J-YW, Chiu C-W, Lin M-F. Coulomb excitations and decays in graphene-related systems. CRC Press; 2019.
[17] Reznik D, Olk C, Neumann D, Copley J. X-ray powder diffraction from carbon nanotubes and nanoparticles. Phys Rev B 1995;52:116.
[18] Li J, Sun JL. Application of X-ray diffraction and electron crystallography for solving complex structure problems. Acc Chem Res 2017;50:2737–45.
[19] Bussolotti F, Chi D, Goh KJ, Huang YL, Wee AT. 2D semiconductor materials and devices. Elsevier; 2020. p. 199–220.
[20] Hafner J. Ab-initio simulations of materials using VASP: density-functional theory and beyond. J Comput Chem 2008;29:2044–78.
[21] Sun L, Li Q, Ren H, Su H, Shi Q, Yang J. Strain effect on electronic structures of graphene nanoribbons: a first-principles study. J Chem Phys 2008;129:074704.
[22] Tran NTT, Nguyen DK, Glukhova OE, Lin M-F. Coverage-dependent essential properties of halogenated graphene: a DFT study. Sci Rep 2017;7:1–13.
[23] Nguyen TDH, Pham HD, Lin S-Y, Lin M-F. Featured properties of Li+-based battery anode: $Li_4Ti_5O_{12}$. RSC Adv 2020;10:14071–9.
[24] Pham HD, Su W-P, Nguyen TDH, Tran NTT, Lin M-F. Rich p-type-doping phenomena in boron-substituted silicene systems. R Soc Open Sci 2020;7:200723.
[25] Khuong Dien V, Thi Han N, Nguyen TDH, Huynh TMD, Pham HD, Lin M-F. Geometric and electronic properties of Li_2GeO_3. Front Mater 2020;7.
[26] Lin C-Y, Ho C-H, Wu J-Y, Do T-N, Shih P-H, Lin S-Y, Lin M-F. Diverse quantization phenomena in layered materials. CRC Press; 2019.
[27] Chen S, Wu J, Lin C, Lin M. Theory of magnetoelectric properties of 2d systems. New J Phys 2016;18:103024.
[28] Galbiati M, Motta N, De Crescenzi M, Camilli L. Group-IV 2D materials beyond graphene on nonmetal substrates: challenges, recent progress, and future perspectives. Appl Phys Rev 2019;6:041310.
[29] Sahin H, Cahangirov S, Topsakal M, Bekaroglu E, Akturk E, Senger RT, Ciraci S. Monolayer honeycomb structures of group-IV elements and III-V binary compounds: first-principles calculations. Phys Rev B 2009;80:155453.

[30] Liu C-C, Feng W, Yao Y. Quantum spin Hall effect in silicene and two-dimensional germanium. Phys Rev Lett 2011;107:076802.
[31] Ridene M, Girard J, Travers L, David C, Ouerghi A. STM/STS investigation of edge structure in epitaxial graphene. Surf Sci 2012;606:1289–92.
[32] Osvath Z, Fulcheri L, Mark GI, Tapaszto L, Gyulai J, Biro LP. STM and STS investigation of few-wall carbon nanotubes containing non-hexagonal carbon rings. In: Nanotechnology: international society for optics and photonics; 2003. p. 587–93.
[33] Kano S, Tada T, Majima Y. Nanoparticle characterization based on STM and STS. Chem Soc Rev 2015;44:970–87.
[34] Spataru CD, Ismail-Beigi S, Capaz RB, Louie SG. Quasiparticle and excitonic effects in the optical response of nanotubes and nanoribbons. Carbon Nanotube 2007:195–227.
[35] Katayama-Yoshida H, Sato K, Fukushima T, Toyoda M, Kizaki H, Dinh V, Dederichs P. Theory of ferromagnetic semiconductors. Physica Status Solidi (A) 2007;204:15–32.
[36] Eliseeva N, Kuzubov A, Ovchinnikov S, Serzhantova M, Tomilin F, Fedorov A. Theoretical study of the magnetic properties of ordered vacancies in 2D hexagonal structures: graphene, 2D-SiC, and h-BN. JETP Lett 2012;95:555–9.
[37] Marty L, Adam E, Albert L, Doyon R, Menard D, Martel R. Exciton formation and annihilation during 1D impact excitation of carbon nanotubes. Phys Rev Lett 2006;96:136803.
[38] Campos-Delgado J, Romo-Herrera JM, Jia X, Cullen DA, Muramatsu H, Kim YA, Hayashi T, Ren Z, Smith DJ, Okuno Y. Bulk production of a new form of sp2 carbon: crystalline graphene nanoribbons. Nano Lett 2008;8:2773–8.
[39] Mehdi Aghaei S, Calizo I. Band gap tuning of armchair silicene nanoribbons using periodic hexagonal holes. J Appl Phys 2015;118:104304.
[40] Han-tang L. Development and application of silicone materials.[J]. Silicone Mater 2006;4.
[41] Cahangirov S, Topsakal M, Ciraci S. Armchair nanoribbons of silicon and germanium honeycomb structures. Phys Rev B 2010;81:195120.
[42] Yuan S, Roldán R, Katsnelson MI. Landau level spectrum of ABA-and ABC-stacked trilayer graphene. Phys Rev B 2011;84:125455.
[43] Do T-N, Lin C-Y, Lin Y-P, Shih P-H, Lin M-F. Configuration-enriched magneto-electronic spectra of AAB-stacked trilayer graphene. Carbon 2015;94:619–32.
[44] Lin C-Y, Huang B-L, Ho C-H, Gumbs G, Lin M-F. Geometry-diversified Coulomb excitations in trilayer AAB stacking graphene. Phys Rev B 2018;98:195442.
[45] Biró L, Gyulai J, Lambin P, Nagy JB, Lazarescu S, Márk G, Fonseca A, Surján P, Szekeres Z, Thiry P. Scanning tunnelling microscopy (STM) imaging of carbon nanotubes. Carbon 1998;36:689–96.
[46] Reetz MT, Helbig W, Quaiser SA, Stimming U, Breuer N, Vogel R. Visualization of surfactants on nanostructured palladium clusters by a combination of STM and high-resolution TEM. Science 1995;267:367–9.
[47] Egerton R, Li P, Malac M. Radiation damage in the TEM and SEM. Micron 2004;35:399–409.
[48] Ryu YK, Frisenda R, Castellanos-Gomez A. Superlattices based on van der Waals 2D materials. Chem Commun 2019;55:11498–510.
[49] Hansson T, Oostenbrink C, van Gunsteren W. Molecular dynamics simulations. Curr Opin Struct Biol 2002;12:190–6.
[50] Birowska M, Milowska K, Majewski J. Van der Waals density functionals for graphene layers and graphite. Acta Phys Pol 2011;120:845–8.
[51] Wong J-H, Wu B-R, Lin M-F. Strain effect on the electronic properties of single layer and bilayer graphene. J Phys Chem C 2012;116:8271–7.
[52] Fu H, Zhang J, Ding Z, Li H, Meng S. Stacking-dependent electronic structure of bilayer silicene. Appl Phys Lett 2014;104:131904.
[53] Chiun-Yan L, Ching-Hong H, Jhao-Ying W, Ming-Fa L. Unusual electronic excitations in ABA trilayer graphene Scientific Reports, vol. 10. Nature Publisher Group; 2020.
[54] Brown L, Hovden R, Huang P, Wojcik M, Muller DA, Park J. Twinning and twisting of tri-and bilayer graphene. Nano Lett 2012;12:1609–15.
[55] Dresselhaus M, Dresselhaus G, Saito R. Physics of carbon nanotubes. Carbon 1995;33:883–91.

[56] Kataura H, Kumazawa Y, Maniwa Y, Umezu I, Suzuki S, Ohtsuka Y, Achiba Y. Optical properties of single-wall carbon nanotubes. Synth Met 1999;103:2555–8.
[57] Cai J, Ruffieux P, Jaafar R, Bieri M, Braun T, Blankenburg S, Muoth M, Seitsonen AP, Saleh M, Feng X. Atomically precise bottom-up fabrication of graphene nanoribbons. Nature 2010;466:470–3.
[58] Simonov KA, Vinogradov NA, Vinogradov AS, Generalov AV, Zagrebina EM, Martensson N, Cafolla AA, Carpy T, Cunniffe JP, Preobrajenski AB. Correction to "effect of substrate chemistry on the bottom-up fabrication of graphene nanoribbons: combined core-level spectroscopy and STM study.". J Phys Chem C 2015;119:880–1.
[59] Chung H-C, Chang C-P, Lin C-Y, Lin M-F. Electronic and optical properties of graphene nanoribbons in external fields. Phys Chem Chem Phys 2016;18:7573–616.
[60] Ding Y, Wang Y. Electronic structures of reconstructed zigzag silicene nanoribbons. Appl Phys Lett 2014;104:083111.
[61] Song Y-L, Zhang Y, Zhang J-M, Lu D-B, Xu K-W. First-principles study of the structural and electronic properties of armchair silicene nanoribbons with vacancies. J Mol Struct 2011;990:75–8.
[62] Dresselhaus MS, Dresselhaus G, Eklund PC. Science of fullerenes and carbon nanotubes: their properties and applications. Elsevier; 1996.
[63] Furuya K, Aikawa Y. Reprocessing of ices in turbulent protoplanetary disks: carbon and nitrogen chemistry. Astrophys J 2014;790:97.
[64] Tomita S, Sakurai T, Ohta H, Fujii M, Hayashi S. Structure and electronic properties of carbon onions. J Chem Phys 2001;114:7477–82.
[65] Chuvilin A, Meyer JC, Algara-Siller G, Kaiser U. From graphene constrictions to single carbon chains. New J Phys 2009;11:083019.
[66] Wang B, Ma X, Caffio M, Schaub R, Li W-X. Size-selective carbon nanoclusters as precursors to the growth of epitaxial graphene. Nano Lett 2011;11:424–30.
[67] Hakansson E. Galvanic corrosion of aluminum/carbon composite systems. 2016.
[68] Song Q, Li K-Z, Li H-L, Li H-J, Ren C. Grafting straight carbon nanotubes radially onto carbon fibers and their effect on the mechanical properties of carbon/carbon composites. Carbon 2012;50:3949–52.
[69] Liu L, Zhao J. Toroidal and coiled carbon nanotubes. Synthes Applicat Carbon Nanotubes Their Compos 2013:257–82.
[70] Yang L, Jiang J, Dong J. Formation mechanism of toroidal carbon nanotubes. Physica Status Solidi (B) 2003;238:115–9.
[71] Lau KT, Lu M, Hui D. Coiled carbon nanotubes: synthesis and their potential applications in advanced composite structures. Compos B Eng 2006;37:437–48.
[72] Alam A, Majumder MK, Kumari A, Kumar VR, Kaushik BK. Performance analysis of single-and multi-walled carbon nanotube based through silicon vias. In: 2015 IEEE 65th electronic components and technology conference (ECTC): IEEE); 2015. p. 1834–9.
[73] Lin M-F, Shyu F-L, Chen R-B. Optical properties of well-aligned multiwalled carbon nanotube bundles. Phys Rev B 2000;61:14114.
[74] Lin C-Y, Wu J-Y, Chang C-P, Lin M-F. Magneto-optical selection rules of curved graphene nanoribbons and carbon nanotubes. Carbon 2014;69:151–61.
[75] Kholmanov I, Cavaliere E, Fanetti M, Cepek C, Gavioli L. Growth of curved graphene sheets on graphite by chemical vapor deposition. Phys Rev B 2009;79:233403.
[76] Schaper AK, Hou H, Wang M, Bando Y, Golberg D. Observations of the electrical behaviour of catalytically grown scrolled graphene. Carbon 2011;49:1821–8.
[77] Yin W-J, Xie Y-E, Liu L-M, Chen Y-P, Wang R-Z, Wei X-L, Zhong J-X, Lau L. Atomic structure and electronic properties of folded graphene nanoribbons: a first-principles study. J Appl Phys 2013;113:173506.
[78] Adachi S. Properties of semiconductor alloys: group-IV, III-V and II-VI semiconductors, vol. 28. John Wiley & Sons; 2009.
[79] Cardona M, Greenaway DL. Optical properties and band structure of group IV-VI and group V materials. Phys Rev 1964;133:A1685.
[80] Lozhnikov VE, Mamonov AV, Borzilov VO, Mamonova MV, Prudnikov PV, Sorokin AA, Baksheev GG. Estimating the performance of ab initio calculation by VASP on openpower high performance system. In: CEUR workshop proceedings; 2019. p. 24–9.

[81] Metois J, Le Lay G. Complementary data obtained on the metal-semiconductor interface by LEED, AES and SEM: Pb/Ge (111). Surf Sci 1983;133:422–42.
[82] Goodenough JB, Park K-S. The Li-ion rechargeable battery: a perspective. J Am Chem Soc 2013;135:1167–76.
[83] Nitta N, Wu F, Lee JT, Yushin G. Li-ion battery materials: present and future. Mater Today 2015;18:252–64.
[84] Li J, Li J, Luo J, Wang L, He X. Recent advances in the $LiFeO_2$-based materials for Li-ion batteries. Int J Electrochem Sci 2011;6:1550–61.
[85] Zhou F, Kang K, Maxisch T, Ceder G, Morgan D. The electronic structure and band gap of $LiFePO_4$ and $LiMnPO_4$. Solid State Commun 2004;132:181–6.
[86] Malik R, Zhou F, Ceder G. Kinetics of non-equilibrium lithium incorporation in $LiFePO_4$. Nat Mater 2011;10:587–90.
[87] Ouyang X, Lei M, Shi S, Luo C, Liu D, Jiang D, Ye Z, Lei M. First-principles studies on surface electronic structure and stability of $LiFePO_4$. J Alloys Compd 2009;476:462–5.
[88] Warren BE. X-Ray diffraction. Courier Corporation; 1990.
[89] Baltes H, Yacoby Y, Pindak R, Clarke R, Pfeiffer L, Berman L. Measurement of the x-ray diffraction phase in a 2D crystal. Phys Rev Lett 1997;79:1285–8.
[90] Speakman SA. Introduction to x-ray powder diffraction data analysis. Center for Materials Science and Engineering at MIT; 2013.
[91] Ahmad S, Miró P, Audiffred M, Heine T. Tuning the electronic structure of graphene through alkali metal and halogen atom intercalation. Solid State Commun 2018;272:22–7.
[92] Charlier J-C, Blase X, Roche S. Electronic and transport properties of nanotubes. Rev Mod Phys 2007;79:677.
[93] Li X, Liang J, Lu Y, Hou Z, Cheng Q, Zhu Y, Qian Y. Sulfur-rich phosphorus sulfide molecules for use in rechargeable lithium batteries. Angew Chem Int Ed 2017;56:2937–41.
[94] Liu J, Lu D, Zheng J, Yan P, Wang B, Sun X, Shao Y, Wang C, Xiao J, Zhang J-G. Minimizing polysulfide shuttle effect in lithium-ion sulfur batteries by anode surface passivation. ACS Appl Mater Interfaces 2018;10:21965–72.
[95] Yuan S, Bao JL, Wang L, Xia Y, Truhlar DG, Wang Y. Graphene-supported nitrogen and boron rich carbon layer for improved performance of lithium–sulfur batteries due to enhanced chemisorption of lithium polysulfides. Adv Energy Mater 2016;6:1501733.
[96] Zhang L, Sun D, Feng J, Cairns EJ, Guo J. Revealing the electrochemical charging mechanism of nanosized Li_2S by in situ and operando X-ray absorption spectroscopy. Nano Lett 2017;17:5084–91.
[97] Zhao Z, Liu L, Yu T, Yang G, Bergara A. Pressure-induced stable Li5P for high-performance lithium-ion batteries. J Phys Chem C 2017;121:21199–205.
[98] Nizamov I y S, Al'metkina LA, Garifzyanova GG, Sergeenko G n G, Batyeva ES. Reactions of phosphorus sulfides (P4S5, P4S7 and P4S10) and 2, 4-BIS (alkylthio)-2, 4-dithioxo-1, 3, 2λ5, 4λ5-dithiadiphosphetanes with dialkyldisulfides and thioacetals in the presence of iodine. Phosphorus, Sulfur, Silicon Relat Elem 1995;102:71–81.
[99] Lepley N, Holzwarth N, Du YA. Structures, Li+ mobilities, and interfacial properties of solid electrolytes Li_3PS_4 and Li_3PO_4 from first principles. Phys Rev B 2013;88:104103.
[100] Neveu A, Pelé V, Jordy C, Pralong V. Exploration of Li–P–S–O composition for solid-state electrolyte materials discovery. J Power Sources 2020;467:228250.
[101] Ding B, Wang J, Fan Z, Chen S, Lin Q, Lu X, et al. Solid-state lithium–sulfur batteries: advances, challenges and perspectives. Mater Today 2020;40:114–31.
[102] Zhao H, Deng N, Yan J, Kang W, Ju J, Ruan Y, Wang X, Zhuang X, Li Q, Cheng B. A review on anode for lithium-sulfur batteries: progress and prospects. Chem Eng J 2018;347:343–65.
[103] Lin SY, Chang SL, Su WP, Lin MF. Structure- and adatom-enriched essential properties of graphene nanoribbons. CRC press; 2018.
[104] Ohara K, Mitsui A, Mori M, Onodera Y, Shiotani S, Koyama Y, Orikasa Y, Murakami M, Shimoda K, Mori K. Structural and electronic features of binary $Li_2SP_2S_5$ glasses. Sci Rep 2016;6:1–9.
[105] Saroja A, Samantaray SS, Sundara R. A room temperature multivalent rechargeable iron ion battery with an ether based electrolyte: a new type of post-lithium ion battery. Chem Commun 2019;55:10416–9.
[106] Wu M, Xu B, Chen L, Ouyang C. Geometry and fast diffusion of AlCl4 cluster intercalated in graphite. Electrochim Acta 2016;195:158–65.
[107] Wang D-Y, Wei C-Y, Lin M-C, Pan C-J, Chou H-L, Chen H-A, Gong M, Wu Y, Yuan C, Angell M. Advanced rechargeable aluminium ion battery with a high-quality natural graphite cathode. Nat Commun 2017;8:1–7.

[108] Sun H, Wang W, Yu Z, Yuan Y, Wang S, Jiao S. A new aluminium-ion battery with high voltage, high safety and low cost. Chem Commun 2015;51:11892–5.

[109] Lin M-C, Gong M, Lu B, Wu Y, Wang D-Y, Guan M, Angell M, Chen C, Yang J, Hwang B-J. An ultrafast rechargeable aluminium-ion battery. Nature 2015;520:324–8.

[110] Kordyuk A. ARPES experiment in fermiology of quasi-2D metals. Low Temp Phys 2014;40:286–96.

[111] Karakachian H, Nguyen TN, Aprojanz J, Zakharov AA, Yakimova R, Rosenzweig P, Polley CM, Balasubramanian T, Tegenkamp C, Power SR. One-dimensional confinement and width-dependent bandgap formation in epitaxial graphene nanoribbons. Nat Commun 2020;11:1–8.

[112] Hipps K. Handbook of applied solid state spectroscopy. Springer; 2006. p. 305–50.

[113] Do T-N, Chang C-P, Shih P-H, Wu J-Y, Lin M-F. Stacking-enriched magneto-transport properties of few-layer graphenes. Phys Chem Chem Phys 2017;19:29525–33.

[114] Unwin PR, Guell AG, Zhang G. Nanoscale electrochemistry of sp2 carbon materials: from graphite and graphene to carbon nanotubes. Acc Chem Res 2016;49:2041–8.

[115] Nicholas RJ, Chuang K-C. Band structure and electron velocity measurement in carbon nanotubes and graphene. Int J Mod Phys B 2009;23:2655–64.

[116] Ye MY, Zhao ZH, Hu ZF, Liu LQ, Ji HM, Shen ZR, Ma TY. /2D heterojunctions of vanadate quantum dots/graphitic carbon nitride nanosheets for enhanced visible-light-driven photocatalysis. Angew Chem Int Ed 2017;56:8407–11.

[117] Arghavan S, Singh A. On the vibrations of single-walled carbon nanotubes. J Sound Vib 2011;330:3102–22.

[118] Spataru CD, Ismail-Beigi S, Benedict LX, Louie SG. Excitonic effects and optical spectra of single-walled carbon nanotubes. Phys Rev Lett 2004;92:077402.

[119] Lefebvre J, Fraser J, Homma Y, Finnie P. Photoluminescence from single-walled carbon nanotubes: a comparison between suspended and micelle-encapsulated nanotubes. Appl Phys A 2004;78:1107–10.

[120] Kimouche A, Ervasti MM, Drost R, Halonen S, Harju A, Joensuu PM, Sainio J, Liljeroth P. Ultra-narrow metallic armchair graphene nanoribbons. Nat Commun 2015;6:1–6.

[121] Tao C, Jiao L, Yazyev OV, Chen Y-C, Feng J, Zhang X, Capaz RB, Tour JM, Zettl A, Louie SG. Spatially resolving edge states of chiral graphene nanoribbons. Nat Phys 2011;7:616–20.

[122] Chung H, Huang Y, Lee M, Chang C, Lin M. Quasi-landau levels in bilayer zigzag graphene nanoribbons. Phys E Low-dimens Syst Nanostruct 2010;42:711–4.

[123] Kortüm G. Reflectance spectroscopy: principles, methods, applications. Springer Science & Business Media; 2012.

[124] Nicodemus FE. Reflectance nomenclature and directional reflectance and emissivity. Appl Opt 1970;9:1474–5.

[125] Woolley JT. Reflectance and transmittance of light by leaves. Plant Physiol 1971;47:656–62.

[126] Shih P-H, Chiu Y-H, Wu J-Y, Shyu F-L, Lin M-F. Coulomb excitations of monolayer germanene. Sci Rep 2017;7:40600.

[127] Yan CH, Wei LF. Time-dependent quantum transports along electronic waveguides with side-branch mesoscopic rings. Phys B Condens Matter 2012;407:4545–9.

[128] Zólyomi V, Wallbank J, Fal'Ko V. Silicane and germanane: tight-binding and first-principles studies. 2D Mater 2014;1:011005.

[129] Yao Y, Ye F, Qi X-L, Zhang S-C, Fang Z. Spin-orbit gap of graphene: first-principles calculations. Phys Rev B 2007;75:041401.

[130] Rapaport DC. The art of molecular dynamics simulation. Cambridge university press; 2004.

[131] Tchernyshyov O, Moessner R, Sondhi SL. Spin-Peierls phases in pyrochlore antiferromagnets. Phys Rev B 2002;66:064403.

[132] Kumar A, Poumirol J, Escoffier W, Goiran M, Raquet B, Broto J. Electronic properties of graphene, few-layer graphene, and bulk graphite under very high magnetic field. Int J Nanosci 2011;10:43–7.

[133] Pham HD, Nguyen TDH, Vo KD, Huynh TMD, Lin M-F. Rich essential properties of boron, carbon, and nitrogen substituted germanenes. App Phys Expr 2020;13:085502.

[134] Jeong TY, Kim H, Choi S-J, Watanabe K, Taniguchi T, Yee KJ, Kim Y-S, Jung S. Spectroscopic studies of atomic defects and bandgap renormalization in semiconducting monolayer transition metal dichalcogenides. Nat Commun 2019;10:1–10.

[135] Estreicher S, Gibbons T, Kang B, Bebek M. Phonons and defects in semiconductors and nanostructures: phonon trapping, phonon scattering, and heat flow at heterojunctions. J Appl Phys 2014;115:012012.

[136] Kozinsky B. Transport in frustrated and disordered solid electrolytes. Handbook Mater Model: Applicat: Curr Emerg Mater 2020:1255–74.

[137] Li Z-P, Mori T, Auchterlonie GJ, Guo Y, Zou J, Drennan J, Miyayama M. Mutual diffusion and microstructure evolution at the electrolyte– anode interface in intermediate temperature solid oxide fuel cell. J Phys Chem C 2011;115:6877–85.

[138] Xiong H, Hong J-S, Luo C-M, Zhong L-L. An ultrathin and broadband metamaterial absorber using multi-layer structures. J Appl Phys 2013;114:064109.

[139] Jia Z, Wang Z, Liang J, Wei B, Wu D. Production of short multi-walled carbon nanotubes. Carbon 1999;37:903–6.

[140] Shaffer M, Koziol K. Polystyrene grafted multi-walled carbon nanotubes. Chem Commun 2002:2074–5.

[141] Kis A, Csanyi G, Salvetat J-P, Lee T-N, Couteau E, Kulik A, Benoit W, Brugger J, Forro L. Reinforcement of single-walled carbon nanotube bundles by intertube bridging. Nat Mater 2004;3:153–7.

[142] Liu JZ, Zheng Q-S, Wang L-F, Jiang Q. Mechanical properties of single-walled carbon nanotube bundles as bulk materials. J Mech Phys Solid 2005;53:123–42.

[143] Ahangari MG, Fereidoon A, Ganji MD, Sharifi N. Density functional theory based molecular dynamics simulation study on the bulk modulus of multi-shell fullerenes. Phys B Condens Matter 2013;423:1–5.

[144] Terrones M, Grobert N, Terrones H. Synthetic routes to nanoscale BxCyNz architectures. Carbon 2002;40:1665–84.

[145] Wang J, Hao J, Liu D, Qin S, Chen C, Yang C, Liu Y, Yang T, Fan Y, Chen Y. Flower stamen-like porous boron carbon nitride nanoscrolls for water cleaning. Nanoscale 2017;9:9787–91.

[146] Li L, Zhu X. Enhanced photocatalytic hydrogen evolution of carbon quantum dot modified 1D protonated nanorods of graphitic carbon nitride. ACS Appl Nano Mater 2018;1:5337–44.

[147] Ahsan MA, He T, Eid K, Abdullah AM, Curry ML, Du A, Puente Santiago AR, Echegoyen L, Noveron JC. Tuning the intermolecular electron transfer of low-dimensional and metal-free BCN/C60 electrocatalysts via interfacial defects for efficient hydrogen and oxygen electrochemistry. J Am Chem Soc 2021;143:1203–15.

[148] Kudu Ö U, Famprikis T, Fleutot B, Braida M-D, Le Mercier T, Islam MS, Masquelier C. A review of structural properties and synthesis methods of solid electrolyte materials in the Li_2S– P_2S_5 binary system. J Power Sources 2018;407:31–43.

[149] Lin S-Y, Chang S-L, Chen H-H, Su S-H, Huang J-C, Lin M-F. Substrate-induced structures of bismuth adsorption on graphene: a first principles study. Phys Chem Chem Phys 2016;18:18978–84.

[150] Stern S, Shah V, Hardy B. Structure-permeability relationships in silicone polymers. J Polym Sci B Polym Phys 1987;25:1263–98.

[151] Tada K, Kitta M, Ozaki H, Tanaka S. A comparative study of $Na_3LiTi_5O_{12}$ and $Li_4Ti_5O_{12}$: geometric and electronic structures obtained by density functional theory calculations. Chem Phys Lett 2019;731:136598.

[152] Liu Y, Lian J, Sun Z, Zhao M, Shi Y, Song H. The first-principles study for the novel optical properties of $LiTi_2O_4$, $Li_4Ti_5O_{12}$, $Li_2Ti_2O_4$ and $Li_7Ti_5O_{12}$. Chem Phys Lett 2017;677:114–9.

[153] Hirsch A. The era of carbon allotropes. Nat Mater 2010;9:868–71.

[154] Falcao EH, Wudl F. Carbon allotropes: beyond graphite and diamond. J Chem Tech Biotechnol: Int Res Process, Environ Clean Technol 2007;82:524–31.

[155] Kishore MA, Varunaa R, Bayani A, Larsson K. Theoretical investigation on BeN_2 monolayer for an efficient bifunctional water splitting catalyst. Sci Rep 2020;10:1–11.

[156] Zhang C, Sun Q. A honeycomb BeN2 sheet with a desirable direct band gap and high carrier mobility. J Phys Chem Lett 2016;7:2664–70.

[157] Hengtao L, Wang H, Yan L, Li X, Chen Y, Wang H. A novel two-dimensional beryllium diphosphide (BeP2) with superconductivity: the first-principles exploration. Phys Chem Chem Phys 2021;23:12834–41.

[158] Zhang H, Holbrook M, Cheng F, Nam H, Liu M, Pan C-R, West D, Zhang S, Chou M-Y, Shih C-K. Epitaxial growth of two-dimensional insulator monolayer honeycomb BeO. ACS Nano 2021;15:2497–505.

[159] Du Y, Zhuang J, Wang J, Li Z, Liu H, Zhao J, Xu X, Feng H, Chen L, Wu K. Quasi-freestanding epitaxial silicene on Ag (111) by oxygen intercalation. Sci Adv 2016;2:e1600067.

[160] Sicolo S, Giordano L, Pacchioni G. Adsorption of late transition metal atoms on MgO/Mo (100) and MgO/Ag (100) ultrathin films: a comparative DFT study. J Phys Chem C 2009;113:16694–701.

[161] Li K, Li Y, Wang Y, He F, Jiao M, Tang H, Wu Z. The oxygen reduction reaction on Pt (111) and Pt (100) surfaces substituted by subsurface Cu: a theoretical perspective. J Mater Chem 2015;3:11444–52.

[162] Mansurov V, Malin T, Galitsyn Y, Zhuravlev K. Graphene-like AlN layer formation on (111) Si surface by ammonia molecular beam epitaxy. J Cryst Growth 2015;428:93–7.
[163] Kumaresan V, Largeau L, Madouri A, Glas F, Zhang H, Oehler F, Cavanna A, Babichev A, Travers L, Gogneau N. Epitaxy of GaN nanowires on graphene. Nano Lett 2016;16:4895–902.
[164] Wu X, Brown L, Kapolnek D, Keller S, Keller B, DenBaars S, Speck J. Defect structure of metal-organic chemical vapor deposition-grown epitaxial (0001) GaN/Al_2O_3. J Appl Phys 1996;80:3228–37.
[165] Dadgar A, Bläsing J, Diez A, Alam A, Heuken M, Krost A. Metalorganic chemical vapor phase epitaxy of crack-free GaN on Si (111) exceeding 1 μm in thickness. Jpn J Appl Phys 2000;39:L1183.
[166] Feltin E, Beaumont B, Laügt M, De Mierry P, Vennégués P, Lahreche H, Leroux M, Gibart P. Stress control in GaN grown on silicon (111) by metalorganic vapor phase epitaxy. Appl Phys Lett 2001;79:3230–2.
[167] Cao X-a, Hu H-t, Ding X-m, Yuan Z-l, Dong Y, Chen X-y, Lai B, Hou X-y. Passivation of the GaAs (100) surface with a vapor-deposited GaS film. J Vac Sci Technol B: Microelectron Nanometer Struct Process, Measure Phenom 1998;16:2656–9.
[168] Si C, Lin Z, Zhou J, Sun Z. Controllable Schottky barrier in GaSe/graphene heterostructure: the role of interface dipole. 2D Mater 2016;4:015027.
[169] Diep NQ, Liu C-W, Wu S-K, Chou W-C, Huynh SH, Chang EY. Screw-dislocation-driven growth mode in two dimensional GaSe on GaAs (001) substrates grown by molecular beam epitaxy. Sci Rep 2019;9:1–8.
[170] Look DC, Molnar RJ. Degenerate layer at GaN/sapphire interface: influence on Hall-effect measurements. Appl Phys Lett 1997;70:3377–9.
[171] Bhattacharya S, Datta A, Chakravorty D. Multiferroic GaN nanofilms grown within Na-4 mica channels. Appl Phys Lett 2010;96:093109.
[172] Brotons-Gisbert M, Andres-Penares D, Martínez-Pastor J, Cros A, Sánchez-Royo J. Optical contrast of 2D InSe on SiO_2/Si and transparent substrates using bandpass filters. Nanotechnology 2017;28:115706.
[173] Desai R, Lakshminarayana D, Patel P, Panchal C. Indium sesquitelluride (In_2Te_3) thin film strain gauge. Sensor Actuator Phys 2005;121:405–9.
[174] Seo J, Hasegawa S, Asahi H. Selective area growth of GaN nanorods on patterned W/SiO_2/Si substrates by RF-MBE. J Cryst Growth 2009;311:4437–41.
[175] Ho I h, Stringfellow G. Solid phase immiscibility in GaInN. Appl Phys Lett 1996;69:2701–3.
[176] Reininger P, Zederbauer T, Schwarz B, Detz H, MacFarland D, Andrews AM, Schrenk W, Strasser G. InAs/AlAsSb based quantum cascade detector. Appl Phys Lett 2015;107:081107.
[177] Lorenz K, Franco N, Alves E, Pereira S, Watson I, Martin R, O'Donnell K. Relaxation of compressively strained AlInN on GaN. J Cryst Growth 2008;310:4058–64.
[178] Guo Q, Tanaka T, Nishio M, Ogawa H. Structural and optical properties of AlInN films grown on sapphire substrates. Jpn J Appl Phys 2008;47:612.
[179] Soga T, Hattori S, Sakai S, Takeyasu M, Umeno M. MOCVD growth of GaAs on Si substrates with AlGaP and strained superlattice layers. Electron Lett 1984;20:916–8.
[180] Ferdous F, Haque A. Energy band structures of strained membrane quantum wires considering the redistribution of elastic strain relaxation. In: 2006 international conference on numerical simulation of semiconductor optoelectronic devices: IEEE); 2006. p. 27–8.
[181] Matsui Y, Murai H, Arahira S, Kutsuzawa S, Ogawa Y. 30-GHz bandwidth 1.55-μm strain-compensated InGaAlAs-InGaAsP MQW laser. IEEE Photon Technol Lett 1997;9:25–7.
[182] Liu L-J, Wu Y-D, Wang Y, An J-M, Hu X-W. Nm long-wavelength vertical-cavity surface emitting lasers. Optoelectron Lett 2018 1 550;14:342–5.
[183] Harris PJ. Carbon nanotube composites. Int Mater Rev 2004;49:31–43.
[184] Wu Z-S, Zhou G, Yin L-C, Ren W, Li F, Cheng H-M. Graphene/metal oxide composite electrode materials for energy storage. Nano Energy 2012;1:107–31.
[185] Thakur VK. Green composites from natural resources. CRC Press; 2013.
[186] Stöberl U, Wurstbauer U, Wegscheider W, Weiss D, Eroms J. Morphology and flexibility of graphene and few-layer graphene on various substrates. Appl Phys Lett 2008;93:051906.
[187] UryuA S, AjikiB H and AndoA T Excitonic two-photon absorption in carbon nanotubes.
[188] Wang X-F, Chakraborty T. Coulomb screening and collective excitations in a graphene bilayer. Phys Rev B 2007;75:041404.
[189] Chi X, Huang Z, Asmara TC, Han K, Yin X, Yu X, Diao C, Yang M, Schmidt D, Yang P. Large enhancement of 2D electron gases mobility induced by interfacial localized electron screening effect. Adv Mater 2018;30:1707428.

[190] Poster Y-YC. Multi-cation thiocyanate-based pseudohalide perovskite solar cells with MASCN additive. Taiwan, Republic of China: National Cheng Kung University; 2018. Yueh-Ya Chiu, Ming-Hsien Li, Peter Chen Department of Photonics, National Cheng Kung University, tainan 701, taiwan, ROC. b Hierarchical Green-energy materials (Hi-GEM) research Center, National Cheng Kung University, 70101 tainan, taiwan, ROC c Division director of Education & research Center for Micro/Nano science and technology (CMNST).
[191] Wang H-Y, Sun JC-Y. Real-time virtual reality co-creation: collective intelligence and consciousness for student engagement and focused attention within online communities. Interact Learn Environ 2021:1–14.
[192] Ho Y, Chang C, Shyu F, Chen R, Chen S, Lin M-F. Electronic and optical properties of double-walled armchair carbon nanotubes. Carbon 2004;42:3159–67.
[193] Lin C-Y, Chen R-B, Ho Y-H, Lin M-F. Electronic and optical properties of graphite-related systems. CRC Press; 2017.
[194] Pang Q, Zhang Y, Zhang J-M, Ji V, Xu K-W. Electronic and magnetic properties of pristine and chemically functionalized germanene nanoribbons. Nanoscale 2011;3:4330–8.
[195] Palczewski AD. Angle-resolved photoemission spectroscopy (ARPES) studies of cuprate superconductors. Ames, IA (United States): Ames Laboratory (AMES); 2010.
[196] Singh J. Optical properties of condensed matter and applications, vol. 6. John Wiley & Sons; 2006.
[197] Roome NJ, Carey JD. Beyond graphene: stable elemental monolayers of silicene and germanene. ACS Appl Mater Interfaces 2014;6:7743–50.
[198] Li Y, Yang Z. The electronic and topological properties of plumbene by first-principles calculations. Bull Am Phys Soc 2020;65.
[199] Guan J, Zhu Z, Tománek D. High stability of faceted nanotubes and fullerenes of multiphase layered phosphorus: a computational study. Phys Rev Lett 2014;113:226801.
[200] Kecik D, Özçelik VO, Durgun E, Ciraci S. Structure dependent optoelectronic properties of monolayer antimonene, bismuthene and their binary compound. Phys Chem Chem Phys 2019;21:7907–17.
[201] Redouani I, Jellal A, Bahaoui A, Bahlouli H. Multibands tunneling in AAA-stacked trilayer graphene. Superlattice Microst 2018;116:44–53.
[202] Lin Y-P, Lin C-Y, Ho Y-H, Do T-N, Lin M-F. Magneto-optical properties of ABC-stacked trilayer graphene. Phys Chem Chem Phys 2015;17:15921–7.
[203] Liu H, Liu Y, Zhu D. Chemical doping of graphene. J Mater Chem 2011;21:3335–45.
[204] Nechaev I, Echenique PM, Chulkov EV. Inelastic decay rate of quasiparticles in a two-dimensional spin-orbit coupled electron system. Phys Rev B 2010;81:195112.
[205] Ohta T, Bostwick A, Seyller T, Horn K, Rotenberg E. Controlling the electronic structure of bilayer graphene. Science 2006;313:951–4.
[206] Malić E, Hirtschulz M, Milde F, Knorr A, Reich S. Analytical approach to optical absorption in carbon nanotubes. Phys Rev B 2006;74:195431.
[207] Charlier A, McRae E, Heyd R, Charlier M, Moretti D. Classification for double-walled carbon nanotubes. Carbon 1999;37:1779–83.
[208] Fan Q, Wei Q, Chai C, Yan H, Zhang M, Lin Z, Zhang Z, Zhang J, Zhang D. Structural, mechanical, and electronic properties of P3m1-BCN. J Phys Chem Solid 2015;79:89–96.
[209] Kumar N, Moses K, Pramoda K, Shirodkar SN, Mishra AK, Waghmare UV, Sundaresan A, Rao C. Borocarbonitrides, BxCyNz. J Mater Chem 2013;1:5806–21.
[210] Foulkes WMC, Haydock R. Tight-binding models and density-functional theory. Phys Rev B 1989;39:12520.
[211] Wang C, Pan B, Tang M, Haas H, Sígalas M, Lee G, Ho K. Environment-dependent tight-binding potential model. MRS Online Proc Library Arch 1997;491.
[212] Ren X 2 the random phase approximation and its applications to real materials.
[213] Ho J-H, Lu C, Hwang C, Chang C, Lin M-F. Coulomb excitations in AA-and AB-stacked bilayer graphites. Phys Rev B 2006;74:085406.
[214] Molle A, Berikaa ER, Pont FM, Bande A. Quantum size effect affecting environment assisted electron capture in quantum confinements. J Chem Phys 2019;150:224105.
[215] Giesbers A, Ponomarenko L, Novoselov K, Geim A, Katsnelson M, Maan J, Zeitler U. Gap opening in the zeroth Landau level of graphene. Phys Rev B 2009;80:201403.
[216] Fromer N, Schüller C, Lai C, Chemla D, Perakis I, Driscoll D, Gossard A. Coulomb correlations in a two-dimensional electron gas in large magnetic fields. Phys Rev B 2002;66:205314.

[217] Do T-N, Gumbs G, Shih P-H, Huang D, Chiu C-W, Chen C-Y, Lin M-F. Peculiar optical properties of bilayer silicene under the influence of external electric and magnetic fields. Sci Rep 2019;9:1–15.
[218] Tang S, Dresselhaus MS. Anisotropic transport for parabolic, non-parabolic, and linear bands of different dimensions. Appl Phys Lett 2014;105:033907.
[219] Tokmachev AM, Averyanov DV, Parfenov OE, Taldenkov AN, Karateev IA, Sokolov IS, Kondratev OA, Storchak VG. Emerging two-dimensional ferromagnetism in silicene materials. Nat Commun 2018;9:1–9.
[220] Mahan GD. Many-particle physics. Springer Science & Business Media; 2013.
[221] Rohlfing M, Louie SG. Electron-hole excitations and optical spectra from first principles. Phys Rev B 2000;62:4927.
[222] Pennington G, Goldsman N. Semiclassical transport and phonon scattering of electrons in semiconducting carbon nanotubes. Phys Rev B 2003;68:045426.
[223] Yang X, Dou X, Rouhanipour A, Zhi L, Räder HJ, Müllen K. Two-dimensional graphene nanoribbons. J Am Chem Soc 2008;130:4216–7.
[224] Chung H-C, Lee M-H, Chang C-P, Huang Y-C, Lin M-F. Effects of transverse electric fields on quasi-landau levels in zigzag graphene nanoribbons. J Phys Soc Jpn 2011;80:044602.
[225] Ren X, Rinke P, Joas C, Scheffler M. Random-phase approximation and its applications in computational chemistry and materials science. J Mater Sci 2012;47:7447–71.
[226] Yan J-A, Xian L, Chou M. Structural and electronic properties of oxidized graphene. Phys Rev Lett 2009;103:086802.
[227] Natori A, Ohno T, Oshiyama A. Work function of alkali-atom adsorbed graphite. J Phys Soc Jpn 1985;54:3042–50.
[228] Li Y, Lu Y, Adelhelm P, Titirici M-M, Hu Y-S. Intercalation chemistry of graphite: alkali metal ions and beyond. Chem Soc Rev 2019;48:4655–87.
[229] Pierce N, Chen G, Rajukumar L P, Chou NH, Koh AL, Sinclair R, Maruyama S, Terrones M, Harutyunyan AR. Intrinsic chirality origination in carbon nanotubes. ACS Nano 2017;11:9941–9.
[230] Sanchez-Valencia JR, Dienel T, Gröning O, Shorubalko I, Mueller A, Jansen M, Amsharov K, Ruffieux P, Fasel R. Controlled synthesis of single-chirality carbon nanotubes. Nature 2014;512:61–4.
[231] Li Y, Chen M, Weinert M, Li L. 1D metallic edge states of oxygen-terminated zigzag graphene edges. 2D Mater 2019;6:025038.
[232] Zuo X, Zhu J, Müller-Buschbaum P, Cheng Y-J. Silicon based lithium-ion battery anodes: a chronicle perspective review. Nano Energy 2017;31:113–43.
[233] Zhang H, Liao Y, Yang G, Zhou X. Theoretical studies on the electronic and optical properties of honeycomb BC3 monolayer: a promising candidate for metal-free photocatalysts. ACS Omega 2018;3:10517–25.
[234] Uddin M, Doan T, Li J, Ziemer KS, Lin J, Jiang H. Electrical transport properties of (BN)-rich hexagonal (BN) C semiconductor alloys. AIP Adv 2014;4:087141.
[235] Bengu E, Marks L. Single-walled BN nanostructures. Phys Rev Lett 2001;86:2385.
[236] Zhang XG, Krstić P, Butler W. Generalized tight-binding approach for molecular electronics modeling. Int J Quant Chem 2003;95:394–403.
[237] Lu C L, Chang C P, Huang Y C, Ho J H, Hwang C C, Lin M F. Electronic properties of AA-and ABC-stacked few-layer graphites. J Phys Soc Jpn 2007;76:024701.
[238] Ho C-H, Tsai S-J, Chen R-B, Chiu Y-H, Lin M-F. Low-energy Landau level spectrum in ABC-stacked trilayer graphene. J Nanosci Nanotechnol 2011;11:4938–47.
[239] Shallcross S, Sharma S, Kandelaki E, Pankratov O. Electronic structure of turbostratic graphene. Phys Rev B 2010;81:165105.
[240] Jackson JD. Classical electrodynamics. John Wiley & Sons; 2007.
[241] Fox M. Optical properties of solids. Oxford University Press; 2001.
[242] Dien VK, Pham HD, Tran NTT, Han NT, Huynh TMD, Nguyen TDH, Fa-Lin M. Orbital-hybridization-created optical excitations in Li_2GeO_3. Sci Rep 2021;11:1–10.
[243] Maier SA, Brongersma ML, Kik PG, Meltzer S, Requicha AA, Atwater HA. Plasmonics—a route to nanoscale optical devices. Adv Mater 2001;13:1501–5.
[244] Hesselmann A. Random-phase-approximation correlation method including exchange interactions. Phys Rev 2012;85:012517.
[245] Hattendorf S, Georgi A, Liebmann M, Morgenstern M. Networks of ABA and ABC stacked graphene on mica observed by scanning tunneling microscopy. Surf Sci 2013;610:53–8.

[246] Chiu C-W, Huang Y-C, Shyu F-L, Lin M-F. Optical absorption spectra in ABC-stacked graphene superlattice. Synth Met 2012;162:800–4.
[247] Mclean RS, Huang X, Khripin C, Jagota A, Zheng M. Controlled two-dimensional pattern of spontaneously aligned carbon nanotubes. Nano Lett 2006;6:55–60.
[248] Liu C, Ding J. Electronic structure of carbon nanotori: the roles of curvature, hybridization, and disorder. J Phys Condens Matter 2006;18:4077.
[249] Guo Y, Molesky S, Hu H, Cortes CL, Jacob Z. Thermal excitation of plasmons for near-field thermophotovoltaics. Appl Phys Lett 2014;105:073903.
[250] Lin C-Y, Do T-N, Huang Y-K, Lin M-F. Optical properties of graphene in magnetic and electric fields. IOP Publishing; 2017.
[251] Thorgrimsson B, McJunkin T, MacQuarrie E, Coppersmith S, Eriksson M. The effect of external electric fields on silicon with superconducting gallium nano-precipitates. J Appl Phys 2020;127:215102.
[252] Khoeini F, Esmaeili L. Effects of electric and magnetic fields on electronic properties of stanene nanoribbons. J Res Many-body Syst 2019;9:61–9.
[253] Li T, Lin M. Electronic properties of bilayer Bernal graphene in a modulated magnetic field. J Magn Magn Mater 2013;348:61–7.
[254] Rappoport TG, Epstein I, Koppens FH, Peres NM. Understanding the electromagnetic response of graphene/metallic nanostructures hybrids of different dimensionality. ACS Photonics 2020;7:2302–8.
[255] Gupta U, Rao C. Advances in the chemistry and physics of materials: Overview of selected topics. World Scientific; 2020. p. 27–54.
[256] Kira M, Koch SW. Many-body correlations and excitonic effects in semiconductor spectroscopy. Prog Quant Electron 2006;30:155–296.
[257] Tassin P, Koschny T, Kafesaki M, Soukoulis CM. A comparison of graphene, superconductors and metals as conductors for metamaterials and plasmonics. Nat Photonics 2012;6:259.
[258] Sáenz PJ, Cristea-Platon T, Bush JW. A hydrodynamic analog of Friedel oscillations. Sci Adv 2020;6:9234.
[259] Hofmann P, Briner B, Doering M, Rust H-P, Plummer E, Bradshaw A. Anisotropic two-dimensional friedel oscillations. Phys Rev Lett 1997;79:265.
[260] Chen S, Wang T, Lee C, Lin M-F. Magneto-electronic properties of graphene nanoribbons in the spatially modulated electric field. Phys Lett A 2008;372:5999–6002.
[261] Lin M-F, Chuang Y-C, Wu J-Y. Electrically tunable plasma excitations in AA-stacked multilayer graphene. Phys Rev B 2012;86:125434.
[262] Kohn W. Image of the fermi surface in the vibration spectrum of a metal. Phys Rev Lett 1959;2:393.
[263] Lu C, Chang C-P, Huang Y-C, Chen R-B, Lin M. Influence of an electric field on the optical properties of few-layer graphene with AB stacking. Phys Rev B 2006;73:144427.
[264] Yankowitz M, Joel I, Wang J, Birdwell AG, Chen Y-A, Watanabe K, Taniguchi T, Jacquod P, San-Jose P, Jarillo-Herrero P. Electric field control of soliton motion and stacking in trilayer graphene. Nat Mater 2014;13:786–9.
[265] Ho Y-H, Wu J-Y, Chen R-B, Chiu Y-H, Lin M-F. Optical transitions between Landau levels: AA-stacked bilayer graphene. Appl Phys Lett 2010;97:101905.
[266] Liu Z, Suenaga K, Harris PJ, Iijima S. Open and closed edges of graphene layers. Phys Rev Lett 2009;102:015501.
[267] Gumbs G, Aĭzin G. Collective excitations in a linear periodic array of cylindrical nanotubes. Phys Rev B 2002;65:195407.
[268] Shemella P, Zhang Y, Mailman M, Ajayan PM, Nayak SK. Energy gaps in zero-dimensional graphene nanoribbons. Appl Phys Lett 2007;91:042101.
[269] Ho Y-H, Chiu Y-H, Lin D-H, Chang C-P, Lin M-F. Magneto-optical selection rules in bilayer Bernal graphene. ACS Nano 2010;4:1465–72.
[270] Ni G, Wang H, Wu J, Fei Z, Goldflam M, Keilmann F, Özyilmaz B, Neto AC, Xie X, Fogler M. Plasmons in graphene moiré superlattices. Nat Mater 2015;14:1217–22.
[271] Lin C-Y, Do T-N, Huang Y-K and Lin M-F AB-stacked graphene.
[272] Dumont G, Boulanger P, Côté M, Ernzerhof M. Peierls instability in carbon nanotubes: a first-principles study. Phys Rev B 2010;82:035419.
[273] Gould T, Ruzsinszky A, Perdew JP. Simple self-interaction correction to random-phase-approximation-like correlation energies. Phys Rev 2019;100:022515.
[274] Schulze H-J, Schuck P, Van Giai N. Two-dimensional electron gas in the random-phase approximation with exchange and self-energy corrections. Phys Rev B 2000;61:8026.

[275] Potthoff M. Self-energy-functional approach to systems of correlated electrons. Eur Phys J B 2003;32:429–36.
[276] Damascelli A. Probing the electronic structure of complex systems by ARPES. Phys Scripta 2004;61. 2004.
[277] Hwang E, Sarma SD. Quasiparticle spectral function in doped graphene: electron-electron interaction effects in ARPES. Phys Rev B 2008;77:081412.
[278] Thess A, Votyakov EV, Kolesnikov Y. Lorentz force velocimetry. Phys Rev Lett 2006;96:164501.
[279] Shyu F-L, Lin M-F. Plasmons and optical properties of semimetal graphite. J Phys Soc Jpn 2000;69:3781–4.
[280] Das DK, Sarkar J, Singh SK. Effect of sample size, temperature and strain velocity on mechanical properties of plumbene by tensile loading along longitudinal direction: a molecular dynamics study. Comput Mater Sci 2018;151:196–203.
[281] Lu C, Chang C-P, Huang Y-C, Lu J, Hwang C-C, Lin M-F. Low-energy electronic properties of the AB-stacked few-layer graphites. J Phys Condens Matter 2006;18:5849.
[282] Yin L-J, Shi L-J, Li S-Y, Zhang Y, Guo Z-H, He L. High-magnetic-field tunneling spectra of A B C-stacked trilayer graphene on graphite. Phys Rev Lett 2019;122:146802.
[283] Chang C, Wang J, Lu C, Huang Y, Lin M-F, Chen R. Optical properties of simple hexagonal and rhombohedral few-layer graphenes in an electric field. J Appl Phys 2008;103:103109.
[284] Meng X, Tongay S, Kang J, Chen Z, Wu F, Li S-S, Xia J-B, Li J, Wu J. Stable p-and n-type doping of few-layer graphene/graphite. Carbon 2013;57:507–14.
[285] Rossi E, Sarma SD. Ground state of graphene in the presence of random charged impurities. Phys Rev Lett 2008;101:166803.
[286] Nakayama T, Shima H. Computing the Kubo formula for large systems. Phys Rev 1998;58:3984.
[287] Lin M, Chuu D, Shung K-K. Thermal conductance and the Peltier coefficient of carbon nanotubes. Phys Rev B 1996;53:11186.
[288] Rao C, Pramoda K. Borocarbonitrides, BxCyNz, 2D nanocomposites with novel properties. Bull Chem Soc Jpn 2019;92:441–68.
[289] Pumera M, Wong CHA. Graphane and hydrogenated graphene. Chem Soc Rev 2013;42:5987–95.
[290] Pontes RB, Mançano RR, da Silva R, Cótica LF, Miwa RH, Padilha JE. Electronic and optical properties of hydrogenated group-IV multilayer materials. Phys Chem Chem Phys 2018;20:8112–8.
[291] Okamoto Y. Density functional theory calculations of alkali metal (Li, Na, and K) graphite intercalation compounds. J Phys Chem C 2014;118:16–9.
[292] Kaneko T, Saito R. First-principles study on interlayer state in alkali and alkaline earth metal atoms intercalated bilayer graphene. Surf Sci 2017;665:1–9.
[293] Rabhi A, Schuck P, Da Providência J. Random phase approximation for the 1D anti-ferromagnetic Heisenberg model. J Phys Condens Matter 2006;18:10249.
[294] Schulz S, Nechaev IA, Güttler M, Poelchen G, Generalov A, Danzenbächer S, Chikina A, Seiro S, Kliemt K, Vyazovskaya AY. Emerging 2D-ferromagnetism and strong spin-orbit coupling at the surface of valence-fluctuating EuIr 2 Si 2. Quant Mater 2019;4:1–8.
[295] Shakil M, Hussain A, Zafar M, Ahmad S, Khan M, Masood MK, Majid A. Ferromagnetism in GaN doped with transition metals and rare-earth elements: a review. Chin J Phys 2018;56:1570–7.
[296] Kou L, Du A, Chen C, Frauenheim T. Strain engineering of selective chemical adsorption on monolayer MoS 2. Nanoscale 2014;6:5156–61.
[297] Pan B, Xing B. Adsorption mechanisms of organic chemicals on carbon nanotubes. Environ Sci Technol 2008;42:9005–13.
[298] Pham HD, Gumbs G, Su W-P, Tran NTT, Lin M-F. Unusual features of nitrogen substitutions in silicene. RSC Adv 2020;10:32193–201.
[299] Lv R, Li Q, Botello-Méndez AR, Hayashi T, Wang B, Berkdemir A, Hao Q, Elías AL, Cruz-Silva R, Gutiérrez HR. Nitrogen-doped graphene: beyond single substitution and enhanced molecular sensing. Sci Rep 2012;2:586.
[300] Santos EJ, Ayuela A, Sánchez-Portal D. First-principles study of substitutional metal impurities in graphene: structural, electronic and magnetic properties. New J Phys 2010;12:053012.
[301] Anastas PT, Zimmerman JB. The periodic table of the elements of green and sustainable chemistry. Green Chem 2019;21:6545–66.
[302] Sivek J, Sahin H, Partoens B, Peeters FM. Adsorption and absorption of boron, nitrogen, aluminum, and phosphorus on silicene: stability and electronic and phonon properties. Phys Rev B 2013;87:085444.

[303] Crespo EA, Ruda M, de Debiaggi SR, Bringa EM, Braschi FU, Bertolino G. Hydrogen absorption in Pd nanoparticles of different shapes. Int J Hydro Ener 2012;37:14831–7.
[304] Nguyen DK, Tran NTT, Chiu Y-H, Lin M-F. Concentration-diversified magnetic and electronic properties of halogen-adsorbed silicene. Sci Rep 2019;9:1–15.
[305] Rafique M, Shuai Y, Hussain N. First-principles study on silicon atom doped monolayer graphene. Phys E Low-dimens Syst Nanostruct 2018;95:94–101.
[306] Margulis V, Muryumin E. Theory of chemisorption of a single hydrogen atom on graphene. Carbon Nanomater Clean Energy Hydro Syst 2008;183.
[307] Boukhvalov D, Katsnelson M, Lichtenstein A. Hydrogen on graphene: electronic structure, total energy, structural distortions and magnetism from first-principles calculations. Phys Rev B 2008;77:035427.
[308] Karlický F, Lepetit B, Lemoine D. Quantum modelling of hydrogen chemisorption on graphene and graphite. J Chem Phys 2014;140:124702.
[309] Ohno S, Wilde M, Fukutani K. Novel insight into the hydrogen absorption mechanism at the Pd (110) surface. J Chem Phys 2014;140:134705.
[310] Tozzini V, Pellegrini V. Prospects for hydrogen storage in graphene. Phys Chem Chem Phys 2013;15:80–9.
[311] Mastail C, Bourennane I, Estève A, Landa G, Rouhani MD, Richard N, Hémeryck A. Oxidation of Germanium and Silicon surfaces (100): a comparative study through DFT methodology. In: IOP conference series: materials science and engineering. IOP Publishing; 2012. p. 012007.
[312] Deal BE, Grove A. General relationship for the thermal oxidation of silicon. J Appl Phys 1965;36:3770–8.
[313] Milošv I, Strehblow HH, Navinšek B, Metikoš-Huković M. Electrochemical and thermal oxidation of TiN coatings studied by XPS. Surf Interface Anal 1995;23:529–39.
[314] Thürmer K, Williams E, Reutt-Robey J. Autocatalytic oxidation of lead crystallite surfaces. Science 2002;297:2033–5.
[315] Sevik C, Bulutay C. Theoretical study of the insulating oxides and nitrides: SiO_2, GeO_2, Al_2O_3, Si_3N_4, and Ge_3N_4. J Mater Sci 2007;42:6555–65.
[316] Sue K, Hakuta Y, Smith RL, Adschiri T, Arai K. Solubility of lead (II) oxide and copper (II) oxide in subcritical and supercritical water. J Chem Eng Data 1999;44:1422–6.
[317] Poulain M. Halide glasses. J Non-Cryst Solids 1983;56:1–14.
[318] Peng L, Hu Y, Xu L. Theoretical and experimental research on the bulk photovoltaic effect in hybrid organic–inorganic perovskites CH3NH3PbI2 X (X= Cl, Br, I). Sci Adv Mater 2016;8:2223–30.
[319] Wang L, Wang S, Ewing R. Amorphization of cubic zirconia by caesium-ion implantation. Phil Mag Lett 2000;80:341–7.
[320] Bhauriyal P, Mahata A, Pathak B. The staging mechanism of $AlCl_4$ intercalation in a graphite electrode for an aluminium-ion battery. Phys Chem Chem Phys 2017;19:7980–9.
[321] Wang Z, Inagaki M. Intercalation of $FeCl_3$ into graphite films in molten salts. Synth Met 1991;44:165–76.
[322] Qi X, Qu J, Zhang H-B, Yang D, Yu Y, Chi C, Yu Z-Z. $FeCl_3$ intercalated few-layer graphene for high lithium-ion storage performance. J Mater Chem 2015;3:15498–504.
[323] Zhan D, Sun L, Ni ZH, Liu L, Fan XF, Wang Y, Yu T, Lam YM, Huang W, Shen ZX. $FeCl_3$-Based few-layer graphene intercalation compounds: single linear dispersion electronic band structure and strong charge transfer doping. Adv Funct Mater 2010;20:3504–9.
[324] Shine DP. Synthesis of polyethers and polysiloxanes. The University of Akron; 1961.
[325] Durgun E, Tongay S, Ciraci S. Silicon and III-V compound nanotubes: structural and electronic properties. Phys Rev B 2005;72:075420.
[326] Su DS, Schlögl R. Nanostructured carbon and carbon nanocomposites for electrochemical energy storage applications. ChemSusChem 2010;3:136–68.
[327] Zhou W, Sit PH-L. First-principles understanding of the staging properties of the graphite intercalation compounds towards dual-ion battery applications. ACS Omega 2020;5:18289–300.

CHAPTER

24

Problems

Shih-Yang Lin[1], Vo Khuong Dien[1], Chi-Hsuan Lee[1], Hsin-Yi Liu[1], Hai Duong Pham[1], Thi My Duyen Huynh[1], Nguyen Thi Han[1], Ngoc Thanh Thuy Tran[2], Thi Dieu Hien Nguyen[1], Wei-Bang Li[1] and Ming-Fa Lin[1]

[1]Department of Physics, National Cheng Kung University, Tainan, Taiwan; [2]Hierarchical Green-Energy Materials (Hi-GEM) Research Center, National Cheng Kung University, Tainan, Taiwan

Problem 24.1: The optimal crystal structures of 1D graphene/silicene/germanene/tinene/plumbene nanoribbons [1–4], as clearly illustrated in Fig. 24.1. The thorough explorations, which are conducted by the first-principles simulations, mainly focus on (a) the total ground state energies, the position-dependent bond lengths, and the buckled geometries [5–7]. Such physical quantities are clearly shown in the related figures. (b) Armchair and zigzag edges, as well as the passivation/absence of hydrogen adatoms, are covered in the full investigations. Certain tables, with the reliable geometric parameters, can provide a very useful information [8]. (c) Discuss the highly non-uniform chemical environments due to the observable fluctuations of bond lengths, indicating the active orbital hybridizations [8]. (d) State certain important similarities and differences of group-IV materials [9]. Moreover, (e) investigate the close relations among quantum confinement, edge structure and buckling, and their critical roles on the fundamental quasiparticle properties [8]. In addition, (f) the similar calculations could be generalized to more complicated crystal structures of bilayer nanoribbons, as a result of various stacking configurations [10,11]. The geometric symmetries will be greatly diversified in few-layer 1D systems [10,11].

Problem 24.2: Double-walled group-IV systems, with the coaxial hollow cylinders, as shown for in Fig. 24.2 for armchair [5–10] carbon nanotubes [12]. (a) The delicate VASP calculations are available in determining the most stable and quasi-stable stacking configurations from the ground state energies [12]. (b) Compared with the perfect cylindrical surfaces of coaxial armchair carbon nanotubes, investigate the buckled cylinders in silicon/germanium/tin/lead nanotubes by characterizing their main features. Very interesting,

https://doi.org/10.1016/B978-0-443-15801-8.00004-9

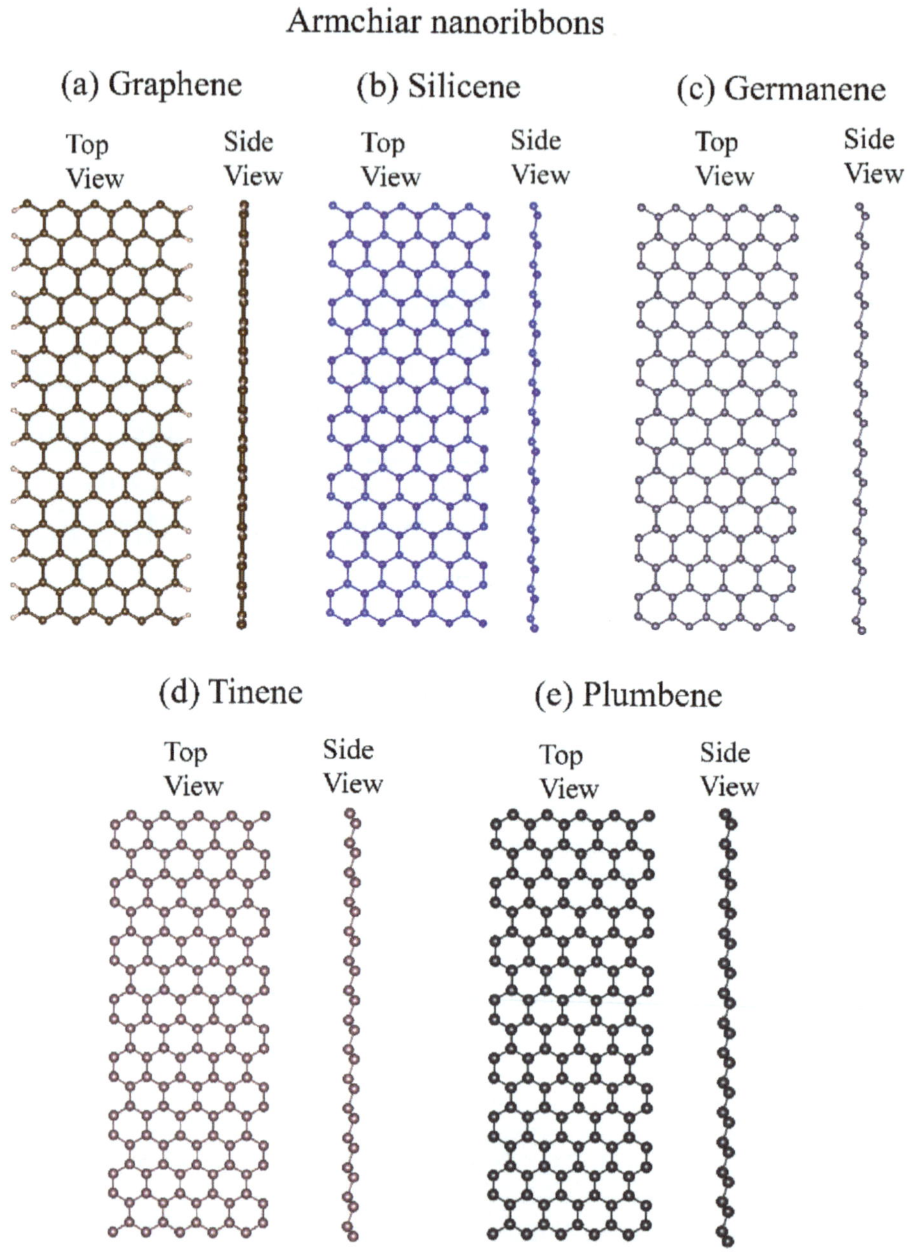

FIG. 24.1 The optimal geometric structures for (A)/(B)/(C)/(D)/(E) 1D armchair graphene/silicene/germanene/tinene/plumbene nanoribbons, with the top and side views (at the left and right hands) in the presence of hydrogen passivation on two open edges.

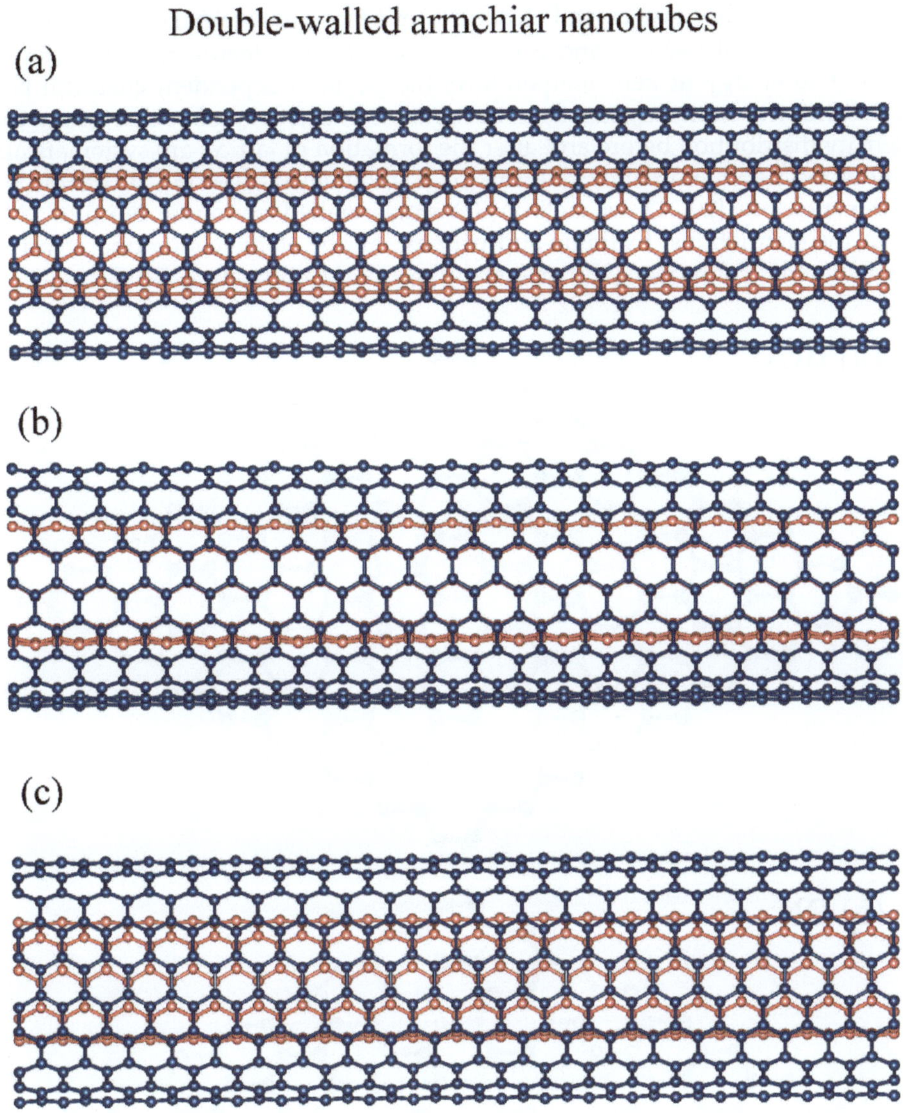

FIG. 24.2 The stable crystal structures of doubled-walled [5–10] carbon nanotubes under the high-symmetry stacking configurations: (A) C_5, (B) D_5h and (C) S_5h.

(c) the latter are expected to present more complicated crystal symmetries because of the strong coupling effects of cylindrical and buckled structures. (d) Clearly show charge density distributions on the coupled and independent armchair nanotubes. (e) Two cases can deduce the drastic changes of orbital hybridizations before/after the significant nanotube coupling. (f) Evaluate whether the interlayer atomic interactions belong to van der Waals ones [13]. (g) The dramatic transformation of lattice symmetry might occur in the increment of atomic number. (h) Compare and achieve certain significant similarities and differences for double-walled and bilayer materials in the main features of geometric properties.

Problem 24.3: The optimal crystal structures of binary and ternary $B_xC_yN_z$-related compounds, e.g., 2D monolayer BC_3 and BN in Fig. 24.3. (a) The featured geometric properties, the ground state energy at zero temperature, the position-dependent chemical bonds, the crystal rotation & translation symmetries, the absence or existence of buckling, and the spatial charge density distribution before and after the formation of lattice, are systematically investigated under the VASP simulations [14,15]. (b) 2D monolayers [16–18], 1D nanoribbons [19], cylindrical nanotubes [20,21], as well as 3D single-walled nanotube bundles [22,23], are included in the simulations. Examine and propose all the stable compound structures with the different atomic component dimensionalities, clearly illustrating B-, C- and N-enriched

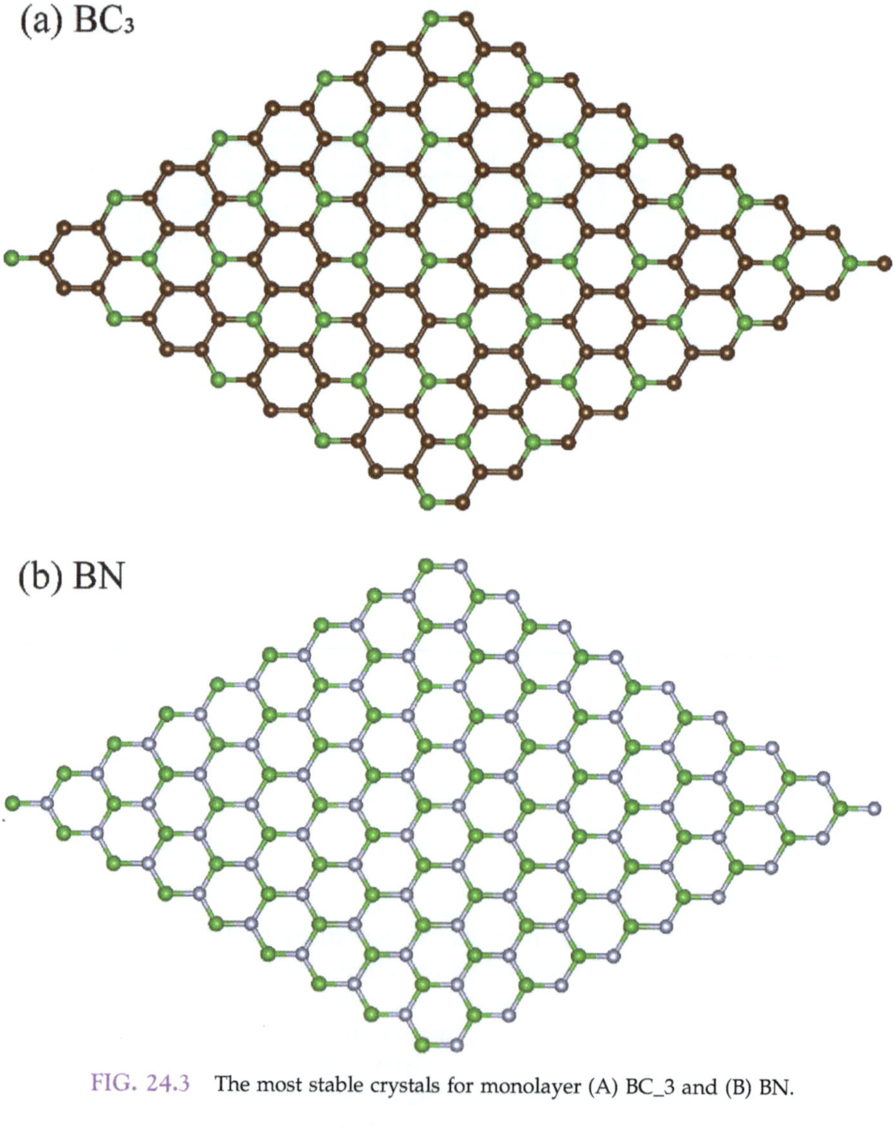

FIG. 24.3 The most stable crystals for monolayer (A) BC_3 and (B) BN.

crystal symmetries [16,24]. (c) The delicate analyses are done by proposing the active and significant multi-orbital hybridizations from the drastic changes of charge densities on the various crystals. (d) The critical mechanisms are further assisted by the reliable calculations on the atom-dominated energy spectra [16,25] and atom- and orbital-decomposed van Hove singularities [18,21].

Problem 24.4: The featured electronic properties, corresponding to the low-dimensional materials in Problems 24.1–24.3, are further explored by the first-principles method. Such quasiparticle behaviors cover (a) band structures & their atom dominances in binary and ternary compounds ($B_xC_yN_z$ in Problem 24.3; [14–16]), as well as (b) the atom- & orbital-decomposed density of states [18,21,26]. (a) Needs to be clearly illustrated for the main features (a direct or indirect band gap, spectrum asymmetry about the Fermi level, energy dispersions with extreme points, subband noncrossings/crossings/anti-crossings and band widths [26]). (c) Discuss the diversified van Hove singularities due to dimensionalities and various critical points (band-edge states in energy-wave-vector space; [8]). (d) The direct combination of (a) and (b), being fully supported by the above-mentioned spatial charge densities, is able to determine the active multi-/single-orbital hybridizations [8]. (e) Thoroughly investigate the relative weights of π, σ and/or sp^3 bondings for the intralayer and interlayer atomic interactions [16], such as, the enhanced or reduced sp^3 hybridizations related to quantum-confinement effects of 1D nanoribbons (open boundary conditions; [8,27]), the existence/absence of the van der Waals interaction in double-walled carbon/silicon/germanium/tin/lead nanotubes, and the covalent, but not metallic bonds in $B_xC_yN_z$ compounds [25].

Problem 24.5: The reliable parameters of the tight-binding model for the emergent materials in Problems 24.1–24.3, being achieved from the VASP simulations on the low-lying valence and conduction subbands [28]. In general, such physical quantities cover the orbital-dependent hopping integrals and on-site Coulomb potentials; [28,29]. (a) It is well known that the carbon-$2p_z$ orbitals of a normal honeycomb lattice are sufficient in characterizing the low-energy essential properties. The π bonding, with hydrogen passivation on two open edges, is assisted by the observable fluctuations of C–C bond lengths (the highly non-uniform chemical environment [30]). This can be solved by the Harrison rule (the strength of orbital hybridization inversely proportional to the neighboring distance [30]). (b) Express the width-dependent Hamiltonian matrix elements of armchair graphene nanoribbons in the absence of the spin-related interactions [31–33]. (c) Calculate band structures and wave functions, discuss their main features, and then classify three types of energy gaps according to the widths of $N_y = 3I$, $3I + 1$, and $3I + 2$ (the dimer number along the transverse direction, I an integer; details in Refs. [34,35]). (d) But for zigzag graphene nanoribbons, the on-site spin-created electron-electron Coulomb interactions need to be included in the Hubbard-like Hamiltonian [36]. (e) By the exact diagonalization, an anti-ferromagnetic configuration across the nanoribbon center, which is accompanied with two ferromagnetic spin distributions on zigzag edges, is examined to survive near the high-symmetry point (the first Brillouin zone boundary [36,37]). (f) Furthermore, the partially flat valence and conduction bands at the Fermi level, with a zero gap, will be dramatically separated into a narrow-gap abnormal energy dispersions [37]. (g) Very interestingly, there exist certain significant differences in 1D graphene nanoribbons and carbon nanotubes in terms of featured energy spectra, wave functions and van Hove singularities, where the former and the latter, respectively, possess the open [37] and periodical [37] boundary conditions. (h) The $2p_z$-orbital-induced orbital hybridizations on a cylindrical surface could be well described an empirical formula. Also, the position-dependent atomic interactions are suitable for the intertube ones [22]. (i) Propose the close relations between 2D graphene and 1D graphene nanoribbon/carbon nanotube [29]/[38]. (j) More complicated Hamiltonian could be built for armchair/zigzag double-walled carbon nanotubes and present the diverse electronic properties due to the rotation and translation symmetries, e.g., band structures of [5, 5]-[10, 10]/[9, 0]-[18.0] [12,39,40].

Problem 24.6: Apparently, non-graphene group-IV systems present the significant sp^3 bondings and buckled geometries [41,42], as thoroughly investigated for 2D honeycomb lattices in this book [43]. The tight-binding model, which is able to provide the concise pictures, needs to satisfy the crystal symmetry of 1D finite-width nanoribbons [44] and cylindrical nanotubes [45]. (a) Determine the enlarged unit cells under the open and periodical boundary conditions, respectively [45]. (b) Test the sp^3- and SOC-dependent Hamiltonian matrix elements for armchair silicene/germanene/stanene/plumbene nanoribbons. Discuss their coupling effects with quantum confinement in terms of the main features of electronic properties, such as, three types of width-dependent band gaps [2], the enhanced or reduced asymmetry of hole and electron spectra [46], various energy dispersions [42], band-edge states [41], and subbband non-crossings/crossings/anticrossings [41]. (c) Add the on-site spin-related Coulomb interactions to evaluate the existence of anti-ferromagnetism and ferromagnetism along the longitudinal and transverse directions, respectively [43]. (d) Clearly illustrate the diverse quasiparticle properties in the increment of atomic number. As for cylindrical nanotubes, the multi-orbital hybridizations of sp^3 bondings might be quite different those of nanoribbons because of the great enhancement of the non-orthogonality between p_z and (s, p_x, p_y) orbitals four orbitals. (e) The hopping integrals of distinct orbitals would become more complicated in a cylindrical surface. A plenty of useful parameters in group-IV armchair/zigzag nanotubes are worthy of the systematic examinations. (f) The delicate analyses are conducted on band structures, wave functions, and orbital- & spin-decomposed van Hove singularities [8,47]. The active orbital hybridizations, corresponding to the specific energy ranges, could account for the featured quasiparticle behaviors.

Problem 24.7: Magnetic quantization phenomena in 2D $B_xC_yN_z$ honeycomb lattices [48]. (a) Plot the primitive unit cells for all the stable crystal structures with integer indices. (b) Establish the multi-orbital hybridizations and on-site Coulomb potentials in C–C, B–C, B–N and C–N bonds. (c) Examine the low-energy band structures to get the reliable parameters in specific materials by comparing with those of VASP simulations [14]. (d) Such systems are present in a uniform perpendicular magnetic field (Fig. 24.4), leading to the quantized Landau levels. The highly degenerate magneto-electronic states mainly arise from the condensation of the 2D neighboring ones, i.e., they correspond to the constant-energy closed wave-vector surface [36,49]. (e) Illustrate the vector-potential-induced Peierls phase and thus the creation of a rectangular Moire superlattice [30]. (f) Calculate the dramatic changes of the nearest-neighbor hopping integrals [49,50]. This is able to build a generalized tight-binding model, in which the magneto-Hamiltonian matrix has a very large dimension. (g) By the exact diagonalization method, the magnetic-field-dependent LL energy spectra and

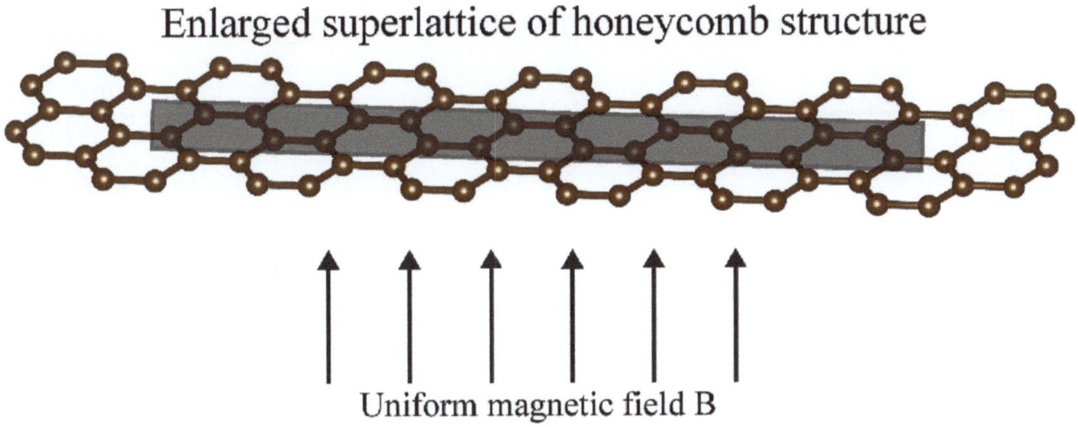

FIG. 24.4 A hexagonal honeycomb crystal in the presence of a uniform perpendicular magnetic field, with an enlarged rectangular Moire superlattice.

localized wave functions are obtained using an efficient way. (i) The well-behaved spatial oscillations, with normal zero-point numbers, are available in defining quantum index. (j) Thoroughly investigate how many subgroups of LLs by plotting the featured oscillations about the spatial probability distributions. In general, this depends on the pairs of valence and conduction energy subbands. (k) Discuss the non-crossing/crossing/anti-crossing behaviors during the variation of magnetic-field strength, especially for the regular or irregular B_z-dependences. (l) Comprehend the dramatic transformation of magneto-wave functions near the critical magnetic fields with anti-crossings. The major and minor modes will present the comparable amplitudes there [30]. (m) The extra quasiparticle phenomena have been successfully predicted in the previous works, e.g., the extraordinary magneto-optical selection rules [30], and quantum Hall conductivities [51].

Problem 24.8: The ballistic transport of 1D materials between two heat reservoirs with bias voltage and different difference (ΔV, and ΔT in Fig. 24.5; [52,53]). Quantum transport properties could be explored by the well-known Landauer-Buttiker formula [54]. The external perturbations are able to drive the quasiparticle transport associated with the electrical and thermal current densities along the longitudinal direction of a nanoribbon or nanotube [55]. (a) Express them in the absence of elastic and inelastic scatterings. (b) Within the linear response [55], well define the significant transport coefficients, the electrical conductance, Peltier coefficient and thermal conductance [56], especially for the specific conditions [53,55]. (c) Propose the concise physical pictures in generating quantum transport structures (plateaus [57]). (d) To clearly examine electronic properties and greatly diversify quantum transports, the gate voltage [58,59], chemical potential and magnetic field perpendicular to nanoribbon honeycomb/parallel to nanotube axis could play the modulation roles. Illustrate the important quasiparticle transport behaviors in graphene/silicene/germanene/stanine/plumbene nanoribbons or nanotubes. The V_g-, E_F- and B_z-dependent transport features are available in distinguishing such materials [60,61]. (e) Delicate explore the complicated relations among the quantum transport coefficients [62], the intrinsic interactions (the single- or multi-orbital hybridizations and spin-orbital couplings [8]), and the external perturbations

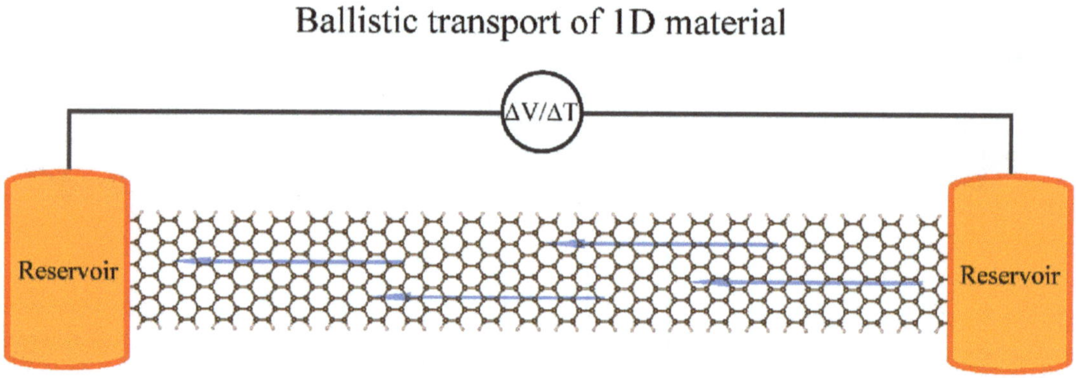

FIG. 24.5 The ballistic transport of 1D materials between two heat reservoirs with bias voltage and different difference (ΔV, ΔT; [52,53]). Quantum transport properties could be explored by the well-known Landauer-Buttiker formula [54].

[63]. (f) Thoroughly discuss how to generalize the similar model calculations for bilayer nanoribbons [10,11] and double-walled coaxial nanotubes in testing the phenomenological methods under the quasiparticle framework [30,33].

Problem 24.9: The binary or ternary $B_xC_yN_z$ compounds, the highly potential candidates in illustrating the diverse quantum Hall effects [64]. The main features magneto-electronic properties will be thoroughly investigated by the generalized tight-binding model (details in Chapter 24 [65,66]). They provide the sufficient information in comprehending the quantized magneto-conductivities. (a) Express the static Kubo formula, with the scatterings of the initial and final Landau level through the electric dipole moment [37,67]. (b) Furthermore, such scattering matrix elements could be evaluated by the gradient approximation (the average momentum approximated by the first derivative of Hamiltonian versus wave vector), as successfully done for graphene-related systems [30,33]. This is the zero-frequency case of the vertical optical transitions [29,60,61,68]. (c) Discuss all the available inter-LL channels without momentum and energy transfers at zero temperatures and determine the existence of magneto-selection rules through the featured Landau levels. (d) By modulating the Fermi level and magnetic-field strength, calculate the magneto-quantum conductivities for all the stable binary and ternary $B_xC_yN_z$ compounds. (e) Investigate the different transport structures in the E_F- and B_z-dependences and propose the critical mechanisms to account for them, such as, the well-like, stair-step and composite quantum structures closely related to the main features of specific Landau levels. (f) Evaluate whether certain materials exhibit the well-behaved magneto-conductivities, being characterized by the analytic formulas. In addition, few-layer graphene systems, with the normal AAA [69], ABA [49,70] and ABC stackings [71,72], present the well-defined dependences. (g) Explore the close relation between the quantum conductivity at the neutral point, Landau-level energies, and intrinsic interactions. This could examine the reliability of hopping integrals and spin-orbital coupling in the generalized tight-binding model [65,66].

Problem 24.10: The diverse optical excitations of graphene-related 3D/2D/1D materials, with the well-behaved stacking configurations and planar honeycomb lattices [70,73]. The direct combinations of the tight-binding model and the dynamic Kubo formula, which are investigated under the intralayer and interlayer atomic interactions of carbon-$2p_z$ single-orbital hybridizations [74], can thoroughly explore the featured absorption spectra. (a) Determine the primitive unit cells and the first Brillouin zones for AA-. AB- & ABC-stated bulk graphites [75], trilayer graphene systems with AAA (Fig. 24.6A [69]), ABA [70], ABC [49] and AAB stackings [76,77], armchair/zigzag/chiral carbon nanotubes [12,78]/[40,79–81], as well as armchair/zigzag/chiral graphene nanoribbons. (b) By using the Fermi golden rule, the spectral function, corresponding to the vertically optical transitions with the zero momentum transfer, covers the joint density of states and the scattering probability of an electric dipole moment. Propose the concise physical pictures for the former and the latter. (c) Derive an analytic formula for the dipole-transition matrix element of monolayer graphene using the gradient approximation, as done in Problem 24.9. The similar calculations could be generalized to all the graphene-based systems. (d) Thoroughly examine the dimension-, stacking-, layer-number- and plane-/cylinder-diversified optical excitations. (e) Fully discuss the various absorption structures due to the dimension-dependent van Hove singularities. (f) Establish a close relation between energy gap and threshold absorption frequency in semiconducting systems. (g) Evaluate the existence of optical selection rules, especially for carbon nanotubes [60] and graphene nanoribbons [29,82]. (h) The above-mentioned optical features are directly reflected in magneto-optical excitations. The continuous frequency dependence will dramatically transform into a lot of prominent absorption peaks, being closely related to 3D & 1D Landau subbands (asymmetric forms [49,67,73,83]), as well as 2D landau levels (delta-function-like ones [36]). (i) Identify and illustrate certain well-defined magneto-optical selection rules by using the symmetric and/or anti-symmetric modes of the initial and final magneto-electronic states [36]. (j) Compare the important differences among all the graphene-related materials. (k) How to utilize the diverse optical excitations in designing the high-performance nanoscale optical devices [84,85].

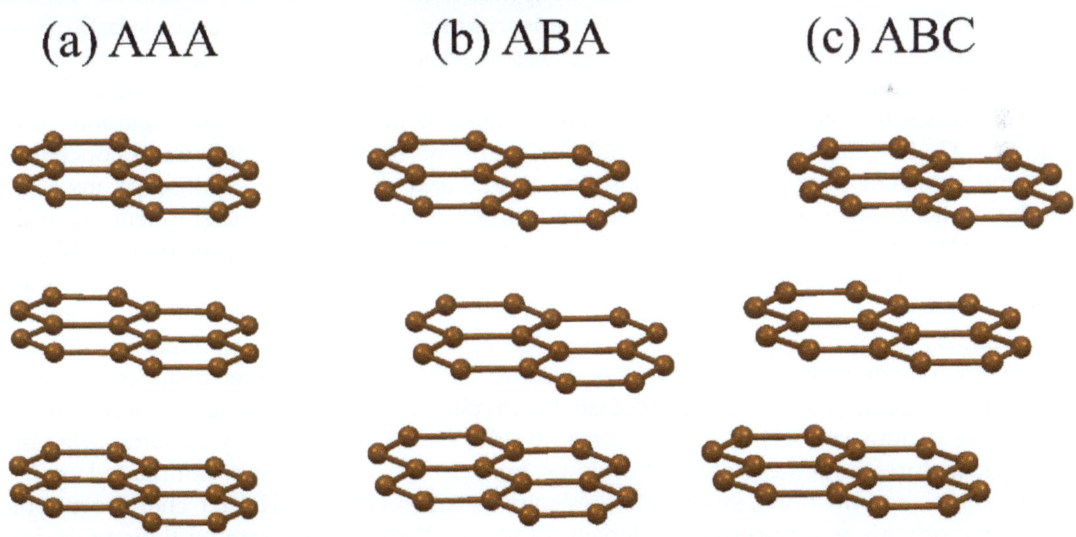

FIG. 24.6 Trilayer graphene systems under the (A) AAA [69], (B) ABA [49,70], (C) ABC [72,131,132] and (D) AAB [76,77] stacking configurations.

Problem 24.11: Excitonic effects in carbon nanotubes/graphene nanoribbons and graphene systems by using the first-principles simulations [63,86–88]. The VASP calculations are available in fully exploring the many-body effects on the vertically optical transitions, while the phenomenological models are not reliable for dealing with quasiparticle behaviors [65]. The precious high-precision absorption measurements [89,90] clearly show the strong evidences of stable/quasi-stable excitons, covering an observable difference (an enough large red shift [91,92]) between threshold absorption frequency and band gap, the great enhancement of the initial excitations, more or modified absorption structures, and strong photoluminescence spectra [93,94]. Calculate all the physical quantities of frequency-dependent optical excitations in the presence/absence of excitonic effects: (a) the real and imaginary parts of the dielectric functions, (b) reflectance spectra, (c) absorption coefficients, and (d) energy loss function at long wave length limits. Thoroughly propose the physical pictures for these quasiparticle properties, such as, their definitions and features. (e) Investigate the many-particle behaviors in armchair and non-armchair single-walled carbon nanotubes (metallic and semiconducting ones [95,96]), e.g., their effects on the frequency, number, form, and intensity of available band-edge excitation spectrum [97,98]. (f) A lot of prominent absorption peaks come to exist because of the transverse periodical boundary condition. Plot the lower-frequency five absorption peaks as a function of radius according three types of geometric structures ($2m + n = 3I$, $3I+1$ and $3I+2$ for $(m.n)$ carbon nanotubes [99,100]). (g) Delicately examine and then identify the close relation between prominent absorption peaks and specific orbital hybridizations. (h) The similar studies could be generalized to finite-width graphene nanoribbons, but only focused on three types of semiconducting armchair nanostructures and zigzag ones (details in Problem 24.5). (i) Explain certain significant differences of 1D graphene nanoribbons and carbon nanotubes and the easier verifications of optical measurements on the former. (j) Check whether semi-metallic bilayer and trilayer graphene systems, with normal stacking configurations (details in Problem 24.9), show the obvious evidence about many-body effects.

Problem 24.12: Chemical modifications on optical excitations of graphene-related systems by the VASP simulations [74,101]. Adatom adsorptions [102] and guest-atom substitutions [103,104] are very efficient in diversifying crystal structures [8,102], electronic properties [102,105,106] and thus optical excitations, (a) Determine the primitive unit cells and their first Brillouin zones for the stage-1/stage-2/stage-3/stage-4 alkali graphite intercalation compounds [107–109], alkali-adsorbed monolayer graphene under the well-defined concentrations and armchair/zigzag distributions (the single- and double-side chemisorptions [74]), and the similar cases inside/on a single-walled carbon nanotube/a planar graphene nanoribbons. (b) Delicately analyze certain similarities and differences, leading to the diversity of other essential quasiparticle properties. (c) While hydrogen, oxygen, halogens, transition metals, and rare-earth atoms serve as active guests, the composite chemical environments might display the drastic changes [101], e.g., the optimal absorption positions, the adatom heights, the planar or buckled carbon-honeycomb lattices, the interlayer distances, as well as the highest-concentration adsorption case. (d) Preparing a table with all the reliable parameters of the various crystals, being supported by the clear figures, is easy to summarize the geometric features from the chemical bondings. (e) Calculate atom- & spin-dominated energy spectra and wave functions, spatial charge/spin densities before and after chemisorptions, and atom-, orbital- & spin-projected density of states. Thoroughly examine their main features and directly link together to determine the critical mechanisms associated with the significant multi-orbital hybridizations in distinct chemical bonds and the ferromagnetic/

antiferromagnetic/nonmagnetic spin distributions (the atom- and orbital-dependent magnetic moments [74]). (f) The atom-, orbital- and spin-dependences, which correspond to the specific valence and conduction energy ranges, are very useful in understanding the prominent absorption structures except for the lower-frequency exciton peaks in semiconducting systems. That is, each strong peak is identified to come from a specific orbital hybridization and spin configuration through the delicate analyses [74]. Moreover, the VASP simulations could be generalized to the substitution cases of guest atoms. The diverse quasiparticle phenomena are expected to reveal in (g) various Moire superlattice [90], (h) complicated band structures & wave functions [110], unique charge/spin density distributions [111], net magnetic moments due to the dominating atoms and orbitals [112], a lot of merged structures in orbital- & spin-decomposed van Hove singularities [74,101], (i) optical gap, excitonic peaks, electron-hole excitations [113], and coherent plasmon modes [107]. Discuss them in detail and propose the concise physical pictures. (j) Finally, three kinds of tables, which cover the significant features of geometric, electronic, and optical properties, could be utilizes to present the full examinations about the development of quasiparticle framework [5].

Problem 24.13: Recently, silicon-related materials might replace the critical anode role of graphite in lithium-ion-based batteries, mainly owing to a great reduce about charging time (Fig. 24.7 [114]). Silicone is built from the bone framework of silicon-oxygen polymers, being surrounded by the various organic functional groups [115]. (a) The diversified Moire

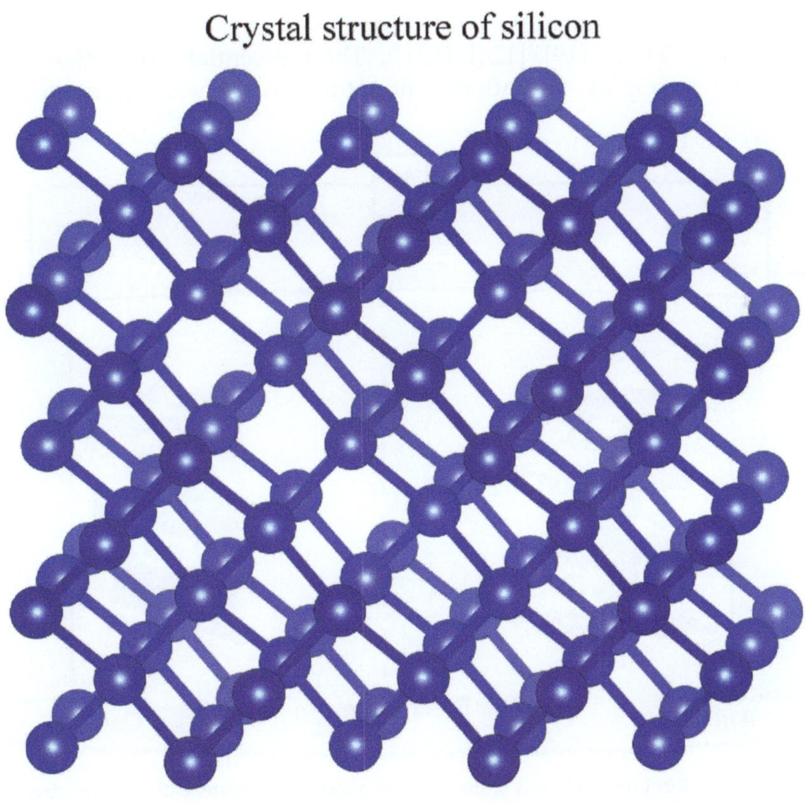

FIG. 24.7 Trilayer graphene systems under the (A) AAA [69], (B) ABA [49,70], (C) ABC [72,131,132] and (D) AAB [76,77] stacking configurations.

superlattices can be reached by modulating the composite components. Plot all the stable configurations. (b) Investigate the ground state energies, the distinct chemical bonds, the highly non-uniform environments, and the delicate 3D atomic positions of large molecules. (c) Tables, with reliable geometric parameters, will be more convenient in comprehending the linear and deformed structures. The concise chemical bondings are suitable in explaining the unusual lattice symmetries. (d) Band structures of long polymer compounds will exhibit a lot of valence and conduction energy subbands with weak wave-vector dependences, as a result of many atoms in a primitive unit cell. Clearly illustrate their main features in semiconducting compounds, such as, the direct or indirect band gap, the high asymmetry of occupied hole and unoccupied electron bands, the well- or undefined band widths, and the subband non-crossings/crossings/anti-crossings, and the band-edge states. (e) Clearly show, examine and identify the active multi-orbital hybridizations in the linear Si—O bone, the functional group roots, and their heterojunctions. That is to say, each silicone has very rich orbital bondings, indicating the various hopping integrals in the tight-binding model [65]. (f) Most important, how to establish the close relations between their unique merits of geometric/electronic properties and dynamic intercalations/de-intercalations about lithium ions is worthy of systematic investigations [114,116], such as, the optimal Li^+-positions, the adatom distribution configurations, and the saturated concentrations, strongly depending on the distinct compounds [114,116].

Problem 24.14: Ternary and quaternary lithium compounds play the critical roles in the anode/electrolyte/cathode materials of Li^+-based batteries [117,118], such as $Li_4Ti_5O_{12}$/Li_2GeO_3/$LiFeO_2$ (Fig. 24.8; [119]/[120]/[121]). Their essential properties could be fully explored by the first-principles simulations, but not the phenomenological models. The latter

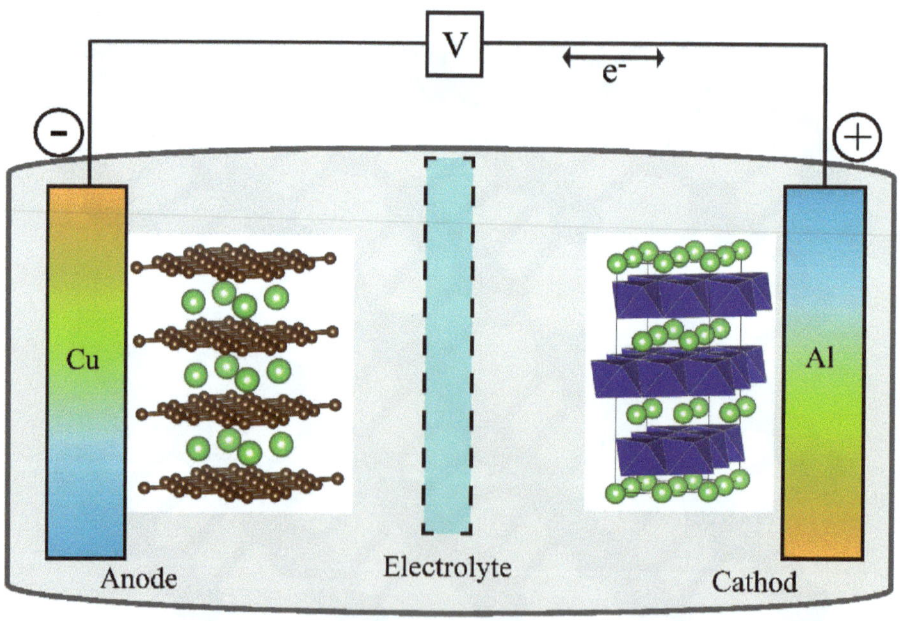

FIG. 24.8 Lithium-ion-based batteries, with anode, electrolyte, cathode and circuit.

are almost absent up to now. (a) Calculate the optimal crystal structures, covering lattice constants, translation & rotation symmetries, position-dependent chemical bonds, and observable fluctuations of bond lengths [119,120]. (b) Plot the spatial charge density distributions and examine the active multi-orbital hybridizations, compared to the isolated atom cases [119,120]. (c) (a) & (b) are utilized to account for the main features of band structures and wave functions. (d) The atom- & orbital-projected van Hove singularities are available in the further identifications of the active orbital bondings, (e) Evaluate whether LiFeO/LiCoO/LiNiO present the obvious ferromagnetism, e.g., the enough magnetic-moment strength from transition metal atoms [121–125]. (f) This is clearly revealed in the spatial spin density distribution. Also noticed that the partial spin configurations might be induced by the neighboring atoms. (g) Test the coexistence of the charge- and spin-co-dominated optical absorption structures. This unusual phenomenon is not predicted/observed in the previous published works. (i) The similar simulations and analyses could be done for another set of Li-, P- and S- related compounds, the critical electrolytes in lithium sulfide batteries [126]. According to the different weights of three core components, there exist a lot of stable crystal structures. (j) This needs to be thoroughly examined by the VASP simulations. The concise pictures are proposed to explain the diverse lattice symmetries [127]. (k) Clearly illustrate that the essential physical/chemical/material properties can provide the enough information about the various potential applications [128]. Also (l) discuss the important differences between $Li_4P_2S_6$/$Li_7P_3S_{11}$/Li_3SP_4 [129] and Li_2SiO_3/Li_2GeO_3 compounds [129]/[130]/[120], especially for the significant multi-orbital hybridizations in various chemical bonds.

References

[1] Yin W-J, Xie Y-E, Liu L-M, Chen Y-P, Wang R-Z, Wei X-L, Zhong J-X, Lau L. Atomic structure and electronic properties of folded graphene nanoribbons: a first-principles study. J Appl Phys 2013;113:173506.

[2] Tchalala MR, Enriquez H, Mayne AJ, Kara A, Dujardin G, Ali MA, Oughaddou H. Atomic structure of silicene nanoribbons on Ag (110). In: Journal of physics: conference series. IOP Publishing; 2014. p. 012002.

[3] Pang Q, Zhang Y, Zhang J-M, Ji V, Xu K-W. Electronic and magnetic properties of pristine and chemically functionalized germanene nanoribbons. Nanoscale 2011;3:4330–8.

[4] Li Y, Yang Z. The electronic and topological properties of plumbene by first-principles calculations. Bull Am Phys Soc 2020;65.

[5] Shih-Yang Lin H-YL, Khanh Nguyen D, Thanh Thuy Tran N, Duong Pham H, Chang S-L, Lin C-Y, Lin M-F. Silicene-based layered materials: essential properties. IOP Publishing; 2020.

[6] Vogt P. Silicene, germanene and other group IV 2D materials. Beilstein-Institut); 2018.

[7] Yuhara J, Le Lay G. Beyond silicene: synthesis of germanene, stanene and plumbene. Jpn J Appl Phys 2020;59:SN0801.

[8] Lin SY, Chang SL, Su WP, Lin MF. Structure- and adatom-enriched essential properties of graphene nanoribbons. CRC press; 2018.

[9] Galbiati M, Motta N, De Crescenzi M, Camilli L. Group-IV 2D materials beyond graphene on nonmetal substrates: challenges, recent progress, and future perspectives. Appl Phys Rev 2019;6:041310.

[10] Zhong X, Pandey R, Karna SP. Stacking dependent electronic structure and transport in bilayer graphene nanoribbons. Carbon 2012;50:784–90.

[11] Morell ES, Vergara R, Pacheco M, Brey L, Chico L. Electronic properties of twisted bilayer nanoribbons. Phys Rev B 2014;89:205405.

[12] Ho Y, Chang C, Shyu F, Chen R, Chen S, Lin M-F. Electronic and optical properties of double-walled armchair carbon nanotubes. Carbon 2004;42:3159–67.

[13] Meunier V, Nardelli MB, Roland C, Bernholc J. Structural and electronic properties of carbon nanotube tapers. Phys Rev B 2001;64:195419.

[14] Kumar N, Moses K, Pramoda K, Shirodkar SN, Mishra AK, Waghmare UV, Sundaresan A, Rao C. Borocarbonitrides, BxCyNz. J Mater Chem 2013;1:5806–21.

[15] Alyörük MM. Piezoelectricity in monolayer BxCyNz structures: a first principles study. Comput Mater Sci 2021;195:110505.

[16] Yadav VK, Chakraborty H, Klein ML, Waghmare UV, Rao C. Defect-enriched tunability of electronic and charge-carrier transport characteristics of 2D borocarbonitride (BCN) monolayers from ab initio calculations. Nanoscale 2019;11:19398–407.

[17] Beniwal S, Hooper J, Miller DP, Costa PS, Chen G, Liu S-Y, Dowben PA, Sykes ECH, Zurek E, Enders A. Graphene-like boron–carbon–nitrogen monolayers. ACS Nano 2017;11:2486–93.

[18] Zhang H, Liao Y, Yang G, Zhou X. Theoretical studies on the electronic and optical properties of honeycomb BC3 monolayer: a promising candidate for metal-free photocatalysts. ACS Omega 2018;3:10517–25.

[19] Chen Z-G, Zou J, Liu G, Li F, Wang Y, Wang L, Yuan X-L, Sekiguchi T, Cheng H-M, Lu GQ. Novel boron nitride hollow nanoribbons. ACS Nano 2008;2:2183–91.

[20] Chen Y, Zou J, Campbell SJ, Le Caer G. Boron nitride nanotubes: pronounced resistance to oxidation. Appl Phys Lett 2004;84:2430–2.

[21] Su W, Chang C, Lin M-F, Li T. Electronic structures and work functions of BC 3 nanotubes: a first-principle study. J Appl Phys 2011;110:014312.

[22] Kis A, Csanyi G, Salvetat J-P, Lee T-N, Couteau E, Kulik A, Benoit W, Brugger J, Forro L. Reinforcement of single-walled carbon nanotube bundles by intertube bridging. Nat Mater 2004;3:153–7.

[23] Liu JZ, Zheng Q-S, Wang L-F, Jiang Q. Mechanical properties of single-walled carbon nanotube bundles as bulk materials. J Mech Phys Solid 2005;53:123–42.

[24] Vaccarini L, Goze C, Henrard L, Hernandez E, Bernier P, Rubio A. Mechanical and electronic properties of carbon and boron–nitride nanotubes. Carbon 2000;38:1681–90.

[25] Terrones M, Grobert N, Terrones H. Synthetic routes to nanoscale BxCyNz architectures. Carbon 2002;40:1665–84.

[26] Fan Q, Wei Q, Chai C, Yan H, Zhang M, Lin Z, Zhang Z, Zhang J, Zhang D. Structural, mechanical, and electronic properties of P3m1-BCN. J Phys Chem Solid 2015;79:89–96.

[27] Sun L, Li Q, Ren H, Su H, Shi Q, Yang J. Strain effect on electronic structures of graphene nanoribbons: a first-principles study. J Chem Phys 2008;129:074704.

[28] Chung H-C, Chang C-P, Lin C-Y, Lin M-F. Electronic and optical properties of graphene nanoribbons in external fields. Phys Chem Chem Phys 2016;18:7573–616.

[29] Li T, Huang Y, Chang S, Chang C, Lin M-F. Magnetoconductance of graphene nanoribbons. Philos Mag A 2009;89:697–709.

[30] Lin C-Y, Ho C-H, Wu J-Y, Do T-N, Shih P-H, Lin S-Y, Lin M-F. Diverse quantization phenomena in layered materials. CRC Press; 2019.

[31] Ho J-H, Lu C, Hwang C, Chang C, Lin M-F. Coulomb excitations in AA-and AB-stacked bilayer graphites. Phys Rev B 2006;74:085406.

[32] Wang X-F, Chakraborty T. Coulomb screening and collective excitations in a graphene bilayer. Phys Rev B 2007;75:041404.

[33] Chiun-Yan Lin J-YW, Chiu C-W, Lin M-F. Coulomb excitations and decays in graphene-related systems. CRC Press; 2019.

[34] Prezzi D, Varsano D, Ruini A, Marini A, Molinari E. Optical properties of graphene nanoribbons: the role of many-body effects. Phys Rev B 2008;77:041404.

[35] Wu J-Y, Chen L-H, Li T-S, Lin M-F. Optical properties of graphene nanoribbon in a spatially modulated magnetic field. Appl Phys Lett 2010;97:031114.

[36] Chung H, Huang Y, Lee M, Chang C, Lin M. Quasi-landau levels in bilayer zigzag graphene nanoribbons. Phys E Low-dimens Syst Nanostruct 2010;42:711–4.

[37] Chung H-C, Lee M-H, Chang C-P, Huang Y-C, Lin M-F. Effects of transverse electric fields on quasi-landau levels in zigzag graphene nanoribbons. J Phys Soc Jpn 2011;80:044602.

[38] Tsai C, Chen S, Shyu F, Chang C, Lin M-F. Magnetization of carbon nanotubes. Phys E Low-dimens Syst Nanostruct 2005;30:86–92.

[39] Shen C, Brozena AH, Wang Y. Double-walled carbon nanotubes: challenges and opportunities. Nanoscale 2011;3:503–18.

[40] Benguediab S, Tounsi A, Zidour M, Semmah A. Chirality and scale effects on mechanical buckling properties of zigzag double-walled carbon nanotubes. Compos B Eng 2014;57:21−4.
[41] Song Y-L, Zhang Y, Zhang J-M, Lu D-B, Xu K-W. First-principles study of the structural and electronic properties of armchair silicene nanoribbons with vacancies. J Mol Struct 2011;990:75−8.
[42] Mehdi Aghaei S, Calizo I. Band gap tuning of armchair silicene nanoribbons using periodic hexagonal holes. J Appl Phys 2015;118:104304.
[43] Cahangirov S, Topsakal M, Ciraci S. Armchair nanoribbons of silicon and germanium honeycomb structures. Phys Rev B 2010;81:195120.
[44] Brey L, Fertig H. Electronic states of graphene nanoribbons studied with the Dirac equation. Phys Rev B 2006;73:235411.
[45] Gumbs G, Aĭzin G. Collective excitations in a linear periodic array of cylindrical nanotubes. Phys Rev B 2002;65:195407.
[46] Ding Y, Wang Y. Electronic structures of reconstructed zigzag silicene nanoribbons. Appl Phys Lett 2014;104:083111.
[47] Tao C, Jiao L, Yazyev OV, Chen Y-C, Feng J, Zhang X, Capaz RB, Tour JM, Zettl A, Louie SG. Spatially resolving edge states of chiral graphene nanoribbons. Nat Phys 2011;7:616−20.
[48] Gupta U, Rao C. Advances in the chemistry and physics of materials: overview of selected topics. World Scientific; 2020. p. 27−54.
[49] Yuan S, Roldán R, Katsnelson MI. Landau level spectrum of ABA-and ABC-stacked trilayer graphene. Phys Rev B 2011;84:125455.
[50] Giesbers A, Ponomarenko L, Novoselov K, Geim A, Katsnelson M, Maan J, Zeitler U. Gap opening in the zeroth Landau level of graphene. Phys Rev B 2009;80:201403.
[51] Dykhne A, Ruzin I. Theory of the fractional quantum Hall effect: the two-phase model. Phys Rev B 1994;50:2369.
[52] Balandin AA. Thermal properties of graphene and nanostructured carbon materials. Nat Mater 2011;10:569−81.
[53] Lin S-Y, Ho Y-H, Shyu F-L, Lin M-F. Electronic thermal property of graphite. J Phys Soc Jpn 2013;82:074603.
[54] Tuovinen R, Perfetto E, Stefanucci G, Van Leeuwen R. Time-dependent Landauer-Büttiker formula: application to transient dynamics in graphene nanoribbons. Phys Rev B 2014;89:085131.
[55] Lin M, Chuu D, Shung K-K. Thermal conductance and the Peltier coefficient of carbon nanotubes. Phys Rev B 1996;53:11186.
[56] Chuu D-S, Lin M-F, Shung KW-K. Thermal conductance and peltier coefficient of carbon nanotubes. In: APS March Meeting Abstracts; 1996. p. 31−108.
[57] Laird EA, Kuemmeth F, Steele GA, Grove-Rasmussen K, Nygård J, Flensberg K, Kouwenhoven LP. Quantum transport in carbon nanotubes. Rev Mod Phys 2015;87:703.
[58] Yoshida M, Popert A, Kato YK. Gate-voltage induced trions in suspended carbon nanotubes. Phys Rev B 2016;93:041402.
[59] Kanda A, Tsukagoshi K, Aoyagi Y, Ootuka Y. Gate-voltage dependence of zero-bias anomalies in multiwall carbon nanotubes. Phys Rev Lett 2004;92:036801.
[60] Lin C-Y, Wu J-Y, Chang C-P, Lin M-F. Magneto-optical selection rules of curved graphene nanoribbons and carbon nanotubes. Carbon 2014;69:151−61.
[61] Sadowski M, Martinez G, Potemski M, Berger C, de Heer WA. Magnetospectroscopy of epitaxial few-layer graphene. Solid State Commun 2007;143:123−5.
[62] Dubois S-M, Zanolli Z, Declerck X, Charlier J-C. Electronic properties and quantum transport in Graphene-based nanostructures. Eur Phys J B 2009;72:1−24.
[63] Yang L, Deslippe J, Park C-H, Cohen ML, Louie SG. Excitonic effects on the optical response of graphene and bilayer graphene. Phys Rev Lett 2009;103:186802.
[64] Uddin M, Doan T, Li J, Ziemer KS, Lin J, Jiang H. Electrical transport properties of (BN)-rich hexagonal (BN) C semiconductor alloys. AIP Adv 2014;4:087141.
[65] Foulkes WMC, Haydock R. Tight-binding models and density-functional theory. Phys Rev B 1989;39:12520.
[66] Konschuh S, Gmitra M, Fabian J. Tight-binding theory of the spin-orbit coupling in graphene. Phys Rev B 2010;82:245412.

[67] Ho C-H, Tsai S-J, Chen R-B, Chiu Y-H, Lin M-F. Low-energy Landau level spectrum in ABC-stacked trilayer graphene. J Nanosci Nanotechnol 2011;11:4938–47.
[68] Ho Y-H, Chiu Y-H, Lin D-H, Chang C-P, Lin M-F. Magneto-optical selection rules in bilayer Bernal graphene. ACS Nano 2010;4:1465–72.
[69] Redouani I, Jellal A, Bahaoui A, Bahlouli H. Multibands tunneling in AAA-stacked trilayer graphene. Superlattice Microst 2018;116:44–53.
[70] Chiun-Yan L, Ching-Hong H, Jhao-Ying W, Ming-Fa L. Unusual electronic excitations in ABA trilayer graphene Scientific Reports, vol. 10. Nature Publisher Group; 2020.
[71] Hattendorf S, Georgi A, Liebmann M, Morgenstern M. Networks of ABA and ABC stacked graphene on mica observed by scanning tunneling microscopy. Surf Sci 2013;610:53–8.
[72] Lu C L, Chang C P, Huang Y C, Ho J H, Hwang C C, Lin M F. Electronic properties of AA-and ABC-stacked few-layer graphites. J Phys Soc Jpn 2007;76:024701.
[73] Ho Y-H, Wu J-Y, Chen R-B, Chiu Y-H, Lin M-F. Optical transitions between Landau levels: AA-stacked bilayer graphene. Appl Phys Lett 2010;97:101905.
[74] Tran NTT, Lin S-Y, Lin C-Y, Lin M-F. Geometric and electronic properties of graphene-related systems: chemical bonding schemes. CRC Press; 2017.
[75] Lin Y-P, Lin C-Y, Ho Y-H, Do T-N, Lin M-F. Magneto-optical properties of ABC-stacked trilayer graphene. Phys Chem Chem Phys 2015;17:15921–7.
[76] Lin C-Y, Huang B-L, Ho C-H, Gumbs G, Lin M-F. Geometry-diversified Coulomb excitations in trilayer AAB stacking graphene. Phys Rev B 2018;98:195442.
[77] Do T-N, Lin C-Y, Lin Y-P, Shih P-H, Lin M-F. Configuration-enriched magneto-electronic spectra of AAB-stacked trilayer graphene. Carbon 2015;94:619–32.
[78] Budyka M, Zyubina T, Ryabenko A, Lin S, Mebel A. Bond lengths and diameters of armchair single wall carbon nanotubes. Chem Phys Lett 2005;407:266–71.
[79] Bulusheva L, Okotrub A, Romanov D, Tomanek D. Electronic structure of (n, 0) zigzag carbon nanotubes: cluster and crystal approach. J Phys Chem 1998;102:975–81.
[80] Sanchez-Valencia JR, Dienel T, Gröning O, Shorubalko I, Mueller A, Jansen M, Amsharov K, Ruffieux P, Fasel R. Controlled synthesis of single-chirality carbon nanotubes. Nature 2014;512:61–4.
[81] Pierce N, Chen G, Rajukumar L P, Chou NH, Koh AL, Sinclair R, Maruyama S, Terrones M, Harutyunyan AR. Intrinsic chirality origination in carbon nanotubes. ACS Nano 2017;11:9941–9.
[82] Chen S, Wang T, Lee C, Lin M-F. Magneto-electronic properties of graphene nanoribbons in the spatially modulated electric field. Phys Lett 2008;372:5999–6002.
[83] Zhang X, Martin I, Jiang H. Landau level anticrossing manifestations in the phase-diagram topology of a two-subband system. Phys Rev B 2006;74:073301.
[84] Cao L, White JS, Park J-S, Schuller JA, Clemens BM, Brongersma ML. Engineering light absorption in semiconductor nanowire devices. Nat Mater 2009;8:643–7.
[85] Atwater HA, Polman A. Plasmonics for improved photovoltaic devices Materials for sustainable energy: a collection of peer-reviewed research and review articles from nature publishing group. 2011. p. 1–11.
[86] Trevisanutto PE, Holzmann M, Côté M, Olevano V. High-energy excitonic effects in graphite and graphene. 2009. arXiv preprint arXiv:0909.1682.
[87] Ichida M, Mizuno S, Tani Y, Saito Y, Nakamura A. Exciton effects of optical transitions in single-wall carbon nanotubes. J Phys Soc 1999;68:3131–3.
[88] Yang L, Cohen ML, Louie SG. Excitonic effects in the optical spectra of graphene nanoribbons. Nano Lett 2007;7:3112–5.
[89] Malić E, Hirtschulz M, Milde F, Knorr A, Reich S. Analytical approach to optical absorption in carbon nanotubes. Phys Rev B 2006;74:195431.
[90] Sivek J, Sahin H, Partoens B, Peeters FM. Adsorption and absorption of boron, nitrogen, aluminum, and phosphorus on silicene: stability and electronic and phonon properties. Phys Rev B 2013;87:085444.
[91] Spataru CD, Ismail-Beigi S, Capaz RB, Louie SG. Quasiparticle and excitonic effects in the optical response of nanotubes and nanoribbons. Carbon Nanotubes 2007:195–227.
[92] Yang L, Park C-H, Son Y-W, Cohen ML, Louie SG. Quasiparticle energies and band gaps in graphene nanoribbons. Phys Rev Lett 2007;99:186801.

[93] Högele A, Galland C, Winger M, Imamoğlu A. Photon antibunching in the photoluminescence spectra of a single carbon nanotube. Phys Rev Lett 2008;100:217401.
[94] Lorchat E, López LEP, Robert C, Lagarde D, Froehlicher G, Taniguchi T, Watanabe K, Marie X, Berciaud S. Filtering the photoluminescence spectra of atomically thin semiconductors with graphene. Nat Nanotechnol 2020;15:283–8.
[95] Ouyang M, Huang J-L, Cheung CL, Lieber CM. Energy gaps in "metallic" single-walled carbon nanotubes. Science 2001;292:702–5.
[96] Odom TW, Huang J-L, Kim P, Lieber CM. Atomic structure and electronic properties of single-walled carbon nanotubes. Nature 1998;391:62–4.
[97] Rohlfing M, Louie SG. Electron-hole excitations and optical spectra from first principles. Phys Rev B 2000;62:4927.
[98] Spataru CD, Ismail-Beigi S, Benedict LX, Louie SG. Excitonic effects and optical spectra of single-walled carbon nanotubes. Phys Rev Lett 2004;92:077402.
[99] Lin M-F, Shung KW-K. Plasmons and optical properties of carbon nanotubes. Phys Rev B 1994;50:17744.
[100] Shyu F-L, Chang C-P, Chen R-B, Chiu C-W, Lin M-F. Magnetoelectronic and optical properties of carbon nanotubes. Phys Rev B 2003;67:045405.
[101] Tran NTT, Nguyen DK, Glukhova OE, Lin M-F. Coverage-dependent essential properties of halogenated graphene: a DFT study. Sci Rep 2017;7:1–13.
[102] Lin S-Y, Chang S-L, Chen H-H, Su S-H, Huang J-C, Lin M-F. Substrate-induced structures of bismuth adsorption on graphene: a first principles study. Phys Chem Chem Phys 2016;18:18978–84.
[103] Rafique M, Shuai Y, Hussain N. First-principles study on silicon atom doped monolayer graphene. Phys E Low-dimens Syst Nanostruct 2018;95:94–101.
[104] Santos EJ, Ayuela A, Sánchez-Portal D. First-principles study of substitutional metal impurities in graphene: structural, electronic and magnetic properties. New J Phys 2010;12:053012.
[105] Sahin H, Ciraci S. Chlorine adsorption on graphene: Chlorographene. J Phys Chem C 2012;116:24075–83.
[106] Pumera M, Wong CHA. Graphane and hydrogenated graphene. Chem Soc Rev 2013;42:5987–95.
[107] Lin M, Huang C, Chuu D. Plasmons in graphite and stage-1 graphite intercalation compounds. Phys Rev B 1997;55:13961.
[108] Li W-B, Lin S-Y, Tran NTT, Lin M-F, Lin K-I. Essential geometric and electronic properties in stage-n graphite alkali-metal-intercalation compounds. RSC Adv 2020;10:23573–81.
[109] Lin W-B, Tran NTT, Lin S-Y. Diverse fundamental properties in stage-n graphite alkali-intercalation compounds: anode materials of Li^{+}-based batteries. 2019. arXiv preprint arXiv:2001.02042.
[110] Boukhvalov D, Katsnelson M, Lichtenstein A. Hydrogen on graphene: electronic structure, total energy, structural distortions and magnetism from first-principles calculations. Phys Rev B 2008;77:035427.
[111] Lin S-Y, Chang S-L, Tran NTT, Yang P-H, Lin M-F. H–Si bonding-induced unusual electronic properties of silicene: a method to identify hydrogen concentration. Phys Chem Chem Phys 2015;17:26443–50.
[112] Pham HD, Gumbs G, Su W-P, Tran NTT, Lin M-F. Unusual features of nitrogen substitutions in silicene. RSC Adv 2020;10:32193–201.
[113] Yang L. Excitonic effects on optical absorption spectra of doped graphene. Nano Lett 2011;11:3844–7.
[114] Zuo X, Zhu J, Müller-Buschbaum P, Cheng Y-J. Silicon based lithium-ion battery anodes: a chronicle perspective review. Nano Energy 2017;31:113–43.
[115] Stern S, Shah V, Hardy B. Structure-permeability relationships in silicone polymers. J Polym Sci B Polym Phys 1987;25:1263–98.
[116] Han-tang L. Development and application of silicone materials.[J]. Silicone Mater 2006;4.
[117] Nitta N, Wu F, Lee JT, Yushin G. Li-ion battery materials: present and future. Mater Today 2015;18:252–64.
[118] Goodenough JB, Park K-S. The Li-ion rechargeable battery: a perspective. J Am Chem Soc 2013;135:1167–76.
[119] Nguyen TDH, Pham HD, Lin SY, Lin MF. Featured properties of Li^{+}-based battery anode: $Li_4Ti_5O_{12}$. RSC Adv 2020;10:14071–9.
[120] Khuong Dien V, Thi Han N, Nguyen TDH, Huynh TMD, Pham HD, Lin M-F. Geometric and electronic properties of Li_2GeO_3. Front Mater 2020;7.
[121] Li J, Li J, Luo J, Wang L, He X. Recent advances in the $LiFeO_2$-based materials for Li-ion batteries. Int J Electrochem Sci 2011;6:1550–61.

[122] Boufelfel A. Electronic structure and magnetism in the layered $LiFeO_2$: DFT + U calculations. J Magn Magn Mater 2013;343:92–8.
[123] Andriyevsky B, Doll K, Jacob T. Electronic and transport properties of $LiCoO_2$. Phys Chem Chem Phys 2014;16:23412–20.
[124] Molenda J, Wilk P, Marzec J. Structural, electrical and electrochemical properties of $LiNiO_2$. Solid State Ionics 2002;146:73–9.
[125] Daré A-M, Hayn R, Richard J-L. Orbital and spin exchange in $LiNiO_2$. EPL 2003;61:803.
[126] Li X, Liang J, Lu Y, Hou Z, Cheng Q, Zhu Y, Qian Y. Sulfur-rich phosphorus sulfide molecules for use in rechargeable lithium batteries. Angew Chem Int Ed 2017;56:2937–41.
[127] Hafner J. Ab-initio simulations of materials using VASP: density-functional theory and beyond. J Comput Chem 2008;29:2044–78.
[128] Yuan S, Bao JL, Wang L, Xia Y, Truhlar DG, Wang Y. Graphene-supported nitrogen and boron rich carbon layer for improved performance of lithium–sulfur batteries due to enhanced chemisorption of lithium polysulfides. Adv Energy Mater 2016;6:1501733.
[129] Kudu Ö U, Famprikis T, Fleutot B, Braida M-D, Le Mercier T, Islam MS, Masquelier C. A review of structural properties and synthesis methods of solid electrolyte materials in the Li2S– P2S5 binary system. J Power Sources 2018;407:31–43.
[130] Han NT, Dien VK, Lin M-F. Excitonic effects in the optical spectra of lithium metasilicate (Li_2SiO_3). 2020. arXiv preprint arXiv:2010.11621.
[131] Chiu C-W, Huang Y-C, Shyu F-L, Lin M-F. Optical absorption spectra in ABC-stacked graphene superlattice. Synth Met 2012;162:800–4.
[132] Yan J-A, Ruan W, Chou M. Enhanced optical conductivity induced by surface states in ABC-stacked few-layer graphene. Phys Rev B 2011;83:245418.

Index

Note: Page numbers followed by '*f*' indicate figures those followed by '*t*' indicate tables.

D

E

F

G

H

Kolibri 360 GmbH
Ferdinand-Jühlke-Straße 7
99095 Erfurt, Deutschland
produktsicherheit@kolibri360.de